Lecture Notes in Statistics 187

Edited by P. Bickel, P. Diggle, S. Fienberg, U. Gather,
I. Olkin, S. Zeger

Patrice Bertail
Paul Doukhan
Philippe Soulier (Editors)

Dependence in Probability and Statistics

With 40 figures

 Springer

Patrice Bertail
CREST
L-S Timbre J340
18 Boulevard Adolphe Pinard
75675 Paris Cedex 14
France
and
Laboratoire MODAL'X
Université Paris X
200 avenue de la République
92000 Nanterre Cedex
France
bertail@ensae.fr

Paul Doukhan
CREST
L-S Timbre J340
18 Boulevard Adolphe Pinard
75675 Paris Cedex 14
France
and
Université de Cergy-Pontoise
33 boulevard du Port
95011 Cergy-Pontoise Cedex
France

Philippe Soulier
Laboratoire MODAL'X
Université Paris X
200 avenue de la République
92000 Nanterre Cedex
France
philippe.soulier@u-paris10.fr

Library of Congress Control Number: 2005938921

ISBN-10: 0-387-31741-4   Printed on acid-free paper.
ISBN-13: 978-0387-31741-0

Printed in the United States of America.   (SBA)

9 8 7 6 5 4 3 2 1

springer.com

Preface

The purpose of this book is to give a detailed account of some recent developments in the field of probability and statistics for dependent data. It covers a wide range of topics from Markov chains theory, weak dependence, dynamical system to strong dependence and their applications. The title of this book has been somehow borrowed from the book "Dependence in Probability and Statistics: a Survey of Recent Result" edited by Ernst Eberlein and Murad S. Taqqu, Birkhäuser (1986), which could serve as an excellent prerequisite for reading this book. We hope that the reader will find it as useful and stimulating as the previous one.

This book was planned during a conference, entitled "STATDEP2005: Statistics for dependent data", organized by the Statistical Laboratory of the CREST (Research Center in Economy and Statistics), in Paris/Malakoff, under the auspices of the French State Statistical Institute, INSEE.

See http://www.crest.fr/pageperso/statdep2005/home.htm for some retrospective informations. However this book is not a conference proceeding. This conference has witnessed the rapid growth of contributions on dependent data in the probabilistic and statistical literature and the need for a book covering recent developments scattered in various probability and statistical journals. To achieve such a goal, we have solicited some participants of the conferences as well as other specialists of the field.

The mathematical level of this book is mixed. Some chapters (chapters 1, 3, 4, 9, 10, part of chapter 12) are general surveys which have been prepared for a broad audience of readership, with good notions in time-series, probability and statistics. Specific statistical procedures with dedicated applications are also presented in the last section and may be of interest to many statisticians. However, due to the innate nature of the field, the mathematical developments are important and some chapters of the book are rather intended to researchers in the field of dependent data.

The book has been organized into three parts: "Weak dependence and related concepts", "Strong dependence" and Statistical estimation and applications".

The first seven chapters consider some recent development in weak dependence, including some recent results for Markov chains as well as some new developments around the notion of weak dependence introduced by Doukhan and Louichi in their seminal paper of 1999. A special emphasis is put on potential applications and developments of these notions in the field of dynamical systems. We believe that this part somehow fills a gap between the statistical literature and the dynamical system literature and that both communities may find it of interest.

The second part, built around 6 chapters, presents some recent or new results on strong dependence with a special emphasis on non-linear processes and random fields currently encountered in applications. Special models exhibiting long range dependence features are also studied in this section. It also proposes some extensions of the notions of weak dependence to anisotropic random fields (chapters 9 and 12) which may motivate new researches in the field.

Finally, the last part considers some general estimation problems ranking from rate of convergence of maximum likelihood estimators, efficient estimation in parametric or non-parametric times series model with an emphasis on applications. Although the important problem of non-stationarity is not specifically addressed (because it covers too large a field), many applications in this section deal with estimations in a non-stationary framework. We hope that these applications will also generate some new theoretical developments in the field of non-stationary time series.

Preparing this book, has been greatly facilitated by the kind cooperation of the authors, which have done their best to follow our recommandations: we would like to thank all of them for their contributions. We would also like to thank the members of the organizing and scientific committees of the Statdep2005 conference, who have kindly accepted to play the role of "associate editors" in the realization of this book. We are in particular grateful to Jérome Dedecker, Youri Golubev, Sylvie Huet, Gabriel Lang, Jose R. León, Eric Moulines, Michael H. Neumann, Emmanuel Rio, Alexandre Tsybakov and Jean-Michel Zakoïan for their help in the refereing processes of the papers. We are also grateful to all the anonymous referees for their great work and all their suggestions.

Malakoff and Nanterre,   *Patrice Bertail*
France   *Paul Doukhan*
December 2005   *Philippe Soulier*

Contents

Part I Weak dependence and related concepts

Regeneration-based statistics for Harris recurrent Markov
chains
Patrice Bertail, Stéphan Clémençon 3

Subgeometric ergodicity of Markov chains
Randal Douc, Eric Moulines, Philippe Soulier 55

Limit Theorems for Dependent U-statistics
Herold Dehling 65

Recent results on weak dependence for causal sequences.
Statistical applications to dynamical systems.
Clémentine Prieur 87

Parametrized Kantorovich-Rubinštein theorem and
application to the coupling of random variables.
Jérôme Dedecker, Clémentine Prieur, Paul Raynaud De Fitte 105

Exponential inequalities and estimation of conditional
probabilities
V. Maume-Deschamps 123

Martingale approximation of non adapted stochastic processes
with nonlinear growth of variance
Dalibor Volný 141

Part II Strong dependence

Almost periodically correlated processes with long memory
Anne Philippe, Donatas Surgailis, Marie-Claude Viano 159

Long memory random fields
Frédéric Lavancier 195

Long Memory in Nonlinear Processes
Rohit Deo, Mengchen Hsieh, Clifford M. Hurvich, Philippe Soulier 221

A LARCH(∞) Vector Valued Process
Paul Doukhan, Gilles Teyssière, Pablo Winant 245

On a Szegö type limit theorem and the asymptotic theory of
random sums, integrals and quadratic forms
Florin Avram, Murad S. Taqqu 259

Aggregation of Doubly Stochastic Interactive Gaussian
Processes and Toeplitz forms of U-Statistics.
Didier Dacunha-Castelle, Lisandro Fermín 287

Part III Statistical Estimation and Applications

On Efficient Inference in GARCH Processes
Christian Francq, Jean-Michel Zakoïan 305

Almost sure rate of convergence of maximum likelihood
estimators for multidimensional diffusions
Dasha Loukianova, Oleg Loukianov 329

Convergence rates for density estimators of weakly dependent
time series
Nicolas Ragache, Olivier Wintenberger 349

Variograms for spatial max-stable random fields
Dan Cooley, Philippe Naveau, Paul Poncet 373

A non-stationary paradigm for the dynamics of multivariate
financial returns
Stefano Herzel, Cătălin Stărică, Reha Tütüncü 391

Multivariate Non-Linear Regression with Applications
Tata Subba Rao and Gyorgy Terdik 431

Nonparametric estimator of a quantile function for the
probability of event with repeated data
Claire Pinçon, Odile Pons 475

Part I

Weak dependence and related concepts

Regeneration-based statistics for Harris recurrent Markov chains

Patrice Bertail[1] and Stéphan Clémençon[2]

[1] CREST-LS, 3, ave Pierre Larousse, 94205 Malakoff, France
   `Patrice.Bertail@ensae.fr`
[2] MODAL'X - Université Paris X Nanterre
   LPMA - UMR CNRS 7599 - Université s Paris VI et Paris VII
   `sclemenc@u-paris10.fr`

1 Introduction

1.1 On describing Markov chains via Renewal processes

Renewal theory plays a key role in the analysis of the asymptotic structure of many kinds of stochastic processes, and especially in the development of asymptotic properties of general irreducible Markov chains. The underlying ground consists in the fact that limit theorems proved for sums of independent random vectors may be easily extended to regenerative random processes, that is to say random processes that may be decomposed at random times, called *regeneration times*, into a sequence of mutually independent blocks of observations, namely *regeneration cycles* (see Smith (1955)). The method based on this principle is traditionally called the *regenerative method*. Harris chains that possess an atom, i.e. a Harris set on which the transition probability kernel is constant, are special cases of regenerative processes and so directly fall into the range of application of the regenerative method (Markov chains with discrete state space as well as many markovian models widely used in operational research for modeling storage or queuing systems are remarkable examples of atomic chains). The theory developed in Nummelin (1978) (and in parallel the closely related concepts introduced in Athreya & Ney (1978)) showed that general Markov chains could all be considered as regenerative in a broader sense (*i.e.* in the sense of the existence of a theoretical regenerative extension for the chain, see § 2.3), as soon as the Harris recurrence property is satisfied. Hence this theory made the regenerative method applicable to the whole class of Harris Markov chains and allowed to carry over many limit theorems to Harris chains such as LLN, CLT, LIL or Edgeworth expansions.

In many cases, parameters of interest for a Harris Markov chain may be thus expressed in terms of regeneration cycles. While, for atomic Markov chains, statistical inference procedures may be then based on a random number of observed regeneration data blocks, in the general Harris recurrent case the regeneration times are theoretical and their occurrence cannot be determined by examination of the data only. Although the *Nummelin splitting technique* for constructing regeneration times has been introduced as a theoretical tool for proving probabilistic results such as limit theorems or probability and moment inequalities in the markovian framework, this article aims to show that it is nevertheless possible to make a practical use of the latter for extending regeneration-based statistical tools. Our proposal consists in an empirical method for building approximatively a realization drawn from a Nummelin extension of the chain with a regeneration set and then recovering "approximate regeneration data blocks". As will be shown further, though the implementation of the latter method requires some prior knowledge about the behaviour of the chain and crucially relies on the computation of a consistent estimate of its transition kernel, this methodology allows for numerous statistical applications.

We finally point out that, alternatively to regeneration-based statistical methods, inference techniques based on data (moving) blocks of fixed length may also be used in our markovian framework. But as will be shown throughout the article, such blocking techniques, introduced for dealing with general time series (in the weakly dependent setting) are less powerful, when applied to Harris Markov chains, than the methods we promote here, which are specifically tailored for (pseudo) regenerative processes.

1.2 Outline

The outline of the paper is as follows. In section 2, notations are set out and key concepts of the Markov chain theory as well as some basic notions about the regenerative method and the Nummelin splitting technique are recalled. Section 3 presents and discusses how to practically construct (approximate) regeneration data blocks, on which statistical procedures we investigate further are based. Sections 4 and 5 mainly survey results established at length in Bertail & Clémençon (2004a,b,c,d). More precisely, the problem of estimating additive functionals of the stationary distribution in the Harris positive recurrent case is considered in section 4. Estimators based on the (pseudo) regenerative blocks, as well as estimates of their asymptotic variance are exhibited, and limit theorems describing the asymptotic behaviour of their bias and their sampling distribution are also displayed. Section 5 is devoted to the study of a specific resampling procedure, which crucially relies on the (approximate) regeneration data blocks. Results proving the asymptotic validity of this particular bootstrap procedure (and its optimality regarding to second order properties in the atomic case) are stated. Section 6 shows how to extend some of the results of sections 4 and 5 to V and U-statistics. A specific notion of robustness for statistics based on the (approximate) regenerative blocks is introduced and investigated in section 7. And asymptotic properties of some regeneration-based statistics related to the extremal behaviour of Markov chains are studied in section 8 in the regenerative case only. Finally, some concluding remarks are collected in section 9 and further lines of research are sketched.

2 Theoretical background

2.1 Notation and definitions

We now set out the notations and recall a few definitions concerning the communication structure and the stochastic stability of Markov chains (for further detail, refer to Revuz (1984) or Meyn & Tweedie (1996)). Let $X = (X_n)_{n \in \mathbb{N}}$ be an aperiodic irreducible Markov chain on a countably generated state space $(E, \mathcal{E})$, with transition probability Π, and initial probability distribution ν. For any $B \in \mathcal{E}$ and any $n \in \mathbb{N}$, we thus have

$$X_0 \sim \nu \text{ and } \mathbb{P}(X_{n+1} \in B \mid X_0, ..., X_n) = \Pi(X_n, B) \text{ a.s.}$$

In what follows, $\mathbb{P}_\nu$ (respectively $\mathbb{P}_x$ for x in E) will denote the probability measure on the underlying probability space such that $X_0 \sim \nu$ (resp. $X_0 = x$), $\mathbb{E}_\nu (.)$ the $\mathbb{P}_\nu$-expectation (resp. $\mathbb{E}_x (.)$ the $\mathbb{P}_x$-expectation), $\mathbb{I}\{\mathcal{A}\}$ will denote the indicator function of the event $\mathcal{A}$ and $\Rightarrow$ the convergence in distribution.

For completeness, recall the following notions. The first one formalizes the idea of communicating structure between specific subsets, while the second one considers the set of time points at which such communication may occur.

- The chain is *irreducible* if there exists a σ-finite measure ψ such that for all set $B \in \mathcal{E}$, when $\psi(B) > 0$, the chain visits B with strictly positive probability, no matter what the starting point.
- Assuming ψ-irreducibility, there is $d' \in \mathbb{N}^*$ and disjoints sets $D_1, ..., D_{d'}$ ($D_{d'+1} = D_1$) weighted by ψ such that $\psi(E \backslash \cup_{1 \leq i \leq d'} D_i) = 0$ and $\forall x \in D_i$, $\Pi(x, D_{i+1}) = 1$. The g.c.d. d of such integers is the *period* of the chain, which is said *aperiodic* if $d = 1$.

A measurable set B is *Harris recurrent* for the chain if for any $x \in B$, $\mathbb{P}_x(\sum_{n=1}^\infty \mathbb{I}\{X_n \in B\} = \infty) = 1$. The chain is said *Harris recurrent* if it is ψ-irreducible and every measurable set B such that $\psi(B) > 0$ is Harris recurrent. When the chain is Harris recurrent, we have the property that $\mathbb{P}_x(\sum_{n=1}^\infty \mathbb{I}\{X_n \in B\} = \infty) = 1$ for any $x \in E$ and any $B \in \mathcal{E}$ such that $\psi(B) > 0$.

A probability measure μ on E is said invariant for the chain when $\mu\Pi = \mu$, where $\mu\Pi(dy) = \int_{x \in E} \mu(dx)\Pi(x, dy)$. An irreducible chain is said *positive recurrent* when it admits an invariant probability (it is then unique).

Now we recall some basics concerning the regenerative method and its application to the analysis of the behaviour of general Harris chains via the Nummelin splitting technique (refer to Nummelin (1984) for further detail).

2.2 Markov chains with an atom

Assume that the chain is ψ-irreducible and possesses an accessible atom, that is to say a measurable set A such that $\psi(A) > 0$ and $\Pi(x,.) = \Pi(y,.)$ for all x, y in A. Denote by $\tau_A = \tau_A(1) = \inf\{n \geq 1,\ X_n \in A\}$ the hitting time on A, by $\tau_A(j) = \inf\{n > \tau_A(j-1),\ X_n \in A\}$ for $j \geq 2$ the successive return times to A and by $\mathbb{E}_A(.)$ the expectation conditioned on $X_0 \in A$. Assume further that the chain is Harris recurrent, the probability of returning infinitely often to the atom A is thus equal to one, no matter what the starting point. Then, it follows from the *strong Markov property* that, for any initial distribution ν, the sample paths of the chain may be divided into i.i.d. blocks of random length corresponding to consecutive visits to A:

$$\mathcal{B}_1 = (X_{\tau_A(1)+1}, ..., \ X_{\tau_A(2)}), ..., \ \mathcal{B}_j = (X_{\tau_A(j)+1}, ..., \ X_{\tau_A(j+1)}), ...$$

taking their values in the torus $\mathbb{T} = \cup_{n=1}^{\infty} E^n$. The sequence $(\tau_A(j))_{j\geq 1}$ defines successive times at which the chain forgets its past, called *regeneration times*. We point out that the class of atomic Markov chains contains not only chains with a countable state space (for the latter, any recurrent state is an accessible atom), but also many specific Markov models arising from the field of operational research (see Asmussen (1987) for regenerative models involved in queuing theory, as well as the examples given in § 4.3). When an accessible atom exists, the *stochastic stability* properties of the chain amount to properties concerning the speed of return time to the atom only. For instance, in this framework, the following result, known as Kac's theorem, holds (*cf* Theorem 10.2.2 in Meyn & Tweedie (1996)).

Theorem 1. *The chain X is positive recurrent iff $\mathbb{E}_A(\tau_A) < \infty$. The (unique) invariant probability distribution μ is then the Pitman's occupation measure given by*

$$\mu(B) = \mathbb{E}_A(\sum_{i=1}^{\tau_A} \mathbb{I}\{X_i \in B\})/\mathbb{E}_A(\tau_A), \ \text{for all } B \in \mathcal{E}.$$

For atomic chains, limit theorems can be derived from the application of the corresponding results to the i.i.d. blocks $(\mathcal{B}_n)_{n\geq 1}$. One may refer for example to Meyn & Tweedie (1996) for the LLN, CLT, LIL, Bolthausen (1980) for the Berry-Esseen theorem, Malinovskii (1985, 87, 89) and Bertail & Clémençon (2004a) for other refinements of the CLT. The same technique can also be applied to establish moment and probability inequalities, which are not asymptotic results (see Clémençon (2001)). As mentioned above, these results are established from hypotheses related to the distribution of the $\mathcal{B}_n$'s. The following assumptions shall be involved throughout the article. Let $\kappa > 0$, $f : E \to \mathbb{R}$ be a measurable function and ν be a probability distribution on $(E, \mathcal{E})$.

Regularity conditions:

$$\mathcal{H}_0(\kappa): \ \mathbb{E}_A(\tau_A^\kappa) < \infty \,,$$
$$\mathcal{H}_0(\kappa, \ \nu): \ \mathbb{E}_\nu(\tau_A^\kappa) < \infty \,.$$

Block-moment conditions:

$$\mathcal{H}_1(\kappa, \ f): \ \mathbb{E}_A((\sum_{i=1}^{\tau_A} |f(X_i)|)^\kappa) < \infty \,,$$
$$\mathcal{H}_1(\kappa, \ \nu, \ f): \ \mathbb{E}_\nu((\sum_{i=1}^{\tau_A} |f(X_i)|)^\kappa) < \infty \,.$$

Remark 1. We point out that conditions $\mathcal{H}_0(\kappa)$ and $\mathcal{H}_1(\kappa, \ f)$ do not depend on the accessible atom chosen : if they hold for a given accessible atom A, they are also fulfilled for any other accessible atom (see Chapter 11 in Meyn & Tweedie (1996)). Besides, the relationship between the "block moment" conditions and the rate of decay of mixing coefficients has been investigated in Bolthausen (1982): for instance, $\mathcal{H}_0(\kappa)$ (as well as $\mathcal{H}_1(\kappa, \ f)$ when f is bounded) is typically fulfilled as soon as the strong mixing coefficients sequence decreases at an arithmetic rate $n^{-\rho}$, for some $\rho > \kappa - 1$.

2.3 General Harris recurrent chains

The Nummelin splitting technique

We now recall the *splitting technique* introduced in Nummelin (1978) for extending the probabilistic structure of the chain in order to construct an artificial regeneration set in the general Harris recurrent case. It relies on the crucial notion of *small set*. Recall that, for a Markov chain valued in a state space $(E, \mathcal{E})$ with transition probability Π, a set $S \in \mathcal{E}$ is said to be *small* if there exist $m \in \mathbb{N}^*$, $\delta > 0$ and a probability measure Γ supported by S such that, for all $x \in S$, $B \in \mathcal{E}$,

$$\Pi^m(x, B) \geq \delta\Gamma(B) \,, \tag{1}$$

denoting by Π^m the m-th iterate of Π. When this holds, we say that the chain satisfies the *minorization condition* $\mathcal{M}(m, S, \delta, \Gamma)$. We emphasize that accessible small sets always exist for ψ-irreducible chains: any set $B \in \mathcal{E}$ such that $\psi(B) > 0$ actually contains such a set (*cf* Jain & Jamison (1967)). Now let us precise how to construct the atomic chain onto which the initial chain X is embedded, from a set on which an iterate Π^m of the transition probability is uniformly bounded below. Suppose that X satisfies $\mathcal{M} = \mathcal{M}(m, S, \delta, \Gamma)$ for $S \in \mathcal{E}$ such that $\psi(S) > 0$. Even if it entails replacing the chain $(X_n)_{n \in \mathbb{N}}$ by the chain $((X_{nm}, ..., X_{n(m+1)-1}))_{n \in \mathbb{N}}$, we suppose $m = 1$. The sample space is expanded so as to define a sequence $(Y_n)_{n\in\mathbb{N}}$ of independent Bernoulli r.v.'s with parameter δ by defining the joint distribution $\mathbb{P}_{\nu,\mathcal{M}}$ whose construction relies on the following randomization of the transition probability Π each time the chain hits S (note that it happens a.s. since the chain is Harris recurrent and $\psi(S) > 0$). If $X_n \in S$ and

- if $Y_n = 1$ (which happens with probability $\delta \in]0,1[$), then X_{n+1} is distributed according to Γ,
- if $Y_n = 0$, (which happens with probability $1 - \delta$), then X_{n+1} is drawn from $(1 - \delta)^{-1}(\Pi(X_{n+1}, .) - \delta\Gamma(.))$.

Set $Ber_\delta(\beta) = \delta\beta + (1-\delta)(1-\beta)$ for $\beta \in \{0,1\}$. We now have constructed a bivariate chain $X^{\mathcal{M}} = ((X_n, Y_n))_{n\in\mathbb{N}}$, called the *split chain*, taking its values in $E \times \{0,1\}$ with transition kernel $\Pi_{\mathcal{M}}$ defined by

- for any $x \notin S$, $B \in \mathcal{E}$, β and β' in $\{0,1\}$,

$$\Pi_{\mathcal{M}}((x,\beta), B \times \{\beta'\}) = \Pi(x, B) \times Ber_\delta(\beta'),$$

- for any $x \in S$, $B \in \mathcal{E}$, β' in $\{0,1\}$,

$$\Pi_{\mathcal{M}}((x,1), B \times \{\beta'\}) = \Gamma(B) \times Ber_\delta(\beta'),$$
$$\Pi_{\mathcal{M}}((x,0), B \times \{\beta'\}) = (1 - \delta)^{-1}(\Pi(x, B) - \delta\Gamma(B)) \times Ber_\delta(\beta').$$

Basic assumptions

The whole point of the construction consists in the fact that $S \times \{1\}$ is an atom for the split chain $X^{\mathcal{M}}$, which inherits all the communication and stochastic stability properties from X (irreducibility, Harris recurrence,...), in particular (for the case $m = 1$ here) the blocks constructed for the split chain are independent. Hence the splitting method enables to extend the regenerative method, and so to establish all of the results known for atomic chains, to general Harris chains. It should be noticed that if the chain X satisfies $\mathcal{M}(m, S, \delta, \Gamma)$ for $m > 1$, the resulting blocks are not independent anymore but 1-dependent, a form of dependence which may be also easily handled. For simplicity 's sake, we suppose in what follows that condition $\mathcal{M}$ is fulfilled with $m = 1$, we shall also omit the subscript $\mathcal{M}$ and abusively denote by $\mathbb{P}_\nu$ the extensions of the underlying probability we consider. The following assumptions, involving the speed of return to the small set S shall be used throughout the article. Let $\kappa > 0$, $f : E \to \mathbb{R}$ be a measurable function and ν be a probability measure on $(E, \mathcal{E})$.

Regularity conditions:

$$\mathcal{H}'_0(\kappa) : \sup_{x\in S} \mathbb{E}_x(\tau_S^\kappa) < \infty,$$
$$\mathcal{H}'_0(\kappa, \nu) : \mathbb{E}_\nu(\tau_S^\kappa) < \infty.$$

Block-moment conditions:

$$\mathcal{H}_1'(\kappa, \ f) \ : \ \sup_{x \in S} \mathbb{E}_x\left(\left(\sum_{i=1}^{\tau_S} |f(X_i)|\right)^\kappa\right) < \infty \, ,$$

$$\mathcal{H}_1'(\kappa, \ f, \ \nu): \ \mathbb{E}_\nu\left(\left(\sum_{i=1}^{\tau_S} |f(X_i)|\right)^\kappa\right) < \infty \, .$$

Remark 2. It is noteworthy that assumptions $\mathcal{H}_0'(\kappa)$ and $\mathcal{H}_1'(\kappa, \ f)$ do not depend on the choice of the small set S (if they are checked for some accessible small set S, they are fulfilled for all accessible small sets *cf* § 11.1 in Meyn & Tweedie (1996)). Note also that in the case when $\mathcal{H}_0'(\kappa)$ (resp., $\mathcal{H}_0'(\kappa, \ \nu)$) is satisfied, $\mathcal{H}_1'(\kappa, \ f)$ (resp., $\mathcal{H}_1'(\kappa, \ f, \ \nu)$) is fulfilled for any bounded f. Moreover, recall that positive recurrence, conditions $\mathcal{H}_1'(\kappa)$ and $\mathcal{H}_1'(\kappa, \ f)$ may be practically checked by using test functions methods (*cf*
Kalashnikov (1978), Tjøstheim (1990)). In particular, it is well known that such block moment assumptions may be replaced by drift criteria of Lyapounov's type (refer to Chapter 11 in Meyn & Tweedie (1996) for further details on such conditions and many illustrating examples, see also Douc *et al.* (2004)).

We recall finally that such assumptions on the initial chain classically imply the desired conditions for the split chain: as soon as X fulfills $\mathcal{H}_0'(\kappa)$ (resp., $\mathcal{H}_0'(\kappa, \ \nu)$, $\mathcal{H}_1'(\kappa, \ f)$, $\mathcal{H}_1'(\kappa, \ f, \ \nu)$), $X^\mathcal{M}$ satisfies $\mathcal{H}_0(\kappa)$ (resp., $\mathcal{H}_0(\kappa, \ \nu)$, $\mathcal{H}_1(\kappa, \ f)$, $\mathcal{H}_1(\kappa, \ f, \ \nu)$).

The distribution of $(Y_1, ..., Y_n)$ conditioned on $(X_1, ..., X_{n+1})$.

As will be shown in the next section, the statistical methodology for Harris chains we propose is based on approximating the conditional distribution of the binary sequence $(Y_1, ..., Y_n)$ given $X^{(n+1)} = (X_1, ..., X_{n+1})$. We thus precise the latter. Let us assume further that the family of the conditional distributions $\{\Pi(x, dy)\}_{x \in E}$ and the initial distribution ν are dominated by a σ-finite measure λ of reference, so that $\nu(dy) = f(y)\lambda(dy)$ and $\Pi(x, dy) = p(x, y)\lambda(dy)$, for all $x \in E$. Notice that the minorization condition entails that Γ is absolutely continuous with respect to λ too, and that

$$p(x, y) \geq \delta\gamma(y), \ \lambda(dy) \text{ a.s.} \tag{2}$$

for any $x \in S$, with $\Gamma(dy) = \gamma(y)dy$. The distribution of $Y^{(n)} = (Y_1, ..., Y_n)$ conditionally to $X^{(n+1)} = (x_1, ..., x_{n+1})$ is then the tensor product of Bernoulli distributions given by: for all $\beta^{(n)} = (\beta_1, ..., \beta_n) \in \{0, 1\}^n$, $x^{(n+1)} = (x_1, ..., x_{n+1}) \in E^{n+1}$,

$$\mathbb{P}_\nu(Y^{(n)} = \beta^{(n)} \mid X^{(n+1)} = x^{(n+1)}) = \prod_{i=1}^{n} \mathbb{P}_\nu(Y_i = \beta_i \mid X_i = x_i, X_{i+1} = x_{i+1}) \, ,$$

with, for $1 \leq i \leq n$,

$$\mathbb{P}_\nu(Y_i = 1 \mid X_i = x_i, \ X_{i+1} = x_{i+1}) = \delta, \text{ if } x_i \notin S \ ,$$

$$\mathbb{P}_\nu(Y_i = 1 \mid X_i = x_i, \ X_{i+1} = x_{i+1}) = \frac{\delta\gamma(x_{i+1})}{p(x_i, x_{i+1})} \ , \text{ if } x_i \in S \ .$$

Roughly speaking, conditioned on $X^{(n+1)}$, from $i = 1$ to n, Y_i is drawn from the Bernoulli distribution with parameter δ, unless X has hit the small set S at time i: in this case Y_i is drawn from the Bernoulli distribution with parameter $\delta\gamma(X_{i+1})/p(X_i, X_{i+1})$. We denote by $\mathcal{L}^{(n)}(p, S, \delta, \gamma, x^{(n+1)})$ this probability distribution.

3 Dividing the sample path into (approximate) regeneration cycles

In the preceding section, we recalled the Nummelin approach for the theoretical construction of regeneration times in the Harris framework. Here we now consider the problem of approximating these random times from data sets in practice and propose a basic preprocessing technique, on which estimation methods we shall discuss further are based.

3.1 Regenerative case

Let us suppose we observed a trajectory $X_1, ..., X_n$ of length n drawn from the chain X. In the regenerative case, when an atom A for the chain is *a priori* known, regeneration blocks are naturally obtained by simply examining the data, as follows.

Algorithm 1 *(Regeneration blocks construction)*

1. *Count the number of visits $l_n = \sum_{i=1}^{n} \mathbb{I}\{X_i \in A\}$ to A up to time n.*
2. *Divide the observed trajectory $X^{(n)} = (X_1, ..., X_n)$ into $l_n + 1$ blocks corresponding to the pieces of the sample path between consecutive visits to the atom A,*

$$\mathcal{B}_0 = (X_1, ..., \ X_{\tau_A(1)}), \ \mathcal{B}_1 = (X_{\tau_A(1)+1}, ..., \ X_{\tau_A(2)}), ...,$$

$$\mathcal{B}_{l_n-1} = (X_{\tau_A(l_n-1)+1}, ..., \ X_{\tau_A(l_n)}), \ \mathcal{B}_{l_n}^{(n)} = (X_{\tau_A(l_n)+1}, ..., \ X_n) \ ,$$

with the convention $\mathcal{B}_{l_n}^{(n)} = \emptyset$ when $\tau_A(l_n) = n$.
3. *Drop the first block $\mathcal{B}_0$, as well as the last one $\mathcal{B}_{l_n}^{(n)}$, when non-regenerative (i.e. when $\tau_A(l_n) < n$).*

The regeneration blocks construction is illustrated by Fig. 1 in the case of a random walk on the half line $\mathbb{R}^+$ with $\{0\}$ as an atom.

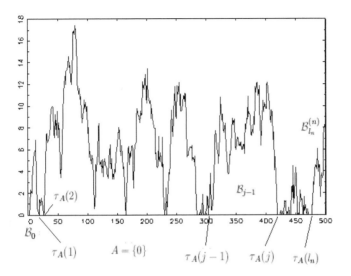

Fig. 1. Dividing the trajectory of a random walk on the half line into regeneration data blocks corresponding to successive visits to $A = 0$

3.2 General Harris case

The principle

Suppose now that observations $X_1, ..., X_{n+1}$ are drawn from a Harris chain X satisfying the assumptions of § 2.3.3 (refer to the latter paragraph for the notations). If we were able to generate binary data $Y_1, ..., Y_n$, so that $X^{\mathcal{M}\,(n)} = ((X_1, Y_1), ..., (X_n, Y_n))$ be a realization of the split chain $X^{\mathcal{M}}$ described in § 2.3, then we could apply the *regeneration blocks construction* procedure to the sample path $X^{\mathcal{M}\,(n)}$. In that case the resulting blocks are still independent since the split chain is atomic. Unfortunately, knowledge of the transition density $p(x, y)$ for $(x, y) \in S^2$ is required to draw practically the Y_i's this way. We propose a method relying on a preliminary estimation of the "nuisance parameter" $p(x, y)$. More precisely, it consists in approximating the splitting construction by computing an estimator $p_n(x, y)$ of $p(x, y)$ using data $X_1, ..., X_{n+1}$, and to generate a random vector $(\widehat{Y}_1, ..., \widehat{Y}_n)$ conditionally to $X^{(n+1)} = (X_1, ..., X_{n+1})$, from distribution $\mathcal{L}^{(n)}(p_n, S, \delta, \gamma, X^{(n+1)})$, which approximates in some sense the conditional distribution $\mathcal{L}^{(n)}(p, S, \delta, \gamma, X^{(n+1)})$ of $(Y_1, ..., Y_n)$ for given $X^{(n+1)}$. Our method, which we call *approximate regeneration blocks construction (ARB construction* in abbreviated form) amounts then to apply the *regeneration blocks construction* procedure to the data $((X_1, \widehat{Y}_1), ..., (X_n, \widehat{Y}_n))$ as if they were drawn from the atomic chain $X^{\mathcal{M}}$. In spite of the necessary consistent transition density estimation step, we shall show in the sequel that many statistical procedures, that would be consistent in the ideal case when they would be based on the regeneration blocks, remain asymptotically valid when implemented from the approximate data blocks. For given parameters $(\delta, \; S, \; \gamma)$ (see § 3.2.2 for a data driven choice of these parameters), the approximate regeneration blocks are constructed as follows.

Algorithm 2 *(Approximate regeneration blocks construction)*

1. *From the data $X^{(n+1)} = (X_1, ..., X_{n+1})$, compute an estimate $p_n(x, y)$ of the transition density such that $p_n(x, y) \geq \delta \gamma(y)$, $\lambda(dy)$ a.s., and $p_n(X_i, X_{i+1}) > 0$, $1 \leq i \leq n$.*
2. *Conditioned on $X^{(n+1)}$, draw a binary vector $(\widehat{Y}_1, ..., \widehat{Y}_n)$ from the distribution estimate $\mathcal{L}^{(n)}(p_n, S, \delta, \gamma, X^{(n+1)})$. It is sufficient in practice to draw the $\widehat{Y}_i$'s at time points i when the chain visits the set S (i.e. when $X_i \in S$), since at these times and at these times only the split chain may regenerate. At such a time point i, draw $\widehat{Y}_i$ according to the Bernoulli distribution with parameter $\delta \gamma(X_{i+1})/p_n(X_i, X_{i+1}))$.*
3. *Count the number of visits $\widehat{l}_n = \sum_{i=1}^{n} \mathbb{I}\{X_i \in S, \widehat{Y}_i = 1)$ to the set $A_{\mathcal{M}} = S \times \{1\}$ up to time n and divide the trajectory $X^{(n+1)}$ into $\widehat{l}_n + 1$ approximate regeneration blocks corresponding to the successive visits of $(X, \widehat{Y})$ to $A_{\mathcal{M}}$,*

$$\widehat{\mathcal{B}}_0 = (X_1, ..., X_{\widehat{\tau}_{A_{\mathcal{M}}}(1)}), \; \widehat{\mathcal{B}}_1 = (X_{\widehat{\tau}_{A_{\mathcal{M}}}(1)+1}, ..., X_{\widehat{\tau}_{A_{\mathcal{M}}}(2)}), ...,$$

$$\widehat{\mathcal{B}}_{\widehat{l}_n - 1} = (X_{\widehat{\tau}_{A_{\mathcal{M}}}(\widehat{l}_n - 1)+1}, ..., X_{\widehat{\tau}_{A_{\mathcal{M}}}(\widehat{l}_n)}), \; \widehat{\mathcal{B}}_{\widehat{l}_n}^{(n)} = (X_{\widehat{\tau}_{A_{\mathcal{M}}}(\widehat{l}_n)+1}, ..., X_{n+1}) \,,$$

where $\widehat{\tau}_{A_{\mathcal{M}}}(1) = \inf\{n \geq 1, \; X_n \in S, \widehat{Y}_n = 1\}$ and $\widehat{\tau}_{A_{\mathcal{M}}}(j + 1) = \inf\{n > \widehat{\tau}_{A_{\mathcal{M}}}(j), \; X_n \in S, \widehat{Y}_n = 1\}$ for $j \geq 1$.
4. *Drop the first block $\widehat{\mathcal{B}}_0$ and the last one $\widehat{\mathcal{B}}_{\widehat{l}_n}^{(n)}$ when $\widehat{\tau}_{A_{\mathcal{M}}}(\widehat{l}_n) < n$.*

Such a division of the sample path is illustrated by Fig. 2 below: from a practical viewpoint the trajectory may only be cut when hitting the small set. At such a point, drawing a Bernoulli r.v. with the estimated parameter indicates whether one should cut here the time series trajectory or not. Of course, due to the dependence induced by the estimated transition density, the resulting blocks are not i.i.d. but, as will be shown later, are close (in some sense) to the one of the true regeneration blocks (which are i.i.d.), provided that the transition estimator is consistent (see assumption $\mathcal{H}_2$ in §1.4.2)

Practical choice of the minorization condition parameters

Because the construction above is highly dependent on the minorization condition parameters chosen, we now discuss how to select the latter with a

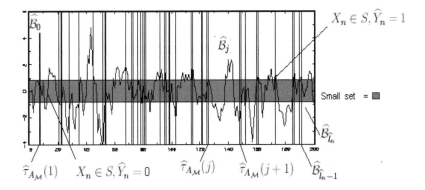

Fig. 2. ARB construction for an AR(1) simulated time-series

data-driven technique so as to construct enough blocks for computing meaningful statistics. As a matter of fact, the rates of convergence of the statistics we shall study in the sequel increase as the mean number of regenerative (or pseudo-regenerative) blocks, which depends on the size of the small set chosen (or more exactly, on how often the chain visits the latter in a trajectory of finite length) and how sharp is the lower bound in the minorization condition: the larger the size of the small set is, the smaller the uniform lower bound for the transition density. This leads us to the following trade-off. Roughly speaking, for a given realization of the trajectory, as one increases the size of the small set S used for the data blocks construction, one naturally increases the number of points of the trajectory that are candidates for determining a block (*i.e.* a cut in the trajectory), but one also decreases the probability of cutting the trajectory (since the uniform lower bound for $\{p(x,y)\}_{(x,y)\in S^2}$ then decreases). This gives an insight into the fact that better numerical results for statistical procedures based on the ARB construction may be obtained in practice for some specific choices of the small set, likely for choices corresponding to a maximum expected number of data blocks given the trajectory, that is

$$N_n(S) = \mathbb{E}_\nu(\sum_{i=1}^{n} \mathbb{I}\{X_i \in S, Y_i = 1\} \,|X^{(n+1)}) \,.$$

Hence, when no prior information about the structure of the chain is available, here is a practical data-driven method for selecting the minorization condition parameters in the case when the chain takes real values. Consider a collection $\mathcal{S}$ of borelian sets S (typically compact intervals) and denote by $\mathcal{U}_S(dy) = \gamma_S(y).\lambda(dy)$ the uniform distribution on S, where $\gamma_S(y) = \mathbb{I}\{y \in S\}/\lambda(S)$ and λ is the Lebesgue measure on $\mathbb{R}$. Now, for any $S \in \mathcal{S}$, set $\delta(S) = \lambda(S). \inf_{(x,y)\in S^2} p(x,y)$. We have for any x, y in S, $p(x,y) \geq \delta(S)\gamma_S(y)$. In the case when $\delta(S) > 0$, the ideal criterion to optimize may be then expressed as

$$N_n(S) = \frac{\delta(S)}{\lambda(S)} \sum_{i=1}^{n} \frac{\mathbb{I}\{(X_i, X_{i+1}) \in S^2\}}{p(X_i, X_{i+1})}. \tag{3}$$

However, as the transition kernel $p(x,y)$ and its minimum over S^2 are unknown, a practical empirical criterion is obtained by replacing $p(x,y)$ by an estimate $p_n(x,y)$ and $\delta(S)$ by a lower bound $\delta_n(S)$ for $\lambda(S).p_n(x,y)$ over S^2 in expression (3). Once $p_n(x,y)$ is computed, calculate $\delta_n(S) = \lambda(S). \inf_{(x,y)\in S^2} p_n(x,y)$ and maximize thus the empirical criterion over $S \in \mathcal{S}$

$$\widehat{N}_n(S) = \frac{\delta_n(S)}{\lambda(S)} \sum_{i=1}^{n} \frac{\mathbb{I}\{(X_i, X_{i+1}) \in S^2\}}{p_n(X_i, X_{i+1})}. \tag{4}$$

More specifically, one may easily check at hand on many examples of real valued chains (see § 4.3 for instance), that any compact interval $V_{x_0}(\varepsilon) = [x_0 - \varepsilon, x_0 + \varepsilon]$ for some well chosen $x_0 \in \mathbb{R}$ and $\varepsilon > 0$ small enough, is a small set, choosing γ as the density of the uniform distribution on $V_{x_0}(\varepsilon)$. For practical purpose, one may fix x_0 and perform the optimization over $\varepsilon > 0$ only (see Bertail & Clémençon (2004c)) but both x_0 and ε may be considered as tuning parameters. A possible numerically feasible selection rule could rely then on searching for (x_0, ε) on a given pre-selected grid $\mathcal{G} = \{(x_0(k), \varepsilon(l)), 1 \leq k \leq K, 1 \leq l \leq L\}$ such that $\inf_{(x,y)\in V_{x_0}(\varepsilon)^2} p_n(x,y) > 0$ for any $(x_0, \varepsilon) \in \mathcal{G}$.

Algorithm 3 *(ARB construction with empirical choice of the small set)*

1. *Compute an estimator $p_n(x,y)$ of $p(x,y)$.*
2. *For any $(x_0, \varepsilon) \in \mathcal{G}$, compute the estimated expected number of pseudo-regenerations:*

$$\widehat{N}_n(x_0, \varepsilon) = \frac{\delta_n(x_0, \varepsilon)}{2\varepsilon} \sum_{i=1}^{n} \frac{\mathbb{I}\{(X_i, X_{i+1}) \in V_{x_0}(\varepsilon)^2\}}{p_n(X_i, X_{i+1})},$$

with $\delta_n(x_0, \varepsilon) = 2\varepsilon. \inf_{(x,y)\in V_{x_0}(\varepsilon)^2} p_n(x,y)$.
3. *Pick (x_0^*, ε^*) in $\mathcal{G}$ maximizing $\widehat{N}_n(x_0, \varepsilon)$ over $\mathcal{G}$, corresponding to the set $S^* = [x_0^* - \varepsilon^*,\ x_0^* + \varepsilon^*]$ and the minorization constant $\delta_n^* = \delta_n(x_0^*, \varepsilon^*)$.*
4. *Apply Algorithm 2 for ARB construction using S^*, δ_n^* and p_n.*

Remark 3. Numerous consistent estimators of the transition density of Harris chains have been proposed in the literature. Refer to Roussas (1969, 91a, 91b), Rosenblatt (1970), Birgé (1983), Doukhan & Ghindès (1983), Prakasa

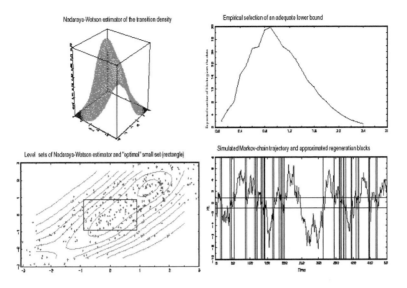

Fig. 3. Illustration of Algorithm 3 : ARB construction with empirical choice of the small set.

Rao (1983), Athreya & Atuncar (1998) or Clémençon (2000) for instance in positive recurrent cases, Karlsen & Tjøstheim (2001) in specific null recurrent cases.

This method is illustrated by Fig. 3 in the case of an $AR(1)$ model: $X_{i+1} = \alpha X_i + \varepsilon_{i+1}$, $i \in \mathbb{N}$, with $\varepsilon_i \overset{i.i.d.}{\sim} \mathcal{N}(0,1)$, $\alpha = 0.95$ and $X_0 = 0$, for a trajectory of length $n = 200$. Taking $x_0 = 0$ and letting ε grow, the expected number regeneration blocks is maximum for ε^* close to 0.9. The true minimum value of $p(x,y)$ over the corresponding square is actually $\delta = 0.118$. The first graphic in this panel shows the *Nadaraya-Watson estimator*

$$p_n(x,y) = \frac{\sum_{i=1}^n K(h^{-1}(x - X_i)) K(h^{-1}(y - X_{i+1}))}{\sum_{i=1}^n K(h^{-1}(x - X_i))} ,$$

computed from the gaussian kernel $K(x) = (2\pi)^{-1} \exp(-x^2/2)$ with an optimal bandwidth h of order $n^{-1/5}$. The second one plots $\widehat{N}_n(\varepsilon)$ as a function of ε. The next one indicates the set S^* corresponding to our empirical selection rule, while the last one displays the "optimal" ARB construction.

Note finally that other approaches may be considered for determining practically small sets and establishing accurate minorization conditions, which conditions do not necessarily involve uniform distributions besides. Refer for instance to Roberts & Rosenthal (1996) for Markov diffusion processes.

A two-split version of the ARB construction

When carrying out the theoretical study of statistical methods based on the ARB construction, one must deal with difficult problems arising from the dependence structure in the set of the resulting data blocks, due to the preliminary estimation step. Such difficulties are somehow similar as the ones that one traditionally faces in a semiparametric framework, even in the i.i.d. setting. The first step of semiparametric methodologies usually consists in a preliminary estimation of some infinite dimensional nuisance parameter (typically a density function or a nonparametric curve), on which the remaining (parametric) steps of the procedure are based. For handling theoretical difficulties related to this dependence problem, a well known method, called the *splitting trick*, amounts to split the data set into two parts, the first subset being used for estimating the nuisance parameter, while the parameter of interest is then estimated from the other subset (using the preliminary estimate). An analogous principle may be implemented in our framework using an additional split of the data in the "middle of the trajectory", for ensuring that a regeneration at least occurs in between with an overwhelming probability (so as to get two independent data subsets, see step 2 in the algorithm below). For this reason, we consider the following variant of the ARB construction. Let $1 < m < n$, $1 \leq p < n - m$.

> **Algorithm 4** *(two-split ARB construction)*
>
> 1. *From the data $X^{(n+1)} = (X_1, ..., X_{n+1})$, keep only the first m observations $X^{(m)}$ for computing an estimate $p_m(x, y)$ of $p(x, y)$ such that $p_m(x, y) \geq \delta \gamma(y)$, $\lambda(dy)$ a.s. and $p_m(X_i, X_{i+1}) > 0$, $1 \leq i \leq n - 1$.*
> 2. *Drop the observations between time $m + 1$ and time $m^* = m + p$ (under standard assumptions, the split chain regenerates once at least between these times with large probability).*
> 3. *From remaining observations $X^{(m^*, n)} = (X_{m^*+1}, ..., X_n)$ and estimate p_m, apply steps 2-4 of **Algorithm 2** (respectively of **Algorithm 3**).*

This procedure is similar to the *2-split method* proposed in Schick (2001), except that here the number of deleted observations is arbitrary and easier to interpret in terms of regeneration. Of course, the more often the split chain regenerates, the smaller p may be chosen. And the main problem consists in picking $m = m_n$ so that $m_n \to \infty$ as $n \to \infty$ for the estimate of the transition kernel to be accurate enough, while keeping enough observation $n - m^*$ for the block construction step: one typically chooses $m = o(n)$ as $n \to \infty$. Further assumptions are required for investigating precisely how to select m. In Bertail & Clémençon (2004d), a choice based on the rate of convergence α_m of the estimator $p_m(x, y)$ (for the MSE when error is measured by the sup-norm over $S \times S$, see assumption $\mathcal{H}_2$ in § 4.2) is proposed: when considering smooth markovian models for instance, estimators with rate $\alpha_m = m^{-1} \log(m)$ may be exhibited and one shows that $m = n^{2/3}$ is then an optimal choice (up to a $\log(n)$). However, one may argue, as in the semiparametric case, that this methodology is motivated by our limitations in the analysis of asymptotic properties of the estimators only, whereas from a practical viewpoint it may deteriorate the finite sample performance of the initial algorithm. To our own experience, it is actually better to construct the estimate $p(x, y)$ from the whole trajectory and the interest of **Algorithm 4** is mainly theoretical.

4 Mean and variance estimation

In this section, we suppose that the chain X is positive recurrent with unknown stationary probability μ and consider the problem of estimating an additive functional of type $\mu(f) = \int f(x)\mu(dx) = \mathbb{E}_\mu(f(X_1))$, where f is a μ-integrable real valued function defined on the state space $(E, \mathcal{E})$. Estimation of additive functionals of type $\mathbb{E}_\mu(F(X_1, ..., X_k))$, for fixed $k \geq 1$, may be investigated in a similar fashion. We set $\overline{f}(x) = f(x) - \mu(f)$.

4.1 Regenerative case

Here we assume further that X admits an *a priori* known accessible atom A. As in the i.i.d. setting, a natural estimator of $\mu(f)$ is the sample mean statistic,

$$\mu'_n(f) = n^{-1}\sum_{i=1}^{n} f(X_i) \ . \tag{5}$$

When the chain is stationary (*i.e.* when $\nu = \mu$), the estimator $\mu'_n(f)$ is zero-bias. However, its bias is significant in all other cases, mainly because of the presence of the first and last (non-regenerative) data blocks $\mathcal{B}_0$ and $\mathcal{B}_{l_n}^{(n)}$ (see Proposition 4.1 below). Besides, by virtue of Theorem 2.1, $\mu(f)$ may be expressed as the mean of the $f(X_i)$'s over a regeneration cycle (renormalized by the mean length of a regeneration cycle)

$$\mu(f) = \mathbb{E}_A(\tau_A)^{-1}\mathbb{E}_A(\sum_{i=1}^{\tau_A} f(X_i)) \ .$$

Because the bias due to the first block depends on the unknown initial distribution (see Proposition 1 below) and thus can not be consistently estimated, we suggest to introduce the following estimators of the mean $\mu(f)$. Define the sample mean based on the observations (eventually) collected after the first regeneration time only by

$$\widetilde{\mu}_n(f) = (n - \tau_A)^{-1} \sum_{i=1+\tau_A}^{n} f(X_i)$$

with the convention $\widetilde{\mu}_n(f) = 0$, when $\tau_A > n$, as well as the sample mean based on the observations collected between the first and last regeneration times before n by

$$\bar{\mu}_n(f) = (\tau_A(l_n) - \tau_A)^{-1} \sum_{i=1+\tau_A}^{\tau_A(l_n)} f(X_i)$$

with $l_n = \sum_{i=1}^{n} \mathbb{I}\{X_i \in A\}$ and the convention $\bar{\mu}_n(f) = 0$, when $l_n \leq 1$ (observe that, by Markov's inequality, $\mathbb{P}_\nu(l_n \leq 1) = O(n^{-1})$ as $n \to \infty$, as soon as $\mathcal{H}_0(1, \nu)$ and $\mathcal{H}_0(2)$ are fulfilled).

Let us introduce some additional notation for the block sums (resp. the block lengths), that shall be used here and throughout. For $j \geq 1$, $n \geq 1$, set

$$L_0 = \tau_A, \ L_j = \tau_A(j+1) - \tau_A(j), \ L_{l_n}^{(n)} = n - \tau_A(l_n)$$

$$f(\mathcal{B}_0) = \sum_{i=1}^{\tau_A} f(X_i), \ f(\mathcal{B}_j) = \sum_{i=1+\tau_A(j)}^{\tau_A(j+1)} f(X_i), \ f(\mathcal{B}_{l_n}^{(n)}) = \sum_{i=1+\tau_A(l_n)}^{n} f(X_i) \ .$$

With these notations, the estimators above may be rewritten as

$$\mu_n'(f) = \frac{f(\mathcal{B}_0) + \sum_{j=1}^{l_n} f(\mathcal{B}_j) + f(\mathcal{B}_{l_n}^{(n)})}{L_0 + \sum_{j=1}^{l_n} L_j + L_{l_n}^{(n)}},$$

$$\widetilde{\mu}_n(f) = \frac{\sum_{j=1}^{l_n} f(\mathcal{B}_j) + f(\mathcal{B}_{l_n}^{(n)})}{\sum_{j=1}^{l_n} L_j + L_{l_n}^{(n)}}, \ \bar{\mu}_n(f) = \frac{\sum_{j=1}^{l_n} f(\mathcal{B}_j)}{\sum_{j=1}^{l_n} L_j} \ .$$

Let $\mu_n(f)$ designs any of the three estimators $\mu_n'(f)$, $\widetilde{\mu}_n(f)$ or $\bar{\mu}_n(f)$. If X fulfills conditions $\mathcal{H}_0(2)$, $\mathcal{H}_0(2, \nu)$, $\mathcal{H}_1(f, 2, A)$, $\mathcal{H}_1(f, 2, \nu)$ then the following CLT holds under $\mathbb{P}_\nu$ (cf Theorem 17.2.2 in Meyn & Tweedie (1996))

$$n^{1/2}\sigma^{-1}(f)(\mu_n(f) - \mu(f)) \Rightarrow \mathcal{N}(0, 1), \text{ as } n \to \infty \ ,$$

with a normalizing constant

$$\sigma^2(f) = \mu(A) \, \mathbb{E}_A((\sum_{i=1}^{\tau_A} f(X_i) - \mu(f)\tau_A)^2) \ . \tag{6}$$

From this expression we propose the following estimator of the asymptotic variance, adopting the usual convention regarding to empty summation,

$$\sigma_n^2(f) = n^{-1} \sum_{j=1}^{l_n-1} (f(\mathcal{B}_j) - \bar{\mu}_n(f)L_j)^2. \tag{7}$$

Notice that the first and last data blocks are not involved in its construction. We could have proposed estimators involving different estimates of $\mu(f)$, but as will be seen later, it is preferable to consider an estimator based on regeneration blocks only. The following quantities shall be involved in the statistical analysis below. Define

$$\alpha = \mathbb{E}_A(\tau_A), \ \beta = \mathbb{E}_A(\tau_A \sum_{i=1}^{\tau_A} \overline{f}(X_i)) = \text{cov}_A(\tau_A, \sum_{i=1}^{\tau_A} \overline{f}(X_i)) \ ,$$

$$\varphi_\nu = \mathbb{E}_\nu(\sum_{i=1}^{\tau_A} \overline{f}(X_i)), \ \gamma = \alpha^{-1} \mathbb{E}_A(\sum_{i=1}^{\tau_A} (\tau_A - i)\overline{f}(X_i)) \ .$$

We also introduce the following technical conditions.

(C1) (*Cramer condition*)

$$\varlimsup_{t \to \infty} | \mathbb{E}_A(\exp(it \sum_{i=1}^{\tau_A} \overline{f}(X_i))) | < 1 \ .$$

(C2) (*Cramer condition*)

$$\varlimsup_{t \to \infty} | \mathbb{E}_A(\exp(it(\sum_{i=1}^{\tau_A} \overline{f}(X_i))^2)) | < 1 \ .$$

(C3) *There exists $N \geq 1$ such that the N-fold convoluted density g^{*N} is bounded, denoting by g the density of the $(\sum_{i=1+\tau_A(1)}^{\tau_A(2)} \overline{f}(X_i) - \alpha^{-1}\beta)^2$'s.*

(C4) *There exists $N \geq 1$ such that the N-fold convoluted density G^{*N} is bounded, denoting by G the density of the $(\sum_{i=1+\tau_A(1)}^{\tau_A(2)} \overline{f}(X_i))^2$'s.*

These two conditions are automatically satisfied if $\sum_{i=1+\tau_A(1)}^{\tau_A(2)} \overline{f}(X_i)$ has a bounded density.

The result below is a straightforward extension of Theorem 1 in Malinovskii (1985) (see also Proposition 3.1 in Bertail & Clémençon (2004a)).

Proposition 1. *Suppose that $\mathcal{H}_0(4)$, $\mathcal{H}_0(2, \nu)$, $\mathcal{H}_1(4, f)$, $\mathcal{H}_1(2, \nu, f)$ and Cramer condition (C1) are satisfied by the chain. Then, as $n \to \infty$, we have*

$$\mathbb{E}_\nu(\mu'_n(f)) = \mu(f) + (\varphi_\nu + \gamma - \beta/\alpha)n^{-1} + O(n^{-3/2}) \ , \qquad (8)$$

$$\mathbb{E}_\nu(\widetilde{\mu}_n(f)) = \mu(f) + (\gamma - \beta/\alpha)n^{-1} + O(n^{-3/2}) \ , \qquad (9)$$

$$\mathbb{E}_\nu(\overline{\mu}_n(f)) = \mu(f) - (\beta/\alpha)n^{-1} + O(n^{-3/2}) \ . \qquad (10)$$

If the Cramer condition (C2) is also fulfilled, then

$$\mathbb{E}_\nu(\sigma_n^2(f)) = \sigma^2(f) + O(n^{-1}) \ , \ \text{as } n \to \infty, \qquad (11)$$

and we have the following CLT under $\mathbb{P}_\nu$,

$$n^{1/2}(\sigma_n^2(f) - \sigma^2(f)) \Rightarrow \mathcal{N}(0, \xi^2(f)) \ , \ \text{as } n \to \infty \ , \qquad (12)$$

with $\xi^2(f) = \mu(A)\text{var}_A((\sum_{i=1}^{\tau_A} \overline{f}(X_i))^2 - 2\alpha^{-1}\beta \sum_{i=1}^{\tau_A} \overline{f}(X_i))$.

Proof. The proof of (8)-(11) is given in Bertail & Clémençon (2004a) and the linearization of $\sigma_n^2(f)$ follows from their Lemma 6.3

$$\sigma_n^2(f) = n^{-1} \sum_{j=1}^{l_n-1} g(\mathcal{B}_j) + r_n , \qquad (13)$$

with $g(\mathcal{B}_j) = \overline{f}(\mathcal{B}_j)^2 - 2\alpha^{-1}\beta\overline{f}(\mathcal{B}_j)$, for $j \geq 1$, and for some $\eta_1 > 0$, $\mathbb{P}_\nu(nr_n > \eta_1 \log(n)) = O(n^{-1})$, as $n \to \infty$. We thus have, as $n \to \infty$,

$$n^{1/2}(\sigma_n^2(f) - \sigma^2(f)) = (l_n/n)^{1/2}l_n^{-1/2} \sum_{j=1}^{l_n-1} (g(\mathcal{B}_j) - \mathbb{E}(g(\mathcal{B}_j)) + o_{\mathbb{P}_\nu}(1) ,$$

and (13) is established with the same argument as for Theorem 17.3.6 in Meyn & Tweedie (1996), as soon as $\mathrm{var}(g(\mathcal{B}_j)) < \infty$, that is ensured by assumption $\mathcal{H}_1(4, f)$.   □

Remark 4. We emphasize that in a non i.i.d. setting, it is generally difficult to construct an accurate (positive) estimator of the asymptotic variance. When no structural assumption, except stationarity and square integrability, is made on the underlying process X, a possible method, currently used in practice, is based on so-called *blocking techniques*. Indeed under some appropriate mixing conditions (which ensure that the following series converge), it can be shown that the variance of $n^{-1/2}\mu_n'(f)$ may be written

$$\mathrm{var}(n^{-1/2}\mu_n'(f)) = \Gamma(0) + 2\sum_{t=1}^{n}(1 - t/n)\Gamma(t)$$

and converges to

$$\sigma^2(f) = \sum_{t=\infty}^{\infty} \Gamma(t) = 2\pi g(0) ,$$

where $g(w) = (2\pi)^{-1}\sum_{t=-\infty}^{\infty} \Gamma(t)\cos(wt)$ and $(\Gamma(t))_{t\geq 0}$ denote respectively the spectral density and the autocovariance sequence of the discrete-time stationary process X. Most of the estimators of $\sigma^2(f)$ that have been proposed in the literature (such as the Bartlett spectral density estimator, the moving-block jackknife/subsampling variance estimator, the overlapping or non-overlapping batch means estimator) may be seen as variants of the basic *moving-block bootstrap estimator*(see Künsch (1989), Liu and Singh(1992))

$$\hat{\sigma}_{M,n}^2 = \frac{M}{Q} \sum_{i=1}^{Q}(\overline{\mu}_{i,M,L} - \mu_n(f))^2 , \qquad (14)$$

where $\overline{\mu}_{i,M,L} = M^{-1}\sum_{t=L(i-1)+1}^{L(i-1)+M} f(X_t)$ is the mean of f on the i-th data block $(X_{L(i-1)+1}, \ldots, X_{L(i-1)+M})$. Here, the size M of the blocks and the amount L of 'lag' or overlap between each block are deterministic (eventually depending on n) and $Q = [\frac{n-M}{L}] + 1$, denoting by $[\cdot]$ the integer part, is the number of blocks that may be constructed from the sample $X_1, ..., X_n$. In the case when $L = M$, there is no overlap between block i and block $i + 1$ (as the original solution considered by Hall (1985), Carlstein (1986)), whereas the case $L = 1$ corresponds to maximum overlap (see Politis & Romano (1992), Politis et al. (2000) for a survey). Under suitable regularity conditions (mixing and moments conditions), it can be shown that if $M \to \infty$ with $M/n \to 0$ and $L/M \to a \in [0, 1]$ as $n \to \infty$, then we have

$$\mathbb{E}(\hat{\sigma}^2_{M,n}) - \sigma^2(f) = O(1/M) + O(\sqrt{M/n}) , \qquad (15)$$

$$\mathrm{var}(\hat{\sigma}^2_{M,n}) = 2c\frac{M}{n}\sigma^4(f) + o(M/n) ,$$

as $n \to \infty$, where c is a constant depending on a, taking its smallest value (namely $c = 2/3$) for $a = 0$. This result shows that the bias of such estimators may be very large. Indeed, by optimizing in M we find the optimal choice $M \sim n^{1/3}$, for which we have $\mathbb{E}(\hat{\sigma}^2_{M,n}) - \sigma^2(f) = O(n^{-1/3})$. Various extrapolation and jackknife techniques or kernel smoothing methods have been suggested to get rid of this large bias (refer to Politis & Romano (1992), Götze & Künsch (1996), Bertail (1997) and Bertail & Politis (2001)). The latter somehow amount to make use of Rosenblatt smoothing kernels of order higher than two (taking some negative values) for estimating the spectral density at 0. However, the main drawback in using these estimators is that they take negative values for some n, and lead consequently to face problems, when dealing with studentized statistics.

In our specific Markovian framework, the estimate $\sigma^2_n(f)$ in the atomic case (or latter $\hat{\sigma}^2_n(f)$ in the general case) is much more natural and allows to avoid these problems. This is particularly important when the matter is to establish Edgeworth expansions at orders higher than two in such a non i.i.d. setting. As a matter of fact, the bias of the variance may completely cancel the accuracy provided by higher order Edgeworth expansions (but also the one of its Bootstrap approximation) in the studentized case, given its explicit role in such expansions (see Götze & Künsch (1996)).

From Proposition 1, we immediately derive that

$$t_n = n^{1/2}\sigma_n^{-1}(f)(\mu_n(f) - \mu(f)) \Rightarrow \mathcal{N}(0, 1) , \text{ as } n \to \infty ,$$

so that asymptotic confidence intervals for $\mu(f)$ are immediately available in the atomic case. This result also shows that using estimators $\widetilde{\mu}_n(f)$ or $\overline{\mu}_n(f)$ instead of $\mu'_n(f)$ allows to eliminate the only quantity depending on the initial distribution ν in the first order term of the bias, which may be interesting for estimation purpose and is crucial when the matter is to deal with an estimator of which variance or sampling distribution may be approximated by a resampling procedure in a nonstationary setting (given the impossibility to approximate the distribution of the "first block sum" $\sum_{i=1}^{\tau_A} f(X_i)$ from one single realization of X starting from ν). For these estimators, it is actually possible to implement specific Bootstrap methodologies, for constructing second order correct confidence intervals for instance (see Bertail & Clémençon (2004b, c) and section 5). Regarding to this, it should be noticed that Edgeworth expansions (E.E. in abbreviated form) may be obtained using the regenerative method by partitioning the state space according to all possible values for the number l_n regeneration times before n and for the sizes of the first and last block as in Malinovskii (1987). Bertail & Clémençon (2004a) proved the validity of an E.E. in the studentized case, of which form is recalled below. Notice that actually (C3) corresponding to their v) in Proposition 3.1 in Bertail & Clémençon (2004a) is not needed in the unstudentized case. Let $\Phi(x)$ denote the distribution function of the standard normal distribution and set $\phi(x) = d\Phi(x)/dx$.

Theorem 2. *Let* $b(f) = \lim_{n\to\infty} n(\mu_n(f) - \mu(f))$ *be the asymptotic bias of* $\mu_n(f)$. *Under conditions* $\mathcal{H}_0(4)$, $\mathcal{H}_0(2, \nu)$ $\mathcal{H}_1(4, f)$, $\mathcal{H}_1(2, \nu, f)$, *(C1), we have the following E.E.,*

$$\sup_{x\in\mathbb{R}} |\mathbb{P}_\nu \left(n^{1/2}\sigma(f)^{-1}(\mu_n(f) - \mu(f)) \le x \right) - E_n^{(2)}(x)| = O(n^{-1}) \,,$$

as $n \to \infty$, *with*

$$E_n^{(2)}(x) = \Phi(x) - n^{-1/2}\frac{k_3(f)}{6}(x^2 - 1)\phi(x) - n^{-1/2}b(f)\phi(x) \,,$$

$$k_3(f) = \alpha^{-1}(M_{3,A} - \frac{3\beta}{\sigma(f)}) \,, \quad M_{3,A} = \frac{\mathbb{E}_A((\sum_{i=1}^{\tau_A} \overline{f}(X_i))^3)}{\sigma(f)^3} \,.$$

A similar limit result holds for the studentized statistic under the further hypothesis that (C2), (C3), $\mathcal{H}_0(s)$ *and* $\mathcal{H}_1(s, f)$ *are fulfilled with* $s = 8 + \varepsilon$ *for some* $\varepsilon > 0$:

$$\sup_{x\in\mathbb{R}} |\mathbb{P}_\nu(n^{1/2}\sigma_n^{-1}(f)(\mu_n(f) - \mu(f)) \le x) - F_n^{(2)}(x)| = O(n^{-1}\log(n)) \,,$$

as $n \to \infty$, *with*

$$F_n^{(2)}(x) = \Phi(x) + n^{-1/2}\frac{1}{6}k_3(f)(2x^2 + 1)\phi(x) - n^{-1/2}b(f)\phi(x) \,.$$

When $\mu_n(f) = \overline{\mu}_n(f)$, *under C4,* $O(n^{-1}\log(n))$ *may be replaced by* $O(n^{-1})$.

This theorem may serve for building accurate confidence intervals for $\mu(f)$ (by E.E. inversion as in Abramovitz & Singh (1983) or Hall (1983)). It also paves the way for studying precisely specific bootstrap methods, as in Bertail & Clémençon (2004c). It should be noted that the skewness $k_3(f)$ is the sum of two terms: the third moment of the recentered block sums and a correlation term between the block sums and the block lengths. The coefficients involved in the E.E. may be directly estimated from the regenerative blocks. Once again by straightforward CLT arguments, we have the following result.

Proposition 2. *For $s \geq 1$, under $\mathcal{H}_1(f, 2s)$, $\mathcal{H}_1(f, 2, \nu)$, $\mathcal{H}_0(2s)$ and $\mathcal{H}_0(2, \nu)$, then $M_{s,A} = \mathbb{E}_A((\sum_{i=1}^{\tau_A} \overline{f}(X_i))^s)$ is well-defined and we have*

$$\widehat{\mu}_{s,n} = n^{-1} \sum_{i=1}^{l_n-1} (f(\mathcal{B}_j) - \overline{\mu}_n(f)L_j)^s = \alpha^{-1} M_{s,A} + O_{\mathbb{P}_\nu}(n^{-1/2}), \text{ as } n \to \infty.$$

4.2 Positive recurrent case

We now turn to the general positive recurrent case (refer to § 2.3 for assumptions and notation). It is noteworthy that, though they may be expressed using the parameters of the minorization condition $\mathcal{M}$, the constants involved in the CLT are independent from these latter. In particular the mean and the asymptotic variance may be written as

$$\mu(f) = \mathbb{E}_{A_\mathcal{M}}(\tau_{A_\mathcal{M}})^{-1} \mathbb{E}_{A_\mathcal{M}}\left(\sum_{i=1}^{\tau_{A_\mathcal{M}}} f(X_i)\right),$$

$$\sigma^2(f) = \mathbb{E}_{A_\mathcal{M}}(\tau_{A_\mathcal{M}})^{-1} \mathbb{E}_{A_\mathcal{M}}\left(\left(\sum_{i=1}^{\tau_{A_\mathcal{M}}} \overline{f}(X_i)\right)^2\right),$$

where $\tau_{A_\mathcal{M}} = \inf\{n \geq 1, (X_n, Y_n) \in S \times \{1\}\}$ and $\mathbb{E}_{A_\mathcal{M}}(.)$ denotes the expectation conditionally to $(X_0, Y_0) \in A_\mathcal{M} = S \times \{1\}$. However, one cannot use the estimators of $\mu(f)$ and $\sigma^2(f)$ defined in the atomic setting, applied to the split chain, since the times when the latter regenerates are unobserved. We thus consider the following estimators based on the *approximate regeneration times* (*i.e.* times i when $(X_i, \widehat{Y}_i) \in S \times \{1\}$), as constructed in § 3.2,

$$\widehat{\mu}_n(f) = \widehat{n}_{A_\mathcal{M}}^{-1} \sum_{j=1}^{\widehat{l}_n-1} f(\widehat{\mathcal{B}}_j) \text{ and } \widehat{\sigma}_n^2(f) = \widehat{n}_{A_\mathcal{M}}^{-1} \sum_{j=1}^{\widehat{l}_n-1} \{f(\widehat{\mathcal{B}}_j) - \widehat{\mu}_n(f)\widehat{L}_j\}^2,$$

with, for $j \geq 1$,

$$f(\widehat{\mathcal{B}}_j) = \sum_{i=1+\widehat{\tau}_{A_\mathcal{M}}(j)}^{\widehat{\tau}_{A_\mathcal{M}}(j+1)} f(X_i), \ \widehat{L}_j = \widehat{\tau}_{A_\mathcal{M}}(j+1) - \widehat{\tau}_{A_\mathcal{M}}(j),$$

$$\widehat{n}_{A_\mathcal{M}} = \widehat{\tau}_{A_\mathcal{M}}(\widehat{l}_n) - \widehat{\tau}_{A_\mathcal{M}}(1) = \sum_{j=1}^{\widehat{l}_n-1} \widehat{L}_j.$$

By convention, $\widehat{\mu}_n(f) = 0$ and $\widehat{\sigma}_n^2(f) = 0$ (resp. $\widehat{n}_{A_\mathcal{M}} = 0$), when $\widehat{l}_n \leq 1$ (resp., when $\widehat{l}_n = 0$). Since the ARB construction involves the use of an estimate $p_n(x, y)$ of the transition kernel $p(x, y)$, we consider conditions on the rate of convergence of this estimator. For a sequence of nonnegative real numbers $(\alpha_n)_{n \in \mathbb{N}}$ converging to 0 as $n \to \infty$,

$\mathcal{H}_2$: $p(x, y)$ *is estimated by* $p_n(x, y)$ *at the rate* α_n *for the MSE when error is measured by the* L^∞ *loss over* $S \times S$:

$$\mathbb{E}_\nu(\sup_{(x,y)\in S\times S} |p_n(x,y) - p(x,y)|^2) = O(\alpha_n) \text{ , as } n \to \infty \text{ .}$$

See Remark 3.1 for references concerning the construction and the study of transition density estimators for positive recurrent chains, estimation rates are usually established under various smoothness assumptions on the density of the joint distribution $\mu(dx)\Pi(x, dy)$ and the one of $\mu(dx)$. For instance, under classical Hölder constraints of order s, the typical rate for the risk in this setup is $\alpha_n \sim (\ln n/n)^{s/(s+1)}$ (refer to Clémençon (2000)).

$\mathcal{H}_3$: *The "minorizing" density* γ *is such that* $\inf_{x\in S} \gamma(x) > 0$.

$\mathcal{H}_4$: *The transition density* $p(x, y)$ *and its estimate* $p_n(x, y)$ *are bounded by a constant* $R < \infty$ *over* S^2.

Some asymptotic properties of these statistics based on the approximate regeneration data blocks are stated in the following theorem (their proof is omitted since it immediately follows from the argument of Theorem 3.2 and Lemma 5.3 in Bertail & Clémençon (2004c)),

Theorem 3. *If assumptions* $\mathcal{H}'_0(2, \nu)$, $\mathcal{H}'_0(8)$, $\mathcal{H}'_1(f, 2, \nu)$, $\mathcal{H}'_1(f, 8)$, $\mathcal{H}_2$, $\mathcal{H}_3$ *and* $\mathcal{H}_4$ *are satisfied by* X, *as well as conditions (C1) and (C2) by the split chain, we have, as* $n \to \infty$,

$$\mathbb{E}_\nu(\widehat{\mu}_n(f)) = \mu(f) - \beta/\alpha \, n^{-1} + O(n^{-1}\alpha_n^{1/2}) \text{ ,}$$
$$\mathbb{E}_\nu(\widehat{\sigma}_n^2(f)) = \sigma^2(f) + O(\alpha_n \vee n^{-1}) \text{ ,}$$

and if $\alpha_n = o(n^{-1/2})$, *then*

$$n^{1/2}(\widehat{\sigma}_n^2(f) - \sigma^2(f)) \Rightarrow \mathcal{N}(0, \xi^2(f))$$

where α, β *and* $\xi^2(f)$ *are the quantities related to the split chain defined in Proposition 4.1* .

Remark 5. The condition $\alpha_n = o(n^{-1/2})$ as $n \to \infty$ may be ensured by smoothness conditions satisfied by the transition kernel $p(x, y)$: under Hölder constraints of order s such rates are achieved as soon as $s > 1$, that is a rather weak assumption.

We also define the *pseudo-regeneration based standardized* (resp., *studentized*) *sample mean by*

$$\widehat{\varsigma}_n = n^{1/2}\sigma^{-1}(f)(\widehat{\mu}_n(f) - \mu(f)) \text{ ,}$$
$$\widehat{t}_n = \widehat{n}_{A_\mathcal{M}}^{1/2} \widehat{\sigma}_n(f)^{-1}(\widehat{\mu}_n(f) - \mu(f)) \text{ .}$$

The following theorem straightforwardly results from Theorem 3.

Theorem 4. *Under the assumptions of Theorem 3, we have as* $n \to \infty$

$$\widehat{\varsigma}_n \Rightarrow \mathcal{N}(0,1) \ and \ \widehat{t}_n \Rightarrow \mathcal{N}(0,1) \ .$$

This shows that from pseudo-regeneration blocks one may easily construct a consistent estimator of the asymptotic variance $\sigma^2(f)$ and asymptotic confidence intervals for $\mu(f)$ in the general positive recurrent case (see Section 5 for more accurate confidence intervals based on a regenerative bootstrap method). In Bertail & Clémençon (2004a), an E.E. is proved for the studentized statistic $\widehat{t}_n$. The main problem consists in handling computational difficulties induced by the dependence structure, that results from the preliminary estimation of the transition density. For partly solving this problem, one may use **Algorithm 4**, involving the *2-split trick*. Under smoothness assumptions for the transition kernel (which are often fulfilled in practice), Bertail & Clémençon (2004d) established the validity of the E.E. up to $O(n^{-5/6} \log(n))$, stated in the result below.

Theorem 5. *Suppose that (C1) is satisfied by the split chain, and that* $\mathcal{H}_0'(\kappa, \ \nu)$, $\mathcal{H}_1'(\kappa, \ f, \ \nu)$, $\mathcal{H}_0'(\kappa)$, $\mathcal{H}_1'(\kappa, \ f)$ *with* $\kappa > 6$, $\mathcal{H}_2$, $\mathcal{H}_3$ *and* $\mathcal{H}_4$ *are fulfilled. Let* m_n *and* p_n *be integer sequences tending to* ∞ *as* $n \to \infty$, *such that* $n^{1/\gamma} \leq p_n \leq m_n$ *and* $m_n = o(n)$ *as* $n \to \infty$. *Then, the following limit result holds for the pseudo-regeneration based standardized sample mean obtained via Algorithm 4*

$$\sup_{x \in \mathbb{R}} |\mathbb{P}_\nu \left(\widehat{\varsigma}_n \leq x \right) - E_n^{(2)}(x)| = O(n^{-1/2} \alpha_{m_n}^{1/2} \vee n^{-3/2} m_n) \ , \ as \ n \to \infty \ ,$$

and if in addition the preceding assumptions with $\kappa > 8$ *and C4) are satisfied, we also have*

$$\sup_{x \in \mathbb{R}} |\mathbb{P}_\nu \left(\widehat{t}_n \leq x \right) - F_n^{(2)}(x)| = O(n^{-1/2} \alpha_{m_n}^{1/2} \vee n^{-3/2} m_n) \ , \ as \ n \to \infty \ ,$$

where $E_n^{(2)}(x)$ *and* $F_n^{(2)}(x)$ *are the expansions defined in Theorem 4.2 related to the split chain. In particular, if* $\alpha_{m_n} = m_n \log(m_n)$, *by picking* $m_n = n^{2/3}$, *these E.E. hold up to* $O(n^{-5/6} \log(n))$.

The conditions stipulated in this result are weaker than the conditions ensuring that the Moving Block Bootstrap is second order correct. More precisely, they are satisfied for a wide range of Markov chains, including nonstationary cases and chains with polynomial decay of $\alpha-$mixing coefficients (*cf* remark 2.1) that do not fall into the validity framework of the MBB methodology. In particular it is worth noticing that these conditions are weaker than Götze & Hipp (1983)'s conditions (in a strong mixing setting).

As stated in the following proposition, the coefficients involved in the E.E.'s above may be estimated from the approximate regeneration blocks.

Proposition 3. *Under* $\mathcal{H}_0'(2s, \nu)$, $\mathcal{H}_1'(2s, \nu, f)$, $\mathcal{H}_0'(2s \vee 8)$, $\mathcal{H}_1'(2s \vee 8, f)$ *with* $s \geq 2$, $\mathcal{H}_2$, $\mathcal{H}_3$ *and* $\mathcal{H}_4$, *the expectation* $M_{s,A_\mathcal{M}} = \mathbb{E}_{A_\mathcal{M}}((\sum_{i=1}^{\tau_{A_\mathcal{M}}} \overline{f}(X_i))^s)$ *is well-defined and we have, as* $n \to \infty$,

$$\widehat{\mu}_{s,n} = n^{-1} \sum_{i=1}^{l_n-1} (f(\widehat{\mathcal{B}}_j) - \widehat{\mu}_n(f)\widehat{L}_j)^s = \mathbb{E}_{A_\mathcal{M}}(\tau_{A_\mathcal{M}})^{-1} M_{s,A_\mathcal{M}} + O_{\mathbb{P}_\nu}(\alpha_{m_n}^{1/2}) \ .$$

4.3 Some illustrative examples

Here we give some examples with the aim to illustrate the wide range of applications of the results previously stated.

Example 1 : countable Markov chains.

Let X be a general irreducible chain with a countable state space E. For such a chain, any recurrent state $a \in E$ is naturally an accessible atom and conditions involved in the limit results presented in § 4.1 may be easily checked at hand. Consider for instance Cramer condition (C1). Denote by Π the transition matrix and set $A = \{a\}$. Assuming that f is μ-centered. We have, for any $k \in \mathbb{N}^*$:

$$\left| \mathbb{E}_A(e^{it \sum_{j=1}^{\tau_A} f(X_j)}) \right| = \left| \sum_{l=1}^{\infty} \mathbb{E}_A(e^{it \sum_{j=1}^{l} f(X_j)} | \tau_A = l) \mathbb{P}_A(\tau_A = l) \right|$$

$$\leq \left| \mathbb{E}_A(e^{it \sum_{j=1}^{k} f(X_j)} | \tau_A = k) \right| \mathbb{P}_A(\tau_A = k) + 1 - \mathbb{P}_A(\tau_A = k) \ .$$

It follows that showing that (C1) holds may boil down to showing the partial conditional Cramer condition

$$\overline{\lim_{t \to \infty}} \left| \mathbb{E}_A(e^{it \sum_{j=1}^{k} f(X_j)} | \tau_A = k) \right| < 1 \ ,$$

for some $k > 0$ such that $\mathbb{P}_A(\tau_A = k) > 0$. In particular, similarly to the i.i.d. case, this condition then holds, as soon as the set $\{f(x)\}_{x \in E}$ is not a point lattice (*i.e.* it is not a regular grid). We point out that the expression obtained in Example 1 of Bertail & Clémençon (2004b) is clearly incorrect (it does not hold at $t = 0$): given that $\forall t \in \mathbb{R}$,

$$\mathbb{E}_A(e^{it \sum_{j=1}^{\tau_A} f(X_j)})$$

$$= \sum_{l=1}^{\infty} \sum_{x_1 \neq a, \dots, x_{l-1} \neq a} e^{it \sum_{j=1}^{l} f(x_j)} \pi(a, x_1)\pi(x_1, x_2)...\pi(x_{l-1}, a) \ ,$$

(C1) does not hold when f maps the state space to a point lattice.

Example 2 : modulated random walk on $\mathbb{R}_+$.

Consider the model

$$X_0 = 0 \text{ and } X_{n+1} = (X_n + W_n)_+ \text{ for } n \in \mathbb{N} , \tag{16}$$

where $x_+ = \max(x, 0)$, (X_n) and (W_n) are sequences of r.v.'s such that, for all $n \in \mathbb{N}$, the distribution of W_n conditionally to $X_0, ..., X_n$ is given by $U(X_n, .)$ where $U(x, w)$ is a transition kernel from $\mathbb{R}_+$ to $\mathbb{R}$. Then, X_n is a Markov chain on $\mathbb{R}_+$ with transition probability kernel $\Pi(x, dy)$ given by

$$\Pi(x, \ \{0\}) = U(x, \ ]-\infty, \ -x]) ,$$
$$\Pi(x, \ ]y, \ \infty[) = U(x, \ ]y - x, \ \infty[) ,$$

for all $x \geq 0$. Observe that the chain Π is δ_0-irreducible when $U(x, .)$ has infinite left tail for all $x \geq 0$ and that $\{0\}$ is then an accessible atom for X. The chain is shown to be positive recurrent iff there exists $b > 0$ and a test function $V : \mathbb{R}_+ \to [0, \infty]$ such that $V(0) < \infty$ and the drift condition below holds for all $x \geq 0$

$$\int \Pi(x, dy)V(y) - V(x) \leq -1 + b\mathbb{I}\{x = 0\} ,$$

(see in Meyn & Tweedie (1996). The times at which X reaches the value 0 are thus regeneration times, and allow to define regeneration blocks dividing the sample path, as shown in Fig. 1. Such a modulated random walk (for which, at each step n, the increasing W_n depends on the actual state $X_n = x$), provides a model for various systems, such as the popular *content-dependent storage process* studied in Harrison & Resnick (1976) (see also Brockwell *et al.* (1982)) or the *work-modulated single server queue* in the context of queuing systems (*cf* Browne & Sigman (1992)). For such atomic chains with continuous state space (refer to Meyn & Tweedie (1996), Feller (1968, 71) and Asmussen (1987) for other examples of such chains), one may easily check conditions used in § 3.1 in many cases. One may show for instance that (C1) is fulfilled as soon as there exists $k \geq 1$ such that $0 < \mathbb{P}_A(\tau_A = k) < 1$ and the distribution of $\sum_{i=1}^{k} f(X_i)$ conditioned on $X_0 \in A$ and $\tau_A = k$ is absolutely continuous. For the regenerative model described above, this sufficient condition is fulfilled with $k = 2$, $f(x) = x$ and $A = \{0\}$, when it is assumed for instance that $U(x, dy)$ is absolutely continuous for all $x \geq 0$ and $\emptyset \neq \text{supp} U(0, dy) \cap \mathbb{R}_+^* \neq \mathbb{R}_+^*$.

Example 3: nonlinear time series.

Consider the heteroskedastic autoregressive model

$$X_{n+1} = m(X_n) + \sigma(X_n)\varepsilon_{n+1}, \ n \in \mathbb{N} ,$$

where $m : \mathbb{R} \to \mathbb{R}$ and $\sigma : \mathbb{R} \to \mathbb{R}_+^*$ are measurable functions, $(\varepsilon_n)_{n \in \mathbb{N}}$ is a i.i.d. sequence of r.v.'s drawn from $g(x)dx$ such that, for all $n \in \mathbb{N}$, ε_{n+1} is independent from the X_k's, $k \leq n$ with $\mathbb{E}(\varepsilon_{n+1}) = 0$ and $\mathbb{E}(\varepsilon_{n+1}^2) = 1$. The transition kernel density of the chain is given by $p(x, y) = \sigma(x)^{-1} g((y - m(x))/\sigma(x))$, $(x, y) \in \mathbb{R}^2$. Assume further that g, m and σ are continuous functions and there exists $x_0 \in \mathbb{R}$ such that $p(x_0, x_0) > 0$. Then, the transition density is uniformly bounded from below over some neighborhood $V_{x_0}(\varepsilon)^2 = [x_0 - \varepsilon, x_0 + \varepsilon]^2$ of (x_0, x_0) in $\mathbb{R}^2$: there exists $\delta = \delta(\varepsilon) \in]0, 1[$ such that,

$$\inf_{(x,y) \in V_{x_0}^2} p(x, y) \geq \delta(2\varepsilon)^{-1} . \tag{17}$$

We thus showed that the chain X satisfies the minorization condition $\mathcal{M}(1, V_{x_0}(\varepsilon), \delta, \mathcal{U}_{V_{x_0}(\varepsilon)})$. Furthermore, block-moment conditions for such time series model may be checked via the practical conditions developed in Douc $et\ al.$ (2004) (see their example 3).

5 Regenerative block-bootstrap

Athreya & Fuh (1989) and Datta & McCormick (1993) proposed a specific bootstrap methodology for atomic Harris positive recurrent Markov chains, which exploits the renewal properties of the latter. The main idea underlying this method consists in resampling a deterministic number of data blocks corresponding to regeneration cycles. However, because of some inadequate standardization, the $regeneration\text{-}based\ bootstrap$ method proposed in Datta & McCormick (1993) is not second order correct when applied to the sample mean problem (its rate is $O_{\mathbb{P}}(n^{-1/2})$ in the stationary case). Prolongating this work, Bertail & Clémençon (2004b) have shown how to modify suitably this resampling procedure to make it second order correct up to $O_{\mathbb{P}}(n^{-1} \log(n))$ in the unstudentized case ($i.e.$ when the variance is known) when the chain is stationary. However this Bootstrap method remains of limited interest from a practical viewpoint, given the necessary modifications (standardization and recentering) and the restrictive stationary framework required to obtain the second order accuracy: it fails to be second order correct in the nonstationary case, as a careful examination of the second order properties of the sample mean statistic of a positive recurrent chain based on its E.E. shows (cf Malinovskii (1987), Bertail & Clémençon (2004a)).

   A powerful alternative, namely the $Regenerative\ Block\text{-}Bootstrap\ (RBB)$, have been thus proposed and studied in Bertail & Clémençon (2004c), that consists in imitating further the renewal structure of the chain by resampling regeneration data blocks, until the length of the reconstructed Bootstrap series is larger than the length n of the original data series, so as to approximate the distribution of the (random) number of regeneration blocks in a series of length n and remove some bias terms (see section 4). Here we survey the asymptotic validity of the RBB for the studentized mean by an adequate estimator of the asymptotic variance. This is the useful version for confidence intervals but also for practical use of the Bootstrap (*cf* Hall (1992)) and for a broad class of Markov chains (including chains with strong mixing coefficients decreasing at a polynomial rate), the accuracy reached by the RBB is proved to be of order $O_{\mathbb{P}}(n^{-1})$ both for the standardized and the studentized sample mean. The rate obtained is thus comparable to the optimal rate of the Bootstrap distribution in the i.i.d. case, contrary to the *Moving Block Bootstrap* (*cf* Götze & Künsch (1996), Lahiri (2003)). The proof relies on the E.E. for the studentized sample mean stated in § 4.1 (see Theorems 4.2, 4.6). In Bertail & Clémençon (2004c) a straightforward extension of the RBB procedure to general Harris chains based on the ARB construction (see § 3.1) is also proposed (it is called *Approximate Regenerative Block-Bootstrap, ARBB* in abbreviated form). Although it is based on the approximate regenerative blocks, it is shown to be still second order correct when the estimate p_n used in the ARB algorithm is consistent. We also emphasize that the principles underlying the (A)RBB may be applied to any (eventually continuous time) regenerative process (and not necessarily markovian) or with a regenerative extension that may be approximated (see Thorisson (2000)).

5.1 The (approximate) regenerative block-bootstrap algorithm.

Once true or approximate regeneration blocks $\widehat{\mathcal{B}}_1, ..., \widehat{\mathcal{B}}_{\widehat{l}_n-1}$ are obtained (by implementing *Algorithm 1, 2, 3* or *4*), the (*approximate*) *regenerative block-bootstrap* algorithm for computing an estimate of the sample distribution of some statistic $T_n = T(\widehat{\mathcal{B}}_1, ..., \widehat{\mathcal{B}}_{\widehat{l}_n-1})$ with standardization $S_n = S(\widehat{\mathcal{B}}_1, ..., \widehat{\mathcal{B}}_{\widehat{l}_n-1})$ is performed in 3 steps as follows.

Algorithm 5 *(Approximate) Regenerative Block-Bootstrap*

1. Draw sequentially bootstrap data blocks $\mathcal{B}_1^*, ..., \mathcal{B}_k^*$ independently from the empirical distribution $\widehat{\mathcal{L}}_n = (\widehat{l}_n - 1)^{-1} \sum_{j=1}^{\widehat{l}_n-1} \delta_{\widehat{\mathcal{B}}_j}$ of the initial blocks $\widehat{\mathcal{B}}_1, ..., \widehat{\mathcal{B}}_{\widehat{l}_n-1}$, until the length of the bootstrap data series $l^*(k) = \sum_{j=1}^{k} l(\mathcal{B}_j^*)$ is larger than n. Let $l_n^* = \inf\{k \geq 1, l^*(k) > n\}$.

2. From the bootstrap data blocks generated at step 1, reconstruct a pseudo-trajectory by binding the blocks together, getting the reconstructed (*A*)*RBB sample path*

$$X^{*(n)} = (\mathcal{B}_1^*, ..., \mathcal{B}_{l_n^*-1}^*) .$$

Then compute the (*A*)*RBB statistic* and its (*A*)*RBB standardization*

$$T_n^* = T(X^{*(n)}) \text{ and } S_n^* = S(X^{*(n)}) .$$

3. The *(A)RBB distribution* is then given by

$$H_{(A)RBB}(x) = \mathbb{P}^*(S_n^{*-1}(T_n^* - T_n) \le x) ,$$

where $\mathbb{P}^*$ denotes the conditional probability given the original data.

Remark 6. A Monte-Carlo approximation to $H_{(A)RBB}(x)$ may be straightfor-wardly computed by repeating independently N times this algorithm.

5.2 Atomic case: second order accuracy of the RBB

In the case of the sample mean, the bootstrap counterparts of the estimators $\bar{\mu}_n(f)$ and $\sigma_n^2(f)$ considered in § 4.1 (using the notation therein) are

$$\mu_n^*(f) = n_A^{*-1} \sum_{j=1}^{l_n^*-1} f(\mathcal{B}_j^*) \text{ and } \sigma_n^{*2}(f) = n_A^{*-1} \sum_{j=1}^{l_n^*-1} \left\{ f(\mathcal{B}_j^*) - \mu_n^*(f)l(\mathcal{B}_j^*) \right\}^2 ,$$

$$(18)$$

with $n_A^* = \sum_{j=1}^{l_n^*-1} l(\mathcal{B}_j^*)$. Let us consider the RBB distribution estimates of the unstandardized and studentized sample means

$$H_{RBB}^U(x) = \mathbb{P}^*(n_A^{1/2}\sigma_n(f)^{-1}\{\mu_n^*(f) - \bar{\mu}_n(f)\} \le x) ,$$
$$H_{RBB}^S(x) = \mathbb{P}^*(n_A^{*-1/2}\sigma_n^{*-1}(f)\{\mu_n^*(f) - \bar{\mu}_n(f)\} \le x) .$$

The following theorem established in Bertail & Clémençon (2004b) shows the RBB is asymptotically valid for the sample mean. Moreover it ensures that the RBB attains the optimal rate of the i.i.d. Bootstrap. The proof of this result crucially relies on the E.E. given in Malinovskii (1987) in the standardized case and its extension to the studentized case proved in Bertail & Clémençon (2004a).

Theorem 6. *Suppose that (C1) is satisfied. Under $\mathcal{H}_0'(2, \nu)$, $\mathcal{H}_1'(2, f, \nu)$, $\mathcal{H}_0'(\kappa)$ and $\mathcal{H}_1(\kappa, f)$ with $\kappa > 6$, the RBB distribution estimate for the un-standardized sample mean is second order accurate in the sense that*

$$\Delta_n^U = \sup_{x \in \mathbb{R}} |H_{RBB}^U(x) - H_\nu^U(x)| = O_{\mathbb{P}_\nu}(n^{-1}) , \text{ as } n \to \infty ,$$

with $H_\nu^U(x) = \mathbb{P}_\nu(n_A^{1/2}\sigma_f^{-1}\{\bar{\mu}_n(f) - \mu(f)\} \le x)$. And if in addition (C4), $\mathcal{H}_0'(\kappa)$ and $\mathcal{H}_1(\kappa, f)$ are checked with $\kappa > 8$, the RBB distribution estimate for the standardized sample mean is also 2nd order correct

$$\Delta_n^S = \sup_{x \in \mathbb{R}} |H_{RBB}^S(x) - H_\nu^S(x)| = O_{\mathbb{P}_\nu}(n^{-1}) , \text{ as } n \to \infty ,$$

with $H_\nu^S(x) = \mathbb{P}_\nu(n_A^{1/2}\sigma_n^{-1}(f)\{\bar{\mu}_n(f) - \mu(f)\} \le x)$.

5.3 Asymptotic validity of the ARBB for general chains

The ARBB counterparts of the statistics $\widehat{\mu}_n(f)$ and $\widehat{\sigma}_n^2(f)$ considered in § 4.2 (using the notation therein) may be expressed as

$$\mu_n^*(f) = n_{A_\mathcal{M}}^{*-1} \sum_{j=1}^{l_n^*-1} f(\mathcal{B}_j^*)$$

and

$$\sigma_n^{*2}(f) = n_{A_\mathcal{M}}^{*-1} \sum_{j=1}^{l_n^*-1} \left\{ f(\mathcal{B}_j^*) - \mu_n^*(f) l(\mathcal{B}_j^*) \right\}^2 \, ,$$

denoting by $n_{A_\mathcal{M}}^* = \sum_{j=1}^{l_n^*-1} l(\mathcal{B}_j^*)$ the length of the ARBB data series. Define the ARBB versions of the pseudo-regeneration based unstudentized and studentized sample means (cf § 4.2) by

$$\widehat{\varsigma}_n^* = n_{A_\mathcal{M}}^{1/2} \frac{\mu_n^*(f) - \widehat{\mu}_n(f)}{\widehat{\sigma}_n(f)} \text{ and } \widehat{t}_n^* = n_{A_\mathcal{M}}^{*1/2} \frac{\mu_n^*(f) - \widehat{\mu}_n(f)}{\sigma_n^*(f)} \, .$$

The unstandardized and studentized version of the ARBB distribution estimates are then given by

$$H_{ARBB}^U(x) = \mathbb{P}^*(\widehat{\varsigma}_n^* \le x \mid X^{(n+1)}) \text{ and } H_{ARBB}^S(x) = \mathbb{P}^*(\widehat{t}_n^* \le x \mid X^{(n+1)}) \, .$$

This is the same construction as in the atomic case, except that one uses the approximate regeneration blocks instead of the exact regenerative ones (cf Theorem 3.3 in Bertail & Clémençon (2004c)).

Theorem 7. *Under the hypotheses of Theorem 4.2, we have the following convergence results in distribution under* $\mathbb{P}_\nu$

$$\Delta_n^U = \sup_{x \in \mathbb{R}} |H_{ARBB}^U(x) - H_\nu^U(x)| \to 0 \, , \text{ as } n \to \infty \, ,$$

$$\Delta_n^S = \sup_{x \in \mathbb{R}} |H_{ARBB}^S(x) - H_\nu^S(x)| \to 0 \, , \text{ as } n \to \infty \, .$$

5.4 Second order properties of the ARBB using the 2-split trick

To bypass the technical difficulties related to the dependence problem induced by the preliminary step estimation, assume now that the pseudo regenerative blocks are constructed according to Algorithm 4 (possibly including the selection rule for the small set of Algorithm 3). It is then easier (at the price of a small loss in the 2nd order term) to get second order results both in the case of standardized and studentized statistics, as stated below (refer to Bertail & Clémençon (2004c) for the technical proof).

Theorem 8. *Suppose that (C1) and (C4) are satisfied by the split chain. Under assumptions* $\mathcal{H}'_0(\kappa,\ \nu)$, $\mathcal{H}'_1(\kappa,\ f,\ \nu)$, $\mathcal{H}'_0(f,\ \kappa)$, $\mathcal{H}'_1(f,\ \kappa)$ *with* $\kappa > 6$, $\mathcal{H}_2$, $\mathcal{H}_3$ *and* $\mathcal{H}_4$, *we have the second order validity of the ARBB distribution both in the standardized and unstandardized case up to order*

$$\Delta^U_n = O_{\mathbb{P}_\nu}(n^{-1/2}\alpha_{m_n}^{1/2} \vee n^{-1/2}n^{-1}m_n\})\ ,\ \text{as}\ n \to \infty\ .$$

And if in addition these assumptions hold with $k > 8$, *we have*

$$\Delta^S_n = O_{\mathbb{P}_\nu}(n^{-1/2}\alpha_{m_n}^{1/2} \vee n^{-1/2}n^{-1}m_n)\ ,\ \text{as}\ n \to \infty\ .$$

In particular if $\alpha_m = m\log(m)$, *by choosing* $m_n = n^{2/3}$, *the ARBB is second order correct up to* $O(n^{-5/6}\log(n))$.

It is worth noticing that the rate that can be attained by the 2-split trick variant of the ARBB for such chains is faster than the optimal rate the MBB may achieve, which is typically of order $O(n^{-3/4})$ under very strong assumptions (see Götze & Künsch (1996), Lahiri (2003)). Other variants of the bootstrap (sieve bootstrap) for time-series may also yield (at least practically) very accurate approximation (see Bühlmann (2002), (1997)). When some specific non-linear structure is assumed for the chain (see our example 3), nonparametric method estimation and residual based resampling methods may also be used : see for instance Franke et al. (2002). However to our knowledge, no rate of convergence is explicitly available for these bootstrap techniques. An empirical comparison of all these recent methods would be certainly of great help but is beyond the scope of this paper.

6 Some extensions to U-statistics

We now turn to extend some of the asymptotic results stated in sections 4 and 5 for sample mean statistics to a wider class of functionals and shall consider statistics of the form $\sum_{1\leq i\neq j\leq n} U(X_i, X_j)$. For the sake of simplicity, we confined the study to U-statistics of degree 2, in the real case only. As will be shown below, asymptotic validity of inference procedures based on such statistics does not straightforwardly follow from results established in the previous sections, even for atomic chains. Furthermore, whereas asymptotic validity of the (approximate) regenerative block-bootstrap for these functionals may be easily obtained, establishing its second order validity and give precise rate is much more difficult from a technical viewpoint and is left to a further study. Besides, arguments presented in the sequel may be easily adapted to V-statistics $\sum_{1\leq i,\ j\leq n} U(X_i, X_j)$.

6.1 Regenerative case

Given a trajectory $X^{(n)} = (X_1, ..., X_n)$ of a Harris positive atomic Markov chain with stationary probability law μ (refer to § 2.2 for assumptions and notation), we shall consider in the following U-statistics of the form

$$T_n = \frac{1}{n(n-1)} \sum_{1 \le i \neq j \le n} U(X_i, X_j) , \qquad (19)$$

where $U : E^2 \to \mathbb{R}$ is a kernel of degree 2. Even if it entails introducing the symmetrized version of T_n, it is assumed throughout the section that the kernel $U(x, y)$ is symmetric. Although such statistics have been mainly used and studied in the case of i.i.d. observations, in dependent settings such as ours, these statistics are also of interest, as shown by the following examples.

• In the case when the chain takes real values and is positive recurrent with stationary distribution μ, the variance of the stationary distribution $s^2 = \mathbb{E}_\mu((X - \mathbb{E}_\mu(X))^2)$, if well defined (note that it differs in general from the asymptotic variance of the mean statistic studied in § 4.1), may be consistently estimated under adequate block moment conditions by

$$\widehat{s}_n^2 = \frac{1}{n-1} \sum_{i=1}^{n} (X_i - \mu_n)^2 = \frac{1}{n(n-1)} \sum_{1 \le i \neq j \le n} (X_i - X_j)^2/2 ,$$

where $\mu_n = n^{-1} \sum_{i=1}^{n} X_i$, which is a U-statistic of degree 2 with symmetric kernel $U(x, y) = (x - y)^2/2$.

• In the case when the chain takes its values in the multidimensional space $\mathbb{R}^p$, endowed with some norm $||.\,||$, many statistics of interest may be written as a U-statistic of the form

$$U_n = \frac{1}{n(n-1)} \sum_{1 \le i \neq j \le n} H(||X_i - X_j||) ,$$

where $H : \mathbb{R} \to \mathbb{R}$ is some measurable function. And in the particular case when $p = 2$, for some fixed t in $\mathbb{R}^2$ and some smooth function h, statistics of type

$$U_n = \frac{1}{n(n-1)} \sum_{1 \le i \neq j \le n} h(t, X_i, X_j)$$

arise in the study of the *correlation dimension* for dynamic systems (see Borovkova *et al.* (1999)). *Depth statistical functions* for spatial data are also particular examples of such statistics (*cf* Serfling & Zuo (2000)).

In what follows, the parameter of interest is

$$\mu(U) = \int_{(x,y) \in E^2} U(x, y)\mu(dx)\mu(dy) , \qquad (20)$$

which quantity we assume to be finite. As in the case of i.i.d. observations, a natural estimator of $\mu(U)$ in our markovian setting is T_n. We shall now study its consistency properties and exhibit an adequate sequence of renormalizing constants for the latter, by using the *regeneration blocks construction* once again. For later use, define $\omega_U : \mathbb{T}^2 \to \mathbb{R}$ by

$$\omega_U(x^{(k)}, y^{(l)}) = \sum_{i=1}^{k} \sum_{j=1}^{l} U(x_i, y_j) \, ,$$

for any $x^{(k)} = (x_1, ..., x_k)$, $y^{(l)} = (y_1, ..., y_l)$ in the torus $\mathbb{T} = \cup_{n=1}^{\infty} E^n$ and observe that ω_U is symmetric, as U.

"Regeneration-based Hoeffding's decomposition"

By the representation of μ as a Pitman's occupation measure (*cf* Theorem 2.1), we have

$$\mu(U) = \alpha^{-2} \mathbb{E}_A \Big(\sum_{i=1}^{\tau_A(1)} \sum_{l=\tau_A(1)+1}^{\tau_A(2)} U(X_i, X_j) \Big)$$
$$= \alpha^{-2} \mathbb{E}(\omega_U(\mathcal{B}_l, \mathcal{B}_k)) \, ,$$

for any integers k, l such that $k \neq l$. In the case of U-statistics based on dependent data, the classical (orthogonal) Hoeffding decomposition (*cf* Serfling (1981)) does not hold anymore. Nevertheless, we may apply the underlying projection principle for establishing the asymptotic normality of T_n by approximatively rewriting it as a U-statistic of degree 2 computed on the regenerative blocks only, in a fashion very similar to the *Bernstein blocks technique* for strongly mixing random fields (*cf* Doukhan (1994)), as follows. As a matter of fact, the estimator T_n may be decomposed as

$$T_n = \frac{(l_n - 1)(l_n - 2)}{n(n-1)} U_{l_n-1} + T_n^{(0)} + T_n^{(n)} + \Delta_n \, , \tag{21}$$

where,

$$U_L = \frac{2}{L(L-1)} \sum_{1 \leq k < l \leq L} \omega_U(\mathcal{B}_k, \mathcal{B}_l) \, ,$$

$$T_n^{(0)} = \frac{2}{n(n-1)} \sum_{1 \leq k \leq l_n-1} \omega_U(\mathcal{B}_k, \mathcal{B}_0) \, ,$$

$$T_n^{(n)} = \frac{2}{n(n-1)} \sum_{0 \leq k \leq l_n-1} \omega_U(\mathcal{B}_k, \mathcal{B}_{l_n}^{(n)}) \, ,$$

$$\Delta_n = \frac{1}{n(n-1)} \Big\{ \sum_{k=0}^{l_n-1} \omega_U(\mathcal{B}_k, \mathcal{B}_k) + \omega_U(\mathcal{B}_{l_n}^{(n)}, \mathcal{B}_{l_n}^{(n)}) - \sum_{i=1}^{n} U(X_i, X_i) \Big\} \, .$$

Observe that the "block diagonal part" of T_n, namely Δ_n, may be straightforwardly shown to converge $\mathbb{P}_\nu$- a.s. to 0 as $n \to \infty$, as well as $T_n^{(0)}$ and $T_n^{(1)}$ by using the same arguments as the ones used in § 4.1 for dealing with sample means, under obvious block moment conditions (see conditions *(ii)-(iii)* below). And, since $l_n/n \to \alpha^{-1}$ $\mathbb{P}_\nu$- a.s. as $n \to \infty$, asymptotic properties of T_n may be derived from the ones of U_{l_n-1}, which statistic depends on the regeneration blocks only. The key point relies in the fact that the theory of U-statistics based on i.i.d. data may be straightforwardly adapted to functionals of the i.i.d. regeneration blocks of the form $\sum_{k<l} \omega_U(\mathcal{B}_k, \mathcal{B}_l)$. Hence, the asymptotic behaviour of the U-statistic U_L as $L \to \infty$ essentially depends on the properties of the linear and quadratic terms appearing in the following variant of *Hoeffding's decomposition*. For k, $l \geq 1$, define

$$\widetilde{\omega}_U(\mathcal{B}_k, \mathcal{B}_l) = \sum_{i=\tau_A(k)+1}^{\tau_A(k+1)} \sum_{j=\tau_A(l)+1}^{\tau_A(l+1)} \{U(X_i, X_j) - \mu(U)\} .$$

(notice that $\mathbb{E}(\widetilde{\omega}_U(\mathcal{B}_k, \mathcal{B}_l)) = 0$ when $k \neq l$) and for $L \geq 1$ write the expansion

$$U_L - \mu(U) = \frac{2}{L} \sum_{k=1}^{L} \omega_U^{(1)}(\mathcal{B}_k) + \frac{2}{L(L-1)} \sum_{1 \leq k < l \leq L} \omega_U^{(2)}(\mathcal{B}_k, \mathcal{B}_l) , \qquad (22)$$

where, for any $b_1 = (x_1, ..., x_l) \in \mathbb{T}$,

$$\omega_U^{(1)}(b_1) = \mathbb{E}(\widetilde{\omega}_U(\mathcal{B}_1, \mathcal{B}_2)|\mathcal{B}_1 = b_1) = \mathbb{E}_A(\sum_{i=1}^{l} \sum_{j=1}^{\tau_A} \widetilde{\omega}_U(x_i, X_j))$$

is the linear term (see also our definition of the *influence function* of the parameter $\mathbb{E}(\omega(\mathcal{B}_1, \mathcal{B}_2))$ in section 7) and for all b_1, b_2 in $\mathbb{T}$,

$$\omega_U^{(2)}(b_1, b_2) = \widetilde{\omega}_U(b_1, b_2) - \widetilde{\omega}_U^{(1)}(b_1) - \widetilde{\omega}_U^{(1)}(b_2)$$

is the quadratic degenerate term (gradient of order 2). Notice that by using the Pitman's occupation measure representation of μ, we have as well, for any $b_1 = (x_1, ..., x_l) \in \mathbb{T}$,

$$(E_A \tau_A)^{-1} \omega_U^{(1)}(b_1) = \sum_{i=1}^{l} \mathbb{E}_\mu(\widetilde{\omega}_U(x_i, X_1)) .$$

For resampling purposes, we also introduce the U-statistic based on the data between the first regeneration time and the last one only:

$$\widetilde{T}_n = \frac{2}{\widetilde{n}(\widetilde{n}-1)} \sum_{1+\tau_A \leq i < j \leq \tau_A(l_n)} U(X_i, X_j) ,$$

with $\widetilde{n} = \tau_A(l_n) - \tau_A$ and $\widetilde{T}_n = 0$ when $l_n \leq 1$ by convention.

Asymptotic normality and asymptotic validity of the RBB

Now suppose that the following conditions, which are involved in the next result, are fulfilled by the chain.

  (i) (Non degeneracy of the U-statistic)

$$0 < \sigma_U^2 = \mathbb{E}(\omega_U^{(1)}(\mathcal{B}_1)^2) < \infty \ .$$

  (ii) (Block-moment conditions: linear part) For some $s \geq 2$,

$$\mathbb{E}(\omega_{|U|}^{(1)}(\mathcal{B}_1)^s) < \infty \text{ and } \mathbb{E}_\nu(\omega_{|U|}^{(1)}(\mathcal{B}_0)^2) < \infty \ .$$

  (iii) (Block-moment conditions: quadratic part) For some $s \geq 2$,

$$\mathbb{E}|\omega_{|U|}(\mathcal{B}_1, \mathcal{B}_2)|^s < \infty \text{ and } \mathbb{E}|\omega_{|U|}(\mathcal{B}_1, \mathcal{B}_1)|^s < \infty \ ,$$
$$\mathbb{E}_\nu|\omega_{|U|}(\mathcal{B}_0, \mathcal{B}_1)|^2 < \infty \text{ and } \mathbb{E}_\nu|\omega_{|U|}(\mathcal{B}_0, \mathcal{B}_0)|^2 < \infty \ .$$

By construction, under *(ii)-(iii)* we have the crucial orthogonality property:

$$\text{cov}(\omega_U^{(1)}(\mathcal{B}_1), \ \omega_U^{(2)}(\mathcal{B}_1, \mathcal{B}_2)) = 0 \ . \tag{23}$$

Now a slight modification of the argument given in Hoeffding (1948) allows to prove straightforwardly that $\sqrt{L}(U_L - \mu(U))$ is asymptotically normal with zero mean and variance $4\sigma_U^2$. Furthermore, by adapting the classical CLT argument for sample means of Markov chains (refer to in Meyn & Tweedie (1996) for instance) and using (23) and $l_n/n \to \alpha^{-1}$ $\mathbb{P}_\nu$-a.s. as $n \to \infty$, one deduces that $\sqrt{n}(T_n - \mu(U)) \Rightarrow \mathcal{N}(0, \Sigma^2)$ as $n \to \infty$ under $\mathbb{P}_\nu$, with $\Sigma^2 = 4\alpha^{-3}\sigma_U^2$.

  Besides, estimating the normalizing constant is important (for constructing confidence intervals or bootstrap counterparts for instance). So we define the natural estimator $\sigma_{U, \, l_n-1}^2$ of σ_U^2 based on the (asymptotically i.i.d.) $l_n - 1$ regeneration data blocks by

$$\sigma_{U, \, L}^2 = (L-1)(L-2)^{-2} \sum_{k=1}^{L} [(L-1)^{-1} \sum_{l=1, k \neq l}^{L} \omega_U(\mathcal{B}_k, \mathcal{B}_l) - U_L]^2 \ ,$$

for $L \geq 1$. The estimate $\sigma_{U, \, L}^2$ is a simple transposition of the *jackknife estimator* considered in Callaert & Veraverbeke (1981) to our setting and may be easily shown to be strongly consistent (by adapting the SLLN for *U*-statistics to this specific functional of the i.i.d regeneration blocks). Furthermore, we derive that $\Sigma_n^2 \to \Sigma^2$ $\mathbb{P}_\nu$-a.s., as $n \to \infty$, where

$$\Sigma_n^2 = 4(l_n/n)^3 \sigma_{U, \, l_n-1}^2 \ .$$

We also consider the regenerative block-bootstrap counterparts T_n^* and Σ_n^{*2} of $\widetilde{T}_n$ and Σ_n^2 respectively, constructed via *Algorithm 5*:

$$T_n^* = \frac{2}{n^*(n^*-1)} \sum_{1 \le i < j \le n^*} U(X_i^*, X_j^*) \, ,$$

$$\Sigma_n^{*2} = 4(l_n^*/n^*)^3 \sigma_{U,\, l_n^*-1}^{*2} \, ,$$

where n^* denotes the length of the RBB data series $X^{*(n)} = (X_1, ..., X_{n^*})$ constructed from the $l_n^* - 1$ bootstrap data blocks, and

$$\sigma_{U,\, l_n^*-1}^{*2} = (l_n^* - 2)(l_n^* - 3)^{-2} \sum_{k=1}^{l_n^*-1} [(l_n^* - 2)^{-1} \sum_{l=1, k \ne l}^{l_n^*-1} \omega_U(\mathcal{B}_k^*, \mathcal{B}_l^*) - U_{l_n^*-1}^*]^2 \, ,$$

(24)

$$U_{l_n^*-1}^* = \frac{2}{(l_n^* - 1)(l_n^* - 2)} \sum_{1 \le k < l \le l_n^*-1} \omega_U(\mathcal{B}_k^*, \mathcal{B}_l^*) \, .$$

We may then state the following result.

Theorem 9. *If conditions (i)-(iii) are fulfilled with $s = 4$, then we have the CLT under $\mathbb{P}_\nu$*

$$\sqrt{n}(T_n - \mu(U))/\Sigma_n \Rightarrow \mathcal{N}(0,1) \, , \text{ as } n \to \infty \, .$$

This limit result also holds for $\widetilde{T}_n$, as well as the asymptotic validity of the RBB distribution: as $n \to \infty$,

$$\sup_{x \in \mathbb{R}} |\mathbb{P}^*(\sqrt{n^*}(T_n^* - \widetilde{T}_n))/\Sigma_n^* \le x) - \mathbb{P}_\nu(\sqrt{n}(\widetilde{T}_n - \mu(U))/\Sigma_n \le x)| \overset{\mathbb{P}_\nu}{\to} 0 \, .$$

Whereas proving the asymptotic validity of the RBB for U-statistics under these assumptions is straightforward (its second order accuracy up to $o(n^{-1/2})$ seems also quite easy to prove by simply adapting the argument used by Helmers (1991) under appropriate Cramer condition on $\omega_U^{(1)}(\mathcal{B}_1)$ and block-moment assumptions), establishing an exact rate, $O(n^{-1})$ for instance as in the case of sample mean statistics, is much more difficult. Even if one tries to reproduce the argument in Bertail & Clémençon (2004a) consisting in partitioning the underlying probability space according to every possible realization of the regeneration times sequence between 0 and n, the problem boils down to control the asymptotic behaviour of the distribution $\mathbb{P}(\sum_{1 \le i \ne j \le m} \omega_U^{(2)}(\mathcal{B}_i, \mathcal{B}_j)/\sigma_{U,\, m}^2 \le y, \sum_{j=1}^m L_j = l)$ as $m \to \infty$, which is a highly difficult technical task (due to the lattice component).

Remark 7. We point out that the approach developed here to deal with the statistic U_L naturally applies to more general functionals of the regeneration blocks $\sum_{k<l} \omega(\mathcal{B}_k, \mathcal{B}_l)$, with $\omega : \mathbb{T}^2 \to \mathbb{R}$ being some measurable function. For instance, the estimator of the asymptotic variance $\widehat{\sigma}_n^2(f)$ proposed in § 4.1 could be derived from such a functional, that may be seen as a U-statistic based on observation blocks with kernel $\omega(\mathcal{B}_k, \mathcal{B}_l) = (f(\mathcal{B}_k) - f(\mathcal{B}_l))^2/2$.

6.2 General case

Suppose now that the observed trajectory $X^{(n+1)} = (X_1, ..., X_{n+1})$ is drawn from a general Harris positive chain with stationary probability μ (see § 2.2 for assumptions and notation). Using the split chain, we have the representation of the parameter $\mu(U)$:

$$\mu(U) = \mathbb{E}_{A_{\mathcal{M}}}(\tau_{A_{\mathcal{M}}})^{-2}\mathbb{E}_{A_{\mathcal{M}}}(\omega_U(\mathcal{B}_1, \mathcal{B}_2)) \ .$$

Using the pseudo-blocks $\widehat{\mathcal{B}}_l$, $1 \leq l \leq \widehat{l}_n - 1$, as constructed in § 3.2, we consider the sequence of renormalizing constants for T_n :

$$\widehat{\Sigma}_n^2 = 4(\widehat{l}_n/n)^3 \widehat{\sigma}_{U, \, \widehat{l}_n - 1}^2 \ , \tag{25}$$

with

$$\widehat{\sigma}_{U, \, \widehat{l}_n - 1}^2 = (\widehat{l}_n - 2)(\widehat{l}_n - 3)^{-2} \sum_{k=1}^{\widehat{l}_n - 1} [(\widehat{l}_n - 2)^{-1} \sum_{l=1, k \neq l}^{\widehat{l}_n - 1} \omega_U(\widehat{\mathcal{B}}_k, \widehat{\mathcal{B}}_l) - \widehat{U}_{\widehat{l}_n - 1}]^2 \ ,$$

$$\widehat{U}_{\widehat{l}_n - 1} = \frac{2}{(\widehat{l}_n - 1)(\widehat{l}_n - 2)} \sum_{1 \leq k < l \leq \widehat{l}_n - 1} \omega_U(\widehat{\mathcal{B}}_k, \widehat{\mathcal{B}}_l) \ .$$

We also introduce the U-statistic computed from the first approximate regeneration time and the last one:

$$\widehat{T}_n = \frac{2}{\widehat{n}(\widehat{n} - 1)} \sum_{1 + \widehat{\tau}_A(1) \leq i < j \leq \widehat{\tau}_A(l_n)} U(X_i, X_j) \ ,$$

with $\widehat{n} = \widehat{\tau}_A(\widehat{l}_n) - \widehat{\tau}_A(1)$. Let us define the bootstrap counterparts T_n^* and Σ_n^* of $\widehat{T}_n$ and $\widehat{\Sigma}_n^2$ constructed from the pseudo-blocks via *Algorithm 5*. Although approximate blocks are used here instead of the (unknown) regenerative ones $\mathcal{B}_l$, $1 \leq l \leq l_n - 1$, asymptotic normality still holds under appropriate assumptions, as shown by the theorem below, which we state in the only case when the kernel U is bounded (with the aim to make the proof simpler).

Theorem 10. *Suppose that the kernel $U(x, y)$ is bounded and that $\mathcal{H}_2$, $\mathcal{H}_3$, $\mathcal{H}_4$ are fulfilled, as well as (i)-(iii) for $s = 4$. Then we have as $n \to \infty$,*

$$\widehat{\Sigma}_n^2 \to \Sigma^2 = 4\mathbb{E}_{A_{\mathcal{M}}}(\tau_{A_{\mathcal{M}}})^{-3}\mathbb{E}_{A_{\mathcal{M}}}(\omega_U^{(1)}(\mathcal{B}_1)^2), \ in \ \mathbb{P}_\nu\text{-}pr.$$

Moreover as $n \to \infty$, under $\mathbb{P}_\nu$ we have the convergence in distribution

$$n^{1/2}\widehat{\Sigma}_n^{-1}(\widehat{T}_n - \mu(U)) \Rightarrow \mathcal{N}(0, 1) \ ,$$

as well as the asymptotic validity of the ARBB counterpart

$$\sup_{x \in \mathbb{R}} |\mathbb{P}^*(\sqrt{n^*}(T_n^* - \widehat{T}_n))/\Sigma_n^* \leq x) - \mathbb{P}_\nu(\sqrt{n}(\widehat{T}_n - \mu(U))/\widehat{\Sigma}_n \leq x)| \xrightarrow[n \to \infty]{\mathbb{P}_\nu} 0 \ .$$

Proof. By applying the results of § 6.1 to the split chain, we get that the variance of the limiting (normal) distribution of $\sqrt{n}(T_n - \mu(U))$ is $\Sigma^2 = 4\mathbb{E}_{A_{\mathcal{M}}}(\tau_{A_{\mathcal{M}}})^{-3}\mathbb{E}_{A_{\mathcal{M}}}(\omega_U^{(1)}(\mathcal{B}_1)^2)$. The key point of the proof consists in considering an appropriate coupling between $(X_i, Y_i)_{1\leq i \leq n}$ and $(X_i, \widehat{Y}_i)_{1 \leq i \leq n}$ (or equivalently between the sequence of the "true" regeneration times between 0 and n and the sequence of approximate ones), so as to control the deviation between functionals constructed from the regeneration blocks and their counterparts based on the approximate ones. The coupling considered here is the same as the one used in the proof of Theorem 3.1 in Bertail & Clémençon (2004c) (refer to the latter article for a detailed construction). We shall now evaluate how $\widehat{\sigma}^2_{U, \, \widehat{l}_n - 1}$ differs from $\sigma^2_{U, \, l_n - 1}$, its counterpart based on the "true" regeneration blocks. Observe first that

$$T_n = \frac{\widehat{n}(\widehat{n} - 1)}{n(n - 1)}\widehat{T}_n + \widehat{T}_n^{(0)} + \widehat{T}_n^{(n)} + \widehat{\Delta}_n \, ,$$

where

$$\widehat{T}_n^{(0)} = \frac{2}{n(n - 1)} \sum_{1 \leq k \leq \widehat{l}_n - 1} \omega_U(\widehat{\mathcal{B}}_k, \widehat{\mathcal{B}}_0) \, ,$$

$$\widehat{T}_n^{(n)} = \frac{2}{n(n - 1)} \sum_{0 \leq k \leq \widehat{l}_n - 1} \omega_U(\widehat{\mathcal{B}}_k, \widehat{\mathcal{B}}_{\widehat{l}_n}^{(n)}) \, ,$$

$$\widehat{\Delta}_n = \frac{1}{n(n - 1)}\{\sum_{k=0}^{\widehat{l}_n - 1} \omega_U(\widehat{\mathcal{B}}_k, \widehat{\mathcal{B}}_k) + \omega_U(\widehat{\mathcal{B}}_{\widehat{l}_n}^{(n)}, \widehat{\mathcal{B}}_{\widehat{l}_n}^{(n)}) - \sum_{i=1}^{n} U(X_i, X_i)\} \, .$$

Now following line by line the proof of lemma 5.2 in Bertail & Clémençon (2004c), we obtain that, as $n \to \infty$, $\widehat{n}/n - 1 = O_{\mathbb{P}_\nu}(1)$, $\widehat{\Delta}_n - \Delta_n$, $\widehat{T}_n^{(0)} - \widehat{T}_n^{(0)}$ and $\widehat{T}_n^{(n)} - \widehat{T}_n^{(n)}$ are $O_{\mathbb{P}_\nu}(n^{-1})$. It follows thus that $\widehat{T}_n = T_n + o_{\mathbb{P}_\nu}(n^{-1/2})$ as $n \to \infty$, and $\sqrt{n}(\widehat{T}_n - \mu(U))$ is asymptotically normal with variance Σ^2. The same limit results is straightforwardly available then for the Bootstrap version by standard regenerative arguments. Furthermore, by Lemma 5.3 in Bertail & Clémençon (2004c) we have $|\widehat{l}_n/n - l_n/n| = O_{\mathbb{P}_\nu}(\alpha_n^{1/2})$ as $n \to \infty$, and thus $\widehat{l}_n/n \to \mathbb{E}_{A_{\mathcal{M}}}(\tau_{A_{\mathcal{M}}})^{-1}$ in $\mathbb{P}_\nu$-pr. as $n \to \infty$. It then follows by simple (especially when U is bounded) but tedious calculations that $\widehat{\Sigma}_n^2 - \Sigma_n^2 = D_n + o_{\mathbb{P}_\nu}(1)$ as $n \to \infty$, with

$$D_n = 4(l_n/n)^3 [\widehat{l}_n^{-1} \sum_{i=1}^{\widehat{l}_n - 1} \{\frac{1}{\widehat{l}_n - 2} \sum_{j=1, j \neq i}^{\widehat{l}_n - 1} \omega_U(\widehat{\mathcal{B}}_i, \widehat{\mathcal{B}}_j)\}^2$$

$$- l_n^{-1} \sum_{i=1}^{l_n - 1} \{\frac{1}{l_n - 2} \sum_{j=1, j \neq i}^{l_n - 1} \omega_U(\mathcal{B}_i, \mathcal{B}_j)\}^2] \, .$$

Now set $\widehat{g}_n(\widehat{\mathcal{B}}_i) = (\widehat{l}_n - 2)^{-1} \sum_{j=1, j \neq i}^{\widehat{l}_n - 1} \omega_U(\widehat{\mathcal{B}}_i, \widehat{\mathcal{B}}_j)$ for $i \in \{1, ..., \widehat{l}_n - 1\}$ and $g_n(\mathcal{B}_i) = (l_n - 2)^{-1} \sum_{j=1, j \neq i}^{l_n - 1} \omega_U(\mathcal{B}_i, \mathcal{B}_j)$ for $i \in \{1, ..., \widehat{l}_n - 1\}$. By standard arguments on U-statistics (see for instance Helmers (1991) and the references therein) and using once again lemma 5.1 and 5.2 in Bertail & Clémençon (2004b), we have uniformly in $i \in \{1, ..., \widehat{l}_n - 1\}$ (resp. in $i \in \{1, ..., \widehat{l}_n - 1\}$), $\widehat{g}_n(\widehat{\mathcal{B}}_i) = \omega_U^{(1)}(\widehat{\mathcal{B}}_i) + o_{\mathbb{P}_\nu}(1)$ (resp. $g_n(\mathcal{B}_i) = \omega_U^{(1)}(\mathcal{B}_i) + o_{\mathbb{P}_\nu}(1)$) as $n \to \infty$. Such uniform bounds are facilitated by the boundedness assumption on U, but one may expect that with refined computations the same results could be established for unbounded kernels.

It follows that as $n \to \infty$,

$$\Delta_n = 4(l_n/n)^3 [\widehat{l}_n^{-1} \sum_{i=1}^{\widehat{l}_n - 1} \{\omega_U^{(1)}(\widehat{\mathcal{B}}_i)\}^2 - l_n^{-1} \sum_{i=1}^{l_n - 1} \{\omega_U^{(1)}(\mathcal{B}_i)\}^2] + o_{\mathbb{P}_\nu}(1) .$$

The first term in the right hand side is also $o_{\mathbb{P}_\nu}(1)$ by lemma 5.2 in Bertail & Clémençon (2004c). The proof of the asymptotic validity of the Bootstrap version is established by following the preceding lines: it may be easily checked by first linearizing and following the proof of Theorem 3.3 in Bertail & Clémençon (2004c). As in the i.i.d case, this asymptotic result essentially boils down then to check that the empirical moments converge to the theoretical ones. This can be done by adapting standard SLLN arguments for U-statistics.   □

7 Robust functional parameter estimation

Extending the notion of *influence function* and/or *robustness* to the framework of general time series is a difficult task (see Künsch (1984) or Martin & Yohai (1986)). Such concepts are important not only to detect *"outliers"* among the data or influential observations but also to generalize the important notion of *efficient estimation* in semiparametric frameworks (see the recent discussion in Bickel & Kwon (2001) for instance). In the markovian setting, a recent proposal based on martingale approximation has been made by Müller *et al.* (2001). Here we propose an alternative definition of the influence function based on the (approximate) regeneration blocks construction, which is easier to manipulate and immediately leads to central limit and convolution theorems.

7.1 Defining the influence function on the torus

The leitmotiv of this paper is that most parameters of interest related to Harris chains are functionals of the distribution $\mathcal{L}$ of the regenerative blocks (observe that $\mathcal{L}$ is a distribution on the torus $\mathbb{T} = \cup_{n \geq 1} E^n$), namely the distribution of $(X_1, ..., X_{\tau_A})$ conditioned on $X_0 \in A$ when the chain possesses an atom A, or the distribution of $(X_1, ..., X_{\tau_{A_\mathcal{M}}})$ conditioned on $(X_0, Y_0) \in A_\mathcal{M}$ in the general case when one considers the split chain (refer to section 2 for assumptions and notation, here we shall omit the subscript A and $\mathcal{M}$ in what follows to make the notation simpler). In view of Theorem 2.1, this is obviously true in the positive recurrent case for any functional of the stationary law μ. But, more generally, the probability distribution $\mathbb{P}_\nu$ of the Markov chain X starting from ν may be decomposed as follows :

$$\mathbb{P}_\nu((X_n)_{n \geq 1}) = \mathcal{L}_\nu((X_1, ..., X_{\tau_{A(1)}})) \prod_{k=1}^{\infty} \mathcal{L}((X_{1+\tau_A(k)}, ..., X_{\tau_A(k+1)})) ,$$

denoting by $\mathcal{L}_\nu$ the distribution of $(X_1, ..., X_{\tau_A})$ conditioned on $X_0 \sim \nu$. Thus any functional of the law of $(X_n)_{n \geq 1}$ may be seen as a functional of $(\mathcal{L}_\nu, \mathcal{L})$. However, pointing out that the distribution of $\mathcal{L}_\nu$ cannot be estimated in most cases encountered in practice, only functionals of $\mathcal{L}$ are of practical interest. The object of this subsection is to propose the following definition of the influence function for such functionals. Let $\mathcal{P}_{\mathbb{T}}$ denote the set of all probability measures on the torus $\mathbb{T}$ and for any $b \in \mathbb{T}$, set $L(b) = k$ if $b \in E^k$, $k \geq 1$. We then have the following natural definition, that straightforwardly extends the classical notion of influence function in the i.i.d. case, with the important novelty that distributions on the torus are considered here.

Definition 1. Let $T : \mathcal{P}_{\mathbb{T}} \to \mathbb{R}$ be a functional on $\mathcal{P}_{\mathbb{T}}$. If for $\mathcal{L}$ in $\mathcal{P}_{\mathbb{T}}$, $t^{-1}(T((1-t)\mathcal{L} + t\delta_b) - T(\mathcal{L}))$ has a finite limit as $t \to 0$ for any $b \in \mathbb{T}$, then the influence function $T^{(1)}$ of the functional T is well defined, and by definition one has for all b in $\mathbb{T}$,

$$T^{(1)}(b, \ \mathcal{L}) = \lim_{t \to 0} \frac{T((1-t)\mathcal{L} + t\delta_b) - T(\mathcal{L})}{t} . \tag{26}$$

7.2 Some examples

The relevance of this definition is illustrated through the following examples, which aim to show how easy it is to adapt known calculations of influence function on $\mathbb{R}$ to this framework.

a) Suppose that X is positive recurrent with stationary distribution μ. Let $f : E \to \mathbb{R}$ be μ-integrable and consider the parameter $\mu_0(f) = \mathbb{E}_\mu(f(X))$. Denote by $\mathcal{B}$ a r.v. valued in $\mathbb{T}$ with distribution $\mathcal{L}$ and observe that $\mu_0(f) = \mathbb{E}_{\mathcal{L}}(f(\mathcal{B}))/\mathbb{E}_{\mathcal{L}}(L(\mathcal{B})) = T(\mathcal{L})$ (recall the notation $f(b) = \sum_{i=1}^{L(b)} f(b_i)$ for any $b \in \mathbb{T}$). A classical calculation for the influence function of ratios yields then

$$T^{(1)}(b, \mathcal{L}) = \frac{d}{dt}(T((1-t)\mathcal{L} + tb)|_{t=0} = \frac{f(b) - \mu(f)L(b)}{\mathbb{E}_{\mathcal{L}}(L(\mathcal{B}))} .$$

Notice that $\mathbb{E}_{\mathcal{L}}(T^{(1)}(\mathcal{B}, \mathcal{L})) = 0$.

b) Let θ be the unique solution of the equation: $\mathbb{E}_\mu(\psi(X, \theta)) = 0$, where $\psi : \mathbb{R}^2 \to \mathbb{R}$ is $\mathcal{C}^2$. Observing that it may be rewritten as $\mathbb{E}_{\mathcal{L}}(\psi(\mathcal{B}, \theta)) = 0$, a similar calculation to the one used in the i.i.d. setting (if differentiating inside the expectation is authorized) gives in this case

$$T_\psi^{(1)}(b, \mathcal{L}) = -\frac{\psi(b, \theta)}{\mathbb{E}_A(\sum_{i=1}^{\tau_A} \frac{\partial \psi(X_i,\theta)}{\partial \theta})} .$$

By definition of θ, we naturally have $\mathbb{E}_{\mathcal{L}}(T_\psi^{(1)}(\mathcal{B}, \mathcal{L})) = 0$.

c) Assuming that the chain takes real values and its stationary law μ has zero mean and finite variance, let ρ be the correlation coefficient between consecutive observations under the stationary distribution:

$$\rho = \frac{\mathbb{E}_\mu(X_n X_{n+1})}{\mathbb{E}_\mu(X_n^2)} = \frac{\mathbb{E}_A(\sum_{n=1}^{\tau_A} X_n X_{n+1})}{\mathbb{E}_A(\sum_{n=1}^{\tau_A} X_n^2)} .$$

For all b in $\mathbb{T}$, the influence function is

$$T_\rho^{(1)}(b, \mathcal{L}) = \frac{\sum_{i=1}^{L(b)} b_i(b_{i+1} - \rho b_i)}{\mathbb{E}_A(\sum_{t=1}^{\tau_A} X_t^2)} ,$$

and one may check that $\mathbb{E}_{\mathcal{L}}(T_\rho^{(1)}(\mathcal{B}, \mathcal{L})) = 0$.

d) It is now possible to reinterpret the results obtained for U-statistics in section 6. With the notation above, the parameter of interest may be rewritten

$$\mu(U) = \mathbb{E}_{\mathcal{L}}(L(\mathcal{B}))^{-2} \mathbb{E}_{\mathcal{L} \times \mathcal{L}}(U(\mathcal{B}_1, \mathcal{B}_2)) ,$$

yielding the influence function: $\forall b \in \mathbb{T}$,

$$\mu^{(1)}(b, \mathcal{L}) = 2\mathbb{E}_{\mathcal{L}}(L(\mathcal{B}))^{-2} \mathbb{E}_{\mathcal{L}}(\tilde{\omega}_U(\mathcal{B}_1, \mathcal{B}_2)|\mathcal{B}_1 = b) .$$

7.3 Main results

In order to lighten the notation, the study is restricted to the case when X takes real values, i.e. $E \subset \mathbb{R}$, but straightforwardly extends to a more general framework. Given an observed trajectory of length n, natural empirical estimates of parameters $T(\mathcal{L})$ are of course the *plug-in estimators* $T(\mathcal{L}_n)$ based on the empirical distribution of the observed regeneration blocks $\mathcal{L}_n = (l_n - 1)^{-1} \sum_{j=1}^{l_n - 1} \delta_{\mathcal{B}_j} \in \mathcal{P}_\mathbb{T}$ in the atomic case, which is defined as soon as $l_n \geq 2$ (notice that $\mathbb{P}_\nu(l_n \leq 1) = O(n^{-1})$ as $n \to \infty$, if $\mathcal{H}_0(1, \nu)$ and $\mathcal{H}_0(2)$ are satisfied). For measuring the closeness between $\mathcal{L}_n$ and $\mathcal{L}$, consider the bounded Lipschitz type metric on $\mathcal{P}_\mathbb{T}$

$$d_{BL}(\mathcal{L}, \mathcal{L}') = \sup_{f \in Lip_\mathbb{T}^1} \{ \int f(b)\mathcal{L}(db) - \int f(b)\mathcal{L}'(db) , \tag{27}$$

for any $\mathcal{L}$, $\mathcal{L}'$ in $\mathcal{P}_\mathbb{T}$, denoting by $Lip_\mathbb{T}^1$ the set of functions $F : \mathbb{T} \to \mathbb{R}$ of type $F(b) = \sum_{i=1}^{L(b)} f(b_i)$, $b \in \mathbb{T}$, where $f : E \to \mathbb{R}$ is such that $\sup_{x \in E} |f(x)| \leq 1$

and is 1-Lipschitz. Other metrics (of Zolotarev type for instance, *cf* Rachev & Ruschendorf (1998)) may be considered. In the general Harris case (refer to § 3.2 for notation), the influence function based on the atom of the split chain, as well as the empirical distribution of the (unobserved) regeneration blocks have to be approximated to be of practical interest. Once again, we shall use the approximate regeneration blocks $\widehat{\mathcal{B}}_1, ..., \widehat{\mathcal{B}}_{\widehat{l}_n-1}$ (using *Algorithm 2, 3*) in the general case and consider

$$\widehat{\mathcal{L}}_n = (\widehat{l}_n - 1) \sum_{j=1}^{\widehat{l}_n-1} \delta_{\widehat{\mathcal{B}}_j} \,,$$

when $\widehat{l}_n \geq 2$. The following theorem provides an asymptotic bound for the error committed by replacing the empirical distribution $\mathcal{L}_n$ of the "true" regeneration blocks by $\widehat{\mathcal{L}}_n$, when measured by d_{BL}.

Theorem 11. *Under* $\mathcal{H}_0'(4), \mathcal{H}_0'(4, \nu), \mathcal{H}_2, \mathcal{H}_3$ *and* $\mathcal{H}_4$, *we have*

$$d_{BL}(\mathcal{L}_n, \widehat{\mathcal{L}}_n) = O(\alpha_n^{1/2}) \,, \text{ as } n \to \infty \,.$$

And if in addition $d_{BL}(\mathcal{L}_n, \mathcal{L}) = O(n^{-1/2})$ *as* $n \to \infty$, *then*

$$d_{BL}(\mathcal{L}_n, \widehat{\mathcal{L}}_n) = O(\alpha_n^{1/2} n^{-1/2}) \,, \text{ as } n \to \infty \,.$$

Proof. With no loss of generality, we assume the X_i's centered. From lemma 5.3 in Bertail & Clémençon (2004c), we have $l_n/\widehat{l}_n - 1 = O_{\mathbb{P}_\nu}(\alpha_n^{1/2})$ as $n \to \infty$. Besides, writing

$$d_{BL}(\mathcal{L}_n, \widehat{\mathcal{L}}_n) \leq (\frac{l_n - 1}{\widehat{l}_n - 1} - 1) \sup_{f \in Lip_{\mathbb{T}}^1} |\frac{1}{l_n - 1} \sum_{j=1}^{l_n-1} f(\mathcal{B}_j)| \,.$$

$$+ \frac{n}{\widehat{l}_n - 1} \sup_{f \in Lip_{\mathbb{T}}^1} |n^{-1} \sum_{j=1}^{l_n-1} f(\mathcal{B}_j) - n^{-1} \sum_{j=1}^{\widehat{l}_n-1} f(\widehat{\mathcal{B}}_j)| \,, \qquad (28)$$

and observing that $\sup_{f \in Lip_{\mathbb{T}}^1} |(l_n-1)^{-1} \sum_{j=1}^{l_n-1} f(\mathcal{B}_j)| \leq 1$, we get that the first term in the right hand side is $O_{\mathbb{P}_\nu}(\alpha_n^{1/2})$ as $n \to \infty$. Now as $\sup_{x \in E} |f(x)| \leq 1$, we have

$$|n^{-1}(\sum_{j=1}^{l_n} f(\mathcal{B}_j) - \sum_{j=1}^{\widehat{l}_n} f(\widehat{\mathcal{B}}_j))| \leq n^{-1}(|\widehat{\tau}_{A_{\mathcal{M}}}(1) - \tau_{A_{\mathcal{M}}}(1)| + |\widehat{\tau}_{A_{\mathcal{M}}}(l_n) - \widehat{\tau}_{A_{\mathcal{M}}}(l_n)|) \,,$$

and from lemma 5.1 in by Bertail & Clémençon (2004b), the term in the right hand side is $o_{\mathbb{P}_\nu}(n^{-1})$ as $n \to \infty$. We thus get

$$d_{BL}(\mathcal{L}_n, \widehat{\mathcal{L}}_n) \leq \alpha_n^{1/2} d_{BL}(\mathcal{L}_n, \mathcal{L}) + o_{\mathbb{P}_\nu}(n^{-1}) \,, \text{ as } n \to \infty \,.$$

And this completes the proof. $\qquad\square$

Given the metric on $\mathcal{P}_{\mathbb{T}}$ defined by d_{BL}, we consider now the *Fréchet differentiability* for functionals $T : \mathcal{P}_{\mathbb{T}} \to \mathbb{R}$.

Definition 2. *We say that T is Fréchet-differentiable at $\mathcal{L}_0 \in \mathcal{P}_{\mathbb{T}}$, if there exists a linear operator $DT^{(1)}_{\mathcal{L}_0}$ and a function $\epsilon^{(1)}(., \mathcal{L}_0) \colon \mathbb{R} \to \mathbb{R}$, continuous at 0 with $\epsilon^{(1)}(0, \mathcal{L}_0) = 0$, such that:*

$$\forall \mathcal{L} \in \mathcal{P}_{\mathbb{T}}, \; T(\mathcal{L}) - T(\mathcal{L}_0) = D^{(1)} T_{\mathcal{L}_0}(\mathcal{L} - \mathcal{L}_0) + R^{(1)}(\mathcal{L}, \mathcal{L}_0) ,$$

with $R^{(1)}(\mathcal{L}, \mathcal{L}_0) = d_{BL}(\mathcal{L}, \mathcal{L}_0) \epsilon^{(1)}(d_{BL}(\mathcal{L}, \mathcal{L}_0), \mathcal{L}_0)$. Moreover, T is said to have a canonical gradient (or influence function) $T^{(1)}(., \mathcal{L}_0)$, if one has the following representation for $DT^{(1)}_{\mathcal{L}_0}$:

$$\forall \mathcal{L} \in \mathcal{P}_{\mathbb{T}}, \; DT^{(1)}_{\mathcal{L}_0}(\mathcal{L} - \mathcal{L}_0) = \int_{\mathbb{T}} T^{(1)}(b, \mathcal{L}_0) \mathcal{L}(db) .$$

Now it is easy to see that from this notion of differentiability on the torus one may directly derive CLT's, provided the distance $d(\mathcal{L}_n, \mathcal{L})$ may be controlled.

Theorem 12. *In the regenerative case, if $T : \mathcal{P}_{\mathbb{T}} \to \mathbb{R}$ is Fréchet differentiable at $\mathcal{L}$ and $d_{BL}(\mathcal{L}_n, \mathcal{L}) = O_{\mathbb{P}_\nu}(n^{-1/2})$ (or $R^{(1)}(\mathcal{L}_n, \mathcal{L}) = o_{\mathbb{P}_\nu}(n^{-1/2})$) as $n \to \infty$, and if $\mathbb{E}_A(\tau_A) < \infty$ and $0 < \mathrm{var}_A(T^{(1)}(\mathcal{B}_1, \mathcal{L})) < \infty$ then under $\mathbb{P}_\nu$,*

$$n^{1/2}(T(\mathcal{L}_n) - T(\mathcal{L})) \Rightarrow \mathcal{N}(0, \mathbb{E}_A(\tau_A) \mathrm{var}_A(T^{(1)}(\mathcal{B}_1, \mathcal{L})) , \quad as \; n \to \infty .$$

In the general Harris case, if the split chain satisfies the assumptions above (with A replaced by $A_\mathcal{M}$), under the assumptions of Theorem 11, as $n \to \infty$ we have under $\mathbb{P}_\nu$,

$$n^{1/2}(T(\widehat{\mathcal{L}}_n) - T(\mathcal{L})) \Rightarrow \mathcal{N}(0, \mathbb{E}_{A_\mathcal{M}}(\tau_{A_\mathcal{M}}) \mathrm{var}_{A_\mathcal{M}}(T^{(1)}(\mathcal{B}_1, \mathcal{L})) .$$

The proof is straightforward and left to the reader. Observe that if one renormalizes by $l_n^{1/2}$ instead of renormalizing by $n^{1/2}$ in the atomic case (resp., by $\widehat{l}_n^{1/2}$ in the general case), the asymptotic distribution would be simply $\mathcal{N}(0, \mathrm{var}_A(T^{(1)}(\mathcal{B}_1, \mathcal{L}))$ (resp., $\mathrm{var}_{A_\mathcal{M}}(T^{(1)}(\mathcal{B}_1, \mathcal{L}))$), which depends on the atom chosen (resp. on the parameters of condition $\mathcal{M}$).

Then going back to the preceding examples, we straightforwardly deduce the following results.

a) Noticing that $n^{1/2}/l_n^{1/2} \to \mathbb{E}_A(\tau_A)^{1/2}$ $\mathbb{P}_\nu$- a.s. as $n \to \infty$, we immediately get that under $\mathbb{P}_\nu$, as $n \to \infty$,

$$n^{1/2}(\mu_n(f) - \mu(f)) \Rightarrow \mathcal{N}(0, \mathbb{E}_A(\tau_A)^{-1} \mathrm{var}_A(\sum_{i=1}^{\tau_A}(f(X_i) - \mu(f)) .$$

b) In a similar fashion, under smoothness assumptions ensuring Fréchet differentiability, the M-estimator $\widehat{\theta}_n$ being the (unique) solution of the block-estimating equation

$$\sum_{i=\tau_A+1}^{\tau_A(l_n)} \psi(X_i,\theta) = \sum_{j=1}^{l_n} \sum_{i=\tau_A(j)+1}^{\tau_A(j+1)} \psi(X_i,\theta) = 0 \;,$$

we formally obtain that, if $\mathbb{E}_A(\sum_{i=1}^{\tau_A} \frac{\partial \psi(X_i,\theta)}{\partial \theta}) \neq 0$ and θ is the true value of the parameter, then under $\mathbb{P}_\nu$, as $n \to \infty$,

$$n^{1/2}(\widehat{\theta}_n - \theta) \Rightarrow \mathcal{N}(0, \ [\frac{\mathbb{E}_A(\sum_{i=1}^{\tau_A} \frac{\partial \psi(X_i,\theta)}{\partial \theta})}{\mathbb{E}_A(\tau_A)}]^{-2} \frac{\mathrm{var}_A(\sum_{i=1}^{\tau_A} \psi(X_i,\theta))}{\mathbb{E}_A(\tau_A)}) \;.$$

Observe that both factors in the variance are independent from the atom A chosen. It is worth noticing that, by writing the asymptotic variance in this way, as a function of the distribution of the blocks, a consistent estimator for the latter is readily available, from the (approximate) regeneration blocks. Examples c) and d) may be treated similarly.

Remark 8. The concepts developed here may also serve as a tool for robustness purpose, for deciding whether a specific data block has an important influence on the value of some given estimate or not, and/or whether it may be considered as "outlier". The concept of robustness we introduce is related to blocks of observations, instead of individual observations. Heuristically, one may consider that, given the regenerative dependency structure of the process, a single suspiciously outlying value at some time point n may have a strong impact on the trajectory, until the (split) chain regenerates again, so that not only this particular observation but the whole "contaminated" segment of observations should be eventually removed. Roughly stated, it turns out that examining (approximate) regeneration blocks as we propose before, allows to identify more accurately outlying data in the sample path, as well as their nature (in the time series context, different type of outliers may occur, such as additive or innovative outliers). By comparing the data blocks (their length, as well as the values of the functional of interest on these blocks) this way, one may detect the ones to remove eventually from further computations.

8 Some extreme values statistics

We now turn to statistics related to the extremal behaviour of functionals of type $f(X_n)$ in the atomic positive Harris recurrent case, where $f : (E, \mathcal{E}) \to \mathbb{R}$ is a given measurable function. More precisely, we shall focus on the limiting distribution of the maximum $M_n(f) = \max_{1 \leq i \leq n} f(X_i)$ over a trajectory of length n, in the case when the chain X possesses an accessible atom A (see Asmussen (1998) and the references therein for various examples of such processes X in the area of queuing systems and a theoretical study of the tail properties of $M_n(f)$ in this setting).

8.1 Submaxima over regeneration blocks

For $j \geq 1$, we define the "submaximum" over the j-th cycle of the sample path:

$$\zeta_j(f) = \max_{1+\tau_A(j) \leq i \leq \tau_A(j+1)} f(X_i) .$$

The $\zeta_j(f)$'s are i.i.d. r.v.'s with common d.f. $G_f(x) = \mathbb{P}(\zeta_1(f) \leq x)$. The following result established by Rootzén (1988) shows that the limiting distribution of the sample maximum of $f(X)$ is entirely determined by the tail behaviour of the df G_f and relies on the crucial observation that the maximum value $M_n(f) = \max_{1 \leq i \leq n} f(X_i)$ over a trajectory of length n, may be expressed in terms of "submaxima" over regeneration blocks as follows

$$M_n(f) = \max(\zeta_0(f), \max_{1 \leq j \leq l_n - 1} \zeta_j(f), \zeta_{l_n}^{(n)}(f)) ,$$

where $\zeta_0(f) = \max_{1 \leq i \leq \tau_A} f(X_i)$ and $\zeta_{l_n}^{(n)}(f) = \max_{1 + \tau_A(l_n) \leq i \leq n} f(X_i)$ denote the maxima over the non regenerative data blocks, and with the usual convention that the maximum over an empty set equals $-\infty$.

Proposition 4. *(Rootzén, 1988) Let $\alpha = \mathbb{E}_A(\tau_A)$ be the mean return time to the atom A. Under the assumption (A1) that the first (non-regenerative) block does not affect the extremal behaviour, i.e. $\mathbb{P}_\nu(\zeta_0(f) > \max_{1 \leq k \leq l} \zeta_k(f)) \to 0$ as $l \to \infty$, we have*

$$\sup_{x \in \mathbb{R}} | \mathbb{P}_\nu(M_n(f) \leq x) - G_f(x)^{n/\alpha} | \to 0 , \quad as \ n \to \infty . \tag{29}$$

Hence, as soon as condition (A1) is fulfilled, the asymptotic behaviour of the sample maximum may be deduced from the tail properties of G_f. In particular, the limiting distribution of $M_n(f)$ (for a suitable normalization) is the extreme df $H_\xi(x)$ of shape parameter $\xi \in \mathbb{R}$ (with $H_\xi(x) = \exp(-x^{-1/\xi})\mathbb{I}\{x > 0\}$ when $\xi > 0$, $H_0(x) = \exp(-\exp(-x))$ and $H_\xi(x) = \exp(-(-x)^{-1/\xi})\mathbb{I}\{x < 0\}$ if $\xi < 0$) iff G_f belongs to the maximum domain of attraction $MDA(H_\xi)$ of the latter df (refer to Resnick (1987) for basics in extreme value theory). Thus, when $G_f \in MDA(H_\xi)$, there are sequences of norming constants a_n and b_n such that $G_f(a_n x + b_n)^n \to H_\xi(x)$ as $n \to \infty$, we then have $\mathbb{P}_\nu(M_n(f) \leq a'_n x + b_n) \to H_\xi(x)$ as $n \to \infty$, with $a'_n = a_n/\alpha^\xi$.

8.2 Tail estimation based on submaxima over regeneration blocks

In the case when assumption (A1) holds, one may straightforwardly derive from (29) estimates of $H_{f, n}(x) = \mathbb{P}_\nu(M_n(f) \leq x)$ as $n \to \infty$ based on the observation of a random number of submaxima $\zeta_j(f)$ over a sample path, as proposed in Glynn & Zeevi (2000):

$$\widehat{H}_{f, n, l}(x) = (\widehat{G}_{f, n}(x))^l ,$$

with $1 \le l \le l_n$ and denoting by $\widehat{G}_{f,\,n}(x) = \frac{1}{l_n - 1}\sum_{i=1}^{l_n - 1}\mathbb{I}\{\zeta_j(f) \le x\}$ the empirical df of the $\zeta_j(f)$'s (with $\widehat{G}_{f,\,n}(x) = 0$ by convention when $l_n \le 1$). We have the following limit result (see also Proposition 3.6 in Glynn & Zeevi (2000) for a different formulation, stipulating the observation of a determin-istic number of regeneration cycles).

Proposition 5. *Let (u_n) be such that $n(1 - G_f(u_n))/\alpha \to \eta < \infty$ as $n \to \infty$. Suppose that assumptions $\mathcal{H}_0(1, \nu)$ and (A1) holds, then $H_{f,\,n}(u_n) \to \exp(-\eta)$ as $\eta \to \infty$. And let $N_n \in \mathbb{N}$ such that $N_n/n^2 \to 0$ as $n \to \infty$, then we have*

$$\widehat{H}_{f,\,N_n,\,l_n}(u_n)/H_{f,n}(u_n) \to 1 \text{ in } \mathbb{P}_\nu - \text{probability, as } n \to \infty . \tag{30}$$

Moreover if $N_n/n^{2+\rho} \to \infty$ as $n \to \infty$ for some $\rho > 0$, this limit result also holds $\mathbb{P}_\nu$- a.s. .

Proof. First, the convergence $H_{f,\,n}(u_n) \to \exp(-\eta)$ as $\eta \to \infty$ straight-forwardly follows from Proposition 8.1. Now we shall show that $l_n(1 - \widehat{G}_{f,\,N_n}(u_n)) \to \eta$ in $\mathbb{P}_\nu$- pr. as $n \to \infty$. As $l_n/n \to \alpha^{-1}$ $\mathbb{P}_\nu$- a.s. as $n \to \infty$ by the SLLN, it thus suffices to prove that

$$n(G_f(u_n) - \widehat{G}_{f,\,N_n}(u_n)) \to 0 \text{ in } \mathbb{P}_\nu - \text{probability as } n \to \infty . \tag{31}$$

Write

$$n(G_f(u_n) - \widehat{G}_{f,\,N_n}(u_n)) = \frac{N_n}{l_{N_n} - 1}\frac{n}{N_n}\sum_{j=1}^{l_{N_n} - 1}\{\mathbb{I}\{\zeta_j(f) \le u_n\} - G_f(u_n)\} ,$$

and observe that $N_n/(l_{N_n} - 1) \to \alpha$, $\mathbb{P}_\nu$- a.s. as $n \to \infty$ by the SLLN again. Besides, from the argument of Theorem 15 in Clémençon (2001), we easily derive that there exist constants C_1 and C_2 such that for all $\varepsilon > 0$, $n \in \mathbb{N}$

$$\mathbb{P}_\nu\left(\left|\sum_{j=1}^{l_{N_n} - 1}\{\mathbb{I}\{\zeta_j(f) \le u_n\} - G_f(u_n)\}\right| \ge \varepsilon\right)$$
$$\le C_1 \exp(-C_2\varepsilon^2/N_n) + \mathbb{P}_\nu\left(\tau_A \ge N_n\right) .$$

From this bound, one immediately establishes (31). And in the case when $N_n = n^{2+\rho}$ for some $\rho > 0$, Borel-Cantelli's lemma, combined with the latter bound shows that the convergence also takes place $\mathbb{P}_\nu$-almost surely.   □

This result indicates that observation of a trajectory of length N_n, with $n^2 = o(N_n)$ as $n \to \infty$, is required for estimating consistently the extremal behaviour of the chain over a trajectory of length n. As shall be shown below, it is nevertheless possible to estimate the tail of the sample maximum $M_n(f)$ from the observation of a sample path of length n only, when assuming some type of behaviour for the latter, namely under maximum domain of attraction hypotheses. As a matter of fact, if one assume that $G_f \in MDA(H_\xi)$ for some $\xi \in \mathbb{R}$, of which sign is *a priori* known, one may implement classical inference procedures (refer to § 6.4 in Embrechts *et al.* (1999) for instance) from the observed submaxima $\zeta_1(f), ..., \zeta_{l_n-1}(f)$ for estimating the shape parameter ξ of the extremal distribution, as well as the norming constants a_n and b_n. We now illustrate this point in the Fréchet case (*i.e.* when $\xi > 0$), through the example of the Hill inference method.

8.3 Heavy-tailed stationary distribution

As shown in Rootzén (1988), when the chain takes real values, assumption (A1) is checked for $f(x) = x$ (for this specific choice, we write $M_n(f) = M_n$, $G_f = G$, and $\zeta_j(f) = \zeta_j$ in what follows) in the particular case when the chain is stationary, i.e. when $\nu = \mu$. Moreover, it is known that when the chain is positive recurrent there exists some index θ, namely the *extremal index* of the sequence $X = (X_n)_{n \in \mathbb{N}}$ (see Leadbetter & Rootzén (1988) for instance), such that

$$\mathbb{P}_\mu(M_n \leq x) \underset{n \to \infty}{\sim} F_\mu(x)^{n\theta} , \tag{32}$$

denoting by $F_\mu(x) = \mu(]-\infty, x]) = \alpha \mathbb{E}_A(\sum_{i=1}^{\tau_A} \mathbb{I}\{X_i \leq x\})$ the stationary df. In this case, as remarked in Rootzén (1988), if (u_n) is such that $n(1 - G(u_n))/\alpha \to \eta < \infty$, we deduce from Proposition 8.1 and (32) that

$$\theta = \lim_{n \to \infty} \frac{\mathbb{P}_A(\max_{1 \leq i \leq \tau_A} X_i > u_n)}{\mathbb{E}_A(\sum_{i=1}^{\tau_A} \mathbb{I}\{X_i > u_n\})} .$$

We may then propose a natural estimate of the extremal index θ based on the observation of a trajectory of length N,

$$\widehat{\theta}_N = \frac{\sum_{j=1}^{l_N - 1} \mathbb{I}\{\zeta_j > u_n\}}{\sum_{i=1}^{N} \mathbb{I}\{X_i > u_n\}} ,$$

which may be shown to be consistent (resp., strongly consistent) under $\mathbb{P}_\mu$ when $N = N_n$ is such that $N_n/n^2 \to \infty$ (resp. $N_n/n^{2+\rho} \to \infty$ for some $\rho > 0$) as $n \to \infty$ and $\mathcal{H}_0(2)$ is fulfilled by reproducing the argument of Proposition 9.2. And Proposition 8.1 combined with (32) also entails that for all ξ in $\mathbb{R}$,

$$G \in MDA(H_\xi) \Leftrightarrow F_\mu \in MDA(H_\xi) .$$

8.4 Regeneration-based Hill estimator

This crucial equivalence holds in particular in the Fréchet case, *i.e.* for $\xi > 0$. Recall that assuming that a df F belongs to $MDA(H_\xi)$ classically amounts then to suppose that it satisfies the tail regularity condition

$$1 - F(x) = L(x)x^{-a} ,$$

where $a = \xi^{-1}$ and L is a slowly varying function, *i.e.* a function L such that $L(tx)/L(x) \to 1$ as $x \to \infty$ for any $t > 0$ (*cf* Theorem 8.13.2 in Bingham *et al.* (1987)). Since the seminal contribution of Hill (1975), numerous papers have been devoted to the development and the study of statistical methods in the i.i.d. setting for estimating the tail index $a > 0$ of a regularly varying df. Various inference methods, mainly based on an increasing sequence of upper order statistics, have been proposed for dealing with this estimation problem, among which the popular *Hill estimator*, relying on a conditional maximum likelihood approach. More precisely, based on i.i.d. observations $X_1, ..., X_n$ drawn from F, the Hill estimator is given by

$$H^X_{k,\,n} = (k^{-1} \sum_{i=1}^{k} \ln \frac{X_{(i)}}{X_{(k+1)}})^{-1}\,, \tag{33}$$

where $X_{(i)}$ denotes the i-th largest order statistic of the sample $X^{(n)} = (X_1, ..., X_n)$, $1 \le i \le n$, $1 \le k < n$. Strong consistency (*cf* Deheuvels *et al.* (1988)) of this estimate has been established when $k = k_n \to \infty$ at a suitable rate, namely for $k_n = o(n)$ and $\ln \ln n = o(k_n)$ as $n \to \infty$, as well as asymptotic normality (see Goldie (1991)) under further conditions on F and k_n, $\sqrt{k_n}(H^X_{k_n,n} - a) \Rightarrow \mathcal{N}(0,\, a^2)$, as $n \to \infty$. Now let us define the *regeneration-based Hill estimator* from the observation of the $l_n - 1$ submaxima $\zeta_1, ..., \zeta_{l_n-1}$, denoting by $\xi_{(j)}$ the j-th largest submaximum,

$$\widehat{a}_{n,\,k} = H^\zeta_{k,\,l_n-1} = \left(k^{-1} \sum_{i=1}^{k} \ln \frac{\zeta_{(i)}}{\zeta_{(k+1)}} \right)^{-1}.$$

Given that $l_n \to \infty$, $\mathbb{P}_\nu$- a.s. as $n \to \infty$, results established in the case of i.i.d. observations straightforwardly extend to our setting (for comparison purpose, see Resnick & Starica (1995) for properties of the classical Hill estimate in dependent settings).

Proposition 6. *Suppose that $F_\mu \in MDA(H_{a^{-1}})$ with $a > 0$. Let (k_n) be an increasing sequence of integers such that $k_n \le n$ for all n, $k_n = o(n)$ and $\ln \ln n = o(k_n)$ as $n \to \infty$. Then the regeneration-based Hill estimator is strongly consistent*

$$\widehat{a}_{n,\,k_{l_n-1}} \to a, \ \mathbb{P}_\nu\text{- a.s. as } n \to \infty\,.$$

Under the further assumption that F_μ satisfies the Von Mises condition and that k_n is chosen accordingly (cf Goldie (1991)), it is moreover asymptotically normal in the sense that

$$\sqrt{k_{l_n-1}}(\widehat{a}_{n,\,k_{l_n-1}} - a) \Rightarrow \mathcal{N}(0,\, a^2) \ \text{under } \mathbb{P}_\nu \text{ as } n \to \infty\,.$$

9 Concluding remarks

Although we are far from having covered the unifying theme of statistics based on (pseudo-) regeneration for Harris Markov chains, an exhaustive treatment of the possible applications of this methodology being naturally beyond the scope of the present survey article, we endeavour to present here enough material to illustrate the power of this method. Most of the results reviewed in this paper are very recent (or new) and this line of research is still in development. Now we conclude by making a few remarks raising several open questions among the topics we focused on, and emphasizing the potential gain that the regeneration-based statistical method could provide in further applications.

• We point out that establishing sharper rates for the 2nd order accuracy of the ARBB when applied to sample mean statistics in the general Harris case presents considerable technical difficulties (at least to us). However, one might expect that this problem could be successfully addressed by refining some of the (rather loose) bounds put forward in the proof. Furthermore, as previously indicated, extending the argument to U-statistics requires to prove preliminary non-uniform limit theorems for U-statistics of random vectors with a lattice component.

• In numerous applications it is relevant to consider null recurrent (eventually regenerative) chains: such chains frequently arise in queuing/network systems, related to teletraffic data for instance (see Resnick (1997) or Glynn & Whitt (1995) for example), with heavy-tailed cycle lengths. Hence, exploring the theoretical properties of the (A)RBB for these specific time series provides thus another subject of further research: as shown by Karlsen & Tjøstheim (1998), consistent estimates of the transition kernel, as well as rates of convergence for the latter, may still be exhibited for β-recurrent null chains (*i.e.* chains for which the return time to an atom is in the domain of attraction of a stable law with $\beta \in]0, 1[$ being the stable index), so that extending the asymptotic validity of the (A)RBB distribution in this case seems conceivable.

• Turning to the statistical study of extremes now (which matters in insurance and finance applications for instance), a thorough investigation of the asymptotic behaviour of extreme value statistics based on the approximate regeneration blocks remains to be carried out in the general Harris case.

We finally mention ongoing work on empirical likelihood estimation in the markovian setting, for which methods based on (pseudo-) regeneration blocks are expected to provide significant results.

References

[AS85] Abramovitz L., Singh K.(1985). Edgeworth Corrected Pivotal Statistics and the Bootstrap, *Ann. Stat.*, **13** ,116-132.

[Asm87] Asmussen, S. (1987). *Applied Probabilities and Queues*. Wiley.

[Asm98] Asmussen, S. (1998). Extremal Value Theory for Queues Via Cycle Maxima. *Extremes*, **1**, No 2, 137-168.

[AA98] Athreya, K.B., Atuncar, G.S. (1998). Kernel estimation for real-valued Markov chains. *Sankhya*, **60**, series A, No 1, 1-17.

[AF89] Athreya, K.B., Fuh, C.D. (1989). Bootstrapping Markov chains: countable case. *Tech. Rep.* B-89-7, Institute of Statistical Science, Academia Sinica, Taipei, Taiwan, ROC.

[AN78] Athreya, K.B., Ney, P. (1978). A new approach to the limit theory of recurrent Markov chains. *Trans. Amer. Math. Soc.,* **245**, 493-501.

[Ber97] Bertail, P. (1997). Second order properties of an extrapolated bootstrap without replacement: the i.i.d. and the strong mixing cases, *Bernoulli*, **3**, 149-179.

[BC04a] Bertail, P., Clémençon, S. (2004a). Edgeworth expansions for suitably normalized sample mean statistics of atomic Markov chains. *Prob. Th. Rel. Fields*, **130**, 388–414 .

[BC04b] Bertail, P., Clémençon, S. (2004b). Note on the regeneration-based bootstrap for atomic Markov chains. *To appear in Test.*

[BC04c] Bertail, P. , Clémençon, S. (2004c). Regenerative Block Bootstrap for Markov Chains. *To appear in Bernoulli.*

[BC04d] Bertail, P. , Clémençon, S. (2004d). Approximate Regenerative Block-Bootstrap for Markov Chains: second-order properties. In *Compstat 2004 Proc.* Physica Verlag.

[BP01] Bertail, P., Politis, D. (2001). Extrapolation of subsampling distribution estimators in the i.i.d. and strong-mixing cases, *Can. J. Stat.*, **29**, 667-680.

[BK01] Bickel, P.J., Kwon, J. (2001). Inference for Semiparametric Models: Some Current Frontiers. *Stat. Sin.*, **11**, No. 4, 863-960.

[BGT89] Bingham N.H., Goldie G.M., Teugels J.L. (1989): *Regular Variation*, Cambridge University Press.

[Bir83] Birgé, L. (1983). Approximation dans les espaces métriques et théorie de l'estimation. *Z. Wahr. verw. Gebiete,* **65**, 181-237.

[BoL80] Bolthausen, E. (1980). The Berry-Esseen Theorem for strongly mixing Harris recurrent Markov Chains. *Z. Wahr. Verw. Gebiete*, **54**, 59-73.

[Bol82] Bolthausen, E. (1982). The Berry-Esseen Theorem for strongly mixing Harris recurrent Markov Chains. *Z. Wahr. Verw. Gebiete*, **60**, 283-289.

[BBD99] Borovkova,S., Burton R., Dehling H. (1999). Consistency of the Takens estimator for the correlation dimension. *Ann. Appl. Prob.*, **9**, No. 2, 376-390.

[BRT82] Brockwell, P.J., Resnick, S.J., Tweedie, R.L. (1982). Storage processes with general release rules and additive inputs. *Adv. Appl. Probab.,* **14**, 392-433.

[BS92] Browne, S., Sigman, K. (1992). Work-modulated queues with applications to storage processes. *J. Appl. Probab.*, **29**, 699-712.

[Bül97] Bühlmann, P. (1997). Sieve Bootstrap for time series. *Bernoulli*, **3**, 123-148.

[Bül02] Bühlmann, P. (2002). Bootstrap for time series. *Stat. Sci.*, **17**, 52-72.

[CV81] Callaert, H., Veraverbeke, N. (1981). The order of the normal approximation for a Studentized statistic. *Ann. Stat.*, **9**, 194-200.

[Car86] Carlstein, E. (1986). The use of subseries values for estimating the variance of a general statistic from a stationary sequence. *Ann. Statist.*, **14**, 1171-1179.

[Clé00] Clémençon, S. (2000). Adaptive estimation of the transition density of a regular Markov chain. *Math. Meth. Stat.,* **9**, No. 4, 323-357.

[Clé01] Clémençon, S. (2001). Moment and probability inequalities for sums of bounded additive functionals of regular Markov chains via the Nummelin splitting technique. *Stat. Prob. Letters*, **55**, 227-238.

[DM93] Datta, S., McCormick W.P. (1993). Regeneration-based bootstrap for Markov chains. *Can. J. Statist.*, **21**, No.2, 181-193.

[DHM88] Deheuvels, P. Häusler, E., Mason, D.M. (1988). Almost sure convergence of the Hill estimator. *Math. Proc. Camb. Philos. Soc.*, **104**, 371-381.

[DFMS04] Douc, R., Fort, G., Moulines, E., Soulier, P. (2004). Practical drift conditions for subgeometric rates of convergence. *Ann. Appl. Prob.*, **14**, No 3, 1353-1377.

[Dou94] Doukhan, P. (1994). *Mixing: Properties and Examples*. Lecture Notes in Statist., 85. Springer, New York.

[DG83] Doukhan, P., Ghindès, M. (1983). Estimation de la transition de probabilité d'une chaîne de Markov Doeblin récurrente. *Stochastic Process. Appl.*, **15**, 271-293.

[EKM01] Embrechts, P., Klüppelberg, C., Mikosch, T. (2001). *Modelling Extremal Events*. Springer-Verlag.

[Fel68] Feller, W. (1968). *An Introduction to Probability Theory and its Applications: vol. I*. John Wiley & Sons, NY, 2nd edition.

[Fel71] Feller, W. (1971). *An Introduction to Probability Theory and its Applications: vol. II*. John Wiley & Sons, NY, 3rd edition

[FKM02] Franke, J. , Kreiss, J. P., Mammen, E. (2002). Bootstrap of kernel smoothing in nonlinear time series. *Bernoulli*, **8**, 1–37.

[GZ00] Glynn, W.P., Zeevi, A. (2000). Estimating Tail Probabilities in Queues via Extremal Statistics. In *Analysis of Communication Networks: Call Centres, Traffic, and Performance* [D.R. McDonald and S.R. Turner, eds.] AMS, Providence, Rhode Island, 135-158.

[GW95] Glynn, W.P., Whitt, W. (1995). Heavy-Traffic Extreme-Value Limits for Queues. Op. Res. Lett. **18**, 107-111.

[Gol91] Goldie, C.M. (1991). Implicit renewal theory and tails of solutions of random equations. *Ann. Appl. Prob.*, **1**, 126-166.

[GH83] Götze, F., Hipp, C. (1983). Asymptotic expansions for sums of weakly dependent random vectors. *Zeit. Wahrschein. verw. Geb.*, **64**, 211-239.

[GK96] Götze, F., Künsch, H.R. (1996). Second order correctness of the blockwise bootstrap for stationary observations. *Ann. Statist.*, **24**, 1914-1933.

[Hal83] Hall P. (1983). Inverting an Edgeworth Expansion. *Ann. Statist.*, **11**, 569-576.

[Hal85] Hall, P. (1985). Resampling a coverage pattern. *Stoch. Process. Applic.*, **20**, 231-246.

[Hal92] Hall, P. (1992). *The Bootstrap and Edgeworth Expansion*. Springer.

[HR76] Harrison, J.M., Resnick, S.J. (1976). The stationary distribution and first exit probabilities of a storage process with general release rule. *Math. Oper. Res.*, **1**, 347-358.

[Hel91] Helmers, R (1991). On the Edgeworth expansion and the bootstrap approximation for a studentized statistics. *Ann. Statist.* ,**19**, 470-484.

[Hoe48] Hoeffding, W. (1948). A class of statistics with asymptotically normal distributions. *Ann. Math. Stat.*, **19**, 293–325.

[JJ67] Jain, J., Jamison, B. (1967). Contributions to Doeblin's theory of Markov processes. *Z. Wahrsch. Verw. Geb.*, **8**, 19-40.

[Kal78] Kalashnikov, V.V. (1978). *The Qualitative Analysis of the Behavior of Complex Systems by the Method of Test Functions*. Nauka, Moscow.

[KT01] Karlsen, H.A., Tjøstheim, D. (2001). Nonparametric estimation in null recurrent time series. *Ann. Statist.,* **29** (2), 372-416.

[Kün84] Künsch, H.R. (1984). Infinitesimal robustness for autoregressive processes. *Ann. Statist.,* **12**, 843-863.

[Kün89] Künsch, H.R. (1989). The jackknife and the bootstrap for general stationary observations. *Ann. Statist.,* **17**, 1217-1241.

[Lah03] Lahiri, S.N. (2003). *Resampling methods for dependent Data,* Springer.

[LR88] Leadbetter, M.R., Rootzén, H. (1988). Extremal Theory for Stochastic Processes. *Ann. Prob.,* **16**, No. 2, 431-478.

[LS92] Liu R., Singh K. (1992). Moving blocks jackknife and bootstrap capture weak dependence. In *Exploring The Limits of The Bootstrap.* Ed. Le Page R. and Billard L., John Wiley, NY.

[Mal85] Malinovskii, V. K. (1985). On some asymptotic relations and identities for Harris recurrent Markov Chains. *In Statistics and Control of Stochastic Processes,* 317-336.

[Mal87] Malinovskii, V. K. (1987). Limit theorems for Harris Markov chains I. *Theory Prob. Appl.,* **31**, 269-285.

[mal89] Malinovskii, V. K. (1989). Limit theorems for Harris Markov chains II. *Theory Prob. Appl.,* **34**, 252-265.

[MY86] Martin, R.D., Yohai, V.J. (1986). Influence functionals for time series. *Ann. Stat.,* **14**, 781-818.

[MT96] Meyn, S.P., Tweedie, R.L., (1996). *Markov chains and stochastic stability.* Springer.

[MSW01] Müller, U.U., Schick, A., Wefelmeyer, W., (2001). Improved estimators for constrained Markov chain models. *Stat. Prob. Lett.,* **54**, 427-435.

[Num78] Nummelin, E. (1978). A splitting technique for Harris recurrent chains. *Z. Wahrsch. Verw. Gebiete,* **43**, 309-318.

[nUM84] Nummelin, E. (1984). *General irreducible Markov chains and non negative operators.* Cambridge University Press, Cambridge.

[PP02] Paparoditis, E. and Politis, D.N. (2002). The local bootstrap for Markov processes. *J. Statist. Plan. Infer.,* **108**, 301–328.

[PR92] Politis, D.N. , Romano, J.P. (1992). A General Resampling Scheme for Triangular Arrays of alpha-mixing Random Variables with Application to the Problem of Spectral Density Estimation, *Ann. Statist.,* **20**, 1985-2007.

[PR94] Politis, D.N., Romano, J.P. (1994). Large sample confidence regions based on subsamples under minimal assumptions. *Ann. Statist.,* **22**, 2031-2050.

[PRW00] Politis, D.N., Romano, J.P., Wolf, T. (2000). *Subsampling.* Springer Series in Statistics, Springer, NY.

[Pol03] Politis, D.N. (2003). The impact of bootstrap methods on time series analysis. *Statistical Science* , **18**, No. 2, 219-230.

[PR83] Prakasa Rao, B.L.S. (1983). *Nonparametric Functional Estimation.* Academic Press, NY.

[RR98] Rachev, S. T., Rüschendorf, L. (1998). *Mass Transportation Problems. Vol. I* and *II.* Springer.

[Res87] Resnick, S. (1987). *Extreme Values, Regular Variation and Point Processes.* Springer, NY.

[Res97] Resnick, S. (1997). Heavy Tail Modeling And Teletraffic Data. *Ann. Stat.,* *25, 1805-1869.*

[RS95] Resnick, S., Starica, C. (1995). Consistency of Hill estimator for dependent data. *J. Appl. Prob.,* **32**, 139-167.

[Rev84] Revuz, D (1984). *Markov chains.* North-Holland, 2nd edition.

[RR96] Roberts, G.O., Rosenthal, J.S. (1996). Quantitative bounds for convergence rates of continuous time Markov processes. *Electr. Journ. Prob.*, **9**, 1-21.

[Ros70] Rosenblatt, M. (1970). Density estimates and Markov sequences. In *Nonparametric Techniques in Statistical Inference,* Ed. M. Puri, 199-210.

[Roo88] Rootzén, H. (1988). Maxima and exceedances of stationary Markov chains. *Adv. Appl. Prob.*, **20**, 371-390.

[Rou69] Roussas, G. (1969). Nonparametric Estimation in Markov Processes. *Ann. Inst. Stat. Math.*, 73-87.

[Rou91a] Roussas, G. (1991a). Estimation of transition distribution function and its quantiles in Markov Processes. In *Nonparametric Functional Estimation and Related Topics,* Ed. G. Roussas, 443-462.

[Rou91b] Roussas, G. (1991b). Recursive estimation of the transition distribution function of a Markov Process. *Stat. Probab. Letters*, **11**, 435-447.

[Ser81] Serfling J. (1981). *Approximation Theorems of Mathematical Statistics,* Wiley, NY.

[SZ00] Serfling, R., Zuo, Y., (2000). General Notions of Statistical Depth Function (in Data Depth). *Ann. Stat.*, **28**, No. 2., 461-482.

[Smi55] Smith, W. L. (1955). Regenerative stochastic processes. *Proc. Royal Stat. Soc.*, A, **232**, 6-31.

[Tjø90] Tjøstheim, D. (1990). Non Linear Time series, *Adv. Appl. Prob.*, **22**, 587-611.

[Tho00] Thorisson, H. (2000). *Coupling, Stationarity and Regeneration.* Springer.

Subgeometric ergodicity of Markov chains

Randal Douc[1], Eric Moulines[2], and Philippe Soulier[3]

[1] CMAP, Ecole Polytechnique, 91128 Palaiseau Cedex, France
   `douc@cmap.polytechnique.fr`
[2] Département TSI, Ecole nationale supérieure des Télécommunications, 46 rue
   Barrault, 75013 Paris, France `moulines@tsi.enst.fr`
[3] Equipe MODAL'X, Université de Paris X Nanterre, 92000 Nanterre, France
   `philippe.soulier@u-paris10.fr`

1 Introduction

Let P be a Markov tranition kernel on a state space X equipped with a countably generated σ-field $\mathcal{X}$. For a control function $f : \mathsf{X} \to [1, \infty)$, the *f-total variation* or *f-norm* of a signed measure μ on $\mathcal{X}$ is defined as

$$\|\mu\|_f := \sup_{|g| \leq f} |\mu(g)| .$$

When $f \equiv 1$, the f-norm is the total variation norm, which is denoted $\|\mu\|_{\mathrm{TV}}$. Assume that P is aperiodic positive Harris recurrent with stationary distribution π. Then the iterated kernels $P^n(x, \cdot)$ converge to π. The rate of convergence of $P^n(x, .)$ to π does not depend on the starting state x, but exact bounds may depend on x. Hence, it is of interest to obtain non uniform or quantitative bounds of the following form

$$\sum_{n=1}^{\infty} r(n) \|P^n(x, \cdot) - \pi\|_f \leq g(x), \quad \text{for all } x \in \mathsf{X} \tag{1}$$

where f is a control function, $\{r(n)\}_{n \geq 0}$ is a non-decreasing sequence, and g is a nonnegative function which can be computed explicitly.

As emphasized in [RR04, section 3.5], quantitative bounds have a substantial history in Markov chain theory. Applications are numerous including convergence analysis of Markov Chain Monte Carlo (MCMC) methods, transient analysis of queueing systems or storage models, etc. With few exception however, these quantitative bounds were derived under conditions which imply geometric convergence, *i.e.* $r(n) = \beta^n$, for some $\beta > 1$ (see for instance [MT94], [Ros95], [RT99], [RR04], and [Bax05]).

Geometric convergence does not hold for many chains of practical interest. Hence it is necessary to derive bounds for chains which converge to the stationary distribution at a rate r which grows to infinity slower than a geometric sequence. These sequences are called subgeometric sequences and are defined in [NT83] as non decreasing sequences r such that $\log r(n)/n \downarrow 0$ as $n \to \infty$. These sequences include among other examples the polynomial sequences $r(n) = n^\gamma$ with $\gamma > 0$ and subgeometric sequences $r(n)e^{cn^\delta}$ with $c > 0$ and $\delta \in (0, 1)$.

The first general results proving subgeometric rates of convergence were obtained by [NT83] and later extended by [TT94], but do not provide computable expressions for the bound in the rhs of (1). A direct route to quantitative bounds for subgeometric sequences has been opened by [Ver97, Ver99], based on coupling techniques. Such techniques were later used in specific contexts by many authors, among others, [FM00] [JR01] [For01] [FM03b].

The goal of this paper is to give a short and self contained proof of general bounds for subgeometric rates of convergence, under practical conditions. This is done in two steps. The first one is Theorem 1 whose proof, based on coupling, provides an intuitive understanding of the results of [NT83] and [TT94]. The second step is the use of a very general drift condition, recently introduced in [DFMS04]. This condition is recalled in Section 2.1 and the bounds it implied are stated in Proposition 1.

This paper complements the works [DFMS04] and [DMS05], to which we refer for applications of the present techniques to practical examples.

2 Explicit bounds for the rate of convergence

The only assumption for our main result is the existence of a small set.

(A1). There exist a set $C \in \mathcal{X}$, a constant $\epsilon > 0$ and a probability measure ν such that, for all $x \in C$, $P(x, \cdot) \geq \epsilon\nu(\cdot)$.

For simplicity, only one-step minorisation is considered in this paper. Adaptations to m-step minorisation can be carried out as in [Ros95] (see also [For01] and [FM03b]).

Let $\check{P}$ be a Markov transition kernel on $\mathsf{X} \times \mathsf{X}$ such that, for all $A \in \mathcal{X}$,

$$\check{P}(x, x', A \times \mathsf{X}) = P(x, A)\mathbb{1}_{(C \times C)^c}(x, x') + Q(x, A)\mathbb{1}_{C \times C}(x, x') \qquad (2)$$

$$\check{P}(x, x', \mathsf{X} \times A) = P(x', A)\mathbb{1}_{(C \times C)^c}(x, x') + Q(x', A)\mathbb{1}_{C \times C}(x, x') \qquad (3)$$

where A^c denotes the complementary of the subset A and Q is the so-called residual kernel defined, for $x \in C$ and $A \in \mathcal{X}$ by

$$Q(x, A) = \begin{cases} (1-\epsilon)^{-1} \left(P(x, A) - \epsilon\nu(A)\right) & 0 < \epsilon < 1 \\ \nu(A) & \epsilon = 1 \end{cases} \qquad (4)$$

One may for example set

$$\check{P}(x, x'; A \times A') =$$
$$P(x, A)P(x', A')\mathbb{1}_{(C \times C)^c}(x, x') + Q(x, A)Q(x', A)\mathbb{1}_{C \times C}(x, x') , \quad (5)$$

but this choice is not always the most suitable; cf. Section 2.2. For $(x, x') \in \mathsf{X} \times \mathsf{X}$, denote by $\check{\mathbb{P}}_{x,x'}$ and $\check{\mathbb{E}}_{x,x'}$ the law and the expectation of a Markov chain with initial distribution $\delta_x \otimes \delta_{x'}$ and transition kernel $\check{P}$.

Theorem 1. *Assume (A1).*

For any sequence $r \in \Lambda$, $\delta > 0$ and all $(x, x') \in \mathsf{X} \times \mathsf{X}$,

$$\sum_{n=1}^{\infty} r(n) \|P^n(x, \cdot) - P^n(x', \cdot)\|_{\mathrm{TV}} \leq (1+\delta)\check{\mathbb{E}}_{x,x'}\left[\sum_{k=0}^{\sigma} r(k)\right] + \frac{1-\epsilon}{\epsilon}M , \quad (6)$$

with $M = (1+\delta)\sup_{n \geq 0}\left\{R^* r(n-1) - \epsilon(1-\epsilon)\delta R(n)/(1+\delta)\right\}_+$ and $R^* = \sup_{(y,y') \in C \times C}\check{\mathbb{E}}_{y,y'}\left[\sum_{k=1}^{\tau} r(k)\right]$.

Let $W : \mathsf{X} \times \mathsf{X} \to [1, \infty)$ and f be a non-negative function f such that $f(x) + f(x') \leq W(x, x')$ for all $(x, x') \in \mathsf{X} \times \mathsf{X}$. Then,

$$\sum_{n=1}^{\infty} \|P^n(x, \cdot) - P^n(x', \cdot)\|_f \leq \check{\mathbb{E}}_{x,x'}\left[\sum_{k=0}^{\sigma} W(X_k, X_k')\right] + \frac{1-\epsilon}{\epsilon}W^* . \quad (7)$$

with $W^* = \sup_{(y,y') \in C \times C}\check{\mathbb{E}}_{y,y'}\left[\sum_{k=1}^{\tau} W(X_k, X_k')\right]$.

Remark 1. Integrating these bounds with respect to $\pi(\mathrm{d}x')$ yields similar bounds for $\|P^n(x, \cdot) - \pi\|_{\mathrm{TV}}$ and $\|P^n(x, \cdot) - \pi\|_f$.

Remark 2. The trade off between the size of the coupling set and the constant ϵ appears clearly: if the small set is big, then the chain returns more often to the small set and the moments of the hitting times can expected to be smaller, but the constant ϵ will be smaller. This trade-off is illustrated numerically in [DMS05, Section 3].

By interpolation, intermediate rates of convergence can be obtained. Let α and β be positive and increasing functions such that, for some $0 \leq \rho \leq 1$,

$$\alpha(u)\beta(v) \leq \rho u + (1-\rho)v , \quad \text{for all } (u, v) \in \mathbb{R}^+ \times \mathbb{R}^+ . \quad (8)$$

Functions satisfying this condition can be obtained from Young's inequality. Let ψ be a real valued, continuous, strictly increasing function on $\mathbb{R}^+$ such that $\psi(0) = 0$; then for all $a, b > 0$,

$$ab \leq \Psi(a) + \Phi(b) , \text{where} \quad \Psi(a) = \int_0^a \psi(x)dx \quad \text{and} \quad \Phi(b) = \int_0^b \psi^{-1}(x)dx ,$$

where ψ^{-1} is the inverse function of ψ. If we set $\alpha(u) = \Psi^{-1}(\rho u)$ and $\beta(v) = \Phi^{-1}((1-\rho)v)$, then the pair (α, β) satisfies (8). A trivial example is obtained by taking $\psi(x) = x^{p-1}$ for some $p \geq 1$, which yields $\alpha(u) = (p\rho u)^{1/p}$ and $\beta(u) = (p(1-\rho)u/(p-1))^{(p-1)/p}$. Other examples are given in Section 2.1.

Corollary 1. *Let α and β be two positive functions satisfying* (8) *for some* $0 \leq \rho \leq 1$. *Then, for any non-negative function f such that $f(x) + f(x') \leq \beta \circ W(x, x')$ and $\delta > 0$, for all $x, x' \in \mathsf{X}$ and $n \geq 1$,*

$$\sum_{n=1}^{\infty} \alpha(r(n)) \| P^n(x, \cdot) - P^n(x', \cdot) \|_f \leq \rho(1 + \delta) \breve{\mathbb{E}}_{x,x'} \left[\sum_{k=0}^{\sigma} r(k) \right]$$

$$+ (1 - \rho) \breve{\mathbb{E}}_{x,x'} \left[\sum_{k=0}^{\sigma} W(X_k, X_k') \right] \frac{1 - \epsilon}{\epsilon} \{ \rho M + (1 - \rho) W^* \} . \quad (9)$$

2.1 Drift Conditions for subgeometric ergodicity

The bounds obtained in Theorem 1 and Corollary 1 are meaningful only if they are finite. Sufficient conditions are given in this section in the form of drift conditions. The most well known drift condition is the so-called Foster-Lyapounov drift condition which not only implies but is actually equivalent to geometric convergence to the stationary distribution, cf. [MT93, Chapter 16]. [JR01], simplifying and generalizing an argument in [FM00], introduced a drift condition which implies polynomial rates of convergence. We consider here the following drift condition, introduced in [DFMS04], which allows to bridge the gap between polynomial and geometric rates of convergence.

Condition $\mathbf{D}(\phi, V, C)$: There exist a function $V : \mathsf{X} \to [1, \infty]$, a concave monotone non decreasing differentiable function $\phi : [1, \infty] \mapsto (0, \infty]$, a measurable set C and a constant $b > 0$ such that

$$PV + \phi \circ V \leq V + b \mathbb{1}_C.$$

If the function ϕ is concave, non decreasing and differentiable, define

$$H_\phi(v) := \int_1^v \frac{dx}{\phi(x)}. \quad (10)$$

Then H_ϕ is a non decreasing concave differentiable function on $[1, \infty)$. Moreover, since ϕ is concave, ϕ' is non increasing. Hence $\phi(v) \leq \phi(1) + \phi'(1)(v - 1)$ for all $v \geq 1$, which implies that H_ϕ increases to infinity. We can thus define its inverse $H_\phi^{-1} : [0, \infty) \to [1, \infty)$, which is also an increasing and differentiable function, with derivative $(H_\phi^{-1})'(x) = \phi \circ H_\phi^{-1}(x)$. For $k \in \mathbb{N}$, $z \geq 0$ and $v \geq 1$, define

$$r_\phi(z) := (H_\phi^{-1})'(z) = \phi \circ H_\phi^{-1}(z) . \quad (11)$$

It is readily checked that if $\lim_{t \to \infty} \phi'(t) = 0$, then $r_\phi \in \Lambda$, cf [DFMS04, Lemma 2.3].

Proposition 2.2 and Theorem 2.3 in [DMS05] show that the drift condition $\mathbf{D}(\phi, V, C)$ implies that the bounds of Theorem 1 are finite. We gather here these results.

Proposition 1. *Assume that Condition $\mathbf{D}(\phi, V, C)$ holds for some small set C and that $\inf_{x \notin C} \phi \circ V(x) > b$. Fix some arbitrary $\lambda \in (0, 1 - b/\inf_{x \notin C} \phi \circ V(x))$ and define $W(x, x') = \lambda \phi(V(x) + V(x') - 1)$. Define also $V^* = (1 - \epsilon)^{-1} \sup_{y \in C} \{PV(y) - \epsilon \nu(V)\}$. Let σ be the hitting time of the set $C \times C$. Then*

$$\check{\mathbb{E}}_{x,x'} \left[\sum_{k=0}^{\sigma} r_\phi(k) \right] \leq 1 + \frac{r_\phi(1)}{\phi(1)} \{V(x) + V(x')\} \mathbb{1}_{(x,x') \notin C \times C} \, ,$$

$$\check{\mathbb{E}}_{x,x'} \left[\sum_{k=0}^{\sigma} W(X_k, X'_k) \right] \leq \sup_{(y,y') \in C \times C} W(y, y') + \{V(x) + V(x')\} \mathbb{1}_{(x,x') \notin C \times C} \, ,$$

$$R^* \leq 1 + \frac{r_\phi(1)}{\phi(1)} \{2V^* - 1\}$$

$$W^* \leq \sup_{(y,y') \in C \times C} W(y, y') + 2V^* - 1 \, .$$

Remark 3. The condition $\inf_{y \notin C} \phi \circ V(y) > b$ may not be fulfilled. If level sets $\{V \leq d\}$ are small, then the set C can be enlarged so that this condition holds. This additional condition may appear rather strong, but can be weakened by using small sets associated to some iterate P^m of the kernel (see *e.g.* [Ros95], [For01] and [FM03b]).

We now give examples of rates that can be obtained by (11).

Polynomial rates

Polynomial rates of convergence are obtained when Condition $\mathbf{D}(\phi, V, C)$ holds with $\phi(v) = cv^\alpha$ for some $\alpha \in [0, 1)$ and $c \in (0, 1]$. The rate of convergence in total variation distance is $r_\phi(n) \propto n^{\alpha/(1-\alpha)}$ and the pairs (r, f) for which (9) holds are of the form $(n^{(1-p)\alpha/(1-\alpha)}, V^{\alpha p})$ for $p \in [0, 1]$, or in other terms, $(n^{\kappa-1}, V^{1-\kappa(1-\alpha)})$ for $1 \leq \kappa \leq 1/(1-\alpha)$, which is Theorem 3.6 of [JR01].

It is possible to extend this result by using more general interpolation functions. For instance, choosing for $b > 0$, $\alpha(x) = (1 \vee \log(x))^b$ and $\beta(x) = x(1 \vee \log(x))^{-b}$ yields the pairs $(n^{(1-p)\alpha/(1-\alpha)} \log^b(n), V^{\alpha p}(1 + \log V)^{-b})$, for $p \in [0, 1]$.

Logarithmic rates of convergence

Rates of convergence slower than any polynomial can be obtained when condition $\mathbf{D}(\phi, V, C)$ holds with a function ϕ that increases to infinity slower than polynomially, for instance $\phi(v) = c(1 + \log(v))^\alpha$ for some $\alpha \geq 0$ and $c \in (0, 1]$. A straightforward calculation shows that

$$r_\phi(n) \asymp \log^\alpha(n) \, .$$

Pairs for which (9) holds are thus of the form $((1 + \log(n))^{(1-p)\alpha}, (1 + \log(V))^{p\alpha})$.

Subexponential rates of convergence

Subexponential rates of convergence faster than any polynomial are obtained when the condition $\mathbf{D}(\phi, V, C)$ holds with ϕ such that $v/\phi(v)$ goes to infinity slower than polynomially. Assume for instance that ϕ is concave and differentiable on $[1, +\infty)$ and that for large v, $\phi(v) = cv/\log^\alpha(v)$ for some $\alpha > 0$ and $c > 0$. A simple calculation yields

$$r_\phi(n) \asymp n^{-\alpha/(1+\alpha)} \exp\left(\{c(1+\alpha)n\}^{1/(1+\alpha)}\right) .$$

Choosing $\alpha(x) = x^{1-p}(1 \vee \log(x))^{-b}$ and $\beta(x) = x^p(1 \vee \log(x))^b$ for $p \in (0,1)$ and $b \in \mathbb{R}$; or $p = 0$ and $b > 0$; or $p = 1$ and $b < -\alpha$ yields the pairs

$$n^{-(\alpha+b)/(1+\alpha)} \exp\left((1-p)\{c(1+\alpha)n\}^{1/(1+\alpha)}\right) , \quad V^p(1 + \log V)^b .$$

2.2 Stochastically monotone chains

Let X be a totally ordered set and let the order relation be denoted by $\preceq$ and for $a \in \mathsf{X}$, let $(-\infty, a]$ denote the set of all $x \in \mathsf{X}$ such that $x \preceq a$. A transition kernel on X is said to be stochastically monotone if $x \preceq y$ implies $P(x, (-\infty, a]) \geq P(y, (-\infty, a])$ for all $a \in \mathsf{X}$. If Assumption (A1) holds, for a small set $C = (-\infty, a_0]$, then instead of defining the kernel $\check{P}$ as in (5), it is convenient to define it, for $x, x' \in \mathsf{X}$ and $A \in \mathcal{X} \otimes \mathcal{X}$, by

$$\check{P}(x, x'; A) = \mathbb{1}_{(x,x') \notin C \times C} \int_0^1 \mathbb{1}_A(P^\leftarrow(x, u), P^\leftarrow(x', u)) \, du$$

$$+ \mathbb{1}_{C \times C}(x, x') \int_0^1 \mathbb{1}_A(Q^\leftarrow(x, u), Q^\leftarrow(x', u)) \, du ,$$

where, for any transition kernel K on X, $K^\leftarrow(x, \cdot)$ is the quantile function of the probability measure $K(x, \cdot)$, and Q is the residual kernel defined in (4). This construction makes the set $\{(x, x') \in \mathsf{X} \times \mathsf{X} : x \preceq x'\}$ absorbing for $\check{P}$. This means that if the chain (X_n, X_n') starts at (x_0, x_0') with $x_0 \preceq x_0'$, then almost surely, $X_n \preceq X_n'$ for all n. Let now σ_C and $\sigma_{C \times C}$ denote the hitting times of the sets C and $C \times C$, respectively. Then, we have the following very simple relations between the moments of the hitting times of the one dimensional chain and that of the bidimensional chain with transition kernel $\check{P}$. For any sequence r and any non negative function V all $x \preceq x'$

$$\check{\mathbb{E}}_{x,x'}\left[\sum_{k=0}^{\sigma_{C \times C}} r(k)V(X_k, X_k')\right] \leq \mathbb{E}_{x'}\left[\sum_{k=0}^{\sigma_C} r(k)V(X_k')\right] .$$

A similar bound obviously holds for the return times. Thus, there only remain to obtain bounds for this quantities, which is very straightforward if moreover condition $\mathbf{D}(\phi, \mathbf{V}, \mathbf{C})$ holds. Examples of stochastically monotone chains with applications to queuing and Monte-Carlo simulation that satisfy condition $\mathbf{D}(\phi, \mathbf{V}, \mathbf{C})$ are given in [DMS05, section 3].

3 Proof of Theorem 1

Define a transition kernel $\tilde{P}$ on the space $\tilde{X} = X \times X \times \{0,1\}$ endowed with the product σ-field $\tilde{\mathcal{X}}$, for any $x, x' \in X$ and $A, A' \in \mathcal{X}$, by

$$\tilde{P}\left((x, x', 0), A \times A' \times \{0\}\right) = \{1 - \epsilon \mathbb{1}_{C \times C}(x, x')\} \check{P}((x, x'), A \times A') , \quad (12)$$

$$\tilde{P}\left((x, x', 0), A \times A' \times \{1\}\right) = \epsilon \mathbb{1}_{C \times C}(x, x') \nu_{x,x'}(A \cap A') , \quad (13)$$

$$\tilde{P}\left((x, x', 1), A \times A' \times \{1\}\right) = P(x, A \cap A') . \quad (14)$$

For any probability measure $\tilde{\mu}$ on $(\tilde{X}, \tilde{\mathcal{X}})$, let $\tilde{\mathbb{P}}_{\tilde{\mu}}$ be the probability measure on the canonical space $(\tilde{X}^{\mathbb{N}}, \tilde{\mathcal{X}}^{\otimes \mathbb{N}})$ such that the coordinate process $\{\tilde{X}_k\}$ is a Markov chain with transition kernel $\tilde{P}$ and initial distribution $\tilde{\mu}$. The corresponding expectation operator is denoted by $\tilde{\mathbb{E}}_{\tilde{\mu}}$.

The transition kernel $\tilde{P}$ can be described algorithmically. Given $\tilde{X}_0 = (X_0, X'_0, d_0) = (x, x', d)$, $\tilde{X}_1 = (X_1, X'_1, d_1)$ is obtained as follows.

- If $d = 1$ then draw X_1 from $P(x, \cdot)$ and set $X'_1 = X_1$, $d_1 = 1$.
- If $d = 0$ and $(x, x') \in C \times C$, flip a coin with probability of heads ϵ.
  - If the coin comes up heads, draw X_1 from $\nu_{x,x'}$ and set $X'_1 = X_1$ and $d_1 = 1$.
  - If the coin comes up tails, draw (X_1, X'_1) from $\check{P}(x, x'; \cdot)$ and set $d_1 = 0$.
- If $d = 0$ and $(x, x') \notin C \times C$, draw (X_1, X'_1) from $\check{P}(x, x'; \cdot)$ and set $d_1 = 0$.

The variable d_n is called the *bell variable*; it indicates whether coupling has occurred by time n ($d_n = 1$) or not ($d_n = 0$). The first index n at which $d_n = 1$ is the coupling time;

$$T = \inf\{k \geq 1 : d_k = 1\}.$$

If $d_n = 1$ then $X_k = X'_k$ for all $k \geq n$. This coupling construction is carried out in such a way that under $\tilde{\mathbb{P}}_{\xi \otimes \xi' \otimes \delta_0}$, $\{X_k\}$ and $\{X'_k\}$ are Markov chains with transition kernel P with initial distributions ξ and ξ' respectively.

The main tool of the proof is the following relation between $\tilde{\mathbb{E}}_{x,x',0}$ and $\check{\mathbb{E}}_{x,x'}$, proved in [DMR04, Lemma 1]. For any non-negative adapted process $(\chi_k)_{k \geq 0}$ and $(x, x') \in X \times X$,

$$\tilde{\mathbb{E}}_{x,x',0}[\chi_n \mathbb{1}_{\{T>n\}}] = \check{\mathbb{E}}_{x,x'}\left[\chi_n (1 - \epsilon)^{N_{n-1}}\right] , \quad (15)$$

where $N_n = \sum_{i=0}^{n} \mathbb{1}_{C \times C}(X_i, X'_i)$ is the number of visits to $C \times C$ before time n.

We now proceed with the proof of Theorem 1.

Step 1 *Lindvall's inequality [Lin79, Lin92]*

$$\sum_{k=0}^{\infty} r(k) \|P^k(x, \cdot) - P^k(x', \cdot)\|_f \leq \tilde{\mathbb{E}}_{x,x',0}\left[\sum_{j=0}^{T-1} r(j) \{f(X_j) + f(X'_j)\}\right] . \quad (16)$$

Proof. For any measurable function ϕ such that $|\phi| \le f$, and for any $(x, x') \in \mathsf{X} \times \mathsf{X}$ it holds that

$$|P^k\phi(x) - P^k\phi(x')| = \left|\tilde{\mathbb{E}}_{x,x',0}[\{\phi(X_k) - \phi(X_k')\}\mathbb{1}_{\{d_k=0\}}]\right|$$
$$\le \tilde{\mathbb{E}}_{x,x',0}[\{f(X_k) + f(X_k')\}\mathbb{1}_{\{T>k\}}] \ .$$

Hence $\|P^k(x, \cdots) - P^k(x', \cdot)\|_f \le \tilde{\mathbb{E}}_{x,x',0}[\{f(X_k) + f(X_k')\}\mathbb{1}_{\{T>k\}}]$. Summing over k yields (16). $\square$

Step 2 *Denote* $W_{r,f}(x, x') = \check{\mathbb{E}}_{x,x'}\left[\sum_{k=0}^{\sigma} r(k)f(X_k, X_k')\right]$ *and* $W^*(r, f) = \sup_{(x,x')\in C\times C}\left[\sum_{k=1}^{\tau} r(k)f(X_k, X_k')\right]/r(0)$. *Then*

$$\tilde{\mathbb{E}}_{x,x',0}\left[\sum_{k=0}^{T-1} r(k)f(X_k, X_k')\right]$$
$$\le W_{r,f}(x, x') + \epsilon^{-1}(1 - \epsilon) W^*_{r,f} \, \tilde{\mathbb{E}}_{x,x',0}[r(T-1)] \ . \quad (17)$$

Proof. Applying (15), we obtain

$$\tilde{\mathbb{E}}_{x,x',0}\left[\sum_{k=0}^{T-1} r(k)f(X_k, X_k')\right] = \sum_{k=0}^{\infty} \tilde{\mathbb{E}}_{x,x',0}\left[r(k)f(X_k, X_k')\mathbb{1}_{\{T>k\}}\right]$$
$$= \sum_{k=0}^{\infty} \check{\mathbb{E}}_{x,x'}\left[r(k)f(X_k, X_k')(1 - \epsilon)^{N_{k-1}}\right]$$
$$= \sum_{j=0}^{\infty}\sum_{k=0}^{\infty}(1 - \epsilon)^j\check{\mathbb{E}}_{x,x'}\left[r(k)f(X_k, X_k')\mathbb{1}_{\{N_{k-1}=j\}}\right]$$
$$= W_{r,f}(x, x') + \sum_{j=1}^{\infty}\sum_{k=0}^{\infty}(1 - \epsilon)^j\check{\mathbb{E}}_{x,x'}\left[r(k)f(X_k, X_k')\mathbb{1}_{\{N_{k-1}=j\}}\right]$$

For $j \ge 0$, let σ_j denote the $(j + 1)$-th visit to $C \times C$. Then $N_{k-1} = j$ iff $\sigma_{j-1} < k \le \sigma_j$. Since r is a subgeometric sequence, $r(n+m) \le r(n)r(m)/r(0)$, thus

$$\sum_{k=0}^{\infty} r(k)f(X_k, X_k')\mathbb{1}_{\{N_{k-1}=j\}} = \sum_{k=\sigma_{j-1}+1}^{\sigma_j} r(k)f(X_k, X_k')$$
$$= \sum_{k=1}^{\tau\circ\theta^{\sigma_{j-1}}} r(\sigma_{j-1} + k)f(X_k, X_k')$$
$$\le \frac{r(\sigma_{j-1})}{r(0)}\left(\sum_{k=1}^{\tau\circ\theta^{\sigma_{j-1}}} r(k)f(X_k, X_k')\right) \circ \theta^{\sigma_{j-1}} \ .$$

Applying the strong Markov property yields

$$\tilde{\mathbb{E}}_{x,x',0}\left[\sum_{k=0}^{T-1} r(k)f(X_k, X'_k)\right] \leq W_{r,f}(x, x')$$

$$+ (1 - \epsilon)W^*(f, g)\sum_{j=0}^{\infty}(1 - \epsilon)^j \tilde{\mathbb{E}}_{x,x'}[r(\sigma_j)] .$$

By similar calculations, (15) yields

$$\tilde{\mathbb{E}}[r(T - 1)] = \epsilon \sum_{j=0}^{\infty}(1 - \epsilon)^j \tilde{\mathbb{E}}[r(\sigma_j)] ,$$

which concludes the proof of (17).   □

Step 3 Applying (17) with $r \equiv 1$ yields (7).

Step 4 If $r \in \Lambda$, then $\lim_{n\to\infty} r(n)/R(k) = 0$, with $R(0) = 1$ and $R(n) = \sum_{k=0}^{n-1} r(k)$, $n \geq 1$. Thus we can define, for $r \in \Lambda$ and $\delta > 0$

$$M_\delta = (1 + \delta)\sup_{n\geq 0}\left\{\epsilon^{-1}(1 - \epsilon)W^*_{r,1}r(n - 1) - \delta R(n)/(1 + \delta)\right\}_+ .$$

M_δ is finite for all $\delta > 0$. This yields

$$\tilde{\mathbb{E}}_{x,x',0}[R(T)] \leq (1 + \delta)W_{r,1}(x, x') + M_\delta .$$

Applying this bound with (16) yields (6).   □

References

[Bax05] Peter H. Baxendale. Renewal theory and computable convergence rates for geometrically ergodic markov chains. *Annals of Applied Probability*, 15(1B):700–738, 2005.

[DFMS04] Randal Douc, Gersende Fort, Eric Moulines, and Philippe Soulier. Practical drift conditions for subgeometric rates of convergence. *Annals of Applied Probability*, 14(3):1353–1377, 2004.

[DMR04] Randal Douc, Eric Moulines, and Jeff Rosenthal. Quantitative bounds for geometric convergence rates of Markov chains. *Annals of Applied Probability*, 14(4):1643–1665, 2004.

[DMS05] Randal Douc, Eric Moulines, and Philippe Soulier. Computable Convergence Rates for Subgeometrically Ergodic Markov Chains Preprint, 2005.

[FM00] Gersende Fort and Eric Moulines. V-subgeometric ergodicity for a Hastings-Metropolis algorithm. *Statistics and Probability Letters* 49(4):401–410, 2000.

[FM03a] Gersende Fort and Eric Moulines. Convergence of the monte carlo expectation maximization for curved exponential families. *Ann. Statist.*, 31(4):1220–1259, 2003.

[FM03b] Gersende Fort and Eric Moulines. Polynomial ergodicity of Markov tran-
         sition kernels,. *Stochastic Processes and Their Applications*, 103:57–99,
         2003.
[For01] Gersende Fort. *Contrôle explicite d'ergodicité de chanes de Markov: ap-
         plications e de convergence de l'algorithme Monte-Carlo EM*. PhD thesis,
         Université de Paris VI, 2001.
[JR01]   Soren Jarner and Gareth Roberts. Polynomial convergence rates of Markov
         chains. *Annals of Applied Probability*, 12(1):224–247, 2001.
[Lin79] Torgny Lindvall. On coupling of discrete renewal sequences. *Z. Wahrsch.
         Verw. Gebiete*, 48(1):57–70, 1979.
[Lin92] Torgny Lindvall. *Lectures on the Coupling Method*. Wiley Series in Prob-
         ability and Mathematical Statistics. John Wiley & Sons, New-York, 1992.
[MT93]   Sean P. Meyn and Robert L. Tweedie. Markov chains and stochastic sta-
         bility. Communications and Control Engineering Series. Springer-Verlag
         London, 1993.
[MT94]   Sean P. Meyn and Robert L. Tweedie. Computable bounds for convergence
         rates of Markov chains. *Annals of Applied Probability*, 4:981–1011, 1994.
[NT83]   Esa Nummelin and Pekka Tuominen. The rate of convergence in Orey's
         theorem for Harris recurrent Markov chains with applications to renewal
         theory. *Stochastic Processes and Their Applications*, 15:295–311, 1983.
[Ros95] Jeffrey S. Rosenthal. Minorization conditions and convergence rates for
         Markov chain Monte Carlo. *Journal American Statistical Association*,
         90:558–566, 1995.
[RR04]   Gareth O. Roberts and Jeffrey S. Rosenthal General state space Markov
         chains and MCMC algorithms. *Probability Suveys*, 1:20–71, 2004
[RT99]   Gareth O. Roberts and Richard L. Tweedie. Bounds on regeneration times
         and convergence rates for Markov chains. *Stochastic Processes and Their
         Applications*, 80:211–229, 1999.
[TT94]   Pekka Tuominen and Richard Tweedie. Subgeometric rates of convergence
         of *f*-ergodic Markov Chains. *Advances in Applied Probability*, 26:775–798,
         1994.
[Ver97] Alexander Veretennikov. On polynomial mixing bounds for stochastic
         differential equations. *Stochastic Process. Appl.*, 70:115–127, 1997.
[Ver99] Alexander Veretennikov. On polynomial mixing and the rate of conver-
         gence for stochastic differential and difference equations. *Theory of prob-
         ability and its applications*, pages 361–374, 1999.

Limit Theorems for Dependent U-statistics

Herold Dehling

Fakultät für Mathematik, Ruhr-Universität Bochum, Universitätsstraße 150, 44780 Bochum, Germany, `herold.dehling@rub.de`

1 Introduction

The asymptotic distribution of U-statistics, and of the related von-Mises-statistics, of independent observations has been investigated for almost 60 years and is rather well understood. All the classical limit theorems for partial sums of independent random variables have a U-statistics counterpart. In this paper we give a survey of some recent progress for U-statistics of weakly dependent observations. We will mostly assume that the observations are generated by functionals of absolutely regular processes. Specifically, we will consider the U-statistic ergodic theorem, the U-statistic central limit theorem and the invariance principle for U-processes. We motivate our investigations by a wide range of examples, e.g. from fractal dimension estimation in time series analysis.

In this paper we will always assume that $(X_n)_{n \in \mathbb{Z}}$ is a stationary ergodic process of $\mathbb{R}^k$-valued random variables. In parts of the paper further restrictions have to be made, e.g. in the form of weak dependence assumptions.

Definition 1. *Let $h \colon \mathbb{R}^k \times \mathbb{R}^k \to \mathbb{R}$ be a measurable symmetric function, i.e. $h(x, y) = h(y, x)$ for all $x, y \in \mathbb{R}^k$. We then define the U-statistic $U_n(h)$ by*

$$U_n(h) = \frac{1}{\binom{n}{2}} \sum_{1 \le i < j \le n} h(X_i, X_j)$$

and the von-Mises-statistic $V_n(h)$ by

$$V_n(h) = \frac{1}{n^2} \sum_{1 \le i,j \le n} h(X_i, X_j) \, .$$

The function h is called the kernel of the U-statistic, respectively von-Mises-statistic.

We can easily extend the definition of U-statistics and von-Mises-statistics to kernels $h : (\mathbb{R}^k)^m \to \mathbb{R}$, in which case $U_n(h)$ and $V_n(h)$ are defined as averages of $h(X_{i_1}, \cdots, X_{i_m})$, $1 \le i_1 < \cdots < i_m \le n$ or $h(X_{i_1}, \cdots, X_{i_m})$, $1 \le i_1, \cdots, i_m \le n$, respectively. In this paper we will restrict attention to the bivariate case, i.e. $m = 2$. Essentially all results remain valid in the general case. Moreover we will mainly consider $\mathbb{R}$-valued U-statistics, again noting that all results also hold for $\mathbb{R}^k$-valued processes.

By symmetry of h, we can rewrite a U-statistics as

$$U_n(h) = \frac{1}{n(n-1)} \sum_{1 \le i \ne j \le n} h(X_i, X_j) \, .$$

The essential difference between U-statistics and von-Mises-statistics thus lies in the fact that the diagonal terms $h(X_i, X_i)$ are included in the von Mises-statistics and excluded in the U-statistics. As

$$n^2 V_n(h) - n(n-1)U_n(h) = \sum_{i=1}^n h(X_i, X_i)$$

and since the asymptotic behavior of the partial sum $\sum_{i=1}^n h(X_i X_i)$ is well understood, one can fairly easily obtain results for U-statistics from corresponding results for von-Mises-statistics and vice versa.

U-statistics have been introduced independently by Halmos (1946) and Hoeffding (1948). Von-Mises-statistics were introduced by von Mises (1947). The motivation in each of these papers was rather different. Halmos was interested in the theory of unbiased estimation, noting that in the case of i.i.d. observations $X_1, \cdots, X_n$

$$\mathbb{E}(U_n(h)) = \mathbb{E}h(X_1, X_2) \, .$$

Hence $U_n(h)$ is an unbiased estimator of the functional $\theta = \theta(F) := \mathbb{E}_F h(X_1, X_2)$, where $\mathbb{E}_F$ indicates that the random variables X_i have marginal distribution F. Moreover

$$U_n(h) = \mathbb{E}(h(X_1, X_2)|X_{(1)}, \cdots, X_{(n)}) \, ,$$

where $X_{(1)} \le \cdots \le X_{(n)}$ denote the order statistics. If the class of possible marginal distributions specified by a given statistical model is rich enough, the order statistic is a complete sufficient statistic and thus $U_n(h)$ is the minimum variance unbiased estimator of $\theta(F)$. Hoeffding (1948) stressed the fact that U-statistics are a generalized mean, namely of the terms $h(X_i, X_j), 1 \le i < j \le n$, and that one could still show asymptotic normality as in the case of ordinary means.

Von-Mises-statistics originated in the theory of differentiable statistical functionals, initiated by von Mises (1947). Suppose we are given a family $\mathcal{P}$ of possible marginal distributions of X_i, where $X_1, \cdots X_n$ is again an i.i.d.

sample. We want to estimate the parameter $\theta = T(F)$ when $T : \mathcal{P} \to \mathbb{R}$ is a given map. A natural estimator for θ is then the plug-in estimator $\hat{\theta} = T(F_n)$, where

$$F_n = \frac{1}{n} \sum_{i=1}^{n} \delta_{X_i}$$

denotes the empirical distribution function. Von Mises proposed to investigate the asymptotic distribution of $\hat{\theta}_n - \theta$ by a Taylor expansion of the operator T in a neighborhood of the true distribution F. Under suitable differentiability assumptions this leads to the expansion

$$T(F_n) - T(F) = D_F T(F_n - F) + \frac{1}{2} D_F^2 T(F_n - F, F_n - F) + \text{higher order terms.}$$

Making use of linearity of the operator $D_F T$, the first order term in this expansion can be rewritten as

$$D_F T(F_n - F) = D_F T \left(\frac{1}{n} \sum_{i=1}^{n} \delta_{X_i} - F \right) = \frac{1}{n} \sum_{i=1}^{n} D_F T \left(\delta_{X_i} - F \right) ,$$

and is thus an average of i.i.d. variables $D_F T(\delta_{X_i} - F)$. As the 2nd order derivative $D_F^2 T$ is a bilinear operator, we obtain

$$D_F^2 T (F_n - F, F_n - F) = D_F^2 T \left(\frac{1}{n} \sum_{i=1}^{n} \delta_{X_i} - F, \frac{1}{n} \sum_{i=1}^{n} \delta_{X_i} - F \right)$$

$$= \frac{1}{n^2} \sum_{1 \le i,j \le n} D_F^2 T \left(\delta_{X_i} - F, \delta_{X_j} - F \right) .$$

Hence $D_F^2 T(F_n - F, F_n - F)$ is a von-Mises-statistic with kernel $h(x,y) = D_F^2 T(\delta_x - F, \delta_y - F)$.

Many sample statistics can be expressed at least approximately as U-statistics or von-Mises-statistics, thus providing a very practical reason for the study of these classes of statistics. Below we list some examples, ranging from standard textbook examples to some recent applications in the area of dimension estimation of distributions with a fractal support.

SAMPLE VARIANCE

The sample variance is defined as

$$s_X^2 := \frac{1}{n-1} \sum_{i=1}^{n} (X_i - \bar{X})^2 ,$$

where $\bar{X} := \frac{1}{n} \sum_{i=1}^{n} X_i$ is the sample mean. Some small calculations show that s_X^2 is a U-statistic with kernel $h(x,y) = \frac{1}{2}(x - y)^2$.

CRAMÉR-VON-MISES STATISTICS

Given a distribution function F_0 and a weight function w we define the kernel

$$h(x, y) := \int \left(\mathbb{1}_{[x,\infty)}(s) - F_0(s) \right) \left(\mathbb{1}_{[y,\infty)}(s) - F_0(s) \right) w(s) \, ds .$$

The associated von-Mises-statistic is

$$V_n(h) = \int \left(F_n(s) - F_0(s) \right)^2 w(s) \, ds .$$

This statistic is known as the Cramér-von Mises-statistic that can be used for testing the hypothesis that F_0 is the underlying distribution of the random variables $X_i, 1 \leq i \leq n$.

χ^2-TEST STATISTIC

The χ^2-statistic for testing goodness-of-fit in models for discrete random variables $X_1, \cdots, X_n$ with possible outcomes $a_1, \ldots, a_K$ is another example of a von-Mises-statistic, arising from the kernel

$$h(x, y) = \sum_{k=1}^{K} \frac{1}{p_k} \left(\mathbb{1}_{\{x=a_k\}} - p_k \right) \left(\mathbb{1}_{\{y=a_k\}} - p_k \right) .$$

In this case, the associated von-Mises-statistic is

$$V_n(h) = \frac{1}{n^2} \sum_{1 \leq i,j \leq n} h(X_i, X_j)$$

$$= \frac{1}{n} \sum_{k=1}^{K} \frac{1}{np_k} \left(N_k - np_k \right)^2 ,$$

where N_k denotes the number of observations among $X_1, \ldots, X_n$ with outcome a_k. Thus, up to a norming constant $\frac{1}{n}$, $V_n(h)$ is the usual χ^2-test statistic.

GRASSBERGER-PROCACCIA ESTIMATOR OF THE CORRELATION DIMENSION

Our last two examples of U-statistics concern the estimation of fractal dimensions. One such notion is the correlation dimension, associated to distributions F on $\mathbb{R}^k$. We first define the correlation integral

$$C(r) = \int_{\mathbb{R}^k} F(B_r(x)) \, dF(x) ,$$

where $B_r(x) := \{ y : \|x - y\| \leq r \}$ denotes the ball of radius r around x. Thus $C(r)$ is the average mass that the distribution F gives to a ball of radius r, averaged with respect to the distribution F. The scaling behavior of $C(r)$ as $r \to 0$ gives information about the dimension of the support of F. If

$$C(r) \sim const. \cdot r^d, \text{ as } r \to 0 , \tag{1}$$

we call d the correlation dimension of F.

Often we do not know the distribution F, but we are only given a finite sample $X_1, \ldots, X_n$ of observations from a stationary process with marginal distribution F. Based on this sample, we want to estimate d. We start by estimating the correlation integral, noting that by Fubini's theorem $C(r)$ can alternatively be expressed as

$$C(r) = \mathbb{P}(\|X - Y\| \leq r),$$

where X, Y are independent random variables with distribution F. Thus, a natural estimator for $C(r)$ is the sample analogue

$$\begin{aligned}
C_n(r) &= \frac{1}{\binom{n}{2}} \# \left\{ 1 \leq i < j \leq n : \|X_i - X_j\| \leq r \right\} \\
&= \frac{1}{\binom{n}{2}} \sum_{1 \leq i < j \leq n} \mathbb{1}_{\{\|X_i - X_j\| \leq r\}} ,
\end{aligned}$$

which is a U−statistic with kernel $h(x, y) = \mathbb{1}_{\{\|x - y\| \leq r\}}$. To get an estimator for d, we take the logarithm on both sides of (1) to obtain

$$\log C(r) \approx const. + d \log r .$$

This suggests to take linear regression of $\log C(r)$ on $\log r$ and to estimate d by the slope of the regression line. This estimator was introduced in 1984 by Grassberger and Procaccia. As the scaling property (1) only holds asymtotically as $r \to 0$, the estimation of d should be based on $C_n(r)$-values in a small region $0 \leq r \leq r$. There is then the usual bias-variance trade-off, as a small region means few observations and thus a larger variance. The choice of r_0 should also depend on n with $r_0 = r_0(n) \to 0$ as $n \to \infty$.

TAKENS' ESTIMATOR

An alternative estimator for the correlation dimension was proposed by Takens (1985). The considerations leading to Takens' estimator start from the assumption that exact scaling holds in a neighborhood of 0, i.e. that

$$C(r) = const \cdot r^d, \ 0 \leq r \leq r_0 ,$$

for some r_0. Suppose moreover that we are given independent random variables $R_1, \ldots, R_K$ with distribution

$$\mathbb{P}(R_k \leq r) = \mathbb{P}(\|X - Y\| \leq r \mid \|X - Y\| \leq r_0) ,$$

where X and Y are independent random variables with distribution F. The Maximum Likelihood estimator of d, based on the observations $R_1, \ldots, R_K$, then becomes

$$\hat{d}_{ML} = \left(-\frac{1}{K} \sum_{k=1}^{K} \log(R_k/r_0) \right)^{-1}.$$

If we replace the independent copies $R_1, \ldots, R_K$ by those dependent pair distances $\|X_i - X_j\|, 1 \leq i, j \leq n$, that satisfy $\|X_i - X_j\| \leq r_0$, we obtain the Takens estimator

$$\hat{d} = \frac{\sum_{1 \leq i < j \leq n} 1\{\|X_i - X_j\| \leq r_0\}}{\sum_{1 \leq i < j \leq n} \log^{-}(\|X_i - X_j\|/r_0)}.$$

Here $\log^{-}(x) = \max(-\log(x), 0)$ denotes the negative part of the logarithm. Thus, the Takens estimator is given by a ratio of two U-statistics, with kernels $1_{\{\|x-y\| \leq r_0\}}$ and $\log^{-}(\|x - y\|/r_0)$, respectively.

Dimension estimation is applied in the analysis of time series arising from a deterministic dynamical system. Let $(\mathcal{X}, \mathcal{F}, \mu)$ be a probability space and $T : \mathcal{X} \to \mathcal{X}$ a measure preserving map, i.e. $\mu(T^{-1}A) = \mu(A)$ for any $A \in \mathcal{F}$. Moreover let $f : \mathcal{X} \to \mathbb{R}$ be a measurable map and consider the process

$$y_n = f(T^n X_0), n \geq 0,$$

where X_0 is a randomly chosen initial value with distribution μ. Though arising from an underlying deterministic system, the process $(y_n)_{n \geq 0}$ may exhibit seemingly random behavior. Information about the underlying dynamical system can be gained from a sample of the so-called reconstruction vectors

$$X_n = (y_n, y_{n-1}, \cdots, y_{n-k+1}).$$

E.g., estimation of the dimension of the distribution of X_n leads to information about the dimension of the attractor of the dynamical system $(\mathcal{X}, \mathcal{F}, \mu, T)$. A theoretical basis for many of these procedures is given by Takens' (1981) reconstruction theorem stating that for generic maps f, the reconstruction map $Rec_k : \mathcal{X} \to \mathbb{R}^k$ defined by

$$Rec_k(x) = (f(x), f(Tx), \cdots, f(T^{k-1}x))$$

is an embedding, provided $k \geq 2d + 1$, where $d = dim(\mathcal{X})$.

2 Independent Limit Theory

In this section we want to give a brief survey of limit theorems for U-statistics when the underlying observations $(X_n)_{n \geq 0}$ form an i.i.d. process of random variables with marginal distribution F. Analogues of the well-known classical limit theorems for partial sums of i.i.d. random variables have been established for U-statistics in the period between 1948 and 1989.

The law of large numbers for U-statistics was proved independently by Hoeffding (1961) and Berk (1966). If $h \in L_1(F \times F)$, we have

$$\lim_{n \to \infty} U_n(h) = \int \int h(x, y) \, dF(x) \, dF(y) \, ,$$

almost surely. One can prove this result using martingale techniques, observing that $(U_n)_{n \geq 2}$ is a backwards martingale.

All further limit theorems make essential use of a technical tool, the so-called Hoeffding decomposition. We define functions $h_1(x)$ and $h_2(x, y)$ by

$$h_1(x) = \mathbb{E}h(x, Y) - \theta \tag{2}$$
$$h_2(x, y) = h(x, y) - h_1(x) - h_1(y) - \theta \, , \tag{3}$$

where $\theta := \mathbb{E}(h(X, Y))$ and where X, Y are independent random variables with distribution F. From (3) we immediately obtain the Hoeffding decomposition of the kernel h, given by

$$h(x, y) = \theta + h_1(x) + h_1(y) + h_2(x, y) \, . \tag{4}$$

In this way, we have written $h(x, y)$ as a sum of a constant term, of two functions of the variables x and y separately, and of a function h_2 of both variables. The functions h_1 and h_2 have special properties, namely

$$\mathbb{E}h_1(X) = 0 \, , \tag{5}$$
$$\mathbb{E}h_2(x, Y) = \mathbb{E}h_2(X, y) = 0, \quad x, y \in \mathbb{R}^k \, , \tag{6}$$

as one can easily show. Property (6) is a crucial property of h_2 and is know as the degeneracy condition. A kernel h satisfying

$$\mathbb{E}h(x, Y) = \mathbb{E}h(X, y) = 0 \quad \forall x, y \in \mathbb{R}^k \, ,$$

is called a degenerate kernel. Sometimes, one calls h degenerate if $\mathbb{E}h(x, Y) = \mathbb{E}h(X, y) \equiv \theta$, in which case $h(x, y) - \theta$ satisfies the above property.

From (4) we obtain via a small computation the Hoeffding decomposition of the U-statistic $U_n(h)$,

$$U_n(h) = \theta + \frac{2}{n} \sum_{i=1}^{n} h_1(X_i) + U_n(h_2) \, .$$

In this way, we have decomposed $U_n(h)$ into a sum of three terms, namely the constant term θ, an average of $h_1(X_i)$ and a U-statistic with a degenerate kernel. By (5) and (6), the terms $h_1(X_i), 1 \leq i \leq n$ and $h_2(X_i, X_j), 1 \leq i < j \leq n$ are all mutually uncorrelated. Thus we get

$$\text{var}\left(\frac{2}{n} \sum_{i=1}^{n} h_1(X_i)\right) = \frac{4}{n} \text{var}\left(h_1(X_1)\right) \, ,$$

$$\text{var}\left(U_n(h_2)\right) = \frac{1}{\binom{n}{2}} \text{var}\left(h_2(X_1, X_2)\right) \, .$$

Thus $U_n(h_2) = O_\mathbb{P}(n^{-1})$ and hence we can obtain the U-statistic central limit theorem from the CLT for partial sums of i.i.d. random variables. The result, due to Hoeffding (1948), is as follows,

$$\sqrt{n}\,(U_n(h) - \theta) \overset{\mathcal{L}}{\to} N\left(0, 4\,\mathrm{var}(h_1(X))\right)\,, \tag{7}$$

provided that $\mathbb{E}(h(X,Y))^2 < \infty$. Using similar arguments, a Donsker type invariance principle and a law of the iterated logarithm for U-statistics have been established.

An interesting special situation occurs when $h_1(x) \equiv 0$, i.e. when h is itself a degenerate kernel. In this case the limit distribution in the CLT is degenerate, and one can apply a different normalization in order to get a non-trivial limit distribution. The result, due to Fillipova (1964), is most conveniently presented when the underlying observations $(X_i)_{i \geq 1}$, are uniformly distributed on $[0,1]$. Invoking the quantile transform technique, one may assume this without loss of generality. Then for any L_2-kernels $h : [0,1]^2 \to \mathbb{R}$ one has

$$\frac{2}{n} \sum_{1 \leq i < j \leq n} h(X_i, X_j) \overset{\mathcal{L}}{\to} \int_0^1 \int_0^1 h(x,y)\,\mathrm{d}W(x)\,\mathrm{d}W(y)\,, \tag{8}$$

where W denotes standard Brownian motion and the stochastic double integral is to be taken in the sense of Itô, i.e. not integrating over the diagonals. The corresponding Donsker invariance principle was established by Denker, Grillenberger and Keller (1985) who could show that in the space $D([0,1])$ equipped with the Skorohod topology

$$\left(\frac{2}{n} \sum_{1 \leq i < j \leq [nt]} h(X_i, X_j)\right)_{0 \leq t \leq 1} \overset{\mathcal{L}}{\to} \left(\int_0^1 \int_0^1 h(x,y)K(t,\mathrm{d}x)K(t,\mathrm{d}y)\right)_{0 \leq t \leq 1}, \tag{9}$$

where $K(t,x)$ denotes the Kiefer-Müller process. The law of the iterated logarithm as well as an almost sure invariance principle for degenerate U-statistics was established in a series of papers by Dehling, Denker, Philipp (1983, 1984) and Dehling (1989a, 1989b), assuming that the kernel h has finite 2nd moments in the bivariate case and finite $(2 + \delta)$th moments in the general case. The latter assumption was weakened to finite 2nd moments also for m-variate U-statistics by Dehling, Utev (1993) and independently by Arcones and Giné (1993). A survey of the limit theory for degenerate U-statistics was given by Dehling (1986).

3 Functionals of Absolutely Regular Processes

In the remaining part of this paper, we will consider U-statistics with dependent observations $(X_n)_{n \in \mathbb{Z}}$. The minimal assumption will be that $(X_n)_{n \in \mathbb{Z}}$ is a stationary ergodic process. For most of our results we will need stronger assumptions concerning the weak dependence of the underlying process. In the context of U-statistic limit theorems, the notion of an absolutely regular process is most suitable.

Definition 2. *(i) Let $(\Omega, \mathcal{A}, \mathbb{P})$ be a probability space, and let $\mathcal{F}, \mathcal{G}$ be sub-σ-fields of $\mathcal{A}$. We define*

$$\beta(\mathcal{F}, \mathcal{G}) := \sup_{F_1, \ldots, F_m, G_1, \ldots, G_n} \sum_{i=1}^{m} \sum_{j=1}^{n} |\mathbb{P}(F_i \cap G_j) - \mathbb{P}(F_i) \cdot \mathbb{P}(G_j)| \ ,$$

where the supremum is taken over all partitions $F_1, \ldots, F_m$ and $G_1, \ldots, G_n$ of Ω into elements of $\mathcal{F}$ and $\mathcal{G}$, respectively.

(ii) Given a stochastic process $(X_n)_{n \in \mathbb{Z}}$ and integers $a \leq b$, we denote by $\mathcal{F}_a^b$ the σ-field generated by the random variables $X_{a+1}, \ldots, X_b$. We define the mixing coefficients of absolute regularity by

$$\beta_k := \sup_{n \in \mathbb{Z}} \beta \left(\mathcal{F}_{-\infty}^n, \mathcal{F}_{n+k}^\infty \right) \ .$$

The process $(X_n)_{n \in \mathbb{Z}}$ is called absolutely regular if $\lim_{k \to \infty} \beta_k = 0$.

Absolutely regular processes were introduced by Volkonskii and Rozanov (1956). Independently they were introduced in ergodic theory by Ornstein under the name weak Bernoulli processes. Comparing with other well-known notions of weak dependence, absolute regularity is weaker than uniform mixing and stronger than strong mixing.

In many applications, e.g. in time series analysis, one encounters stochastic processes that do not satisfy any weak dependence condition but that can be represented as a functional of a weakly dependent process. If $(X_n)_{n \geq 0}$ is the orbit of a dynamical system given by the measure preserving map $T : \mathcal{X} \to \mathcal{X}$ and the initial value X_0, i.e. $X_n = T(X_{n-1})$, then $(X_n)_{n \geq 0}$ does not satisfy any of the weak dependence properties of probability theory. However, in some cases one can still express $(X_n)_{n \geq 0}$ as a functional of an absolutely regular process. This was e.g. established by Hofbauer and Keller (1982) for piecewise monotone expanding maps of $[0, 1]$.

Definition 3. *Let $(\Omega, \mathcal{F}, \mathbb{P})$ be a probability space and let $(Z_n)_{n \in \mathbb{Z}}$ be a stationary stochastic process. We say that the process $(X_n)_{n \in \mathbb{Z}}$ is a functional of $(Z_n)_{n \in \mathbb{Z}}$ if there exists a measurable function $f : \mathbb{R}^{\mathbb{Z}} \to \mathbb{R}$ such that*

$$X_n = f \left((Z_{n+k})_{k \in \mathbb{Z}} \right) \ .$$

Similarly. we say that the process $(X_n)_{n \in \mathbb{N}}$ is a one-sided functional of $(Z_n)_{n \in \mathbb{N}}$ if $X_n = f((Z_{n+k})_{k \geq 0})$, for some function $f : \mathbb{R}^{\mathbb{N}} \to \mathbb{R}$.

Example 1. Consider the transformation $T : [0,1] \to [0,1]$, defined by $T(x) = 2x \,[\mathrm{mod}\,1]$ and the process $X_n = T^n(X_0)$, where X_0 is a uniformly $[0,1]$-distributed random variable. Let $Z_n = \mathbb{1}_{\{X_n \geq \frac{1}{2}\}}, n \geq 0$. Then $(Z_n)_{n \geq 0}$ is a sequence of i.i.d. symmetric Bernoulli random variables and

$$X_n = \sum_{k=0}^{\infty} Z_{n+k} \frac{1}{2^{k+1}} \ .$$

Thus, the deterministic sequence $(X_n)_{n \geq 0}$ can be represented as a functional of the i.i.d. process $(Z_n)_{n \geq 0}$.

The concept of a functional of a mixing process is one way to treat processes that are 'almost' weakly dependent, which has been used already by Billingsley (1968) and by Ibragimov and Linnik (1971). An alternative approach has recently been developed by Doukhan and Louhichi (1999).

Limit theorems for functionals of absolutely regular processes require some form of continuity of the functional $f : \mathbb{R}^{\mathbb{Z}} \to \mathbb{R}$. Below we formulate two continuity properties, the r-approximation condition and the Lipschitz condition.

Definition 4. *(i) Let $(X_n)_{n \in \mathbb{Z}}$ be a functional of $(Z_n)_{n \in \mathbb{Z}}$, and let $(a_l)_{l \geq 0}$ be a sequence of non-negative constants satisfying $\lim_{l \to \infty} a_l = 0$. We say that the process $(X_n)_{n \in \mathbb{Z}}$ is an r-approximating functional of $(Z_n)_{n \in \mathbb{Z}}$ with constants $(a_l)_{l \geq 0}$ if*

$$\mathbb{E} \,|X_0 - \mathbb{E}(X_0 | Z_{-l}, \cdots, Z_l)|^r \leq a_l \ . \tag{10}$$

(ii) We say that $(X_n)_{n \in \mathbb{Z}}$ is a Lipschitz functional of $(Z_n)_{n \in \mathbb{Z}}$ if $X_n = f((Z_{n+k})_{k \in \mathbb{Z}})$ and if there exists $\alpha \in [0,1)$ such that $f : \mathbb{R}^{\mathbb{Z}} \to \mathbb{R}$ satisfies

$$|f((z_i)_{i \in \mathbb{Z}}) - f((z_i')_{i \in \mathbb{Z}})| \leq const. \, \alpha^n, \tag{11}$$

for all $(z_i)_{i \in \mathbb{Z}}$ and $(z_i')_{i \in \mathbb{Z}}$ such that $z_{-n} = z_{-n}', \ldots, z_n = z_n'$.

The r-approximation condition is weaker than the Lipschitz condition and is satisfied by many examples, see e.g. Borovkova, Burton and Dehling (2001).

An important role in the treatment of absolutely regular processes and their functionals is played by coupling techniques. For a stationary and absolutely regular process $(Z_n)_{n \in \mathbb{Z}}$ with mixing coefficients $(\beta_k)_{k \geq 0}$, Berbee (1979) showed that one can find a copy $(Z_n')_{n \in \mathbb{Z}}$ with the same joint distribution as the original process $(Z_n)_{n \in \mathbb{Z}}$ such that

$$(Z_n')_{n < 0} \text{ is independent of } (Z_n)_{n \in \mathbb{N}} \tag{12}$$

$$\mathbb{P}(Z_j = Z_j', \ \forall j \geq k) \geq 1 - \beta_k, \text{ for all } k \in \mathbb{N} \ . \tag{13}$$

Thus one can find a stochastic process $(Z_n')_{n \in \mathbb{Z}}$ whose development until time $n = 0$ is independent of $(Z_n)_{n \in \mathbb{Z}}$ and that from time $n = 1$ on couples to the development of $(Z_n)_{n \in \mathbb{Z}}$ in such a way that the two processes become close with large probability.

The following lemma gives a coupling for functionals of absolutely regular processes. First we introduce some notation. For a stochastic process $(X_n)_{n \in \mathbb{Z}}$ and an index set $I = \{i_1 < \cdots < i_m\} \subset \mathbb{Z}$ we write

$$X_I = (X_{i_1}, \ldots, X_{i_m}) \, .$$

Lemma 1. *Let $(X_k)_{k \in \mathbb{Z}}$ be an r-approximating functional with constants $(a_l)_{l \geq 0}$ of an absolutely regular process with mixing coefficients $(\beta_k)_{k \geq 0}$. Let $i_1 < \ldots i_j < i_{j+1} < \ldots < i_k$ be integers and define $I_1 := (i_1, \ldots, i_j)$ and $I_2 := (i_{j+1}, \ldots, i_k)$. Then there exist copies (X'_{I_1}, X'_{I_2}) and (X''_{I_1}, X''_{I_2}) of (X_{I_1}, X_{I_2}) with the following properties*

$$(X''_{I_1}, X''_{I_2}) \text{ is independent of } (X_{I_1}, X_{I_2}) \tag{14}$$

$$\mathbb{P}(\|X''_{I_1} - X'_{I_1}\| \geq \epsilon) \leq 2^k \left(\frac{2}{\epsilon}\right)^r a_m \tag{15}$$

$$\mathbb{P}(\|X'_{I_2} - X_{I_2}\| \geq \epsilon) \leq 2^k \left(\frac{2}{\epsilon}\right)^r a_m + \beta_{i_{j+1}-i_j-2m} \tag{16}$$

where $m \in \mathbb{N}_0$ and $\epsilon > 0$ are given.

Proof. By the Berbee coupling method, we can find copies $(Z'_n)_{n \in \mathbb{Z}}$ and $(Z''_n)_{n \in \mathbb{Z}}$ of the underlying absolutely regular process $(Z_n)_{n \in \mathbb{Z}}$ such that

$$(Z'_n)_{n \leq i_j + m} \text{ is independent of } (Z_n)_{n \leq i_j + m}$$

$$\mathbb{P}\left(\bigcup_{n=i_{j+1}-m}^{\infty} \{Z'_n \neq Z_n\}\right) \leq \beta_{i_{j+1}-i_j-2m}$$

$$(Z''_n)_{n \leq i_j + m} = (Z'_n)_{n \leq i_j + m}$$

$$(Z''_n)_n \text{ is independent of } (Z_n) \, .$$

Let $(X'_n)_{n \in \mathbb{Z}}$ and $(X''_n)_{n \in \mathbb{Z}}$ denote the corresponding functionals of $(Z'_n)_{n \in \mathbb{Z}}$ and $(Z''_n)_{n \in \mathbb{Z}}$, respectively. Define

$$f_{I_1}((u_i)_{i \in \mathbb{Z}}) := \big(f((u_{i_1+i})_{i \in \mathbb{Z}}), \ldots, f((u_{i_j+i})_{i \in \mathbb{Z}})\big)$$
$$f_m(u_{-m}, \ldots, u_m) := \mathbb{E}(f(Z_i)_{i \in \mathbb{Z}} | Z_{-m} = u_{-m}, \ldots, Z_m = u_m)$$
$$f_{I_1,m}((u_i)_{i \in \mathbb{Z}}) := \big(f_m(u_{i_1-m}, \ldots, u_{i_1+m}), \ldots, f_m(u_{i_j-m}, \ldots, u_{i_j+m})\big)$$

and analogously f_{I_2} and $f_{I_2,m}$. Thus we have $X_{I_1} = f_{I_1}((Z_i)_{i \in \mathbb{Z}})$, $X_{I_2} = f_{I_2}((Z_i)_{i \in \mathbb{Z}})$. Observe that $f_{I_1,m}((Z_i)_{i \in \mathbb{Z}})$ is a function of $(Z_i)_{i \leq i_j+m}$ only, and that $f_{I_2,m}((Z_i)_{i \in \mathbb{Z}})$ is a function of $(Z_i)_{i \geq i_{j+1}-m}$, only. Now we get

$$\mathbb{P}\left(\|X'_{I_2} - X_{I_2}\| \geq \epsilon\right) \leq \mathbb{P}\left(\|f_{I_2}((Z'_i)_{i \in \mathbb{Z}}) - f_{I_2,m}((Z'_i)_{i \in \mathbb{Z}})\| \geq \epsilon/2\right)$$

$$+\mathbb{P}\left(\|f_{I_2}(Z_i)_{i \in \mathbb{Z}} - f_{I_2,m}((Z_i)_{i \in \mathbb{Z}})\| \geq \epsilon/2\right)$$

$$+\mathbb{P}\left(f_{I_2,m}((Z'_i)_{i \in \mathbb{Z}}) \neq f_{I_2,m}((Z_i)_{i \in \mathbb{Z}})\right)$$

$$\leq 2\left(\frac{2}{\epsilon}\right)^r \mathbb{E}\|f_{I_2}((Z_i)_{i \in \mathbb{Z}}) - f_{I_2,m}((Z_i)_{i \in \mathbb{Z}})\|^r$$

$$+\beta_{i_{j+1}-i_j-m}$$

$$\leq 2^k\left(\frac{2}{\epsilon}\right)^r a_m + \beta_{i_{j+1}-i_j-m}\ .$$

Similarly, we can prove (15). (14) is a direct consequence of the independence of $(Z_i)_{i \in \mathbb{Z}}$ and $(Z'_i)_{i \in \mathbb{Z}}$. $\qquad\square$

4 U-statistic ergodic theorem

Birkhoff's (1930) ergodic theorem states that for a stationary ergodic process $(X_i)_{i \geq 0}$ with one-dimensional marginal distribution F and an $L_1(F)$-function $g : \mathbb{R} \to \mathbb{R}$ we have

$$\frac{1}{n}\sum_{i=0}^{n-1} g(X_i) \to \int g(x)\,\mathrm{d}F(x)\ ,$$

almost surely. In view of this result and the law of large numbers for U-statistics, we can ask whether

$$\lim_{n \to \infty} \frac{1}{\binom{n}{2}} \sum_{1 \leq i < j \leq n} h(X_i, X_j) = \int\int h(x, y)\,\mathrm{d}F(x)\,\mathrm{d}F(y) \qquad (17)$$

holds, almost surely. Aaronson, Burton, Dehling, Gilat, Hill and Weiss (1996) have shown that this is generally not the case, as the following example shows.

Example 2. Let $T : [0,1] \to [0,1]$ be defined by $Tx = 2x \mod 1$, as in Example 1, and define again $X_n = T^n X_0$, where X_0 is a uniformly $[0,1]$-distributed random variable. Consider the kernel

$$h(x, y) = \mathbb{1}_G(x, y) + \mathbb{1}_G(y, x)\ ,$$

where $G := \cup_{n=0}^{\infty}\{(x, T^n x) : 0 \leq x \leq 1\}$, i.e. G is the union of graphs of the iterates of T. Note that $h(X_i, X_j) \equiv 1$ for all $i < j$ and that G is a null set in $[0,1]^2$. Thus we get that the left hand side of (17) equals 1, whereas the right hand side equals 0, i.e. the U-statistics ergodic theorem fails.

Several more counterexamples to the U-statistic ergodic theorem have been given by Borovkova, Burton and Dehling (1999). Investigating the above example somewhat closer, we can see that poor mixing properties of the process $(X_n)_{n \geq 0}$ together with extreme discontinuity of h make the ergodic theorem fail. This motivates the conditions in the following theorem that gives sufficient conditions for the U-statistic ergodic theorem.

Theorem 1. (Aaronson et al, 1996) *Let $(X_i)_{i \geq 0}$ be a stationary ergodic process and let $h : \mathbb{R}^2 \to \mathbb{R}$ be a symmetric kernel. Assume moreover that one of the following two conditions is satisfied*

(i) h is bounded and $F \times F$- almost surely continuous

(ii) h is bounded and $(X_i)_{i \geq 0}$ is an absolutely regular process.

Then

$$\lim_{n \to \infty} \frac{1}{\binom{n}{2}} \sum_{1 \leq i < j \leq n} h(X_i, X_j) = \int \int h(x, y) \, dF(x) \, dF(y) \, ,$$

i.e. the U-statistic ergodic theorem holds.

Remark 1. (i) Condition (i) is satisfied by the kernel arising in the sample correlation integral, $h(x, y) = \mathbb{1}_{\{\|x - y\| \leq r\}}$, provided that

$$F \times F(\{(x, y) : \|y - x\| = r\}) = 0 \, .$$

This holds e.g. for atom-free F, and hence we have

$$\lim_{n \to \infty} C_n(r) = C(r) \, .$$

This special case was proved independently by Serinko (1996).

(ii) Borovkova, Burton and Dehling (1999, 2002) have given sufficient conditions for the law of large numbers for U-statistics with possibly unbonded kernels.

5 *U*-statistic CLT

The Central Limit Theorem for U-statistics with absolutely regular observations was proved first by Yoshihara (1976) and later sharpened by Denker and Keller (1983). We define

$$\sigma^2 = \mathbb{E}(h_1(X_1))^2 + 2 \sum_{j=1}^{\infty} \text{cov}(h_1(X_1), h_1(X_j)) \, ; \tag{18}$$

note that $\sigma^2 = \lim_{n \to \infty} \frac{1}{n} \text{var}(\sum_{j=1}^{n} h_1(X_j))$.

Theorem 2. (Denker and Keller, 1983) *Let $(X_i)_{i \geq 1}$ be an absolutely regular process with mixing coefficients satisfying $\sum_{k=1}^{\infty} \beta_k^{\delta/(2+\delta)} < \infty$ and let $h : \mathbb{R}^2 \to \mathbb{R}$ be a symmetric kernel satisfying*

$$\sup_{i < j} \mathbb{E}|h(X_i, X_j)|^{2+\delta} < \infty \, .$$

Then $\sqrt{n} \, (U_n(h) - \theta) \xrightarrow{\mathcal{L}} N(0, 4\sigma^2)$.

The proof of Theorem 2 makes use of the Hoeffding decomposition. As in the case of i.i.d. observations, we have

$$\sqrt{n}\left(U_n(h) - \theta\right) = \frac{2}{\sqrt{n}}\sum_{i=1}^{n} h_1(X_i) + \sqrt{n}\, U_n(h_2)\,, \tag{19}$$

where h_1 and h_2 are defined as in (5) and (6). In order to prove weak convergence of $\sqrt{n}\left(U_n(h) - \theta\right)$, we prove that $\frac{1}{\sqrt{n}}\sum_{i=1}^{n} h_1(X_i) \xrightarrow{\mathcal{L}} N(0,\sigma^2)$ and that $\sqrt{n}\, U_n(h_2) \to 0$ in probability. The first part can be achieved by applying one of the well-known central limit theorems for partial sums of weakly dependent random variables, see e.g. Bradley (2005), Doukhan (1994) or Rio (2000). The second part will follow if we can prove that $\mathbb{E}(\sqrt{n}\, U_n(h_2))^2 \to 0$. Note that

$$\mathbb{E}\left(\sqrt{n}\, U_n(h_2)\right)^2$$
$$= \frac{4}{n(n-1)^2}\sum_{1 \le i_1 < j_1 \le n, 1 \le i_2 < j_2 \le n} \mathbb{E}\left(h_2(X_{i_1}, X_{j_1})h_2(X_{i_2}, X_{j_2})\right)\,. \tag{20}$$

In the case of i.i.d. observations, the terms $h(X_i, X_j)$, $1 \le i < j \le n$, are mutually uncorrelated and thus the right hand side of (20) equals $\frac{4}{n(n-1)^2}\binom{n}{2} = O(\frac{1}{n})$. For weakly dependent observations, we have to find upper bounds on $\mathbb{E}(h_2(X_{i_1}, X_{j_1})h_2(X_{i_2}, X_{j_2}))$ in terms of the maximal spacing among the indices. We formulate the required result in a more general setting. Let $(X_n)_{n\in\mathbb{Z}}$ be a stationary process and let $i_1 < \ldots < i_j < i_{j+1} < \ldots < i_k$ be integers. Denote for abbreviation $I_1 = (i_1, \ldots, i_j)$, $I_2 = (i_{j+1}, \ldots, i_k)$ and the random vectors $X_{I_1} = (X_{i_1}, \ldots, X_{i_j})$, $X_{I_2} = (X_{i_{j+1}}, \ldots, X_{i_k})$. Finally, let $\mathbb{E}_{X_{I_1}}$ be the expectation operator with respect to the random variables X_{I_1}, keeping the remaining random variables fixed. I.e., we have for any measurable function $g : \mathbb{R}^k \to \mathbb{R}$

$$\mathbb{E}_{X_{I_1}} g(X_{i_1}, \ldots, X_{i_k}) = \int_{\mathbb{R}^j} g(x_1, \ldots, x_j, X_{i_{j+1}}, \ldots, X_{i_k})\, d\mathbb{P}_{X_{I_1}}(x_1, \ldots, x_j)\,.$$

In the same way we define the expectation operator $\mathbb{E}_{X_{I_2}}$.

Lemma 2. (Yoshihara, 1976) *Let $(X_n)_{n\in\mathbb{Z}}$ be an absolutely regular process with mixing coefficients $(\beta_k)_{k\ge0}$, let $i_1 < \ldots < i_j < i_{j+1} < \ldots < i_k$ be integers. Let $r, s > 0$ satisfy $\frac{1}{r} + \frac{1}{s} = 1$ and let $g : \mathbb{R}^k \to \mathbb{R}$ be a measurable function. Then*

$$\left|\mathbb{E}\left(g(X_{i_1}, \ldots, X_{i_k})\right) - \mathbb{E}_{X_{I_1}}\left(\mathbb{E}_{X_{I_2}}\left(g(X_{i_1}, \ldots, X_{i_k})\right)\right)\right| \tag{21}$$
$$\le 4\max\left\{\left(\mathbb{E}|g(X_{i_1}, \ldots, X_{i_k})|^r\right)^{1/r}, \left(\mathbb{E}_{X_{I_1}}\mathbb{E}_{X_{I_2}}|g(X_{i_1}, \ldots, X_{i_k})|^r\right)^{1/r}\right\}\beta_{i_{j+1}-i_j}^{1/s}\,,$$

where $I_1 = (i_1, \ldots, i_j)$ and $I_2 = (i_{j+1}, \ldots, i_k)$.

With the help of this inequality one can now show that $\mathbb{E}(\sqrt{n}U_n(h_2))^2 \to 0$. Inequalities of the type (21) are natural generalizations of correlation inequalities that give upper bounds on $\mathrm{cov}(\xi, \eta)$ for random variables ξ and η that are $\mathcal{F}^n_{-\infty}$, respectively $\mathcal{F}^\infty_{n+k}$ measurable. We thus call (21) a generalized correlation inequality. They form a crucial ingredient in the study of U-statistics of weakly dependent observations, in the same way as standard correlation inequalities do for partial sums.

Central limit theorems for U-statistics of functionals of an absolutely regular process require some form of continuity of the kernel h. Below we formulate two different conditions that were introduced in the U-statistic context by Denker and Keller (1986).

Definition 5. *Let $h : \mathbb{R}^2 \to \mathbb{R}$ be a symmetric kernel.*
(a) We say that h satisfies the Lipschitz condition if there exist $L, \rho > 0$ and $r \geq 0$ such that

$$|h(x_1, x_2) - h(y_1, y_2)| \leq L \left(|x_1 - y_1|^\rho + |x_2 - y_2|^\rho\right)$$
$$\cdot \left(1 + |x_1|^r + |x_2|^r + |y_1|^r + |y_2|^r\right) .$$

(b) We say that h satisfies the oscillation condition if

$$\int osc(h, \epsilon, (x_1, x_2)) \, dF(x_1) \, dF(x_2) = O(\epsilon^r)$$

as $\epsilon \to 0$, where

$$osc(h, \epsilon, (x_1, x_2)) := \sup_{y_i, y_i' \in \mathbb{R}, |y_i - x_i| < \epsilon, |y_i' - x_i| < \epsilon, i = 1, 2} |h(y_1, y_2) - h(y_1', y_2')|$$

denotes the ϵ-oscillation of h in (x_1, x_2).

Theorem 3. (Denker and Keller, 1986) *Let $(X_n)_{n \in \mathbb{Z}}$ be a Lipschitz functional of the absolutely regular process $(Z_n)_{n \in \mathbb{Z}}$ with mixing coefficients satisfying*

$$\beta_n^{\delta/(2+\delta)} = O(n^{-2-\epsilon})$$

for some $\epsilon, \delta > 0$. Moreover let $h : \mathbb{R}^2 \to \mathbb{R}$ be a symmetric kernel satisfying the Lipschitz condition or the oscillation condition. Then $\sqrt{n}(U_n(h) - \theta) \xrightarrow{\mathcal{L}} N(0, 4\sigma^2)$.

In their proof of Theorem 3, Denker and Keller follow the usual pattern of proof of limit theorems for functionals of absolutely regular processes, or generally of any kind of mixing processes. Introduce the following finite block functionals X_n^m as approximations to $X_n = f((Z_{n+k})_{k \in \mathbb{Z}})$,

$$X_n^m = \mathbb{E}(X_n \mid Z_{n-m}, \ldots, Z_{n+m}) .$$

Then $(X_n^m)_{n \in \mathbb{Z}}$ is still an absolutely regular process and hence Theorem 2 applies. By the Lipschitz property of the functional f and by the Lipschitz or oscillation condition satisfied by the kernel h, one can show that the original U-statistic $U_n(h)$ is closely approximated by

$$U_n^m(h) = \frac{1}{\binom{n}{2}} \sum_{1 \leq i < j \leq n} h(X_i^m, X_j^m) \,,$$

as $m \to \infty$. With some technical effort one can then show that the CLT holds for $U_n(h)$.

Borovkova, Burton and Dehling (2001) proposed a different approach to the study of functionals of an absolutely regular process, treating the process $(X_n)_{n \in \mathbb{Z}}$ directly, without approximation by finite block functionals. In this way they could derive limit theorems such as U-process invariance principles. In what follows we will outline some ingredients of their technique in connection with the U-statistic CLT. Again, one has to require some form of continuity of the kernel h.

Definition 6. *Let $(X_n)_{n \in \mathbb{N}}$ be a stationary stochastic process and let $h : \mathbb{R}^2 \to \mathbb{R}$ be a symmetric kernel. We say that h satisfies the p-continuity condition if there exists $\phi : [0, \infty) \to [0, \infty)$ satisfying $\phi(\epsilon) = o(1)$ as $\epsilon \to 0$ such that*

$$\mathbb{E}\left(|h(X, Y) - h(X', Y)|^p \mathbb{1}_{\{|X - X'| \leq \epsilon\}}\right) \leq \phi(\epsilon) \,,$$

for all triples of random variables X, X', Y with marginal distribution F and such that (X, Y) either has distribution $F \times F$ or $\mathbb{P}_{X_0, X_k}$ for some $k \in \mathbb{N}_0$.

Remark 2. (i) Lipschitz continuous kernels are p-continuous under a moment assumption on the variables X_k. We get namely

$$|h(X, Y) - h(X', Y)|^p \mathbb{1}_{\{|X - X'| \leq \epsilon\}} \leq \epsilon^p {}^p (1 + |X|^r + |X'|^r + 2|Y|^r)^p$$

and thus p-continuity holds with $\phi(\epsilon) = O(\epsilon^p {}^p)$ provided $\mathbb{E}|X_0|^{rp} < \infty$.

(ii) p-continuity is close in spirit to the oscillation condition, in that it also requires continuity in some average sense. The main difference is that in the oscillation condition only the product measure $F \times F$ enters whereas for p-continuity we need to consider averages with respect to the joint distribution of (X_i, X_j) for all pairs $i < j$.

Example 3. We consider the kernel that was used in the definition of the sample correlation integral, i.e. $h(x, y) = \mathbb{1}_{\{|x - y| \leq t\}}$. Observing that

$$\left|\mathbb{1}_{\{|x - y| \leq t\}} - \mathbb{1}_{\{|x' - y| \leq t\}}\right|^p \mathbb{1}_{\{|x - x'| \leq \epsilon\}} \leq \mathbb{1}_{\{t - \epsilon \leq |x - y| \leq t + \epsilon\}} \,,$$

we get

$$\mathbb{E}\left|\mathbb{1}_{\{|X_0 - X_k| \leq t\}} - \mathbb{1}_{\{|X_0' - X_k| \leq t\}}\right|^p \mathbb{1}_{\{|X_0 - X_0'| \leq \epsilon\}} \leq \mathbb{P}(t - \epsilon \leq |X_0 - X_k| \leq t + \epsilon) \,.$$

Thus the kernel $\mathbb{1}_{\{|x - y| \leq t\}}$ is p-continuous provided that $\sup_k \mathbb{P}(t - \epsilon \leq |X_0 - X_k| \leq t + \epsilon) = o(1)$, as $\epsilon \to 0$ and similarly $\mathbb{P}(t - \epsilon \leq |X - Y| \leq t + \epsilon) = o(1)$, where X and Y are independent with the same distribution F. These conditions specify equicontinuity of the family of distribution functions of $|X_0 - X_k|$ at t as well as continuity of the distribution function of $|X - Y|$.

As shown in the remarks about the proof of Theorem 2 we are often led to expressions of the type $\mathbb{E}(h(X_{i_1}, X_{j_1})h(X_{i_2}, X_{j_2}))$ with a given kernel $h : \mathbb{R}^2 \to \mathbb{R}$. For the treatment of such expectations we introduce a generalization of the notion of p-continuity to functions $g : \mathbb{R}^k \to \mathbb{R}$.

Definition 7. *Let $(X_n)_{n\in\mathbb{Z}}$ be a stationary stochastic process. A function $g : \mathbb{R}^k \to \mathbb{R}$ is called p-continuous if there exists a function $\phi : \mathbb{R}_+ \to \mathbb{R}_+$ with $\phi(\epsilon) = o(1)$ as $\epsilon \to 0$ such that for all index sets $I = \{i_1 < \ldots < i_k\} \subset \mathbb{Z}$ and all non-empty disjoint sets I_1, I_2 with $I_1 \cup I_2 = I$ we have*

$$\mathbb{E}\left|g(\xi, \eta) - g(\xi, \eta')\right|^p \mathbb{1}_{\{\|\eta-\eta'\|\leq\epsilon\}} \leq \phi(\epsilon), \tag{22}$$

for all random variables $\xi : \Omega \to \mathbb{R}^{|I_1|}$, $\eta, \eta' : \Omega \to \mathbb{R}^{|I_2|}$ such that (ξ, η) has distribution $\mathbb{P}_{X_{I_1}, X_{I_2}}$ or $\mathbb{P}_{X_{I_1}} \times \mathbb{P}_{X_{I_2}}$.

Lemma 3. *(Borovkova et al, 2001) Let $h : \mathbb{R}^2 \to \mathbb{R}$ be a symmetric p-continuous kernel and define*

$$g(x_1, x_2, x_3, x_4) = h(x_1, x_2)h(x_3, x_4). \tag{23}$$

(i) If h is bounded, then g is also p-continuous.
(ii) If $\sup_k \mathbb{E}|h(X_0, X_k)|^p < \infty$ and $\mathbb{E}_{X_0}\mathbb{E}_{X_k}|h(X_0, X_k)|^p < \infty$, then g is p/2-continuous.

Lemma 4. *(Borovkova et al, 2001) Let $h : \mathbb{R}^2 \to \mathbb{R}$ be a symmetric p-continuous kernel. Then the terms h_1 and h_2 of the Hoeffding decomposition are also p-continuous.*

The following proposition gives a generalized correlation inequality for functionals of absolutely regular processes. Such inequalities were first proved by Borovkova et al (2001). We improve their Lemma 4.3 by replacing the constants $\alpha_k = \sqrt{2\sum_{i=k}^\infty a_i}$ by $\sqrt{a_k}$.

Proposition 1. *Let $(X_n)_{n\in\mathbb{Z}}$ be an absolutely regular process with mixing coefficients $(\beta_k)_{k\geq 0}$, let $i_1 < \ldots < i_j < i_{j+1} < \ldots < i_k$ be integers. Let $r, s > 0$ satisfy $\frac{1}{r} + \frac{1}{s} = 1$ and let $g : \mathbb{R}^k \to \mathbb{R}$ be a measurable function. Then*

$$\left|\mathbb{E}\left(g(X_{i_1}, \ldots, X_{i_k})\right) - \mathbb{E}_{X_{I_1}}\left(\mathbb{E}_{X_{I_2}}\left(g(X_{i_1}, \ldots, X_{i_k})\right)\right)\right| \tag{24}$$

$$\leq 4\max\left\{\left(\mathbb{E}|g(X_{i_1}, \ldots, X_{i_k})|^r\right)^{1/r}, \left(\mathbb{E}_{X_{I_1}}\mathbb{E}_{X_{I_2}}|g(X_{i_1}, \ldots, X_{i_k})|^r\right)^{1/r}\right\}$$

$$\cdot \left(\beta_{i_{j+1}-i_j-2m} + a_m^{1/2}\right)^{1/s} + 2\phi(a_m^{1/2}).$$

where $I_1 = (i_1, \ldots, i_j)$, $I_2 = (i_{j+1}, \ldots, i_k)$.

Proof. Let (X'_{I_1}, X'_{I_2}) and (X''_{I_1}, X''_{I_2}) be copies of (X_{I_1}, X_{I_2}) as defined in Lemma 1. By independence of (X'_{I_1}, X'_{I_2}) and (X''_{I_1}, X''_{I_2}), we get

$$\left| \mathbb{E} \left(g(X_{i_1}, \dots, X_{i_k}) \right) - \mathbb{E}_{X_{I_1}} \left(\mathbb{E}_{X_{I_2}} \left(g(X_{i_1}, \dots, X_{i_k}) \right) \right) \right|$$

$$= \left| \mathbb{E} g \left(X''_{I_1}, X_{I_2} \right) - \mathbb{E} g \left(X'_{I_1}, X'_{I_2} \right) \right|$$

$$\leq \left| \mathbb{E} g \left(X''_{I_1}, X_{I_2} \right) - \mathbb{E} g \left(X''_{I_1}, X'_{I_2} \right) \right| + \left| \mathbb{E} g \left(X''_{I_1}, X'_{I_2} \right) - \mathbb{E} g \left(X'_{I_1}, X'_{I_2} \right) \right| \quad (25)$$

We now define the events $B := \{ \| X_{I_2} - X'_{I_2} \| \leq a_m^{1/2} \}$ and $D := \{ \| X'_{I_1} - X''_{I_1} \| \leq a_m^{1/2} \}$. By (15) and (16) we get $\mathbb{P}(D^c) \leq 4 a_m^{1/2}$ and $\mathbb{P}(B^c) \leq 4 a_m^{1/2} + \beta_{i_{j+1} - i_j - 2m}$. Now we can bound the two terms on the right hand side of (25) separately. By the 1-continuity property of g we get

$$\mathbb{E} \left| g \left(X''_{I_1}, X_{I_2} \right) - g \left(X''_{I_1}, X'_{I_2} \right) \right| \mathbb{1}_B \leq \phi(a_m^{1/2}) .$$

Using Hölder's inequality we find

$$\mathbb{E} \left| g \left(X''_{I_1}, X_{I_2} \right) - g \left(X''_{I_1}, X'_{I_2} \right) \right| \mathbb{1}_{B^c} \leq 2 M (\mathbb{P}(B^c))^{1/s}$$

$$\leq 2 M \left(a_m^{1/2} + \beta_{i_{j+1} - i_j - 2m} \right) .$$

Similarly we obtain the following bounds for the second term on the right hand side of (25),

$$\left| \mathbb{E} g \left(X''_{I_1}, X'_{I_2} \right) - \mathbb{E} g \left(X'_{I_1}, X'_{I_2} \right) \right| \mathbb{1}_D \leq \phi(a_m^{1/2})$$

$$\left| \mathbb{E} g \left(X''_{I_1}, X'_{I_2} \right) - \mathbb{E} g \left(X'_{I_1}, X'_{I_2} \right) \right| \mathbb{1}_{D^c} \leq 2 M a_m^{1/2s} .$$

Putting the last four inequalities together we obtain the statement of the proposition. $\qquad \square$

Theorem 4. *Let* $(X_n)_{n \in \mathbb{Z}}$ *be a 1-approximating functional with constants* $(a_l)_{l \geq 0}$ *of an absolutely regular process with mixing coefficients* $(\beta_k)_{k \geq 0}$. *Let* h *be a bounded 1-continuous kernel and suppose that*

$$\sum_{k=1}^{\infty} k^2 (\beta_k + a_k^{1/2} + \phi(a_k^{1/2})) < \infty .$$

Then, as $n \to \infty$

$$\sqrt{n} \left(U_n(h) - \theta \right) \xrightarrow{\mathcal{L}} N(0, \sigma^2) ,$$

where σ^2 *is defined as in (18).*

6 Empirical U-processes

Recall that one of the examples motivating the study of dependent U-statistics was the sample correlation integral

$$C_n(r) = \frac{1}{\binom{n}{2}} \sum_{1 \leq i < j \leq n} \mathbb{1}_{\{ \| X_i - X_j \| \leq r \}} .$$

For fixed r, this is simply a U-statistic. Considered as a stochastic process indexed by $r \in [0, R]$, however, $C_n(r)$ becomes the empirical distribution function of the observations $\|X_i - X_j\|$, $1 \leq i < j \leq n$.

More generally, one can consider the empirical distribution function of $h(X_i, X_j)$, $1 \leq i < j \leq n$, given by

$$U_n(t) := \frac{1}{\binom{n}{2}} \sum_{1 \leq i < j \leq n} \mathbb{1}_{\{h(X_i, X_j) \leq t\}} = \frac{1}{\binom{n}{2}} \sum_{1 \leq i < j \leq n} h_t(X_i, X_j), \qquad (26)$$

where we have defined $h_t(x, y) := \mathbb{1}_{\{h(x,y) \leq t\}}$. Moreover, we define the empirical U-process

$$\sqrt{n}(U_n(t) - U(t)), \ 0 \leq t \leq t_0 \,,$$

where $U(t) = \mathbb{P}(h(X, Y) \leq t)$.

Weak convergence of the empirical U-process to an appropriate Gaussian process was established by Silverman (1983) and by Serfling (1984) in the case of i.i.d. observations and by Borovkova (1995) and Arcones and Yu (1994) for absolutely regular processes. Borovkova et al (2001) could establish the same result for functionals of absolutely regular processes.

Theorem 5. *Let $(X_n)_{n \in \mathbb{Z}}$ be a 1-approximating functional with constants $(a_l)_{l \geq 0}$ of an absolutely regular process with mixing coefficients $(\beta_k)_{k \geq 0}$. Assume that $h_t(x, y)$ are 1-continuous with $\phi_t \equiv \phi$ and that*

$$\sum_{k=1}^{\infty} k^2 (a_k^{1/2} + \beta_k + \phi(a_k^{1/2}))^{1/3 - \epsilon} < \infty$$

for some $\epsilon > 0$. Moreover assume that $|U(t) - U(s)| \leq C|t - s|$ and that $|\mathbb{E}h_t(X_0, X_k) - \mathbb{E}h_s(X_0, X_k)| \leq C\,|t - s|$ for all k. Then

$$\sqrt{n}\,(U_n(t) - U(t)) \to W(t) \,,$$

where $(W(t))_{0 \leq t \leq 1}$ is a mean-zero Gaussian process with

$$\mathbb{E}(W(s)W(t)) = 4\mathrm{cov}(h_{s,1}(X_1), h_{t,1}(X_1)) + 4\sum_{k=2}^{\infty} \mathrm{cov}(h_{s,1}(X_1), h_{t,1}(X_k))$$

$$+ 4\sum_{k=2}^{\infty} \mathrm{cov}(h_{s,1}(X_k), h_{t,1}(X_1)) \,.$$

The proof of this theorem uses again the Hoeffding decomposition. For each fixed t we get

$$\sqrt{n}(U_n(t) - U(t)) = \frac{2}{\sqrt{n}} \sum_{i=1}^{n} h_{t,1}(X_i) + \frac{2}{\sqrt{n}(n-1)} \sum_{1 \leq i < j \leq n} h_{t,2}(X_i, X_j) \,.$$

The first term on the right hand side is an ordinary empirical process indexed by the class of functions $h_{t,1}$ and converges to $(W(t))_{0 \leq t \leq 1}$ by standard empirical process theory for dependent samples. The second term converges to 0 in probability uniformly in t. The proof of this fact uses the chaining technique in combination with the generalized correlation inequality; details are given in Borovkova et al (2001).

Acknowledgement. The author would like to thank Tim Branzke as well as an anonymous referee for carefully reading the manuscript. Their comments and suggestions helped immensely to improve the presentation and especially to reduce the number of misprints.

References

[ABDGH96] J. AARONSON, R.M. BURTON, H.G. DEHLING, D. GILAT, T. HILL and B. WEISS (1996). Strong laws for L- and U-statistics, *Transactions of the American Mathematical Society* **348**, 2845–2865.

[AG95] M.A. ARCONES and E. GINÉ (1995). On the law of the iterated logarithm for canonical U-statistics. *Stochastic Processes and their Applications* **58**, 217–245.

[AY94] M.A. ARCONES and B. YU (1994). Central limit theorems for empirical processes and U-processes of stationary mixing sequences, *Journal of Theoretical Probability* **7**, 47-71.

[Ber79] H.C.P. BERBEE (1979). *Random walks with stationary increments and renewal theory*. Mathematical Centre Tracts **112**, Mathematisch Centrum, Amsterdam.

[Ber66] R.H. BERK (1966) Limiting behavior of posterior distributions when the model is incorrect. *Annals of Mathematical Statistics* **37**, 51–58.

[BP79] I. BERKES and W. PHILIPP (1979). Approximation theorems for independent and weakly dependent random vectors, *Annals of Probability* **7**, 29-54.

[Bil68] P. BILLINGSLEY (1968). *Convergence of Probability Measures*. John Wiley, New York.

[Bor95] S.A. BOROVKOVA (1995). Weak convergence of the empirical process of U-statistics structure for dependent observations. *Theory of Stochastic Processes* **2 (18)**, 115-124.

[BB99] S.A. BOROVKOVA, R.M. BURTON and H.G. DEHLING (1999). Consistency of the Takens estimator for the correlation dimension. *Annals of Applied Probability* **9**, 376-390.

[BBD01] S. BOROVKOVA, R. BURTON and H. DEHLING (2001). Limit theorems for functionals of mixing processes with applications to U-statistics and dimension estimation. *Transactions of the American Mathematical Society* **353**, 4261–4318.

[BBD02] S. BOROVKOVA, R. BURTON and H. DEHLING (2002). From Dimension Estimation to Asymptotics of Dependent U-Statistics. In: *Limit Theorems in Probability and Statistics I* (I. Berkes, E. Csáki, M. Csörgő, eds.), Budapest 2002, 201–234.

[Bra05] R.C. BRADLEY (2005). *Introduction to Strong Mixing Conditions*, Volume 1 – 3. Technical Report, Department of Mathematics, Indiana University, Bloomington. Custom Publishing of I.U., Bloomington.

[Deh86] H. DEHLING (1986) Almost sure approximations for U-statistics. in: *Dependence in Probability and Statistics* E. Eberlein, M. .S Taqqu (eds.), p. 119-135, Birkhäuser, Boston

[Deh89a] H. DEHLING (1989a) The functional law of the iterated logarithm for von-Mises functionals and multiple Wiener integrals. *Journal of Multivariate Analysis* **28**, 177-189.

[Deh89b] H. DEHLING (1989b) Complete convergence of triangular arrays and the law of the iterated logarithm for U-statistics. *Statistics and Probability Letters* **7**, 319-321.

[DDP83] H. DEHLING, M. DENKER and W. PHILIPP (1984) Invariance principles for von Mises and U-Statistics. *Zeitschrift für Wahrscheinlichkeitstheorie und verwandte Gebiete* **67**, 139–167.

[DDP84] H. DEHLING, M. DENKER and W. PHILIPP (1986) A bounded law of the iterated logarithm for Hilbert space valued martingales and its application to U-statistics. *Probability Theory and Related Fields* **72**, 111–131.

[DU96] H. DEHLING and S.A. UTEV (1996) The Law of the Iterated Logarithm for Degenerate U-Statistics. in: *Probability Theory and Mathematical Statistics*, I. A. Ibragimov and A. Yu. Zaitsev (eds.), p. 19–28, Gordon and Breach Publishers.

[DK83] M. DENKER and G. KELLER (1983). On *U*-Statistics and v. Mises' Statistics for Weakly Dependent Processes, *Zeitschrift für Wahrscheinlichkeitstheorie und verwandte Gebiete* **64**, 505–522.

[DK86] M. DENKER and G. KELLER (1986). Rigorous statistical procedures for data from dynamical systems, *Journal of Statistical Physics* **44**, 67–93.

[Dou94] P. DOUKHAN (1994). *Mixing: Properties and Examples*. Springer Verlag, New York.

[DL99] P. DOUKHAN and S. LOUHICHI (1999). A new weak dependence condition and applications to moment inequalities. *Stochastic Processes and their Applications* **84**, 313–343.

[DMR95] P. DOUKHAN, P. MASSART and E. RIO (1995). Invariance principles for absolutely regular empirical processes, *Annales de l'Institut Henri Poincaré. Probabilités et Statistique* **31**, 393–427.

[Fil62] A. A. FILLIPOVA (1962). Mises' theorem on the asymptotic behavior of functionals of empirical distribution functions and its statistical applications. *Theory of Probability and its Applications* **7**, 24–57.

[GP83] P. GRASSBERGER and I. PROCACCIA (1983). Characterization of strange attractors, *Physical Review Letters* **50**, 346–349.

[Hal46] P. HALMOS (1946). The theory of unbiased estimation, *Annals of Mathematical Statistics* **17**, 34–43.

[HK82] F. HOFBAUER and G. KELLER (1982). Equilibrium states for piecewise monotonic transformations, *Ergodic Theory and Dynamical Systems* **2**, 23–43.

[Hoe48] W. HOEFFDING (1948). A class of statistics with asymptotically normal distribution, *Annals of Mathematical Statistics* **19**, 293–325.

[Hoe61] W. HOEFFDING (1961). The strong law of large numbers for *U*-statistics. *University of North Carolina Mimeo Report* No. 302.

[IL71] I.A. IBRAGIMOV and YU.V. LINNIK (1971). *Independent and stationary sequences of random variables,* Wolters-Noordhoff Publishing, Groningen.

[Von47] R. VON MISES (1947). On the asymptotic distribution of differentiable statistical functions, Annals of Mathematical Statistics **18**, 309–348.

[Rio00] E. RIO (2000). *Théorie asymptotique des processus aléatoires faiblement dépendants,* Springer Verlag, Berlin.

[Ser84] R. SERFLING (1984). Generalized *L*-, *M*- and *R*-statistics, *Annals of Statistics* **12**, 76–86.

[Ser96] R.J. SERINKO (1996). Ergodic theorems arising in correlation dimension estimation. *Journal of Statistical Physics* **85**, 25–40.

[Sil83] B. SILVERMAN (1983). Convergence of a class of empirical distribution functions of dependent random variables, *Annals of Probability* **11**, 745–751.

[Tak81] F. TAKENS (1981). Detecting strange attractors in turbulence, In: *Dynamical systems and turbulence.* Lecture Notes in Mathematics **898**, 336–381. Springer-Verlag.

[Tak85] F. TAKENS (1985). On the numerical determination of the dimension of the attractor, In: *Dynamical Systems and Bifurcations.* Lecture Notes in Mathematics **1125**, 99–106. Springer-Verlag.

[VR59] V.A. VOLKONSKII and YU.A. ROZANOV (1959). Some limit theorems for random functions I. *Theory of Probability and Its Aplications* **4**, 178–197.

[Yos76] K. YOSHIHARA (1976). Limiting behaviour of *U*-statistics for stationary, absolutely regular processes, *Zeitschrift für Wahrscheinlichkeitstheorie und Verwandte Gebiete* **35**, 237-252.

Recent results on weak dependence for causal sequences. Statistical applications to dynamical systems.

Clémentine Prieur

INSA Toulouse, Laboratoire de Statistique et Probabilités, Université Paul Sabatier, 118 route de Narbonne, 31062 Toulouse cedex 4, France. E-mail: prieur@cict.fr

1 Introduction

There exists a wide literature on limit theorems under various classical mixing conditions such as strong mixing condition ($\alpha-$mixing), absolute regularity ($\beta-$mixing), or $\Phi-$mixing. For recent and complete results on the properties of these coefficients, we refer to the monographs by Doukhan [Dou94], Rio [Rio00a] and Bradley [Bra02]. However, many commonly used models do not satisfy these mixing conditions. For example, Andrews [AND84] proved that if $(\varepsilon_i)_{i\geq 1}$ is i.i.d. with marginal $\mathcal{B}(1/2)$, then the stationary solution $(X_i)_{i\geq 0}$ of the equation

$$X_n = \frac{1}{2}(X_{n-1} + \varepsilon_n), \quad X_0 \text{ independent of } (\varepsilon_i)_{i\geq 1}$$

is not $\alpha-$mixing. Many authors have therefore introduced modifications of these various mixing coefficients. Let $(\Omega, \mathcal{A}, \mathbb{P})$ be a probability space, X a real-valued random variable with law $\mathbb{P}_X$ and $\mathcal{M}$ a $\sigma-$algebra of $\mathcal{A}$. Let us first recall the definition of the usual mixing coefficients between $\mathcal{M}$ and $\sigma(X)$, introduced respectively by Rosenblatt [Ros56], Volkonskii and Rozanov [RV59] and Ibragimov [Inr62]:

$$\alpha(\mathcal{M}, \sigma(X)) = \frac{1}{2} \sup_{A \in \mathcal{B}(\mathbb{R})} \|\mathbb{P}_{X|\mathcal{M}}(A) - \mathbb{P}_X(A)\|_1 ,$$

$$\beta(\mathcal{M}, \sigma(X)) = \| \sup_{A \in \mathcal{B}(\mathbb{R})} |\mathbb{P}_{X|\mathcal{M}}(A) - \mathbb{P}_X(A)| \|_1 ,$$

$$\phi(\mathcal{M}, \sigma(X)) = \sup_{A \in \mathcal{B}(\mathbb{R})} \|\mathbb{P}_{X|\mathcal{M}}(A) - \mathbb{P}_X(A)\|_\infty .$$

We refer to the book of Doukhan [Dou94] for the properties of these coefficients. However, to derive limit theorems for random processes modeling real-world phenomena which do not satisfy classical mixing conditions, it is useful to weaken the definitions above. Of course the goal is to catch more examples without losing too much of the nice properties of classical mixing processes. Following an idea of Rosenblatt [Ros56], we can look at what happens when considering coarser sets than $\mathcal{M}$ or $\mathcal{B}(\mathbb{R})$. Changing $\mathcal{M}$ leads to coefficients which behave quite differently from the usual mixing coefficients. Let us cite for instance the work of Doukhan and Louhichi [DL99]. One different approach is to consider instead of $\mathcal{B}(\mathbb{R})$ the coarser set $\{]-\infty,t], t \in \mathbb{R}\}$. This has been done by Rio [Rio00a] and later by Peligrad [Pel02] for the strong mixing coefficient. Dedecker and Prieur [DP05] have broadened and systematically developed this approach. The coefficients thus obtained measure the difference between the conditional distribution function $F_{X|\mathcal{M}}$ of $\mathbb{P}_{X|\mathcal{M}}$ and the distribution function F_X of $\mathbb{P}_X$. We define below the four dependence coefficients introduced in [DP05].

Definition 1.
$\tau(\mathcal{M}, X) = \int \|F_{X|\mathcal{M}}(t) - F_X(t)\|_1 \, dt$,
$\alpha(\mathcal{M}, X) = \sup_{t \in \mathbb{R}} \|F_{X|\mathcal{M}}(t) - F_X(t)\|_1$,
$\beta(\mathcal{M}, X) = \|\sup_{t \in \mathbb{R}} |F_{X|\mathcal{M}}(t) - F_X(t)|\|_1$,
$\phi(\mathcal{M}, X) = \sup_{t \in \mathbb{R}} \|F_{X|\mathcal{M}}(t) - F_X(t)\|_\infty$.

The coefficient $\alpha(\mathcal{M}, X)$ was first introduced by Rio ([Rio00a], equation 1.10 c). It has then been used by Peligrad [Pel02], while $\tau(\mathcal{M}, X)$ was introduced in the current form by Dedecker and Prieur [DP03]. However, Rüschendorf first introduced the coefficient τ, in a "dual" form (equation 10 in [Rüs85]). These coefficients are smaller than their corresponding mixing coefficients, but they are in many situations easier to compute. It is worth of interest to notice that the coefficients $\tau(\mathcal{M}, X)$, $\alpha(\mathcal{M}, X)$, $\beta(\mathcal{M}, X)$ and $\phi(\mathcal{M}, X)$, as other measures of dependence, can be defined as supremum over some family of functions. For this, we first need to introduce some definitions and notations.

Definition 2. *A σ-finite signed measure is the difference of two positive σ-finite measures, one of them at least being finite. We say that a function h from $\mathbb{R}$ to $\mathbb{R}$ is σ-BV if there exists a σ-finite signed measure dh such that $h(x) = h(0) + dh([0, x[)$ if $x \geq 0$ and $h(x) = h(0) - dh([x, 0[)$ if $x \leq 0$ (h is left continuous). The function h is BV if the signed measure dh is finite. Recall also the Hahn-Jordan decomposition: for any σ-finite signed measure μ, there is a set D such that $\mu_+(A) = \mu(A \cap D) \geq 0$ and $-\mu_-(A) = \mu(A \setminus D) \leq 0$. μ_+ and μ_- are singular, one of them at least is finite and $\mu = \mu_+ - \mu_-$. The measure $|\mu| = \mu_+ + \mu_-$ is called the total variation measure for μ. Denote by $\|\mu\| = |\mu|(\mathbb{R})$.*

We then get the result written in Lemma 1 below (compare to Theorem 4.4 in Bradley [Bra02] for usual mixing coefficients).

Lemma 1. *[DP05] Let $(\Omega, \mathcal{A}, \mathbb{P})$ be a probability space, X a real-valued random variable and $\mathcal{M}$ a σ-algebra of $\mathcal{A}$. Let Λ_1 be the space of 1-Lipschitz functions from $\mathbb{R}$ to $\mathbb{R}$, and BV_1 be the space of BV functions h such that $\|dh\| \leq 1$. We have*

(A1). $\tau(\mathcal{M}, X) = \left\| \sup\left\{ \left| \int f(x)\mathbb{P}_{X|\mathcal{M}}(dx) - \int f(x)\mathbb{P}_X(dx) \right|, f \in \Lambda_1 \right\} \right\|_1.$

(A2). $\alpha(\mathcal{M}, X) = \sup\{ \|\mathbb{E}(f(X)|\mathcal{M}) - \mathbb{E}(f(X))\|_1, f \in BV_1 \}.$

(A3). $\beta(\mathcal{M}, X) = \left\| \sup\left\{ \left| \int f(x)\mathbb{P}_{X|\mathcal{M}}(dx) - \int f(x)\mathbb{P}_X(dx) \right|, f \in BV_1 \right\} \right\|_1.$

(A4). $\phi(\mathcal{M}, X) = \sup\{ \|\mathbb{E}(f(X)|\mathcal{M}) - \mathbb{E}(f(X))\|_\infty, f \in BV_1 \}.$

This paper is a survey of recent results on dependence in the causal frame. For the non-causal frame, we refer e.g. to Doukhan and al or Ango Nze and Doukhan [Dou03, AND02]. In Section 2 we state useful tools in limit theory for dependent sequences. In Section 3 we give a way to compare the different coefficients of dependence. Section 4 is devoted to statistical applications. In Section 5, we give exponential inequalites. We then see in Section 6 how to extend some of the coefficients of Definition 1 to the multidimensionnal case. To conclude, we give in Section 7 some results for particular classes of dynamical systems on $[0, 1]$.

2 Main tools for statistical applications

2.1 Covariance inequalities

For statistical applications, it was made clear by Viennet [Vie97] that covariance inequalities in the style of Delyon [Del90] are more efficient than the usual covariance inequalities. Viennet's result applies to linear estimators and provides optimal results for the mean integrated square error.

In Proposition 1 below, we give two covariance inequalities. Inequality (1) is a weak version of that of Delyon [Del90] in which appear two random variables $b_1(\sigma(Y), \sigma(X))$ and $b_2(\sigma(X), \sigma(Y))$ each having mean $\beta(\sigma(Y), \sigma(X))$. Inequality (2) is a weak version of that of Peligrad [Pel83], where the dependence coefficients are $\phi(\sigma(Y), \sigma(X))$ and $\phi(\sigma(X), \sigma(Y))$.

Proposition 1. *[Ded04] Let X and Y be two real-valued random variables on the probability space $(\Omega, \mathcal{A}, \mathbb{P})$. Let $F_{X|Y} = \{t \to P_{X|Y}(] - \infty, t])\}$ be a distribution function of X given Y and let F_X be the distribution function of X. Define the random variable $b(\sigma(Y), X) = \sup_{x \in \mathbb{R}} |F_{X|Y}(x) - F_X(x)|$. For any conjugate exponents p and q, we have the inequalities*

$$|\text{cov}(Y, X)| \leq 2\{\mathbb{E}(|X|^p b(\sigma(X), Y))\}^{\frac{1}{p}} \{\mathbb{E}(|Y|^q b(\sigma(Y), X))\}^{\frac{1}{q}} \tag{1}$$

$$\leq 2\phi(\sigma(X), Y)^{\frac{1}{p}} \phi(\sigma(Y), X)^{\frac{1}{q}} \|X\|_p \|Y\|_q . \tag{2}$$

In order to derive the MISE of the unknown marginal density of a stationary sequence (Section 4), we need Corollary 1 below.

Corollary 1. *[Ded04] Let f_1, f_2, g_1, g_2 be four increasing functions, and let $f = f_1 - f_2$ and $g = g_1 - g_2$. For any random variable Z, let $\Delta_p(Z) = \inf_{a \in \mathbb{R}} \|Z - a\|_p$ and $\Delta_{p,\sigma(X),Y}(Z) = \inf_{a \in \mathbb{R}} (\mathbb{E}(|Z - a|^p b(\sigma(X), Y)))^{1/p}$. For any conjugate exponents p and q, we have the inequalities*

$$|\text{cov}(g(Y), f(X))| \leq 2\{\Delta_{p,\sigma(X),Y}(f_1(X)) + \Delta_{p,\sigma(X),Y}(f_2(X))\}$$
$$\{\Delta_{q,\sigma(Y),X}(g_1(Y)) + \Delta_{q,\sigma(Y),X}(g_2(Y))\} ,$$

$$|\text{cov}(g(Y), f(X))| \leq 2\phi(\sigma(X), Y)^{\frac{1}{p}} \phi(\sigma(Y), X)^{\frac{1}{q}} \{\Delta_p(f_1(X)) + \Delta_p(f_2(X))\}$$
$$\{\Delta_q(g_1(Y)) + \Delta_q(g_2(Y))\} .$$

In particular, if μ is a signed measure with total variation $\|\mu\|$ and $f(x) = \mu(] - \infty, x])$, we have

$$|\text{cov}(Y, f(X))| \leq \|\mu\| \mathbb{E}(|Y| b(\sigma(Y), X)) \leq \phi(\sigma(Y), X) \|\mu\| \|Y\|_1 . \tag{3}$$

Two different proofs of inequality (3) above can be found in [DP05] and [Ded04].

2.2 Coupling

Coupling is another popular and useful method to obtain limit theorems for sequences of dependent random variables. The coupling result stated in this section just concerns real valued sequences. Thanks to conditional quantile transformation [Maj78], we get an explicit formula for the coupled variable. A more general result in higher dimension has been stated by Rüschendorf [Rüs85] (Section 6.2).

Lemma 2. *[Rüs85, DP03] Let $(\Omega, \mathcal{A}, \mathbb{P})$ be a probability space, X an integrable real-valued random variable, and $\mathcal{M}$ a σ-algebra of $\mathcal{A}$. Assume that there exists a random variable δ uniformly distributed over $[0,1]$, independent of the σ-algebra generated by X and $\mathcal{M}$. Then there exists a random variable X^*, measurable with respect to $\mathcal{M} \vee \sigma(X) \vee \sigma(\delta)$, independent of $\mathcal{M}$ and distributed as X, such that*

$$\|X - X^*\|_1 = \tau(\mathcal{M}, X) . \tag{4}$$

Let $U = F_{X|\mathcal{M}}(X - 0) + \delta \left(F_{X|\mathcal{M}}(X) - F_{X|\mathcal{M}}(X - 0) \right)$, where $F_{X|\mathcal{M}}(t - 0) = \sup_{s < t} F_{X|\mathcal{M}}(s)$. The random variable U is uniformly distributed over $[0,1]$ and independent of $\mathcal{M}$. An explicit solution for the coupling is then given by $X^* = F_X^{-1}(U)$.

3 Comparison of coefficients

The four coefficients introduced in Definition 1 can be compared to each other. It is the purpose of Proposition 2 below. Comparing coefficients to each other is useful for example when deriving upper bounds for $\tau(\mathcal{M}, X)$, $\alpha(\mathcal{M}, X)$, $\beta(\mathcal{M}, X)$ and $\phi(\mathcal{M}, X)$.

Proposition 2. *[DP05] Let $(\Omega, \mathcal{A}, \mathbb{P})$ be a probability space, X a real-valued random variable and $\mathcal{M}$ a σ-algebra of $\mathcal{A}$.*

(A1). We have the inequalities $\alpha(\mathcal{M}, X) \leq \beta(\mathcal{M}, X) \leq \phi(\mathcal{M}, X)$.
(A2). Let Q_X be the generalized inverse of the tail function $t \to \mathbb{P}(|X| > t)$: if $u \in]0, 1[$, $Q_X(u) = \inf\{t \in \mathbb{R} : \mathbb{P}(|X| > t) \leq u\}$. We have the inequality

$$\tau(\mathcal{M}, X) \leq 2 \int_0^{\alpha(\mathcal{M}, X)} Q_X(u) du .$$

(A3). Assume moreover that X has a continuous distribution function F with modulus of continuity w. Define the function g by $g(x) = xw(x)$. Then

$$\beta(\mathcal{M}, X) \leq \frac{2\tau(\mathcal{M}, X)}{g^{-1}(\tau(\mathcal{M}, X))} . \tag{5}$$

In particular, if F is Hölder, that is there exist $C > 0$ and $\alpha \in]0, 1]$ such that for all (x, y) $|F(x) - F(y)| \leq C|x - y|^{\alpha}$, then

$$\beta(\mathcal{M}, X) \leq 2C^{1/(\alpha+1)} (\tau(\mathcal{M}, X))^{\alpha/(\alpha+1)} .$$

When X has a density bounded by K, we obtain the bound

$$\beta(\mathcal{M}, X) \leq 2\sqrt{K\tau(\mathcal{M}, X)} . \tag{6}$$

4 Application to mean integrated square error

We deal in this section with the problem of estimating the unknown marginal density f from the observations $(X_1, \ldots, X_n)$ of a stationary sequence $(X_i)_{i \geq 0}$. There exist many works on density estimation under various mixing conditions. The results of Proposition 3 are comparable to results of Mokkadem (Theorem 1.2 in [Rio00a], and also [Mok90]). The ones of Proposition 4 are to be compared to Theorem 1.3 (a) in [Rio00a]. These results of Mokkadem and Rio are obtained for the strong mixing coefficients $\alpha(\sigma(X_0), \sigma(X_i))$. Viennet [Vie97] proved, under the minimal assumption on the β−mixing coefficients

$$\sum_{k>0} \beta(\sigma(X_0), \sigma(X_k)) < +\infty , \tag{7}$$

that the mean integrated square error (MISE) is of the same order than in the i.i.d. case. Let us now state the results we obtain when working with our weaker coefficient β defined in Definition 1. We first define the coefficients $\tau(i)$, $\alpha(i)$, $\beta(i)$, and $\phi(i)$ related to a sequence of real-valued random variables.

Definition 3. *Let $(\Omega, \mathcal{A}, \mathbb{P})$ be a probability space. Let $(X_i)_{i \geq 0}$ be a sequence of integrable real-valued random variables and $(\mathcal{M}_i)_{i \geq 0}$ be a sequence of σ-algebras of $\mathcal{A}$. The sequence of coefficients $\tau(i)$ is then defined by*

$$\tau(i) = \sup_{k \geq 0} \tau(\mathcal{M}_k, X_{i+k}) . \tag{8}$$

The coefficients $\alpha(i)$, $\beta(i)$ and $\phi(i)$ are defined in the same way.

According to Definition 3 above and to the stationarity of $(X_i)_{i \geq 0}$, we let $\beta(i) = \beta(\sigma(X_0), X_i)$. Both Propositions 3 and 4 give upper bounds for the variance of estimators of the marginal density f.

Proposition 3. *[DP05] Let K be any BV function such that $\int |K(x)| dx$ is finite. Let $(X_i)_{i \geq 0}$ be a stationary sequence, and define*

$$Y_{k,n} = h^{-1} K(h^{-1}(x - X_k)) \quad and \quad f_n(x) = \frac{1}{n} \sum_{k=1}^{n} Y_{k,n} . \tag{9}$$

The following inequality holds

$$nh \int \mathrm{var}(f_n(x)) dx \leq \int (K(x))^2 dx + 2\left(\sum_{k=1}^{n-1} \beta(k)\right) \|dK\| \int |K(x)| dx .$$

Proposition 4. *[DP05] Let $(\varphi_i)_{1 \leq i \leq n}$ be an orthonormal system of $\mathbb{L}^2(\mathbb{R}, \lambda)$ (λ is the Lebesgue measure) and assume that each φ_i is BV. Let $(X_i)_{i \geq 0}$ be a stationary sequence, and define*

$$Y_{j,n} = \frac{1}{n} \sum_{k=1}^{n} \varphi_j(X_k) \quad and \quad f_n = \sum_{j=1}^{m} Y_{j,n} \varphi_j . \tag{10}$$

The following inequality holds

$$n \int \mathrm{var}(f_n(x)) dx \leq \sup_{x \in \mathbb{R}}\left(\sum_{j=1}^{m} \varphi_j^2(x)\right) + 2\left(\sum_{k=1}^{n-1} \beta(k)\right) \sup_{x \in \mathbb{R}}\left(\sum_{j=1}^{m} \|d\varphi_j\| \, |\varphi_j(x)|\right) .$$

These upper bounds allow deriving some quite sharp rates of convergence for some estimators of the unknown density f. We can write indeed $f_n(x) - f(x) = (f_n(x) - \mathbb{E}f_n(x)) + (\mathbb{E}f_n(x) - f(x))$. The bias term $\mathcal{BIAS}_n(x) = \mathbb{E}f_n(x) - f(x)$ does not depend on the dependence properties of the stationary sequence $(X_i)_{i \geq 0}$, but only on the regularity of the marginal density f. We refer to Prieur [Pri01] page 63 for a detailed study of this term.

For kernel density estimators, optimal results can be obtained for the MISE under the condition

$$\sum_{k>0} \beta(\sigma(X_0), X_k) < +\infty$$

as far as we assume that the kernel K is BV and Lebesgue integrable. This condition is weaker than condition (7). For projection estimators, it does not work in the general case. We have to assume moreover that the basis is well localized, because the variance inequality of Proposition 4 is less precise than that of Viennet. It is due to the fact that the covariance inequality (3) of Corollary 1 is not symetric contrary to Delyon's covariance inequality [Del90] used in Viennet [Vie97]. We refer to [DP05] for more detailed results as far as for the proofs.

5 Exponential inequalities

In this section, we state Hoeffding-type inequalities (Propositions 5 and 6) for partial sums under $\phi-$mixing conditions. The coefficients $\phi(k)$ are defined as in Definition 3.

Proposition 5. *[DP05] Let $(X_i)_{i\geq 0}$ be a sequence of centered and square integrable random variables and $\mathcal{M}_i = \sigma(X_j, 1 \leq j \leq i)$. For any BV function h, define*

$$S_n(h) = \sum_{i=1}^{n} h(X_i) \quad and \quad b_{i,n} = \Big(\sum_{k=0}^{n-i} \phi(k)\Big) \|dh\| \, \|h(X_i) - \mathbb{E}(h(X_i))\|_{p/2} \, .$$

For any $p \geq 2$ we have the inequality

$$\|S_n(h) - \mathbb{E}(S_n(h))\|_p \leq \Big(2p \sum_{i=1}^{n} b_{i,n}\Big)^{1/2} \leq \|dh\| \Big(2p \sum_{k=0}^{n-1} (n-k)\phi(k)\Big)^{1/2} . \quad (11)$$

We also have that

$$\mathbb{P}(|S_n(h) - \mathbb{E}(S_n(h))| > x) \leq e^{1/e} \exp\left(\frac{-x^2}{4e\|dh\|^2 \sum_{k=0}^{n-1}(n-k)\phi(k)}\right) . \quad (12)$$

Proposition 6. *Let $(X_i)_{i\geq 0}$ be a sequence of centered and square integrable random variables and $\mathcal{M}_i = \sigma(X_j, 1 \leq j \leq i)$. For any BV function h, define $S_n(h) = \sum_{i=1}^{n} h(X_i)$. For any $p \geq 2$ we have the inequality*

$$\mathbb{P}(|S_n(h) - \mathbb{E}(S_n(h))| > x) \leq 2 \exp\left(\frac{-x^2}{2\|dh\|^2 \sum_{i=1}^{n}\big(1 + 2\sum_{k=1}^{n-i+1} \phi(k)\big)^2}\right) . \quad (13)$$

Inequalities (12) and (13) are of the same type as soon as $\sum_{k>0} \phi(k)$ is finite. Proposition 5 applies to obtain an empirical central limit theorem for classes of BV functions. We refer to Dedecker and Prieur [DP05] for statement and proof of the empirical central limit theorem as far as for applications.

6 The higher dimension case

6.1 Extension of the coefficients

To get more precise inequalities and limit theorems, it is often necessary to consider the dependence between a past σ−algebra and several points in the future of the sequence. Therefore we focus in this section on the problems arising when extending coefficients based on the conditional distribution function to dimension greater than 2. This is a rather complicated problem. One way to proceed is to start with the functional definition (Definition 2) of the coefficients. Even by doing so, it remains difficult to extend $\alpha(\mathcal{M}, X)$, $\beta(\mathcal{M}, X)$ and $\phi(\mathcal{M}, X)$ because the notion of bounded variation is delicate as soon as we are in dimension greater or equal to 2. But dealing with τ is more efficient. We get the following immediate extension, whose "dual" form has been introduced by Rüschendorf [Rüs85].

Let $(\Omega, \mathcal{A}, \mathbb{P})$ be a probability space, $\mathcal{M}$ a σ-agebra of $\mathcal{A}$ and X a random variable with values in a Polish space $(\mathcal{X}, d)$. As in $\mathbb{R}$ there exists a conditional distribution $\mathbb{P}_{X|\mathcal{M}}$ of X given $\mathcal{M}$ (Theorem 10.2.2 in [Dud89]). Let $\Lambda_1(\mathcal{X})$ be the space of 1-lipschitz functions from $\mathcal{X}$ to $\mathbb{R}$. Assume that $\int d(0, x)\mathbb{P}_X(dx)$ is finite and define

$$\tau(\mathcal{M}, X) = \left\|\sup\left\{\left|\int f(x)\mathbb{P}_{X|\mathcal{M}}(dx) - \int f(x)\mathbb{P}_X(dx)\right|, f \in \Lambda_1(\mathcal{X})\right\}\right\|_1. \quad (14)$$

If $d(0, X)$ is bounded, we can also define the uniform version of τ, which was first introduced by Rio [Rio86]:

$$\varphi(\mathcal{M}, X) = \sup\{\|\mathbb{E}(f(X)|\mathcal{M}) - \mathbb{E}(f(X))\|_\infty, f \in \Lambda_1(\mathcal{X})\}.$$

Let us just mention here that Rio's definition of φ [Rio86] is slightly different from the definition above. He defines the class $\Lambda_1(\mathcal{X})$ as the set of 1−Lipschitz functions from $\mathcal{X}$ to $[0, 1]$.

Thanks to such definitions in spaces of higher dimension, we now define the dependence between two sequences $(X_i)_{i\geq0}$ and $(\mathcal{M}_i)_{i\geq0}$ by considering k−tuples in the future and not only a single variable.

In the following, if $(\mathcal{X}, d)$ is some Polish space, we put on $\mathcal{X}^k$ the distance

$$d_1(x, y) = d(x_1, y_1) + \cdots + d(x_k, y_k). \quad (15)$$

We then define

$$\tau_k(i) = \max_{1\leq l\leq k} \frac{1}{l} \sup\{\tau(\mathcal{M}_p, (X_{j_1}, \ldots, X_{j_l})), p + i \leq j_1 < \cdots < j_l\} \quad (16)$$

$$\text{and } \tau_\infty(i) = \sup_{k>0} \tau_k(i).$$

The coefficients φ_k and φ_∞ are defined in the same way.

6.2 Coupling

A coupling result for the one-dimensional case and for the coefficient τ has already been stated in Lemma 2 of Section 2.2. In order to extend this result, let us first give some definitions.

Definition 4. *In the following,*

- $(\mathcal{X}, d)$ *is a Polish space,*
- $(\Omega, \mathcal{A}, \mathbb{P})$ *is a given probability space,*
- $c : \mathcal{X} \times \mathcal{X} \to \mathbb{R}_+$ *is a cost function satisfying*

$$c(x, y) = \sup_{u \in Lip_{\mathcal{X}}^{(c)}} |u(x) - u(y)| , \qquad (17)$$

where $Lip_{\mathcal{X}}^{(c)}$ is the class of continuous bounded functions u on $\mathcal{X}$ such that $|u(x) - u(y)| \leq c(x, y)$.

We assume that $\int c(x, x_0)\mathbb{P}_X(dx)$ is finite for some (and therefore any) x_0 in $\mathcal{X}$. We then introduce the following generalization of the coefficient τ defined by (14)

$$\tau_c(\mathcal{M}, X) = \left\| \sup \left\{ \left| \int f(x)\mathbb{P}_{X|\mathcal{M}}(dx) - \int f(x)\mathbb{P}_X(dx) \right|, f \in Lip_{\mathcal{X}}^{(c)} \right\} \right\|_1 . \quad (18)$$

Let us notice that if $Lip_{\mathcal{X}}^{(c)}$ is a separating class, this coefficient measures the dependence between $\mathcal{M}$ and X ($\tau_c(\mathcal{M}, X) = 0$ if and only if X is independent of $\mathcal{M}$).

We are now in position to state a general coupling result.

Lemma 3. *[Rüs85, DPR05] Let X be a random variable with values in $(\mathcal{X}, d)$. Assume that there exists a random variable δ uniformly distributed over $[0, 1]$, independent of the σ-algebra generated by X and $\mathcal{M}$. Then there exists a random variable X^*, measurable with respect to $\mathcal{M} \vee \sigma(X) \vee \sigma(\delta)$, independent of $\mathcal{M}$ and distributed as X, such that*

$$\tau_c(\mathcal{M}, X) = \mathbb{E}(c(X, X^*)) . \qquad (19)$$

Remark 1.
Lemma 3 above can be extended to the case where $\mathcal{X}$ is not a Polish space [DPR05].
If $c(x, y) = \mathbb{I}_{x \neq y}$, (19) was first written by Berbee [Ber79]. In the case where c is a distance for which $\mathcal{X}$ is a Polish space, the result has been proved by Rüschendorf [Rüs85] in the particular case where Ω is Polish. The proof of Lemma 3 can be found in [Rüs85, DPR05]. It mainly relies on a parametrized version of the Kantorovich-Rubinštein Theorem (Proposition 4 in [Rüs85] and, for the duality, Theorem 2.1 in [DPR05]).

6.3 Exponential inequalities and statistical results

Working with the coefficients (16), we obtain further results for τ−dependent sequences. We sum up these results below.

Proposition 7. *[DP03] Let $(X_i)_{i>0}$ be a sequence of real-valued random variables bounded by M, and $\mathcal{M}_i = \sigma(X_k, 1 \leq k \leq i)$. Let $S_k = \sum_{i=1}^{k}(X_i - \mathbb{E}(X_i))$ and $\overline{S}_n = \max_{1 \leq k \leq n} |S_k|$. Let q be some positive integer, v_q some nonnegative number such that*

$$v_q \geq \|X_{q[n/q]+1} + \cdots + X_n\|_2^2 + \sum_{i=1}^{[n/q]} \|X_{(i-1)q+1} + \cdots + X_{iq}\|_2^2 .$$

and h the function defined by $h(x) = (1+x)\ln(1+x) - x$.

(A1). For any positive λ, $\mathbb{P}(|S_n| \geq 3\lambda) \leq 4 \exp\left(-\frac{v_q}{(qM)^2} h\left(\frac{\lambda qM}{v_q}\right)\right) + \frac{n}{\lambda}\tau_q(q+ 1)$.

(A2). For any $\lambda \geq Mq$, $\mathbb{P}(\overline{S}_n \geq (\mathbb{1}_{q>1} + 3)\lambda) \leq 4 \exp\left(-\frac{v_q}{(qM)^2} h\left(\frac{\lambda qM}{v_q}\right)\right) + \frac{n}{\lambda}\tau_q(q+1)$.

Proposition 7 above extends Bennett's inequality for independent sequences. Starting from the second inequality of Proposition 7 and from the coupling property of Lemma 2, we can prove a functional law of the iterated logarithm (Theorem 1 below). We need some preliminary notations. Let $(X_i)_{i\in\mathbb{Z}}$ be a stationary sequence of real-valued random variables. Let $Q = Q_{X_0}$ be defined as in Proposition 2 and let G be the inverse of $x \to \int_0^x Q(u)du$. Let $\mathcal{S}$ be the subset of $C([0,1])$ consisting of all absolutely continuous functions with respect to the Lebesgue measure such that $h(0) = 0$ and $\int_0^1 (h'(t))^2 dt \leq 1$.

Theorem 1. *[DP03] Let $(X_i)_{i\in\mathbb{Z}}$ be a stationary sequence of real-valued zero-mean square integrable random variables, and $\mathcal{M}_i = \sigma(X_j, j \leq i)$. Let $S_n = X_1 + \cdots + X_n$ and define the partial sum process $S_n(t) = S_{[nt]} + (nt - [nt])X_{[nt]+1}$. If*

$$\sum_{k=1}^{\infty} \int_0^{\tau_\infty(k)} Q \circ G(u)\, du < \infty \qquad (20)$$

then $\mathrm{var}(S_n)$ converges to $\sigma^2 = \sum_{k\in\mathbb{Z}} \mathrm{cov}(X_0, X_k)$. If furthermore $\sigma > 0$ then the process
$\{\sigma^{-1}(2n\ln\ln n)^{-1/2}S_n(t) : t \in [0,1]\}$ is almost surely relatively compact in $C([0,1])$ with limit set $\mathcal{S}$.

To conclude with the statistical applications, we would like to mention two results: the first one is a concentration inequality which is a straightforward consequence of Theorem 1 in Rio [Rio00b], and the second one is a Berry-Esseen inequality due to Rio [Rio86].

The space $\Lambda_1(\mathcal{X}^n)$ is the space of 1-Lipschitz functions from $\mathcal{X}^n$ to $\mathbb{R}$ with respect to d_1 defined by (15).

Theorem 2. *[DP05] Let $(X_1, \ldots X_n)$ be a sequence of random variables with values in a Polish space $(\mathcal{X}, d)$ and $\mathcal{M}_i = \sigma(X_1, \ldots, X_i)$.*
Let $\Delta_i = \inf\{2\|d(X_i, x)\|_\infty, \ x \in \mathcal{X}\}$ and define

$$B_n = \Delta_n \quad \text{and for } 1 \leq i < n, \quad B_i = \Delta_i + 2\varphi(\mathcal{M}_i, (X_{i+1}, \ldots, X_n)) \, .$$

For any f in $\Lambda_1(\mathcal{X}^n)$, we have that

$$\mathbb{P}(f(X_1, \ldots, X_n) - \mathbb{E}(f(X_1, \ldots, X_n)) \geq x) \leq \exp\left(\frac{-2x^2}{B_1^2 + \cdots + B_n^2}\right) .$$

Theorem 3. *[Rio86] Let $(X_i)_{i \in \mathbb{Z}}$ be a stationary sequence of real-valued bounded and centered random variables and $\mathcal{M}_i = \sigma(X_j, j \leq i)$. Let $S_n = X_1 + \cdots + X_n$ and $\sigma_n = \|S_n\|_2$. If $\limsup_{n \to \infty} \sigma_n = \infty$ and*

$$\sum_{n > 0} n\varphi_3(n) < \infty \, , \tag{21}$$

then σ_n^2 converges to $\sigma^2 = \sum_{k \in \mathbb{Z}} \operatorname{cov}(X_0, X_k)$. Moreover $\sigma > 0$ and

$$\sup_{x \in \mathbb{R}} \left| \mathbb{P}(S_n \leq x\sigma_n) - \frac{1}{\sqrt{2\pi}} \int_{-\infty}^{x} \exp(-x^2/2)dx \right| \leq \frac{C}{\sqrt{n}} \, ,$$

where C depends only on $\|X_0\|_\infty$, $(\varphi_3(k))_{k \geq 0}$ and σ.

7 Application to dynamical systems on $[0, 1]$

The mixing conditions studied in this paper are similar to the usual mixing conditions. However, they are weaker and therefore applicable to substantially broader classes of processes. In this section, we are interested in classes of dynamical systems on $[0, 1]$. We refer to [DP03, DP05] for other classes of examples.

The statistical study of dynamical systems is really important because even very simple and determinist dynamical systems may behave in an unpredictable way. Indeed, if we consider two close initial conditions, one can obtain after some finite time two orbits with quite different behaviours. In the litterature on dynamical systems, mixing in the ergodic-theoric sense (MES) is different from mixing in the sense of Rosenblatt [Ros56]. A dynamical system (T^n, μ) is said to be MES if for any sets A and B in $\mathcal{B}(\mathbb{R})$, the sequence $D_n(A, B, \mu, T) = |\mu(A \cap T^{-n}(B)) - \mu(A)\mu(B)|$ converges to zero. For such dynamical systems it is easy to see that strong mixing is a uniform version of MES. As MES only gives a non uniform control of $D_n(A, B, \mu, T)$, it is not sufficient in general to obtain functional limit theorems or deviation inequalities for large classes of functions. There exist number of works on the statistical properties of dynamical systems. Of course, we do not pretend to give a complete account of the literature. One way to derive covariance inequalities for dynamical systems is to study the spectral properties of the associated Perron-Frobenius operator in some well chosen Banach space ([HK82, LY74, Mor94] for example). More recently Pène obtained rates of convergence in the central limit theorem for two-dimensional dispersive billiards [Pen02] and also rate of convergence in multidimensional limit theorems for the Prokhorov metric [Pen04]. Chazottes, Collet, Martinez and Schmitt gave exponential inequalities for expanding maps of the interval [CMS02], Devroye inequalities [CCS04] and statistical applications [CCS05] for a class of non-uniformly hyperbolic dynamical systems. A first approach to study dynamical systems using tools of the theory of weak dependence, in the sense of Doukhan and Louhichi [DL99], can be found in Prieur [Pri01]. Some more recent works of Dedecker and Prieur [DP05] prove that furhter results can be obtained by working with the coefficient Φ defined in Definition 1 of Section 1 and with the Markov chain associated to the dynamical system (Section 7.1 for a precise definition of this Markov chain).

We now introduce the model (Section 7.1), and then apply some of the results of Sections 4 and 6 (Sections 7.2, 7.3 below) to this model.

7.1 Introduction of the model

Let $I = [0,1]$, T be a map from I to I and define $X_i = T^i$. If μ is invariant by T, the sequence $(X_i)_{i \geq 0}$ of random variables from (I, μ) to I is strictly stationary.

For any finite measure ν on I, we use the notations $\nu(h) = \int_I h(x)\nu(dx)$. For any finite signed measure ν on I, let $\|\nu\| = |\nu|(I)$ be the total variation of ν. Denote by $\|g\|_{1,\lambda}$ the $\mathbb{L}^1$-norm with respect to the Lebesgue measure λ on I.

Covariance inequalities. In many interesting cases, one can prove that, for any BV function h and any k in $\mathbb{L}^1(I, \mu)$,

$$|\mathrm{cov}(h(X_0), k(X_n))| \leq a_n \|k(X_n)\|_1 (\|h\|_{1,\lambda} + \|dh\|) , \qquad (22)$$

for some nonincreasing sequence a_n tending to zero as n tends to infinity. Note that if (22) holds, then

$$
\begin{aligned}
|\mathrm{cov}(h(X_0), k(X_n))| &= |\mathrm{cov}(h(X_0) - h(0), k(X_n))| \\
&\leq a_n \|k(X_n)\|_1 (\|h - h(0)\|_{1,\lambda} + \|dh\|) .
\end{aligned}
$$

Since $\|h - h(0)\|_{1,\lambda} \leq \|dh\|$, we obtain that

$$|\mathrm{cov}(h(X_0), k(X_n))| \leq 2a_n \|k(X_n)\|_1 \|dh\| . \qquad (23)$$

Inequality (23) is similar to the second inequality in Proposition 1 item 2, with $X = X_0$ and $Y = k(X_n)$, and one can wonder if $\phi(\sigma(X_n), X_0) \leq 2a_n$. The answer is positive, due to the following Lemma.

Lemma 4. *[DP05] Let $(\Omega, \mathcal{A}, \mathbb{P})$ be a probability space, X a real-valued random variable and $\mathcal{M}$ a σ-algebra of $\mathcal{A}$. We have the equality*

$$\phi(\mathcal{M}, X) =$$
$$\sup\{|\mathrm{cov}(Y, h(X))| \ : \ Y \text{ is } \mathcal{M}\text{-measurable}, \|Y\|_1 \leq 1 \text{ and } h \in BV_1\}.$$

Hence, we obtain an easy way to prove that a dynamical system $(T^i)_{i \geq 0}$ is ϕ-dependent:

$$\text{If (22) holds, then } \phi(\sigma(X_n), X_0) \leq 2a_n. \tag{24}$$

In many cases, Inequality (22) follows from the spectral properties of the Markov operator associated to T. In these cases, due to the underlying Markovian structure, (24) holds with $\mathcal{M}_n = \sigma(X_i, i \geq n)$ instead of $\sigma(X_n)$.

Proof of Lemma 4. Write first $|\mathrm{cov}(Y, h(X))| = |\mathbb{E}(Y(\mathbb{E}(h(X)|\mathcal{M}) - \mathbb{E}(h(X))))|$. For any positive ε, there exists A_ε in $\mathcal{M}$ such that $\mathbb{P}(A_\varepsilon) > 0$ and for any ω in A_ε,

$$|\mathbb{E}(h(X)|\mathcal{M})(\omega) - \mathbb{E}(h(X))| > \|\mathbb{E}(h(X)|\mathcal{M}) - \mathbb{E}(h(X))\|_\infty - \varepsilon.$$

Define the random variable

$$Y_\varepsilon := \frac{\mathbb{1}_{A_\varepsilon}}{\mathbb{P}(A_\varepsilon)} \, \mathrm{sign}\left(\mathbb{E}(h(X)|\mathcal{M}) - \mathbb{E}(h(X))\right).$$

Then Y_ε is $\mathcal{M}$-measurable, $\mathbb{E}|Y_\varepsilon| = 1$ and $|\mathrm{cov}(Y_\varepsilon, h(X))| \geq \|\mathbb{E}(h(X)|\mathcal{M}) - \mathbb{E}(h(X))\|_\infty - \varepsilon$. This being true for any positive ε, we infer from Lemma 1 that

$$\phi(\mathcal{M}, X) \leq$$
$$\sup\{|\mathrm{cov}(Y, h(X))| \ : \ Y \text{ is } \mathcal{M}\text{-measurable}, \|Y\|_1 \leq 1 \text{ and } h \in BV_1\}.$$

The converse inequality follows immediately from Lemma 1 of Section 1.

Spectral gap. Define the operator $\mathcal{L}$ from $\mathbb{L}^1(I, \lambda)$ to $\mathbb{L}^1(I, \lambda)$ *via* the equality

$$\int_0^1 \mathcal{L}(h)(x)k(x)d\lambda(x) = \int_0^1 h(x)(k \circ T)(x)d\lambda(x),$$

where $h \in \mathbb{L}^1(I, \lambda)$ and $k \in \mathbb{L}^\infty(I, \lambda)$.

The operator $\mathcal{L}$ is called the Perron-Frobenius operator of T. In many interesting cases, the spectral analysis of $\mathcal{L}$ in the Banach space of BV-functions equiped with the norm $\|h\|_v = \|dh\| + \|h\|_{1,\lambda}$ can be done by using the Theorem of Ionescu-Tulcea and Marinescu [LY74, HK82]. Assume that 1 is a simple eigenvalue of $\mathcal{L}$ and that the rest of the spectrum is contained in a closed disk of radius strictly smaller than one. Then there exists a unique T-invariant absolutely continuous probability μ whose density f_μ is BV, and

$$\mathcal{L}^n(h) = \lambda(h)f_\mu + \Psi^n(h) \quad \text{with} \quad \|\Psi^n(h)\|_v \leq K\rho^n \|h\|_v. \tag{25}$$

for some $0 \leq \rho < 1$ and $K > 0$. Assume moreover that:

$I_* = \{f_\mu \neq 0\}$ is an interval, and

$$\text{there exists } \gamma > 0 \text{ such that } f_\mu > \gamma^{-1} \text{ on } I_*. \tag{26}$$

Without loss of generality assume that $I_* = I$ (otherwise, take the restriction to I_* in what follows). Define now the Markov kernel associated to T by

$$P(h)(x) = \frac{\mathcal{L}(f_\mu h)(x)}{f_\mu(x)}. \tag{27}$$

It is easy to check (for instance [BGR00]) that $(X_0, X_1, \ldots, X_n)$ has the same distribution as $(Y_n, Y_{n-1}, \ldots, Y_0)$ where $(Y_i)_{i \geq 0}$ is a stationary Markov chain with invariant distribution μ and transition kernel P. Since $\|fg\|_\infty \leq \|fg\|_v \leq 2\|f\|_v\|g\|_v$, we infer that, taking $C = 2K\gamma(\|df_\mu\| + 1)$,

$$P^n(h) = \mu(h) + g_n \quad \text{with} \quad \|g_n\|_\infty \leq C\rho^n \|h\|_v. \tag{28}$$

This estimate implies (22) with $a_n = C\rho^n$. Indeed,

$$\begin{aligned}
|\text{cov}(h(X_0), k(X_n))| &= |\text{cov}(h(Y_n), k(Y_0))| \\
&\leq \|k(Y_0)(\mathbb{E}(h(Y_n)|\sigma(Y_0)) - \mathbb{E}(h(Y_n)))\|_1 \\
&\leq \|k(Y_0)\|_1 \|P^n(h) - \mu(h)\|_\infty \\
&\leq C\rho^n \|k(Y_0)\|_1 (\|dh\| + \|h\|_{1,\lambda}) .
\end{aligned}$$

Collecting the above facts, we infer that $\phi(\sigma(X_n), X_0) \leq 2C\rho^n$. Moreover, using the Markov property we obtain that

$$\begin{aligned}
\phi(\sigma(X_n, \ldots, X_{m+n}), X_0) &= \phi(\sigma(Y_0, \ldots Y_m), Y_{n+m}) \\
&= \phi(\sigma(Y_m), Y_{n+m}) \\
&= \phi(\sigma(X_n), X_0) .
\end{aligned}$$

This being true for any integer m, it holds for $\mathcal{M}_n = \sigma(X_i, i \geq n)$. We conclude that if (25) and (26) hold then there exists $C > 0$ and $0 \leq \rho < 1$ such that

$$\phi(\sigma(X_i, i \geq n), X_0) \leq 2C\rho^n . \tag{29}$$

Application: Expanding maps. Let $([a_i, a_{i+1}[)_{1 \leq i \leq N}$ be a finite partition of $[0, 1[$. We make the same assumptions on T as in Collet *et al* [CMS02].

(A1). For each $1 \leq j \leq N$, the restriction T_j of T to $]a_j, a_{j+1}[$ is strictly monotonic and can be extented to a function $\overline{T}_j$ in $C^2([a_j, a_{j+1}])$.

(A2). Let I_n be the set where $(T^n)'$ is defined. There exists $A > 0$ and $s > 1$ such that $\inf_{x \in I_n} |(T^n)'(x)| > As^n$.

(A3). The map T is topologically mixing: for any two nonempty open sets U, V, there exists $n_0 \geq 1$ such that $T^{-n}(U) \cap V \neq \emptyset$ for all $n \geq n_0$.

If T satisfies 1. 2. and 3. then (25) holds. If furthermore (26) holds ([Mor94] for sufficient conditions), then (29) holds.

7.2 MISE

Let us now specify the results concerning the MISE in the particular case of dynamical systems described in Section 7.1 above. We know via Proposition 2 that $\beta(\mathcal{M}, X) \leq \phi(\mathcal{M}, X)$. Hence, Propositions 3 and 4 apply to dynamical systems satisfying (22) with $2\sum_{i=1}^{n-1} a_k$ instead of $\sum_{i=1}^{n-1} \beta(k)$. For kernel estimators such a result can also be deduced from a variance estimate [Pri01].

7.3 Exponential inequalities

This section is devoted to exponential inequalities which can be obtained for dynamical systems. We first need to give the dependence properties of the Markov chain associated to our dynamical system. Let T be a map from $[0, 1]$ to $[0, 1]$ satisfying Conditions 1. 2. and 3. of Section 7.1. Assume moreover that the density f_μ of the invariant probability μ satisfies (26). Let $X_i = T^i$ and define P as in (27). We know from Section 7.1 that on $([0, 1], \mu)$, the sequence $(X_n, \ldots, X_0)$ has the same distribution as $(Y_0, \ldots, Y_n)$ where $(Y_i)_{i \geq 0}$ is the stationary Markov chain with Markov Kernel P. Consequently

$$\varphi(\sigma(X_j, j \geq i + k), (X_0, \ldots, X_k)) = \varphi(\sigma(Y_0), (Y_i, \ldots, Y_{i+k})) . \qquad (30)$$

To bound $\varphi(\sigma(Y_0), (Y_i, \ldots, Y_{i+k}))$, the first step is to compute $\mathbb{E}(f(Y_0, \ldots, Y_k)|Y_0 = x)$. As for P, define the operator Q_k by

$$\int_0^1 Q_k(f)(x)g(x)f_\mu(x)d\lambda(x) =$$
$$\int_0^1 f(T^k(x), T^{k-1}(x), \ldots, x)g(T^k(x))f_\mu(x)d\lambda(x) .$$

Clearly $\mathbb{E}(f(Y_0, \ldots, Y_k)|Y_0 = x) = Q_k(f)(x)$ and by definition

$$\varphi(\sigma(Y_0), (Y_i, \ldots, Y_{i+k})) = \sup_{f \in \Lambda_1(\mathbb{R}^{k+1})} \|P^i \circ Q_k(f) - \mu(Q_k(f))\|_\infty$$
$$= \sup_{f \in \Lambda_1(\mathbb{R}^{k+1})} \|(P^i - \mu) \circ (Q_k(f) - Q_k(f)(0))\|_\infty \qquad (31)$$

Here, we use a recent result of Collet *et al.* [CMS02]. Denote by $\Lambda_{L_1, \ldots, L_n}$ the set of functions f from $\mathbb{R}^n$ to $\mathbb{R}$ such that

$$|f(x_1, \ldots, x_n) - f(y_1, \ldots, y_n)| \leq L_1|x_1 - y_1| + \cdots + L_n|x_n - y_n| . \qquad (32)$$

Adapting Lasota-Yorke's approach to higher dimension Collet *et al.* prove (page 312 line 6 [CMS02]) that there exist $K > 0$ and $0 \leq \sigma < 1$ such that, for any f in $\Lambda_{L_1, \ldots, L_{k+1}}$,

$$\|dQ_k(f)\| \leq K \sum_{i=0}^k \sigma^i L_{i+1} . \qquad (33)$$

Applying (28), we infer from (31) and (33) that

$$\varphi(\sigma(Y_0), (Y_i, \ldots, Y_{i+k})) \le C\rho^i \|Q_k(f) - Q_k(f)(0)\|_v \le C\rho^i 2\|dQ_k(f)\|$$
$$\le C\rho^i 2K \sum_{j=0}^{k} \sigma^j .$$

Moreover, according to (30), the same bound holds for $\varphi(\sigma(X_j, j \ge i + k), (X_0, \ldots, X_k))$. For the Markov chain $(Y_i)_{i\ge0}$ and the σ-algebras $\mathcal{M}_i = \sigma(Y_j, j \le i)$ we obtain from (33) that

$$\varphi_\infty(i) \le \Big(2CK \sum_{j\ge0} \sigma^j\Big)\rho^i .$$

Several exponential inequalities for dynamical systems are derived from dependence properties of the associated Markov chain. Exponential inequalities of Propositions 5 and 6 in Section 4 can be applied to the sequence $(X_i - \mathbb{E}X_i)_{i\in\mathbb{N}}$. In this section, we would like to focus on the concentration inequality for Lipschitz functions stated in Theorem 2. Starting from (33) and (28) we get that, for any function f belonging to $\Lambda_{L_1, \ldots, L_n}$,

$$\mathbb{P}(f(Y_1, \ldots, Y_n) - \mathbb{E}(f(Y_1, \ldots, Y_n)) \ge x) \le \exp\Big(\frac{-2x^2}{M_1^2 + \cdots + M_n^2}\Big) . \quad (34)$$

with

$M_n = L_n \Delta_0$ and for $1 \le i < n$,
$$M_i = \Delta_0 L_i + 4CK\rho(L_{i+1} + \cdots + L_n \sigma^{n-i-1}) .$$

Since $(X_1, \ldots, X_n)$ has the same distribution as $(Y_n, \ldots, Y_1)$, the bound (34) holds for $f(X_1, \ldots, X_n)$ with

$M_n = L_1 \Delta_0$ and for $1 \le i < n$,
$$M_i = \Delta_0 L_{n-i+1} + 4CK\rho(L_{n-i} + \cdots + L_1 \sigma^{n-i+1}) .$$

Remark 2. Assume that (34) holds for $M_i = \delta_0 L_i + \delta_i L_{i+1} + \cdots + \delta_{n-i} L_n$ (which is the case in the four examples studied above) and let $C_n = \delta_0 + \cdots + \delta_{n-1}$. Applying Cauchy-Schwarz's inequality, we obtain the bound $M_i^2 \le C_n \sum_{j=i}^{n} \delta_{j-i} L_i^2$, and consequently $\sum_{i=1}^{n} M_i^2 \le C_n^2 \sum_{i=1}^{n} L_i^2$. Hence, (34) yields the upper bound

$$\mathbb{P}(f(X_1, \ldots, X_n) - \mathbb{E}(f(X_1, \ldots, X_n)) \ge x) \le \exp\Big(\frac{-2x^2}{C_n^2(L_1^2 + \cdots + L_n^2)}\Big) . \quad (35)$$

For expanding maps, (35) has been proved by Collet *et al* [CMS02].

Aknowledgment. I would like to thank Jérôme Dedecker for suggestions concerning the organization of the paper.

References

[AND84] Andrews, D. W. K.: Nonstrong mixing autoregressive processes. J. Appl. Probab., **21**, 930-934 (1984).

[AND02] Ango Nze, P., Doukhan, P.: Weak Dependence: Models and Applications. In: Empirical process, Techniques for dependent data. Birkhäuser (2002).

[BGR00] Barbour, A.D., Gerrard, R.M., Reinert, G.: Iterates of expanding maps. Probab. Theory Relat. Fields, **116**, 151-180 (2000).

[Ber79] Berbee, H.C.P.: Random walks with stationary increments and renewal theory. Math. Cent. Tracts. Amsterdam (1979).

[Bra02] Bradley, R.C.: Introduction to Strong Mixing Conditions, Volume 1. Technical Report, Department of Mathematics, I. U. Bloomington (2002).

[CCS04] Chazottes, J.R., Collet, P., Schmitt B.: Devroye Inequality for a Class of Non-Uniformly Hyperbolic Dynamical Systems. Preprint (2004). Available on http://arxiv.org/abs/math/0412166.

[CCS05] Chazottes, J.R., Collet, P., Schmitt B.: Statistical Consequences of Devroye Inequality for Processes. Applications to a Class of Non-Uniformly Hyperbolic Dynamical Systems. Preprint (2005). Available on http://arxiv.org/abs/math/0412167.

[CMS02] Collet, P., Martinez, S., Schmitt, B.: Exponential inequalities for dynamical measures of expanding maps of the interval. Probab. Theory. Relat. Fields, **123**, 301-322 (2002).

[Ded04] Dedecker, J.: Inégalités de covariance. C. R. Acad. Sci. Paris, Ser I, **339**, 503-506 (2004).

[DP03] Dedecker, J., Prieur, C.: Coupling for τ-dependent sequences and applications. J. Theoret. Probab., **17 no 4**, 861-885 (2004).

[DP05] Dedecker, J., Prieur, C.: New dependence coefficients. Examples and applications to statistics. To appear in Probab. Theory Relat. Fields (2005).

[DPR05] Dedecker, J., Prieur, C., Raynaud De Fitte, P.: Parametrized Kantorovitch-Rubinštein Theorem and coupling for the minimal distance (2005).

[Del90] Delyon, B.: Limit theorems for mixing processes. Tech. Report 546 IRISA, Rennes I (1990).

[Dou94] Doukhan, P.: Mixing: properties and examples. Lecture Notes in Statist., **85**, Springer-Verlag (1994).

[Dou03] Doukhan, P.: Models inequalities and limit theorems for stationary sequences. In: P. Doukhan, G. Oppenheim and M.S. Taqqu (eds.), Theory and applications of long-range dependence, Birkhäuser (2003).

[DL99] Doukhan, P., Louhichi, S.: A new weak dependence condition and applications to moment inequalities. Stochastic Process. Appl., **84**, 313-342 (1999).

[Dud89] Dudley, R.M.: Real analysis and probability. Wadworsth Inc., Belmont, California (1989).

[HK82] Hofbauer, F., Keller, G.: Ergodic properties of invariant measures for piecewise monotonic transformations. Math. Z., **180**, 119-140 (1982).

[Inr62] Ibragimov, I.A.: Some limit theorems for stationary processes. Theory Probab. Appl., **7**, 349-382 (1962).

[LY74] Lasota, A., Yorke, J.A.: On the existence of invariant measures for piecewise monotonic transformations. Trans. Amer. Math. Soc., **186**, 481-488 (1974).

[Maj78] Major, P.: On the invariance principle for sums of identically distributed random variables. J. Multivariate Anal., **8**, 487-517 (1978).

[Mok90] Mokkadem, A.: Study of risks of kernel estimators. Teor. Veroyatnost. i Primenen, **35**, no. 3, 531-538 (1990); translation in Theory Probab. Appl., **35** (1990), no. 3, 478-486 (1991).

[Mor94] Morita, T.: Local limit theorem and distribution of periodic orbits of Lasota-Yorke transformations with infinite Markov partition. J. Math. Soc. Japan, **46**, 309-343 (1994).

[Pel83] Peligrad, M.: A note on two measures of dependence and mixing sequences. Adv. Appl. Probab., **15**, 461-464 (1983).

[Pel02] Peligrad, M.: Some remarks on coupling of dependent random variables. Stat. Prob. Letters, **60**, 201-209 (2002).

[Pen02] Pène, F.: Rates of convergence in the CLT for two-dimensional dispersive billiards. Comm. Math. Phys. **225**, no. 1, 91-119 (2002).

[Pen04] Pène, F.: Multiple decorrelation and rate of convergence in multidimensional limit theorems for the Prokhorov metric. Ann. Probab. **32**, no. 3B, 2477-2525 (2004).

[Pri01] Prieur, C.: Density Estimation For One-Dimensional Dynamical Systems. ESAIM, Probab. & Statist., WWW.emath.fr/ps, **5**, 51-76 (2001).

[Rio86] Rio, E.: Sur le théorème de Berry-Esseen pour les suites faiblement dépendantes. Probab. Theory Relat. Fields, **104**, 255-282 (1996).

[Rio00a] Rio, E.: Théorie asymptotique des processus aléatoires faiblement dépendants. Collection Mathématiques & Applications **31**. Springer, Berlin (2000).

[Rio00b] Rio, E.: Inégalités de Hoeffding pour les fonctions lipschitziennes de suites dépendantes. C. R. Acad. Sci. Paris Série I, **330**, 905-908 (2000).

[Ros56] Rosenblatt, M.: A central limit theorem and a strong mixing condition. Proc. Nat. Acad. Sci. U. S. A., **42**, 43-47 (1956).

[RV59] Rozanov, Y.A., Volkonskii, V.A.: Some limit theorems for random functions I. Theory Probab. Appl., **4**, 178-197 (1959).

[Rüs85] Rüschendorf, L.: The Wasserstein Distance and Approximation Theorems. Z. Wahrscheinlichkeitstheorie verw. Gebiete, **70**, 117-129 (1985).

[Vie97] Viennet, G.: Inequalities for absolutely regular sequences: application to density estimation. Probab. Theory Relat. Fields, **107**, 467-492 (1997).

Starting from (6) (see [Ber79, Theorem 4.3.2]), Berbee obtained the following coupling result ([Ber79, Theorem 4.4.7]): let $X = (X_k)_{k \geq 1}$ be a $\mathbb{S}_1^\infty$-valued random variable and let $X_{(i)} = (X_k)_{k \geq i}$. If Ω is rich enough, there exists X^* distributed as X and independant of $\mathcal{M}$ such that, for any $i \geq 1$,

$$\frac{1}{2}\mathbb{E}(\|\mathbb{P}_{X_{(i)}|\mathcal{M}} - \mathbb{P}_{X_{(i)}}\|_v) = \mathbb{P}(X_{(i)} \neq X_{(i)}^*) \,, \tag{7}$$

where $\mathbb{P}_{X_{(i)}}$ is the distribution of $X_{(i)}$ and $\mathbb{P}_{X_{(i)}|\mathcal{M}}$ is a regular distribution of $X_{(i)}$ given $\mathcal{M}$. If $(X_k)_{k \in \mathbb{Z}}$ is a strictly stationary sequence of $\mathbb{M}$-valued random variables and $\mathcal{M} = \sigma(X_i, i \leq 0)$, the sequences for which $\mathbb{P}(X_{(i)} \neq X_{(i)}^*)$ converges to zero as i tends to infinity are called β-mixing or absolutely regular sequences. The property (7) is very powerful (see [Rio98] and [BBD01] for recent applications).

In Section 4, we shall see that, contrary to (1), the property (6) is characteristic of the discrete metric. Hence, no analogue of (7) is possible if the underlying cost function is not proportional to the discrete metric.

Preliminary notations

For any topological space $\mathfrak{T}$, we denote by $\mathcal{B}_{\mathfrak{T}}$ the Borel σ–algebra of $\mathfrak{T}$ and by $\mathcal{P}(\mathfrak{T})$ the space of probability laws on $(\mathfrak{T}, \mathcal{B}_{\mathfrak{T}})$, endowed with the narrow topology, that is, for every mapping $\varphi : \mathfrak{T} \to [0, 1]$, the mapping $\mu \mapsto \int_{\mathfrak{T}} \varphi \, d\mu$ is l.s.c. if and only if φ is l.s.c.

Throughout, $\mathbb{S}$ is a given completely regular topological space and $(\Omega, \mathcal{A}, \mathbb{P})$ a given probability space. Note that in [Rüs85], both Ω and $\mathbb{S}$ were assumed to be Polish. However the results are valid in much more general spaces, without significant changes in the proofs. *The reader who is not interested by this level of generality may assume as well in the sequel that all topological spaces we consider are Polish.* On the other hand, we give in appendix some definitions and references which might be useful for a complete reading.

2 Parametrized Kantorovich–Rubinštein theorem

Most of the ideas of this Section are contained in [Rüs85], except for the duality part of point 2 of Theorem 1, which draws inspiration from [CRV04, §3.4].

For any $\mu, \nu \in \mathcal{P}(\mathbb{S})$, let $D(\mu, \nu)$ be the set of probability laws π on $(\mathbb{S} \times \mathbb{S}, \mathcal{B}_{\mathbb{S} \times \mathbb{S}})$ with marginals μ and ν, that is, $\pi(A \times \mathbb{S}) = \mu(A)$ and $\pi(\mathbb{S} \times A) = \nu(A)$ for every $A \in \mathcal{B}_{\mathbb{S}}$. Let us recall the

Kantorovich–Rubinštein duality theorem [Lev84], [RR98, Theorem 4.6.6] *Assume that $\mathbb{S}$ is a completely regular pre-Radon space[4], that is, ev-*

---

[4] In [Lev84] and [RR98, Theorem 4.6.6], the space $\mathbb{S}$ is assumed to be a universally measurable subset of some compact space. But this amounts to assume that it is completely regular and pre-Radon: see [RR98, Lemma 4.5.17] and [GP84, Corollary 11.8].

ery finite τ-additive Borel measure on $\mathbb{S}$ is inner regular with respect to the compact subsets of $\mathbb{S}$. Let $c : \mathbb{S} \times \mathbb{S} \to [0, +\infty]$ be a universally measurable mapping. For every $(\mu, \nu) \in \mathcal{P}(\mathbb{S}) \times \mathcal{P}(\mathbb{S})$, let us denote

$$\Delta_{\mathrm{KR}}^{(c)}(\mu, \nu) := \inf_{\pi \in D(\mu,\nu)} \int_{\mathbb{S} \times \mathbb{S}} c(x, y) \, \mathrm{d}\pi(x, y) \,,$$

$$\Delta_{\mathrm{L}}^{(c)}(\mu, \nu) := \sup_{f \in \mathrm{Lip}_{\mathbb{S}}^{(c)}} (\mu(f) - \nu(f)) \,,$$

where $\mathrm{Lip}_{\mathbb{S}}^{(c)} = \{u \in C_b(\mathbb{S}) ; \forall x, y \in \mathbb{S} \quad |u(x) - u(y)| \leq c(x, y)\}$. Then the equality $\Delta_{\mathrm{KR}}^{(c)}(\mu, \nu) = \Delta_{\mathrm{L}}^{(c)}(\mu, \nu)$ holds for all $(\mu, \nu) \in \mathcal{P}(\mathbb{S}) \times \mathcal{P}(\mathbb{S})$ if and only if (4) holds.

Note that, if c satifies (4), it is the supremum of a set of continuous functions, thus it is l.s.c. Every continuous metric c on $\mathbb{S}$ satisfies (4) (see [RR98, Corollary 4.5.7]), and, if $\mathbb{S}$ is compact, every l.s.c. metric c on $\mathbb{S}$ satisfies (4) (see [RR98, Remark 4.5.6]).

Denote

$$\mathcal{Y}(\Omega, \mathcal{A}, \mathbb{P}; \mathbb{S}) = \{\mu \in \mathcal{P}(\Omega \times \mathbb{S}, \mathcal{A} \otimes \mathcal{B}_{\mathbb{S}}); \forall A \in \mathcal{A} \quad \mu(A \times \mathbb{S}) = \mathbb{P}(A)\} \,.$$

When no confusion can arise, we omit some part of the information, and use notations such as $\mathcal{Y}(\mathcal{A})$ or simply $\mathcal{Y}$ (same remark for the set $\mathcal{Y}^{c,1}(\Omega, \mathcal{A}, \mathbb{P}; \mathbb{S})$ defined below). If $\mathbb{S}$ is a Radon space, every $\mu \in \mathcal{Y}$ is *disintegrable*, that is, there exists a (unique, up to $\mathbb{P}$-a.e. equality) $\mathcal{A}_{\mu}^*$-measurable mapping $\omega \mapsto \mu_{\omega}$, $\Omega \to \mathcal{P}(\mathbb{S})$, such that

$$\mu(f) = \int_{\Omega} \int_{\mathbb{S}} f(\omega, x) \, \mathrm{d}\mu_{\omega}(x) \, \mathrm{d}\mathbb{P}(\omega) \,,$$

for every measurable $f : \Omega \times \mathbb{S} \to [0, +\infty]$ (see [Val73]). If furthermore the compact subsets of $\mathbb{S}$ are metrizable, the mapping $\omega \mapsto \mu_{\omega}$ can be chosen $\mathcal{A}$-measurable, see the Appendix.

Let c satisfy (4). We denote

$$\mathcal{Y}^{c,1}(\Omega, \mathcal{A}, \mathbb{P}; \mathbb{S}) = \{\mu \in \mathcal{Y}; \int_{\Omega \times \mathbb{S}} c(x, x_0) \, \mathrm{d}\mu(\omega, x) < +\infty\}$$

where x_0 is some fixed element of $\mathbb{S}$ (this definition is independent of the choice of x_0). For any $\mu, \nu \in \mathcal{Y}$, let $\underline{D}(\mu, \nu)$ be the set of probability laws π on $\Omega \times \mathbb{S} \times \mathbb{S}$ such that $\pi(. \times . \times \mathbb{S}) = \mu$ and $\pi(. \times \mathbb{S} \times .) = \nu$. We now define the parametrized versions of $\Delta_{\mathrm{KR}}^{(c)}$ and $\Delta_{\mathrm{L}}^{(c)}$. Set, for $\mu, \nu \in \mathcal{Y}^{c,1}$,

$$\underline{\Delta}_{\mathrm{KR}}^{(c)}(\mu, \nu) = \inf_{\pi \in \underline{D}(\mu,\nu)} \int_{\Omega \times \mathbb{S} \times \mathbb{S}} c(x, y) \, \mathrm{d}\pi(\omega, x, y) \,.$$

Let also $\underline{\mathrm{Lip}}^{(c)}$ denote the set of measurable integrands $f : \Omega \times \mathbb{S} \to \mathbb{R}$ such that $f(\omega, .) \in \mathrm{Lip}_{\mathbb{S}}^{(c)}$ for every $\omega \in \Omega$. We denote

$$\Delta_{\mathrm{L}}^{(c)}(\mu,\nu) = \sup_{f\in\underline{\mathrm{Lip}}^{(c)}} (\mu(f) - \nu(f)) \ .$$

Theorem 1 (Parametrized Kantorovich–Rubinštein theorem). *Assume that $\mathbb{S}$ is a completely regular Radon space and that the compact subsets of $\mathbb{S}$ are metrizable (e.g. $\mathbb{S}$ is a regular Suslin space). Let $c : \mathbb{S} \times \mathbb{S} \to [0, +\infty[$ satisfy (4). Let $\mu, \nu \in \mathcal{Y}^{c,1}$ and let $\omega \mapsto \mu_\omega$ and $\omega \mapsto \nu_\omega$ be disintegrations of μ and ν respectively.*

(A1). Let $G : \omega \mapsto \Delta_{\mathrm{KR}}^{(c)}(\mu_\omega, \nu_\omega) = \Delta_{\mathrm{L}}^{(c)}(\mu_\omega, \nu_\omega)$ and let $\mathcal{A}^*$ be the universal completion of $\mathcal{A}$. There exists an $\mathcal{A}^*$–measurable mapping $\omega \mapsto \lambda_\omega$ from Ω to $\mathcal{P}(\mathbb{S} \times \mathbb{S})$ such that λ_ω belongs to $D(\mu_\omega, \nu_\omega)$ and

$$G(\omega) = \int_{\mathbb{S}\times\mathbb{S}} c(x,y) \, \mathrm{d}\lambda_\omega(x,y) \ .$$

(A2). The following equalities hold:

$$\Delta_{\mathrm{KR}}^{(c)}(\mu,\nu) = \int_{\Omega\times\mathbb{S}\times\mathbb{S}} c(x,y) \, \mathrm{d}\lambda(\omega,x,y) = \Delta_{\mathrm{L}}^{(c)}(\mu,\nu) \ ,$$

where λ is the element of $\mathcal{Y}(\Omega, \mathcal{A}, \mathbb{P}; \mathbb{S} \times \mathbb{S})$ defined by $\lambda(A \times B \times C) = \int_A \lambda_\omega(B \times C) \, \mathrm{d}\mathbb{P}(\omega)$ for any A in $\mathcal{A}$, B and C in $\mathcal{B}_\mathbb{S}$. In particular, λ belongs to $\underline{D}(\mu,\nu)$, and the infimum in the definition of $\Delta_{\mathrm{KR}}^{(c)}(\mu,\nu)$ is attained for this λ.

Remark 1. In the case where both Ω and $\mathbb{S}$ are Polish spaces, point 1 and the first equality in point 2 of Theorem 1 are contained in Proposition 4 of Rüschendorf [Rüs85]. The proof we give below follows that of Proposition 4 in [Rüs85] and of Theorem 3.4.1 in [CRV04]. As in [Rüs85], the main argument is a measurable selection lemma given in [CV77].

The set of compact subsets of a topological space $\mathfrak{T}$ is denoted by $\mathcal{K}(\mathfrak{T})$.

Lemma 1 (A measurable selection lemma). *Assume that $\mathbb{S}$ is a Suslin space. Let $c : \mathbb{S} \times \mathbb{S} \to [0, +\infty]$ be an l.s.c. mapping. Let $\mathcal{B}^*$ be the universal completion of the σ–algebra $\mathcal{B}_{\mathcal{P}(\mathbb{S})\times\mathcal{P}(\mathbb{S})}$. For any $\mu, \nu \in \mathcal{P}(\mathbb{S})$, let*

$$r(\mu,\nu) = \inf_{\pi\in D(\mu,\nu)} \int c(x,y) \, \mathrm{d}\pi(x,y) \in [0, +\infty] \ .$$

The function r is $\mathcal{B}^*$–measurable. Furthermore, the multifunction

$$K : \begin{cases} \mathcal{P}(\mathbb{S}) \times \mathcal{P}(\mathbb{S}) \to \mathcal{K}\left(\mathcal{P}(\mathbb{S} \times \mathbb{S})\right) \\ (\mu,\nu) \qquad \mapsto \left\{\pi \in D(\mu,\nu); \int c(x,y) \, \mathrm{d}\pi(x,y) = r(\mu,\nu)\right\} \end{cases}$$

has a $\mathcal{B}^*$–measurable selection, that is, there exists a $\mathcal{B}^*$–measurable mapping $\lambda : (\mu,\nu) \mapsto \lambda_{\mu,\nu}$ defined on $\mathcal{P}(\mathbb{S}) \times \mathcal{P}(\mathbb{S})$ with values in $\mathcal{K}\left(\mathcal{P}(\mathbb{S} \times \mathbb{S})\right)$, such that $\lambda_{\mu,\nu} \in K(\mu,\nu)$ for all $\mu, \nu \in \mathcal{P}(\mathbb{S})$.

Proof. Observe first that the mapping r can be defined as

$$r : (\mu, \nu) \mapsto \inf \{\psi(\pi); \pi \in D(\mu, \nu)\} \ ,$$

with

$$\psi : \begin{cases} \mathcal{P}(\mathbb{S} \times \mathbb{S}) \to [0, +\infty] \\ \pi \qquad\quad \mapsto \int_{\mathbb{S} \times \mathbb{S}} c(x, y) \, \mathrm{d}\pi(x, y) \ . \end{cases}$$

The mapping ψ is l.s.c. because it is the supremum of the l.s.c. mappings $\pi \mapsto \pi(c \wedge n)$, $n \in \mathcal{N}$ (if c is bounded and continuous, ψ is continuous). Furthermore, we have $D = \Phi^{-1}$, where Φ is the continuous mapping

$$\Phi : \begin{cases} \mathcal{P}(\mathbb{S} \times \mathbb{S}) \to \mathcal{P}(\mathbb{S}) \times \mathcal{P}(\mathbb{S}) \\ \lambda \qquad\quad \mapsto (\lambda(\cdot \times \mathbb{S}), \lambda(\mathbb{S} \times \cdot)) \ . \end{cases}$$

(Recall that $D(\mu, \nu)$ is the set of probability laws π on $\mathbb{S} \times \mathbb{S}$ with marginals μ and ν.) Therefore, the graph $\mathrm{gph}\,(D)$ of D is a closed subset of the Suslin space $\mathbb{X} = \big(\mathcal{P}(\mathbb{S}) \times \mathcal{P}(\mathbb{S})\big) \times \mathcal{P}(\mathbb{S} \times \mathbb{S})$. Applying Lemma III.39 of [CV77] as done in [Rüs85], we infer that r is $\mathcal{B}^*$–measurable. Now the fact that K has a $\mathcal{B}^*$–measurable selection follows from the application of Lemma III.39 given in paragraph 39 of [CV77]. □

Proof (Proof of Theorem 1). By the Radon property, the probability measures $\mu(\Omega \times .)$ and $\nu(\Omega \times .)$ are tight, that is, for every integer $n \geq 1$, there exists a compact subset K_n of $\mathbb{S}$ such that $\mu(\Omega \times (\mathbb{S} \backslash K_n)) \leq 1/n$ and $\nu(\Omega \times (\mathbb{S} \backslash K_n)) \leq 1/n$. Now, we can clearly replace $\mathbb{S}$ in the statements of Theorem 1 by the smaller space $\cup_{n \geq 1} K_n$. But $\cup_{n \geq 1} K_n$ is Suslin (and even Lusin), so we can assume without loss of generality that $\mathbb{S}$ is a regular Suslin space. We easily have

$$\underline{\Delta}_{\mathrm{L}}^{(c)}(\mu, \nu) = \sup_{f \in \mathrm{Lip}^{(c)}} \int_\Omega \int_{\mathbb{S}} \int_{\mathbb{S}} (f(\omega, x) - f(\omega, y)) \, \mathrm{d}\mu_\omega(x) \, \mathrm{d}\nu_\omega(y) \, \mathrm{d}\mathbb{P}(\omega)$$

$$\leq \int_\Omega \int_{\mathbb{S}} \int_{\mathbb{S}} c(x, y) \, \mathrm{d}\mu_\omega(x) \, \mathrm{d}\nu_\omega(y) \, \mathrm{d}\mathbb{P}(\omega) \leq \underline{\Delta}_{\mathrm{KR}}^{(c)}(\mu, \nu) \ . \qquad (8)$$

So, to prove Theorem 1, we only need to prove that $\underline{\Delta}_{\mathrm{KR}}^{(c)}(\mu, \nu) \leq \underline{\Delta}_{\mathrm{L}}^{(c)}(\mu, \nu)$ and that the minimum in the definition of $\underline{\Delta}_{\mathrm{KR}}^{(c)}(\mu, \nu)$ is attained.

Using the notations of Lemma 1, we have $G(\omega) = r(\mu_\omega, \nu_\omega)$, thus G is $\mathcal{A}^*$–measurable (indeed, the mapping $\omega \mapsto (\mu_\omega, \nu_\omega)$ is measurable for $\mathcal{A}^*$ and $\mathcal{B}^*$ because it is measurable for $\mathcal{A}$ and $\mathcal{B}_{\mathcal{P}(\mathbb{S}) \times \mathcal{P}(\mathbb{S})}$). From Lemma 1, the multifunction $\omega \mapsto D(\mu_\omega, \nu_\omega)$ has an $\mathcal{A}^*$–measurable selection $\omega \mapsto \lambda_\omega$ such that, for every $\omega \in \Omega$, $G(\omega) = \int_{\mathbb{S} \times \mathbb{S}} c(x, y) \, \mathrm{d}\lambda_\omega(x, y)$. We thus have

$$\underline{\Delta}_{\mathrm{KR}}^{(c)}(\mu, \nu) \leq \int_{\Omega \times \mathbb{S} \times \mathbb{S}} c(x, y) \, \mathrm{d}\lambda(\omega, x, y) = \int_\Omega G(\omega) \, \mathrm{d}\mathbb{P}(\omega) \ . \qquad (9)$$

Furthermore, since $\mu, \nu \in \mathcal{Y}^{c,1}$, we have $G(\omega) < +\infty$ a.e. Let Ω_0 be the almost sure set on which $G(\omega) < +\infty$. Fix an element x_0 in $\mathbb{S}$. We have, for every $\omega \in \Omega_0$,

$$G(\omega) = \sup_{g \in \mathrm{Lip}_{\mathbb{S}}^{(c)}} (\mu_\omega(g) - \nu_\omega(g)) = \sup_{g \in \mathrm{Lip}_{\mathbb{S}}^{(c)},\ g(x_0)=0} (\mu_\omega(g) - \nu_\omega(g)) .$$

Let $\epsilon > 0$. Let $\widetilde{\mu}$ and $\widetilde{\nu}$ be the finite measures on $\mathbb{S}$ defined by

$$\widetilde{\mu}(B) = \int_{\Omega \times B} c(x_0, x)\, d\mu(\omega, x) \quad \text{and} \quad \widetilde{\nu}(B) = \int_{\Omega \times B} c(x_0, x)\, d\nu(\omega, x)$$

for any $B \in \mathcal{B}_{\mathbb{S}}$. Let $\mathbb{S}_0$ be a compact subset of $\mathbb{S}$ containing x_0 such that $\widetilde{\mu}(\mathbb{S} \setminus \mathbb{S}_0) \leq \epsilon$ and $\widetilde{\nu}(\mathbb{S} \setminus \mathbb{S}_0) \leq \epsilon$. For any $f \in \underline{\mathrm{Lip}}^{(c)}$, we have

$$\left| \int_\Omega (\mu_\omega - \nu_\omega)(f(\omega, .))\, d\mathbb{P}(\omega) - \int_\Omega (\mu_\omega - \nu_\omega)(f(\omega, .) \mathbb{1}_{\mathbb{S}_0})\, d\mathbb{P}(\omega) \right|$$

$$= \left| \int_\Omega (\mu_\omega - \nu_\omega)(f(\omega, .) \mathbb{1}_{\mathbb{S} \setminus \mathbb{S}_0})\, d\mathbb{P}(\omega) \right| \leq 2\epsilon . \quad (10)$$

Set, for all $\omega \in \Omega_0$,

$$G'(\omega) = \sup_{g \in \mathrm{Lip}_{\mathbb{S}}^{(c)},\ g(x_0)=0} (\mu_\omega - \nu_\omega)(g \mathbb{1}_{\mathbb{S}_0}) .$$

We thus have

$$\left| \int_{\Omega_0} G\, d\mathbb{P} - \int_{\Omega_0} G'\, d\mathbb{P} \right| \leq 2\epsilon . \quad (11)$$

Let $\mathrm{Lip}_{\mathbb{S}}^{(c)}\big|_{\mathbb{S}_0}$ denote the set of restrictions to $\mathbb{S}_0$ of elements of $\mathrm{Lip}_{\mathbb{S}}^{(c)}$. The set $\mathbb{S}_0$ is metrizable, thus $C_b(\mathbb{S}_0)$ (endowed with the topology of uniform convergence) is metrizable separable, thus its subspace $\mathrm{Lip}_{\mathbb{S}}^{(c)}\big|_{\mathbb{S}_0}$ is also metrizable separable. We can thus find a dense countable subset $D = \{u_n;\ n \in \mathcal{N}\}$ of $\mathrm{Lip}_{\mathbb{S}}^{(c)}$ for the seminorm $\|u\|_{C_b(\mathbb{S}_0)} := \sup_{x \in \mathbb{S}_0} |u(x)|$. Set, for all $(\omega, x) \in \Omega_0 \times \mathbb{S}$,

$$N(\omega) = \min\left\{ n \in \mathcal{N};\ \int_{\mathbb{S}} u_n(x)\, d(\mu_\omega - \nu_\omega)(x) \geq \Delta_{\mathrm{L}}^{(c)}(\mu_\omega, \nu_\omega) - \epsilon G'(\omega) - \epsilon \right\} ,$$

and $f(\omega, x) = u_{N(\omega)}(x)$. We then have, using (10) and (11),

$$\Delta_{\mathrm{L}}^{(c)}(\mu, \nu) \geq \int_{\Omega_0 \times \mathbb{S}} f\, d(\mu - \nu) \geq \int_{\Omega_0 \times \mathbb{S}_0} f\, d(\mu - \nu) - 2\epsilon$$

$$\geq \int_{\Omega_0} G'\, d\mathbb{P} - 3\epsilon \geq \int_{\Omega_0} G\, d\mathbb{P} - 5\epsilon .$$

Thus, in view of (8) and (9),

$$\Delta_{\mathrm{KR}}^{(c)}(\mu, \nu) = \int_{\Omega \times \mathbb{S} \times \mathbb{S}} c(x, y)\, d\lambda(\omega, x, y) = \Delta_{\mathrm{L}}^{(c)}(\mu, \nu) .$$

$\square$

3 Application: coupling for the minimal distance

In this section $\mathbb{S}$ is a completely regular Radon space with metrizable compact subsets, $c : \mathbb{S} \times \mathbb{S} \to [0, +\infty]$ is a mapping satisfying (4) and $\mathcal{M}$ is a sub-σ-algebra of $\mathcal{A}$. Let X be a random variable with values in $\mathbb{S}$, let $\mathbb{P}_X$ be the distribution of X, and let $\mathbb{P}_{X|\mathcal{M}}$ be a regular conditional distribution of X given $\mathcal{M}$ (see Section 5 for the existence). We assume that $\int c(x, x_0) \mathbb{P}_X(dx)$ is finite for some (and therefore any) x_0 in $\mathbb{S}$ (which means exactly that the unique measure of $\mathcal{Y}(\mathcal{M})$ with disintegration $\mathbb{P}_{X|\mathcal{M}}(\cdot, \omega)$ belongs to $\mathcal{Y}^{c,1}(\mathcal{M})$). The proof of the following result is comparable to that of Corollary 4.2.5 in [Ber79] and of Proposition 5 in [Rüs85].

Theorem 2 (A general coupling theorem). *Assume that Ω is rich enough, that is, there exists a random variable U from $(\Omega, \mathcal{A})$ to $([0, 1], \mathcal{B}([0, 1]))$, independent of $\sigma(X) \vee \mathcal{M}$ and uniformly distributed over $[0, 1]$. Let Q be any element of $\mathcal{Y}^{c,1}(\mathcal{M})$. There exists a $\sigma(U) \vee \sigma(X) \vee \mathcal{M}$-measurable random variable Y, such that $Q_{\cdot}$ is a regular conditional probability of Y given $\mathcal{M}$, and*

$$\mathbb{E}\left(c(X, Y) | \mathcal{M}\right) = \sup_{f \in \mathrm{Lip}_{\mathbb{S}}^{(c)}} \left| \int f(x) \mathbb{P}_{X|\mathcal{M}}(dx) - \int f(x) Q_{\cdot}(dx) \right| \quad \mathbb{P}\text{-}a.s. \quad (12)$$

Proof. We apply Theorem 1 to the probability space $(\Omega, \mathcal{M}, \mathbb{P})$ and to the disintegrated measures $\mu_\omega(\cdot) = \mathbb{P}_{X|\mathcal{M}}(\cdot, \omega)$ and $\nu_\omega = Q_\omega$. As in the proof of Theorem 1, we assume without loss of generality that $\mathbb{S}$ is Lusin regular. From point 1 of Theorem 1 we infer that there exists a mapping $\omega \mapsto \lambda_\omega$ from Ω to $\mathcal{P}(\mathbb{S} \times \mathbb{S})$, measurable for $\mathcal{M}^*$ and $\mathcal{B}_{\mathcal{P}(\mathbb{S} \times \mathbb{S})}$, such that λ_ω belongs to $D(\mathbb{P}_{X|\mathcal{M}}(\cdot, \omega), Q_\omega)$ and $G(\omega) = \int_{\mathbb{S} \times \mathbb{S}} c(x, y) \lambda_\omega(dx, dy)$.

On the measurable space $(\mathbb{M}, \mathcal{T}) = (\Omega \times \mathbb{S} \times \mathbb{S}, \mathcal{M}^* \otimes \mathcal{B}_{\mathbb{S}} \otimes \mathcal{B}_{\mathbb{S}})$ we put the probability

$$\pi(A \times B \times C) = \int_A \lambda_\omega(B \times C) \mathbb{P}(d\omega).$$

If $I = (I_1, I_2, I_3)$ is the identity on $\mathbb{M}$, we see that a regular conditional distribution of (I_2, I_3) given I_1 is given by $\mathbb{P}_{(I_2, I_3)|I_1 = \omega} = \lambda_\omega$. Since $\mathbb{P}_{X|\mathcal{M}}(\cdot, \omega)$ is the first marginal of λ_ω, a regular conditional probability of I_2 given I_1 is given by $\mathbb{P}_{I_2|I_1 = \omega}(\cdot) = \mathbb{P}_{X|\mathcal{M}}(\cdot, \omega)$. Let $\lambda_{\omega, x} = \mathbb{P}_{I_3|I_1 = \omega, I_2 = x}$ be a regular conditional distribution of I_3 given (I_1, I_2), so that $(\omega, x) \mapsto \lambda_{\omega, x}$ is measurable for $\mathcal{M}^* \otimes \mathcal{B}_{\mathbb{S}}$ and $\mathcal{B}_{\mathcal{P}(\mathbb{S})}$. From the uniqueness (up to $\mathbb{P}$-a.s. equality) of regular conditional probabilities, it follows that

$$\lambda_\omega(B \times C) = \int_B \lambda_{\omega, x}(C) \mathbb{P}_{X|\mathcal{M}}(dx, \omega) \quad \mathbb{P}\text{-}a.s. \quad (13)$$

Assume that we can find a random variable $\tilde{Y}$ from Ω to $\mathbb{S}$, measurable for $\sigma(U) \vee \sigma(X) \vee \mathcal{M}^*$ and $\mathcal{B}_{\mathbb{S}}$, such that $\mathbb{P}_{\tilde{Y}|\sigma(X) \vee \mathcal{M}^*}(\cdot, \omega) = \lambda_{\omega, X(\omega)}(\cdot)$. Since $\omega \mapsto \mathbb{P}_{X|\mathcal{M}}(\cdot, \omega)$ is measurable for $\mathcal{M}^*$ and $\mathcal{B}_{\mathcal{P}(\mathbb{S})}$, one can check that $\mathbb{P}_{X|\mathcal{M}}$

is a regular conditional probability of X given $\mathcal{M}^*$. For A in $\mathcal{M}^*$, B and C in $\mathcal{B}_\mathbb{S}$, we thus have

$$
\begin{aligned}
\mathbb{E}\left(\mathbb{1}_A \mathbb{1}_{X \in B} \mathbb{1}_{\tilde{Y} \in C}\right) &= \mathbb{E}\left(\mathbb{1}_A \mathbb{E}\left(\mathbb{1}_{X \in B} \mathbb{E}\left(\mathbb{1}_{\tilde{Y} \in C} | \sigma(X) \vee \mathcal{M}^*\right) | \mathcal{M}^*\right)\right) \\
&= \int_A \left(\int_B \lambda_{\omega,x}(C) \mathbb{P}_{X|\mathcal{M}}(\mathrm{d}x, \omega)\right) \mathbb{P}(\mathrm{d}\omega) \\
&= \int_A \lambda_\omega(B \times C) \mathbb{P}(\mathrm{d}\omega) .
\end{aligned}
$$

We infer that λ_ω is a regular conditional probability of $(X, \tilde{Y})$ given $\mathcal{M}^*$. By definition of λ_ω, we obtain that

$$
\mathbb{E}\left(c(X, \tilde{Y})|\mathcal{M}^*\right) = \sup_{f \in \mathrm{Lip}_\mathbb{S}^{(c)}} \left| \int f(x) \mathbb{P}_{X|\mathcal{M}}(\mathrm{d}x) - \int f(x) Q_\cdot(\mathrm{d}x) \right| \quad \mathbb{P}\text{-a.s.}
$$
(14)

Since $\mathbb{S}$ is Lusin, it is standard Borel (see Section 5). Applying Lemma 2, there exists a $\sigma(U) \vee \sigma(X) \vee \mathcal{M}$-measurable modification Y of $\tilde{Y}$, so that (14) still holds for $\mathbb{E}(c(X,Y)|\mathcal{M}^*)$. We obtain (12) by noting that $\mathbb{E}(c(X,Y)|\mathcal{M}^*) = \mathbb{E}(c(X,Y)|\mathcal{M})$ $\mathbb{P}$-a.s.

It remains to build $\tilde{Y}$. Since $\mathbb{S}$ is standard Borel, there exists a one to one map f from $\mathbb{S}$ to a Borel subset of $[0,1]$, such that f and f^{-1} are measurable for $\mathcal{B}([0,1])$ and $\mathcal{B}_\mathbb{S}$. Define $F(t,\omega) = \lambda_{\omega,X(\omega)}(f^{-1}(]-\infty, t]))$. The map $F(\cdot, \omega)$ is a distribution function with càdlàg inverse $F^{-1}(\cdot, \omega)$. One can see that the map $(u, \omega) \to F^{-1}(u, \omega)$ is $\mathcal{B}([0,1]) \otimes \mathcal{M}^* \vee \sigma(X)$-measurable. We now use the fact that Ω is rich enough: the existence of the random variable U uniformly distributed over $[0,1]$ and independent of $\sigma(X) \vee \mathcal{M}$ allows some independent randomization. Let $T(\omega) = F^{-1}(U(\omega), \omega)$ and $\tilde{Y} = f^{-1}(T)$. It remains to see that $\mathbb{P}_{\tilde{Y}|\sigma(X) \vee \mathcal{M}^*}(\cdot, \omega) = \lambda_{\omega, X(\omega)}(\cdot)$. For any A in $\mathcal{M}^*$, B in $\mathcal{B}_\mathbb{S}$ and t in $\mathbb{R}$, we have

$$
\mathbb{E}\left(\mathbb{1}_A \mathbb{1}_{X \in B} \mathbb{1}_{\tilde{Y} \in f^{-1}(]-\infty, t])}\right) = \int_A \mathbb{1}_{X(\omega) \in B} \mathbb{1}_{U(\omega) \le F(t, \omega)} \mathbb{P}(\mathrm{d}\omega) .
$$

Since U is independent of $\sigma(X) \vee \mathcal{M}$, it is also independent of $\sigma(X) \vee \mathcal{M}^*$. Hence

$$
\begin{aligned}
\mathbb{E}\left(\mathbb{1}_A \mathbb{1}_{X \in B} \mathbb{1}_{\tilde{Y} \in f^{-1}(]-\infty, t])}\right) &= \int_A \mathbb{1}_{X(\omega) \in B} F(t, \omega) \mathbb{P}(\mathrm{d}\omega) \\
&= \int_A \mathbb{1}_{X(\omega) \in B} \lambda_{\omega, X(\omega)}(f^{-1}(]-\infty, t])) \mathbb{P}(\mathrm{d}\omega) .
\end{aligned}
$$

Since $\{f^{-1}(]-\infty, t]), t \in [0,1]\}$ is a separating class, the result follows. □

Coupling and dependence coefficients

Define the coefficient

$$\tau_c(\mathcal{M}, X) = \left\| \sup_{f \in \mathrm{Lip}_{\mathbb{S}}^{(c)}} \left| \int f(x) \mathbb{P}_{X|\mathcal{M}}(\mathrm{d}x) - \int f(x) \mathbb{P}_X(\mathrm{d}x) \right| \right\|_1 . \qquad (15)$$

If $\mathrm{Lip}_{\mathbb{S}}^{(c)}$ is a separating class, this coefficient measures the dependence between $\mathcal{M}$ and X ($\tau_c(\mathcal{M}, X) = 0$ if and only if X is independent of $\mathcal{M}$). From point 2 of Theorem 1, we see that an equivalent definition is

$$\tau_c(\mathcal{M}, X) = \sup_{f \in \mathrm{Lip}_{\mathbb{S},\mathcal{M}}^{(c)}} \int f(\omega, X(\omega)) \mathbb{P}(\mathrm{d}\omega) - \int \left(\int f(\omega, x) \mathbb{P}_X(\mathrm{d}x) \right) \mathbb{P}(\mathrm{d}\omega) .$$

where $\mathrm{Lip}_{\mathbb{S},\mathcal{M}}^{(c)}$ is the set of integrands f from $\Omega \times \mathbb{S} \to \mathbb{R}$, measurable for $\mathcal{M} \otimes \mathcal{B}_{\mathbb{S}}$, such that $f(\omega, .)$ belongs to $\mathrm{Lip}_{\mathbb{S}}^{(c)}$ for any $\omega \in \Omega$.

Let $c(x, y) = \mathbb{1}_{x \neq y}$ be the discrete metric and let $\| \cdot \|_v$ be the variation norm. From the Riesz-Alexandroff representation theorem (see [Whe83, Theorem 5.1]), we infer that for any (μ, ν) in $\mathcal{P}(\mathbb{S}) \times \mathcal{P}(\mathbb{S})$,

$$\sup_{f \in \mathrm{Lip}_{\mathbb{S}}^{(c)}} |\mu(f) - \nu(f)| = \frac{1}{2} \|\mu - \nu\|_v .$$

Hence, for the discrete metric $\tau_c(\mathcal{M}, X) = \beta(\mathcal{M}, \sigma(X))$ is the β-mixing coefficient between $\mathcal{M}$ and $\sigma(X)$ introduced in [RV59]. If c is a distance for which $\mathbb{S}$ is Polish, $\tau_c(\mathcal{M}, X)$ has been introduced in [Rüs85, Inequality (10)] in its "dual" form, and in [DP04], [DP05] in its present from (obviously the reference to [Rüs85] is missing in these two papers).

Applying Theorem 2 with $Q = \mathbb{P} \otimes \mathbb{P}_X$, we see that this coefficient has a characteristic property which is often called the *coupling* or *reconstruction* property.

Corollary 1 (reconstruction property). *If Ω is rich enough (see Theorem 2), there exists a $\sigma(U) \vee \sigma(X) \vee \mathcal{M}$-measurable random variable X^*, independent of $\mathcal{M}$ and distributed as X, such that*

$$\tau_c(\mathcal{M}, X) = \mathbb{E}\left(c(X, X^*)\right) . \qquad (16)$$

If $c(x, y) = \mathbb{1}_{x \neq y}$, (16) is given in [Ber79, Corollary 4.2.5] (note that in Berbee's corollary, $\mathbb{S}$ is assumed to be standard Borel. For other proofs of Berbee's coupling, see [Bry82], [Rüs85, Proposition 5 and Remark 2 page 123] and [Rio00, Section 5.3]). If c is a distance for which $\mathbb{S}$ is a Polish space, (16) has been proved in [Rüs85, Proposition 6] (in [Rüs85] a more general result for sequences is given, in the spirit of [BP79]. For an other proof of (16) when $(\mathbb{S}, c)$ is Polish, see [DP04]).

Coupling is a very useful property in the area of limit theorems and statistics. Many authors have used Berbee's coupling to prove various limit theorems (see for instance the review paper [MP02] and the references therein) as well as exponential inequalities (see for instance the paper [DMR95] for

Bernstein-type inequalities and applications to empirical central limit theorems). Unfortunately, these results apply only to β-mixing sequences, but this property is very hard to check and many simple processes (such as iterates of maps or many non-irreducible Markov chains) are not β-mixing. In many cases however, this difficulty may be overcome by considering another distance c, more adapted to the problem than the discrete metric (typically c is a norm for which $\mathbb{S}$ is a separable Banach space). The case $\mathbb{S} = \mathbb{R}$ and $c(x, y) = |x - y|$, is studied in the paper [DP04], where many non β-mixing examples are given. In this paper the authors used the coefficients τ_c to prove Bernstein-type inequalities and a strong invariance principle for partial sums. In the paper [DP05, Section 4.4] the same authors show that if T is an uniformly expanding map preserving a probability μ on $[0, 1]$, then $\tau_c(\sigma(T^n), T) = O(a^n)$ for $c(x, y) = |x - y|$ and some a in $[0, 1[$.

The following inequality (which can be deduced from [MP02, page 174]) shows clearly that $\beta(\mathcal{M}, \sigma(X))$ is in some sense the more restrictive coefficient among all the $\tau_c(\mathcal{M}, X)$: for any x in $\mathbb{S}$, we have that

$$\tau_c(\mathcal{M}, X) \leq 2 \int_0^{\beta(\mathcal{M}, \sigma(X))} Q_{c(X,x)}(u) du \, , \qquad (17)$$

where $Q_{c(X,x)}$ is the generalized inverse of the function $t \mapsto \mathbb{P}(c(X, x) > t)$. In particular, if c is bounded by M, $\tau_c(\mathcal{M}, X) \leq 2M\beta(\mathcal{M}, \sigma(X))$.

A simple example

Let $(X_i)_{i \geq 0}$ be a stationary Markov chain with values in a Polish space $\mathbb{S}$, satisfying the equation $X_{n+1} = F(X_n, \xi_{n+1})$, where $(\xi_i)_{i>0}$ is a sequence of independent and identically distributed random variables with values in some measurable space $\mathbb{M}$ and independent of X_0, and F is a measurable function from $\mathbb{S} \times \mathbb{M}$ to $\mathbb{S}$. Let X_0^* be a random variable distributed as X_0 and independent of $(X_0, (\xi_i)_{i>0})$, and let $X_{n+1}^* = F(X_n^*, \xi_{n+1})$. The sequence $(X_i^*)_{i \geq 0}$ is independent of X_0 and distributed as $(X_i)_{i \geq 0}$. From the definition (15) of τ_c, we easily infer that

$$\tau_c(\sigma(X_0), X_k) \leq \mathbb{E}(c(X_k, X_k^*)) \, .$$

Let μ be the distribution of X_0 and $(X_n^{(x)})_{n \geq 0}$ the chain starting from $X_0^{(x)} = x$. With these notations, we have that

$$\mathbb{E}(c(X_k, X_k^*)) = \iint \mathbb{E}(c(X_k^{(x)}, X_k^{(y)})) \mu(dx) \mu(dy) \, .$$

If there exists a sequence $(\delta_i)_{i \geq 0}$ of nonnegative numbers such that

$$\mathbb{E}(c(X_k^{(x)}, X_k^{(y)})) \leq \delta_k c(x, y) \, ,$$

then

$$\tau_c(\sigma(X_0), X_k) \leq \delta_k \mathbb{E}(c(X_0, X_0^*)) \ .$$

For instance, in the case where $\mathbb{E}(c(F(x, \xi_0), F(y, \xi_0))) \leq \kappa c(x, y)$ for some $\kappa < 1$, we can take $\delta_k = \kappa^k$. An important example is the case where $\mathbb{S} = \mathbb{M}$ is a separable Banach space and $X_{n+1} = f(X_n) + \xi_{n+1}$ for some κ lipschitz function f with respect to c.

Let us consider the well known example $2X_{n+1} = X_n + \xi_{n+1}$, where X_0 has uniform distribution λ over $[0, 1]$ and ξ_1 is Bernoulli distributed with parameter $1/2$. If $c(x, y) = |x - y|$, it follows from our preceding remarks that $\tau_c(\sigma(X_0), X_k) \leq 2^{-k}$. However, it is well known that this chain is not β mixing. Indeed, it is a stationary Markov chain with invariant distribution λ and transition kernel

$$K(x, \cdot) = \frac{1}{2}(\delta_{x/2} + \delta_{(x+1)/2}) \ ,$$

so that $\|K^k(x, .) - \lambda\|_v = 2$. Consequently $\beta(\sigma(X_0), \sigma(X_k)) = 1$ for any $k \geq 0$.

A simple application

Let $(X_i)_{i \in \mathbb{Z}}$ be a stationary sequence of real-valued random variables with common distribution function F. Let $\mathcal{M}_0 = \sigma(X_k, k \leq 0)$, and let $F_{X_k | \mathcal{M}_0}$ be a conditional distribution function of X_k given $\mathcal{M}_0$. Let $F_n = n^{-1} \sum_{i=1}^{n} \mathbb{1}_{X_i \leq t}$ be the empirical distribution function. Let μ be a finite measure on $(\mathbb{R}, \mathcal{B}(\mathbb{R}))$. In [DM03, Example 2, Section 2.2], it is proved that the process $\{t \mapsto \sqrt{n}(F_n(t) - F(t))\}$ converges weakly in $\mathbb{L}^2(\mu)$ to a mixture of $\mathbb{L}^2(\mu)$-valued Gaussian random variables as soon as

$$\sum_{k>0} \mathbb{E}\left(\int |F_{X_k|\mathcal{M}_0}(t) - F(t)|^2 \mu(dt) \right)^{1/2} < \infty \ . \tag{18}$$

Let X_k^* be a random variable distributed as X_k and independent of $\mathcal{M}_0$, and let $F_\mu(x) = \mu(]-\infty, x[)$. Since $F = F_{X_k^*|\mathcal{M}_0}$, it follows that

$$\mathbb{E}\left(\int |F_{X_k|\mathcal{M}_0}(t) - F(t)|^2 \mu(dt) \right)^{1/2} \leq \mathbb{E}\left(\sqrt{|F_\mu(X_k) - F_\mu(X_k^*)|} \right) \ .$$

Let $d_\mu(x, y) = \sqrt{|F_\mu(x) - F_\mu(y)|}$. From (16) it follows that one can choose X_k^* such that

$$\mathbb{E}\left(\sqrt{|F_\mu(X_k) - F_\mu(X_k^*)|} \right) = \tau_{d_\mu}(\mathcal{M}_0, X_k) \ .$$

Consequently (18) holds as soon as $\sum_{k>0} \tau_{d_\mu}(\mathcal{M}_0, X_k) < \infty$. This is an example where the natural cost function d_μ is not the discrete metric $c(x, y) = \mathbb{1}_{x \neq y}$ nor the usual norm $c(x, y) = |x - y|$.

4 A counter example to maximal coupling

In this section we prove that no analogue of Goldstein's maximal coupling (see [Gol79]) is possible if the cost function is not proportional to the discrete metric.

More generally, we consider the following problem. Let $\mathbb{M}$ be a Polish space and $\mathbb{S} = \mathbb{M} \times \mathbb{M}$. Let c be any symmetric measurable function from $\mathbb{M} \times \mathbb{M}$ to $\mathbb{R}^+$, such that $c(x,y) = 0$ if and only if $x = y$. Let $\mathcal{F}$ be the class of symmetric measurable functions φ from $\mathbb{R}^+ \times \mathbb{R}^+$ to $\mathbb{R}^+$, such that $x \mapsto \varphi(0, x)$ is increasing. For $\varphi \in \mathcal{F}$, we define the cost function $c_\varphi((x_1, x_2), (y_1, y_2)) = \varphi(c(x_1, y_1), c(x_2, y_2))$ on $\mathbb{S} \times \mathbb{S}$.

The question $\mathbf{Q}$ is the following. For which couples (φ, c) do we have the property: for any probability measures μ, ν on $\mathbb{S}$ with marginals $\mu_{(2)}(A) = \mu(\mathbb{M} \times A)$ and $\nu_{(2)}(A) = \nu(\mathbb{M} \times A)$, there exists a probability measure λ in $D(\mu, \nu)$ with marginal $\lambda_{(2)}(A \times B) = \lambda(\mathbb{M} \times A \times \mathbb{M} \times B)$, such that

$$\Delta_{\mathrm{KR}}^{(c_\varphi)}(\mu, \nu) \quad = \quad \int \varphi(c(x_1, y_1), c(x_2, y_2)) \, \lambda(\mathrm{d}x_1, \mathrm{d}x_2, \mathrm{d}y_1, \mathrm{d}y_2) \, , \quad (19)$$

$$\Delta_{\mathrm{KR}}^{(c)}(\mu_{(2)}, \nu_{(2)}) = \int c(x_2, y_2) \lambda_{(2)}(\mathrm{d}x_2, \mathrm{d}y_2) \, . \quad (20)$$

From Goldstein's result we know that the couple $(\varphi(x, y) = x \vee y, c(x, y) = \mathbb{1}_{x \neq y})$ is a solution to $\mathbf{Q}$. The following proposition shows that, if c is not proportional to the discrete metric, no couple (φ, c) can be a solution to $\mathbf{Q}$.

Proposition 1. *Suppose that c is not proportional to the discrete metric. There exist a_1, b_1, a_2, b_2 in $\mathbb{M}$ such that $a_1 \neq b_1$ and $a_2 \neq b_2$ and two probabilities μ and ν on $\{(a_1, a_2), (a_1, b_2), (b_1, a_2), (b_1, b_2)\}$ for which, for any $\varphi \in \mathcal{F}$, there is no λ in $D(\mu, \nu)$ satisfying (19) and (20) simultaneously.*

Proof. Since c is not proportional to the discrete metric, there exist at least two points (a_1, b_1) and (a_2, b_2) in $\mathbb{M} \times \mathbb{M}$ such that $a_1 \neq b_1$, $a_2 \neq b_2$ and $c(a_1, b_1) > c(a_2, b_2) > 0$. Define the probabilities μ and ν by

$$\begin{array}{ll}
\mu(a_1, a_2) = \tfrac{1}{2} \, , & \nu(a_1, a_2) = 0 \, , \\
\mu(a_1, b_2) = 0 \, , & \nu(a_1, b_2) = \tfrac{1}{2} \, , \\
\mu(b_1, a_2) = 0 \, , & \nu(b_1, a_2) = \tfrac{1}{2} \, , \\
\mu(b_1, b_2) = \tfrac{1}{2} \, , & \nu(b_1, b_2) = 0 \, .
\end{array}$$

The set $D(\mu, \nu)$ is the set of probabilities λ_α such that $\lambda_\alpha(a_1, a_2, a_1, b_2) = \lambda_\alpha(b_1, b_2, b_1, a_2) = \alpha$, $\lambda_\alpha(a_1, a_2, b_1, a_2) = \lambda_\alpha(b_1, b_2, a_1, b_2) = 1/2 - \alpha$, for α in $[0, 1/2]$. Consequently, for any φ in $\mathcal{F}$,

$$\int \varphi(c(x_1, y_1), c(x_2, y_2)) \lambda_\alpha(\mathrm{d}x_1, \mathrm{d}x_2, \mathrm{d}y_1, \mathrm{d}y_2)$$

$$= 2\alpha \, \varphi(0, c(a_2, b_2)) + (1 - 2\alpha) \, \varphi(c(a_1, b_1), 0) \, . \quad (21)$$

Since $c(a_1, b_1) > c(a_2, b_2)$, since φ is symmetric, and since $x \mapsto \varphi(0, x)$ is increasing, $\varphi(c(a_1, b_1), 0) > \varphi(0, c(a_2, b_2))$. Therefore, the unique solution to (19) is $\lambda_{1/2}$. Now

$$\int c(x_2, y_2)\lambda_{1/2}(\mathrm{d}x_1, \mathrm{d}x_2, \mathrm{d}y_1, \mathrm{d}y_2) = c(a_2, b_2) > 0 \ .$$

Since $\mu_{(2)} = \nu_{(2)}$, $\Delta_{\mathrm{KR}}^{(c)}(\mu_{(2)}, \nu_{(2)}) = 0$. Hence $\lambda_{1/2}$ does not satisfy (20).   □

Remark 2. If now c is the discrete metric $c(x, y) = \mathbb{1}_{x \neq y}$, the right hand term in equality (21) is $\varphi(c(a_1, b_1), 0)$. Consequently, any λ_α is solution to (19) and λ_0 is solution to both (19) and (20). We conjecture that if c is the discrete metric, then any couple (φ, c), $\varphi \in \mathcal{F}$, is a solution to **Q**.

5 Appendix: topological and measure-theoretical complements

Topological spaces

Let us recall some definitions (see [ScH73, GP84] for complements on Radon and Suslin spaces). A topological space $\mathbb{S}$ is said to be

- *regular* if, for any $x \in \mathbb{S}$ and any closed subset F of $\mathbb{S}$ which does not contain x, there exist two disjoint open subsets U and V such that $x \in U$ and $F \subset V$,
- *completely regular* if, for any $x \in \mathbb{S}$ and any closed subset F of $\mathbb{S}$ which does not contain x, there exists a continuous function $f : \mathbb{S} \to [0, 1]$ such that $f(x) = 0$ and $f = 1$ on F (equivalently, $\mathbb{S}$ is *uniformizable*, that is, the topology of $\mathbb{S}$ can be defined by a set of semidistances),
- *pre-Radon* if every finite τ–additive Borel measure on $\mathbb{S}$ is inner regular with respect to the compact subsets of $\mathbb{S}$ (a Borel measure μ on $\mathbb{S}$ is τ–additive if, for any family $(F_\alpha)_{\alpha \in A}$ of closed subsets of $\mathbb{S}$ such that $\forall \alpha, \beta \in A \ \exists \gamma \in A \ F_\gamma \subset F_\alpha \cap F_\beta$, we have $\mu(\cap_{\alpha \in A} F_\alpha) = \inf_{\alpha \in A} \mu(F_\alpha)$),
- *Radon* if every finite Borel measure on $\mathbb{S}$ is inner regular with respect to the compact subsets of $\mathbb{S}$,
- *Suslin*, or *analytic*, if there exists a continuous mapping from some Polish space onto $\mathbb{S}$,
- *Lusin* if there exists a continuous injective mapping from some Polish space onto $\mathbb{S}$. Equivalently, $\mathbb{S}$ is Lusin if there exists a Polish topology on $\mathbb{S}$ which is finer than the given topology of $\mathbb{S}$.

Obviously, every Lusin space is Suslin and every Radon space is pre-Radon. Much less obviously, every Suslin space is Radon. Every regular Suslin space is completely regular.

Many usual spaces of Analysis are Lusin: besides all separable Banach spaces (e.g. L^p ($1 \leq p < +\infty$), or the Sobolev spaces $\mathrm{W}^{s,p}(\Omega)$ ($0 < s < 1$

and $1 \leq p < +\infty)$), the spaces of distributions $\mathcal{E}'$, $\mathcal{S}'$, $\mathcal{D}'$, the space $\mathcal{H}(\mathcal{C})$ of holomorphic functions, or the topological dual of a Banach space, endowed with its weak*–topology are Lusin. See [ScH73, pages 112–117] for many more examples.

Standard Borel spaces

A measurable space $(\mathbb{M}, \mathcal{M})$ is said to be *standard Borel* if it is Borel-isomorphic with some Polish space $\mathbb{T}$, that is, there exists a mapping $f : \mathbb{T} \to \mathbb{M}$ which is one-one and onto, such that f and f^{-1} are measurable for $\mathcal{B}_{\mathbb{T}}$ and $\mathcal{M}$. We say that a topological space $\mathbb{S}$ is *standard Borel* if $(\mathbb{S}, \mathcal{B}_{\mathbb{S}})$ is standard Borel.

If τ_1 and τ_2 are two comparable Suslin topologies on $\mathbb{S}$, they share the same Borel sets. In particular, every Lusin space is standard Borel.

A useful property of standard Borel spaces is that every standard space $\mathbb{S}$ is Borel-isomorphic with a Borel subset of $[0, 1]$. This a consequence of e.g. [Kec95, Theorem 15.6 and Corollary 6.5], see also [Sko76] or [DM75, Théorème III.20]. (Actually, we have more: every standard Borel space is countable or Borel-isomorphic with $[0, 1]$. Thus, for standard Borel spaces, the Continuum Hypothesis holds true!)

Another useful property of standard Borel spaces is that, if $\mathbb{S}$ is a standard Borel space, if $X : \Omega \mapsto \mathbb{S}$ is a measurable mapping, and if $\mathcal{M}$ is a sub-σ-algebra of $\mathcal{A}$, there exists a regular conditional distribution $\mathbb{P}_{X|\mathcal{M}}$ (see e.g. [Dud02, Theorem 10.2.2] for the Polish case, which immediately extends to standard Borel spaces from their definition). Note that, if $\mathbb{S}$ is radon, then the distribution $\mathbb{P}_X$ of X is tight, that is, for every integer $n \geq 1$, there exists a compact subset K_n of $\mathbb{S}$ such that $\mathbb{P}_X(\mathbb{S} \setminus K_n) \geq 1/n$. Hence one can assume without loss of generality that X takes its values in $\cup_{n \geq 1} K_n$. If moreover $\mathbb{S}$ has metrizable compact subsets, then $\cup_{n \geq 1} K_n$ is Lusin (and hence standard Borel), and there exists a regular conditional distribution $\mathbb{P}_{X|\mathcal{M}}$. Thus, if $\mathbb{S}$ is Radon with metrizable compact subsets, every element μ of $\mathcal{Y}$ has an $\mathcal{A}$-measurable disintegration. Indeed, denoting $\mathcal{A}' = \mathcal{A} \otimes \{\emptyset, \mathbb{S}\}$, one only needs to consider the conditional distribution $\mathbb{P}_{X|\mathcal{A}'}$ of the random variable $X : (\omega, x) \mapsto x$ defined on the probability space $(\Omega \times \mathbb{S}, \mathcal{A} \otimes \mathcal{B}_{\mathbb{S}}, \mu)$.

For any σ–algebra $\mathcal{M}$ on a set $\mathbb{M}$, the *universal completion* of $\mathcal{M}$ is the σ-algebra $\mathcal{M}^* = \cap_{\mu} \mathcal{M}_{\mu}^*$, where μ runs over all finite nonegative measures on $\mathcal{M}$ and $\mathcal{M}_{\mu}^*$ is the μ–completion of $\mathcal{M}$. A subset of a topological space $\mathbb{S}$ is said to be *universally measurable* if it belongs to $\mathcal{B}_{\mathbb{S}}^*$. The following lemma can be deduced from e.g. [VW96, Exercise 10 page 14] and the Borel-isomorphism theorem.

Lemma 2. *Assume that $\mathbb{S}$ is a standard Borel space. Let $X : \Omega \to \mathbb{S}$ be $\mathcal{A}^*$–measurable. Then there exists an $\mathcal{A}$–measurable modification $Y : \Omega \to \mathbb{S}$ of X, that is, Y is $\mathcal{A}$–measurable and satisfies $Y = X$ a.e.*

References

[Ber79]   Henry C. P. Berbee, *Random walks with stationary increments and renewal theory*, Mathematical Centre Tracts, vol. 112, Mathematisch Centrum, Amsterdam, 1979.

[BP79]   I. Berkes and W. Philipp, *Approximation theorems for independent and weakly dependent random vectors*, Ann. Probab. **7** (1979), 29-54.

[BBD01]   S. Borovkova, R. Burton and H. Dehling, *Limit theorems for functionals of mixing processes with application to U-statistics and dimension estimation*, Trans. Amer. Math. Soc. **353** (2001) 4261-4318.

[Bry82]   W. Bryc, *On the approximation theorem of I. Berkes and W. Philipp*, Demonstratio Mathematica **15** (1982), no. 3, 807–816.

[CRV04]   Charles Castaing, Paul Raynaud de Fitte, and Michel Valadier, *Young measures on topological spaces. With applications in control theory and probability theory*, Kluwer Academic Publishers, Dordrecht, 2004.

[CV77]   Charles Castaing and Michel Valadier, *Convex analysis and measurable multifunctions*, Lecture Notes in Math., no. 580, Springer Verlag, Berlin, 1977.

[DM03]   Jérôme Dedecker and Florence Merlevède, *The conditional central limit theorem in Hilbert spaces*, Stochastic Process. Appl. **108** (2003), 229-262.

[DP04]   Jérôme Dedecker and Clémentine Prieur, *Couplage pour la distance minimale*, C. R. Math. Acad. Sci. Paris **338** (2004), no. 10, 805–808.

[DP04]   Jérôme Dedecker and Clémentine Prieur, *Coupling for τ-dependent sequences and applications*, J. Theor. Probab **17** (2004), no. 4, 861-885.

[DP05]   Jérôme Dedecker and Clémentine Prieur, *New dependence coefficients. Examples and applications to statistics*, Probab. Theory Relat. Fields **132** (2005), 203-236.

[DM75]   Claude Dellacherie and Paul André Meyer, *Probabilités et potentiel. Chapitres I à IV*, Hermann, Paris, 1975.

[Dob70]   R. L. Dobrušin, *Prescribing a system[] of random variables by conditional distributions*, Theor. Probability Appl. **15** (1970), 458-486.

[DMR95]   Paul Doukhan, Pascal Massart, and Emmanuel Rio, *Invariance principles for absolutely regular empirical processes*, Annales Inst. H. Poincaré Probab. Statist. **31** (1995), 393–427.

[Dud02]   R. M. Dudley, *Real analysis and probability*, Cambridge University Press, Cambridge, 2002.

[GP84]   R. J. Gardner and W. F. Pfeffer, *Borel measures*, Handbook of set-theoretic topology, North-Holland, Amsterdam, 1984, pp. 961–1043.

[Gol79]   S. Goldstein, *Maximal coupling*, Z. Wahrsch. Verw. Gebiete **46** (1979), 193-204.

[Kec95]   Alexander S. Kechris, *Classical descriptive set theory*, Graduate Texts in Mathematics, no. 156, Springer Verlag, New York, 1995.

[Lev84]   V. L. Levin, *The problem of mass transfer in a topological space and probability measures with given marginal measures on the product of two spaces*, Dokl. Akad. Nauk SSSR **276** (1984), no. 5, 1059–1064, English translation: Soviet Math. Dokl. 29 (1984), no. 3, 638–643.

[MP02]   Florence Merlevède and Magda Peligrad, *On the coupling of dependent random variables and applications*, Empirical process techniques for dependent data, Birkhäuser, 2002, pp. 171–193.

[RR98]   S. T. Rachev and L. Rüschendorf, *Mass transportation problems. Volume I: Theory*, Probability and its Applications, Springer Verlag, New York, Berlin, 1998.

[Rio98]   Emmanuel Rio, *Processus empiriques absolument réguliers et entropie universelle*, Probab. Theory Relat. Fields **111** (1998), no. 4, 585–608.

[Rio00]   Emmanuel Rio, *Théorie asymptotique des processus aléatoires faiblement dépendants*, Mathématiques et Applications, no. 31, Springer, Berlin, Heidelberg, 2000.

[Rüs85]   L. Rüschendorf, *The Wasserstein Distance and Approximation Theorems*, Z. Wahrsch. Verw. Gebiete **70** (1985), 117-129.

[RV59]   Y. A. Rozanov and V. A. Volkonskii, *Some limit theorems for random functions I*, Teor. Verojatnost. i Primenen. **4** (1959), 186–207.

[ScH73]   Laurent Schwartz, *Radon measures on arbitrary topological spaces and cylindrical measures*, Tata Institute of Fundamental Research Studies in Mathematics, Oxford University Press, London, 1973.

[Sko76]   A. V. Skorohod, *On a representation of random variables*, Teor. Verojatnost. i Primenen. **21** (1976), no. 3, 645–648, English translation: Theor. Probability Appl. 21 (1976), no. 3, 628–632 (1977).

[Val73]   Michel Valadier, *Désintégration d'une mesure sur un produit*, C. R. Acad. Sci. Paris Sér. I **276** (1973), A33–A35.

[VW96]   Aad W. van der Vaart and Jon A. Wellner, *Weak convergence and empirical processes. With applications to statistics*, Springer Series in Statistics, Springer Verlag, Berlin, 1996.

[Whe83]   Robert F. Wheeler, *A survey of Baire measures and strict topologies*, Exposition. Math. **1** (1983), no 2, 97-190.

Exponential inequalities and estimation of conditional probabilities

V. Maume-Deschamps

Université de Bourgogne B.P. 47870 21078 Dijon Cedex FRANCE
vmaume@u-bourgogne.fr

1 Introduction

This paper deals with the problems of typicality and conditional typicality of "empirical probabilities" for stochastic process and the estimation of potential functions for Gibbs measures and dynamical systems. The questions of typicality have been studied in [FKT88] for independent sequences, in [BRY98, Ris89] for Markov chains. In order to prove the consistency of estimators of transition probability for Markov chains of unknown order, results on typicality and conditional typicality for some (Ψ)-mixing process where obtained in [CsS, Csi02]. Unfortunately, lots of natural mixing process do not satisfy this Ψ-mixing condition (see [DP05]). We consider a class of mixing process inspired from [DP05]. For this class, we prove strong typicality and strong conditional typicality. In the particular case of Gibbs measures (or complete connexions chains) and for certain dynamical systems, from the typicality results we derive an estimation of the potential as well as a procedure to test the nullity of the asymptotic variance of the process.
More formally, we consider $X_0, \ldots., X_n, \ldots$ a stochastic process taking values on an complete set Σ and a sequence of countable partitions of Σ, $(\mathcal{P}_k)_{k \in \mathbb{N}}$ such that if $P \in \mathcal{P}_k$ then there exists a unique $\widetilde{P} \in \mathcal{P}_{k-1}$ such that almost surely, $X_j \in P$ implies $X_{j-1} \in \widetilde{P}$. Our aim is to obtain empirical estimates of the probabilities:

$$\mathbb{P}(X_j \in P), \ P \in \mathcal{P}_k \ ,$$

the conditional probabilities:

$$\mathbb{P}(X_j \in P \mid X_{j-1} \in \widetilde{P}), \ P \in \mathcal{P}_k$$

and the limit when $k \to \infty$ when it makes sense.
We shall define a notion of mixing with respect to a class of functions. Let $\mathcal{C}$ be a Banach space of real **bounded** functions endowed with a norm of the form:

$$\|f\|_C = C(f) + \|f\| \, ,$$

where $C(f)$ is a semi-norm (i.e. $\forall f \in C$, $C(f) \geq 0$, $C(\lambda f) = |\lambda| C(f)$ for $\lambda \in \mathbb{R}$, $C(f + g) \leq C(f) + C(g)$) and $\| \ \|$ is a norm on C. We will denote by C_1 the subset of functions in C such that $C(f) \leq 1$.

Particular choices of C may be the space BV of functions of bounded variation on Σ if it is totally ordered or the space of Hölder (or piecewise Hölder) functions. Recall that a function f on Σ is of bounded variation if it is bounded and

$$\bigvee f := \sup \sum_{i=0}^{n} |f(x_i) - f(x_{i+1})| < \infty \, ,$$

where the sup is taken over all finite sequences $x_1 < \cdots < x_n$ of elements of Σ. The space BV endowed with the norm $\|f\| = \bigvee f + \|f\|_\infty$ is a Banach space.

Inspired from [DP05], we define the Φ_C-mixing coefficients.

Definition 1. *For $i \in \mathbb{N}$, let $\mathcal{M}_i$ be the sigma algebra generated by X_1, ..., X_i. For $k \in \mathbb{N}$,*

$$\Phi_C(k) = \sup\{|\mathbb{E}(Yf(X_{i+k})) - \mathbb{E}(Y)\mathbb{E}(f(X_{i+k}))| \, , i \in \mathbb{N} \, ,$$
$$Y \ is \ \mathcal{M}_i - measurable \ with \ \|Y\|_1 \leq 1, \ f \in C_1\} \, . \qquad (*)$$

Our main assumption on the process is the following.

Assumption 1

$$\sum_{k=0}^{n-1} (n - k)\Phi_C(k) = O(n) \, .$$

Remark 1. Assumption 1 is equivalent to $(\Phi_C(k))_{k \in \mathbb{N}}$ summable. We prefer to formulate it in the above form because it appears more naturally in our context.

Our definition is inspired from Csiszár's (which is Ψ-mixing for variables taking values in a finite alphabet) and Dedecker-Prieur. It covers lots of natural systems (see Section 3 for an example with dynamical systems and [DP05] for further examples). Our definition extends Csiszár's which was for random variables on a finite alphabet.

We consider a sequence $(\mathcal{P}_k)_{k \in \mathbb{N}}$ of countable partitions of Σ such that: almost surely, for all $j, k \in \mathbb{N}$, we have

for any $P \in \mathcal{P}_k$, there exists $\widetilde{P} \in \mathcal{P}_{k-1}$, $X_j \in P \Rightarrow X_{j-1} \in \widetilde{P}$. $\quad (**)$

For $i, \ell \in \mathbb{N}$, for $P \in \mathcal{P}_k$, consider the random variable:

$$N_i^\ell(P) = \sum_{j=i}^{\ell+i-1} \mathbb{1}_P(X_j) \, .$$

Our aim is to have quantitative informations on how close are the empirical probabilities $\frac{N_i^{i+n}(P)}{n}$ to the expected value $Q_i^{i+n}(P) := \mathbb{E}\left(\frac{N_i^{i+n}(P)}{n}\right)$. We are especially interested in "large scale typicality": k will grow with n. We wonder also about "conditional typicality", for $P \in \mathcal{P}_k$, let

$$\hat{g}_n(P) = \frac{N_1^{n+1}(P)}{N_0^{n-1}(\widetilde{P})} \frac{n-1}{n}.$$

Our main result is that $\hat{g}_n(P)$ is a consistent estimator of the conditional probabilities $Q_n(P|\widetilde{P}) := \frac{Q_1^{n+1}(P)}{Q_0^{n-1}(\widetilde{P})}$. This follows from an exponential inequality (see Theorem 1).

If the conditional probabilities $Q_n(P|\widetilde{P})$ converge when $k \to \infty$, we may obtain an estimator of the limit function. This is the case for certain dynamical systems (see Section 3) and g-measures (see Section 4). In these settings, we obtain a consistent estimator of the potential function. This may leads to a way of testing the nullity of the asymptotic variance of the system (see Section 5 for details).

Section 2 contains general results on typicality and conditional typicality for some weak-dependant sequences. In Section 3, we apply these results to expanding dynamical systems of the interval. Section 4 is devoted to Gibbs measures and chains with complete connections. Finally, in Section 5 we sketch an attempt to test the nullity of the asymptotic variance of the system.

2 Typicality and conditional typicality via exponential inequalities

Following Csiszár, we wonder about typicality that is: how close are the "empirical probabilities" $\frac{N_i^{n+i}(P)}{n}$ to the expected probability $Q_i^{n+i}(P)$? This is done via a "Hoeffding-type" inequality for partial sums.

The following Proposition has been obtained in [DP05], we sketch here the proof because our context is a bit different.

Proposition 1. *Let (X_i) be a sequence a random variables. Let the coefficients $\Phi_{\mathcal{C}}(k)$ be defined by (*). For $\varphi \in \mathcal{C}$, $p \geq 2$, define*

$$S_n(\varphi) = \sum_{i=1}^{n} \varphi(X_i)$$

and

$$b_{i,n} = \left(\sum_{k=0}^{n-i} \Phi(k)\right) \|\varphi(X_i) - \mathbb{E}(\varphi(X_i))\|_{\frac{p}{2}} C(\varphi).$$

For any $p \geq 2$, we have the inequality:

$$\|S_n(\varphi) - \mathbb{E}(S_n(\varphi))\|_p \leq \left(2p \sum_{i=1}^{n} b_{i,n} \right)^{\frac{1}{2}}$$

$$\leq C(\varphi) \left(2p \sum_{k=0}^{n-1} (n-k)\Phi_\mathcal{C}(k) \right)^{\frac{1}{2}} . \qquad (1)$$

As a consequence, we obtain

$$\mathbb{P}\left(|S_n(\varphi) - \mathbb{E}(S_n(\varphi))| > t\right) \leq \mathrm{e}^{\frac{1}{e}} \exp\left(\frac{-t^2}{2\mathrm{e}(C(\varphi))^2 \sum_{k=0}^{n-1} (n-k)\Phi_\mathcal{C}(k)} \right) . \qquad (2)$$

Proof (Sketch of proof). There are two ingredients to get (1). Firstly we need a counterpart to Lemma 4 in [DP05].

Lemma 1.

$$\Phi_\mathcal{C}(k) = \sup\left\{ \|\mathbb{E}(\varphi(X_{i+k})|\mathcal{M}_i) - \mathbb{E}(\varphi(X_{i+k}))\|_\infty \ , \ \varphi \in \mathcal{C}_1 \right\} .$$

We postpone the proof of Lemma 1 to the end of the proof of the proposition. Secondly, we apply Proposition 4 in [DD03] to get: (let $Y_i = \varphi(X_i) - \mathbb{E}(\varphi(X_i))$)

$$\|S_n(\varphi) - \mathbb{E}(S_n(\varphi))\|_p \leq \left(2p \sum_{i=1}^{n} \max_{i \leq \ell \leq n} \left\| Y_i \sum_{k=i}^{\ell} \mathbb{E}(Y_k|\mathcal{M}_i) \right\|_{\frac{p}{2}} \right)^{\frac{1}{2}}$$

$$\leq \left(2p \sum_{i=1}^{n} \|Y_i\|_{\frac{p}{2}} \sum_{k=i}^{n} \|\mathbb{E}(Y_k|\mathcal{M}_i)\|_\infty \right)^{\frac{1}{2}} \leq \left(2p \sum_{i=1}^{n} b_{i,n} \right)^{\frac{1}{2}} .$$

We have used that by Lemma 1, $\|\mathbb{E}(Y_{k+i}|\mathcal{M}_i)\|_\infty \leq C(\varphi)\Phi_\mathcal{C}(k)$. To obtain the second part of inequality (2), use $\|Y_i\|_{\frac{p}{2}} \leq \|Y_i\|_\infty \leq C(\varphi)\Phi_\mathcal{C}(0))$. The second inequality (2) follows from (1) as in [DP05]. □

Proof (Proof of Lemma 1). We write

$$\mathbb{E}(Yf(X_{i+k})) - \mathbb{E}(Y)\mathbb{E}(f(X_{i+k})) = \mathbb{E}(Y[\mathbb{E}(f(X_{i+k})|\mathcal{M}_i) - \mathbb{E}(f(X_{i+k}))])$$
$$\leq \|\mathbb{E}(f(X_{i+k})|\mathcal{M}_i) - \mathbb{E}(f(X_{i+k}))\|_\infty.$$

To prove the converse inequality, for $\varepsilon > 0$, consider an event A_ε such that for $\omega \in A_\varepsilon$,

$$|\mathbb{E}(f(X_{i+k})|\mathcal{M}_i)(\omega) - \mathbb{E}(f(X_{i+k}))| \geq \|\mathbb{E}(f(X_{i+k})|\mathcal{M}_i) - \mathbb{E}(f(X_{i+k}))\|_\infty - \varepsilon,$$

and consider the random variable

$$Y_\varepsilon = \frac{\mathbb{1}_{A_\varepsilon}}{\mathbb{P}(A_\varepsilon)} \mathrm{sign}(\mathbb{E}(h(X_{i+k})|\mathcal{M}_i)(\omega) - \mathbb{E}(f(X_{i+k}))) .$$

Y_ε is $\mathcal{M}_i$-measurable, $\|Y_\varepsilon\|_1 \leq 1$ and

$$\mathbb{E}(Y_\varepsilon f(X_{i+k})) - \mathbb{E}(Y_\varepsilon)\mathbb{E}(f(X_{i+k})) \geq \|\mathbb{E}(f(X_{i+k})|\mathcal{M}_i) - \mathbb{E}(f(X_{i+k}))\|_\infty - \varepsilon .$$

Thus, the lemma is proved. □

We shall apply inequality (2) to the function $\varphi = \mathbb{1}_P$, $P \in \mathcal{P}_k$.

Corollary 1. *If the process $(X_1, \ldots, X_n, \ldots)$ satisfies Assumption 1, if the sequence of partitions $(\mathcal{P}_k)_{k \in \mathbb{N}}$ satisfies (**) and for all $P \in \mathcal{P}_k$, $\mathbb{1}_P \in \mathcal{C}$, then, there exists a constant $C > 0$ such that for all $k \in \mathbb{N}$, for all $P \in \mathcal{P}_k$, for any $t \in \mathbb{R}$, for all $i, n \in \mathbb{N}$,*

$$\mathbb{P}\left(\left|\frac{N_i^{n+i}(P)}{n} - Q_i^{n+i}(P)\right| > t\right) \leq e^{\frac{1}{e}} e^{\left(-\frac{Ct^2 n}{C(\mathbb{1}_P)^2}\right)}. \tag{3}$$

Proof. It follows directly from (2) applied to $\varphi = \mathbb{1}_P$ and Assumption 1. □

Let us denote by $\hat{\mathbb{P}}_i^{n+i}(P) = \frac{N_i^{n+i}(P)}{n}$. The following corollary is a counterpart to Csiszár's result (Theorem 1 in [Csi02]) in our context.

Corollary 2. *There exists $C > 0$ such that for all $P \in \mathcal{P}_k$ for which $\left(\frac{Q_i^{n+i}(P)}{C(\mathbb{1}_P)}\right)^2 n \geq \ln^2 n$, we have:*

$$\mathbb{P}\left(\left|\frac{\hat{\mathbb{P}}_i^{n+i}(P)}{Q_i^{n+i}(P)} - 1\right| > t\right) \leq e^{\frac{1}{e}} e^{\left(-Ct^2 \ln^2 n\right)}.$$

Proof. We apply Corollary 1 with $t \cdot Q_i^{n+i}(P)$ instead of t. We get:

$$\mathbb{P}\left(\left|\frac{\hat{\mathbb{P}}_i^{n+i}(P)}{Q_i^{n+i}(P)} - 1\right| > t\right) \leq e^{\frac{1}{e}} \exp\left(-\frac{Ct^2 (Q_i^{n+i}(P))^2 n}{(C(\mathbb{1}_P))^2}\right).$$

The result follows.   □

Remark 2. Let us consider the case where $\mathcal{C} = BV$. If the partition $\mathcal{P}_k$ is a partition into interval, then for all $P \in \mathcal{P}_k$, $C(\mathbb{1}_P) = 2$.

We are now in position to prove our theorem on conditional typicality. Recall that

$$\hat{g}_n(P) = \frac{n-1}{n} \frac{N_1^{n+1}(P)}{N_0^{n-1}(\tilde{P})}.$$

Theorem 1. *Let the process $(X_p)_{p \in \mathbb{N}}$ satisfy Assumption 1, let the sequence of partitions $(\mathcal{P}_k)_{k \in \mathbb{N}}$ satisfy (**) and assume that if $P \in \mathcal{P}_k$ then $\mathbb{1}_P \in \mathcal{C}$. There exists $K > 0$ such that for all $\varepsilon < 1$, for all $P \in \mathcal{P}_k$ for which*

$$\frac{Q_0^{n-1}(\tilde{P})}{C(\mathbb{1}_P)} \quad and \quad \frac{Q_0^{n-1}(\tilde{P})}{C(\mathbb{1}_{\tilde{P}})} \geq n^{-\frac{\varepsilon}{2}},$$

we have

$$\mathbb{P}\left(\left|\hat{g}_n(P) - Q_n(P|\tilde{P})\right| > t\right) \leq 4e^{-Kt^2 n^{1-\varepsilon}} + 2e^{-Kn^{1-\varepsilon}}.$$

If the sequence is stationary, the result may be rewritten as:

$$\mathbb{P}\left(\left|\hat{g}_n(P) - \mathbb{P}(X_1 \in P \mid X_0 \in \tilde{P})\right| > t\right) \leq 4e^{-Kt^2 n^{1-\varepsilon}} + 2e^{-Kn^{1-\varepsilon}}.$$

Proof. Fix $R > 0$, let us bound the probability

$$\mathbb{P}\left(\left|\hat{g}_n(P) - Q_n(P|\widetilde{P})\right| > t\right)$$

with the sum of the probabilities:

$$(1) = \mathbb{P}\left(\left|\hat{\mathbb{P}}_1^{n+1}(P) - Q_1^{n+1}(P)\right| > \frac{t \cdot Q_0^{n-1}(\widetilde{P})}{2}\right),$$

$$(2) = \mathbb{P}\left(\left|\hat{\mathbb{P}}_0^{n-1}(\widetilde{P}) - Q_0^{n-1}(\widetilde{P})\right| > \frac{tQ_0^{n-1}(\widetilde{P})R}{2}\right),$$

$$(3) = \mathbb{P}\left(\frac{\hat{\mathbb{P}}_0^{n-1}(\widetilde{P})}{\hat{\mathbb{P}}_1^{n+1}(P)} < R\right).$$

The terms (1) and (2) are easily bounded using Corollary 1: we get

$$(1) \le e^{\frac{1}{e}} \exp\left(-\frac{Ct^2 n^{1-\varepsilon}}{4}\right) \quad (2) \le e^{\frac{1}{e}} \exp\left(-\frac{Ct^2 R^2 (n-1)^{1-\varepsilon}}{4}\right).$$

It remains to bound the term (3). We have (recall that almost surely, $X_j \in P \Rightarrow X_{j-1} \in \widetilde{P}$):

$$\frac{\hat{\mathbb{P}}_1^{n+1}(P)}{\hat{\mathbb{P}}_0^{n-1}(\widetilde{P})} \le \frac{n-1}{n}\left(1 + \frac{\mathbb{1}_{\{X_n \in P\}}}{N_0^{n-1}(\widetilde{P})}\right).$$

So we have that $\dfrac{\hat{\mathbb{P}}_1^{n+1}(P)}{\hat{\mathbb{P}}_0^{n-1}(\widetilde{P})} < 2$ unless if $N_0^{n-1}(\widetilde{P}) = 0$. Take $R = \frac{1}{2}$, we have:

$$(3) \le \mathbb{P}(N_0^{n-1}(\widetilde{P}) = 0)$$

and

$$\mathbb{P}(N_0^{n-1}(\widetilde{P}) = 0) \le \mathbb{P}\left(\hat{\mathbb{P}}_0^{n-1}(\widetilde{P}) \le \frac{Q_0^{n-1}(\widetilde{P})}{2}\right).$$

Apply Corollary 1 with $t = \frac{Q_0^{n-1}(\widetilde{P})}{2}$ (of course our hypothesis imply that $Q_0^{n-1}(\widetilde{P}) > 0$) to get

$$(3) \le e^{\frac{1}{e}} e^{-Cn^{1-\varepsilon}}.$$

These three bounds give the result (we have bounded $e^{\frac{1}{e}}$ by 2). $\qquad\square$

3 Applications to dynamical systems

We turn now to our main motivation: dynamical systems. Consider a dynamical system (Σ, T, μ). Σ is a complete space, $T : \Sigma \to \Sigma$ is a measurable map, μ is a T-invariant probability measure on Σ. As before, $\mathcal{C}$ is a Banach space of bounded functions on Σ (typically, $\mathcal{C}$ will be the space of function of bounded variations or a space of piecewise Hölder functions, see examples in Section 3.1). It is endowed with a norm of the form:

$$\|f\|_{\mathcal{C}} = C(f) + \|f\| \,,$$

where $C(f)$ is a semi-norm (i.e. $\forall f \in \mathcal{C}$, $C(f) \geq 0$, $C(\lambda f) = |\lambda| C(f)$ for $\lambda \in \mathbb{R}$, $C(f + g) \leq C(f) + C(g)$) and $\| \ \|$ is a norm on $\mathcal{C}$. In addition, we assume that the norm $\| \ \|$ on $\mathcal{C}$ is such that for any $\varphi \in \mathcal{C}$, there exists a real number $R(\varphi)$ such that $\|\varphi + R(\varphi)\| \leq C(\varphi)$ (for example, this is the case if $\| \ \| = \| \ \|_\infty$ and $C(\varphi) = \bigvee(\varphi)$ or $\| \ \| = \| \ \|_\infty$ and $C(\varphi)$ is the Hölder constant). We assume that the dynamical system satisfy the following mixing property: for all $\varphi \in L^1(\mu)$, $\psi \in \mathcal{C}$,

$$\left| \int_\Sigma \psi \cdot \varphi \circ T^n \, d\mu - \int_\Sigma \psi \, d\mu \int_\Sigma \varphi \, d\mu \right| \leq \Phi(n) \|\varphi\|_1 \|\psi\|_{\mathcal{C}} \,, \qquad (4)$$

with $\Phi(n)$ summable.

Consider a countable partition $A_1, \ldots, A_p, \ldots$ of Σ. Denote by $\mathcal{P}_k$ the countable partition of Σ whose atoms are defined by: for $i_0, \ldots, i_{k-1}$, denote

$$A_{i_0, \ldots, i_{k-1}} = \{ x \in \Sigma \ / \text{ for } j = 0, \ldots, k-1, \ T^j(x) \in A_{i_j} \} \,.$$

We assume that for all $i_0, \ldots, i_{k-1}$, $f = \mathbb{1}_{A_{i_0,\ldots,i_{k-1}}} \in \mathcal{C}$ and let $C(i_0, \ldots, i_{k-1})$ be denoted by $C(f)$. Consider the process taking values into Σ: $X_j(x) = T^j(x)$, $j \in \mathbb{N}$, $x \in \Sigma$. Clearly if $X_j \in A_{i_0,\ldots,i_{k-1}}$ then $X_{j+1} \in A_{i_1,\ldots,i_{k-1}}$. That is for any $P \in \mathcal{P}_k$, there exists a unique $\widetilde{P} \in \mathcal{P}_{k-1}$ such that $X_j \in P \Rightarrow X_{j+1} \in \widetilde{P}$.

Condition (4) may be rewritten as: for all $\varphi \in L^1(\mu)$, $\psi \in \mathcal{C}$,

$$|\mathrm{Cov}(\psi(X_0), \varphi(X_n))| \leq \Phi(n) \|\varphi\|_1 \|\psi\|_{\mathcal{C}} \,.$$

Moreover, we assume that for $\psi \in \mathcal{C}$, there exists a real number $R(\psi)$ such that $\|\psi + R(\psi)\| \leq C(\psi)$. We have:

$$
\begin{aligned}
|\mathrm{Cov}(\psi(X_0), \varphi(X_n))| &= |\mathrm{Cov}([\psi(X_0) + R(\psi)], \varphi(X_n))| \\
&\leq \Phi(n) \|\varphi\|_1 \|\psi + R(\psi)\|_{\mathcal{C}} \leq \Phi(n) \|\varphi\|_1 C(\psi) \,. \quad (5)
\end{aligned}
$$

Using the stationarity of the sequence (X_j), we have for all $i \in \mathbb{N}$, for $\psi \in \mathcal{C}_1$, $\varphi \in L^1$, $\|\varphi\|_1 \leq 1$,

$$|\mathrm{Cov}(\psi(X_i), \varphi(X_{n+i}))| \leq 2\Phi(n) . \tag{6}$$

So, our Assumptions 1 and (**) are satisfied for a "time reversed" process: consider a process $(Y_n)_{n\in\mathbb{N}}$ such that $(Y_n, \cdots, Y_0)$ as the same law as $(X_0, \cdots, X_n)$, then $\mathrm{Cov}(\psi(X_i), \varphi(X_{n+i})) = \mathrm{Cov}(\psi(Y_{i+n}), \varphi(Y_i))$ and the process $(Y_n)_{n\in\mathbb{N}}$ satisfies our Assumptions 1. Using the stationarity, it satisfies also(**), see [BGR00] and [DP05] for more developments on this "trick". Applying Theorem 1 to the process $(Y_n)_{n\in\mathbb{N}}$ and using that

$$\sum_{j=1}^{n} \mathbb{1}_P(Y_j) \overset{\mathrm{Law}}{=} \sum_{0}^{n-1} \mathbb{1}_P(X_j)$$

and

$$\sum_{j=0}^{n-2} \mathbb{1}_{\tilde{P}}(Y_j) \overset{\mathrm{Law}}{=} \sum_{j=0}^{n-2} \mathbb{1}_{\tilde{P}}(X_j) ,$$

we obtain the following result.

Theorem 2. *There exists a constant $C > 0$, such that for all $k, n \in \mathbb{N}$, for any sequence $i_0, \ldots, i_{k-1}$, for all $t \in \mathbb{R}$,*

$$\mathbb{P}\left(\left|\frac{N_0^n(A_{i_0,\ldots,i_{k-1}})}{n} - \mu(A_{i_0,\ldots,i_{k-1}})\right| > t\right) \leq e^{\frac{1}{e}}e^{-\frac{Ct^2n}{C(i_0,\ldots,i_{k-1})^2}} .$$

Let $\hat{g}_n(A_{i_0,\ldots,i_{k-1}}) = \frac{N_0^n(A_{i_0,\ldots,i_{k-1}})}{N_0^{n-1}(A_{i_1,\ldots,i_{k-1}})}\frac{n-1}{n}$, there exists $K > 0$ such that for all $\varepsilon < 1$, if

$$\frac{\mu(A_{i_1,\ldots,i_{k-1}})}{C(i_0,\ldots,i_{k-1})} \text{ and } \frac{\mu(A_{i_1,\ldots,i_{k-1}})}{C(i_1,\ldots,i_{k-1})} \geq n^{-\frac{\varepsilon}{2}} ,$$

then we have:

$$\mathbb{P}\left(\left|\hat{g}_n(A_{i_0,\ldots,i_{k-1}}) - \mathbb{P}(X_0 \in A_{i_0}|X_1 \in A_{i_1},\ldots,X_{k-1} \in A_{i_{k-1}})\right| > t\right)$$
$$\leq 4e^{-Kt^2n^{1-\varepsilon}} + 2e^{-Kn^{1-\varepsilon}} .$$

Let us end this section with a lemma stating that the elements $P \in \mathcal{P}_k$ are exponentially small. It indicates that we might not expect to take k of order greater than $\ln n$ in the above theorem.

Lemma 2. *Assume that $C_{\max} = \max\limits_{j=1,\ldots,} C(\mathbb{1}_{A_j}) < \infty$. There exists $0 < \gamma < 1$ such that for all $P \in \mathcal{P}_k$, we have*

$$\mu(P) \leq \gamma^k .$$

Proof. The proof of Lemma 2 follows from the mixing property. It is inspired from [Pac00]. Let $n_0 \in \mathbb{N}$ to be fixed later. Let $P \in \mathcal{P}_k$, for some indices $i_0, \ldots, i_{k-1}$, we have that

$$P = \{x \in A_{i_0}, \ldots, T^{k-1}x \in A_{i_{k-1}}\} .$$

Then, let $\ell = [\frac{k}{n_0}]$,

$$\mu(P) = \mathbb{P}(X_0 \in A_{i_0}, \ldots, X_{k-1} \in A_{i_{k-1}})$$
$$\leq \mathbb{P}(X_0 \in A_{i_0}, X_{n_0} \in A_{i_{n_0}}, \ldots, X_{\ell n_0} \in A_{i_{\ell n_0}}) .$$

The random variable

$$Y = \frac{\mathbb{1}_{A_{i_{n_0}}}(X_{n_0}) \cdots \mathbb{1}_{A_{i_{\ell n_0}}}(X_{\ell n_0})}{\mathbb{P}(X_{n_0} \in A_{i_{n_0}}, \ldots, X_{\ell n_0} \in A_{i_{\ell n_0}})}$$

is $\mathcal{M}_{\ell n_0}$-measurable with L^1 norm less than 1 and $\frac{\mathbb{1}_{A_{i_0}}}{C_{\max}}$ is in $\mathcal{C}_1$. From the mixing property (6), we get: (let $s = \sup_{j=1,\ldots} \mu(A_j) < 1$)

$$\mathbb{P}(X_0 \in A_{i_0}, X_{n_0} \in A_{i_{n_0}}, \ldots, X_{\ell n_0} \in A_{i_{\ell n_0}})$$
$$\leq \mathbb{P}(X_{n_0} \in A_{i_{n_0}}, \ldots, X_{\ell n_0} \in A_{i_{\ell n_0}}) \cdot (\Phi_C(n_0)C_{\max} + s) .$$

Choosing n_0 such that $\Phi_C(n_0)C_{\max} + s < 1$, we obtain the result by induction.
□

3.1 Expanding maps of the interval

In this section, we consider piecewise expanding maps on the interval $I = [0, 1]$. That is, T is a piecewise expanding map, defined on a finite partition into intervals $A_1, \ldots, A_\ell$. $\mathcal{P}_k$ is the partition of I with atoms: $A_{i_0} \cap T^{-1}A_{i_1} \cap \cdots \cap T^{-(k-1)}A_{i_{k-1}}$. If for all $j = 1, \ldots, \ell$, $T(A_j)$ is a union of the A_p's, T is said to be a *Markov map*. For $x \in I$, let $C_k(x)$ be the atom of the partition $\mathcal{P}_k$ containing x. Under an assumption of aperiodicity in the Markov case or covering in general, the map T admits a unique invariant measure absolutely continuous with respect to the Lebesgue measure m. Let h be the invariant density. The potential of the system is $g = \frac{h}{|T'| \cdot h \circ T}$, we have also that g^{-1} is the Radon-Nikodym derivative of $\mu \circ T$ with respect to μ (if $\mu = hm$). We shall prove that $g(x)$ may be estimated by $\hat{g}_{n,k}(x) := \hat{g}_n(C_k(x))$ for $k = \Theta(\ln n)$. Formally the assumptions on the system are the following.

Assumption 2(A1). *the restriction of T to each $\overline{A_j}$ is a C^2 one-to-one map from $\overline{A_j}$ to $T(\overline{A_j}) =: B_j$.*

(A2). T is expanding: there exists $1 < \theta^{-1}$ such that for all $x \in I$, $\theta^{-1} \leq |T'(x)|$.

(A3). If T is a Markov map, we assume that it is aperiodic: there exists $N \in \mathbb{N}$ such that for all $i, j = 1, \ldots, \ell$, for all $n \geq N$,

$$T^{-n} A_i \cap A_j \neq \emptyset .$$

(A4). If T is not Markov, we assume that it satisfies the covering property: for all $k \in \mathbb{N}$, there exists $N(k)$ such that for all $P \in \mathcal{P}_k$,

$$T^{N(k)} P = [0, 1] .$$

The above conditions are sufficient to ensure existence and uniqueness of an absolutely continuous invariant measure as well as an estimation of the speed of mixing (see for example [Sch96] for the Markov case and [CoL96], [Liv95] for the general case). Under more technical assumptions, these results on existence and uniqueness of an absolutely continuous invariant measure as well as an estimation of the speed of mixing remain valid, with an infinite countable partition ([Bro96], [L,S,V], [Mau01]).

Theorem 3. *([Sch96], [CoL96], [Liv95]) Let $\mathcal{C}$ be the space of functions on $[0,1]$ of bounded variations. Let T satisfy the assumptions 2. Then we have the following mixing property: there exists $C > 0$, $0 < \xi < 1$ such that for all $\varphi \in L^1(\mu)$, $\psi \in \mathcal{C}$,*

$$\left| \int_\Sigma \psi \cdot \varphi \circ T^n d\mu - \int_\Sigma \psi d\mu \int_\Sigma \varphi d\mu \right| \leq C\xi^n \|\varphi\|_1 \|\psi\|_\mathcal{C} .$$

Moreover, we have that the invariant density h belongs to BV and $0 < \inf h \leq \sup h < \infty$. If the map is Markov, then h is C^1 on each B_j.

In other words, our system satisfy (4) for bounded variation functions. Moreover, for any $k \in \mathbb{N}$, the element P of $\mathcal{P}_k$ are subintervals, so the indicators $\mathbb{1}_P$ belong to BV and $C(\mathbb{1}_P) = \bigvee(\mathbb{1}_P) = 2$. So, we shall apply Theorem 2, this will lead to the announced estimation of the potential g.

Let us also introduce a very useful tool in dynamical systems: the transfer operator. For $f \in BV$, let

$$\mathcal{L}(f)(x) = \sum_{y/T(y)=x} g(y)f(y) .$$

We have $\mathcal{L}(\mathbb{1}) = \mathbb{1}$, for all $f_1 \in BV$, $f_2 \in L^1(\mu)$,

$$\int_I \mathcal{L}(f_1) \cdot f_2 d\mu = \int_I f_1 \cdot f_2 \circ T d\mu .$$

The process $(Y_n)_{n \in \mathbb{N}}$ introduced after Lemma 2 is a Markov process with kernel $\mathcal{L}$ (see [BGR00]). The following three lemmas are the last needed bricks between Theorem 2 and the estimation of the potential g.

Lemma 3. *Assume that T satisfies Assumption 2 and is a Markov map, let γ be given by Lemma 2. There exists $K > 0$ such that for all $k \in \mathbb{N}$, for all $x \in I$,*

$$(1 - K\gamma^k)g(x) \leq \frac{\mu(C_k(x))}{\mu(C_{k-1}(Tx))} \leq (1 + K\gamma^k)g(x) . \tag{7}$$

Proof. Because the map is Markov, for all $x \in I$, $T(C_k(x)) = C_{k-1}(Tx)$. We have:

$$\mu(T(C_k(x))) = \int \frac{1}{g} \mathbb{1}_{C_k(x)} d\mu,$$

$$\min_{y \in C_k(x)} \frac{1}{g(y)} \int \mathbb{1}_{C_k(x)} d\mu \le \int \frac{1}{g} \mathbb{1}_{C_k(x)} d\mu \le \max_{y \in C_k(x)} \frac{1}{g(y)} \int \mathbb{1}_{C_k(x)} d\mu .$$

Since the map is Markov, h and $h \circ T$ are C^1 on each $C_k(x)$, so g is C^1 on $C_k(x)$ and since T is expanding, we conclude that

$$\max_{y \in C_k(x)} \frac{1}{g(y)} \le (1 + K\gamma^k) \frac{1}{g(x)}$$

and

$$\min_{y \in C_k(x)} \frac{1}{g(y)} \ge (1 - K\gamma^k) \frac{1}{g(x)} .$$

The result follows.   □

If the map T is not Markov, we shall prove a result not so strong (but sufficient for our purpose). To deal with non Markov maps, we have to modify the above proof at two points: firstly, we have not $T(C_k(x)) = C_{k-1}(Tx)$ for all x (but for lots of them) ; secondly, $g = \frac{h}{|T'| h \circ T}$ is not smooth (due to h). The following lemma shows that we control the irregularity of h.

Lemma 4. *Let $a = \bigvee h$, for any interval P, let $\bigvee_P h$ be the variation of h on P. For all $k \ge 1$, for all $u_k > 0$,*

$$\mu\{x \in [0,1] \ / \bigvee_{C_k(x)} h \ge u_k\} \le \frac{\gamma^k}{u_k a} .$$

Proof. We have:

$$\mu\{x \in [0,1] \ / \bigvee_{C_k(x)} h \ge u_k\} = \sum_{\substack{P \in \mathcal{P}_k \\ \bigvee_P h \ge u_k}} \mu(P) ,$$

$$a = \bigvee h \ge \sum_{P \in \mathcal{P}_k} \bigvee_P h \ge \#\{P \in \mathcal{P}_k \ / \bigvee_P h \ge u_k\} u_k .$$

In other words, $\#\{P \in \mathcal{P}_k \ / \bigvee_P h \ge u_k\} \le \frac{a}{u_k}$. Using Lemma 2, we get:

$$\mu\{x \in [0,1] \ / \bigvee_{C_k(x)} h \ge u_k\} \le \#\{P \in \mathcal{P}_k \ / \bigvee_P h \ge u_k\} \gamma^k \le \frac{\gamma^k}{u_k a} .$$

                                                                                    □

Corollary 3. *For all $\kappa > \gamma$, there exists a constant $K > 0$ and for all $k \in \mathbb{N}^*$, a set B_k such that $\mu(B_k) \le \frac{\gamma^k}{\kappa^k a}$ and if $x \notin B_k$, $y \in C_k(x)$,*

$$(1 - K\kappa^k) \le \frac{g(x)}{g(y)} \le (1 + K\kappa^k) . \tag{8}$$

Proof. Recall that $g = \frac{h}{|T'|h \circ T}$. Because T is piecewise C^2 and expanding, $\frac{1}{|T'|}$ satisfies an equation of the type (8) for all $x \in [0,1]$, for $\kappa = \gamma$. We just have to prove that h satisfies such an inequality. Fix $\kappa > \gamma$, let

$$B_k = \left\{ x \in [0,1] \ / \ \bigvee_{C_k(x)} h \geq \kappa^k \right\}.$$

Let $x \notin B_k$ and $y \in C_k(x)$.

$$|h(x) - h(y)| \leq \bigvee_{C_k(x)} h \leq \kappa^k .$$

Now, $\frac{h(x)}{h(y)} = 1 + \frac{h(x) - h(y)}{h(y)}$, thus

$$1 - \frac{1}{\sup h}\kappa^k \leq \frac{h(x)}{h(y)} \leq 1 + \frac{1}{\inf h}\kappa^k .$$

Of course, the same equation holds for $h \circ T$ by replacing k with $k - 1$, combining this equations (for h, $h \circ T$ and $|T'|$) gives the result. $\square$

Lemma 5. *Assume that T satisfies Assumption 2 and is not necessary a Markov map. There exists $K > 0$ such that for all $k \in \mathbb{N}$, for all $\kappa > \gamma$,*

$$\mu \left\{ x \in I \ / \ (1 - K\kappa^k)g(x) \leq \frac{\mu(C_k(x))}{\mu(C_{k-1}(Tx))} \leq (1 + K\kappa^k)g(x) \right\}$$

$$\geq 1 - \left(2\ell\gamma^k + a \left(\frac{\gamma}{\kappa} \right)^k \right) .$$

Proof. We begin with a simple remark. Let us denote $\partial \mathcal{P}$ the union of the boundaries of the A_j's. For $x \in [0,1]$, if $\overline{C_k(x)} \cap \partial \mathcal{P} = \emptyset$ then $T(C_k(x)) = C_{k-1}(Tx)$, otherwise, $T(C_k(x))$ is strictly included into $C_{k-1}(Tx)$. This elementary remark is very useful in the study of non Markov maps. The points x such that $T(C_k(x)) = C_{k-1}(Tx)$ will be called k-Markov points. If the map is Markov then all points are k-Markov for all $k \in \mathbb{N}$. For k-Markov points, we may rewrite the proof of Lemma 3 to get the inequalities:

$$\min_{y \in C_k(x)} \frac{1}{g(y)}\mu(C_k(x)) \leq \mu(C_{k-1}(Tx)) \leq \max_{y \in C_k(x)} \frac{1}{g(y)}\mu(C_k(x)) .$$

Now, we use Corollary 3 and we have that if x is a k-Markov point that do not belong to B_k then

$$(1 - K\kappa^k)g(x) \leq \frac{\mu(C_k(x))}{\mu(C_{k-1}(Tx))} \leq (1 + K\kappa^k)g(x) . \tag{9}$$

So, we have that the set D_k of points not satisfying 9 for one k is included into the set of points x such that $\overline{C_k(x)} \cap \partial \mathcal{P} \neq \emptyset$ or in B_k (given by Corollary 3). Clearly, there are at most 2ℓ elements P of $\mathcal{P}_k$ such that $\overline{P} \cap \partial \mathcal{P} \neq \emptyset$, moreover, by Lemma 2, we have for $P \in \mathcal{P}_k$, $\mu(P) \leq \gamma^k$. We have proved that $\mu(D_k) \leq 2\ell\gamma^k + \frac{\gamma^k}{\kappa^k a}$. $\square$

We are now in position to prove that $\hat{g}_{n,k}(x)$ is a consistent estimator of the potential $g(x)$.

Theorem 4. *For all $\kappa > \gamma$, there exists D_k and E_k finite union of elements of $\mathcal{P}_k$ satisfying $\mu(D_k) \le 2\ell\gamma^k + a\left(\frac{\gamma}{\kappa}\right)^k$, $\mu(E_k) \le \gamma^k$ and there exists $L > 0$ such that if*

- $x \notin D_k \cup E_k$,
- $\dfrac{\ln(\frac{t}{2K})}{\ln(\kappa)} \le k \le \dfrac{\varepsilon \ln 2n}{2 \ln(\frac{\ell}{\gamma})}$

then

$$\mathbb{P}(|\hat{g}_{n,k}(x) - g(x)| > t) \le 4e^{-Lt^2 n^{1-\varepsilon}} + 2e^{-Ln^{1-\varepsilon}} .$$

Proof. Fix $\kappa > \gamma$, let D_k be given by Lemma 5: if $x \notin D_k$ then

$$(1 - K\kappa^k)g(x) \le \frac{\mu(C_k(x))}{\mu(C_{k-1}(Tx))} \le (1 + K\kappa^k)g(x),$$

let E_k be the set of points x such that $\mu(C_k(x)) \le \frac{\gamma^k}{\ell^k}$. Clearly, if $x \in D_k$ then $C_k(x) \subset D_k$ and if $x \in E_k$ then $C_k(x) \subset E_k$, so D_k and E_k are finite union of elements of $\mathcal{P}_k$.

Let $x \notin D_k \cup E_k$, then $\mu(C_k(x)) > \frac{\gamma^k}{\ell^k}$. If $k \le \frac{\varepsilon}{2} \frac{\ln 2n}{\ln(\frac{\ell}{\gamma})}$ then $\mu(C_k(x)) \ge 2n^{-\frac{\varepsilon}{2}}$. Since $C_k(x)$ is an interval, we have $C(\mathbb{1}_{C_k(x)}) = \bigvee(\mathbb{1}_{C_k(x)}) = 2$ and then

$$\frac{\mu(C_{k-1}(Tx))}{C(\mathbb{1}_{C_k(x)})} = \frac{\mu(C_{k-1}(Tx))}{C(\mathbb{1}_{C_{k-1}(Tx)})} \ge \frac{\mu(C_k(x))}{2} \ge n^{-\frac{\varepsilon}{2}} .$$

We shall use Theorem 2.

$$\mathbb{P}(|\hat{g}_{n,k}(x) - g(x)| > t)$$

$$\le \mathbb{P}(|\hat{g}_{n,k}(x) - \frac{\mu(C_k(x))}{\mu(C_{k-1}(Tx))}| > t - |\frac{\mu(C_k(x))}{\mu(C_{k-1}(Tx))} - g(x)|)$$

$$\le \mathbb{P}(|\hat{g}_{n,k}(x) - \frac{\mu(C_k(x))}{\mu(C_{k-1}(Tx))}| > t - K\kappa^k) \text{ (because } x \notin D_k)$$

$$\le 4e^{-L(t-K\kappa^k)^2 n^{1-\varepsilon}} + 2e^{-Ln^{1-\varepsilon}} \text{ (where we have used Theorem 2) .}$$

If $\ln(t/2)/\ln(1/\kappa) \le k$, we conclude

$$\mathbb{P}(|\hat{g}_{n,k}(x) - g(x)| > t) \le 4e^{-Lt^2 n^{1-\varepsilon}} + 2e^{-Ln^{1-\varepsilon}} .$$

$\square$

We derive the following corollary. Fix $\kappa > \gamma$.

Corollary 4. *Let $\alpha = c/\{2(1+c)\}$ with $c = \ln(1/\kappa)/\ln(l/\gamma)$ and $k(n)$ be an increasing sequence such that*

$$\frac{\ln\left(\frac{1}{2Kn^\alpha}\right)}{\ln(\kappa)} \leq k(n) \leq \frac{\varepsilon}{2} \frac{\ln 2n}{\ln(\frac{\ell}{\gamma})} .$$

Let $\hat{g}_n = \hat{g}_{n,k(n)}$, then $|\hat{g}_n(x) - g(x)| = O_\mathbb{P}(n^{-\alpha})$.

Proof. It suffices to prove that:

$$\lim_{M\to\infty} \limsup_{n\to\infty} \mathbb{P}(n^\alpha|\hat{g}_n(x) - g(x)| > M) = 0 .$$

We chose $t = n^{-\alpha}$ in Theorem 4 and obtain:

$$\mathbb{P}(n^\alpha|\hat{g}_n(x) - g(x)| > M) \leq \mathbb{P}(|\hat{g}_n(x) - g(x)| > \frac{1}{n^\alpha}) \leq 4e^{-Ln^{1-\varepsilon-2\alpha}} + o(1) .$$

The best rate is obtained for $\alpha = c/\{2(1+c)\}$ with $c = \ln(1/\kappa)/\ln(l/\gamma)$.   □

Remark 3. In [CMS02], an exponential inequality is proven for Lipschitz functions of several variables for expanding dynamical systems of the interval. We can not use such a result here because characteristic functions of intervals are not Lipschitz, the result could maybe be improved to take into consideration piecewise Lipschitz functions. The Lipchitz constant enter in the bound of the exponential inequality and any kind of piecewise Lipschitz constant would be exponentially big for $\mathbb{1}_P$, $P \in \mathcal{P}_k$. Nevertheless, such a result for functions of several variables could be interesting to estimate the conditional probabilities and potential g: we could construct an estimator by replacing $N_j^\ell(A_{i_0,...,i_{k-1}})$ with

$$\widetilde{N_j^n}(A_{i_0,...,i_{k-1}}) = \left|\{p \in \{j, ..., n+j-k\} \ / \ X_j \in A_{i_0}, ..., X_{j+k-1} \in A_{i_{k-1}}\}\right| .$$

4 Gibbs measures and chains with complete connections

In this section, we state our results in the particular setting of Gibbs measures or chains with complete connections. Gibbs measures and chains with complete connections are two different point of view of the same thing - consider a stationary process $(X_i)_{i\in\mathbb{N}}$ or $\mathbb{Z}$ taking values into a finite set A satisfying: for all $a_0, ..., a_k, ...$ in A. If $\mathbb{P}(X_0 = a_0, ..., X_k = a_k) \neq 0$ for all k, then

$$\lim_{k\to\infty} \mathbb{P}(X_0 = a_0|X_1 = a_1, ..., X_{k-1} = a_{k-1}) = \mathbb{P}(X_0 = a_0|X_i = a_i, \ i \geq 1) ,$$

exists. Moreover, there exists a summable sequence $\gamma_k > 0$ such that if $a_0 = b_0$, ..., $a_k = b_k$,

$$\left|\frac{\mathbb{P}(X_0 = a_0|X_i = a_i, \ i \geq 1)}{\mathbb{P}(X_0 = b_0|X_i = b_i, \ i \geq 1)} - 1\right| \leq \gamma_k . \tag{10}$$

Define $\Sigma \subset A^{\mathbb{N}}$ be the set of admissible sequences:

$$\Sigma = \{x = (x_0, \ldots, x_k, \ldots,) \in A^{\mathbb{N}} \mid$$
$$\text{for all } k \geq 0, \ \mathbb{P}(X_0 = x_0, \ldots, X_k = x_k) \neq 0\} .$$

Σ is compact for the product topology and is invariant by the shift map σ: $\sigma(x_0, x_1, \ldots) = (x_1, \ldots)$. We denote by μ the image measure of the X_i's. We assume that the process is mixing: there exists $N > 0$ such that for all $i, j \in A$, for all $n > N$,

$$\mathbb{P}(X_0 = i \text{ and } X_n = j) \neq 0 .$$

We shall denote by

$$A_j = \{x \in \Sigma \ / \ x_0 = j\} \text{ and } A_{i_0, \ldots, i_{k-1}} = \{x \in \Sigma \ / \ x_j = i_j \ j = 0, \ldots k - 1\} .$$

As before, $\mathcal{P}_k$ is the partition of Σ whose atoms are the $A_{i_0, \ldots, i_{k-1}}$'s and $C_k(x)$ is the atom of $\mathcal{P}_k$ containing x.
We assume also that the process has a *Markov structure*: for $x = (x_0, \ldots,) \in \Sigma$, $ax = (a, x_0, \ldots) \in \Sigma$ if and only if $ay \in \Sigma$ for all $y \in A_{x_0}$.
For $x \in \Sigma$, let $g(x) = \mathbb{P}(X_0 = x_0 | X_i = x_i, \ i \geq 1)$. We shall prove that $\hat{g}_{n,k}$ is a consistent estimator of g.
It is known (see [KMS97], [Mau98], [BGF99], [Pol00]) that such a process is mixing for suitable functions.
Let $\gamma_n^\star = \sum_{k \geq n} \gamma_k$, define a distance on Σ by $d(x, y) = \gamma_n^\star$ if and only if $x_j = y_j$ for $j = 0, \ldots, n - 1$ and $x_n \neq y_n$. Let L be the space of Lipschitz functions for this distance, endowed with the norm $\|\psi\| = \sup |\psi| + L(\psi)$ where $L(\psi)$ is the Lipschitz constant of ψ.

Theorem 5. *([KMS97], [Mau98], [BGF99], [Pol00]) A process satisfying (10), being mixing and having a Markov structure is mixing for functions in L in the sense that equation (4) is verified for $\varphi \in L^1(\mu)$ and $\psi \in L$ with $\Phi(n) \overset{n \to \infty}{\longrightarrow} 0$. If $\gamma_n^\star$ is summable, so is $\Phi(n)$.*

In what follows, we assume that $\gamma_n^\star$ is summable. For any $\psi \in L$, let $R = -\inf \psi$ then $\sup |\psi + R| \leq L(\psi)$, then we have (6) for the process $(X_i)_{i \in \mathbb{N}}$ and $\psi \in L$ such that $L(\psi) \leq 1$ and Theorem 2 is satisfied.
We have that

$$L(\mathbb{1}_{A_j}) \leq \frac{1}{\gamma_0^\star} \text{ and } L\left(\mathbb{1}_{A_{i_0, \ldots, i_{k-1}}}\right) \leq \frac{1}{\gamma_k^\star} .$$

Equation (10) gives the following lemma which will be used instead of Corollary 3.

Lemma 6. *For all $x \in \Sigma$, for all $k \in \mathbb{N}$, $y \in C_k(x)$,*

$$1 - \gamma_k \leq \frac{g(x)}{g(y)} \leq 1 + \gamma_k .$$

Following the proof of Lemma 3, we get: for all $x \in \Sigma$, for $k \in \mathbb{N}$,

$$(1 - \gamma_k)g(x) \leq \frac{\mu(C_k(x))}{\mu(C_{k-1}(T(x)))} \leq (1 + \gamma_k)g(x) . \tag{11}$$

Let $\gamma < 1$ be given by Lemma 2, let $\ell = |A|$.

Theorem 6. *Assume that $\gamma_k^\star$ is summable, and that the process satisfy (10), is mixing and has a Markov structure. Then there exists $L > 0$ such that if:*

(A1). $\mu(C_k(x)) \geq \frac{\gamma^k}{\ell^k}$,

(A2). $\left(\frac{\gamma}{\ell}\right)^k \gamma_k^\star \geq n^{-\frac{\varepsilon}{2}}$,

(A3). $\gamma_k \leq \frac{t}{2}$.

we have

$$\mathbb{P}(|\hat{g}_{n,k}(x) - g(x)| > t) \leq 4e^{-Lt^2 n^{1-\varepsilon}} + 2e^{-Ln^{1-\varepsilon}} .$$

Moreover,

(A1). $\mu\{x \in \Sigma \ / \ \mu(C_k(x)) < \frac{\gamma^k}{\ell^k}\} \leq \gamma^k$,

(A2). $\left(\frac{\gamma}{\ell}\right)^k \gamma_k^\star \geq n^{-\frac{\varepsilon}{2}}$ *if $k \leq a \ln n$ for suitable $a > 0$,*

(A3). $\gamma_k \leq \frac{t}{2}$ *if $k \geq bt^{-\frac{1}{2}}$ for suitable $b > 0$.*

Proof. The proof follows the proof of Theorem 4 using Lemma 6 instead of Lemma 5. The estimates on k are obtained by noting that since $\gamma_k^\star$ is summable then $\gamma_k = o(\frac{1}{k^2})$ and $\gamma_k^\star = o(\frac{1}{k})$. Of course, better estimates may be obtained if γ_k decreases faster. $\qquad\square$

As in Section 3.1, we derive the following corollary.

Corollary 5. *For $k = \Theta(\ln n)$, there exists $\alpha > 0$ such that $\hat{g}_{n,k}$ goes to $g(x)$ in probability at rate $n^{-\alpha}$.*

5 Testing if the asymptotic variance is zero: the complete case

In this section, we study the problem of testing whether the asymptotic variance of the process is zero. This is motivated by the fact that for the process studied in the previous sections, we may prove a central limit theorem provided the asymptotic variance is not zero (see [Bro96], [Val01] for examples). We are concerned with a process $(X_j)_{j \in \mathbb{N}}$ satisfying Conditions of Section 3.1 or Section 4. We assume moreover that the system is complete: $T(A_i) = I$ for all i if we are in the context of Section 3.1 or $\sigma(A_i) = \Sigma$ if we are in the context of Section 4. Our arguments should probably be generalized to non complete situations. In what follows, we shall denote T for $T : I \to I$ as well as $\sigma : \Sigma \to \Sigma$.

Definition 2. *([Bro96]) Let*

$$S_n = \sum_{j=0}^{n-1}(X_j - \mathbb{E}(X_0)) \text{ and } M_n = \int \left(\frac{S_n}{\sqrt{n}}\right)^2 d\mathbb{P}.$$

The sequence M_n converges to V which we shall call the asymptotic variance.

Proposition 2. *([Bro96], [CM04]) The asymptotic variance V is zero if and only if the potential $\log g$ is a cohomologous to a constant: $\log g = \log a + u - u \circ T$, with $a > 0$, $u \in BV$ or $u \in L$.*

Because we are in a stationary setting, we have that the asymptotic variance is zero if and only if g is indeed constant (the fact that the system is complete is here very important). We deduce a way of testing if the asymptotic variance is zero. Using Theorem 4 or Theorem 6, we have that if g is constant,

$$\mathbb{P}(|\sup \hat{g}_{n,k} - \inf \hat{g}_{n,k}| > t) \le 2 \cdot (4e^{-Lt^2 n^{1-\varepsilon}} + 2e^{-Ln^{1-\varepsilon}}) + \gamma^k.$$

To use such a result, we have to compute $\sup \hat{g}_{n,k}$ and $\inf \hat{g}_{n,k}$, so we have ℓ^k computations to make with $k = \Omega(\ln n)$. A priori, all the constants in the above inequality, may be specified. In theory, for $t > 0$, we may find k, n satisfying the hypothesis of Theorem 4 or Theorem 6 so that $\mathbb{P}(|\sup \hat{g}_{n,k} - \inf \hat{g}_{n,k}| > t)$ is smaller than a specified value. If the computed values of $\sup \hat{g}_{n,k}$ and $\inf \hat{g}_{n,k}$ agree with this estimation this will indicates that g is probably constant so that the asymptotic variance is probably 0.

Acknowledgement

Many ideas of the paper have been discussed with Bernard Schmitt, it is a pleasure to thank him. I am grateful to Jérôme Dedecker who patiently answered my questions on weak-dependence coefficients. I whish also to thank the referees of this paper for many interesting remarks and for the suggestion of the proof of Corollary 4.

References

[BGR00] A.D. Barbour, R. Gerrard, G. Reinert, *Iterates of expanding maps.* Probab. Theory Related Fields 116 (2000), no. 2, 151–180.

[BRY98] A. Barron, J. Rissanen, B. Yu, *The minimum description length principle in coding and modeling.* Information theory: 1948–1998. IEEE Trans. Inform. Theory 44 (1998), no. 6, 2743–2760.

[BGF99] X Bressaud, R. Fernàndez, A. Galves, *Decay of correlations for non-Hölderian dynamics. A coupling approach.* Electron. J. Probab. 4 (1999), no. 3, 19 pp.

[Bro96] A. Broise, *Transformations dilatantes de l'intervalle et théorèmes limites.* Études spectrales d'opérateurs de transfert et applications. Astérisque 1996, no. 238, 1–109.

[CM04] F. Chazal, V. Maume-Deschamps, *Statistical properties of Markov dynamical sources: applications to information theory.* Discrete Math. Theor. Comput. Sci. 6 (2004), no. 2, 283–314

[CoL96] P. Collet, *Some ergodic properties of maps of the interval.* Dynamical systems (Temuco, 1991/1992), 55–91, Travaux en Cours, 52, Hermann, Paris, 1996.

[CMS02] P. Collet, S. Martinez, B. Schmitt, *Exponential inequalities for dynamical measures of expanding maps of the interval.* Probab. Theory Related Fields 123 (2002), no. 3, 301–322.

[Csi02] I. Csiszár, *Large-scale typicality of Markov sample paths and consistency of MDL order estimators.* Special issue on Shannon theory: perspective, trends, and applications. IEEE Trans. Inform. Theory 48 (2002), no. 6, 1616–1628.

[CsS] I. Csiszár, P. C. Shields, *The consistency of the BIC Markov order estimator.* Ann. Statist. 28 (2000), no. 6, 1601–1619.

[DD03] J. Dedecker, P. Doukhan, *A new covariance inequality and applications.* Stochastic Process. Appl. 106 (2003), no. 1, 63–80.

[DP05] J. Dedecker, C. Prieur, *New dependence coefficients. Examples and applications to statistics.* Probab. Theory and Relat. Fields 132 (2005), 203–236.

[FKT88] P. Flajolet, P. Kirschenhofer, R.F. Tichy, *Deviations from uniformity in random strings.* Probab. Theory Related Fields 80 (1988), no. 1, 139–150.

[KMS97] A. Kondah, V. Maume, B. Schmitt, *Vitesse de convergence vers l'état d'équlibre pour des dynamiques markoviennes non holdériennes.* Ann. Inst. H. Poincaré Probab. Statist. 33 (1997), no. 6, 675–695.

[Liv95] C. Liverani, *Decay of correlations for piecewise expanding maps.* J. Statist. Phys. 78 (1995), no. 3-4, 1111–1129.

[L,S,V] C. Liverani, B. Saussol, S. Vaienti, *Conformal measure and decay of correlation for covering weighted systems.* Ergodic Theory Dynam. Systems 18 (1998), no. 6, 1399–1420.

[Mau01] V. Maume-Deschamps, *Correlation decay for Markov maps on a countable state space.* Ergodic Theory Dynam. Systems 21 (2001), no. 1, 165–196.

[Mau98] V Maume-Deschamps, *Propriétés de mélange pour des sytèmes dynamiques markoviens.* Thèse de l'Université de Bourgogne, available at `http://math.u-bourgogne.fr/IMB/maume`.

[Pac00] F. Paccaut *Propriétés statistiques de systèmes dynamiques non markoviens* Ph D Thesis Université de Bourgogne (2000). Available at `http://www.lamfa.u-picardie.fr/paccaut/publi.html`

[Pol00] M. Policott *Rates of mixing for potentials of summable variation.* Trans. Amer. Math. Soc. 352 (2000), no. 2, 843–853.

[Ris89] J. Rissanen, non k-Markov points.*Stochastic complexity in statistical inquiry.* World Scientific Series in Computer Science, 15. World Scientific Publishing Co., Inc., Teaneck, NJ, 1989. vi+178 pp.

[Sch96] B. Schmitt, *Ergodic theory and thermodynamic of one-dimensional Markov expanding endomorphisms.* Dynamical systems (Temuco, 1991/1992), 93–123, Travaux en Cours, 52, Hermann, Paris, 1996.

[Val01] B. Vallée *Dynamical sources in information theory: fundamental intervals and word prefixes* Algorithmica, **29**, 262-306, (2001).

Martingale approximation of non adapted stochastic processes with nonlinear growth of variance

Dalibor Volný

Laboratoire de Mathématiques Raphaël Salem, Université de Rouen, Avenue de l'Université BP 12, 76801 Saint Etienne du Rouvray
dalibor.volny@univ-rouen.fr

1 Introduction

Let $(\Omega, \mathcal{A}, \mu)$ be a probability space, T a measure preserving bijective and bimeasurable mapping of Ω onto itself. By $\mathcal{I}$ we denote the σ-algebra of $A \in \mathcal{A}$ for which $T^{-1}A = A$; if all elements of $\mathcal{I}$ are of measure 0 or 1 we say that μ is ergodic. A sequence of $X_i = f \circ T^i$, $i \in \mathbf{Z}$, where f is a measurable function, is strictly stationary and any strictly stationary sequence of random variables X_k can be represented in this way. By a filtration we shall mean a sequence of σ-fields $(\mathcal{F}_k)_k$ where $\mathcal{F}_k = T^{-k}\mathcal{F}_0$, $\mathcal{F}_k \subset \mathcal{F}_{k+1}$.

One of the tools for proving central limit theorems for stationary sequences of random variables has been approximating the partial sums $S_n(f) = \sum_{i=0}^{n-1} f \circ T^i$ by a martingale, thus reducing the original problem to a study of limit theorems for martingale differences. In most of the known results, the variances of the partial sums $S_n(f)$ grow linearly. The two methods shown here admit a nonlinear growth. In [WW04], one of the most general conditions for the existence of such a martingale approximation has been given. In the original version, the result was formulated for additive functionals of stationary Markov chains but in fact (as the authors remarked) it holds true for stationary sequences in general. Here we shall show a generalization to the case when the filtration $(\mathcal{F}_k)_k$ is not the natural filtration of the process (X_k). The proof is presented in a way avoiding the language of Markov chains.

Another approach giving central limit theorems for processes with non-linear growth of variances is given by a sequence of martingale-coboundary representations.

In the last chapter, a few related results concerning the conditional central limit theorem and the choice of filtration are announced.

2 On an approximation of Wu and Woodroofe

Let $(\mathcal{F}_k)_k$ be a filtration. For X integrable and $k \in \mathbf{Z}$ let us denote

$$Q_k(X) = \mathbb{E}(X|\mathcal{F}_k), \quad R_k(X) = X - \mathbb{E}(X|\mathcal{F}_k).$$

Remark that

$$X = Q_k(X) + R_k(X),$$
$$Q_k Q_{k-1}(X) = Q_{k-1} Q_k(X) = Q_{k-1}(X),$$
$$R_k R_{k-1}(X) = R_{k-1} R_k(X) = R_k(X),$$
$$R_{k-1}(X) - R_k(X) = Q_k(X) - Q_{k-1}(X) = \mathbb{E}(X|\mathcal{F}_k) - \mathbb{E}(X|\mathcal{F}_{k-1}). \quad (1)$$

For a sequence $(X_k)_k$ we denote $S_n = \sum_{i=1}^{n} X_i$; if, moreover, $X_k \in L^2$ and $\mathbb{E}(X_k \,|\, \mathcal{I}) = 0$, we denote $\sigma_n^2 = \mathbb{E}(S_n^2)$. We shall abuse the notation by omitting to write a.s. and by writing L^2 instead of $\mathcal{L}^2$.

In [WW04] Wu and Woodroofe proved

Theorem 1. *Let $(X_k)_k$ be a stationary and ergodic sequence adapted to the filtration $(\mathcal{F}_k)$, $X_k \in L^2$, $\mathbb{E}(X_k) = 0$. Then there exists a stationary martingale difference array $(D_{n,i})$ (adapted to the filtration $(\mathcal{F}_k)_k$ for each n) such that for $M_{n,k} = \sum_{i=1}^{k} D_{n,i}$,*

$$\max_{1 \le k \le n} \|S_k - M_{n,k}\|_2 = o(\sigma_n)$$

if and only if $\|Q_0(S_n)\|_2 = o(\sigma_n)$.

In such a case

$$\sigma_n = \ell(n)\sqrt{n}$$

where $\ell(n)$ is a slowly varying function.

We shall show an analogue of Theorem 1 as well as analogs of other limit theorems for noncausal sequences.

Let us suppose that all random variables we work with belong to the Hilbert space H of $X \in L^2$ such that $\mathbb{E}(X|\mathcal{F}_{-\infty}) = 0$ and $\mathbb{E}(X|\mathcal{F}_\infty) = X$. This assumption implies $\mathbb{E}(X|\mathcal{I}) = 0$ (cf. e.g. [Vol87], Theorem 2). Let us denote

$$UX = X \circ T, \quad X \in H.$$

For every $X \in H$ and any $k \in \mathbf{Z}$, we have by (1) that $X = Q_k(X) + R_k(X)$. For a process $(X_k)_k$, the sequence of $Q_k(X_k)$ is adapted to the filtration $(\mathcal{F}_k)$. As we shall see the martingale approximations of the sequence $(R_k(X_k))_k$ can be studied in the same way as approximations of adapted processes.

Let H^- be the range of Q_0, H^{--} the range of Q_{-1}; $H^+ = H \ominus H^-$. Define

$$P_k X = \mathbb{E}(X \,|\, \mathcal{F}_k) - \mathbb{E}(X \,|\, \mathcal{F}_{k-1})$$
$$= R_{k-1}(X) - R_k(X) = Q_k(X) - Q_{k-1}(X), \quad k \in \mathbf{Z}.$$

Recall that for any σ-algebra $\mathcal{F} \subset \mathcal{A}$ and for $f \in L^2$ we have $U\mathbb{E}(f \mid \mathcal{F}) = \mathbb{E}(Uf \mid T^{-1}\mathcal{F})$, hence

$$UQ_k(f) = Q_{k+1}(Uf), \quad UR_k(f) = R_{k+1}U(f), \quad UP_i f = P_{i+1}Uf.$$

For $f \in H^+$ define

$$Vf = \sum_{i=1}^{\infty} U^{-i} P_0 U^{-i} f.$$

V is an isomorphism of the Hilbert space H^+ onto the Hilbert space H^{--}. To see this, it is sufficient to realise that for each $i > 0$ V is an isomorphism of $H_i = L^2(\mathcal{F}_i) \ominus L^2(\mathcal{F}_{i-1})$ onto H_{-i}: we have $UP_i = P_{i+1}U$ hence for $f \in H_i$, $Vf = U^{-2i}f$. H^+ is the direct sum of H_i, $i > 0$, while H^{--} is the direct sum of H_{-i}, $i > 0$ and U is a unitary operator.

Remark that one can easily extend the definition of V to the whole space H by defining

$$Vf = \sum_{i \in \mathbb{Z}} U^{-i} P_0 U^{-i} f.$$

Proposition 1. *For every $k \in \mathbb{Z}$, $n \geq 0$ and $f \in H^+$ we have*

$$VU^k f = U^{-k} Vf, \tag{2}$$

$$VR_n(f) = Q_{-n-1}(Vf). \tag{3}$$

Proof. Because $f = \sum_{i=1}^{\infty} P_i f$ we have $P_0 U^{k-i} f = 0$ for $i \leq k$ hence

$$VU^k f = \sum_{i=k+1}^{\infty} U^{-i} P_0 U^{k-i} f = \sum_{i=1}^{\infty} U^{-k} U^{-i} P_0 U^{-i} f = U^{-k} Vf,$$

which proves (2).

For any $f \in H$, $R_n f = \sum_{i=n+1}^{\infty} P_i f$ hence

$$VR_n(f) = \sum_{i=n+1}^{\infty} U^{-i} P_0 U^{-i} f = Q_{-n-1}(Vf),$$

which proves (3). $\qquad\square$

Corollary 1. *Let $f \in H^+$, $X_i = U^i f$, $i \in \mathbb{Z}$. Define $Z_i = U^i Vf$, $i \in \mathbb{Z}$, $S_n = \sum_{i=1}^{n} X_i$, $S'_n = \sum_{i=1}^{n} Z_i$, $n \geq 1$. Then the process (Z_i) is adapted to the filtration $(\mathcal{F}_{i-1})$ and*

i. $\mathbb{E}(S_n^2) = \mathbb{E}(S'^2_n)$,

ii. $\|R_{n-1}(S_n)\|_2 = \left\| Q_0 \left(\sum_{i=0}^{n-1} Z_i \right) \right\|_2$.

Proof. By (2) we have $\|S_n'\|_2 = \left\|\sum_{i=1}^n U^{-i} V f\right\|_2 = \left\|V\sum_{i=1}^n X_i\right\|_2 = \|S_n\|_2$. This proves (i).

By (3) and (2), $V R_{n-1}(S_n) = Q_{-n}(\sum_{k=1}^n V U^k f) = Q_{-n}(\sum_{i=1}^n Z_{-i})$ hence

$$\|R_{n-1}(S_n)\|_2 = \|V R_{n-1}(S_n)\|_2 = \left\|U^n Q_{-n}\left(\sum_{i=1}^n Z_{-i}\right)\right\|_2 = \left\|Q_0\left(\sum_{i=1}^n Z_{n-i}\right)\right\|_2 .$$

This proves (ii). $\square$

Theorem 2. *Let $(X_k = U^k X)_k$ be a stationary sequence,*
  $X \in L^2(\mathcal{F}_\infty) \ominus L^2(\mathcal{F}_{-\infty})$, $\sigma_n = \|S_n\|_2 \to \infty$. *The following are equivalent:*

a. $\|Q_0(S_n)\|_2 = o(\sigma_n)$, $\|R_n(S_n)\|_2 = o(\sigma_n)$;
b. there exists a stationary martingale difference array $(D_{n,i})$ (adapted to the filtration $(\mathcal{F}_k)_k$) such that for $M_{n,k} = \sum_{i=1}^k D_{n,i}$,

$$\max_{1 \le k \le n} \|S_k - M_{n,k}\|_2 = o(\sigma_n) ;$$

c. there exists a stationary adapted sequence $(Y_k = U^k Y)_k$ such that $\|S_n(X - Y)\|_2 = o(\sigma_n)$ and $\|Q_0(S_n(Y))\|_2 = o(\sigma_n)$.

If (a) or (b) or (c) holds then $\sigma_n = \ell(n)\sqrt{n}$ where $\ell(n)$ is a slowly varying function.

Proof. Let us suppose (a). Notice that under this assumption

$$\|Q_1(S_n)\|_2 = o(\sigma_n) . \tag{4}$$

To prove this let $\epsilon > 0$. Then there exists N such that $\|Q_0(S_k)\|_2 < \epsilon \sigma_k$ for all $k \ge N$. Note that $Q_1(S_n) = Q_1(X_1) + Q_1(S_n - X_1)$. By stationarity $\|Q_1(S_n - X_1)\|_2 = \|Q_0(S_{n-1})\|_2$ from which it follows $\|Q_1(S_n)\|_2 \le \|Q_0(S_{n-1})\|_2 + \|X_1\|_2$. $\|Q_0(S_n)\|_2 = o(\sigma_n)$ and $\sigma_n \to \infty$ hence $\|Q_1(S_k)\|_2 < \epsilon \sigma_{k-1} + \|X_1\|_2 < 2\epsilon \sigma_{k-1}$ for all k sufficiently big. Because $\sigma_n \to \infty$ and $|\sigma_n - \sigma_{n-1}| \le \|X_1\|_2$ we have $\sigma_n / \sigma_{n-1} \to 1$. This finishes the proof of (4).

We shall prove that $\sigma_n = \ell(n)\sqrt{n}$ where $\ell(n)$ is a slowly varying function. For positive integers n, m we have

$$S_n = Q_0(S_n) + (Q_n - Q_0)(S_n) + R_n(S_n) ,$$
$$S_{n+m} - S_n = Q_n(S_{n+m} - S_n) + (Q_{n+m} - Q_n)(S_{n+m} - S_n)$$
$$+ R_{n+m}(S_{n+m} - S_n) ,$$

hence by orthogonality

$$\mathbb{E}[S_n(S_{n+m} - S_n)] = \mathbb{E}[Q_0(S_n)Q_n(S_{n+m} - S_n)]$$
$$+ \mathbb{E}[(Q_n - Q_0)(S_n)Q_n(S_{n+m} - S_n)]$$
$$+ \mathbb{E}[R_n(S_n)(Q_{n+m} - Q_n)(S_{n+m} - S_n)]$$
$$+ \mathbb{E}[R_n(S_n)R_{n+m}(S_{n+m} - S_n)] .$$

From the Schwartz inequality and (a) it follows

$$|\mathbb{E}[S_n(S_{n+m} - S_n)]| \leq o(\sigma_n)o(\sigma_m) + \sigma_n o(\sigma_m) + o(\sigma_n)\sigma_m + o(\sigma_n)o(\sigma_m) .$$

By the remark in [IL71], p.330 (cf. [WW04], proof of Lemma 1) it follows that $\sigma_n^2 = \ell'(n)n$ where $\ell'(n)$ is a slowly varying function. This proves that $\sigma_n = \ell(n)\sqrt{n}$ where $\ell(n)$ is a slowly varying function.

By [Res87], Theorem 06, we have $\sum_{k=1}^{n} k^{1/2}\ell(k) \approx (2/3)n^{3/2}\ell(n)$ hence

$$\sup_n \frac{1}{n\sigma_n} \sum_{k=1}^{n} \sigma_k < \infty. \tag{5}$$

We define

$$G_{n,k} = G_{n,k}(X) = \frac{1}{n} \sum_{j=1}^{n} \sum_{i=0}^{j-1} Q_k(X_{k+i}) ,$$

$$H_{n,k} = H_{n,k}(X) = \frac{1}{n} \sum_{j=1}^{n} \sum_{i=0}^{j-1} R_k(X_{k-i}) .$$

Then

$$Q_k(X_k) = G_{n,k} - Q_k(G_{n,k+1}) + \frac{1}{n} \sum_{j=1}^{n} Q_k(X_{k+j}) ,$$

$$R_k(X_k) = H_{n,k} - R_k(H_{n,k-1}) + \frac{1}{n} \sum_{j=1}^{n} R_k(X_{k-j}) .$$

Denote

$$D_{n,k}^{(1)} = D_{n,k}^{(1)}(X) = G_{n,k} - Q_{k-1}(G_{n,k}) ,$$

$$D_{n,k}^{(2)} = D_{n,k}^{(2)}(X) = H_{n,k-1} - R_k(H_{n,k-1}) ,$$

$$D_{n,k} = D_{n,k}(X) = D_{n,k}^{(1)} + D_{n,k}^{(2)} .$$

Then

$$D_{n,k}^{(1)} = \frac{1}{n} \sum_{j=1}^{n} \sum_{i=0}^{j-1} P_k X_{k+i} = \sum_{i=0}^{n-1} \frac{n-i}{n} P_k X_{k+i} ,$$

$$D_{n,k}^{(2)} = \frac{1}{n} \sum_{j=1}^{n} \sum_{i=0}^{j-1} P_k X_{k-i-1} = \sum_{i=1}^{n} \frac{n-i+1}{n} P_k X_{k-i} ,$$

$$\sum_{k=1}^{n} Q_k(X_k) = \sum_{k=2}^{n} D_{n,k}^{(1)} + G_{n,1} - Q_n(G_{n,n+1}) + \frac{1}{n} \sum_{k=1}^{n} \sum_{j=1}^{n} Q_k(X_{k+j}) ,$$

$$\sum_{k=1}^{n} R_k(X_k) = \sum_{k=2}^{n} D_{n,k}^{(2)} + H_{n,n} - R_1(H_{n,0}) + \frac{1}{n} \sum_{k=1}^{n} \sum_{j=1}^{n} R_k(X_{k-j}) ,$$

$$\tag{6}$$

therefore

$$\sum_{k=1}^{n} X_k = \sum_{k=2}^{n} D_{n,k} + G_{n,1} - Q_n(G_{n,n+1})$$

$$+ \frac{1}{n} \sum_{k=1}^{n} \sum_{j=1}^{n} Q_k(X_{k+j}) + H_{n,n} - R_1(H_{n,0}) + \frac{1}{n} \sum_{k=1}^{n} \sum_{j=1}^{n} R_k(X_{k-j}) .$$

It remains to be proved that all the terms $D_{n,1}$, $G_{n,1}$, $G_{n,n+1}$, $H_{n,n}$, $H_{n,0}$, $\frac{1}{n} \sum_{k=1}^{n} \sum_{j=1}^{n} Q_k(X_{k+j})$, $\frac{1}{n} \sum_{k=1}^{n} \sum_{j=1}^{n} R_k(X_{k-j})$ are $o(\sigma_n)$. Let us show this for $G_{n,1}$; the other proofs are similar.

By (4), for each $\epsilon > 0$ there exists N such that $\|Q_1(S_k)\|_2 < \epsilon \sigma_k$ for all $k \geq N$. For $n \geq N$,

$$\|G_{n,1}\|_2 \leq \frac{1}{n} \sum_{k=1}^{N-1} \|Q_1(S_k)\|_2 + \frac{1}{n} \sum_{k=N}^{n} \|Q_1(S_k)\|_2 \leq \frac{1}{n} \sum_{k=1}^{N-1} \sigma_k + \frac{\epsilon}{n} \sum_{k=N}^{n} \sigma_k .$$

We have

$$\frac{1}{n} \sum_{k=N}^{n} \sigma_k \leq \sigma_n \sup_l (1/(l\sigma_l)) \sum_{k=1}^{l} \sigma_k .$$

By (5), $\sup_l (1/(l\sigma_l)) \sum_{k=1}^{l} \sigma_k < \infty$ and for N fixed, $n \to \infty$, the term $\frac{1}{n} \sum_{k=1}^{N-1} \sigma_k$ is going to zero.

We have proved that $\max_{1 \leq k \leq n} \|S_k - M_{n,k}\|_2 = o(\sigma_n)$. Notice that we actually proved

$$\max_{1 \leq k \leq n} \| \sum_{j=1}^{k} Q_k(X_k) - \sum_{j=1}^{k} D_{n,j}^{(1)} \|_2 = o(\sigma_n) ,$$

$$\max_{1 \leq k \leq n} \| \sum_{j=1}^{k} R_k(X_k) - \sum_{j=1}^{k} D_{n,j}^{(2)} \|_2 = o(\sigma_n) . \tag{7}$$

The proof that (b) implies $\|Q_0(S_n)\|_2 = o(\sigma_n)$, $\|R_n(S_n)\|_2 = o(\sigma_n)$ follows from $Q_0(M_{n,n}) = 0 = R_n(M_{n,n})$.

It remains to prove that (a), (b) are equivalent to (c). Let us suppose (a), (b). We have $X = Q_0(X_0) + R_0(X_0) = X' + X''$; define

$$Y = Q_0(X_0) + V R_0(X_0) = Y' + Y'' , \quad Y_k = U^k Y = Y_k' + Y_k'' , \quad k \in \mathbb{Z} .$$

We shall denote $S_n(Y) = \sum_{i=1}^{n} Y_i$. Recall that by (1)

$$Q_k Q_{k-1}(X) = Q_{k-1} Q_k(X) = Q_{k-1}(X) ,$$
$$R_k R_{k-1}(X) = R_{k-1} R_k(X) = R_k(X)$$

and notice that

$$X'_k = Y'_k = U^k Q_0(X_0) = Q_k(X_k) \ .$$

Therefore $Q_0(\sum_{i=1}^n Y'_i) = Q_0(\sum_{i=1}^n Q_i(X_i)) = Q_0(\sum_{i=1}^n X_i) = Q_0(S_n)$ hence

$$\|Q_0(S_n(Y'))\|_2 = o(\sigma_n) \ .$$

To show

$$\|Q_0(S_n(Y))\|_2 = o(\sigma_n) \tag{8}$$

it remains to prove

$$\|Q_0(S_n(Y''))\|_2 = o(\sigma_n) \ . \tag{9}$$

We have

$$Q_0(S_n(Y'')) = Q_0(Y''_n) - Q_0(Y''_0) + Q_0\Big(\sum_{i=0}^{n-1} Y''_i\Big)$$

and by Corollary 1(ii) $\|Q_0(\sum_{i=0}^{n-1} Y''_i)\|_2 = \|R_{n-1}(S_n(X''))\|_2$ hence

$$\big| \|R_{n-1}(S_n(X''))\|_2 - \|Q_0(S_n(Y''))\|_2 \big| \leq 2\|X_0\|_2.$$

$R_{n-1}(S_n(X'')) = R_{n-1}(U^n X'') + R_{n-1}(S_{n-1}(X''))$, therefore

$$\big| \|R_{n-1}(S_n(X''))\|_2 - \|R_{n-1}(S_{n-1}(X''))\|_2 \big| \leq \|X''_0\|_2 \leq \|X_0\|_2.$$

From $X''_k = U^k R_0(X) = R_k(X_k)$ it follows
$\|R_{n-1}(S_{n-1}(X''))\|_2 = \|R_{n-1}(S_{n-1}(X))\|_2$. We thus have

$$\big| \|Q_0(S_n(Y''))\|_2 - \|R_{n-1}(S_{n-1}(X))\|_2 \big| \leq 3\|X_0\|_2.$$

Because $|\sigma_n - \sigma_{n-1}| \leq \|X_0\|_2$, using (a), we deduce

$$\|Q_0(S_n(Y''))\|_2 = o(\sigma_n). \tag{10}$$

Let us show $\|S_n(X - Y)\|_2 = o(\sigma_n)$.
We denote

$$D_{n,k}(Y') = G_{n,k}(Y') - Q_{k-1} G_{n,k}(Y') = \sum_{i=0}^{n-1} \frac{n-i}{n} P_k Y'_{k+i},$$

$$D_{n,k}(Y'') = G_{n,k}(Y'') - Q_{k-1} G_{n,k}(Y'') = \sum_{i=0}^{n-1} \frac{n-i}{n} P_k Y''_{k+i},$$

$$D_{n,k}(Y) = D_{n,k}(Y') + D_{n,k}(Y'').$$

We have $Y = \sum_{i=0}^{\infty} P_{-i} Y$. For $i \geq 0$, $P_0 Y'_i = P_0 X_i$ and $P_0 Y''_0 = 0$. For $i \geq 1$, $P_0 Y''_i = P_0 U^i \sum_{j=1}^{\infty} U^{-j} P_0 U^{-j} X = P_0 \sum_{j=1}^{\infty} P_{i-j} U^{i-2j} X = P_0 U^{-i} X$. Using stationarity we deduce

$$\sum_{i=0}^{n-1} \frac{n-i}{n} P_k Y'_{k+i} = \sum_{i=0}^{n-1} \frac{n-i}{n} P_k X_{k+i},$$

$$\sum_{i=0}^{n-1} \frac{n-i}{n} P_k Y''_{k+i} = \sum_{i=1}^{n-1} \frac{n-i}{n} P_k X_{k-i},$$

therefore

$$D_{n,k} - D_{n,k}(Y) = \frac{1}{n} \sum_{i=1}^{n} P_k X_{k-i} = \frac{1}{n} R_{k-1}\Big(\sum_{i=1}^{n} X_{k-i}\Big) - \frac{1}{n} R_k\Big(\sum_{i=1}^{n} X_{k-i}\Big),$$

$1 \le k \le n$. By (a) $\|D_{n,k} - D_{n,k}(Y)\|_2 = o(\sigma_n/n)$, hence $\|M_{n,n} - M_{n,n}(Y)\|_2 = o(\sigma_n)$.

Next we prove
$$\|S_n(Y) - M_{n,n}(Y)\|_2 = o(\sigma_n).$$

By (7), $\|S_n(Y') - M_{n,n}(Y')\|_2 = \|S_n(X') - M_{n,n}(X')\|_2 = o(\sigma_n)$ where $M_{n,n}(X') = \sum_{k=1}^{n} D_{n,k}^{(1)}$. By (6),

$$\| S_n(Y'') - M_{n,n}(Y'')\|_2 \le$$

$$\| D_{n,1}(Y'')\|_2 + \|G_{n,1}(Y'')\|_2 - \|Q_n(G_{n,n+1}(Y''))\|_2 + \|\frac{1}{n} \sum_{k=1}^{n} \sum_{j=1}^{n} Q_k(Y''_{k+j})\|_2.$$

We show $\|G_{n,1}(Y'')\|_2 = o(\sigma_n)$; for the other terms on the right it can be done in the same manner.

$$\|G_{n,1}(Y'')\|_2 \le \frac{1}{n} \sum_{k=1}^{n} \|Q_1(S_k(Y''))\|_2.$$

From (4) and (8) it follows $\|Q_1(S_k(Y''))\|_2 = o(\sigma_k)$ hence we can prove $\|G_{n,1}(Y'')\|_2 = o(\sigma_n)$ in the same way as we proved $\|G_{n,1}^{(1)}\|_2 = o(\sigma_n)$.

Therefore,

$$\|S_n(Y) - S_n(X)\|_2$$
$$\le \|S_n(Y) - M_{n,n}(Y)\|_2 + \|M_{n,n}(Y) - M_{n,n}\|_2 + \|M_{n,n} - S_n(X)\|_2 = o(\sigma_n).$$

This finishes the proof of (c).

Eventually we prove that (c) implies (a), (b). Suppose (c). We get

$$\|Q_0(S_n(Y))\|_2 = o(\|(S_n(Y))\|_2)$$

so that (a) is satisfied for Y. By the equivalence of (a) and (b) we get (b) for Y and from $\|S_n(X - Y)\|_2 = o(\sigma_n)$ it follows (b) for X. Therefore, (a) holds for X as well. $\square$

Proposition 2. *In Theorem 2, we can take the following martingale approximation:*

$$M_{n,k} = \sum_{j=1}^{k} D_{n,j}$$

where

$$D_{n,k} = \sum_{j=0}^{\infty} \left(\frac{n}{n+1}\right)^{j+1} P_k X_{k+j} + \sum_{j=0}^{\infty} \left(\frac{n}{n+1}\right)^{j+1} P_k X_{k-1-j}.$$

Proof. Define

$$Y_{n,k}^+ = \sum_{j=0}^{\infty} \left(\frac{n}{n+1}\right)^{j+1} \mathbb{E}(X_{k+j} \mid \mathcal{F}_k),$$

$$Y_{n,k}^- = \sum_{j=0}^{\infty} \left(\frac{n}{n+1}\right)^{j+1} [X_{k-j} - \mathbb{E}(X_{k-j} \mid \mathcal{F}_k)],$$

$$Y_{n,k} = Y_{n,k}^+ + Y_{n,k}^-.$$

We then have

$$\left(1 + \frac{1}{n}\right) Y_{n,k}^+ = \mathbb{E}(X_k \mid \mathcal{F}_k) + Q_k(Y_{n,k+1}^+)$$

and

$$\left(1 + \frac{1}{n}\right) Y_{n,k}^- = [X_k - \mathbb{E}(X_k \mid \mathcal{F}_k)] + R_k(Y_{n,k-1}^-)$$

hence

$$\left(1 + \frac{1}{n}\right) Y_{n,k} = X_k + Q_k(Y_{n,k+1}^+) + R_k(Y_{n,k-1}^-),$$

$$\sum_{k=1}^{n} X_k = \sum_{k=1}^{n} [Y_{n,k} - Q_k(Y_{n,k+1}^+) - R_k(Y_{n,k-1}^-)] + \frac{1}{n} \sum_{k=1}^{n} Y_{n,k}.$$

We have

$$\sum_{k=1}^{n} [Y_{n,k} - Q_k(Y_{n,k+1}^+) - R_k(Y_{n,k-1}^-)]$$

$$= \sum_{k=1}^{n} [Y_{n,k}^+ - Q_k(Y_{n,k+1}^+)] + \sum_{k=1}^{n} [Y_{n,k}^- - R_k(Y_{n,k-1}^-)]$$

$$= \sum_{k=2}^{n} [Y_{n,k}^+ - Q_{k-1}(Y_{n,k}^+)] + Y_{n,1}^+ - Q_n(Y_{n,n+1}^+)$$

$$+ \sum_{k=2}^{n} [Y_{n,k-1}^- - R_k(Y_{n,k-1}^-)] + Y_{n,n}^- - R_1(Y_{n,0}^-).$$

Notice that $Q_k(Y_{n,k}^+) = Y_{n,k}^+$, $R_k(Y_{n,k}^-) = Y_{n,k}^-$. By (1) we thus get

$$Y_{n,k}^+ - Q_{k-1}(Y_{n,k}^+) = \sum_{j=0}^{\infty} \left(1 + \frac{1}{n}\right)^{-(j+1)} P_k X_{k+j}$$

and

$$Y_{n,k-1}^- - R_k(Y_{n,k-1}^-) = \sum_{j=0}^{\infty} \left(1 + \frac{1}{n}\right)^{-(j+1)} P_k X_{k-1-j},$$

hence for

$$D_{n,k} = \sum_{j=0}^{\infty} \left(\frac{n}{n+1}\right)^{j+1} P_k(X_{k+j}) + \sum_{j=0}^{\infty} \left(\frac{n}{n+1}\right)^{j+1} P_k(X_{k-1-j})$$

it is

$$\sum_{k=1}^{n} X_k = \sum_{k=2}^{n} D_{n,k} + Y_{n,1}^+ - Q_n(Y_{n,n+1}^+) + Y_{n,n}^- - R_1(Y_{n,0}^-) + \frac{1}{n} \sum_{k=1}^{n} Y_{n,k}.$$

In the same way as in the proof of Corollary 1 in [WW04] we prove that $\|Y_{n,k}\|_2 = o(\sigma_n)$. $\qquad\square$

The optimal approximation was studied in [Rüs85]. In [KV05] it is shown that for $\sigma_n' = \|\sum_{k=1}^{n} X_k'\|_2$ and $\sigma_n'' = \|\sum_{k=1}^{n} X_k''\|_2$ we can have $\limsup_{n\to\infty} \sigma_n'/\sigma_n'' = \infty$, $\liminf_{n\to\infty} \sigma_n'/\sigma_n'' = 0$, $\lim_{n\to\infty} \sigma_n/\sigma_n' = 0 = \lim_{n\to\infty} \sigma_n/\sigma_n''$.

The approximation from Theorems 1, 2 alone does not imply the CLT. For the CLT it is needed also a limit theorem for the triangular array $(D_{n,i})$ ([WW04]). On the other hand, it is possible to generalize to nonadapted processes the CLT of Maxwell and Woodroofe ([MW00], Theorem 1):

Theorem 3. *Let*

$$\sum_{n=1}^{\infty} n^{-3/2}\|Q_0(S_n)\|_2 < \infty, \quad \sum_{n=1}^{\infty} n^{-3/2}\|R_n(S_n)\|_2 < \infty.$$

Then there exists a martingale (M_n) with strictly stationary increments such that $\|S_n - M_n\|_2 = o(\sqrt{n})$ as $n \to \infty$.

Proof. We use the representation $X_k = X_k' + X_k''$ where $X_k' = Q_k(X_k) = U^k Q_0(X_0)$ and $X_k'' = R_k(X_k) = U^k R_0(X_0)$, $k \in \mathbb{Z}$. We denote $S_n(X') = \sum_{i=1}^{n} X_i'$, $S_n(X'') = \sum_{i=1}^{n} X_i''$. The sequence (X_k') satisfies the assumption $\sum_{n=1}^{\infty} n^{-3/2}\|Q_0(S_n(X'))\|_2 < \infty$ of [MW00], Theorem 1 (cf. [PI05], Theorem 1, for the nonergodic version), hence there exists a martingale (M_n') such that $\|M_n' - \sum_{i=1}^{n} X_i'\|_2 = o(\sqrt{n})$.

As in Corollary 1(ii) we define $Z_k = U^k V X_0''$. The process (Z_k) is adapted to the filtration $(\mathcal{F}_k)$. By Corollary 1,

$$\left\|Q_0\left(\sum_{k=0}^{n-1} Z_k\right)\right\|_2 = \|R_{n-1}(\sum_{k=1}^{n} X_k'')\|_2$$
$$= \|R_{n-1}(S_n) - \mathbb{E}(X_n \mid \mathcal{F}_n) + \mathbb{E}(X_n \mid \mathcal{F}_{n-1})\|_2$$
$$\leq \|R_{n-1}(S_n)\|_2 + 2\|X_0\|_2 .$$

From this, $\|R_n(S_n) - R_{n-1}(S_n)\|_2 \leq \|R_n(S_n) - R_{n-1}(S_{n-1})\|_2 + \|R_{n-1}(S_{n-1}) - R_{n-1}(S_n)\|_2$ and the assumption $\sum_{n=1}^{\infty} n^{-3/2}\|R_n(S_n)\|_2 < \infty$ it follows that

$$\sum_{n=1}^{\infty} n^{-3/2}\left\|Q_0\left(\sum_{k=0}^{n-1} Z_k\right)\right\|_2 < \infty .$$

Hence [MW00], Theorem 1 ([PI05], Theorem 1 in the nonergodic case), can be applied again and there exist martingale differences D_k such that $D_{k+1} = UD_k$, $P_k D_k = D_k$ and $\left\|\sum_{i=1}^{n}(Z_i - D_i)\right\|_2 = o(\sqrt{n})$. For $k \geq 1$ we have $V^{-1}Z_{-k} = X_k''$ (by (2), $VX_k'' = VR_k(X_k) = VU^k X_0'' = U^{-k}VX_0'' = Z_{-k}$) and $V^{-1}D_{-k} = U^{2k}U^{-k}D_0 = D_k$ (recall that for $X = P_k X$, $VX = U^{-2k}X$) hence

$$\left\|\sum_{k=1}^{n}(X_k'' - D_k)\right\|_2 = \left\|\sum_{k=1}^{n}(Z_{-k} - D_{-k})\right\|_2 = o(\sqrt{n}) .$$

$M_n = M_n' + \sum_{i=1}^{n} D_i$ give the martingale approximation. □

Using different methods, a result similar to that of Maxwell and Woodroofe was proved by Derriennic and Lin [DL01]. For adapted sequences, Peligrad and Utev [PI05] proved under the same assumptions as Maxwell and Woodroofe the Donsker invariance principle. In [Wu05] the strong laws (strong laws of large numbers, laws of the iterated logarithm, strong invariance principles) are studied.

3 Successive martingale-coboundary representations

A classical result ([Gor69], Theorem 1, [HH80]) establishes the CLT for stationary sequences $(f \circ T^i)$ for which there exists a certain class G of functions h such that $(1/\sqrt{n}) \limsup_{n \to \infty} \|S_n(f - h)\|_2 = 0$; for each $h \in G$ there exist $g, m \in L^2$ such that $h = m + g - g \circ T$ and $(m \circ T^i)$ is a martingale difference sequence. All the martingale difference sequences have the same filtration. In [Vol93] it is shown that Gordin's condition is equivalent to the existence of a single square integrable function m such that $(m \circ T^i)$ is a martingale difference sequence and $\limsup_{n \to \infty} \|S_n(f - m)\|_2 = 0$. The existence of the decomposition $f = m + g - g \circ T$ with $m, g \in L^2$ and $(m \circ T^i)$ a martingale difference sequence implies the Gordin's condition; the converse, however, is not true ([Vol93]).

Let us show a more general case. Let $\mathcal{M} \subset T^{-1}\mathcal{M}$ be a sub-σ-algebra of $\mathcal{A}$, $P_i f = \mathbb{E}(f|T^{-i-1}\mathcal{M}) - \mathbb{E}(f|T^{-i}\mathcal{M})$. For each n let $0 \leq k_1(n) \leq k_2(n) < \infty$,

$$f_n = \sum_{i=-k_1(n)}^{k_2(n)} P_i f , \quad m_n = \sum_{i=-k_1(n)}^{k_2(n)} P_0 U^i f .$$

Then,

$$f_n = m_n + g_n - U g_n$$

where

$$g_n = \sum_{i=1}^{k_1(n)} \sum_{k=0}^{k_1(n)-i} P_{-i} U^k f_n - \sum_{i=0}^{k_2(n)} \sum_{k=1}^{k_2(n)-i} P_i U^{-k} f_n .$$

As an immediate consequence of [Gor69], Theorem 1, we get:

Theorem 4. *If*

i.

$$\lim_{n\to\infty} \frac{1}{n\|m_n\|_2^2} \|S_n(f - f_n)\|_2^2 = 0 ,$$

ii.

$$\lim_{n\to\infty} \frac{\|g_n\|_2^2}{n\|m_n\|_2^2} = 0 ,$$

iii.

$$\frac{1}{\|m_n\|_2\sqrt{n}} S_n(m_n) \xrightarrow[\mathcal{D}]{} \nu ,$$

then the distributions of $(1/\|m_n\|_2\sqrt{n})S_n(f)$ *weakly converge to the law* ν.

Proof. From (i) and (ii) it follows that

$$\lim_{n\to\infty} \frac{1}{\sqrt{n}\|m_n\|_2} \|S_n(f - m_n)\|_2 = 0$$

hence the weak limit of the distributions of $(1/\sqrt{n}\|m_n\|_2)S_n(f)$ is the same as the weak limit of the distributions of $(1/\sqrt{n}\|m_n\|_2)S_n(m_n)$. By (iii) this limit equals ν.   □

Using the same idea as in Theorem 4 we can prove the central limit theorem for linear processes with divergent series of a_n.

Let (Z_i) be a martingale difference sequence with $Z_i \in L^2$, (a_i) a sequence of random variables, independent of (Z_i), such that $\sum_{i\in\mathbb{Z}} \|a_i\|_2^2 < \infty$. We define

$$X_k = \sum_{i\in\mathbb{Z}} a_{k-i} Z_i , \quad k \in \mathbb{Z} .$$

Theorem 5. *Let* (Z_i) *be a stationary martingale difference sequence with* $\|Z_i\|_2 = 1$ *for all* $i \in \mathbb{Z}$, $\sum_{i\in\mathbb{Z}} \|a_i\|_2^2 < \infty$ *and*

$$X_k = \sum_{i\in\mathbb{Z}} a_{k-i} Z_i , \quad Y_{n,k} = \sum_{i=-n+1}^{n-1} a_i Z_k , \quad k \in \mathbb{Z} , \quad n = 1, 2, \ldots$$

$$s_n = \left\| \sum_{i=0}^{n-1} a_i \right\|_2, \quad s_{-n} = \left\| \sum_{i=-n+1}^{0} a_i \right\|_2, \quad s(n) = \left\| \sum_{i=-n+1}^{n-1} a_i \right\|_2 .$$

Let

i.

$$\sup_{n \geq 0} \max_{0 \leq k \leq n} \left\| \sum_{i=k}^{n-1} a_i \right\|_2 / s_n < \infty, \quad \sup_{n \geq 0} \max_{0 \leq k \leq n} \left\| \sum_{i=k}^{n-1} a_{-i} \right\|_2 / s_{-n} < \infty ,$$

ii. for every $\epsilon > 0$,

$$\max_{\epsilon n \leq k \leq n} \left\| \sum_{i=k}^{n-1} a_i \right\|_2 / s_n \to 0, \quad \max_{\epsilon n \leq k \leq n} \left\| \sum_{i=k}^{n-1} a_{-i} \right\|_2 / s_{-n} \to 0 ,$$

iii.

$$\frac{n}{s_n^2} \sum_{i=n}^{\infty} \|a_i\|_2^2 \to 0, \quad \frac{n}{s_{-n}^2} \sum_{i=n}^{\infty} \|a_{-i}\|_2^2 \to 0 .$$

If $\liminf_{n\to\infty} s(n) > 0$, $\liminf_{n\to\infty} s(n)/s_n > 0$, $\liminf_{n\to\infty} s(n)/s_{-n} > 0$, *then*

$$\frac{1}{s(n)\sqrt{n}} \left\| \sum_{k=0}^{n-1} (X_k - Y_{n,k}) \right\|_2 \to 0 \quad as \quad n \to \infty .$$

If $n^{-1/2} \sum_{k=0}^{n-1} Z_k$ *converge in distribution to* $N(0,1)$ *and* $s(n)^{-1} \sum_{i=-n+1}^{n-1} a_i$ *converge in distribution to a random variable* η, *then* $s(n)^{-1} n^{-1/2} \sum_{k=0}^{n-1} Y_{n,k}$ *converge in distribution to a law with the characteristic function* $\varphi(t) = \mathbb{E}[\exp(-\frac{1}{2}\eta^2 t^2)]$.

Proof. For $l < k$, define $\sum_{j=k}^{l} a_j = 0$. We have

$$\sum_{k=0}^{n-1} (X_k - Y_{n,k}) = \sum_{k=0}^{n-1} \left(\sum_{i=-\infty}^{\infty} a_{k-i} Z_i - \sum_{i=-n+1}^{n-1} a_i Z_k \right)$$

$$= \sum_{i=-\infty}^{-1} Z_i \sum_{k=0}^{n-1} a_{k-i} + \sum_{i=n}^{\infty} Z_i \sum_{k=0}^{n-1} a_{k-i} + \sum_{i=0}^{n-1} Z_i \left(\sum_{k=0}^{n-1} a_{k-i} - \sum_{j=-n+1}^{n-1} a_j \right)$$

$$= \sum_{i=-\infty}^{-1} Z_i \sum_{k=0}^{n-1} a_{k-i} + \sum_{i=n}^{\infty} Z_i \sum_{k=0}^{n-1} a_{k-i} - \sum_{i=0}^{n-1} Z_i \left(\sum_{j=i+1}^{n-1} a_{-j} + \sum_{j=n-i}^{n-1} a_j \right) .$$

By mutual orthogonality of Z_i we thus get

$$\left\| \sum_{k=0}^{n-1} (X_k - Y_{n,k}) \right\|_2^2 = \sum_{i=-\infty}^{-1} \left\| \sum_{k=0}^{n-1} a_{k-i} \right\|_2^2$$

$$+ \sum_{i=n}^{\infty} \left\| \sum_{k=0}^{n-1} a_{k-i} \right\|_2^2 + \sum_{i=0}^{n-1} \left\| \sum_{j=i+1}^{n-1} a_{-j} \right\|_2^2 + \sum_{i=0}^{n-1} \left\| \sum_{j=n-i}^{n-1} a_j \right\|_2^2 .$$

By (i) and (ii) we have

$$\frac{1}{ns_{-n}^2} \sum_{i=0}^{n-1} \left\| \sum_{j=i+1}^{n-1} a_{-j} \right\|_2^2 \to 0 \,, \quad \frac{1}{ns_n^2} \sum_{i=0}^{n-1} \left\| \sum_{j=n-i}^{n-1} a_j \right\|_2^2 \to 0 \quad \text{as} \quad n \to \infty \,.$$

It remains to show

$$\frac{1}{ns_n^2} \sum_{i=-\infty}^{-1} \left\| \sum_{k=0}^{n-1} a_{k-i} \right\|_2^2 \to 0 \,, \quad \frac{1}{ns_{-n}^2} \sum_{i=n}^{\infty} \left\| \sum_{k=0}^{n-1} a_{k-i} \right\|_2^2 \to 0 \quad \text{as} \quad n \to \infty \,.$$

We shall prove the first convergence, the other one follows from the same idea by symmetry.

$$\sum_{i=-\infty}^{-1} \left\| \sum_{k=0}^{n-1} a_{k-i} \right\|_2^2 = \sum_{i=1}^{\infty} \left\| \sum_{j=0}^{n-1} a_{i+j} \right\|_2^2 = \sum_{i=1}^{n-1} \left\| \sum_{j=0}^{n-1} a_{i+j} \right\|_2^2 + \sum_{i=n}^{\infty} \left\| \sum_{j=0}^{n-1} a_{i+j} \right\|_2^2 \,.$$

By (i) and (ii)

$$\frac{1}{ns_n^2} \sum_{i=1}^{n-1} \left\| \sum_{j=0}^{n-1} a_{i+j} \right\|_2^2 \to 0 \quad \text{as} \quad n \to \infty \,.$$

By the Cauchy-Schwarz inequality $\sum_{i=n}^{\infty} \left\| \sum_{j=0}^{n-1} a_{i+j} \right\|_2^2 \le n^2 \sum_{i=n}^{\infty} \|a_i\|_2^2$ and using (iii) we deduce

$$\frac{1}{ns_n^2} \sum_{i=n}^{\infty} \left\| \sum_{j=0}^{n-1} a_{i+j} \right\|_2^2 \le \frac{n}{s_n^2} \sum_{i=n}^{\infty} \|a_i\|_2^2 \to 0 \quad \text{as} \quad n \to \infty \,.$$

$\square$

A direct calculation shows that the assumptions of Theorem 5 are verified for the following linear process with nonrandom coefficients a_i.

Corollary 2. *Let (Z_i) be a stationary martingale difference sequence with $\|Z_i\|_2 = 1$ for all $i \in \mathbb{Z}$, $\sum_{i \in \mathbb{Z}} a_i^2 < \infty$ and*

$$X_k = \sum_{i \in \mathbb{Z}} a_{k-i} Z_i \,, \quad Y_{n,k} = \sum_{i=-n+1}^{n-1} a_i Z_k \,, \quad k \in \mathbb{Z} \ n = 1, 2, \dots$$

where $a_n = \log^a |n|/|n|$ for $|n| \ge 1$, $a < 1$. If $\frac{1}{\sqrt{n}} \sum_{k=0}^{n-1} Z_k$ converge in distribution to $N(0,1)$ then $\frac{1}{s(n)\sqrt{n}} \sum_{k=0}^{n-1} Y_{n,k}$, $s_n = \sum_{i=0}^{n-1} a_i$, converge in distribution to $N(0,1)$ as well.

For $a_k = k^{-\beta}$ with $1/2 < \beta < 1$ the assumptions of Theorem 5 are not satisfied and as noticed in [WW04], the approximation in the sense of Theorem 2 does not exist either; nevertheless, S_n/σ_n converge to a normal law. The results for stationary linear processes presented here partially overlap with those from [WW04] and [Yok95].

4 Concluding remarks

The central limit theorems proved by martingale approximation have their conditional versions. Let $(\mathcal{F}_k)_k$ be a given filtration, Δ a suitable distance between probability distributions (for example the Lévy distance). Let ν be a probability law and $\nu_{n,\omega}$ be the conditional laws of S_n/σ_n given $\mathcal{F}_0$. We say that the sequence (S_n/σ_n) converges to ν conditionally given $\mathcal{F}_0$ if

$$\lim_{n\to\infty} \int \Delta(\nu, \nu_{n,\omega})\,\mu(d\omega) = 0 \ .$$

The conditional convergence was studied by Dedecker and Merlevède [DM02]. It implies the weak convergence but in general, the implication does not hold vice versa. Dedecker and Merlevède showed, for example ([DM02], Proposition 1), that for adapted sequences satisfying the assumptions of [Gor69], Theorem 1, the convergence is conditional. Wu and Woodroofe proved ([WW04]) that for a sequence (X_k) adapted to the filtration, S_n/σ_n converge conditionally to the standard normal law given $\mathcal{F}_0$ if and only if the approximation in the sense of Theorem 2 exists and the approximating array of martingale differences satisfies conditions of Gänssler and Häeusler ([GH79], Theorem 2, see also conditions (11) and (12) in [WW04]).

In [OV05] (cf. also [Vol88]) it is shown that in the case of a nonergodic measure μ the conditional convergence can take place while for the ergodic components the sequence of laws of S_n/σ_n does not weakly converge (has different limit points). For several central limit theorems this situation cannot happen, e.g. for [MW00], Theorem 1.

The existence of the martingale approximation depends on the choice of the filtration. Even for the stationary linear process $X_k = \sum_{i\in\mathbb{Z}} a_{k-i}Z_i$, $k \in \mathbb{Z}$ it can happen that there exists no aproximation by martingales adapted to the filtration $(\mathcal{F}_k)_k$ generated by the sequence of Z_i but there exists an approximation with respect to another filtration $(\mathcal{F}'_k)_k$ ([Vol05]). The sequence (S_n/σ_n) then can converge conditionally given $\mathcal{F}'_0$ but not given $\mathcal{F}_0$.

References

[DM02] Jérôme Dedecker and Florence Merlevède. Necessary and sufficient conditions for the conditional central limit theorem. *Annals of Probabability*, 30(3):1044–1081, 2002.

[DL01] Yves Derriennic and Michael Lin. Fractional Poisson equations and ergodic theorems for fractional coboundaries. *Israel Journal of Mathematics*, 123:93–130, 2001.

[GH79] Peter Gänssler and Erich Häusler. Remarks on the functional central limit theorem for martingales. *Zeitschrift für Wahrscheinligkeits Theorie und Verwandte Gebiete*, 50(3):237–243, 1979.

[Gor69] M. I. Gordin. The central limit theorem for stationary processes. *Doklady Akademii Nauk SSSR*, 188:739–741, 1969.

[HH80] Peter Hall and Chris C. Heyde. *Martingale limit theory and its application.* Academic Press, Harcourt Brace Jovanovich Publishers, New York, 1980. Probability and Mathematical Statistics.

[IL71] Idlar A. Ibragimov and Yu. V. Linnik. *Independent and stationary sequences of random variables.* Wolters-Noordhoff Publishing, Groningen, 1971. With a supplementary chapter by I. A. Ibragimov and V. V. Petrov, Translation from the Russian edited by J. F. C. Kingman.

[KV05] J. Klicnarová and Dalibor Volný. On optimal martingale approximation. Preprint, 2005.

[MW00] Michael Maxwell and Michael Woodroofe. Central limit theorems for additive functionals of Markov chains. *Annals of Probability,* 28(2):713–724, 2000.

[OV05] L. Ouchti and Dalibor Volný. A conditional CLT which fails in all ergodic components Preprint, 2005.

[PI05] Magda Peligrad and Sergey Utev. A new maximal inequality and invariance principle for stationary sequences. *Annals of Probability,* 33(2):798–815, 2005.

[Res87] Sidney I. Resnick. *Extreme values, regular variation, and point processes,* volume 4 of *Applied Probability. A Series of the Applied Probability Trust.* Springer-Verlag, New York, 1987.

[Rüs85] Ludger Rüschendorf. The Wasserstein distance and approximation theorems. *Zeitschrift für Wahrscheinligkeits Theorie und Verwandte Gebiete,* 70(1):117–129, 1985.

[Vol87] Dalibor Volný. Martingale decompositions of stationary processes. *Yokohama Mathematical Journal,* 35(1-2):113–121, 1987.

[Vol88] Dalibor Volný. Counter examples to the central limit problem for stationary dependent random variables. *Yokohama Mathematical Journal,* 36(1):69–78, 1988.

[Vol93] Dalibor Volný. Approximating martingales and the central limit theorem for strictly stationary processes. *Stochastic Processes and Their Applications,* 44(1):41–74, 1993.

[Vol05] Dalibor Volný. Martingale approximations of stochastic processes depending on the choice of the filtration Preprint, 2005.

[Wu05] Wei Biao Wu. A strong convergence theory for stationary processes. *Preprint,* 2005.

[WW04] Wei Biao Wu and Michael Woodroofe. Martingale approximations for sums of stationary processes. *Annals of Probability,* 32(2):1674–1690, 2004.

[Yok95] Ryōzō Yokoyama. On the central limit theorem and law of the iterated logarithm for stationary processes with applications to linear processes. *Stochastic Processes and Their Applications,* 59(2):343–351, 1995.

Part II

Strong dependence

Almost periodically correlated processes with long memory

Anne Philippe[1], Donatas Surgailis[2], and Marie-Claude Viano[3]

[1] Laboratoire de Mathématiques Jean Leray, UMR CNRS 6629,
   Université de Nantes,
   2 rue de la Houssinière - BP 92208,
   44322 Nantes Cedex 3, France `Anne.Philippe@math.univ-nantes.fr`
[2] Institute of Mathematics and Informatics, 2600 Vilnius, Lithuania
   `sdonatas@ktl.mii.lt`
[3] Laboratoire Paul Painlevé UMR CNRS 8524,
   UFR de Mathématiques – Bat M2
   Université de Lille 1,
   Villeneuve d'Ascq, 59655 Cedex, France `Marie-Claude.Viano@univ-lille1.fr`

1 Introduction

Since their introduction by [Gla61, Gla63] much attention has been given to periodically correlated (PC) or almost periodically correlated (APC) processes, mainly because of their potential use in modeling of cyclical phenomena appearing in hydrology, climatology and econometrics. Following the pioneer work of [Gla63], an important part of the literature has been devoted to APC continuous time processes. The reader can refer to [DH94] for a review including spectral analysis.

In the present paper, we focus on discrete time. A discrete time process is PC when there exists a non zero integer T such that

$$\mathbb{E}(X_{t+T}) \equiv \mathbb{E}(X_t) \quad \text{and} \quad \mathrm{cov}(X_{t+T}, X_{s+T}) \equiv \mathrm{cov}(X_t, X_s) .$$

Usually, as it is the case throughout the present paper, $\mathbb{E}(X_t)$ is supposed to be zero, the attention being focused on the second order periodicity. A review on PC discrete time processes is proposed in [LB99]. A large part of the literature on this topic is devoted to the so-called PARMA (periodic ARMA) models, processes having representation of the form

$$X_{tT+j} - \sum_{k=1}^{p(j)} \phi_k(j) X_{tT+j-k} = \sum_{k=0}^{q(j)} \theta_k(j) \varepsilon_{tT+j-k} , \quad t \geq 0 , j = 0, \ldots, T-1 ,$$

$$(1)$$

where (ε_t) is a zero-mean white noise with unit variance. See for example [BL01, LB00, BLS04] for existence of a solution, statistical developments and forecasting methods. [BH94] give invertibility conditions for periodic moving averages. [HMM02] provide conditions for the existence of L^2 bounded solutions of AR(1) models with almost periodic coefficients.

All these models are short-memory ones. Now, it is well known that in the above mentioned scientific fields many data sets presenting some periodicity also exhibit long range dependence. Such phenomena can be modeled via stationary processes: the so called seasonal fractional models presented and studied among others by [GZW89, VDO95, OOV00] belong to this category. Another idea could be to turn to non stationary models and build PC or APC processes allowing for some strong dependence. To our knowledge, the only attempt to mix together in such a way periodicity and long memory is in [HL95] who propose a 2-PC process consisting in fact in two independent fractional long memory components based on two independent white noises

$$X_{2t} = (I - B)^{-d_1} \varepsilon_t^{(1)} \quad \text{and} \quad X_{2t+1} = (I - B)^{-d_2} \varepsilon_t^{(2)} , \tag{2}$$

where B denotes the backshift operator.

The aim of this article is to propose models of APC long memory processes and to investigate their second order properties and the convergence of their partial sums.

Theoretically, in order to build a long memory PC process (X_t) with period T, it is enough to adjust a long memory stationary model to the T-variate process $(X_{Tt}, \cdots, X_{Tt+T-1})'$. In other words, examples of PC long memory processes are nothing else than examples of long memory stationary vector processes (see for example [Arc94, Mar05] for a study of multivariate long memory). However models of this sort may be not easy to generalize in the direction of almost periodicity, so we prefer to keep to a few particular constructions.

Among the different definitions of almost periodicity we adopt in this paper the following one : a sequence $\mathbf{d} = (d_t, t \in \mathbb{Z}) \in \ell^\infty(\mathbb{Z})$ is *almost periodic* if it is the limit in $\ell^\infty(\mathbb{Z})$ of periodic elements
i.e. for any $\epsilon > 0$ there exists a p_ϵ-periodic sequence $\mathbf{d}^{(\epsilon)}$ such that

$$\sup_{t \in \mathbb{Z}} |d_t - d_t^{(\epsilon)}| \le \epsilon .$$

Recall that an almost periodic sequence is bounded and averageable, in the sense that the following limit exits, uniformly with respect to s

$$(t - s)^{-1} \sum_{u=s}^{t} d_u \to \overline{d} , \qquad \text{as} \quad t - s \to +\infty . \tag{3}$$

The limit $\overline{d}$ is called the mean value of the sequence $\mathbf{d}$. We define APC processes as in [Gla63]:

Definition 1. *A zero mean second order process is APC if for fixed h, the sequence* $(\mathrm{cov}(X_t, X_{t+h}))_t$ *is almost periodic.*

In the present paper we focus on APC processes with long memory. Our starting point are Definitions 2 and 3 suggested by the corresponding definitions in the theory of stationary processes [Ber94] and the theory of point processes [DV00, DV97].

Definition 2. *A second order APC process* (X_t) *has long memory if*

$$\limsup_{N \to \infty} N^{-1} \mathrm{var}\Big(\sum_{t=1}^{N} X_t \Big) = \infty \,.$$

Definition 3. *The Hurst index* H *of APC process* (X_t) *is defined by*

$$H = \inf \Big\{ h : \limsup_{N \to \infty} N^{-2h} \mathrm{var}\Big(\sum_{t=1}^{N} X_t \Big) < \infty \Big\} \,.$$

Since the covariance function of an APC process is bounded, the Hurst index is well-defined and takes values in $[0, 1]$. For example, when

$$\mathrm{var}\Big(\sum_{t=1}^{N} X_t \Big) = L(N) N^{2\alpha} \,,$$

for some $0 < \alpha < 1$ and some positive $L(\cdot)$ slowly varying at infinity, the Hurst index is $H = \alpha$.

Remark 1. Definitions 2 and 3 refer to partial sums in the interval $t \in [1, 2, \ldots, N]$. It is not difficult to show that this interval can be replaced by any interval $[s + 1, \ldots, s + N]$ and so,

$$H = \inf \Big\{ h : \limsup_{N \to \infty} N^{-2h} \mathrm{var}\Big(\sum_{t=s+1}^{s+N} X_t \Big) < \infty, \quad \forall s \in \mathbb{Z} \Big\} \,.$$

In the stationary case, the Hurst index is closely related to the decay rate of the autocovariance $\mathrm{cov}(X_t, X_{t+j})$, roughly meaning that $\mathrm{cov}(X_t, X_{t+j})$ decays as j^{2H-2} when the lag j increases. For an APC process (X_t), the mean value of the almost periodic sequence $\mathrm{cov}(X_t, X_{t+j})$ denoted by

$$\rho(j) := \lim_{N \to \infty} N^{-1} \sum_{t=1}^{N} \mathrm{cov}(X_t, X_{t+j}) \,, \tag{4}$$

takes the place of the autocovariance, as shall be seen in Proposition 1 below. In the sequel we call $\rho(j)$ the *averaged autocovariance* of the APC process (X_t).

The paper is organized as follows. Section 2 links the averaged covariance (4) to the Hurst index of an APC process, and provides the expression and the asymptotic behavior of $\rho(j)$ for most of the models introduced in Section 3, underscoring their long memory characteristics. In Section 3 we present four families of PC or APC long memory models. The three first ones are directly deduced from the classical fractional integrated stationary model by modulating the variance, the time index or the memory exponent, while the last one is newer and less simple. In Section 4, we study the invertibility and the covariance structure of the models of this new family. Section 5 provides, for the processes defined in Section 3, the convergence of the Donsker lines in the Skorohod space $D([0,1])$, showing that, even in the case of modulated memory, the asymptotic behavior is the same as what is obtained in the case of stationary long memory. Section 6, is devoted to simulations, illustrating and completing the theoretical results. Technical proofs are relegated to the Appendix.

2 Hurst index and the averaged autocovariance

The following result points out the link between decay rate of $\rho(j)$, the long memory of the process and the value of its Hurst parameter.

Proposition 1. *Let $\rho(j)$ be the averaged autocovariance of an APC process (X_t). Assume that*

$$\rho(j) = s(j)j^{2H-2}L(j) , \tag{5}$$

where $1/2 < H < 1$, L is slowly varying at infinity, and where $s(j)$ is bounded and Cesaro summable with mean value $\bar{s} \neq 0$:

$$\lim_{k \to \infty} k^{-1} \sum_{j=1}^{k} s(j) =: \bar{s} \neq 0 . \tag{6}$$

Moreover, assume that the convergence towards the mean value of the sequence $(\mathrm{cov}(X_t, X_{t+j}))_t$ is uniform with respect to j:

$$\lim_{N \to \infty} \sup_{j \in \mathbb{Z}} \left| \frac{1}{\rho(j)N} \sum_{t=1}^{N} \mathrm{cov}(X_t, X_{t+j}) - 1 \right| = 0 . \tag{7}$$

Then

$$\mathrm{var}\left(\sum_{t=1}^{N} X_t \right) = L_1(N)N^{2H} ,$$

where $L_1(N)$ is a slowly varying function such that

$$\lim_{N \to \infty} \frac{L_1(N)}{L(N)} = \frac{\bar{s}}{H(2H-1)} .$$

In particular, (X_t) has long memory and its Hurst index is H.

Proof. We have

$$\mathrm{var}\Big(\sum_{t=1}^{N} X_t\Big) = 2 \sum_{j=1}^{N-1}\sum_{t=1}^{N-j} \mathbb{E} X_t X_{t+j} + \sum_{t=1}^{N} \mathbb{E} X_t^2 =: 2\Sigma_1 + \Sigma_2 . \tag{8}$$

Write

$$\frac{\Sigma_1}{L(N)N^{2H}} = N^{1-2H} \sum_{j=1}^{N-1} \frac{L(j)s(j)}{L(N)} j^{2H-2}\theta_{j,N} ,$$

where

$$\theta_{j,N} := \frac{1}{\rho(j)N} \sum_{t=1}^{N-j} \mathbb{E} X_t X_{t+j} = \frac{N-j}{N} + \theta'_{j,N},$$

$$\theta'_{j,N} := \Big(\frac{N-j}{N}\Big)\Big(\frac{1}{\rho(j)(N-j)} \sum_{t=1}^{N-j} \mathbb{E} X_t X_{t+j} - 1\Big) .$$

Now, we need the following technical lemma whose proof is in the Appendix.

Lemma 1. *Let $H > 1/2$. Let $L(\cdot)$ be slowly varying at infinity and $s(j), j \geq 1$ be a bounded sequence summable in the Cesaro sense to $\bar{s} \neq 0$. Then*

$$\lim_{N\to\infty} \frac{N^{1-2H}}{L(N)} \sum_{j=1}^{N} j^{2H-2}L(j)s(j) = \frac{\bar{s}}{2H-1} .$$

¿From this lemma it immediately follows that

$$N^{1-2H} \sum_{j=1}^{N-1} \frac{L(j)s(j)}{L(N)} j^{2H-2}\Big(1 - \frac{j}{N}\Big) \sim \frac{\bar{s}}{2H(2H-1)} .$$

On the other hand, from (7), for arbitrary small $\epsilon > 0$ there exists $K < \infty$ such that $\sup_{0\leq j\leq N-K} |\theta'_{j,N}| < \epsilon$ and therefore

$$N^{1-2H}\Big| \sum_{j=1}^{N-K} \frac{L(j)s(j)}{L(N)} j^{2H-2}\theta'_{j,N}\Big| \leq C\epsilon N^{1-2H} \sum_{j=1}^{N-1} \frac{L(j)}{L(N)} j^{2H-2} < C\epsilon , \tag{9}$$

with C independent of N. Finally, as $N \to \infty$,

$$N^{1-2H}\Big| \sum_{j=N-K+1}^{N-1} \frac{L(j)}{L(N)} j^{2H-2}\theta'_{j,N}\Big| < CK^2 N^{-2H} L^{-1}(N) \to 0 ,$$

by definition of $\theta'_{j,N}$ and the boundedness of $\mathbb{E} X_t X_{t+j}$. This proves

$$\frac{\Sigma_1}{L(N)N^{2H}} \to \frac{\bar{s}}{H(2H-1)} .$$

Finally, Σ_2 defined in (8) satisfies $\Sigma_2 = O(N) = o(L(N)N^{2H})$.   □

For many families of processes the memory and the Hurst parameter can be deduced from a direct inspection of the covariance, as stated in the next proposition which shall be used in Section 3.

Proposition 2. *Let (X_t) be an APC process such that there exists $1/2 < H < 1$ and an almost periodic function $A_t, t \in \mathbb{Z}$ with mean value $\bar{A} \neq 0$ such that*

$$\mathrm{cov}(X_s, X_t) \sim A_s A_t |t - s|^{2H-2} \qquad \text{uniformly as} \quad t - s \to \infty . \tag{10}$$

Then:

(A1). the averaged covariance $\rho(j)$ has the form

$$\rho(j) = (s(j) + o(1))j^{2H-2} ,$$

where $s(j)$ is the mean value of the almost periodic sequence $(A_t A_{t+j})_t$
(A2). the process (X_t) has long memory with Hurst parameter H.

Proof. (i) From (10),

$$\lim_{N\to\infty} N^{-1} \sum_{t=1}^{N} \left(\mathrm{cov}(X_t, X_{t+j})j^{2-2H} - A_t A_{t+j} \right) = 0 ,$$

leading to

$$\rho(j) = \lim_{N\to\infty} N^{-1} \sum_{t=1}^{N} \mathrm{cov}(X_t, X_{t+j}) = s(j)j^{2H-2} .$$

Condition (6), with $\bar{s} = \bar{A}^2$, follows from almost periodicity.
(ii) It suffices to show that

$$\mathrm{var}\left(\sum_{t=1}^{N} X_t \right) \sim \frac{\bar{A}^2}{H(2H-1)} N^{2H} , \qquad \text{as} \quad N \to \infty . \tag{11}$$

As condition (7) is not necessarily satisfied, (11) cannot be deduced from Proposition 1 and has to be proved directly. By condition (10), the proof reduces to $\sum_{t\neq s=1}^{N} A_s A_t |t - s|^{2H-2} \sim CN^{2H}$, which in turn follows from

$$\sum_{1\leq t\neq s\leq N} (A_s A_t - \bar{A}^2)|t - s|^{2H-2} = o(N^{2H}) .$$

Let $\tilde{A}_t := A_t - \bar{A}$, then

$$\sum_{1\leq t\neq s\leq N} (A_s A_t - \bar{A}^2)|t - s|^{2H-2} = \sum_{1\leq t\neq s\leq N} A_s \tilde{A}_t |t - s|^{2H-2}$$

$$+ \bar{A} \sum_{1\leq t\neq s\leq N} \tilde{A}_t |t - s|^{2H-2} .$$

¿From (47), we have

$$\sum_{1 \leq t \neq s \leq N} \tilde{A}_t |t - s|^{2H-2} = o(N^{2H}) .$$

It remains to show that

$$\sum_{1 \leq t \neq s \leq N} A_s \tilde{A}_t |t - s|^{2H-2} = \sum_{1 \leq s < t \leq N} + \sum_{1 < t < s \leq N} =: Q_{1,N} + Q_{2,N} = O(N^{2H}) .$$

Then,

$$Q_{1,N} = \sum_{s=1}^{N} A_s \sum_{j=1}^{N-s} j^{2H-2} \tilde{A}_{s+j}$$

$$= \sum_{s=1}^{N} A_s \sum_{j=1}^{N-s} j^{2H-2} (\alpha_{s,s+j} - \alpha_{s,s+j-1}) ,$$

where $\alpha_{s,s+j} := \sum_{i=1}^{j} \tilde{A}_{s+i}$ and $j^{-1} \alpha_{s,s+j} \to 0 (j \to \infty)$ uniformly in s. Rewrite the sum over j as

$$(N-s)^{2H-2} \alpha_{s,N} - \sum_{j=1}^{N-s-1} \left(j^{2H-2} - (j+1)^{2H-2} \right) \alpha_{s,s+j} .$$

We obtain

$$\sum_{s=1}^{N} A_s (N-s)^{2H-2} \alpha_{s,N} = \sum_{s=1}^{N} A_s (N-s)^{2H-1} \frac{1}{N-s} \sum_{i=1}^{N-s} \tilde{A}_{s+i}$$

$$= o \left(\sum_{s=1}^{N} (N-s)^{2H-1} \right) = o(N^{2H}) .$$

Also,

$$\sum_{j=1}^{N-s-1} j^{2H-1} \left(1 - \left(1 + \frac{1}{j} \right)^{2H-2} \right) (j^{-1} \alpha_{s,s+j}) = o \left(\sum_{j=1}^{N-s} j^{2H-2} \right)$$

$$= o((N-s)^{2H-1}) ,$$

implies a similar result for the remaining sum, yielding $Q_{1,N} = o(N^{2H})$. The sums $Q_{2,N}$ is treated similarly. This proves part (ii) of the proposition. $\square$

3 Examples

Here and throughout the paper, ε_t is a zero-mean white noise with unit variance.

The four classes below are more or less directly built from the classical FARIMA $(0, d, 0)$

$$Y_t = (I - B)^{-d}\varepsilon_t = \varepsilon_t + \sum_{j \geq 1} \psi_j(d)\varepsilon_{t-j} \, , \tag{12}$$

with

$$\psi_j(d) = \frac{\Gamma(d+j)}{\Gamma(d)\Gamma(j+1)} = \prod_{k=1}^{j} \frac{k-1+d}{k} \, , \quad \forall j \neq 0 \, ,$$

where $d \in (0, 1/2)$.

Recall that the covariance of (Y_t) is

$$\Gamma_d(h) := \text{cov}(Y_0, Y_h) = \frac{\Gamma(h+d)\Gamma(1-2d)}{\Gamma(h-d+1)\Gamma(d)\Gamma(1-d)} \, , \tag{13}$$

and behaves as a power of h as $h \to \infty$:

$$\Gamma_d(h) \sim \frac{\Gamma(1-2d)}{\Gamma(d)\Gamma(1-d)}|h|^{2d-1} =: c(d)|h|^{2d-1} \, . \tag{14}$$

The first two examples are obtained by amplitude modulation and phase modulation.

3.1 Amplitude modulation

Let Y_t be the FARIMA $(0, d, 0)$ defined in (12) and S_t an almost periodic deterministic sequence. The process defined by

$$X_t^{\text{AM}} = S_t Y_t \, , \tag{15}$$

is a APC process since its covariance is

$$\text{cov}(X_s^{\text{AM}}, X_t^{\text{AM}}) = S_s S_t \Gamma_d(t-s) \sim c(d) S_s S_t |t-s|^{2d-1} \quad \text{as} \quad |t-s| \to \infty \, . \tag{16}$$

¿From Proposition 2 the process (X_t^{AM}) has long memory with Hurst parameter $H^{\text{AM}} = d + (1/2)$. Moreover, the averaged covariance is

$$\rho^{\text{AM}}(j) = (s(j) + o(1))j^{2H-2} \, ,$$

where $s(j)$ is the mean value of the sequence $(S_t S_{t+j})_t$.

Remark 2. As noticed by [LB99], this model is rather simplistic in so far as, at least if $S_t > 0$ for every t, the correlation $\text{cor}(X_s^{\text{AM}}, X_t^{\text{AM}}) = \text{cor}(Y_t, Y_s)$ is stationary.

3.2 Phase modulation

With the same ingredients as above, assuming that S_t takes only integer values, consider

$$X_t^{\mathrm{PM}} = Y_{t+S_t} \ . \tag{17}$$

Being integer valued, the sequence S_t is necessarily periodic and the process is PC with covariance

$$\mathrm{cov}(X_s^{\mathrm{PM}}, X_t^{\mathrm{PM}}) = \Gamma_d(|t - s + S_t - S_s|) \sim c(d)|s - t|^{2d-1} \quad \text{as} \quad |s - t| \to \infty \ . \tag{18}$$

¿From Proposition 2, the process (X_t^{PM}) has long memory and its Hurst parameter is $H^{\mathrm{PM}} = d + (1/2)$ and $\rho^{\mathrm{PM}}(j) \sim j^{2H-2}$.

Remark 3. Contrary to what happens in amplitude modulation, $\mathrm{var}(X_s^{\mathrm{PM}})$ does not depend on s. Moreover the proper periodicity of the covariance disappears as the lag $|s - t|$ tends to infinity, giving rise to an asymptotic stationarity.

In this model, as in the previous one when S_t is periodic, the memory is the same for the T stationary components $(X_{Tt+h})_{t\in\mathbb{Z}}$ and for the cross correlations, and is characterized by the parameter d of the underlying FARIMA. It is no more the case in the following example.

3.3 Modulated memory

Here the periodicity is directly obtained by modulating the memory exponent d. Consider $\mathbf{d} = (d_t, t \in \mathbb{Z})$ an almost periodic sequence such that for all t, $d_t \in (0, 1/2)$ and that $d^+ := \sup\{d_t : t \in \mathbb{Z}\} < 1/2$. Let ε_t be a zero-mean white noise and X_t^{MM} defined by

$$X_t^{\mathrm{MM}} = (I - B)^{-d_t} \epsilon_t \qquad \forall t \in \mathbb{Z} \ , \tag{19}$$

this expression merely meaning that

$$X_t^{\mathrm{MM}} = \sum_{j=0}^{\infty} \psi_j(d_t)\varepsilon_{t-j} \ , \quad \text{with} \quad \psi_j(d_t) = \frac{\Gamma(d_t + j)}{\Gamma(d_t)\Gamma(j+1)} \ . \tag{20}$$

It is easy to check that, as $t - s \to \infty$,

$$\mathrm{cov}(X_s^{\mathrm{MM}}, X_t^{\mathrm{MM}}) = \sum_{j=0}^{\infty} \psi_j(d_s)\psi_{j+t-s}(d_t) \sim \frac{\Gamma(1 - d_s - d_t)}{\Gamma(d_t)\Gamma(1 - d_t)}(t - s)^{d_s+d_t-1} \ , \tag{21}$$

showing that the memory exponent itself is almost periodic.

Proposition 3 *(A1). The process (X_t^{MM}) is APC. It has long memory and its Hurst parameter is given by $H^{\mathrm{MM}} = (1/2) + d^+$.*

(A2). Assume that **d** *is T-periodic. Then, as* $j \to \infty$,

$$\rho^{\mathrm{MM}}(j) = c(d^+)(s(j) + o(1))j^{2d^+ - 1} ,$$

where

$$s(j) = \#\{1 \leq t \leq T : t \in D^+, t + j \in D^+\} , \qquad (22)$$

with

$$D^+ := \{t \in \mathbb{Z} : d_t = d^+\} .$$

Proof. (1) Let $\mathbf{d}^{(\varepsilon)}$ be a $T^{(\varepsilon)}$-periodic sequence such that $\sup |d_t - d_t^{(\varepsilon)}| < \varepsilon/2$. Since for any t

$$|\psi_k(d_t) - \psi_k(d_t^{(\varepsilon)})| \leq C\varepsilon\psi_k(d^+) , \qquad (23)$$

where $C < \infty$ is independent of t, it is easy to obtain

$$\sup_{t,s \in \mathbb{Z}} \left| \mathrm{cov}(X_s^{\mathrm{MM}}, X_t^{\mathrm{MM}}) - \sum_{j=0}^{\infty} \psi_j(d_s^{(\varepsilon)})\psi_{j+t-s}(d_t^{(\varepsilon)}) \right| < C\varepsilon \sum_{j=0}^{\infty} \psi_j(d^+)^2 ,$$

implying that (X_t^{MM}) is APC.

Observe that $0 \leq \mathrm{cov}(X_s^{\mathrm{MM}}, X_t^{\mathrm{MM}}) \leq C(|t - s| \vee 1)^{2d^+ - 1}$ (this is obvious by (20), (21) and the monotonicity of $\psi_j(d)$ with respect to d), and that one can find $C > 0$ and $p \in \{0, \ldots, T^{(\varepsilon)} - 1\}$, such that

$$\mathrm{cov}(X_s^{\mathrm{MM}}, X_t^{\mathrm{MM}}) \geq C(t - s)^{2d^+ - 1 - 2\epsilon} , \qquad t \neq s \in \{p + kT^{(\epsilon)} , \quad k \in \mathbb{Z}\} . \qquad (24)$$

Indeed, let $D_\epsilon^+ := \{t \in \mathbb{Z} : d_t^{(\epsilon)} = d_\epsilon^+\}$. Then for $t \in D_\epsilon^+$,

$$d_t > d_t^{(\epsilon)} - \epsilon/2 = d_\epsilon^+ - \epsilon/2 > d^+ - \epsilon ,$$

implying, for any $t \neq s \in D_\epsilon^+$,

$$\mathrm{cov}(X_s^{\mathrm{MM}}, X_t^{\mathrm{MM}}) \geq \sum_{k=0}^{\infty} \psi_k(d^+ - \epsilon)\psi_{k+t-s}(d^+ - \epsilon) \geq C(t - s)^{2d^+ - 1 - 2\epsilon} ,$$

whence (24). We thus obtain

$$C_1 N^{2d^+ + 1 - 2\epsilon} \leq \mathrm{var}\left(\sum_{t=1}^{N} X_t^{\mathrm{MM}}\right) \leq C_2 N^{2d^+ + 1} ,$$

for any $\epsilon > 0$ and N large enough, meaning that the Hurst index of $(X_t^{\mathrm{MM}})_t$ is $H^{\mathrm{MM}} = d^+ + (1/2)$.

(2) Write

$$\rho_N(j) = N^{-1} \sum_{t=1}^{N} \operatorname{cov}(X_t, X_{t+j})$$

$$= \frac{1}{T} \sum_{1 \le t \le T : t \in D^+} \operatorname{cov}(X_t, X_{t+j}) + \frac{1}{T} \sum_{1 \le t \le T : t \notin D^+} \operatorname{cov}(X_t, X_{t+j}) + O(N^{-1})$$

$$=: \rho_1(j) + \rho_2(j) + O(N^{-1}),$$

where

$$\rho_2(j) = o(j^{2d^+ - 1}) = o(j^{2H-2})$$
$$\rho_1(j) \sim c(d^+)s(j)j^{2d^+ - 1}, \quad \text{if } t + j \in D^+ \text{for some} t = 1, \ldots, T,$$
$$\rho_1(j) = o(j^{2d^+ - 1}), \quad \text{if } t + j \notin D^+, t = 1, \ldots, T.$$

and where $s(j)$ is defined in (22). This proves part (ii).   □

Remark 4. Consider the periodic case. Note that in (22) s is periodic and that $s(j) = 0$ is quite possible for a sub-sequence j_n (see for example MODEL (A3) in Section 6 where $s(2) = s(3) = 0$). Then $\rho^{\text{MM}}(j_n) = o(j_n^{2d^+ - 1})$, implying that the averaged covariance presents abrupt changes of memory. Nevertheless, the asymptotic behavior of $\operatorname{var}\left(\sum_{t=1}^{N} X_t^{\text{MM}}\right)$ is roughly the same as that of the stationary FARIMA$(0, d^+, 0)$ process.

3.4 Modulated coefficients

Despite the fact that it allows for some memory variability, a major drawback of the previous model is that, apart when **d** is a constant sequence, nothing is clear concerning its invertibility. At any case, in general, we have

$$\varepsilon_t \neq (I - B)^{d_t} X_t^{\text{MM}}.$$

So, we present another extension of the standard FARIMA model (12) based on the ideas of [PSV04] about fractional processes with time varying parameter.

Let $\mathbf{d} = (d_t, t \in \mathbb{Z})$ be an almost periodic sequence. Consider the time-varying fractionally integrated model (TVFI, in the sequel)

$$X_t^{\text{CM}} = \sum_{j=0}^{\infty} a_j(t)\varepsilon_{t-j}, \tag{25}$$

where $(\varepsilon_t)_{t \in \mathbb{Z}}$ is a zero-mean L^2 white noise and where the coefficients $a_j(t)$, directly built from formula (13) of the the FARIMA coefficients, are defined by $a_0(t) \equiv 1$ and

$$a_j(t) = \prod_{k=1}^{j} \frac{k - 1 + d_{t-k}}{k} = \frac{d_{t-1}}{1} \cdots \frac{d_{t-j} + j - 1}{j}, \quad \forall j \neq 0. \tag{26}$$

As it is the case for the model (19), the standard fractional process (12) is recovered when $d_t \equiv d \in (0, 1/2)$.

Under suitable conditions on the sequence $\mathbf{d}$, the next section establishes that the series in (25) converges in the L^2 sense. The induced process is APC and bounded in the L^2 sense and an explicit expansion of ε_t with respect to the (X_j^{CM})'s is available. Moreover, the covariance of this process behaves as

$$\mathrm{cov}(X_s^{\mathrm{CM}}, X_t^{\mathrm{CM}}) \sim c(\bar{d})\gamma(s)\gamma(t)(t-s)^{2\bar{d}-1} , \qquad t-s \to \infty ,$$

for a suitable $\bar{d} \in (0, 1/2)$ to be defined, and where $\gamma(t)$ is almost periodic (see (33) and Proposition 6).

According to Proposition 2, the process (X^{CM}) exhibits long memory and its Hurst parameter is equal to $H^{\mathrm{CM}} = 1/2 + \bar{d}$. Moreover, the averaged covariance is given by

$$\rho^{\mathrm{CM}}(j) = c(\bar{d}) \, (s(j) + o(1)) \, j^{2\bar{d}-1} \qquad \text{as} \quad j \to \infty .$$

$s(j)$ being the mean value of the sequence $(\gamma(\ell)\gamma(\ell+j))_\ell$.

Remark 5. This model is richer than the the amplitude modulated one: comparing with (16) shows that both covariances have the same asymptotic behavior, but it is clear that the correlation of X^{CM} is only asymptotically stationary.

3.5 Note on stationary seasonal memory

It may be interesting to compare the above models to some particular stationary ones. Stationary processes presenting seasonal long memory are well known (see for example [OOV00, LV00] for references, properties and simulations). As an example, consider the fractional ARMA process

$$X_t^{\mathrm{S}} = (I - B)^{-d}(I - 2B\cos(\frac{2\pi}{T}) + B^2))^{-d}\epsilon_t , \tag{27}$$

where $d \in (0, 1/2)$ and T a positive period. This process is stationary and, if the period T is an integer, its covariance has the same kind of asymptotic behavior as the averaged covariance $\rho(j)$ for the processes X^{AM}, X^{MM} and X^{CM}: precisely, as j tends to infinity

$$\mathrm{cov}(X_t^{\mathrm{S}}, X_{t+j}^{\mathrm{S}}) = \left(c_1 + c_2 \cos\left(\frac{2\pi}{T}j\right) + o(1) \right) j^{2d-1} ,$$

where c_1 and c_2 are non zero constants (see [LV00]).

Such processes are not properly PC processes, but they have many similar properties.

4 Existence, invertibility and covariance structure of the TVFI APC processes

4.1 Existence and invertibility

For these questions, we only give the results, referring to [PSV04] for the proofs.

Consider the operators

$$A(\mathbf{d})x_t = \sum_{j=0}^{\infty} a_j(t)x_{t-j} \quad \text{and} \quad B(\mathbf{d})x_t = \sum_{j=0}^{\infty} b_j(t)x_{t-j} , \quad (28)$$

where the coefficients a_j are defined in (26), and where the b_j's are defined by $b_0(t) \equiv 1$ and

$$b_j(t) = d_{t-1} \prod_{k=2}^{j} \frac{k-1+d_{t-j+k-2}}{k} , \quad j \geq 1 . \quad (29)$$

First, $B(-\mathbf{d})$ is the inverse of $A(\mathbf{d})$, since, at least formally,

$$B(-\mathbf{d})A(\mathbf{d})x_t = A(\mathbf{d})B(-\mathbf{d})x_t = x_t \quad \forall t . \quad (30)$$

Now, [PSV04] prove that if the sequence $\mathbf{d}$ satisfies the condition

$[H_D^{(a)}]$: for all integers $s < t$ such that $t - s$ is larger than some $K > 0$

$$\sum_{s<u<t} \frac{d_u - D}{t - u} \leq 0 ,$$

the coefficients of the time varying operator $A(\mathbf{d})$ are bounded above by the coefficients $\psi_j(D)$, i.e. there exists $C > 0$ such that

$$|a_j(t)| \leq C\psi_j(D) , \quad \forall t \in \mathbb{Z} , \quad \forall j \in \mathbb{N} . \quad (31)$$

(see [PSV04] for the proof)

Taking into account the well know properties of the FARIMA$(0,d,0)$ filter, inequalities (31) immediately imply that

(A1). if $[H_D^{(a)}]$ holds with $D < 0$, then $A(\mathbf{d})$ boundedly operates from $\mathcal{L}^1$ to $\mathcal{L}^1$

(A2). if $[H_D^{(a)}]$ holds with $D < 1/2$, then $A(\mathbf{d})$ boundedly operates from $\mathcal{L}_0^2$ to $\mathcal{L}^2$,

where $\mathcal{L}^p$ $(p = 1, 2)$ denotes the class of all real-valued random processes $\{x_t, t \in \mathbb{Z}\}$ such that $\sup_t \mathbb{E}|x_t|^p < \infty$ and $\mathcal{L}_o^2 \subset \mathcal{L}^2$ the subclass of all zero mean *orthogonal* sequences.

Similar results hold for the second operator $B(\mathbf{d})$ if condition $[H_D^{(a)}]$ is replaced by

$[H_D^{(b)}]$: there exists $K < \infty$ such that for all integers $s < t$ such that $t - s > K$

$$\sum_{s<u<t} \frac{d_u - D}{u - s} \leq 0 .$$

The following lemma, whose proof is in the Appendix, shows that conditions $[H_D^{(a)}]$ and $[H_D^{(b)}]$ are well adapted to almost periodic sequences.

Lemma 2. *Let* **d** *be an almost periodic sequence with mean value* $\overline{d}$ *defined in (3). Conditions $[H_D^{(a)}]$ and $[H_D^{(b)}]$ hold with any* $D > \overline{d}$.

Finally, gathering all these results and remarks allows to specify the convergence of (25) and to obtain an inverse expansion.

Proposition 4. *Let* **d** *be an almost periodic sequence with mean value* $\overline{d} \in (0, 1/2)$. *Consider the operators* $A(\mathbf{d})$ *and* $B(\mathbf{d})$ *defined in (28), (26) and (29). Let* ε *be a zero-mean* L^2 *white noise. Then the process*

$$X_t^{\mathrm{CM}} = A(\mathbf{d})\varepsilon_t = \sum_{j=0}^{\infty} a_j(t)\varepsilon_{t-j} , \quad t \in \mathbb{Z} ,$$

is bounded in L^2 *and the inversion formula is*

$$\varepsilon_t = B(-\mathbf{d})X_t^{\mathrm{CM}} = \sum_{j=0}^{\infty} b_j^-(t)X_{t-j}^{\mathrm{CM}} , \quad \forall \, t \in \mathbb{Z} .$$

where the $b_j^-(t)$*'s denote the coefficient of the linear operator* $B(-\mathbf{d})$.

Remark 6. The condition $\overline{d} \in (0, 1/2)$ does not imply that the sequence itself is in this domain. Excursions of the sequence outside the classical stability interval $(0, 1/2)$ of the FARIMA are allowed, producing interesting sample paths as shall be seen in Section 6.

4.2 Covariance structure

Next propositions state almost periodicity and the slow decay of the covariances.

Proposition 5. *Under conditions of Proposition 4, the process* (X_t^{CM}) *defined in (25) is almost periodically correlated.*

Proof. The proof is in the Appendix. □

As a preliminary towards the asymptotic study of the covariances, we get the asymptotic behavior of the coefficients of the time varying filter. For this purpose we need to assume that there exists $\delta > 0$ and $C > 0$ such that

$$\sup_{s<t} \left| \frac{1}{t-s} \sum_{u=s}^{t} (d_u - \bar{d}) \right| \le C|t-s|^{-\delta} . \tag{32}$$

The next lemma points out that the coefficients of the time varying filter are asymptotically close to $\gamma(v)\psi_{v-u}(\bar{d})$ where the $(\psi_k(\bar{d}))$'s are the coefficients of the FARIMA $(0,\bar{d},0)$ and where γ is the almost periodic sequence (33).

Lemma 3. *Under the assumption (32) on the almost periodic sequence* **d** *and if* $0 < \bar{d} < 1/2$,
 (i) the infinite product

$$\gamma(v) := \prod_{u<v} \left(1 + \frac{d_u - \bar{d}}{\bar{d} + v - u - 1} \right) , \tag{33}$$

converges for any $v \in \mathbb{Z}$ *and the sequence* $\gamma(v)$ *is almost periodic.*
 (ii) the coefficients defined in (26) can be rewritten

$$a_{v-u}(v) = \gamma(v)\psi_{v-u}(\bar{d})\theta_{u,v} , \tag{34}$$

with

$$|\theta_{u,v} - 1| = O(|v-u|^{-\delta}) \ as \ v - u \to +\infty .$$

The proof is postponed in the Appendix.

This result is now used to show that the covariance (35) asymptotically behaves as the product of the almost periodic sequences $\gamma(s)\gamma(t)$ and $(t-s)^{2\bar{d}-1}$. This last term is the asymptotic covariance of a stationary FARIMA $(0,\bar{d},0)$, showing a kind of asymptotic averaging of the memory.

Proposition 6. *Let* $(\epsilon_t)_{t\in\mathbb{Z}}$ *is a weak white noise. Under the assumption (32) on the almost periodic sequence* **d** *and if* $0 < \bar{d} < 1/2$, *the covariance function of the process* X^{CM} *defined in (25),* $X_t^{\mathrm{CM}} = A(\mathbf{d})\epsilon_t$ *satisfies, as* $t - s \to \infty$

$$\mathrm{cov}(X_s^{\mathrm{CM}}, X_t^{\mathrm{CM}}) \sim c(\bar{d})\gamma(s)\gamma(t)(t-s)^{2\bar{d}-1} , \tag{35}$$

uniformly in $s < t$.

Proof. Let $s < t$. The covariance is equal to

$$\mathrm{cov}(X_s^{\mathrm{CM}}, X_t^{\mathrm{CM}}) = \sum_{u\le s} a_{s-u} a_{t-u} = \gamma(s)\gamma(t) \sum_{u\le s} \psi_{s-u}(\bar{d})\psi_{t-u}(\bar{d}) + R_A(s,t) .$$

Firstly, we have [BD91] (Chap. 13)

$$\sum_{u\le s} \psi_{s-u}(\bar{d})\psi_{t-u}(\bar{d}) \sim \frac{\Gamma(1-2\bar{d})}{\Gamma(\bar{d})\Gamma(1-\bar{d})} (t-s)^{2\bar{d}-1} \ (t-s\to\infty) .$$

Secondly, using Lemma 3,

$$|R_A(s,t)| = \left| \gamma(s)\gamma(t) \sum_{u \leq s} \psi_{s-u}(\overline{d})\psi_{t-u}(\overline{d})(\theta_{s,u}\theta_{t,u} - 1) \right|$$

$$\leq C \sum_{u \leq s} \psi_{s-u}(\overline{d})\psi_{t-u}(\overline{d})(1 + s - u)^{-\delta} = o((t-s)^{2\overline{d}-1}),$$

uniformly in $s < t$. This proves (35). $\qquad\qquad\qquad\qquad\qquad\qquad\qquad\square$

5 Convergence of the partial sums

In this section, we turn to the asymptotic behavior of partial sums processes for each of the four classes of models of Section 3.

First consider the FARIMA (0,d,0) process (12): it is now well known that its suitably normalized partial sums converge in the Skorohod space towards a fractional Brownian motion (see for instance [Taq03] or [Dav70]):

$$N^{-d-(1/2)} \sum_{t=1}^{[N\tau]} Y_t \xrightarrow{D[0,1]} \sigma(d) B_{d+1/2}(\tau), \tag{36}$$

where $\sigma(d)$ is a positive constant such that

$$\sigma(d)^2 = \frac{c(d)}{d(2d+1)} = \frac{1}{d(2d+1)} \frac{\Gamma(1-2d)}{\Gamma(d)\Gamma(1-d)}. \tag{37}$$

We shall see that this result still holds, with an adapted parameter d and a modified constant factor, for all the models presented in Section 3. Almost periodicity of the covariance has no effect on the partial sums. This is not very surprising since a PC process can be viewed as a multivariate stationary one. Returning to stationary seasonal long memory, it is worth noticing that the partial sums of the process X^S defined in (27) have exactly the same asymptotic behavior (see for example [OV03]). Indeed, for a stationary sequence, the partial sums are asymptotically insensitive to unboundedness of the spectrum at non zero frequencies.

Consider the case of phase modulation (processes X_t^{PM}).

Theorem 1. Let X^{PM} be a process defined in (17). Then, as $N \to \infty$

$$N^{-d-(1/2)} \sum_{t=1}^{[N\tau]} X_t^{PM} \xrightarrow{D[0,1]} \sigma(d) B_{d+1/2}(\tau), \tag{38}$$

where $B_{d+1/2}(\tau)$ is a fractional Brownian motion with Hurst parameter $d + 1/2$.

Proof. By definition of X^{PM} we have

$$Z_t := X_t^{\mathrm{PM}} - Y_t = Y_{t+S_t} - Y_t$$

$$= \sum_{j=0}^{\infty} (\psi_{j+S_t}(d) - \psi_j(d))\varepsilon_{t-j} + \sum_{t<u\leq t+S_t} \psi_{t+S_t-u}(d)\varepsilon_u := Z'_t + Z''_t .$$

As S_t is bounded, the process (Z''_t) is finitely dependent and therefore $\mathbb{E}\big(\sum_{t=s+1}^{N+s} Z''_t\big)^2 \leq CN$, with C independent of s and of N. Also, since $|\psi_{j+p}(d) - \psi_j(d)| \leq Cj^{-1}\psi_j(d) \leq Cj^{d-2}$ for all $j \geq 0, 0 \leq p \leq K$ and some $C = C(K) < \infty$, we obtain for any $s \leq t$

$$\mathbb{E}Z'_s Z'_t = \sum_{j=0}^{\infty} j^{d-2}(j+t-s)_+^{d-2} \leq C(t-s)^{d-2} ,$$

and hence $\mathbb{E}\big(\sum_{t=s+1}^{N+s} Z'_t\big)^2 \leq CN$. Therefore

$$\sum_{t=1}^{[N\tau]} X_t^{\mathrm{PM}} = \sum_{t=1}^{[N\tau]} Y_t + O_p(N^{1/2}) ,$$

proving the convergence of fidis distributions in (38). From the above bounds, for any $0 \leq \tau < \tau + h \leq 1, Nh \geq 1$

$$\mathbb{E}\Big(\sum_{t=[N\tau]}^{[N(\tau+h)]} X_t^{\mathrm{PM}} \Big)^2 \leq 2\mathbb{E}\Big(\sum_{t=[N\tau]}^{[N(\tau+h)]} Y_t \Big)^2 + 2\mathbb{E}\Big(\sum_{t=[N\tau]}^{[N(\tau+h)]} Z_t \Big)^2$$

$$\leq C((Nh)^{1+2d} + (Nh)) \leq C(Nh)^{1+2d} ,$$

implying tightness of the partial sums of (X_t^{PM}) in $D[0,1]$ [Bil68] (Theorem 15.6). □

Remark 7. Since the process X^{PM} is periodically correlated, the above result could be viewed as a consequence of functional limit theorems on multivariate stationary processes. In fact it is a consequence of Theorem 6 in [Arc94]. Indeed this theorem works if the same memory parameter governs all the cross-covariances of the vector process $(X_{Tt}, \ldots, X_{T(t+1)} - 1))$, which is the case for the process X_t^{PM}. However the expression of the limiting process in [Arc94] is rather involved, we preferred to give a more direct proof.

Now, we consider the modulated memory processes in the particular case of a T-periodic sequence **d**.

Remark 8. Despite the fact that we restrict the study to the periodic case, Theorem 6 in [Arc94] does not apply to processes X^{MM} since different memory parameters appear in the T components of the associated vector process.

Theorem 2. *Consider the process X^{MM} defined in (19) from a T-periodic sequence **d**. Denote by $d^+ \in (0, 1/2)$ the maximum of **d** on a period. Then, as $N \to \infty$*

$$N^{-d^+ - (1/2)} \sum_{t=1}^{[N\tau]} X_t^{MM} \xrightarrow{D[0,1]} \frac{\kappa\sigma(d^+)}{T} B_{d^+ + 1/2}(\tau) .$$

where $\kappa = \#\{1 \leq t \leq T : d_t = d^+\}$

Proof. Tightness is obvious by $|\mathrm{Cov}(X_s, X_t)| \leq C|t - s|^{2d^+ - 1}$. To prove the fidi convergence, let $D^+ = \{t \in \mathbb{Z} : d_t = d^+\}$. Without loss of generality assume for simplicity $kT \in D^+, k \in \mathbb{Z}_+$. We have

$$\sum_{t=1}^{N} X_t^{MM} = \sum_{k=1}^{[N/T]} \sum_{(k-1)T < t \leq kT, t \in D^+} X_t^{MM} + o_p(N^{2d^+ + 1})$$

$$= \kappa \sum_{k=1}^{[N/T]} Z_{kT} + \sum_{k=1}^{[N/T]} Z_k' + o_p(N^{2d^+ + 1}) ,$$

where

$$Z_k' := \sum_{(k-1)T < t \leq kT, t \in D^+} X_t^{MM} - \kappa X_{kT}^{MM} ,$$

and $Z_t = \sum_{j=0}^{\infty} \psi_j(d^+)\varepsilon_{t-j}$ is the FARIMA $(0, d^+, 0)$. Similarly as in the proof of Theorem 1, we get that $\mathbb{E}\big(\sum_{k=1}^{[N/T]} Z_k'\big)^2 = O(N)$. Then, the remaining details are standard. $\square$

Hence, at least in the purely periodic case, the partial sums of the model with modulated memory (17) behave exactly as if the process was governed only by the strongest memory component. The result still holds (up to a multiplicative constant) if the maximum d^+ of the sequence is achieved at several points of the period. Whether it holds for other almost periodic sequences is an open question.

The following theorem permits to treat in the same time the cases of amplitude modulation (processes X_t^{AM}) and of modulated coefficients (processes X_t^{CM}).

Theorem 3. *Let $\mathbf{q}$ be an almost periodic sequence and d a real number in $(0, 1/2)$. Consider a zero-mean white noise (ε_t) and the second order process*

$$X_t = \sum_{s \leq t} a_{t-s}(t)\varepsilon_s , \qquad \text{for all } t \in \mathbb{Z}$$

where the time varying coefficients $a_{t-s}(t)$ satisfy

$$a_{t-s}(t) = q(t)\psi_{t-s}(d)\theta_{t,s} , \tag{39}$$

with, for some $\delta > 0$,

$$|\theta_{t,s} - 1| < C|t - s|^{-\delta} \quad \forall s < t . \tag{40}$$

Then, as $N \to \infty$,

$$N^{-d-(1/2)} \sum_{t=1}^{[N\tau]} X_t \xrightarrow{D[0,1]} \sigma(d)\, \bar{q}\, B_{d+1/2}(\tau) \,, \tag{41}$$

where $B_{d+1/2}(\tau)$ is a fractional Brownian motion with Hurst parameter equal to $d + 1/2$. The constant $\sigma(d) > 0$ is defined in (37) and $\bar{q}$ is the mean value of $\mathbf{q}$.

Proof. We follow the scheme of discrete stochastic integrals (see [Sur03, Sur83]). For simplicity, we only consider the one-dimensional convergence at $\tau = 1$.

The left hand side of (41) can be expressed as

$$N^{-d-(1/2)} \sum_{t=1}^{N} X_t = \int f_N \mathrm{d}Z_N := I_N \,,$$

where

$$f_N(x) := N^{-d} \begin{cases} \sum_{t=1}^{N} a_{t-s}(t) \,, & x \in ((s-1)/N, s/N], x \in (-\infty, 1] \,, \\ 0 \,, & \text{otherwise} \,, \end{cases}$$

and Z_N is defined on finite intervals $(x', x'']$, $x' < x''$ by

$$Z_N((x', x'']) := N^{-1/2} \sum_{x' < s/N \le x''} \epsilon_s \,.$$

According to the central limit theorem, for any $m < \infty$ and any disjoint intervals $(x'_i, x''_i], i = 1, \ldots, m$,

$$(Z_N((x'_1, x''_1]), \ldots, Z_N((x'_m, x''_m])) \xrightarrow{\text{law}} (Z((x'_1, x''_1]), \ldots, Z((x'_m, x''_m]) \,, \tag{42}$$

where $Z(\mathrm{d}x)$ is a standard Gaussian noise with mean zero and variance $\mathrm{d}x$.

On the other hand, the limit process in (41) can be written $\int f\mathrm{d}Z$ with

$$f(x) := \frac{\bar{q}\sigma(d)}{\nu(d)} \begin{cases} \int_0^1 (t-x)_+^{d-1}\mathrm{d}t \,, & x \le 1 \,, \\ 0 \,, & \text{otherwise} \,. \end{cases}$$

See [Taq03] for details on the representation of B_H as a stochastic integral and an explicit form of the positive constant $\nu(d)$ such that

$$\nu(d)^2 = \int_{-\infty}^{1} \left(\int_0^1 (t-x)_+^{d-1}\mathrm{d}t \right)^2 \mathrm{d}x \,.$$

As proved in [Sur03, Sur83], the convergence $I_N \xrightarrow{\text{law}} I$ follows from the convergence $\|f_N - f\|_2 \to 0$.

By using (40) and (39), it suffices to show $\|\tilde{f}_N - f\|_2 \to 0$, where

$$\tilde{f}_N(x) := N^{-d} \begin{cases} \sum_{t=1}^{N} q(t)\psi_{t-s}(d) , & x \in (\frac{s-1}{N}, \frac{s}{N}] \cap (-\infty, 1] , \\ 0 , & \text{otherwise} . \end{cases}$$

Next, split $\tilde{f}_N = f'_N + f''_N$, where

$$f'_N(x) := \begin{cases} N^{-d}\bar{q}\sum_{t=1}^{N} \psi_{t-s}(d) , & \text{for } x \in (\frac{s-1}{N}, \frac{s}{N}] \cap (-\infty, 1] \\ 0 , & \text{otherwise} , \end{cases}$$

and

$$f''_N(x) := \begin{cases} N^{-d}\sum_{t=1}^{N}(q(t) - \bar{q})\psi_{t-s}(\bar{d}) , & \text{for } x \in (\frac{s-1}{N}, \frac{s}{N}] \cap (-\infty, 1] , \\ 0 , & \text{otherwise} . \end{cases}$$

It is easily seen that $\|f'_N - f\|_2 \to 0$, hence it suffices to show that

$$\|f''_N\|_2 \to 0 , \tag{43}$$

which is equivalent to

$$R_N := \sum_{s=-\infty}^{N} \left(\sum_{t=1}^{N}(q(t) - \bar{q})\psi_{t-s}(d)\right)^2 = o(N^{1+2d}) .$$

We split the sum $R_N = \sum_{s=-\infty}^{0} \cdots + \sum_{s=1}^{N} \cdots =: R'_N + R''_N$ and put

$$G(t) := \sum_{u=1}^{t}(q(u) - \bar{q}) .$$

Then $\epsilon(t) := |G(t)/t| \to 0 (t \to \infty)$ since $q(\cdot) - \bar{q}$ is almost periodic with zero mean value.

Fix $s \le 0$. Using summation by parts we obtain

$$\sum_{t=1}^{N}(q(t) - \bar{q})\psi_{t-s}(d) = G(N)\psi_{N-s}(d) + \sum_{t=1}^{N-1} G(t)(\psi_{t+1-s}(d) - \psi_{t-s}(d)) .$$

Let us recall that the FARIMA coefficients satisfy

$$|\psi_{N-s}(d)| \le C(N - s)^{d-1} \quad \text{and} \quad |\psi_{t+1-s}(d) - \psi_{t-s}(d))| \le C|t - s|^{d-2} .$$

we obtain

$$R'_N \le C\epsilon^2(N)N^2 \sum_{s=0}^{\infty}(N + s)^{2d-2} + C\sum_{s=0}^{\infty}\left(\sum_{t=1}^{N} \epsilon(t)t(t + s)^{d-2}\right)^2 .$$

Here, $N^2 \sum_{s=0}^{\infty} (N+s)^{2d-2} = O(N^{1+2d})$ while the second sum on the r.h.s. above is dominated by

$$CN^{2d+1} \int_0^{\infty} \left(\int_0^1 \epsilon(t/N)t(t+s)^{d-2}dt \right)^2 ds .$$

Since

$$\int_0^{\infty} \left(\int_0^1 t(t+s)^{d-2}dt \right)^2 ds < \infty ,$$

the above inequalities and the dominated convergence theorem imply

$$R'_N = o(N^{1+2d}) .$$

The relation $R''_N = o(N^{1+2d})$ follows similarly, with $G(t)$ replaced by $G(t) - G(s)$. This proves (43) and the finite dimensional convergence in (41).

Tightness in $D[0,1]$ is obtained by the criterion given in [Bil68] (Theorem 15.6). Using Lemma 3 and boundedness of $q_A(t)$, the proof is standard. $\square$

Returning now to amplitude modulation and to coefficients modulation processes presented in Subsections 3.1 and 3.4, it is clear that the assumptions of Theorem 3 are satisfied by processes X_t^{AM} with $q(t) = S_t$, $\theta_{s,t} \equiv 1$ and by processes X_t^{CM} with $q(t) = \gamma(t)$, $d = \bar{d}$, $\theta_{s,t}$ defined in (34), under the conditions of Lemma 3.

Corollary 1.

(A1). Consider the process X_t^{AM} defined in (15) from the FARIMA (0,d,0) with $d \in (0,1/2)$ and the almost periodic sequence S_t with mean value $\bar{S}$. As $N \to \infty$,

$$N^{-d-(1/2)} \sum_{t=1}^{[N\tau]} X_t^{AM} \xrightarrow{D[0,1]} \sigma(d) \, \bar{S} B_{d+1/2}(\tau) .$$

(A2). Consider the process X_t^{CM} defined in (25) and (26) from the almost periodic sequence $\mathbf{d}$ submitted to the conditions of Lemma 3. As $N \to \infty$,

$$N^{-\bar{d}-(1/2)} \sum_{t=1}^{[N\tau]} X_t^{CM} \xrightarrow{D[0,1]} \sigma(\bar{d}) \, \bar{q} B_{\bar{d}+1/2}(\tau) .$$

6 Simulation

In this section, we provide some numerical examples to illustrate the properties of models of Section 3. We only consider periodic sequences.

6.1 Comparison of the models via simulation

The period is $T = 5$ for all models, and the memory parameters are chosen in such a way that they provide the same limiting behaviour of the partial sums. Namely

(A1). (X_t^{AM}) : $d = 0.35$ and $S_t = \gamma(t)$ where γ is the infinite product (33) associated to the sequence $\mathbf{d}$ of the model (X_t^{CM}) just below,

(A2). (X_t^{PM}) : $d = 0.35$ and $(S_1, \ldots, S_5) = (1, 1, 5, 5, 5)$

(A3). (X_t^{MM}) : $(d_1, \ldots, d_5) = (0.25, 0.25, 0.35, 0.35, 0.35)$

(A4). (X_t^{CM}) : $(d_1, \ldots, d_5) = (-0.25, -0.25, 0.75, 0.75, 0.75)$, this implies that $\bar{d} = 0.35$.

Moreover, the periodic sequence driving X_t^{AM} is chosen to be $S_t = \gamma(t)$ where γ is the infinite product (33) associated to model X_t^{CM} with the above preassigned parameters d_j. Hence, the theoretical covariances (16) and (35) have exactly the same asymptotic behavior.

The models are built from a Gaussian zero-mean white noise with unit variance, and for each one a sample path of length $n = 10^4$ is simulated.

Estimation of the averaged autocovariance.

Let

$$\hat{\rho}_N(j) = \frac{1}{N} \sum_{t=1}^{N-j} (X_t - \bar{X})(X_{t+j} - \bar{X}), \qquad \bar{X} := \frac{1}{N} \sum_{t=1}^{N} X_t,$$

be the sample autocovariance and the sample mean. Next proposition proves that, under rather weak assumptions (for example an assumption on the fourth cumulants which is trivially satisfied for Gaussian processes), the empirical autocovariance is an L^2 convergent estimator of the averaged covariance (4). Denote by $\chi(t_1, t_2, t_3, t_4)$ the 4th order joint cumulant of $X_{t_1}, X_{t_2}, X_{t_3}, X_{t_4}$.

Proposition 7. *Let* (X_t) *be a zero mean APC process with averaged covariance* ρ *such that* $\sup_{t \in \mathbb{Z}} \mathbb{E} X_t^4 < \infty$. *Moreover, assume*

$$\lim_{h \to \infty} \sup_{t \in \mathbb{Z}} |\mathbb{E}(X_t X_{t+h})| = 0, \quad and$$

$$\lim_{h \to \infty} \sup_{t_1 \leq t_2 \leq t_3 \leq t_1 + h} |\chi(X_{t_1}, X_{t_2}, X_{t_3}, X_{t_1+h})| = 0. \quad (44)$$

Then for any $j \geq 0$

$$\mathbb{E}(\hat{\rho}_N(j) - \rho(j))^2 \to 0 \qquad (N \to \infty).$$

Proof. Write

$$\rho(j) - \hat{\rho}_N(j) =$$

$$\rho(j) - \frac{1}{N} \sum_{t=1}^{N} (X_t - \bar{X})(X_{t+j} - \bar{X}) + \frac{1}{N} \sum_{t=N-j+1}^{N} (X_t - \bar{X})(X_{t+j} - \bar{X}),$$

where it is clear that the variance of the last term tends to zero as $N \to \infty$. Now,

$$\rho(j) - \frac{1}{N} \sum_{t=1}^{N} (X_t - \bar{X})(X_{t+j} - \bar{X}) = \sum_{i=1}^{3} \delta_i(j),$$

with

$$\delta_1(j) := \rho(j) - N^{-1} \sum_{t=1}^{N} \mathbb{E}(X_t X_{t+j}),$$

$$\delta_2(j) := -N^{-1} \sum_{t=1}^{N} (X_t X_{t+j} - \mathbb{E}(X_t X_{t+j})),$$

$$\delta_3(j) := \bar{X}^2.$$

By the definition of $\rho(j)$, $\delta_1(j) \to 0$. Writing

$$\mathbb{E}(X_t X_{t+j} X_s X_{s+j}) = \chi(t, t+j, s, s+j)$$
$$+ \mathbb{E}(X_t X_{t+j})\mathbb{E}(X_s X_{s+j})) + \mathbb{E}(X_s X_{t+j})\mathbb{E}(X_t X_{s+j}))$$
$$+ \mathbb{E}(X_t X_s)\mathbb{E}(X_{t+j} X_{s+j})),$$

and using assumptions (44) easily leads to the convergence to zero of $\mathbb{E}(\delta_2(j))^2$ and $\mathbb{E}(\delta_3(j))^2$. □

Figures 1 show the empirical autocovariances of the four sample paths. From Proposition 7, they should mimic the theoretical sequences $\rho(j)$ whose behavior is specified, for each model, in Section 3.

In our situation, given the choice of the parameters, $\rho^{AM}(j)$ and $\rho^{CM}(j)$ have exactly the same asymptotic behavior: a 5-periodic sequence damped by $j^{-0.3}$. In $\rho^{PM}(j)$, the periodic effect is present for small values of j but asymptotically disappears, the behavior being then exactly as $j^{-0.3}$. As for $\rho^{MM}(j)$, the highest value of this maximum is achieved for $h = 5k$, so that the convergence to zero of $\rho^{MM}(5k)$ is slower than the convergence of the other subsequences $\rho^{MM}(5k+\ell)$. As expected, those remarks are well illustrated by the shapes of the corresponding empirical autocovariances of Figure 1.

The associated T-variate processes and estimation of the memory

Consider the four 5-variate stationary processes $Z_t = (X_{5t}, \cdots, X_{5t+4})'$. Their correlations are estimated from the corresponding subsequences of the

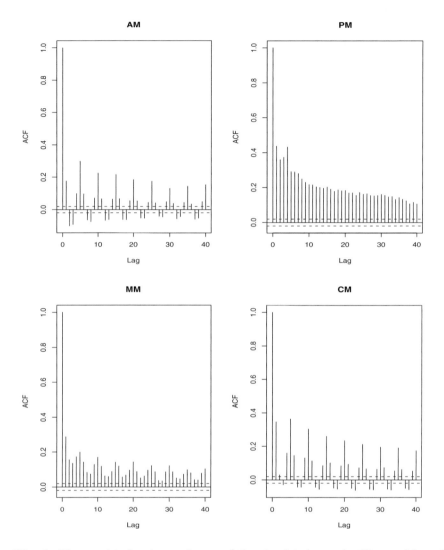

Fig. 1. The empirical autocovariances of the simulated sample. The models and their parameters are precised in Section 6.1

four sample paths and represented in Figures 2-3-4-5. To each model corresponds 25 graphics. The graphic (j, k) represents $\hat{\rho}_{j,k}(h)$, an estimation of $\text{cor}(X_{5t+j}, X_{5(t+h)+k})$. For all the models, long memory is well visible in the autocorrelations $\hat{\rho}_{j,j}(h)$. Comparing what happens for models X_t^{AM}, X_t^{CM} is particularly interesting since for these two models, the empirical autocovariances are quite similar, giving the wrong impression that these models are close to each other. Here, we see that, for the process X_t^{AM}, the graphs $\hat{\rho}_{j,k}(h)$

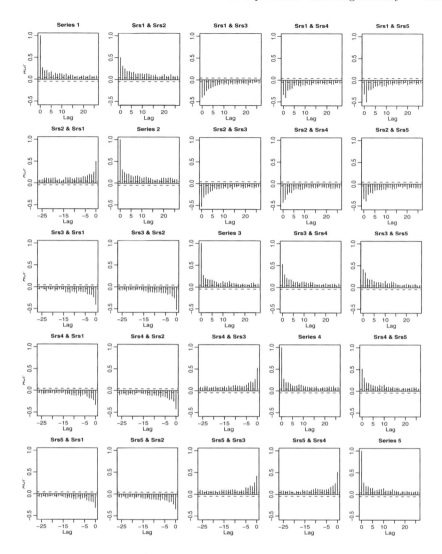

Fig. 2. The empirical autocorrelation of the 5 variate simulated sample for the processes X^{AM}. The parameters are given in Section 6.1, MODEL (A1).

remain the same (except for the sign) when $j - k$ is kept constant, which is not surprising since

$$\mathrm{cor}(X^{\mathrm{AM}}_{5t+j}, X^{\mathrm{AM}}_{5(t+h)+k}) = \mathrm{cor}(Y_{5t+j}, Y_{5(t+h)+k}) = \frac{\Gamma_d(5h + k - j)}{\Gamma_d(0)} \ .$$

This is not at all the case for the process X^{CM}_t.

According to the asymptotic behavior of the covariance function, we know that for X^{CM}, X^{AM} and X^{PM}, the sub processes of $(X_{5t+\ell})_t$ $\ell \in \{0, \ldots, 4\}$

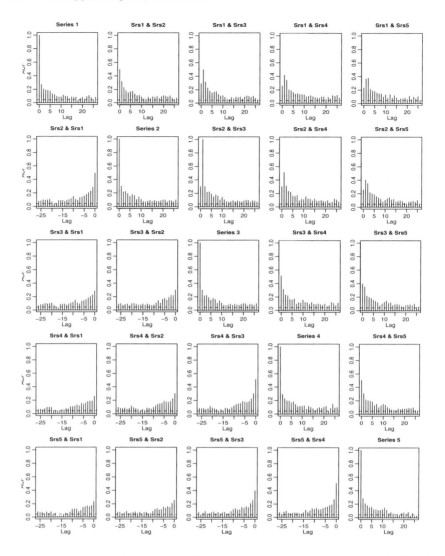

Fig. 3. The empirical autocorrelation of the 5 variate simulated sample for the processes X^{PM}. The parameters are given in Section 6.1, MODEL (A2).

have the same long memory parameter. This is visible in Figures 2-3 but less evident in the two first columns of Figure 5, surely because for small values of the lag, $\mathrm{cor}(X_s^{CM}, X_t^{CM})$ can be very far from the correlation of the FARIMA $(0, \bar{d}, 0)$. As seen in Figure 4, the situation for X^{MM} is different. For example the memory coefficient of X_{5t}^{MM} and X_{5t+1}^{MM} is $d_1 = 0$ whereas it is $d_3 = 0.35$ for the 3 other sub processes. Notice also that, for model X^{PM}, the cross correlations can take their maximum values for non a zero lag. For instance,

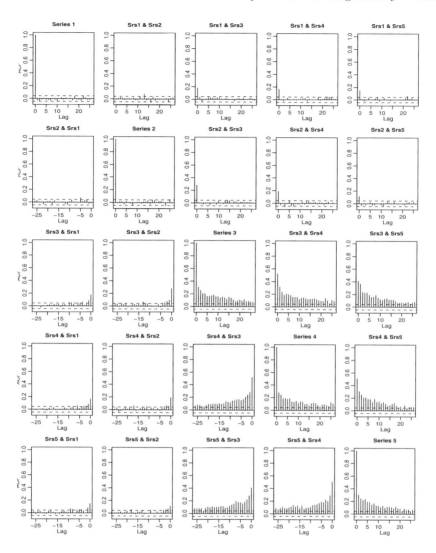

Fig. 4. The empirical autocorrelation of the 5 variate simulated sample for the processes X^{MM}. The parameters are given in Section 6.1, MODEL (A3).

processes X_{5k+2} and X_{5k+3}, have a cross correlation equal to 1 at lag 1, as easily seen on Figure 3. This is due to the fact that the sequence $t + S_t$ is not monotonic.

Table 1 provides estimations of the memory parameters. These estimations are calculated from each of the five sub samples $X_{5t+\ell}$ ($\ell = 1, \ldots, 5$), using the local Whittle estimate (see for instance [BLO03]). For the models X^{AM} and X^{PM} the estimated parameter is d, with value 0.35. For model X^{MM}, each

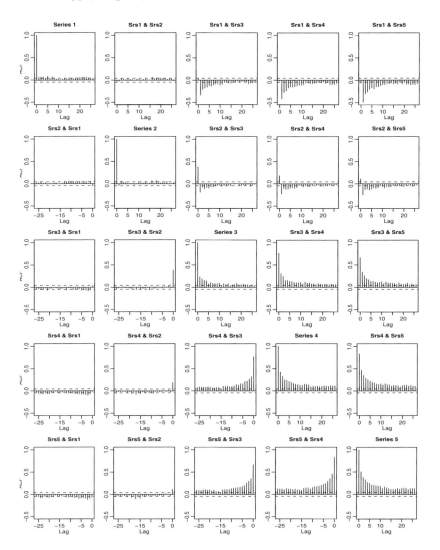

Fig. 5. The empirical autocorrelation of the 5 variate simulated sample for the processes X^{CM}. The parameters are given in Section 6.1, MODEL (A4).

ℓ	1	2	3	4	5
X^{AM}	0.34	0.35	0.30	0.34	0.34
X^{PM}	0.35	0.39	0.39	0.36	0.36
X^{MM}	-0.08	0.12	0.28	0.35	0.27
X^{CM}	0.30	0.34	0.29	0.35	0.37

Table 1. Estimation of long memory parameters calculated on each sub processes. The sample size is 10^4.

sub sample gives an estimation of its own parameter, namely $d_1 = d_2 = 0$ and $d_3 = d_4 = d_5 = 0.35$. For model X^{CM} the estimated parameter is the mean value $\bar{d}$, with value 0.35. It can be noticed that, despite the fact that the memory is not at all visible on Figure 5 for the two first sub samples, they nevertheless furnish not so bad estimations of $\bar{d}$. This is not surprising, since the Whittle estimator is only based on low frequencies of the periodogram, ignoring the small lags in the covariance.

6.2 Other example of X^{CM} processes

As noticed before, the sequence $\mathbf{d}$ on which is based a X^{CM} process can leave the stability domain $]0, 1/2[$, the only constraint concerning its mean value (see Remark 6). We present here a model for which this phenomenon is particularly marked, causing important bursts in the sample paths.

We consider the following sequence $\mathbf{d}$ with period $T = 150$ and mean value $\bar{d} = 0.3$

$$d_t = \begin{cases} -.4 & \text{if } t \in \{1, \ldots, 50\}, \\ .4 & \text{if } t \in \{51, \ldots, 100\}, \\ .9 & \text{if } t \in \{101, \ldots, 150\}. \end{cases} \tag{45}$$

Figure 6 well illustrates that the values of $(d_t)_{t \in \mathbb{Z}}$ outside the interval $]0, 1/2[$ create strong local nonstationarities, suggesting that models X^{CM} could be used to model non linearities.

7 Appendix

7.1 Proof of Lemma 1

Proof. Write

$$\sum_{j=1}^{N-1} \frac{L(j)s(j)}{L(N)} j^{2H-2} = \sum_{j=1}^{N-1} s(j) j^{2H-2} + \sum_{j=1}^{N-1} s(j) \left(\frac{L(j)}{L(N)} - 1 \right) j^{2H-2}$$

$$=: R_1 + R_2 .$$

Firstly, the term R_2 is negligible, i.e.

$$R_2 = o(N^{2H-1}) . \tag{46}$$

Indeed,

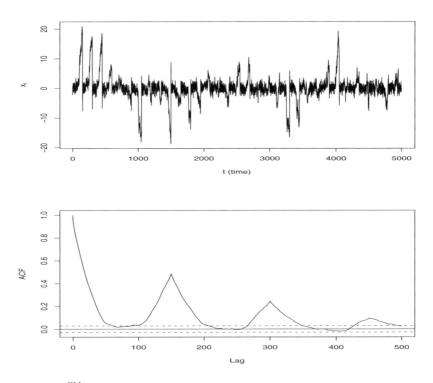

Fig. 6. X^{CM} process associated to the $150-$periodic sequence $(d_t)_{t\in\mathbb{Z}}$ defined in (45). *[top]* A simulated sample path of length 10^4. *[Bottom]* The empirical autocorrelations of the simulated sample

$$|R_2| \le C \sum_{j=1}^{N} \left|\frac{L(j)}{L(N)} - 1\right| j^{2H-2}$$

$$= C \left(\sum_{1\le j<N/2} \left|\frac{L(j)}{L(N)} - 1\right| j^{2H-2} + \sum_{N/2\le j\le N} \left|\frac{L(j)}{L(N)} - 1\right| j^{2H-2} \right)$$

$$=: C(R_{21} + R_{22}) \,.$$

Then,

$$R_{22} \le \sup_{x\in[1/2,\infty)} x^{-\kappa} \left|\frac{L(Nx)}{L(N)} - 1\right| \sum_{N/2\le j\le N} j^{2H-2}(j/N)^{\kappa} \,,$$

where κ is any positive number. Consequently,

$$\sum_{N/2\le j<N} j^{2H-2}(j/N)^{\kappa} \le N^{-\kappa} \sum_{1\le j<N} j^{2H-2+\kappa} = O(N^{2H-1}) \,,$$

and $R_{22} = o(N^{2H-1})$ follows from the well known property of slowly varying functions:

$$\sup_{x \in [x_0, \infty)} \frac{1}{x^\kappa} \left| \frac{L(Nx)}{L(N)} - 1 \right| \to 0, \qquad \text{for all } x_0, \kappa > 0 \,.$$

Relation $R_{21} = o(N^{2H-1})$ follows similarly by

$$\sup_{x \in (0, x_0]} x^\kappa \left| \frac{L(Nx)}{L(N)} - 1 \right| \to 0, \qquad \text{for all } x_0, \kappa > 0 \,.$$

This proves (46).

Let us now show that $N^{1-2H} R_1$ converges to $\frac{\bar{s}}{2H-1}$. Clearly, the statement follows from the fact that

$$\lim_{N \to \infty} N^{1-2H} \sum_{j=1}^{N} j^{2H-2} U(j) = 0 \,, \tag{47}$$

for any bounded sequence $U(j)$ such that $T(j) := j^{-1} \sum_{k=1}^{j} U(k)$ tends to zero as $j \to \infty$. To prove (47), first notice that since $T(j) \to 0$

$$N^{1-2H} \sum_{j=1}^{N} j^{2H-2} T(j-1) \to 0, \text{ as } N \to \infty \,.$$

Hence,

$$N^{1-2H} \sum_{j=1}^{N} j^{2H-2} U(j) = N^{1-2H} \Big(\sum_{j=1}^{N} j^{2H-1} (T(j) - T(j-1)) + \sum_{j=1}^{N} j^{2H-2} T(j-1) \Big)$$

$$= N^{1-2H} \sum_{j=1}^{N} j^{2H-1} (T(j) - T(j-1)) + o(1) \,.$$

A summation by parts leads to

$$N^{1-2H} \sum_{j=1}^{N} j^{2H-1} (T(j) - T(j-1)) =$$

$$T(N) + N^{1-2H} \sum_{j=1}^{N-1} (j^{2H-1} - (j+1)^{2H-1}) T(j) \,.$$

Then, as $N \to \infty$, $T(N) \to 0$ and

$$N^{1-2H} \sum_{j=1}^{N-1} |j^{2H-1} - (j+1)^{2H-1}| |T(j)| \le c N^{1-2H} \sum_{j=1}^{N-1} j^{2H-2} |T(j)| \to 0 \,.$$

Therefore (47) is proved. $\qquad\qquad \square$

7.2 Proof of Lemma 2

For any $\delta > 0$, there exists $K < \infty$ such that for all $s < t$, $t - s > K$

$$\alpha_{s,t} := \sum_{s<u<t} (d_u - D') < 0 \quad \text{with} \quad D' = \delta + \bar{d} \,. \tag{48}$$

Let $D = \bar{d} + 2\delta$. Then, for all $s < t, t - s > K$ we have

$$\sum_{s<u<t} \frac{d_u - D}{t - u} = -\delta \sum_{s<u<t} \frac{1}{t - u} + \sum_{s<u<t} \frac{d_u - D'}{t - u} =: -\delta J'_{s,t} + J''_{s,t} \,,$$

where $J'_{s,t} \to \infty$ as $t - s \to \infty$. Therefore, it suffices to prove that $J''_{s,t}$ is bounded above. Using (48) and summation by parts,

$$J''_{s,t} = \frac{h_{s,t}}{t - s - 1} + \sum_{i=1}^{t-s-2} h_{t-i-1,t}\left(\frac{1}{i} - \frac{1}{i+1}\right) < \sum_{i=1}^{t-s-2} \frac{h_{t-i-1,t}}{i(i+1)} =: I_{s,t} \,.$$

Then, for fixed K,

$$I_{s,t} = \sum_{i=1}^{K} \frac{h_{t-i-1,i}}{i(i+1)} + \sum_{i=K+1}^{t-s-2} \frac{h_{t-i-1,i}}{i(i+1)} \leq \sum_{i=1}^{K} \frac{h_{t-i-1,i}}{i(i+1)} \,,$$

which is bounded, uniformly in t. This proves $[H_D^{(a)}]$ with any $D > \bar{d}$. Similarly, we obtain that $[H_D^{(b)}]$ with any $D > \bar{d}$.

7.3 Proof of Proposition 5

Let $\mathbf{d}_\varepsilon$ be a periodic sequence such that $\|\mathbf{d} - \mathbf{d}_\varepsilon\|_\infty \leq \varepsilon$. Denote

$$\alpha_{s-j}(s) = d_{s-1} \ldots (d_j + s - j - 1) \,,$$

and $\alpha_{s-j}^{(\varepsilon)}(s)$ the corresponding sequence built from $\mathbf{d}_\varepsilon$. The familiar inequality

$$\left| \prod_{j=1}^{k} a_j - \prod_{j=1}^{k} b_j \right| \leq$$

$$|a_k - b_k| \prod_{j=1}^{k-1} |a_j| + |a_{k-1} - b_{k-1}| \prod_{j=1}^{k-2} |a_j||b_k| + \ldots + |a_1 - b_1| \prod_{j=2}^{k} |b_j| \,,$$

leads to

$$\Delta_{s-j}^{(\varepsilon)}(s) := |\alpha_{s-j}^{(\varepsilon)}(s) - \alpha_{s-j}(s)| \leq (s - j)\varepsilon(2\|\mathbf{d}\|_\infty)^{s-j-1} \,. \tag{49}$$

Let us now compare $\text{cov}(X_s^{\text{CM}}, X_t^{\text{CM}})$ to the periodic sequence

$$\sigma^{(\varepsilon)}(s,t) = \sum_{j<s} \frac{\alpha^{(\varepsilon)}_{s-j}(s)\alpha^{(\varepsilon)}_{t-j}(t)}{(s-j)!(t-j)!} \, .$$

¿From (49),

$$|\text{cov}(X^{\text{CM}}_s, X^{\text{CM}}_t) - \sigma^{(\varepsilon)}(s,t)| \leq \sum_{j<s} \frac{|\alpha^{(\varepsilon)}_{s-j}(s)\alpha^{(\varepsilon)}_{t-j}(t) - \alpha_{s-j}(s)\alpha_{t-j}(t)|}{(s-j)!(t-j)!}$$

$$\leq \sum_{j<s} \frac{|\alpha_{s-j}(s)||\Delta^{(\varepsilon)}_{s-j}(s)| + |\alpha^{(\varepsilon)}_{t-j}(t)||\Delta^{(\varepsilon)}_{s-j}(s)|}{(s-j)!(t-j)!}$$

$$\leq C\varepsilon \sum_{j<s} \frac{(2\|\mathbf{d}\|_\infty)^{s+t-2j}}{(s-j)!(t-j)!}$$

$$\leq C\varepsilon \left(\sup_{l\geq 0} \frac{(2\|\mathbf{d}\|_\infty)^l}{l!} \right) \sum_{j\geq 1} \frac{(2\|\mathbf{d}\|_\infty)^j}{j!} \, ,$$

with some constant C, and the proof is over.

7.4 Proof of Lemma 3

Proof of (i). For fixed $n \geq 1$, we define the sequence γ^n by

$$\gamma^n(v) := \prod_{p=1}^n \left(1 + \frac{d_{v-p} - \overline{d}}{\overline{d} + p - 1} \right) =: \prod_{p=1}^n (1 + \beta_p(v)) \, ,$$

The sequence β_p is almost periodic for any $p \geq 1$, hence so is γ^n for any $n \geq 1$. It is well-known that uniform limit of almost periodic functions is almost periodic. The statement of the lemma thus follows from

$$\sup_{v\in\mathbb{Z}} |\gamma^n(v) - \gamma^m(v)| \to 0 \, , \quad \text{as} \quad m, n \to \infty \, .$$

Since $\mathbf{d}$ is bounded, there exists n_0 such that $|\beta_p(v)| < 1/2$ holds for any $p \geq n_0, v \in \mathbb{Z}$. Now, uniform convergence of γ^n is equivalent to the uniform convergence of γ^n/γ^{n_0}, and we can suppose in the sequel that $n_0 = 1$, or that $|\beta_p(v)| < 1/2$ holds for all $p \geq 1$.

Then $\gamma^n(v) > 0 \ \forall n \geq 1$, and

$$|\gamma^n(v) - \gamma^m(v)| \leq \gamma^n(v) \left| \prod_{p=n+1}^m (1 + \beta_p(v)) - 1 \right| \quad (n < m) \, .$$

Using the inequalities $e^{x-x^2} \leq 1 + x \leq e^x$ for $|x| < 1/2$ we obtain

$$\gamma^n(v) \leq \exp\left\{\sum_{p=1}^{n} \beta_p(v)\right\},$$

$$\exp\left\{\sum_{p=n+1}^{m} \beta_p(v) - \sum_{p=n+1}^{m} \beta_p^2(v)\right\} \leq \prod_{p=n+1}^{m} (1 + \beta_p(v)) \leq \exp\left\{\sum_{p=n+1}^{m} \beta_p(v)\right\}.$$

Here, $\sum_{p=n+1}^{m} \beta_p^2(v) \leq Cn^{-1}$, due to $\beta_p^2(v) \leq Cp^{-2}$.

Let $D_{n,m}(v) := \sum_{p=n+1}^{m} (d_{v-p} - \bar{d})$. A summation by parts leads to

$$\sum_{p=n+1}^{m} \beta_p(v) = \frac{D_{n,m}(v)}{\bar{d} + m - 1} + \sum_{p=n+1}^{m-1} D_{n,p}(v)\left(\frac{1}{\bar{d} + p - 1} - \frac{1}{\bar{d} + p}\right). \quad (50)$$

By (32),

$$|D_{n,m}(v)| \leq C|n - m|^{1-\delta}$$

where the constant C does not depend on v.

The first term on the r.h.s. of (50) does not exceed

$$C|n - m|^{1-\delta} m^{-1} \leq Cn^{-\delta},$$

while

$$\sum_{p=n+1}^{m-1} |D_{n,p}(v)|\left|\frac{1}{\bar{d} + p - 1} - \frac{1}{\bar{d} + p}\right| \leq C \sum_{p=n+1}^{\infty} |p - n|^{1-\delta} p^{-2} \leq Cn^{-\delta}.$$

Therefore, for n sufficiently large,

$$e^{-Cn^{-\delta}} \leq \prod_{p=n+1}^{m} (1 + \beta_p(v)) \leq e^{Cn^{-\delta}}, \quad (51)$$

for some constant $C > 0$ independent of v, n and $m > n$. It also follows that $\sup_{n \geq 1, v \in \mathbb{Z}} \gamma_A^n(v) < \infty$, and (i) is proved.

Now let us prove (ii). By definition,

$$\theta_{u,v} = a_{v-u}/\gamma(v)\psi_{v-u}(\bar{d}) = \prod_{p=v-u}^{\infty} (1 + \beta_p(v))^{-1},$$

Then from (51), $e^{-C(v-u)^{-\delta}} \leq \theta_{u,v} \leq e^{C(v-u)^{-\delta}}$ for all $v - u$ sufficiently large, yielding $|\theta_{u,v} - 1| = O(|v - u|^{-\delta})$ and (ii) is proved.

References

[Arc94]   Arcones, M. (1994). Limit theorems for nonlinear functionals of a stationary sequence of vectors. *The Annals of Probability*, 22:2242–2274.

[BLO03]   Bardet, J., Lang, G., Oppenheim, G., Philippe, A., Stoev, S., and Taqqu, M. (2003). Semi-parametric estimation of the long range dependent processes : A survey. In Doukhan, P., Oppenheim, G., and Taqqu, M., editors, *Theory and Applications of Long Range Dependence*. Birkhäuser, Boston.

[BL01]   Basawa, I. and Lund, R. (2001). Large sample properties of parameter estimates for periodic ARMA models. *Journal of Time Series Analysis*, 22(6):651–666.

[BLS04]   Basawa, I., Lund, R., and Shao, Q. (2004). First-order seasonal autoregressive processes with periodically varying parameters. *Stat. Probab. Lett.*, 67(4):299–306.

[BH94]   Bentarzi, M. and Hallin, M. (1994). On the invertibility of periodic moving-average models. *Journal of Time Series Analysis*, 15(3):263–268.

[Ber94]   Beran, J. (1994). *Statistics for long memory processes*, volume 61 of *Monographs on Statistics and Applied Probability*. Chapman and Hall, New York.

[Bil68]   Billingsley, P. (1968). *Convergence of probability measures*. John Wiley & Sons Inc., New York.

[BD91]   Brockwell, P. J. and Davis, R. A. (1991). *Time Series: Theory and Methods*. Springer-Verlag, New York, 2nd edition.

[DV97]   Daley, D. and Vesilo, R. (1997). Long range dependence of point processes, with queueing examples. *Stochastic Processes and their Applications*, 70(2):265–282.

[DV00]   Daley, D. and Vesilo, R. (2000). Long range dependence of inputs and outputs of some classical queues. In *McDonald, David R. (ed.) et al., Analysis of communication networks: call centres, traffic and performance*, pages 179–186.

[Dav70]   Davydov, Y. (1970). The invariance principle for stationary processes. *Theory of Probability and its Applications*, 15:487–498.

[DH94]   Dehay, D. and Hurd, H. L. (1994). Representation and estimation for periodically and almost periodically correlated random processes. In *Cyclostationarity in communications and signal processing.*, pages 295–326. New York, IEEE.

[Gla61]   Gladyshev, E. G. (1961). Periodically correlated random sequences. *Soviet mathematics*, 2.

[Gla63]   Gladyshev, E. G. (1963). Periodically and almost PC random processes with continuous time parameter. *Theory of Probability and its Applications* , 8.

[GZW89]   Gray, H. L., Zhang, N.-F., and Woodward, W. (1989). On generalized fractional processes. *Journal of Time Series Analysis*, 10:233–258.

[HL95]   Hui, Y. and Li, W. (1995). On fractionally differenced periodic processes. *Sankhya, The Indian Journal of Statistics. Series B*, 57(1):19–31.

[HMM02]   Hurd, H., Makagon, A., and Miamee, A. G. (2002). On AR(1) models with periodic and almost periodic coefficients. *Stochastic Process. Appl.*, 100:167–185.

[LV00]   Leipus, R. and Viano, M.-C. (2000). Modelling long memory time series with finite or infinite variance: a general approach. *Journal of Time Series Analysis*, 21(1):61–74.

[LB00]   Lund, R. and Basawa, I. (2000). Recursive prediction and likelihood evaluation for periodic ARMA models. *Journal of Time Series Analysis*, 21(1):75–93.

[LB99]   Lund, R. B. and Basawa, I. V. (1999). Modeling and inference for periodically correlated time series. In *Asymptotics, nonparametrics, and time series*, volume 158 of *Statist. Textbooks Monogr.*, pages 37–62. Dekker, New York.

[Mar05]   Marinucci, D. (2005). The empirical process for bivariate sequences with long memory. *Statistical Inference for Stochastic Processes*, 8(2):205–223.

[OOV00]   Oppenheim, G., Ould Haye, M., and Viano, M.-C. (2000). Long memory with seasonal effects. *Statistical Inference for Stochastic Processes*, 3:53–68.

[OV03]   Ould Haye, M. and Viano, M.-C. (2003). Limit theorems under seasonal long memory. In Doukhan, P., Oppenheim, G., and Taqqu, M., editors, *Theory and Applications of Long Range Dependence*. Birkhäuser, Boston.

[PSV04]   Philippe, A., Surgailis, D., and Viano, M.-C. (2004). Time-varying fractionally integrated processes with nonstationary long memory. Technical report, Pub. IRMA Lille, 61(9).

[Sur83]   Surgailis, D. (1983). Domains of attraction of self-similar multiple integrals. *Lithuanian Mathematical Journal*, 22(3):327–340.

[Sur03]   Surgailis, D. (2003). Non-CLTs: U-statistics, multinomial formula and approximations of multiple Itô-Wiener integrals. In Doukhan, P., Oppenheim, G., and Taqqu, M., editors, *Theory and Applications of Long Range Dependence*. Birkhäuser, Boston.

[Taq03]   Taqqu, M. S. (2003). Fractional Brownian motion and long range dependence. In Doukhan, P., Oppenheim, G., and Taqqu, M., editors, *Theory and Applications of Long Range Dependence*. Birkhäuser, Boston.

[VDO95]   Viano, M.-C., Deniau, C., and Oppenheim, G. (1995). Long range dependence and mixing for discrete time fractional processes. *Journal of Time Series Analysis*, 16(3):323–338.

Long memory random fields

Frédéric Lavancier

LS-CREST, ENSAE, 3 avenue Pierre Larousse, 92 245 Malakoff, France and
Laboratoire Paul Painlevé, UMR CNRS 8424, 59655 Villeneuve d'Ascq, France
`lavancier@ensae.fr`

1 Introduction

A random field $X = (X_n)_{n \in \mathbb{Z}^d}$ is usually said to exhibit long memory, or
strong dependence, or long-range dependence, when its covariance function
$r(n)$, $n \in \mathbb{Z}^d$, is not absolutely summable : $\sum_{n \in \mathbb{Z}^d} |r(n)| = \infty$. An alternative
definition involves spectral properties : a random field is said to be strongly
dependent if its spectral density is unbounded. These two points of view are
closely related but not equivalent.

Generalizing a hypothesis widely used in dimension 1, most studies on
long-range dependent random fields assume that the covariance function behaves at infinity as

$$r(h) \underset{h \to \infty}{\sim} |h|^{-\alpha} L(|h|) \, b\left(\frac{h}{|h|}\right) , \qquad 0 < \alpha < d , \tag{1}$$

where L is slowly varying at infinity and where b is continuous on the unit
sphere of $\mathbb{R}^d$, $|.|$ denoting the l^1-norm on $\mathbb{R}^d$.

Even if the form (1) is not exactly isotropic because of the presence of the
function b defined on the unit sphere, the long memory is due to the term
$|h|^{-\alpha}$ which depends only on the norm. So we will call isotropic this kind of
long-range dependence. Let us focus on the spectral domain to precise this
notion of isotropy.

Definition 1. *A stationary random field exhibits isotropic long memory if it
admits a spectral density which is continuous everywhere except at 0 where*

$$f(x) \sim |x|^{\alpha - d} L\left(\frac{1}{|x|}\right) b\left(\frac{x}{|x|}\right) , \qquad 0 < \alpha < d , \tag{2}$$

*where L is slowly varying at infinity and where b is continuous on the unit
sphere in $\mathbb{R}^d$.*

Conditions (1) and (2) are linked by a result of [Wai65] who proved that if the covariance of a random field satisfies (1) and if its spectral density is continuous outside 0, then this random field exhibits isotropic long memory according to definition 1.

Conditions (1) and (2) are regular ways for a random field to be strongly dependent. Now, it is easy to build long memory random fields which fail to satisfy these conditions, either by filtering white noises through unbounded filters like some special AR filters or by aggregating random parameters short memory random fields. Besides, non-isotropic long memory fields naturally arise in statistical mechanics in relatively simple situations of phase transition.

So, the aim of the paper is to give a presentation as complete as possible of isotropic or non-isotropic long memory random fields.

In the first section, we present families of models presenting different kinds of long memory with special glance to Ising model and Gaussian systems in the more specific domain of statistical mechanics.

In the second section, we present a review of the available limit theorems. The first part is devoted to the convergence of partial sums and the second part to the empirical process. We present some well-known results concerning the isotropic long-memory setting : the asymptotic behaviour of the partial sums investigated by [DM79] for Gaussian subordinated fields and by [Sur82] for functionals of linear fields ; the convergence of the empirical process for linear fields proved in [DLS02]. We also give the asymptotic behaviour of the partial sums and of the empirical process in some non-isotropic long memory cases. For these new results, we explain the scheme of proof, based on a spectral convergence theorem. In both situations of isotropic and non-isotropic strong dependence, we observe, like in dimension $d = 1$, a non standard rate of convergence and a non standard limiting process.

2 Modeling long memory stationary random fields

We present two classes of long-memory stationary random fields. The first class is a straightforward generalization of models now widely used for random processes ($d = 1$). The second one comes from mechanical statistics and is for that reason specifically adapted to dimensions $d > 1$.

2.1 Filtering and aggregation

Filtering white noises through unbounded filters or aggregating random coefficients ARMA processes are the two main ways leading to long-memory processes. Since the pioneer works of [GJ80], [Gra80] and [Hos81], these methods have been generalized and improved, providing large families of long-memory one-dimensional processes. See for instance [BD91] for filtered processes and [OV04] for aggregation schemes. These methods are easily extended to dimensions $d > 1$. In fact they lead to rather close covariance structures, but the aggregation method produces only Gaussian random fields. Both provide useful simulation methods.

Filtering

Let us consider a zero-mean white noise $(\epsilon_n)_{n \in \mathbb{Z}^d}$ with spectral representation

$$\epsilon_n = \int_{[-\pi,\pi]^d} e^{i<n,\lambda>} dZ(\lambda) \, ,$$

where the control measure of Z has constant density $\sigma^2/(2\pi)^d$ on $[-\pi,\pi]^d$, and the random field X obtained from ϵ by the filtering operation

$$X_n = \int_{[-\pi,\pi]^d} e^{i<n,\lambda>} a(\lambda) dZ(\lambda) \, , \tag{3}$$

where $a \in L^2([-\pi,\pi]^d)$.

The spectral density of the induced field is

$$f_X(\lambda) = \frac{\sigma^2}{(2\pi)^d} |a(\lambda)|^2 \, , \quad \forall \lambda \in [-\pi,\pi]^d \, , \tag{4}$$

and long-memory is achieved when a is unbounded at certain frequencies.

Example 1 (Long memory ARMA fields). ARMA fields are obtained when $a(\lambda) = \frac{Q}{P}(e^{i\lambda})$ where P and Q are polynomial functions. Denoting by L_j the lag operator for index j , i.e.

$$L_j X_{n_1,n_2\ldots,n_d} = X_{n_1,\ldots,n_{j-1},n_j-1,n_{j+1},\ldots,n_d} \, ,$$

we can write an ARMA field in the most popular way

$$P(L_1,\ldots,L_d) X_{n_1,\ldots,n_d} = Q(L_1,\ldots,L_d) \epsilon_{n_1,\ldots,n_d} \, . \tag{5}$$

If $P(e^{i\lambda}) \neq 0$ for all $\lambda \in [-\pi,\pi]^d$, (5) admits a unique stationary solution (cf. for instance [Ros85] and [Guy93]).

But contrary to the one dimensional case, this condition is not necessary when $d > 1$, and there exist stationary fields having an ARMA representation (5) with $P(e^{i\lambda}) = 0$ at some frequencies λ. In this case, the induced field X exhibits long memory since its spectral density, given by (4), is unbounded.

The following ARMA representation in dimension $d = 5$ is a trivial example of this phenomena :

$$X_{n_1,\ldots,n_5} - \frac{1}{5}(X_{n_1-1,n_2,\ldots,n_5} + X_{n_1,n_2-1,n_3,n_4,n_5} + \cdots + X_{n_1,\ldots,n_5-1}) = \epsilon_{n_1,\ldots,n_5} \, .$$

This representation admits a stationary solution since the filter $a(\lambda_1,\ldots,\lambda_5) = \left(1 - \frac{1}{5}(e^{i\lambda_1} + \cdots + e^{i\lambda_5})\right)^{-1}$ is in $L^2([-\pi,\pi]^5)$, and the induced field X is strongly dependent because its spectral density is unbounded at $\lambda = 0$.

Example 2 (Fractional integration). Generalizing the FARIMA processes defined by

$$(I - L)^\alpha X_n = \epsilon_n ,$$

we consider random fields of the form

$$(P(L_1, \dots, L_d))^\alpha X_{n_1, \dots, n_d} = \epsilon_{n_1, \dots, n_d} ,$$

where P is a polynomial having roots on the unit circle and where $\alpha > 0$ is chosen such that $a(\lambda) = \left(P(e^{i\lambda})\right)^{-\alpha} \in L^2([-\pi, \pi]^d)$.

As an example, consider, for a fixed positive integer k, the model

$$(I - L_1 L_2^k)^\alpha X_{n_1, n_2} = \epsilon_{n_1, n_2} ,$$

where $0 < \alpha < 1/2$. The spectral density of X is

$$f_X(\lambda_1, \lambda_2) = \frac{\sigma^2}{4\pi^2} \left| 1 - e^{i(\lambda_1 + k\lambda_2)} \right|^{-2\alpha} ,$$

where σ^2 is the variance of the white noise ϵ. The field X exhibits non-isotropic long memory since f_X is unbounded all over the line $\lambda_1 + k\lambda_2 = 0$ and fails to satisfy (2). Using well known results on FARIMA processes (cf [BD91]) easily leads to:

$$\begin{cases} \rho(h, kh) = \prod_{0 < j \le h} \frac{j - 1 + \alpha}{j - \alpha} & h = \pm 1, \pm 2, \dots \\ \rho(h, l) = 0 & \text{if } l \ne kh , \end{cases}$$

where ρ denotes the correlation function of X. The field X has a non summable correlation function in the direction $l = kh$ since $\rho(h, kh)$ is asymptotically proportional to $h^{2\alpha - 1}$. Compared to (1), this confirms that X is a non-isotropic long memory random field.

Aggregation

Let us consider a sequence $(X^{(q)})_{q \ge 1}$ of independent copies of the field

$$P(L_1, \dots, L_d) X_{n_1, \dots, n_d} = \epsilon_{n_1, \dots, n_d} , \tag{6}$$

where P is a polynomial function with random coefficients such that P has almost surely no roots on the unit sphere and $(\epsilon_n)_{n \in \mathbb{Z}^d}$ is a zero-mean white noise with variance σ^2.

The representation (6) admits almost surely the solution :

$$X_n = \sum_{j \in \mathbb{Z}^d} c_j \epsilon_{n-j} , \tag{7}$$

where $(c_j)_{j \in \mathbb{Z}^d}$ are the coefficients of the Laurent expansion of P^{-1}. The field X given by (7) belongs to L^2 if and only if

$$\sum_{j \in \mathbb{Z}^d} \mathbb{E}(|c_j|^2) < \infty , \tag{8}$$

and its spectral density is

$$f(\lambda) = \frac{\sigma^2}{(2\pi)^d} \mathbb{E} \left| P^{-1}\left(e^{i\lambda}\right) \right|^2 . \tag{9}$$

Now, from the central limit theorem, the finite dimensional distributions of $N^{-1/2} \sum_{q=0}^{N} X_n^{(q)}$ converge as $N \to \infty$ to the so-called aggregated field Z

$$Z_n = \lim_{N \to \infty} \frac{1}{\sqrt{N}} \sum_{q=0}^{N} X_n^{(q)} , \quad n \in \mathbb{Z}^d .$$

This process is Gaussian and has the same second order characteristics as the $X^{(q)}$'s. In particular, its spectral density is (9) and long memory is obtained when $\mathbb{E} \left| P^{-1}\left(e^{i\lambda}\right) \right|^2$ is unbounded.

Example 3. Let us consider, in dimension $d = 2$, the AR representation

$$X_{n,m} - aX_{n-1,m} - bX_{n,m-1} + abX_{n-1,m-1} = \epsilon_{n,m} , \tag{10}$$

where a and b are independent and where a (resp. b) has on $[0,1]$ the density

$$(1-x)^\alpha \Phi_1(x) , \quad (\text{resp. } (1-x)^\beta \Phi_2(x)) , \tag{11}$$

where $0 < \alpha, \beta < 1$ and where Φ_j, $j = 1,2$ are bounded, continuous at $x = 1$, with $\Phi_1(1)\Phi_2(1) \neq 0$.

It is easily checked that the above random parameters AR fields satisfy all the required conditions to lead to an aggregated random field with long memory (see for instance [OV04]). The spectral density of the aggregated field Z is a tensorial product and

$$f(\lambda_1, \lambda_2) \sim c|\lambda_1|^{\alpha-1}|\lambda_2|^{\beta-1} \quad \text{when } \lambda \to 0 ,$$

where c is a positive constant. Therefore Z exhibits long memory.

Example 4. Consider the AR representation

$$X_{n,m} - aX_{n+k,m-1} = \epsilon_{n,m} , \tag{12}$$

where $k \in \mathbb{Z}$ is fixed and where a is a random parameter on $[0,1]$ with density (11).

The spectral density of the induced aggregated field Z is unbounded on the line $\lambda_2 = k\lambda_1$ since

$$f(\lambda_1, \lambda_2) \sim c\,|\lambda_2 - k\lambda_1|^{\alpha-1} , \quad \text{as } \lambda_2 - k\lambda_1 \to 0$$

where c is a positive constant. Hence the long-memory is non-isotropic.

We present two 2-dimensional models produced by aggregating $N = 1000$ autoregressive fields with random parameters. The first one (figure 1) is constructed according to the scheme of example 3, the parameters a and b having the same density $\frac{3}{2}\sqrt{1-x}$.

The second model (figure 2) is constructed as in example 4 with $k = -1$ and where a has the same density as above. For both models, an image of size 100×100 has been obtained where, at each point, the realization of the random variable is represented by a level of gray.

Anisotropy clearly appears in figure 2. The strong dependence only occurs in one direction and the long memory is non-isotropic. Its periodogram is unbounded all over the line $\lambda_2 + \lambda_1 = 0$ and fails to follow (2). Moreover, its covariance function decays slowly in only one direction and is not of the form (1). In contrast, in the first model, strong dependence occurs along two directions, with the same intensity. This is the reason why the phenomena of anysotropy is less visible in figure 1.

2.2 Long memory in statistical mechanics

Statistical mechanics explains the macroscopic behaviour of systems of particles by their microscopic properties and provides interpretations of thermodynamic or magnetic phenomena like phase transition. There is phase transition when a system is unstable. For instance, it is the case during the liquid-vapour transition of a gas or when a magnetic material is in transition between the ferromagnetic and the paramagnetic phase. A rigorous mathematical formalism of statistical mechanics can be found for example in [Geo88]. Our aim is to underline the strong dependence properties of some systems in phase transition by focusing on the Ising model and on systems based on quadratic interactions.

Let us consider a system of particles on the lattice $\mathbb{Z}^d$. The state of a particle located on $j \in \mathbb{Z}^d$ is described by the spin x_j, a random variable with values in a polish space X. The pair potential $\Phi = (\Phi_{i,j})_{i,j \in \mathbb{Z}^d}$ gives the interactions between the pairs of particles.

A system configuration is an element $\omega = (x_i)_{i \in \mathbb{Z}^d}$ of the space $\Omega = X^{\mathbb{Z}^d}$. The energy on each finite set Λ of $\mathbb{Z}^d$ involves not only the energy quantity inside the set Λ but also the edges interactions:

$$E_\Lambda(\omega) = \sum_{\{i,j\} \subset \Lambda} \Phi_{i,j}(x_i, x_j) + \sum_{\substack{i \in \Lambda \\ j \in \Lambda^c}} \Phi_{i,j}(x_i, x_j) \ . \tag{13}$$

Now, consider on Ω an a priori measure $\rho = \otimes_{i \in \mathbb{Z}^d} \rho_i$ (typically ρ_i is the Lebesgue measure when $X = \mathbb{R}$ or a Bernoulli measure when $X = \{\pm 1\}$). A measure μ on Ω is called a Gibbs measure associated with the potential Φ with respect to ρ if, for every finite set Λ, ω_Λ and ω_{Λ^c} denoting the restriction of ω to Λ and to its complementary set,

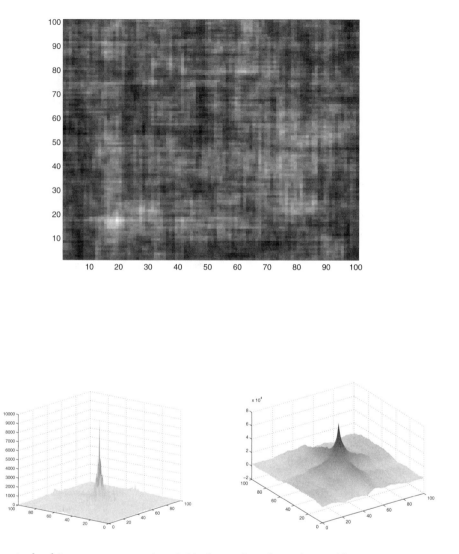

Fig. 1. *[top]* Long memory random field of a product form obtained by aggregating random parameters AR fields of the form (10) with $\alpha = \beta = 0.5$ *[bottom-left]* Its periodogram *[bottom-right]* Its covariance function

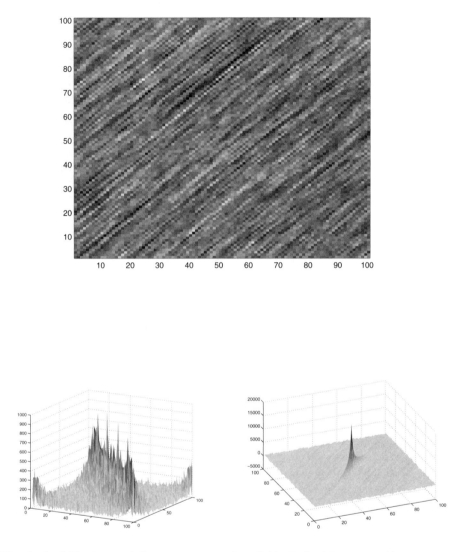

Fig. 2. *[top]* Non-isotropic long memory random field obtained by aggregating random parameters AR fields of the form (12) with $k = -1$ *[bottom-left]* Its periodogram *[bottom-right]* Its covariance function

$$\mu\left(d\omega_\Lambda | \omega_{\Lambda^c}\right) = \frac{1}{Z_\Lambda(\omega_{\Lambda^c})} e^{-E_\Lambda(\omega)} \rho(d\omega) , \qquad (14)$$

where $Z_\Lambda(\omega_{\Lambda^c})$ is a normalizing constant.

A Gibbs measure is locally characterized by (14). This formalism, attributed to Dobrushin, Landford and Ruelle, guarantees the coherence of the conditional distributions.

For a given system, a fundamental question is whether a Gibbs measure exists or not. Phase transition occurs when there exists several Gibbs measures. The set of all Gibbs measures is a convex set whose extreme elements are the pure phases, the other Gibbs measures being mixtures of the pure phases.

Now, consider the spins' system equipped with the Gibbs measure as a random field. When the second order moments exist, we can measure the memory of the spins' system via the covariance between two sites i and j, $r(i,j) = \text{cov}(x_i, x_j)$. In the following examples the field is stationary ($r(h) = \text{cov}(x_i, x_{i+h})$) and presents long-range dependence properties.

The Ising model

The well known Ising model has been introduced to study magnetism and fluid dynamic. The state space is $X = \{-1, 1\}$, the a priori measure is the Bernoulli measure $1/2(\delta_{-1} + \delta_1)$ and the potential is restricted to the nearest neighbors:

$$\Phi_{i,j}(x_i, x_j) = \begin{cases} \beta x_i x_j & \text{if } |i - j| = 1 \\ 0 & \text{otherwise,} \end{cases}$$

where $\beta > 0$ is a constant representing the inverse temperature.

In dimension $d = 1$, there exists a unique Gibbs measure for any β, therefore the system is never in phase transition. In dimension $d \geq 2$, Gibbs measures exist and phase transition takes place if β is greater than a critical value β_c depending on the dimension d (see [Ons44] in dimension $d = 2$ and [Dob65] in any dimension). When $d = 2$, $\beta_c = \frac{1}{2}\ln(1 + \sqrt{2}) \approx 0.441$.

Let us consider the covariance function. In their physical approach of the Ising model, [KO49] and [Fis64] obtain the asymptotic behaviour of r. When $\beta \neq \beta_c$, the covariance function decays exponentially but when $\beta = \beta_c$ the rate of decay is slow and the covariance is not summable. We have

$$r(h) \underset{h \to \infty}{\sim} \begin{cases} |h|^{-1} e^{-\kappa|h|} & \text{if } \beta \neq \beta_c \\ |h|^{-(d-2+\mu)} & \text{if } \beta = \beta_c , \end{cases}$$

where $\kappa > 0$ is the Boltzmann's constant and $\mu \in [0, 2]$ is a critical parameter which is $1/4$ in case $d = 2$. The strong dependence at the critical point is isotropic.

Remark 1. The long-range dependence structure of the Ising model at the critical point was pointed out in [CJL78] where one can also find others models exhibiting long memory.

Remark 2. There exist some models, slightly more complex than the Ising model, which exhibit long-range dependence without being in phase transition. This is the case for the XY model and for the Heiseinberg model : they are never in phase transition when $d \leq 2$ but their covariance function in dimension $d = 2$ is not summable all over an interval of low temperatures (see [KT78]).

Homogeneous Gaussian models

The state space is $X = \mathbb{R}$, the a priori measure ρ is the Lebesgue measure and the potential is

$$\Phi_{i,j}(x_i, x_j) = \begin{cases} \beta \left(\frac{1}{2} J(0) x_i^2 + ex_i \right) & \text{if } i = j \\ \beta J(i - j) x_i x_j & \text{if } i \neq j, \end{cases}$$

where β and e are constants representing respectively the inverse temperature and an external magnetic field and where $(J(i))_{i \in \mathbb{Z}^d}$ is a positive definite real sequence with $J(i) = J(-i)$ for every i and $\sum_{i \in \mathbb{Z}^d} J(i) < \infty$. We suppose for simplicity that $e = 0$. Contrary to the Ising model, the temperature has no influence on the appearance of phase transition. The main parameter is the sequence J, improperly named potential.

This system was studied by [Kün80] and [Dob80]. All the results can be found in [Geo88]. The pure phases are Gaussian and their characteristics are directly linked to the potential J via its Fourier transform

$$\hat{J}(\lambda) = \sum_{n \in \mathbb{Z}^d} J(n) e^{i<n,\lambda>} , \quad \lambda \in [-\pi, \pi]^d .$$

Theorem 1 (Künsch, Dobrushin). *Under the above hypotheses on J and in the case $e = 0$, the set of Gibbs measures is non empty if and only if*

$$\int_{[-\pi,\pi]^d} \hat{J}^{-1}(\lambda) d\lambda < \infty .$$

In this case, the pure phases are the Gaussian measures with covariance function

$$r(h) = \int_{[-\pi,\pi]^d} \hat{J}^{-1}(\lambda) e^{i<h,\lambda>} d\lambda \tag{15}$$

and with mean vector a sequence $(u_n)_{n \in \mathbb{Z}^d}$ such that, for all $k \in \mathbb{Z}^d$,

$$\sum J(n) u_{k+n} = 0 .$$

Remark 3. In the case $e \neq 0$, further hypotheses are needed for the existence of a Gibbs measure.

The occurrence of phase transition in the particular case $e = 0$ can be deduced from Theorem 1 and is given in the following corollary. Note that, despite the fact that the pure phases are Gaussian, all Gibbs measures are not necessarily so. Phase transition can take place with one or several measures without second moment. Insofar as we are interested in the covariance function, the corollary is stated in the L^2 setting:

Corollary 1 (Künsch). *Under the hypotheses of Theorem 1, there exist several Gibbs measures with finite second moments if and only if $\hat{J}$ has at least one root in $[-\pi, \pi]^d$.*

Therefore, when the system is in phase transition, every Gibbs measure having a finite second moment is strongly dependent. Indeed $\hat{J}^{-1}$, which is the spectral density of the pure phases, according to (15), is unbounded if there is phase transition.

Example 5. In dimension $d \geq 3$, the harmonic potential is a simple example of finite range interaction leading to long-memory random fields. The potential is defined by:

$$J(n) = \begin{cases} -\frac{1}{2d} & \text{if } |n| = 1 \\ 1 & \text{if } n = 0 \\ 0 & \text{otherwise} \end{cases}$$

and we have

$$\hat{J}(\lambda) = 1 - \sum_{|n|=1} \frac{1}{2d} e^{i<n,\lambda>} = 1 - \frac{1}{d} \sum_{k=1}^{d} \cos(\lambda_k)$$

whose inverse is integrable on $[-\pi, \pi]^d$ since $d \geq 3$. Hence, Theorem 1 guarantees the existence of a Gibbs measure associated with this potential. Moreover $\hat{J}(0) = 0$ and according to Corollary 1, the system is in phase transition and the second order Gibbs measures exhibit long memory. The long-range dependence is isotropic in the sense of definition 1.

Example 6. In dimension $d = 2$, consider the potential :

$$J(k, l) = \begin{cases} \prod_{0 < j \leq k} \frac{j - 1 - \alpha}{j + \alpha} & \text{if } l = pk, \ |k| > 1 \\ 1 & \text{if } k = l = 0 \\ 0 & \text{otherwise} \end{cases}$$

where p is a non null fixed integer and $\alpha \in]0, 1/2[$.

The sequence $J(k, pk)$ corresponds to the autocorrelation function of an integrated stationary process of order α, from which (see [BD91])

$$J(k, pk) \sim \frac{\Gamma(1 + \alpha)}{\Gamma(-\alpha)} k^{-2\alpha - 1}, \quad \text{when} \quad k \to \infty .$$

This shows the summability of J. Moreover, using the well known properties of the FARIMA processes,

$$\sum_{k \in \mathbb{Z}} J(k, pk) e^{ik\lambda} \frac{\Gamma^2(1+\alpha)}{\Gamma(1+2\alpha)} \left| 2 \sin\left(\frac{\lambda}{2}\right) \right|^{2\alpha}.$$

Finally

$$\hat{J}(\lambda_1, \lambda_2) = \sum_{k,l \in \mathbb{Z}^2} J(k, l) e^{i(k\lambda_1 + l\lambda_2)} = \sum_{k \in \mathbb{Z}} J(k, pk) e^{ik(\lambda_1 + p\lambda_2)}$$

$$= \frac{\Gamma^2(1+\alpha)}{\Gamma(1+2\alpha)} \left| 2 \sin\left(\frac{\lambda_1 + p\lambda_2}{2}\right) \right|^{2\alpha}.$$

Since $\alpha \in]0, 1/2[$, $\hat{J}^{-1}$ is integrable on $[-\pi, \pi]^2$ and the existence of a Gibbs measure is guaranteed by Theorem 1. In addition, $\hat{J}$ vanishes all along the line $\lambda_1 + p\lambda_2 = 0$ which shows that the system is in phase transition according to Corollary 1 and that the Gibbs measures exhibit non-isotropic long memory.

3 Limit theorems under isotropic and non-isotropic strong dependence

We present some limit theorems for the partial sums process and the doubly indexed empirical process of long memory random fields.

3.1 Partial sums of long memory random fields

Since the results for isotropic long-memory fields are nearly classical while those related to non-isotropic long memory are newer and still incomplete, we split this section in two parts according to the regularity of the strong dependence. The first one is devoted to isotropic long memory: available results concern Gaussian subordinated fields and some particular functionals of linear fields. In the second part, related to non-isotropic long memory, we first present the spectral convergence theorem on which is based the convergence of the partial sums. Then we apply it to some non-isotropic long memory fields.

In a third part, we give a tightness criterion for partial sums and we apply it to situations needed for the doubly-indexed empirical process treated in the next section.

In the sequel we adopt the notation $A_n = \{1, \ldots, n\}^d$ and $\overset{fidi}{\Longrightarrow}$ for the convergence of the finite dimensional distributions.

Convergence of partial sums under isotropic long memory

The first study of partial sums is due to [DM79] who considered Gaussian subordinated fields presenting isotropic long memory. Then the same results for some functional of linear fields are obtained in [Sur82] and [AT87].

Let us first introduce the so-called Hermite process Z_m of order m which is the limiting process we shall encounter here.

$$Z_m(t) = \int_{\mathbb{R}^{md}} \prod_{j=1}^{d} \frac{e^{it_j\left(x_j^{(1)}+\cdots+x_j^{(m)}\right)} - 1}{i\left(x_j^{(1)}+\cdots+x_j^{(m)}\right)} Z_{G_0}(dx^{(1)})\ldots Z_{G_0}(dx^{(m)}) \quad (16)$$

where Z_{G_0} is the random Gaussian spectral field with control measure G_0. The spectral measure G_0 depends on a parameter α and a function b continuous on the unit sphere in $\mathbb{R}^d$ and it is given by

$$2^d \int_{\mathbb{R}^d} e^{i<t,x>} \prod_{j=1}^{d} \frac{1 - \cos(x_j)}{x_j^2} G_0(dx) = \int_{[-1,1]^d} \frac{b\left(\frac{x+t}{|x+t|}\right)}{|x+t|^\alpha} \prod_{j=1}^{d}(1 - |x_j|)dx .$$

$$(17)$$

When $d = 1$ (16) simplifies because G_0 admits a density proportional to $|x|^{\alpha-1}$ and in this case

$$Z_m(t) = \kappa^{-k/2} \int_{\mathbb{R}^m} \frac{e^{it(x^{(1)}+\cdots+x^{(m)})} - 1}{i(x^{(1)}+\cdots+x^{(m)})} \prod_{k=1}^{m} \left|x^{(k)}\right|^{\frac{\alpha-1}{2}} dW(x^{(k)}) ,$$

where W is the Gaussian white noise spectral field and where $\kappa = \int_{\mathbb{R}} e^{ix}|x|^{\alpha-1}$.

Let us now summarize the convergence results.

Theorem 2. *[[DM79]] Let $(X_n)_{n \in \mathbb{Z}^d}$ be a zero-mean, stationary, Gaussian random field. Let H be a measurable function such that*

$$\int_{\mathbb{R}} H(x)e^{\frac{-x^2}{2}} dx = 0 \quad and \quad \int_{\mathbb{R}} H^2(x)e^{\frac{-x^2}{2}} dx < \infty .$$

Denote by m its Hermite rank.

We suppose that (X_n) admits the following covariance function

$$r(k) = |k|^{-\alpha} L(|k|) b\left(\frac{k}{|k|}\right) ,$$

with $r(0) = 1$, where $0 < m\alpha < d$ and where L is a slowly varying function at infinity and b is a continuous function on the unit sphere in $\mathbb{R}^d$.

Then

$$\frac{1}{N^{d-m\alpha/2}(L(N))^{m/2}} \sum_{k \in A_{[Nt]}} H(X_k) \overset{fidi}{\Longrightarrow} c_m Z_m(t) , \quad (18)$$

where Z_m is the Hermite process of order m defined in (16) and where c_m is the coefficient of rank m in the Hermite expansion of H.

The following theorem concerns linear fields. The class of functions H is restricted to the Appell polynomials.

Theorem 3. *[[Sur82] and [AT87]] Let $(\epsilon_n)_{n\in\mathbb{Z}^d}$ be a sequence of zero-mean i.i.d random fields with variance 1 and finite moments of any order. Let $(X_n)_{n\in\mathbb{Z}^d}$ be the linear field*

$$X_n = \sum_{k\in\mathbb{Z}^d} a_k \epsilon_{n-k} \ ,$$

where

$$a_k = |k|^{-\beta} L(|k|) a\left(\frac{k}{|k|}\right) \ , \qquad d < 2\beta < d\left(1 + 1/m\right) \ , \tag{19}$$

where L is a slowly varying function at infinity and a is a continuous function on the unit sphere in $\mathbb{R}^d$.

Let P_m be the mth Appell polynomial associated with the distribution of X_0. Then

$$\frac{1}{N^{d-m(\beta-\frac{d}{2})}} \sum_{k\in A_{[Nt]}} P_m(X_k) \overset{fidi}{\Longrightarrow} Z_m(t) \ ,$$

where Z_m is the Hermite process of order m defined by (16) and (17) in which $\alpha = 2\beta - d$ and

$$b(t) = \int_{\mathbb{R}^d} a\left(\frac{s}{|s|}\right) a\left(\frac{s-t}{|s-t|}\right) |s|^{-\beta} |t-s|^{-\beta} ds \ .$$

Remark 4. Theorem 3 relates to isotropic long memory since condition (19) implies that the covariance function of X has asymptotically the form (1).

Remark 5. One can find a presentation of the tools for proving Theorems 2 and 3 in [DOT03].

Convergence of partial sums under non-isotropic long memory

The proofs of Theorem 2 and 3 rely on the convergence of multiple stochastic integrals. This method fails to work under non-isotropic long memory. So we turn to a method based on convergence of spectral measures.

Starting from a filter $a \in L^2([-\pi,\pi]^d)$ and a zero-mean random field ξ having a spectral density f_ξ, we consider the linear field

$$X_n = \sum_{k\in\mathbb{Z}^d} a_k \xi_{n-k} \ , \qquad n \in \mathbb{Z}^d \tag{20}$$

where a_k are the Fourier coefficients of a:

$$a(\lambda) = \sum_{k\in\mathbb{Z}^d} a_k e^{-i<k,\lambda>} \ .$$

The filter a is directly linked to the spectral density f_X of X by the relation :

$$f_X(\lambda) = f_\xi(\lambda)|a(\lambda)|^2 \ .$$

First, the partial sums are rewritten using the spectral field W of ξ. Since

$$\xi_k = \int_{[-\pi,\ \pi]^d} e^{i<k,\lambda>} dW(\lambda) \ , \tag{21}$$

if the random measure W_n on $[-n\pi;\ n\pi]^d$ is defined for all Borel set A by

$$W_n(A) = n^{d/2}W(n^{-1}A) \ ,$$

we have

$$n^{-d/2} \sum_{k \in A_{[nt]}} X_k = \int_{[-n\pi,n\pi]^d} a\left(\frac{\lambda}{n}\right) \prod_{j=1}^d \frac{e^{i\lambda_j[t_j n]/n} - 1}{n(e^{i\lambda_j/n} - 1)} dW_n(\lambda) \ , \tag{22}$$

where $[nt] = ([nt_1], \ldots, [nt_d])$.

Hence, in order to investigate the convergence of the partial sums (22), it suffices to handle stochastic integrals of the form $\int \Phi_n dW_n$ where $\Phi_n \in L^2(\mathbb{R}^d)$. This is made possible by the spectral convergence theorem.

The spectral convergence theorem

Let $(\xi_k)_{k \in \mathbb{Z}^d}$ be a real stationary random field. We work under the following assumptions :

H1 : The zero-mean stationary random field $(\xi_k)_{k \in \mathbb{Z}^d}$ has a spectral density f_ξ bounded above by $M > 0$. Moreover, the sequence of partial sums of the noise

$$S_n^\xi(t) = n^{-d/2} \sum_{k \in A_{[nt]}} \xi_k, \quad t \in [0,1]^d \ , \tag{23}$$

converges in the finite dimensional distributions sense to a field B.

Theorem 4. *Under* **H1**, *there exists a linear application I_0 from $L^2(\mathbb{R}^d)$ into $L^2(\Omega, \mathcal{A}, \mathbb{P})$ which has the following properties :*

(i) $\forall \Phi \in L^2(\mathbb{R}^d) \quad \mathbb{E}\left(I_0(\Phi)\right)^2 \leq (2\pi)^d M ||\Phi||_2^2$

(ii) $I_0\left(\prod_{j=1}^d \frac{e^{it_j\lambda_j}-1}{i\lambda_j}\right) = B(t_1, \ldots, t_d)$

(iii) *If the sequence Φ_n converges in $L^2(\mathbb{R}^d)$ to Φ, then $\int \Phi_n(x) dW_n(x)$ converges in law to $I_0(\Phi)$.*

(iv)   *If ξ is i.i.d, then $\forall \Phi \in L^2(\mathbb{R}^d)$ $I_0(\Phi) = \int \Phi dW_0$, where W_0 is the Gaussian white noise spectral field.*

Remark 6. When ξ is i.i.d, B is the Brownian sheet, property (ii) corresponding to its harmonisable representation

$$B(t) = \int \prod_{j=1}^d \frac{e^{it_j\lambda_j} - 1}{i\lambda_j} dW_0(\lambda) \ ,$$

and I_0 becomes in this case an isometry from $L^2(\mathbb{R}^d)$ into $L^2(\Omega, \mathcal{A}, \mathbb{P})$ which can then be considered as the stochastic integral with respect to W_0.

In the general case, point (i) shows that I_0 might not be an isometry so that I_0 cannot be always viewed as a stochastic integral.

Remark 7. Although our purpose is only to investigate the convergence of the partial sums, Theorem 4 appears to be a useful tool to obtain the asymptotic properties of any linear statistic writable in the form $\int \Phi_n dW_n$.

Proof. The theorem is proved in [LS00] in dimension $d = 1$. The details of the generalization to the context of random fields can be found in [Lav05a], so we only give a sketch of the proof. Let us consider the field

$$B_n(t) = \int_{[-n\pi,\ n\pi]^d} \prod_{j=1}^{d} \frac{e^{it_j\lambda_j} - 1}{i\lambda_j} dW_n(\lambda) . \tag{24}$$

Denoting $\hat{\Phi}$ the Fourier transform of Φ, we prove after some integrations by parts that

$$\int_{[-n\pi,\ n\pi]^d} \hat{\Phi}(x)dW_n(x) = \frac{(-1)^d}{(2\pi)^{d/2}} \int_{\mathbb{R}^d} \frac{\partial\Phi(t_1,\ldots,t_d)}{\partial t_1 \ldots \partial t_d} B_n(t_1,\ldots,t_d)dt_1 \ldots dt_d . \tag{25}$$

Besides, $B_n - S_n^\xi$ converges to 0 in L^2, which leads to the finite dimensional convergence of B_n to B. Then, extending to $d > 1$ a theorem of [Gri76] leads to the convergence in law of (25) to

$$I_B(\Phi) = (-1)^d \int_{\mathbb{R}^d} \frac{\partial\Phi(t)}{\partial t_1 \ldots \partial t_d} B(t)dt .$$

Finally the linear application I_0 of the theorem is defined by

$$I_0(\Phi) = I_B(\check{\Phi}) , \tag{26}$$

where $\check{\Phi}$ is the inverse Fourier transform of Φ in $L^2(\mathbb{R}^d)$ and we have

$$\mathbb{E}\left(I_0(\Phi)\right)^2 = \mathbb{E}\left(I_B(\check{\Phi})\right)^2$$

$$\leq \varliminf \mathbb{E}\left((2\pi)^{d/2} \int_{[-n\pi,\ n\pi]^d} \hat{\check{\Phi}}dW_n\right)^2 \leq (2\pi)^d M||\Phi||_2^2 ,$$

which is (i) of Theorem 4.

Theorem 4.2 in [Bil68] implies that $\int \Phi dW_n$ converges to $I_0(\Phi)$. Hence $\int \Phi_n dW_n$ converges to $I_0(\Phi)$ as soon as Φ_n goes to Φ in $L^2(\mathbb{R}^d)$. This proves (iii).

The particular choice $\check{\Phi} = \mathbb{1}_{[0,t_1] \times \cdots \times [0,t_d]}$ in (26) leads to (ii).

Convergence of partial sums

In view of the spectral representation (22) and of Theorem 4, for proving the convergence of the partial sums it is sufficient to check the L^2-convergence of $a(x/n)$. This leads to several types of proofs according to the form of a.

The following propositions focus on filters which lead to non-isotropic long memory random fields. Their proofs can be found in [Lav05a].

The first result concerns the simplest situation of a tensorial product.

Proposition 1. *Let $(\xi_k)_{k \in \mathbb{Z}^d}$ be a noise satisfying* **H 1**. *Let $(X_k)_{k \in \mathbb{Z}^d}$ be the random field defined by (20), constructed by filtering ξ through a filter of the form :*

$$a(\lambda_1, \dots, \lambda_d) = \prod_{j=1}^d a_j(\lambda_j) , \tag{27}$$

where the a_j's satisfy:

$$a_j(\lambda_j) \sim |\lambda_j|^{-\alpha_j} \quad \text{when } \lambda_j \to 0 ,$$

with $0 < \alpha_j < 1/2$. Then

$$\frac{1}{n^{d/2-(\sum_{j=1}^d \alpha_j)}} \sum_{k \in A_{[nt]}} X_k \stackrel{fidi}{\Longrightarrow} I_0\left(\prod_{j=1}^d \frac{e^{it_j\lambda_j}-1}{i\lambda_j|\lambda_j|^{\alpha_j}}\right) , \tag{28}$$

where I_0 is the linear application defined in Theorem 4.

Remark 8. When ξ is i.i.d, the limiting field (28) is the Fractional Brownian sheet with parameters $(\alpha_j, \quad j = 1, \dots, d)$.

It is well known that, in dimension $d = 1$, only the spectral behaviour at 0 determines the asymptotic of the partial sums. This result still holds for $d = 2$, as stated in the next proposition.

Proposition 2. *Let $(\xi_k)_{k \in \mathbb{Z}^d}$ be a stationary random field satisfying* **H 1**. *Let $(X_k)_{k \in \mathbb{Z}^d}$ be defined by (20), constructed by filtering ξ through a.*

(i) If the filter $a \in L^2([-\pi, \pi]^d)$ is continuous at the origin with $a(0) \neq 0$, then, for $d \leq 2$,

$$\frac{1}{n^{d/2}} \sum_{k \in A_{[nt]}} X_k \stackrel{fidi}{\Longrightarrow} a(0)B(t) , \tag{29}$$

where B is the limit of the partial sums of ξ introduced in hypotheses **H 1**.

(ii) If the filter a is equivalent at 0 to a homogeneous function $\tilde{a}$, i.e. for all c, $\tilde{a}(c\lambda) = |c|^{-\alpha} \tilde{a}(\lambda)$, with degree $\alpha \in]0, 1[$ such that $a \in L^2([-\pi, \pi]^d)$, then, for $d \leq 2$,

$$\frac{1}{n^{d/2+\alpha}} \sum_{k \in A_{[nt]}} X_k \stackrel{fidi}{\Longrightarrow} I_0\left(\tilde{a}(\lambda) \prod_{j=1}^d \frac{e^{it_j\lambda_j}-1}{i\lambda_j}\right) , \tag{30}$$

where I_0 is the linear application defined in Theorem 4.

Remark 9. When ξ is i.i.d, the limiting process can be written as a stochastic integral with respect to a Gaussian white noise measure (cf Remark 6).

Remark 10. Filtering a white noise through a filter satisfying the hypotheses in (i) can produce a weakly dependent random field, for instance if a is continuous on $[-\pi, \pi]^d$. It produces non-isotropic long memory when a is unbounded since the covariance function is then not absolutely summable. This memory involves only non-zero singularities of the spectral density and, as expected, does not modify the limit obtained under weak dependence.

Condition (ii) of Theorem 2 can be satisfied with isotropic as well as with non isotropic long-memory. The memory is non-isotropic for instance when the filter is $a(\lambda_1, \lambda_2) = |\lambda_1 + \theta\lambda_2|^{-\alpha}$, where $0 < \alpha < 1/2$ and $\theta \in \mathbb{R}$, $\theta \neq 0$.

Unfortunately, probably due to the spectral method, these results cannot be extended in dimension $d \geq 3$ without further assumptions. We only give an example of filters unbounded all over a linear subspace of $[-\pi, \pi]^d$.

Proposition 3. *Let $(\xi_k)_{k \in \mathbb{Z}^d}$ be a stationary random field satisfying* **H1**. *Let $(X_k)_{k \in \mathbb{Z}^d}$ be the random field defined by (20).*

Suppose that a has the following form :

$$a(\lambda) = \left| \sum_{i=1}^{d} c_i \lambda_i \right|^{-\alpha} ,$$

where $0 < \alpha < 1/2$ and the c_i's are real constants.

Then, as long as

$$0 < 2\alpha < \frac{1}{(d-2) \vee 1} , \tag{31}$$

we have

$$\frac{1}{n^{d/2+\alpha}} \sum_{k \in A_{[nt]}} X_k \overset{fidi}{\Longrightarrow} I_0 \left(a(\lambda) \prod_{j=1}^{d} \frac{e^{it_j \lambda_j} - 1}{i\lambda_j} \right) , \tag{32}$$

where I_0 is the linear application defined in Theorem 4.

Remark 11. The condition (31) on α is a restriction only when $d \geq 4$.

Tightness criteria for partial sums

So far, only the convergence of the finite-dimensional distributions of the partial sums has been stated. In dimension $d = 1$, a convenient criterion for tightness is given in [Taq75] from which the convergence in $\mathcal{D}([0, 1])$ follows easily.

General conditions for tightness in $\mathcal{D}([0, 1]^d)$ of a sequence of random fields are given in [BW71]. The following lemma, a corollary of Theorems 2 and 3 in [BW71], is very useful for proving tightness of the partial sums of strongly dependent fields.

Lemma 1. *Let us consider a stationary random field* $(X_k)_{k \in \mathbb{Z}^d}$ *and its normalized partial sum process*

$$S_n(t) = d_n^{-1} \sum_{k_1=0}^{[nt_1]} \cdots \sum_{k_d=0}^{[nt_d]} X_{k_1,\ldots,k_d} \,, \quad t \in [0,1]^d \,.$$

If the finite-dimensional distributions of S_n *converge to those of* X *and if there exist* $c > 0$ *and* $\beta > 1$ *such that for all* $p_1, \ldots, p_d \in \{1, \ldots, n\}$

$$\mathbb{E} \left(d_n^{-1} \sum_{k_1=0}^{p_1} \cdots \sum_{k_d=0}^{p_d} X_{k_1,\ldots,k_d} \right)^2 \le c \left(\prod_{i=1}^{d} \frac{p_i}{n} \right)^{\beta} \,, \tag{33}$$

then

$$S_n \stackrel{\mathcal{D}([0,1]^d)}{\Longrightarrow} X \,.$$

Moreover the field X *admits a continuous version.*

The details of the proof can be found in [Lav05b].

In the next section, we study the doubly-indexed empirical process of long memory random fields and we investigate its asymptotic behaviour for the long memory Gaussian subordinated fields of Theorem 2 and for the non-isotropic long memory situation of Proposition 3. For this, we need the convergence of the partial sums in $\mathcal{D}([0,1]^d)$ in both settings. Since the convergence of their finite-dimensional distributions has already been stated, only tightness is missing, which is the subject of the next results. Their proofs, based on the tightness criterion presented in Lemma 1, can be found in [Lav05b].

Proposition 4. *Under the hypothesis of Theorem 2, the partial sums process*

$$\frac{1}{N^{d-m\alpha/2}(L(N))^{m/2}} \sum_{k \in A_{[Nt]}} H(X_k)$$

is tight and convergence (18) takes place in $\mathcal{D}([0,1]^d)$.

Proposition 5. *Under the hypothesis of Proposition 3, the partial sums process*

$$\frac{1}{N^{d/2+\alpha}} \sum_{k \in A_{[Nt]}} X_k$$

is tight and convergence (32) takes place in $\mathcal{D}([0,1]^d)$.

3.2 Empirical Process of long memory random fields

We study the asymptotic behaviour of the empirical process

$$\sum_{j \in A_{[nt]}} \left[\mathbb{1}_{\{G(X_j) \leq x\}} - F(x) \right] \; , \tag{34}$$

where G is a measurable function and where F is the cumulative distribution function of $G(X_1)$, $(X_k)_{k \in \mathbb{Z}^d}$ being a long-range dependent stationary random field.

Our presentation relates to Gaussian subordinated random fields and to (non necessarily Gaussian) linear random fields.

In the first situation, we prove a uniform weak reduction principle and apply it to different situations of strong dependence. We present the convergence of (34) in $\mathcal{D}(\overline{\mathbb{R}} \times [0,1]^d)$ when X is Gaussian with isotropic long-range dependence, generalizing in dimension $d > 1$ the result of [DT89]. In the non-isotropic long memory setting, we give the convergence of (34) in $\mathcal{D}(\overline{\mathbb{R}} \times [0,1]^d)$ when the random field X is linear, Gaussian, and when the Hermite rank of $\mathbb{1}_{\{G(X_j) \leq x\}} - F(x)$ is 1.

In the situation of (non necessarily Gaussian) linear random fields a uniform weak reduction principle is more difficult to obtain. The only available results are those proved in [DLS02] where the authors obtain the convergence of (34) for $t = 1$, when G is the identity function, and in the situation of isotropic long-memory.

In each situation described above, the limiting process is degenerated insofar as it has the form $f(x)Z(t)$ where f is a deterministic function and Z a random field. This asymptotic behaviour of the empirical process is a characteristic property of strong dependence in dimension $d = 1$. It seems to be also the case with random fields even if the strong dependence is anisotropic such as in Corollary 3 below.

Empirical process of Gaussian subordinated fields

The main tool to obtain the convergence of the empirical process is the uniform weak reduction principle introduced in [DT89] which allows to replace in most cases the empirical process by the first term in its expansion on the Hermite basis. We present an inequality generalizing this principle to dimension $d > 1$. Then we specify the dependence structure of the random field in two corollaries. The first one refers to the isotropic long-range dependent Gaussian fields of Theorem 2. The second one relates to non-isotropic long memory. It focuses on the random field of Proposition 3 which is in addition supposed here to be Gaussian. The proofs of this section are detailed in [Lav05b].

Let $(X_n)_{n \in \mathbb{Z}^d}$ be a stationary Gaussian random field with covariance function r such that $r(0) = 1$.

Let G be a measurable function. We consider the following expansion on the Hermite basis :

$$\mathbb{1}_{\{G(X_j) \leq x\}} - F(x) = \sum_{q=m}^{\infty} \frac{J_q(x)}{q!} H_q(X_j) \; ,$$

where $F(x) = \mathbb{P}(G(X_1) \leq x)$. H_q is the Hermite polynomial of degree q and

$$J_q(x) = \mathbb{E}\left[\mathbb{1}_{\{G(X_1) \leq x\}} H_q(X_1)\right] \ .$$

Let

$$S_n(x) = \sum_{j \in A_n} \left[\mathbb{1}_{\{G(X_j) \leq x\}} - F(x) - \frac{J_m(x)}{m!} H_m(X_j)\right] \ .$$

Now, we formulate the inequality leading to the uniform weak reduction principle. Its proof follows the same lines as in [DT89].

Theorem 5. *Let*

$$d_N^2 = \mathrm{var}\left(\sum_{j \in A_N} H_m(X_j)\right) = m! \sum_{j,k \in A_N^2} r^m(k-j) \ .$$

If $d_N \longrightarrow \infty$, we have, for all η, $\delta > 0$ and for all $n \leq N$,

$$\mathbb{P}\left(\sup_x d_N^{-1} |S_n(x)| > \eta\right) \leq C N^\delta d_N^{-2} \sum_{j,k \in A_N^2} |r(k-j)|^{m+1} + \frac{d_n^2}{N^{2d}} \ , \qquad (35)$$

where C is a positive constant depending only on η.

If the limit of $d_N^{-1} \sum_{j \in A_{[Nt]}} H_m(X_j)$ is known, inequality (35) provides the asymptotic behaviour of the empirical process (34) if the upper bound in (35) vanishes when N goes to infinity.

The first corollary below relates to the Gaussian subordinated fields of Theorem 2.

Corollary 2. *Under the above notations, we suppose that the Gaussian field $(X_n)_{n \in \mathbb{Z}^d}$ admits the covariance function*

$$r(k) = |k|^{-\alpha} L(|k|) b\left(\frac{k}{|k|}\right) \ , \qquad r(0) = 1 \ , \qquad (36)$$

where $0 < m\alpha < d$, where L is slowly varying at infinity and where b is continuous on the unit sphere in $\mathbb{R}^d$.

Then

$$\frac{1}{N^{d-m\alpha/2}(L(N))^{m/2}} \sum_{j \in A_{[Nt]}} \left[\mathbb{1}_{\{G(X_j) \leq x\}} - F(x)\right] \overset{\mathcal{D}(\bar{\mathbb{R}} \times [0,1]^d)}{\Longrightarrow} \frac{J_m(x)}{m!} Z_m(t) \ ,$$

where the convergence takes place in $\mathcal{D}(\bar{\mathbb{R}} \times [0,1]^d)$ endowed with the uniform topology and the σ-field generated by the open balls and where Z_m, defined in (16), is the Hermite process of order m.

Proof (Sketch of proof). From (36), as $N \to \infty$

$$d_N^2 \sim N^{2d-m\alpha}(L(N))^m \ ,$$

and

$$\sum_{j,k \in A_N} |r(k-j)|^{m+1} = O(N^{2d-(m+1)\alpha}L(N)^{m+1}) + O(N^d) \ .$$

Hence the upper bound in (35) goes to zero for small values of δ.

Moreover Theorem 2 gives the convergence of $d_N^{-1} \sum_{j \in A_{[Nt]}} H_m(X_j)$ to the Hermite process, this convergence taking place in $\mathcal{D}([0,1]^d)$ from Proposition 4. Now, J_m is bounded and so :

$$J_m(x)d_N^{-1} \sum_{j \in A_{[Nt]}} H_m(X_j) \overset{\mathcal{D}(\bar{\mathbb{R}} \times [0,1]^d)}{\Longrightarrow} J_m(x)Z_m(t) \ . \tag{37}$$

The measurability of the empirical process is obtained if $\mathcal{D}(\bar{\mathbb{R}} \times [0,1]^d)$, endowed with the uniform topology, is equipped with the σ-field generated by the open balls. Finally (37) and (35) give the convergence claimed in the corollary.

The next corollary focuses on the non-isotropic random field of Proposition 3 based on Gaussian noise. Since this Proposition only gives the limit distribution of $d_N^{-1} \sum_{j \in A_{[Nt]}} X_j$, we restrict ourselves to functions G such that the Hermite rank of (34) is 1.

Corollary 3. *Let $(\epsilon_n)_{n \in \mathbb{Z}^d}$ be a stationary Gaussian field with a bounded spectral density. We consider the linear field*

$$X_n = \sum_{k \in \mathbb{Z}^d} a_k \epsilon_{n-k} \ , \tag{38}$$

where the (a_k)'s are, up to a normalisation providing $\mathrm{var}(X_1) = 1$, the Fourier coefficients of

$$a(\lambda) = \left| \sum_{i=1}^{d} c_i \lambda_i \right|^{-\alpha} \ , \quad 0 < \alpha < 1/2 \ , \tag{39}$$

where $(c_1, \ldots, c_d)$ are real valued parameters.

We suppose that the Hermite rank of $\mathbb{1}_{\{G(X_n) \leq x\}} - F(x)$ is 1.

If

$$0 < 2\alpha < \frac{1}{(d-2) \vee 1} \ , \tag{40}$$

then

$$\frac{1}{n^{d/2+\alpha}} \sum_{j \in A_{[nt]}} \left(\mathbb{1}_{\{G(X_j) \leq x\}} - F(x) \right) \overset{\mathcal{D}(\bar{\mathbb{R}} \times [0,1]^d)}{\Longrightarrow} J_1(x)R(t) \ ,$$

where $J_1(x) = \mathbb{E}[\mathbb{1}_{\{G(X_1) \leq x\}} X_1]$, *and where the convergence takes place in* $\mathcal{D}(\bar{\mathbb{R}} \times [0,1]^d)$ *endowed with the uniform topology and the σ-field generated by the open balls.*

When ϵ *is a white noise, the limiting field is defined by*

$$R(t) = \int_{\mathbb{R}^d} a(u) \prod_{j=1}^{d} \frac{e^{it_j u_j} - 1}{iu_j} dW_0(u) ,$$

where W_0 is the Gaussian white noise spectral field.

Remark 12. As in Proposition 3, the condition (40) is not a restriction when $d \leq 3$.

Proof (Sketch of proof). From (39), $d_n^2 \sim n^{d+2\alpha}$ when $n \to \infty$ and

$$\text{if } 0 < 2\alpha < 1/2 , \quad \sum_{j,k \in A_n^2} r^2(k-j) = O(n^d) ,$$

$$\text{if } 1/2 < 2\alpha < 1 , \quad \sum_{j,k \in A_n^2} r^2(k-j) = O(n^{d-1+4\alpha}) .$$

Therefore the upper bound in (35) tends to zero if δ is small enough. Since Proposition 3 and Proposition 5 prove the convergence of the partial sums of X in $\mathcal{D}([0,1]^d)$, the convergence of the empirical process follows.

Empirical process of long memory linear fields

Without the Gaussian assumption, a general uniform weak reduction principle as in Theorem 5 is not yet available. This has been done in [DLS02] in the particular case of the isotropic long memory linear random fields of Theorem 3. These authors obtain the convergence of the empirical process (34) for $t = 1$ and when G is the identity function.

Theorem 6 ([DLS02]). *Let ϵ be a zero-mean i.i.d random field with variance 1. Assume that there exist positive constants C and δ such that*

$$\left| \mathbb{E} e^{ia\epsilon_0} \right| \leq C(1 + |a|)^{-\delta} , \quad a \in \mathbb{R} ,$$

and

$$\mathbb{E} |\epsilon_0|^{2+\delta} < \infty .$$

Let X be the linear field defined by

$$X_n = \sum_{k \in \mathbb{Z}^d} a_k \epsilon_{n-k} , \quad n \in \mathbb{Z}^d$$

with

$$a_k = |k|^{-\alpha} b\left(\frac{k}{|k|}\right) , \quad k \in \mathbb{Z}^d ,$$

where $d/2 < \alpha < d$ and where b is continuous on the unit sphere in $\mathbb{R}^d$. Then, with $Z \sim \mathcal{N}(0,1)$ a standard Gaussian variable,

$$\frac{1}{n^{3d/2-\alpha}} \sum_{k \in A_n} \left[\mathbb{1}_{\{X_k \leq x\}} - F(x)\right] \stackrel{\mathcal{D}(\mathbb{R})}{\Longrightarrow} cf(x)Z ,$$

where c is a positive constant, F denoting the cumulative distribution function of X_1 and $f = F'$.

Remark 13. In [DLS02], the authors actually studied the convergence of the weighted empirical process

$$\sum_{k \in A_n} \gamma_{n,k} \mathbb{1}_{\{X_k \leq x + \xi_{n,k}\}} ,$$

where $\sup_n \max_{k \in A_n}(|\xi_{n,k}| + |\gamma_{n,k}|) = O(1)$. They obtain the same result.

4 Conclusion

All the above results confirm some specificities of the long memory compared with the short one : particularly a non standard normalisation and a degenerated limit for the empirical process. However, the study is far from being complete and should be extended for instance in the direction of seasonal phenomena, as it is done in dimension $d = 1$ ([OH02]), where the correct approximation of the empirical process might not be based on the first term of the Hermite expansion.

Finally, all results on the empirical process are a first step towards the study of U-statistics, Cramer Von Mises or Kolmogorov Smirnov statistics, and of M and L-statistics. They are the object of a current work.

References

[AT87]   F. Avram and M. Taqqu. Noncentral limit theorems and appell polynomials. *The Annals of Probability*, 15:767–775, 1987.

[BD91]   P.J. Brockwell and R.A. Davis. *Time Series: Theory and Methods*. Springer Series in Statistics. Springer-Verlag, 1991.

[Bil68]   P. Billingsley. *Convergence of probability measures*. John Wiley and Sons, 1968.

[BW71]   P. J. Bickel and M. J. Wichura. Convergence criteria for multiparameters stochastic processes and some applications. *The Annals of Mathematical Statistics*, 42(5):1656–1670, 1971.

[CJL78]   M. Cassandro and G. Jona-Lasinio. Critical point behaviour and probability theory. *Advances in Physics*, 27(6):913–941, 1978.

[DLS02]   P. Doukhan, G. Lang, and D. Surgailis. Asymptotics of weighted empirical processes of linear fields with long-rang dependence. *Annales de l'Institut Henri Poincaré*, 6:879–896, 2002.

[DM79]   R. L. Dobrushin and P. Major. Non central limit theorems for non-linear functionals of gaussian fields. *Zeitschrift für Warscheinligkeitstheorie verwande Gebiete*, 50:27–52, 1979.

[Dob65]   R. L. Dobrushin. Existence of a phase transition in two and three dimensional ising models. *Theory of Probability and Applications*, 10:193–213, 1965.

[Dob80]   R. L. Dobrushin. Gaussian random fields - gibbsian point of view. In R. L. Dobrushin and Ya. G. Sinai, editors, *Multicomponent random systems*, volume 6 of *Advances in Probability and Related Topics*. New York: Dekker, 1980.

[DOT03]   P. Doukhan, G. Oppenheim, and M. Taqqu, editors. *Theory and applications of long-range dependence.* Birkhäuser, 2003.

[DT89]   H. Dehling and M. S. Taqqu. The empirical process of some long-range dependent sequences with an application to u-statistics. *The Annals of Statistics*, 4:1767–1783, 1989.

[Fis64]   M. E. Fisher. Correlation functions and the critical region of simple fluids. *Journal of Mathematical Physics*, 5(7):944–962, 1964.

[Geo88]   H. O. Georgii. *Gibbs measure and phase transitions.* De Gruyter, 1988.

[GJ80]   C.W.J. Granger and R. Joyeux. An introduction to long memory time series and fractional differencing. *Journal of Time Series Analysis*, 1:15–30, 1980.

[Gra80]   C. W. Granger. Long memory relationships and the aggregation of dynamic models. *Journal of Econometrics*, 14:227–238, 1980.

[Gri76]   L Grinblatt. A limit theorem for measurable random processes and its applications. *Proceedings of the American Mathematical Society*, 61(2):371–376, 1976.

[Guy93]   X. Guyon. *Champs aléatoires sur un réseau.* Masson, 1993.

[Hos81]   J.R.M Hosking. Fractional differencing. *Biometrika*, 68:165–176, 1981.

[Kün80]   H. Künsch. *Reellwertige Zufallsfelder auf einem Gitter : Interpolationsprobleme, Variationsprinzip und statistische Analyse.* PhD thesis, ETH Zürich, 1980.

[KO49]   B. Kaufman and L. Onsager. Crystal statistics III: Short-range order in a binary ising lattice. *Physical Review*, 76:1244–1252, 1949.

[KT78]   J. M. Kosterlitz and D. J. Thouless. Two-dimensional physics. In *Progess in Low Temperature Physics*, volume VIIB, page 371. North-Holland, Amsterdam, 1978.

[Lav05a]   F. Lavancier. Invariance principles for non-isotropic long memory random fields. *preprint. Available at http://math.univ-lille1.fr/ lavancier.*, 2005.

[Lav05b]   F. Lavancier. Processus empirique de fonctionnelles de champs gaussiens à longue mémoire. *preprint 63, IX, IRMA, Lille. Available at http://math.univ-lille1.fr/ lavancier.*, 2005.

[LS00]   G. Lang and Ph. Soulier.   Convergence de mesures spectrales
         aléatoires et applications à des principes d'invariance. *Stat. Inf. for
         Stoch. Proc.*, 3:41–51, 2000.

[OH02]   M. Ould Haye. Asymptotical behavior of the empirical process for
         seasonal long-memory data. *ESAIM*, 6:293–309, 2002.

[Ons44] L. Onsager. Crystal statistics I : A two dimensional model with
         order-disorder transition. *Physical Review*, 65:117–149, 1944.

[OV04]   G. Oppenheim and M.-C. Viano. Aggregation of random parame-
         ter Ornstein-Uhlenbeck or AR processes: some convergence results.
         Journal of Time Series Analysis., 25(3):335–350, 2004.

[Ros85] M. Rosenblatt. *Stationary sequences and random fields*. Birkhäuser,
         1985.

[Sur82] D. Surgailis. Zones of attraction of self-similar multiple integrals.
         Lithuanian Mathematics Journal, 22:327–340, 1982.

[Taq75] M. S. Taqqu. Weak convergence to fractional brownian motion and
         to the rosenblatt process. *Zeitschrift für Wahrscheinlichkeitstheorie
         und verwande Gebiete*, 31:287–302, 1975.

[Wai65] S. Wainger. *Special trigonometric series in k-dimensions*. Number 59.
         AMS, 1965.

Long Memory in Nonlinear Processes

Rohit Deo[1], Mengchen Hsieh[1], Clifford M. Hurvich[1], and Philippe Soulier[2]

[1] New York University, 44 W. 4'th Street, New York NY 10012, USA
  {rdeo,mhsieh,churvich}@stern.nyu.edu
[2] Université Paris X, 200 avenue de la République, 92001 Nanterre cedex, France
  philippe.soulier@u-paris10.fr

1 Introduction

It is generally accepted that many time series of practical interest exhibit strong dependence, i.e., long memory. For such series, the sample autocorrelations decay slowly and log-log periodogram plots indicate a straight-line relationship. This necessitates a class of models for describing such behavior. A popular class of such models is the autoregressive fractionally integrated moving average (ARFIMA) (see [Ade74], [GJ80]), [Hos81], which is a linear process. However, there is also a need for nonlinear long memory models. For example, series of returns on financial assets typically tend to show zero correlation, whereas their squares or absolute values exhibit long memory. See, e.g., [DGE93]. Furthermore, the search for a realistic mechanism for generating long memory has led to the development of other nonlinear long memory models. In this chapter, we will present several nonlinear long memory models, and discuss the properties of the models, as well as associated parametric and semiparametric estimators.

Long memory has no universally accepted definition; nevertheless, the most commonly accepted definition of long memory for a weakly stationary process $X = \{X_t,\ t \in \mathbb{Z}\}$ is the regular variation of the autocovariance function: there exist $H \in (1/2, 1)$ and a slowly varying function L such that

$$\mathrm{cov}(X_0, X_t) = L(t)|t|^{2H-2} . \tag{1}$$

Under this condition, it holds that:

$$\lim_{n \to \infty} n^{-2H} L(n)^{-1} \mathrm{var}\left(\sum_{t=1}^{n} X_t\right) = 1/(2H(2H-1)). \tag{2}$$

The condition (2) does not imply (1). Nevertheless, we will take (2) as an alternate definition of long memory. In both cases, the index H will be referred to as the *Hurst index* of the process X. This definition can be expressed in terms of the parameter $d = H - 1/2$, which we will refer to as the *memory parameter*. The most famous long memory processes are fractional Gaussian noise and the $ARFIMA(p, d, q)$ process, whose memory parameter is d and Hurst index is $H = 1/2 + d$. See for instance [Taq03] for a definition of these processes.

The second-order properties of a stationary process are not sufficient to characterize it, unless it is a Gaussian process. Processes which are linear with respect to an i.i.d. sequence (strict sense linear processes) are also relatively well characterized by their second-order structure. In particular, weak convergence of the partial sum process of a Gaussian or strict sense linear long memory processes $\{X_t\}$ with Hurst index H can be easily derived. Define $S_n(t) = \sum_{k=1}^{[nt]} (X_k - \mathbb{E}[X_k])$ in discrete time or $S_n(t) = \int_0^{nt} (X_s - \mathbb{E}[X_s]) \mathrm{d}s$ in continuous time. Then $\mathrm{var}(S_n(1))^{-1/2} S_n(t)$ converges in distribution to a constant times the fractional Brownian motion with Hurst index H, that is the Gaussian process B_H with covariance function

$$\mathrm{cov}(B_H(s), B_H(t)) = \frac{1}{2} \{|s|^{2H} - |t - s|^{2H} + t^{2H}\} .$$

In this paper, we will introduce nonlinear long memory processes, whose second order structure is similar to that of Gaussian or linear processes, but which may differ greatly from these processes in many other aspects. In Section 2, we will present these models and their second-order properties, and the weak convergence of their partial sum process. These models include conditionally heteroscedastic processes (Section 2.1) and models related to point processes (Section 2.2). In Section 3, we will consider the problem of estimating the Hurst index or memory parameter of these processes.

2 Models

2.1 Conditionally heteroscedastic models

These models are defined by

$$X_t = \sigma_t v_t , \tag{3}$$

where $\{v_t\}$ is an independent identically distributed series with finite variance and σ_t^2 is the so-called volatility. We now give examples.

LMSV and LMSD

The Long Memory Stochastic Volatility (LMSV) and Long Memory Stochastic Duration (LMSD) models are defined by Equation (3), where $\sigma_t^2 = \exp(h_t)$ and $\{h_t\}$ is an unobservable Gaussian long memory process with memory parameter $d \in (0, 1/2)$, independent of $\{v_t\}$. The multiplicative innovation series

$\{v_t\}$ is assumed to have zero mean in the LMSV model, and positive support with unit mean in the LMSD model. The LMSV model was first introduced by [BCdL98] and [Har98] to describe returns on financial assets, while the LMSD model was proposed by [DHH05] to describe durations between transactions on stocks.

Using the moment generating function of a Gaussian distribution, it can be shown (see [Har98]) for the LMSV/LMSD model that for any real s such that $\mathbb{E}[|v_t|^s] < \infty$,

$$\rho_s(j) \sim C_s j^{2d-1} \qquad j \to \infty,$$

where $\rho_s(j)$ denotes the autocorrelation of $\{|x_t|^s\}$ at lag j, with the convention that $s = 0$ corresponds to the logarithmic transformation. As shown in [SV02], the same result holds under more general conditions without the requirement that $\{h_t\}$ be Gaussian.

In the LMSV model, assuming that $\{h_t\}$ and $\{v_t\}$ are functions of a multivariate Gaussian process, [Rob01] obtained similar results on the autocorrelations of $\{|X_t|^s\}$ with $s > 0$ even if $\{h_t\}$ is not independent of $\{v_t\}$. Similar results were obtained in [SV02], allowing for dependence between $\{h_t\}$ and $\{v_t\}$.

The LMSV process is an uncorrelated sequence, but powers of LMSV or LMSD may exhibit long memory. [SV02] proved the convergence of the centered and renormalized partial sums of any absolute power of these processes to fractional Brownian motion with Hurst index $1/2$ in the case where they have short memory.

FIEGARCH

The weakly stationary FIEGARCH model was proposed by [BM96]. The FIEGARCH model, which is observation-driven, is a long-memory extension of the EGARCH (exponential GARCH) model of [Nel91]. The FIEGARCH model for returns $\{X_t\}$ takes the form 2.1 innovation series $\{v_t\}$ are i.i.d. with zero mean and a symmetric distribution, and

$$\log \sigma_t^2 = \omega + \sum_{j=1}^{\infty} a_j g(v_{t-j}) \tag{4}$$

with $g(x) = \theta x + \gamma(|x| - \mathbb{E}|v_t|)$, $\omega > 0$, $\theta \in \mathbb{R}$, $\gamma \in \mathbb{R}$, and real constants a_j such that the process $\{\log \sigma_t^2\}$ has long memory with memory parameter $d \in (0, 1/2)$. If θ is nonzero, the model allows for a so-called leverage effect, whereby the sign of the current return may have some bearing on the future volatility. In the original formulation of [BM96], the $\{a_j\}$ are the $AR(\infty)$ coefficients of an $ARFIMA(p, d, q)$ process.

As was the case for the LMSV model, here we can once again express the log squared returns as in (18) with $\mu = \mathbb{E}[\log v_t^2] + \omega$, $u_t = \log v_t^2 - \mathbb{E}[\log v_t^2]$, and $h_t = \log \sigma_t^2 - \omega$. Here, however, the processes $\{h_t\}$ and $\{u_t\}$ are not mutually independent. The results of [SV02] also apply here, and in particular, the processes $\{|X_t|^u\}$, $\{\log(X_t^2)\}$ and $\{\sigma_t\}$ have the same memory parameter d.

ARCH(∞) and FIGARCH

In ARCH(∞) models, the innovation series $\{v_t\}$ is assumed to have zero mean and unit variance, and the conditional variance is taken to be a weighted sum of present and past squared returns:

$$\sigma_t^2 = \omega + \sum_{k=1}^{\infty} a_j X_{t-j}^2 \,, \tag{5}$$

where $\omega, a_j, j = 1, 2, \ldots$ are nonnegative constants. The general framework leading to (3) and (5) was introduced by [Rob91]. [KL03] have shown that $\sum_{j=1}^{\infty} a_j \leq 1$ is a necessary condition for existence of a strictly stationary solution to equations (3), (5), while [GKL00] showed that $\sum_{j=1}^{\infty} a_j < 1$ is a sufficient condition for the existence of a strictly stationary solution. If $\sum_{j=1}^{\infty} a_j = 1$, the existence of a strictly stationary solution has been proved by [KL03] only in the case where the coefficients a_j decay exponentially fast. In any case, if a stationary solution exists, its variance, if finite, must be equal to $\omega(1 - \sum_{k=1}^{\infty} a_k)^{-1}$, so that it cannot be finite if $\sum_{k=1}^{\infty} a_k = 1$ and $\omega > 0$. If $\omega = 0$, then the process which is identically equal to zero is a solution, but it is not known whether a nontrivial solution exists.

In spite of a huge literature on the subject, the existence of a strictly or weakly stationary solution to (3), (5) such that $\{\sigma_t^2\}$, $\{|X_t|^u\}$ or $\{\log(X_t^2)\}$ has long memory is still an open question. If $\sum_{j=1}^{\infty} a_j < 1$, and the coefficients a_j decay sufficiently slowly, [GKL00] found that it is possible in such a model to get hyperbolic decay in the autocorrelations $\{\rho_r\}$ of the squares, though the rates of decay they were able to obtain were proportional to $r^{-\theta}$ with $\theta > 1$. Such autocorrelations are summable, unlike the autocorrelations of a long-memory process with positive memory parameter. For instance, if the weights $\{a_j\}$ are proportional to those given by the $AR(\infty)$ representation of an ARFIMA(p, d, q) model, then $\theta = -1 - d$. If $\sum_{j=1}^{\infty} a_j = 1$, then the process has infinite variance so long memory as defined here is irrelevant.

Let us mention for historical interest the FIGARCH (fractionally integrated GARCH) model which appeared first in [BBM96]. In the FIGARCH model, the weights $\{a_j\}$ are given by the $AR(\infty)$ representation of an ARFIMA(p, d, q) model, with $d \in (0, 1/2)$, which implies that $\sum_{j=1}^{\infty} a_j = 1$, hence the very existence of FIGARCH series is an open question, and in any case, if it exists, it cannot be weakly stationary. The lack of weak stationarity of the FIGARCH model was pointed out by [BBM96]. Once again, at the time of writing this paper, we are not aware of any rigorous result on this process or on any ARCH(∞) process with long memory.

LARCH

Since the ARCH structure (appearently) fails to produce long memory, an alternative definition of heteroskedasticity has been considered in which long memory can be proved rigorously. [GS02] considered models which satisfy the equation $X_t = \zeta_t A_t + B_t$, where $\{\zeta_t\}$ is a sequence of i.i.d. centered random variables with unit variance and A_t and B_t are linear in $\{X_t\}$ instead of quadratic as in the ARCH specification. This model nests the LARCH model introduced by [Rob91], obtained for $B_t \equiv 0$. The advantage of this model is that it can exhibit long memory in the conditional mean B_t and/or in the conditional variance A_t, possibly with different memory parameters. See [GS02, Corollary 4.4]. The process $\{X_t\}$ also exhibits long memory with a memory parameter depending on the memory parameters of the mean and the conditional variance [GS02, Theorem 5.4]. If the conditional mean exhibits long memory, then the partial sum process converges to the fractional Brownian motion, and it converges to the standard Brownian motion otherwise. See [GS02, Theorem 6.2]. The squares $\{X_t^2\}$ may also exhibit long memory, and their partial sum process converge either to the fractional Brownian motion or to a non Gaussian self-similar process. This family of processes is thus very flexible. An extension to the multivariate case is given in [DTW05].

We conclude this section by the following remark. Even though these processes are very different from Gaussian or linear processes, they share with weakly dependent processes the Gaussian limit and the fact that weak limits and L^2 limits have consistent normalisations, in the sense that, if ξ_n denotes one of the usual statistics computed on a time series, there exists a sequence v_n such that $v_n \xi_n$ converges weakly to a non degenerate distribution and $v_n^2 \mathbb{E}[\xi_n^2]$ converges to a positive limit (which is the variance of the asymptotic distribution). In the next subsection, we introduce models for which this is no longer true.

2.2 Shot noise processes

General forms of the shot-noise process have been considered for a long time; see for instance [Tak54], [Dal71]. Long memory shot noise processes have been introduced more recently; an early reference seems to be [GMS93]. We present some examples of processes related to shot noise which may exhibit long memory. For simplicity and brevity, we consider only stationary processes.

Let $\{t_j, \ j \in \mathbb{Z}\}$ be the points of a stationary point process on the line, numbered for instance in such a way that $t_{-1} < 0 \leq t_0$, and for $t \geq 0$, let $N(t) = \sum_{j \geq 0} \mathbb{1}_{\{t_j \leq t\}}$ be the number of points between time zero and t. Define then

$$X_t = \sum_{j \in \mathbb{Z}} \epsilon_j \mathbb{1}_{\{t_j \leq t < t_j + \eta_j\}}, \quad t \geq 0. \tag{6}$$

In this model, the shocks $\{\epsilon_j\}$ are an i.i.d. sequence; they are generated at birth times $\{t_j\}$ and have durations $\{\eta_j\}$. The observation at time t is the sum of all surviving present and past shocks. In model (6), we can take time to be continuous, $t \in \mathbb{R}$ or discrete, $t \in \mathbb{Z}$. This will be made precise later for each model considered. We now describe several well known special cases of model (6).

(A1). Renewal-reward process; [TL86], [Liu00].

The durations are exactly the interarrival times of the renewal process: $\eta_0 = t_0$, $\eta_j = t_{j+1} - t_j$, and the shocks are independent of their birth times. Then there is exactly one surviving shock at time t:

$$X_t = \epsilon_{N(t)}. \tag{7}$$

(A2). ON-OFF model; [TWS97].

This process consists of alternating ON and OFF periods with independent durations. Let $\{\eta_k\}_{\geq 1}$ and $\{\zeta_k\}_{k \geq 1}$ be two independent i.i.d. sequences of positive random variables with finite mean. Let t_0 be independent of these sequences and define $t_j = t_0 + \sum_{k=1}^{j}(\eta_k + \zeta_k)$. The shocks ϵ_j are deterministic and equal to 1. Their duration is η_j. The η_js are the ON periods and the ζ_js are the OFF periods. The first interval t_0 can also be split into two successive ON and OFF periods η_0 and ζ_0. The process X can be expressed as

$$X_t = \mathbb{1}_{\{t_{N(t)} \leq t < t_{N(t)} + \eta_{N(t)}\}}. \tag{8}$$

(A3). Error duration process; [Par99].

This process was introduced to model some macroeconomic data. The birth times are deterministic, namely $t_j = j$, the durations $\{\eta_j\}$ are i.i.d. with finite mean and

$$X_t = \sum_{j \leq t} \epsilon_j \mathbb{1}_{\{t < j + \eta_j\}}. \tag{9}$$

(A4). Infinite Source Poisson model.

If the t_j are the points of a homogeneous Poisson process, the durations $\{\eta_j\}$ are i.i.d. with finite mean and $\epsilon_j \equiv 1$, we obtain the infinite source Poisson model or M/G/∞ input model considered among others in [MRRS02].

[MRR02] have considered a variant of this process where the shocks (referred to as transmission rates in this context) are random, and possibly contemporaneously dependent with durations.

In the first two models, the durations satisfy $\eta_j \leq t_{j+1} - t_j$, hence are not independent of the point process of arrivals (which is here a renewal process). Nevertheless η_j is independent of the past points $\{t_k, \ k \leq j\}$. The process can be defined for all $t \geq 0$ without considering negative birth times and shocks. In the last two models, the shocks and durations are independent of the renewal process, and any past shock may contribute to the value of the process at time t.

Stationarity and second order properties

• The renewal-reward process (7) is strictly stationary since the renewal process is stationary and the shocks are i.i.d. It is moroever weakly stationary if the shocks have finite variance. Then $\mathbb{E}[X_t] = \mathbb{E}[\epsilon_1]$ and

$$\text{cov}(X_0, X_t) = \mathbb{E}[\epsilon^2]\, \mathbb{P}(\eta_0 > t) = \lambda \mathbb{E}[\epsilon_1^2]\, \mathbb{E}[(\eta_1 - t)_+]\,, \tag{10}$$

where η_0 is the delay distribution and $\lambda = \mathbb{E}[(t_1 - t_0)]^{-1}$ is intensity of the stationary renewal process. Note that this relation would be true for a general stationary point process. Cf. for instance [TL86] or [HHS04].

• The stationary version of the ON-OFF was studied in [HRS98]. The first On and OFF period η_0 and ζ_0 can be defined in such a way that the process X is stationary. Let F_{on} and F_{off} be the distribution functions of the ON and OFF periods η_1 and ζ_1. [HRS98, Theorem 4.3] show that if $1 - F_{\text{on}}$ is regularly varying with index $\alpha \in (1, 2)$ and $1 - F_{\text{off}}(t) = o(1 - F_{\text{on}}(t))$ as $t \to \infty$, then

$$\text{cov}(X_0, X_t) \sim c\mathbb{P}(\eta_0 > t) = c\lambda \mathbb{E}[(\eta_1 - t)_+]\,, \tag{11}$$

• Consider now the case when the durations are independent of the birth times. To be precise, assume that $\{(\eta_j, \epsilon_j)\}$ is an i.i.d. sequence of random vectors, independent of the stationary point process of points $\{t_j\}$. Then the process $\{X_t\}$ is strictly stationary as long as $\mathbb{E}[\eta_1] < \infty$, and has finite variance if $\mathbb{E}[\epsilon_1^2 \eta_1] < \infty$. Then $\mathbb{E}[X_t] = \lambda \mathbb{E}[\epsilon_1 \eta_1]$ and

$$\begin{aligned}
\text{cov}(X_0, X_t) = {}& \lambda \mathbb{E}[\epsilon_1^2 (\eta_1 - t)_+] \\
& + \{\text{cov}(\epsilon_1 N(-\eta_1, 0], \epsilon_2 N(t - \eta_2, t]) - \lambda \mathbb{E}[\epsilon_1 \epsilon_2 (\eta_1 \wedge (\eta_2 - t)_+]\}\,,
\end{aligned}$$

where λ is the intensity of the stationary point process, i.e. $\lambda^{-1} = \mathbb{E}[t_0]$. The last term has no known general expression for a general point process, but it vanishes in two particular cases:

- if N is a homogeneous Poisson point process;
- if ϵ_1 is centered and independent of η_1.

In the latter case (10) holds, and in the former case, we obtain a formula which generalizes (10):

$$\text{cov}(X_0, X_t) = \lambda \mathbb{E}[\epsilon_1^2 (\eta_1 - t)_+]\,. \tag{12}$$

We now see that second order long memory can be obtained if (10) holds and the durations have regularly varying tails with index $\alpha \in (1, 2)$ or,

$$\mathbb{E}[\epsilon_1^2 \mathbb{1}_{\{\eta_1 > t\}}] = \ell(t)t^{-\alpha}\,. \tag{13}$$

Thus, if (13) and either (11) or (12) hold, then X has long memory with Hurst index $H = (3 - \alpha)/2$ since

$$\text{cov}(X_0, X_t) \sim \frac{\lambda}{\alpha - 1} \, \ell(t) t^{1-\alpha} \, . \tag{14}$$

Examples of interest in teletraffic modeling where ϵ_1 and η_1 are not independent but (13) holds are provided in [MRR02] and [FRS05].

We conjecture that (14) holds in a more general framework, at least if the interarrival times of the point process have finite variance.

Weak convergence of partial sums

This class of long memory process exhibits a very distinguishing feature. Instead of converging weakly to a process with finite variance, dependent stationary increments such as the fractional Brownian motion, the partial sums of some of these processes have been shown to converge to an α-stable Levy process, that is, an α-stable process with independent and stationary increment. Here again there is no general result, but such a convergence is easy to prove under restrictive assumptions. Define

$$S_T(t) = \int_0^{Tt} \{X_s - \mathbb{E}[X_s]\} \, ds \, .$$

Then it is known in the particular cases described above that the finite dimensional distributions of the process $\ell(T)T^{-1/\alpha}S_T$ (for some slowly varying function ℓ) converge weakly to those of an α-stable process. This was proved in [TL86] for the renewal reward process, in [MRRS02] for the ON-OFF and infinite source Poisson processes when the shocks are constant. A particular case of dependent shocks and durations is considered in [MRR02]. [HHS04] proved the result in discrete time for the error duration process; the adaptation to the continuous time framework is straightforward. It is also probable that such a convergence holds when the underlying point process is more general.

Thus, these processes are examples of second order long memory process with Hurst index $H \in (1/2, 1)$ such that $T^{-H}S_T(t)$ converges in probability to zero. This behaviour is very surprising and might be problematic in statistical applications, as illustrated in Section 3.

It must also be noted that convergence does not hold in the space $\mathcal{D}$ of right-continuous, left-limited functions endowed with the J_1 topology, since a sequence of processes with continuous path which converge in distribution in this sense must converge to a process with continuous paths. It was proved in [RvdB00, Theorem 4.1] that this convergence holds in the M_1 topology for the infinite source Poisson process. For a definition and application of the M_1 topology in queuing theory, see [Whi02].

Slow growth and fast growth

Another striking feature of these processes is the slow growth versus fast growth phenomenon, first noticed by [TL86] for the renewal-reward process and more rigorously investigated by [MRRS02] for the ON-OFF and infinite source Poisson process[3]. Consider M independent copies $X^{(i)}$, $1, \leq i \leq M$ of these processes and denote

$$A_{M,T}(t) = \sum_{i=1}^{M} \int_{0}^{Tt} \{X_{s}^{(i)} - \mathbb{E}[X_{s}]\} \, ds \, .$$

If M depends on T, then, according to the rate growth of M with respect to T, a stable or Gaussian limit can be obtained. More precisely, the slow growth and fast growth conditions are, up to slowly varying functions $MT^{1-\alpha} \to 0$ and $MT^{1-\alpha} \to \infty$, respectively. In other terms, the slow and fast growth conditions are characterized by $\mathrm{var}(A_{M,T}(1)) \ll b(MT)$ and $\mathrm{var}(A_{M,T}(1)) \gg b(MT)$, respectively, where b is the inverse of the quantile function of the durations.

Under the slow growth condition, the finite dimensional distributions of $L(MT)(MT)^{-1/\alpha}A_{M,T}$ converge to those of a Levy α-stable process, where L is a slowly varying function. Under the fast growth condition, the sequence of processes $T^{-H}\ell^{-1/2}(T)M^{-1/2}A_{M,T}$ converges, in the space $\mathcal{D}(\mathbb{R}_{+})$ endowed with the J_1 topology, to the fractional Brownian motion with Hurst index $H = (3 - \alpha)/2$. It is thus seen that under the fast growth condition, the behaviour of a Gaussian long memory process with Hurst index H is recovered.

Non stationary versions

If the sum defining the process X in (6) is limited to non negative indices j, then the sum has always a finite number of terms and there is no restriction on the distribution of the interarrival times $t_{j+1} - t_j$ and the durations η_j. These models can then be nonstationary in two ways: either because of initialisation, in which case a suitable choice of the initial distribution can make the process stationary; or because these processes are non stable and have no stationary distribution. The latter case arises when the interarrival times and/or the durations have infinite mean. These models were studied by [RR00] and [MR04] in the case where the point process of arrivals is a renewal process. Contrary to the stationary case, where heavy tailed durations imply non Gaussian limits, the limiting process of the partial sums has non stationary increments and can be Gaussian in some cases.

2.3 Long Memory in Counts

The time series of counts of the number of transactions in a given fixed interval of time is of interest in financial econometrics. Empirical work suggests that such series may possess long memory. See [DHH05]. Since the counts are

---

[3] Actually, in the case of the Infinite Source Poisson process, [MRRS02] consider a single process but with an increasing rate λ depending on T, rather than superposition of independent copies. The results obtained are nevertheless of the same nature.

induced by the durations between transactions, it is of interest to study the properties of durations, how these properties generate long memory in counts, and whether there is a connection between potential long memory in durations and long memory in counts.

The event times determine a counting process $N(t)$ = Number of events in $(0, t]$. Given any fixed clock-time spacing $\Delta t > 0$, we can form the time series $\{\Delta N_{t'}\} = \{N(t'\Delta t) - N[(t' - 1)\Delta t]\}$ for $t' = 1, 2, \ldots$, which counts the number of events in the corresponding clock-time intervals of width Δt. We will refer to the $\{\Delta N_{t'}\}$ as the *counts*. Let $\tau_k > 0$ denote the waiting time (duration) between the $k - 1$'st and the k'th transaction.

We give some preliminary definitions taken from [DVJ03].

Definition 1. *A point process* $N(t) = N(0, t]$ *is stationary if for every* $r = 1, 2, \ldots$ *and all bounded Borel sets* $A_1, \ldots, A_r$, *the joint distribution of* $\{N(A_1 + t), \ldots, N(A_r + t)\}$ *does not depend on* $t \in [0, \infty)$.

A second order stationary point process is long-range count dependent (LRcD) if

$$\lim_{t \to \infty} \frac{\text{var}(N(t))}{t} = \infty .$$

A second order stationary point process $N(t)$ *which is LRcD has Hurst index* $H \in (1/2, 1)$ *given by*

$$H = \sup\{h : \limsup_{t \to \infty} \frac{\text{var}(N(t))}{t^{2h}} = \infty\} .$$

Thus if the counts $\{\Delta N_{t'}\}_{t'=-\infty}^{\infty}$ on intervals of any fixed width $\Delta t > 0$ are LRD with memory parameter d then the counting process $N(t)$ must be LRcD with Hurst index $H = d + 1/2$. Conversely, if $N(t)$ is an LRcD process with Hurst index H, then $\{\Delta N_{t'}\}$ cannot have exponentially decaying autocorrelations, and under the additional assumption of a power law decay of these autocorrelations, $\{\Delta N_{t'}\}$ is LRD with memory parameter $d = H - 1/2$.

There exists a probability measure P^0 under which the doubly infinite sequence of durations $\{\tau_k\}_{k=-\infty}^{\infty}$ are a stationary time series, i.e., the joint distribution of any subcollection of the $\{\tau_k\}$ depends only on the lags between the entries. On the other hand, the point process N on the real line is stationary under the measure P. A fundamental fact about point processes is that in general (a notable exception is the Poisson process) there is no single measure under which both the point process N and the durations $\{\tau_k\}$ are stationary, i.e., in general P and P^0 are not the same. Nevertheless, there is a one-to-one correspondence between the class of measures P^0 that determine a stationary duration sequence and the class of measures P that determine a stationary point process. The measure P^0 corresponding to P is called the *Palm distribution*. The counts are stationary under P, while the durations are stationary under P^0.

We now present an important theoretical result obtained by [Dal99].

Theorem 1. *A stationary **renewal** point process is LRcD and has Hurst index $H = (1/2)(3-\alpha)$ under P if the interarrival time has tail index $1 < \alpha < 2$ under P^0.*

Theorem 1 establishes a connection between the tail index of a duration process and the persistence of the counting process. According to the theorem, the counting process will be LRcD if the duration process is *iid* with infinite variance. Here, the memory parameter of the counts is completely determined by the tail index of the durations.

This prompts the question as to whether long memory in the counts can be generated solely by dependence in finite-variance durations. An answer in the affirmative was given by [DRV00], who provide an example outside of the framework of the popular econometric models. We now present a theorem on the long-memory properties of counts generated by durations following the LMSD model. The theorem is a special case of a result proved in [DHSW05], who give sufficient conditions on durations to imply long memory in counts.

Theorem 2. *If the durations $\{\tau_k\}$ are generated by the LMSD process with memory parameter d, then the induced counting process $N(t)$ has Hurst index $H = 1/2 + d$, i.e. satisfies $\mathrm{var}(N(t)) \sim Ct^{2d+1}$ under P as $t \to \infty$ where $C > 0$.*

3 Estimation of the Hurst index or memory parameter

A weakly stationary process with autocovariance function satisfying (1) has a spectral density f defined by

$$f(x) = \frac{1}{2\pi} \sum_{t \in \mathbb{Z}} \gamma(t) e^{itx} . \tag{15}$$

This series converges uniformly on the compact subsets of $[-\pi, \pi] \setminus \{0\}$ and in $L^1([-\pi, \pi], dx)$. Under some strengthening of condition (1), the behaviour of the function f at zero is related to the rate of decay of γ. For instance, if we assume in addition that L is ultimately monotone, we obtain the following Tauberian result [Taq03, Proposition 4.1], with $d = H - 1/2$.

$$\lim_{x \to 0} L(x)^{-1} x^{2d} f(x) = \pi^{-1} \Gamma(2d) \cos(\pi d). \tag{16}$$

Thus, a natural idea is to estimate the spectral density in order to estimate the memory paramter d. The statistical tools are the discrete Fourier transform (DFT) and the periodogram, defined for a sample $U_1, \ldots, U_n$, as

$$J_{n,j}^U = (2\pi n)^{-1/2} \sum_{t=1}^{n} U_t e^{itw_j}, \quad I_U(\omega_j) = |J_{n,j}^U|^2,$$

where $\omega_j = 2j\pi/n$, $1 \le j < n/2$ are the so-called Fourier frequencies. (Note that for clarity the index n is omitted from the notation). In the classical weakly stationary short memory case (when the autocovariance function is absolutely summable), it is well known that the periodogram is an asymptotically unbiased estimator of the spectral density f_U defined in (15). This is no longer true for second order long memory processes. [HB93] showed (in the case where the function L is continuous at zero but the extension is straightforward) that for any fixed positive integer j, there exists a positive constant $c(k, H)$ such that

$$\lim_{n \to \infty} \mathbb{E}[I_U(\omega_j)/f_U(\omega_j)] = c(j, H).$$

The previous results are true for any second order long memory process. Nevertheless, spectral method of estimation of the Hurst parameter, based on the heuristic (but incorrect) assumption that the renormalised DFTs $f_U^{-1/2}(\omega_j)J_{n,j}^U$ are i.i.d. standard complex Gaussian have been proposed and theoretically justifed in some cases. The most well known is the GPH estimator of the Hurst index, introduced by [GPH83] and proved consistent and asymptotically Gaussian for Gaussian long memory processes by [Rob95b] and for a restricted class of linear processes by [Vel00]. Another estimator, often referred to as the local Whittle or GSE estimator was introduced by [Kün87] and again proved consistent asymptotically Gaussian by [Rob95a] for linear long memory processes.

These estimators are built on the m first log-periodogram ordinates, where m is an intermediate sequence, i.e. $1/m + m/n \to 0$ as $n \to \infty$. The choice of m is irrelevant to consistency of the estimator but has an influence on the bias. The rate of convergence of these estimators, when known, is typically slower than $\sqrt{n}$. Trimming of the lowest frequencies, which means taking the ℓ first frequencies out is sometimes used, but there is no theoretical need for this practice, at least in the Gaussian case. See [HDB98]. For nonlinear series, we are not sure yet if trimming may be needed in general.

In the following subsections, we review what is known, both theoretically and empirically, about these and related methods for the different types of nonlinear processes described previsoulsy.

We start by describing the behaviour of the renormalized DFTs at low frequencies, that is, when the index j of the frequency ω_j remains fixed as $n \to \infty$.

3.1 Low-Frequency DFTs of Counts from Infinite-Variance Durations

To the best of our knowledge there is no model in the literature for long memory processes of counts. Hence the question of parametric estimation has not arisen so far in this context. However, one may still be interested in semiparametric estimation of long memory in counts. We present the following result on the behavior of the Discrete Fourier Transforms (DFTs) of processes of counts induced by infinite-variance durations that will be of relevance to us in understanding the behavior of the GPH estimator. Let n denote the number of observations on the counts, $\omega_j = 2\pi j/n$, and define

$$J_{n,j}^{\Delta N} = \frac{1}{\sqrt{2\pi n}} \sum_{t'=1}^{n} \Delta N_{t'} e^{it'\omega_j} .$$

Assume that the distribution of the durations satisfies

$$P(\tau_k \geq x) \backsim \ell(x)x^{-\alpha} \qquad x \to \infty \qquad (17)$$

where $\ell(x)$ is a slowly varying function with $\lim_{x\to\infty} \frac{\ell(kx)}{\ell(x)} = 1 \ \forall k > 0$ and $\ell(x)$ is ultimately monotone at ∞.

Theorem 3. *Let $\{\tau_k\}$ be i.i.d. random variables which satisfy (17) with $\alpha \in (1,2)$ and mean μ_τ. Then for each fixed j, $\ell(n)^{-1}n^{1/2-1/\alpha}J_{n,j}^{\Delta N}$ converges in distribution to a complex α-stable distribution. Moreover, for each fixed j, $\omega_j^d J_{n,j}^{\Delta N} \xrightarrow{p} 0$, where $d = 1 - \alpha/2$.*

The theorem implies that when j is fixed, the normalized periodogram of the counts, $\omega_j^{2d} I_{\Delta N}(\omega_j)$ converges in probability to zero. The degeneracy of the limiting distribution of the normalized DFTs of the counts suggests that the inclusion of the very low frequencies may induce negative finite-sample bias in semiparametric estimators. In addition, the fact that the suitably normalized DFT has an asymptotic stable distribution could further degrade the finite-sample behavior of semiparametric estimators, more so perhaps for the Whittle-likelihood-based estimators than for the GPH estimator since the latter uses the logarithmic transformation.

By contrast, for linear long-memory processes, the normalized periodogram has a nondegenerate positive limiting distribution. See, for example, [TH94].

3.2 Low-Frequency DFTs of Counts from LMSD Durations

We now study the behavior of the low-frequency DFTs of counts generated from finite-variance LMSD durations.

Theorem 4. *Let the durations $\{\tau_k\}$ follow an LMSD model with memory parameter d. Then for each fixed j, $\omega_j^d J_{n,j}^{\Delta N}$, converges in distribution to a zero-mean Gaussian random variable.*

This result is identical to what would be obtained if the counts were a linear long-memory process, and stands in stark contrast to Theorem 3. The discrepancy between these two theorems suggests that the low frequencies will contribute far more bias to semiparametric estimates of d based on counts if the counts are generated by infinite-variance durations than if they were generated from LMSD durations.

3.3 Low and High Frequency DFTs of Shot-Noise Processes

Let X be either the renewal-reward process defined in (7) or the error duration process (9). [HHS04], Theorem 4.1, have proved that Theorem 3 still holds, i.e. $n^{1/2-1/\alpha} J_{n,j}^X$ converges in distribution to an α-stable law, where α is the tail index of the duration. This result can probably be extended to all the shot-noise process for which convergence in distribution of the partial sum process can be proved.

The DFTs of these processes have an interesting feature, related to the slow growth/fast growth phenomenon. The high frequency DFTs, i. e. the DFT $J_{n,j}^X$ computed at a frequency ω_j whose index j increases as n^ρ for some $\rho > 1-1/\alpha$, renormalized by the square root of the spectral density computed at ω_j, have a Gaussian weak limit. This is proved in Theorem 4.2 of [HHS04].

3.4 Estimation of the memory parameter of the LMSV and LMSD models

We now discuss parametric and semiparametric estimation of the memory parameter for the LMSV/LMSD models. Note that in both the LMSV and LMSD models, $\log X_t^2$ can be expressed as the sum of a long memory signal and *iid* noise. Specifically, we have

$$\log X_t^2 = \mu + h_t + u_t, \qquad (18)$$

where $\mu = E\left(\log v_t^2\right)$ and $u_t = \log v_t^2 - E\left(\log v_t^2\right)$ is a zero-mean *iid* series independent of $\{h_t\}$. Since all the existant methodology for estimation for the LMSV model exploits only the above signal plus noise representation, the methodology continues to hold for the LMSD model.

Assuming that $\{h_t\}$ is Gaussian, [DH01] derived asymptotic theory for the log-periodogram regression estimator (GPH; [GPH83]) of d based on $\{\log X_t^2\}$. This provides some justification for the use of GPH for estimating long memory in volatility. Nevertheless, it can also be seen from Theorem 1 of [DH01] that the presence of the noise term $\{u_t\}$ induces a negative bias in the GPH estimator, which in turn limits the number m of Fourier frequencies which can be used in the estimator while still guaranteeing $\sqrt{m}$-consistency and asymptotic normality. This upper bound, $m = o[n^{4d/(4d+1)}]$, where n is the sample size, becomes increasingly stringent as d approaches zero. The results in [DH01] assume that $d > 0$ and hence rule out valid tests for the presence of long memory in $\{h_t\}$. Such a test based on the GPH estimator was provided and justified theoretically by [HS02].

[SP03] proposed a nonlinear log-periodogram regression estimator $\hat{d}_{\mathrm{NLP}}$ of d, using Fourier frequencies $1, \ldots, m$. They partially account for the noise term $\{u_t\}$ through a first-order Taylor expansion about zero of the spectral density of the observations, $\{\log X_t^2\}$. They establish the asymptotic normality of $m^{1/2}(\hat{d}_{\mathrm{NLP}} - d)$ under assumptions including $n^{-4d} m^{4d+1/2} \to \mathrm{Const}$. Thus,

$\hat{d}_{\text{NLP}}$, with a variance of order $n^{-4d/(4d+1/2)}$, converges faster than the GPH estimator, but still arbitrarily slowly if d is sufficiently close to zero. [SP03] also assumed that the noise and signal are Gaussian. This rules out most LMSV/LMSD models, since $\{\log v_t^2\}$ is typically non-Gaussian.

For the LMSV/LMSD model, results analogous to those of [DH01] were obtained by [Art04] for the GSE estimator, based once again on $\{\log X_t^2\}$. The use of GSE instead of GPH allows the assumption that $\{h_t\}$ is Gaussian to be weakened to linearity in a Martingale difference sequence. [Art04] requires the same restriction on m as in [DH01]. A test for the presence of long memory in $\{h_t\}$ based on the GSE estimator was provided by [HMS05].

[HR03] proposed a local Whittle estimator of d, based on log squared returns in the LMSV model. The local Whittle estimator, which may be viewed as a generalized version of the GSE estimator, includes an additional term in the Whittle criterion function to account for the contribution of the noise term $\{u_t\}$ to the low frequency behavior of the spectral density of $\{\log X_t^2\}$. The estimator is obtained from numerical optimization of the criterion function. It was found in the simulation study of [HR03] that the local Whittle estimator can strongly outperform GPH, especially in terms of bias when m is large.

Asymptotic properties of the local Whittle estimator were obtained by [HMS05], who allowed $\{h_t\}$ to be a long-memory process, linear in a Martingale difference sequence, with potential nonzero correlation with $\{u_t\}$. Under suitable regularity conditions on the spectral density of $\{h_t\}$, [HMS05] established the $\sqrt{m}$-consistency and asymptotic normality of the local Whittle estimator, under certain conditions on m. If we assume that the short memory component of the spectral density of $\{h_t\}$ is sufficiently smooth, then their condition on m reduces to

$$\lim_{n\to\infty} \left(m^{-4d-1+\delta}n^{4d} + n^{-4}m^5\log^2(m)\right) = 0 \tag{19}$$

for some arbitrarily small $\delta > 0$.

The first term in (19) imposes a lower bound on the allowable value of m, requiring that m tend to ∞ faster than $n^{4d/(4d+1)}$. It is interesting that [DH01], under similar smoothness assumptions, found that for $m^{1/2}(\hat{d}_{GPH} - d)$ to be asymptotically normal with mean zero, where $\hat{d}_{GPH}$ is the GPH estimator, the bandwidth m must tend to ∞ at a rate *slower* than $n^{4d/(4d+1)}$. Thus for any given d, the optimal rate of convergence for the local Whittle estimator is faster than that for the GPH estimator.

Fully parametric estimation in LMSV/LMSD models once again is based on $\{\log X_t^2\}$ and exploits the signal plus noise representation (18). When $\{h_t\}$ and $\{u_t\}$ are independent, the spectral density of $\{\log X_t^2\}$ is simply the sum of the spectral densities of $\{h_t\}$ and $\{u_t\}$, viz.

$$f_{\log X^2}(\lambda) = f_h(\lambda) + \sigma_u^2/(2\pi), \tag{20}$$

where $f_{\log X^2}$ is the spectral density of $\{\log X_t^2\}$, f_h is the spectral density of $\{h_t\}$ and $\sigma_u^2 = \text{var}(u_t)$, all determined by the assumed parametric model. This representation suggests the possibility of estimating the model parameters in the frequency domain using the Whittle likelihood. Indeed, [Hos97] claims that the resulting estimator is $\sqrt{n}$-consistent and asymptotically normal. We believe that though the result provided in [Hos97] is correct, the proof is flawed. [Deo95] has shown that the quasi-maximum likelihood estimator obtained by maximizing the Gaussian likelihood of $\{\log X_t^2\}$ in the time domain is $\sqrt{n}$-consistent and asymptotically normal.

One drawback of the latent-variable LMSV/LMSD models is that it is difficult to derive the optimal predictor of $|X_t|^s$. In the LMSV model, $\{|X_t|^s\}$ for $s > 0$ serves as a proxy for volatility, while in the LMSD model, $\{X_t\}$ represents durations. A computationally efficient algorithm for optimal linear prediction of such series was proposed in [DHL05], exploiting the Preconditioned Conjugate Gradient (PCG) algorithm. In [CHL05], it is shown that the computational cost of this algorithm is $O(n \log^{5/2} n)$, in contrast to the much more expensive Levinson algorithm, which has cost of $O(n^2)$.

3.5 Simulations on the GPH Estimator for Counts

We simulated i.i.d. durations from a positive stable distribution with tail index $\alpha = 1.5$, with an implied d for the counts of .25. We also simulated durations from an LMSD $(1, d, 0)$ model with Weibull innovations, $AR(1)$ parameter of $-.42$, and $d = .3545$, as was estimated from actual tick-by-tick durations in [DHH05]. The stable durations were multiplied by a constant $c = 1.21$ so that the mean duration matches that found in actual data. For the LMSD durations, we used $c = 1$. One unit in the rescaled durations is taken to represent one second. Tables 1 and 2, for the stable and LMSD cases respectively, present the GPH estimates based on the resulting counts for different values of Δt, using $n = 10,000$, $m = n^{0.5}$ and $m = n^{0.8}$. For the stable case, the bias was far more strongly negative for the smaller value of m, whereas for the LMSD case, the bias did not change dramatically with m. This is consistent with the discussion in Section 3.2, and also with the averaged $\log - \log$ periodogram plots presented in Figure 1, where the averaging is taken over a large number of replications, and all positive Fourier frequencies are considered, $j = 1, \ldots, n/2$. The plot for the stable durations (upper panel) shows a flat slope at the low frequencies. For this process, using more frequencies in the regression seems to mitigate the negative bias induced by the flatness in the lower frequencies as indicated by the less biased estimates of d when $m = n^{0.8}$.

For the LMSD process, by Theorem 2 the counts have the same memory parameter as the durations, $d = .3545$. We did not find severe negative bias in the GPH estimators on the counts, though the estimate of d seems to increase with Δt in the case when $m = n^{0.5}$. The averaged $\log - \log$ periodogram plot presented in the lower panel of Figure 1 shows a near-perfect straight line across all frequencies, which is quite different from the pattern we observed in the case of counts based on stable durations. The straight-line relationship here is consistent with the bias results in our LMSD simulations, and with the discussion in Section 3.2.

Statistical properties of $\hat{d}_{GPH}$ and the choice of m for Gaussian long-memory time series have been discussed in recent literature. [Rob95b] showed for Gaussian processes that the GPH estimator is $m^{1/2}$-consistent and asymptotically normal if an increasing number of low frequencies L is trimmed from the regression of the log periodogram on log frequency. [HDB98] showed that trimming can be avoided for Gaussian processes. In our simulations, we did not use any trimming. There is as yet no theoretical justification for the GPH estimator in the current context since the counts are clearly non-Gaussian, and presumably constitute a nonlinear process. It is not clear whether trimming would be required for such a theory, but our simulations and theoretical results suggest that in some situations trimming may be helpful, while in others it may not be needed.

Table 1. Mean of GPH estimators for counts with different $\triangle t$. Counts generated from *iid* stable durations with skewness parameter $\beta = 0.8$ and tail index $\alpha = 1.5$. The corresponding memory parameter for counts is $d = .25$. We generated 500 replications each with sample size $n = 10,000$. The number of frequencies in the log periodogram regression was $m = n^{0.8} = 1585$ and $m = \sqrt{n} = 100$. t-values marked with $*$ reject the null hypothesis, $d = 0.25$ in favor of $d < 0.25$.

$\triangle t$	$m = n^{0.5}$		$m = n^{0.8}$	
$c = 1.21$	$Mean(\hat{d}_{GPH})$	t-Value	$Mean(\hat{d}_{GPH})$	t-Value
5 min	0.1059	-17.65^*	0.2328	-5.77^*
10 min	0.0744	-23.08^*	0.2212	-8.31^*
20 min	0.0715	-23.23^*	0.2186	-7.75^*

Table 2. Mean of the GPH estimators for counts with different Δt. Counts generated from LMSD durations with Weibull $(1, \gamma)$ shocks. The number of frequencies in the log periodogram regression was $m = \sqrt{n}$ and $m = n^{0.8}$. We used $d = .3545$ and $\gamma = 1.3376$ for our simulations. We simulated 200 replications of the counts, each with sample size $n = 10,000$. t-values marked with $*$ reject the null hypothesis, $d = 0.3545$ in favor of $d < 0.3545$.

$\triangle t$	$m = n^{0.5}$		$m = n^{0.8}$	
$c = 1$	$Mean(\hat{d}_{GPH})$	t-Value	$Mean(\hat{d}_{GPH})$	t-Value
5 min	0.3458	-1.76^*	0.3471	-6.49^*
30 min	0.3873	3.45^*	0.3469	-3.59^*
60 min	0.3923	4.05^*	0.3478	-3.20^*

Fig. 1. Averaged $\log - \log$ periodogram plots for the counts generated from *iid* Stable and LMSD durations.

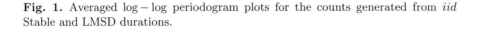

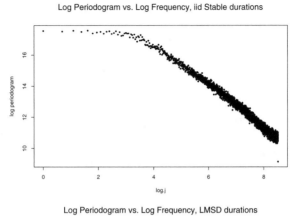

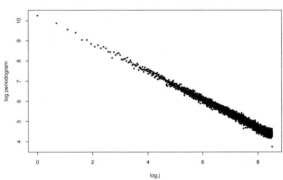

3.6 Estimation of the memory parameter of the Infinite Source Poisson process

Due to the underlying Poisson point process, the Infinite Poisson Source process is a very mathematically tractable model. Computations are very easy and in particular, convenient formulas for cumulants of integrals along paths of the process are available. This allows to derive the theoretical properties of estimators of the Hurst index or memory parameter. [FRS05] have defined an estimator of the Hurst index of the Infinite Poisson source process (with random transmission rate) related to the GSE and proved its consistency and rate of convergence. Instead of using the DFTs of the process, so-called wavelets coefficients are defined as follows. Let ψ be a measurable compactly supported function on $\mathbb{R}$ such that $\int \psi(s)\,\mathrm{d}s = 0$. For $j \in \mathbb{N}$ and $k = 0, \ldots, 2^j - 1$, define

$$w_{j,k} = 2^{j/2} \int \psi(2^{-j}s - k) X_s\,\mathrm{d}s .$$

If (13) holds, then $\mathbb{E}[w_{j,k}] = 0$ and $\mathrm{var}(w_{j,k}) = L(2^j)2^{(2-\alpha)j} = L(2^j)2^{2dj}$, where α is the tail index of the durations, $d = 1 - \alpha/2$ is the memory parameter and L is a slowly varying function at infinity. This scaling property makes it natural to define a contrast function

$$\hat{W}(d') = \log\left(\sum\nolimits_{(j,k)\in\Delta} 2^{-2d'j}w_{j,k}^2\right) + \delta d' \log(2) ,$$

where Δ is the admissible set of coefficients, which depends on the interval of observation and the support of the function ψ. The estimator of d is then $\hat{d} = \arg\min_{d'\in(0,1/2)} W(d')$. [FRS05] have proved under some additional technical assumptions that this estimator is consistent. The rate of convergence can be obtained, but the asymptotic distribution is not known, though it is conjectured to be Gaussian, if the set Δ is properly chosen.

Note in passing that here again, the slow growth/fast growth phenomenon arises. It can be shown, if the shocks and durations are independent, that for fixed k, $2^{(1-\alpha)j/2}w_{j,k}$ converges to an α-stable distribution, but if k tends to infinity at a suitable rate, $2^{-dj}w_{jk}$ converges to a complex Gaussian distribution. This slow growth/fast growth phenomenon is certainly a very important property of these processes that should be understood more deeply.

References

[Ade74]   Rolf K. Adenstedt. On large-sample estimation for the mean of a stationary random sequence. *The Annals of Statistics*, 2:1095–1107, 1974.

[Art04]   Josu Arteche. Gaussian semiparametric estimation in long memory in stochastic volatility and signal plus noise models. *Journal of Econometrics*, 119(1):131–154, 2004.

[BBM96]   Richard T. Baillie, Tim Bollerslev, and Hans Ole Mikkelsen. Fraction-
           ally integrated generalized autoregressive conditional heteroskedasticity.
           Journal of Econometrics, 74(1):3–30, 1996.
[BCdL98]   F. Jay Breidt, Nuno Crato, and Pedro de Lima. The detection and esti-
           mation of long memory in stochastic volatility. *Journal of Econometrics*,
           83(1-2):325–348, 1998.
[BM96]   Tim Bollerslev and Hans Ole Mikkelsen. Modeling and pricing long
           memory in stock market volatility. *Journal of Econometrics*, 73(1):151–
           184, 1996.
[CHL05]   Willa Chen, Clifford M. Hurvich, and Yi Lu. On the correlation matrix of
           the discrete fourier transform and the fast solution of large toeplitz sys-
           tems for long-memory time series. To appear in Journal of the American
           Statistical Association, 2005.
[Dal71]   Daryl J. Daley. Weakly stationary point processes and random measures.
           Journal of the Royal Statistical Society. Series B. Methodological, 33:406–
           428, 1971.
[Dal99]   Daryl J. Daley. The Hurst index of long-range dependent renewal pro-
           cesses. *The Annals of Probability*, 27(4):2035–2041, 1999.
[Deo95]   Rohit Deo. On GMM and QML estimation for the long memory stochas-
           tic volatility model. Working paper, 1995.
[DGE93]   Zhuanxin Ding, Clive W.J. Granger, and Robert F. Engle. A long mem-
           ory property of stock market returns and a new model. *Journal of Em-
           pirical Finance*, 1:83–106, 1993.
[DH01]   Rohit Deo and Clifford M. Hurvich. On the log periodogram regression
           estimator of the memory parameter in long memory stochastic volatility
           models. *Econometric Theory*, 17(4):686–710, 2001.
[DHH05]   Rohit Deo, Mengchen Hsieh, and Clifford M. Hurvich. Tracing the source
           of memory in volatility. Preprint, 2005.
[DHL05]   Rohit Deo, Clifford M. Hurvich, and Yi Lu. Forecasting realized volatility
           using a long-memory stochastic volatility model: estimation, prediction
           and seasonal adjustment. To appear in *Journal of Econometrics*, 2005.
[DHSW05]   Rohit Deo, Clifford M. Hurvich, Philippe Soulier, and Yi Wang. Propa-
           gation of memory parameter from durations to counts. Preprint, available
           on http://www.tsi.enst.fr/ soulier/dhsw.pdf, 2005.
[DRV00]   Daryl J. Daley, Tomasz Rolski, and Rein Vesilo. Long-range dependent
           point processes and their Palm-Khinchin distributions. *Advances in Ap-
           plied Probability*, 32(4):1051–1063, 2000.
[DTW05]   Paul Doukhan, Gilles Teyssiere, and Pablo Winant. A larch(∞) vec-
           tor valued process. In Patrice Bertail, Paul Doukhan and Philippe
           Soulier (eds), *Dependence in Probability and Statistics*. Springer, New
           York, 2005.
[DVJ03]   Daryl J. Daley and David Vere-Jones. *An introduction to the theory of
           point processes. Vol. I: Elementary theory and methods. 2nd ed.* Proba-
           bility and Its Applications. New York, NY: Springer., 2003.
[EKM97]   Paul Embrechts, Claudia Klüppelberg, and Thomas Mikosch. *Modelling
           Extremal Events for Insurance and Finance*. Number 33 in Stochastic
           modelling and applied probability. Berlin: Springer, 1997.
[FRS05]   Gilles Faÿ, François Roueff, and Philippe Soulier. Estimation of the
           memory parameter of the infinite source poisson process. Prépublication
           05-12 de l'Université Paris 10, 2005.

[GJ80]   Clive W.J. Granger and Roselyne Joyeux. An introduction to long mem-
           ory time series and fractional differencing. *Journal of Time Series Anal-
           ysis*, 1:15–30, 1980.

[GKL00]   Liudas Giraitis, Piotr Kokoszka, and Remigijus Leipus.   Stationary
           ARCH models: dependence structure and central limit theorem. *Econo-
           metric Theory*, 16(1):3–22, 2000.

[GMS93]   Liudas Giraitis, Stanislas A. Molchanov, and Donatas Surgailis. Long
           memory shot noises and limit theorems with application to Burgers'
           equation. In *New directions in time series analysis, Part II*, volume 46
           of *IMA Vol. Math. Appl.*, pages 153–176. Springer, New York, 1993.

[GPH83]   John Geweke and Susan Porter-Hudak. The estimation and application
           of long memory time series models. *Journal of Time Series Analysis*,
           4(4):221–238, 1983.

[GS02]   Liudas Giraitis and Donatas Surgailis.   ARCH-type bilinear models
           with double long memory. *Stochastic Processes and their Applications*,
           100:275–300, 2002.

[Har98]   Andrew C. Harvey.   Long memory in stochastic volatility.   In J.
           Knight and S. Satchell (eds), *Forecasting volatility in financial markets*.
           Butterworth-Heinemann, London, 1998.

[HB93]   Clifford M. Hurvich and Kaizô I. Beltrão. Asymptotics for the low-
           frequency ordinates of the periodogram of a long-memory time series.
           Journal of Time Series Analysis, 14(5):455–472, 1993.

[HDB98]   Clifford M. Hurvich, Rohit Deo, and Julia Brodsky. The mean squared
           error of Geweke and Porter-Hudak's estimator of the memory parameter
           of a long-memory time series. *Journal of Time Series Analysis*, 19, 1998.

[HHS04]   Mengchen Hsieh, Clifford M. Hurvich, and Philippe Soulier. Asymp-
           totics for duration driven long memory processes.   Available on
           http://www.tsi.enst.fr/ soulier, 2004.

[HMS05]   Clifford M. Hurvich, Eric Moulines, and Philippe Soulier. Estimating
           long memory in volatility. *Econometrica*, 73(4):1283–1328, 2005.

[Hos81]   J. R. M. Hosking. Fractional differencing. *Biometrika*, 60:165–176, 1981.

[Hos97]   Yuzo Hosoya. A limit theory for long-range dependence and statistical
           inference on related models. *The Annals of Statistics*, 25(1):105–137,
           1997.

[HR03]   Clifford M. Hurvich and Bonnie K. Ray. The local whittle estimator of
           long memory stochastic volatility. *Journal of Financial Econometrics*,
           1:445–470, 2003.

[HRS98]   David Heath, Sidney Resnick, and Gennady Samorodnitsky. Heavy tails
           and long range dependence in ON/OFF processes and associated fluid
           models. *Mathematics of Operations Research*, 23(1):145–165, 1998.

[HS02]   Clifford M. Hurvich and Philippe Soulier. Testing for long memory in
           volatility. *Econometric Theory*, 18(6):1291–1308, 2002.

[IW71]   D. L. Iglehart and Ward Whitt. The equivalence of central limit the-
           orems for counting processes and associated partial sums. *Annals of
           Mathematical Statistics*, 42:1372–1378, 1971.

[KL03]   Vytautas Kazakevičius and Remigijus Leipus. A new theorem on the
           existence of invariant distributions with applications to ARCH processes.
           Journal of Applied Probability, 40(1):147–162, 2003.

[Kün87]   H. R. Künsch. Statistical aspects of self-similar processes. In Yu.A. Prohorov and V.V. Sazonov (eds), *Proceedings of the first World Congres of the Bernoulli Society*, volume 1, pages 67–74. Utrecht, VNU Science Press, 1987.

[Liu00]   Ming Liu. Modeling long memory in stock market volatility. *Journal of Econometrics*, 99:139–171, 2000.

[MR04]   Thomas Mikosch and Sidney Resnick. Activity rates with very heavy tails. Technical Report 1411, Cornell University; to appear in Stochastic Processes and Their Applications, 2004.

[MRR02]   Krishanu Maulik, Sidney Resnick, and Holger Rootzén. Asymptotic independence and a network traffic model. *Journal of Applied Probability*, 39(4):671–699, 2002.

[MRRS02] Thomas Mikosch, Sidney Resnick, Holger Rootzén, and Alwin Stegeman. Is network traffic approximated by stable Lévy motion or fractional Brownian motion? *The Annals of Applied Probability*, 12(1):23–68, 2002.

[Nel91]   Daniel B. Nelson. Conditional heteroskedasticity in asset returns: a new approach. *Econometrica*, 59(2):347–370, 1991.

[Par99]   William R. Parke. What is fractional integration? *Review of Economics and Statistics*, pages 632–638, 1999.

[Rob91]   Peter M. Robinson. Testing for strong serial correlation and dynamic conditional heteroskedasticity in multiple regression. *Journal of Econometrics*, 47(1):67–84, 1991.

[Rob95a] P. M. Robinson. Gaussian semiparametric estimation of long range dependence. *The Annals of Statistics*, 23(5):1630–1661, 1995.

[Rob95b] P. M. Robinson. Log-periodogram regression of time series with long range dependence. *The Annals of Statistics*, 23(3):1048–1072, 1995.

[Rob01]   P. M. Robinson. The memory of stochastic volatility models. *Journal of Econometrics*, 101(2):195–218, 2001.

[RR00]   Sidney Resnick and Holger Rootzén. Self-similar communication models and very heavy tails. *The Annals of Applied Probability*, 10(3):753–778, 2000.

[RvdB00] Sidney Resnick and Eric van den Berg. Weak convergence of high-speed network traffic models. *Journal of Applied Probability*, 37(2):575–597, 2000.

[SP03]   Yixiao Sun and Peter C. B. Phillips. Nonlinear log-periodogram regression for perturbed fractional processes. *Journal of Econometrics*, 115(2):355–389, 2003.

[SV02]   Donatas Surgailis and Marie-Claude Viano. Long memory properties and covariance structure of the EGARCH model. *ESAIM. Probability and Statistics*, 6:311–329, 2002.

[Tak54]   Lajos Takács. On secondary processes derived from a Poisson process and their physical applications. With an appendix by Alfréd Rényi. *Magyar Tud. Akad. Mat. Fiz. Oszt. Közl.*, 4, 1954.

[Taq03]   Murad Taqqu. Fractional brownian motion and long-range dependence. In Paul Doukhan, Georges Oppenheim and Murad S. Taqqu (eds), *Theory and applications of Long-Range Dependence*. Boston, Birkhäuser, 2003.

[TH94]   Norma Terrin and Clifford M. Hurvich. An asymptotic Wiener-Itô representation for the low frequency ordinates of the periodogram of a long memory time series. *Stochastic Processes and their Applications*, 54(2):297–307, 1994.

[TL86]   Murad S. Taqqu and Joshua M. Levy. Using renewal processes to gen-
           erate long range dependence and high variability. In E. Eberlein and
           M.S. Taqqu (eds), *Dependence in Probability and Statistics*. Boston,
           Birkhäuser, 1986.

[TWS97]   Murad Taqqu, Walter Willinger, and Robert Sherman. Proof of a funda-
           mental result in self-similar traffic modeling. *Computer Communication
           Review.*, 27, 1997.

[Vel00]   Carlos Velasco. Non-Gaussian log-periodogram regression. *Econometric
           Theory*, 16(1):44–79, 2000.

[Whi02]   Ward Whitt. *Stochastic-process limits*. Springer Series in Operations
           Research. Springer-Verlag, New York, 2002.

Appendix

Proof (of Theorem 3). For simplicity, we set the clock-time spacing $\Delta t = 1$.
Define

$$S_{\tau,n}(\theta) = \sum_{k=1}^{\lfloor n\theta \rfloor} \tau_k \qquad 0 \leq \theta \leq 1 \, ,$$

$$S_{\Delta N,n}(\theta) = \sum_{t'=1}^{\lfloor n\theta \rfloor} \Delta N_{t'} \qquad 0 \leq \theta \leq 1 \, .$$

Since $\alpha < 2$ and $\{\tau_k\}$ is an i.i.d. sequence, by the fonctional central limit
theorem (FCLT) for random variables in the domain of attraction of a stable
law (see [EKM97, Theorem 2.4.10]), $l(n)n^{-1/\alpha}\{S_{\tau,n}(\theta) - \lfloor n\theta \rfloor \mu_\tau\}$ converges
weakly in $\mathcal{D}(0,1)$ to an α-stable motion, for some slowly varying function l.
Now define

$$U_n(\theta) = (2\pi)^{-1/2}l(n)n^{-1/\alpha}\{S_{\Delta N,n}(\theta) - \lfloor n\theta \rfloor / \mu_\tau\} \, .$$

By the equivalence of FCLTs for the counting process and its associated partial
sums of duration process (see [IW71]), U_n also converges weakly in $\mathcal{D}([0,1])$
to an α-stable motion, say S. Summation by parts yields, for any nonzero
Fourier frequency ω_j (with fixed $j > 0$)

$$l(n)n^{1/2-1/\alpha}J_{n,j}^{\Delta N} = (2\pi)^{-1/2}l(n)n^{-1/\alpha}\sum_{t'=1}^{n}\{\Delta N_{t'} - 1/\mu_\tau\}e^{it'\omega_j}$$

$$= \sum_{t'=1}^{n}\{U_n(t'/n) - U_n((t'-1)/n)\}e^{it'\omega_j} = \int_0^1 e^{2ij\pi x}\,dU_n(x) \, .$$

Hence by the continuous mapping theorem

$$\sqrt{2\pi}\, l(n)n^{1/2-1/\alpha} J_{n,j}^{\Delta N} \xrightarrow{d} \int_0^1 e^{2i\pi jx}\, dS(x)$$

which is a stochastic integral with respect to a stable motion, hence has a stable law.

To prove the second statement of the theorem, note that for fixed j and as $n \to \infty$, $f(\omega_j) \sim l_1(n)\omega_j^{-2d}$ for some slowly varying function l_1, so

$$f^{-1/2}(\omega_j) J_{n,j}^{\Delta N} = \frac{l(n)n^{1/\alpha-1/2}}{f^{1/2}(\omega_j)} \frac{J_{n,j}^{\Delta N}}{l(n)n^{1/\alpha-1/2}}$$

$$\sim C_1 l(n)n^{1/\alpha+\alpha/2-3/2} \frac{J_{n,j}^{\Delta N}}{\mu_\tau^{-1-1/\alpha} l(n)n^{1/\alpha-1/2}}. \quad (21)$$

Since $1/\alpha+\alpha/2-3/2 < 0$, we have $l(n)n^{1/\alpha+\alpha/2-3/2} \to 0$. Hence by Slutsky's Theorem, (21) converges to zero. □

Proof (of Theorem 4). Let $S_n(t) = n^{-H} \sum_{k=1}^{[nt]} (\tau_k - \mathbb{E}[\tau_k])$, $t \in (0,1)$. It is shown in Surgailis and Viano (2002) that $S_n(t) \xRightarrow{d} B_H(t)$ in $\mathcal{D}([0,1])$ where $B_H(t)$ is fractional Brownian motion with Hurst parameter $H = d + 1/2$. Thus, by Iglehart and Whitt (1971), it follows that $t^{-H}N \to AB_H$ in $\mathcal{D}([0,1])$, where A is a nonzero constant. The result follows as above by the continuous mapping theorem and summation by parts. □

A LARCH(∞) Vector Valued Process

Paul Doukhan[1], Gilles Teyssière[2] and Pablo Winant[3]

[1] Laboratoire de Statistique, CREST. Timbre J340, 3, avenue Pierre Larousse, 92240 Malakoff, France. doukhan@ensae.fr
[2] Statistique Appliquée et MOdélisation Stochastique, Université Paris 1, Centre Pierre Mendès France, 90 rue de Tolbiac, F-75634 Paris Cedex 13. France. stats@gillesteyssiere.net
[3] ENS Lyon, 46 allée d'Italie, 69384 Lyon, France. pablo.winant@ens-lyon.org

1 Introduction

The purpose of this chapter is to propose a unified framework for the study of ARCH(∞) processes that are commonly used in the financial econometrics literature. We extend the study, based on Volterra expansions, of univariate ARCH(∞) processes by Giraitis et al. [GKL00] and Giraitis and Surgailis [GS02] to the multi-dimensional case.

Let $\{\xi_t\}_{t\in\mathbb{Z}}$ be a sequence of real valued random matrices independent and identically distributed of size $d \times m$, $\{a_j\}_{j\in\mathbb{N}^*}$ be a sequence of real matrices $m \times d$, and a be a real vector of dimension m. The vector ARCH(∞) process is defined as the solution to the recurrence equation:

$$X_t = \xi_t \left(a + \sum_{j=1}^{\infty} a_j X_{t-j} \right). \tag{1}$$

The following section 2 displays a chaotic expansion solution to this equation; we also consider a random fields extension of this model. Some approximations of this solutions are listed in the next section 3, where we consider approximations by m-dependent sequences, coupling results and approximations by Markov sequences. Section 4 details the weak dependence properties of the model and section 5 provides an existence and uniqueness condition for the solution of the previous equation; in that case, long range dependence may occur. The end of this section is dedicated to review examples of this vector valued model.

The vector ARCH(∞) model nests a large variety of models, the two first extensions being obvious:

(A1). The univariate linear ARCH(∞) (LARCH) model, where the X_t and a_j are scalar,

(A2). The bilinear model, with

$$X_t = \zeta_t \left(\alpha + \sum_{j=1}^{\infty} \alpha_j X_{t-j} \right) + \beta + \sum_{j=1}^{\infty} \beta_j X_{t-j} \, ,$$

where all variables are scalar, and ζ_t are iid centered innovations. We set

$$\xi_t = (\zeta_t, 1) \, , \quad a = \begin{pmatrix} \alpha \\ \beta \end{pmatrix} , \quad a_j = \begin{pmatrix} \alpha_j \\ \beta_j \end{pmatrix} \, .$$

In that case, the expansion (3) is the same as the one used by Giraitis and Surgailis [GS02].

(A3). With a suitable re-parameterization, this vector ARCH(∞) nests the standard GARCH–type processes used in the financial econometrics literature for modeling the non-linear structure of the conditional second moments. The GARCH(p, q) model is defined as

$$r_t = \sigma_t \varepsilon_t \, ,$$

$$\sigma_t^2 = \sum_{j=1}^{p} \beta_j \sigma_{t-j}^2 + \gamma_0 + \sum_{j=1}^{q} \gamma_j r_{t-j}^2 \, , \quad \gamma_0 > 0 \, , \quad \gamma_j \geq 0 \, , \quad \beta_i \geq 0 \, ,$$

where the ε_t are centered and iid. This model is nested in the class of bilinear models with the following re-parameterization

$$\alpha_0 = \frac{\gamma_0}{1 - \sum \beta_i} \, , \quad \sum \alpha_i z^i = \frac{\sum \gamma_i z^i}{1 - \sum \beta_i z^i} \, ,$$

see Giraitis *et al.* [GLS05]. The covariance function of the sequence $\{r_t^2\}$ has an exponential decay, which is implied by the exponential decay of the sequence of weights α_j; see Giraitis *et al.* [GKL00].

(A4). The ARCH(∞) model, where the sequence of weights β_j might have either a exponential decay or a hyperbolic decay.

$$r_t = \sigma_t \varepsilon_t \, , \quad \sigma_t^2 = \beta_0 + \sum_{j=1}^{\infty} \beta_j r_{t-j}^2 \, ,$$

with the following parameterization

$$X_t = r_t^2 \, , \quad \xi_t = \left(\frac{\varepsilon_t^2 - \lambda_1}{\kappa}, 1 \right) \, , \quad a = \begin{pmatrix} \kappa \beta_0 \\ \lambda_1 \beta_0 \end{pmatrix} \, , \quad a_j = \begin{pmatrix} \kappa \beta_j \\ \lambda_1 \beta_j \end{pmatrix} \, ,$$

where the ε are centered and iid, $\lambda_1 = \mathbb{E}(\varepsilon_0^2)$, and $\kappa^2 = \mathrm{var}(\varepsilon_0^2)$. Note that the first coordinate of ξ_0 is thus a centered random variable. Conditions for stationarity of the unidimensional ARCH(∞) model have been derived using Volterra expansions by Giraitis *et al.* [GKL00] and Giraitis and Surgailis [GS02]. The present paper is a multidimensional generalization of these previous works.

(A5). We can consider models with several innovations and variables such as:

$$Z_t = \zeta_{1,t} \left(\alpha + \sum_{j=1}^{\infty} \alpha_j^1 Z_{t-j} \right) + \mu_{1,t} \left(\beta + \sum_{j=1}^{\infty} \beta_j^1 Y_{t-j} \right) + \gamma + \sum_{j=1}^{\infty} \gamma_j^1 Z_{t-j} ,$$

$$Y_t = \zeta_{2,t} \left(\alpha + \sum_{j=1}^{\infty} \alpha_j^2 Y_{t-j} \right) + \mu_{2,t} \left(\beta + \sum_{j=1}^{\infty} \beta_j^2 Z_{t-j} \right) + \gamma + \sum_{j=1}^{\infty} \gamma_j^2 Y_{t-j} .$$

This model is straightforwardly described through equation (1) with $d = 2$ and $m = 3$. Here $\xi_t = \begin{pmatrix} \zeta_{1,t} & \mu_{1,t} & 1 \\ \zeta_{2,t} & \mu_{2,t} & 1 \end{pmatrix}$ is a 2×3 iid sequence, $a_j = \begin{pmatrix} \alpha_j^1 & \alpha_j^2 \\ \beta_j^1 & \beta_j^2 \\ \gamma_j^1 & \gamma_j^2 \end{pmatrix}$ is a 3×2 matrix and $a = \begin{pmatrix} \alpha \\ \beta \\ \gamma \end{pmatrix}$ is a vector in $\mathbb{R}^3$ and the process $X_t = \begin{pmatrix} Z_t \\ Y_t \end{pmatrix}$ is a vector of dimension 2. Dimensions $m = 3$ and $d = 2$ are only set here for simplicity. Replacing $m = 3$ by $m = 6$ would allow to consider different coefficients α, β and γ for both lines in this system of two coupled equations.

This generalizes the class of multivariate ARCH(∞) processes, defined in the p-dimensional case as:

$$R_t = \Sigma_t^{\frac{1}{2}} \varepsilon_t ,$$

where R_t is a p–dimensional vector, Σ_t is a $p \times p$ positive definite matrix, and ε_t is a p–dimensional vector. Those models are formally investigated by Farid Boussama in [Bou98]; published references include [Bou00] and [EK96].

This model is of interest in financial econometrics as the volatility of asset prices of linked markets, e.g., major currencies in the Foreign Exchange (FX) market, are correlated, and in some cases display a common strong dependence structure; see [Tey97]. This common dependence structure can be modeled with the assumption that the innovations $\varepsilon_1, \ldots, \varepsilon_p$ are correlated. An (empirically) interesting case for the bivariate model (X_t, Y_t) is obtained with the assumption that the $(\zeta_{1,t}, \zeta_{2,t})$ are cross-correlated.

2 Existence and Uniqueness in L^p

In the sequel, we set $A(x) = \sum_{j \geq x} \|a_j\|$, $A = A(1)$, where $\| \cdot \|$ denotes the matrix norm.

Theorem 1. *Let $p > 0$, we denote*

$$\varphi = \sum_{j \geq 1} \|a_j\|^{p \wedge 1} \left(\mathbb{E}\|\xi_0\|^p\right)^{\frac{1}{p \wedge 1}} . \tag{2}$$

If $\varphi < 1$, then a stationary solution in L^p to equation (1) is given by:

$$X_t = \xi_t \left(a + \sum_{k=1}^{\infty} \sum_{j_1,\ldots,j_k \geq 1} a_{j_1} \xi_{t-j_1} \cdots a_{j_k} \xi_{t-j_1-\cdots-j_k} \cdot a\right) . \tag{3}$$

Proof. The norm used for the matrices is any multiplicative norm. We have to show that expression (3) is well defined under the conditions stated above, converges absolutely in L^p, and that it satisfies equation (1).

Step 1. We first show that expression (3) is well defined (after the second line we omit to precise the norms). For $p \geq 1$, we have

$$\sum_{j_1,\ldots,j_k \geq 1} \|a_{j_1} \xi_{t-j_1} \cdots a_{j_k} \xi_{t-j_1-\cdots-j_k}\|_{m \times m}$$

$$\leq \sum_{j_1,\ldots,j_k \geq 1} \|a_{j_1}\|_{m \times d} \cdots \|a_{j_k}\|_{m \times d} \|\xi_{t-j_1}\|_{d \times m} \cdots \|\xi_{t-j_1-\cdots-j_k}\|_{d \times m} .$$

The series thus converges in norm L^p because

$$\sum_{k=1}^{\infty} \sum_{j_1,\ldots,j_k \geq 1} \left(\mathbb{E}\|a_{j_1} \xi_{t-j_1} \cdots a_{j_k} \xi_{t-j_1-\cdots-j_k}\|^p\right)^{1/p}$$

$$\leq \sum_{k=1}^{\infty} \sum_{j_1,\ldots,j_k \geq 1} \|a_{j_1}\| \cdots \|a_{j_k}\| \left(\mathbb{E}\|\xi_{t-j_1}\|^p\right)^{1/p} \cdots \left(\mathbb{E}\|\xi_{t-j_1-\cdots-j_k}\|^p\right)^{1/p}$$

$$\leq \sum_{k=1}^{\infty} \sum_{j_1,\ldots,j_k \geq 1} \|a_{j_1}\| \cdots \|a_{j_k}\| \left(\mathbb{E}\|\xi_0\|^p\right)^{\frac{k}{p}} \leq \sum_{k=1}^{\infty} \varphi^k .$$

The series $\sum_{k=1}^{\infty} \varphi^k$ is finite since $\varphi < 1$, hence the series (3) converges in L^p.

For $p < 1$, the convergence is defined through the metric $d_p(U, V) = \mathbb{E}\|U - V\|^p$ between vector valued L^p random variables U, V and we start from

$$\left(\sum_{j_1,\ldots,j_k \geq 1} \|a_{j_1} \xi_{t-j_1} \cdots a_{j_k} \xi_{t-j_1-\cdots-j_k}\|\right)^p$$

$$\leq \sum_{j_1,\ldots,j_k \geq 1} \|a_{j_1} \xi_{t-j_1} \cdots a_{j_k} \xi_{t-j_1-\cdots-j_k}\|^p ,$$

and we use the same arguments as for $p = 1$.

Step 2. We now show that equation (3) is solution to equation (1):

$$X_t = \xi_t \left(a + \sum_{k=1}^{\infty} \sum_{j_1,\ldots,j_k \geq 1} a_{j_1} \xi_{t-j_1} \cdots a_{j_k} \xi_{t-j_1-\cdots-j_k} \cdot a\right)$$

$$= \xi_t \left(a + \sum_{j \geq 1} a_j \xi_{t-j} + \right.$$

$$\left. + \sum_{k=2}^{\infty} \sum_{j_1 \geq 1} a_j \xi_{t-j} \sum_{j_2, \ldots, j_k \geq 1} a_{j_2} \xi_{t-j-j_2} \cdots a_{j_k} \xi_{t-j-j_2-\cdots-j_k} \cdot a \right)$$

$$= \xi_t \left(a + \sum_{j \geq 1} a_j \times \right.$$

$$\left. \xi_{t-j} \left\{ a + \sum_{k=2}^{\infty} \sum_{j_2, \ldots, j_k \geq 1} a_{j_2} \xi_{(t-j)-j_2} \cdots a_{j_k} \xi_{(t-j)-j_2-\cdots-j_k} \cdot a \right\} \right)$$

$$= \xi_t \left(a + \sum_{j \geq 1} a_j X_{t-j} \right).$$

Remark 1. The uniqueness of this solution is not demonstrated without additional condition; see Theorem 2 and section 5 below.

Theorem 2. *Assume that $p \geq 1$ then from (2), $\varphi = \sum_j \|a_j\| \|\xi_0\|_p$. Assume $\varphi < 1$. If a stationary solution $(Y_t)_{t \in \mathbb{Z}}$ to equation (1) exists (a.s.), if Y_t is independent of the sigma-algebra generated by $\{\xi_s; s > t\}$, for each $t \in \mathbb{Z}$, then this solution is also in L^p and it is (a.s.) equal to the previous solution $(X_t)_{t \in \mathbb{Z}}$ defined by equation (3).*

Proof. Step 1. We first prove that $\|Y_0\|_p < \infty$. From equation (1) and from $\{Y_t\}_{t \in \mathbb{Z}}$'s stationarity, we derive

$$\|Y_0\|_p \leq \|\xi_0\|_p \left(\|a\| + \sum_{j=1}^{\infty} \|a_j\| \|Y_0\|_p \right) < \infty,$$

hence, the first point in the theorem follows from:

$$\|Y_0\|_p \leq \frac{\|\xi_0\|_p \|a\|}{1 - \varphi} < \infty.$$

Step 2. As in [GKL00] we write recursively $Y_t = \xi_t \left(a + \sum_{j \geq 1} a_j Y_{t-j} \right) = X_t^m + S_t^m$, with

$$X_t^m = \xi_t \left(a + \sum_{k=1}^{m} \sum_{j_1, \cdots, j_k \geq 1} a_{j_1} \xi_{t-j_1} \cdots a_{j_k} \xi_{t-j_1-\cdots-j_k} a \right),$$

$$S_t^m = \xi_t \left(\sum_{j_1, \cdots, j_{m+1} \geq 1} a_{j_1} \xi_{t-j_1} \cdots a_{j_m} \xi_{t-j_1-\cdots-j_m} a_{j_{m+1}} Y_{t-j_1-\cdots-j_m} \right).$$

We have

$$\|S_t^m\|_p \leq \|\xi\|_p \sum_{j_1,\cdots,j_{m+1}\geq 1} \|a_{j_1}\|\cdots\|a_{j_{m+1}}\|\|\xi\|_p^m\|Y_0\|_p = \|Y_0\|_p\varphi^{m+1} .$$

We recall the additive decomposition of the chaotic expansion X_t in equation (3) as a finite expansion plus a negligible remainder that can be controlled $X_t = X_t^m + R_t^m$ where

$$R_t^m = \xi_t\left(\sum_{k>m}\sum_{j_1,\cdots,j_k\geq 1} a_{j_1}\xi_{t-j_1}\cdots a_{j_k}\xi_{t-j_1-\cdots-j_k}a\right) ,$$

satisfies

$$\|R_t^m\|_p \leq \|a\|\|\xi_0\|_p \sum_{k>m} \varphi^k \leq \|a\|\|\xi_0\|_p\frac{\varphi^m}{1-\varphi} \to 0 .$$

Then, the difference between those two solutions is controlled as a function of m with $X_t - Y_t = R_t^m - S_t^m$, hence

$$\|X_t - Y_t\|_p \leq \|R_t^m\|_p + \|S_t^m\|_p$$

$$\leq \frac{\varphi^m}{1-\varphi}\|a\|\|\xi_0\|_p + \|Y_0\|_p\varphi^m \leq 2\frac{\varphi^m}{1-\varphi}\|a\|\|\xi_0\|_p ,$$

and thus $Y_t = X_t$ a.s.

We also consider the following extension of equation (1) to random fields $\{X_t\}_{t\in\mathbb{Z}^D}$:

Lemma 1. *Assume that a_j are $m \times d$-matrices now defined for each $j \in \mathbb{Z}^D \setminus \{0\}$. Fix an arbitrary norm $\|\cdot\|$ on $\mathbb{Z}^D$. We extend the previous function A to $A(x) = \sum_{\|j\|\geq x}\|a_j\|$, $A = A(1)$ and we suppose with $p = \infty$ that $\varphi = A\|\xi_0\|_\infty < 1$. Then the random field*

$$X_t = \xi_t\left(a + \sum_{k=1}^{\infty}\sum_{j_1\neq 0}\cdots\sum_{j_k\neq 0} a_{j_1}\xi_{t-j_1}\cdots a_{j_k}\xi_{t-j_1-\cdots-j_k}a\right) \qquad (4)$$

is a solution to the recursive equation:

$$X_t = \xi_t\left(a + \sum_{j\neq 0} a_j X_{t-j}\right) , \qquad t\in\mathbb{Z}^D . \qquad (5)$$

Moreover, each stationary solution to this equation is also bounded and equals X_t, a.s.

The proof is the same as before, we first prove that any solution is bounded and we expand it as the sum of the first terms in this chaotic expansion, up to a small remainder (wrt to sup norm); the only important modification follows from the fact that now $j_1 + \cdots + j_\ell$ may really vanish for nonzero j_i's which entails that the bound with expectation has to be replaced by upper bounds.

Remark 2. In the previous lemma, the independence of the ξ's does not play a role. We may have stated it for arbitrary random fields $\{\xi_t\}$ such that $\|\xi_t\|_\infty \leq M$ for each $t \in \mathbb{Z}^D$; such models with dependent inputs are interesting but assumptions on the innovations are indeed very strong. This means that such models are heteroscedastic but with bounded innovations: according to [MH04], this restriction excludes extreme phenomena like crashes and bubbles. Mandelbrot school has shown from the seminal paper [Man63] that asset prices returns do not have a Gaussian distribution as the number of extreme deviations, the so–called "Noah effects", of asset returns is far greater than what is allowed by the Normal distribution, even with ARCH–type effects. It is the reason why this extension is not pursued in the present paper.

3 Approximations

This section is aimed to approximate a sequence $\{X_t\}$ given by (3), solution to eqn. (1) by a sequence $\{\tilde{X}_t\}$. We shall prove that we can control the approximation error $\mathbb{E}\|X_t - \tilde{X}_t\|$ within reasonable small bounds.

3.1 Approximation by Independence

The purpose is to approximate X_t by a random variable independent of X_0. We set

$$\tilde{X}_t = \xi_t \left(a + \sum_{k=1}^\infty \sum_{j_1 + \cdots + j_k < t} a_{j_1} \xi_{t-j_1} \cdots a_{j_k} \xi_{t-j_1-\cdots-j_k} a \right) .$$

Proposition 1. *Define φ from* (2). *A bound for the error is given by:*

$$\mathbb{E}\|X_t - \tilde{X}_t\| \leq \mathbb{E}\|\xi_0\| \left(\mathbb{E}\|\xi_0\| \sum_{k=1}^{t-1} k\varphi^{k-1} A\left(\frac{t}{k}\right) + \frac{\varphi^t}{1-\varphi} \right) \|a\| .$$

Furthermore, we have as particular results that if b, $C > 0$ and $q \in [0,1)$, then for a suitable choice of constants K, K':

$$\mathbb{E}\|X_t - \tilde{X}_t\| \leq \begin{cases} K \frac{(\log(t))^{b \vee 1}}{t^b} , & \text{for Riemannian decays } A(x) \leq Cx^{-b} , \\ K'(q \vee \varphi)^{\sqrt{t}} , & \text{for geometric decays } A(x) \leq Cq^x . \end{cases}$$

Remark 3. Note that in the first case this decay is essentially the same Riemannian one while it is sub-geometric (like $t \mapsto e^{-c\sqrt{t}}$) when the decay of the coefficients is geometric.

Remark 4. In the paper Riemannian or Geometric decays always refer to the previous relations.

Idea of the Proof. A careful study of the terms in X_t's expansion which do not appear in $\tilde{X}_t$ entails the following bound with the triangular inequality. For this, quote that if $j_1 + \cdots + j_k \geq t$ for some $k \geq 1$ then, at least, one of the indices $j_1, \ldots,$ or j_k is larger than t/k. The additional term corresponds to those terms with indices $k > t$ in the expansion (3).

The following extension to the case of the random fields determined in lemma 1 is immediate by setting

$$\tilde{X}_t = \xi_t \left(a + \sum_{k=1}^{\infty} \sum_{\substack{j_1, \ldots, j_k \neq 0 \\ \|j_1\| + \cdots + \|j_k\| < \|t\|}} a_{j_1} \xi_{t-j_1} \cdots a_{j_k} \xi_{t-j_1-\cdots-j_k} a \right).$$

Proposition 2. *The random field* $(X_t)_{t \in \mathbb{Z}^D}$ *defined in lemma 1 satisfies:*

$$\mathbb{E}\|X_t - \tilde{X}_t\| \leq \mathbb{E}\|\xi_0\| \left(\|\xi_0\|_\infty \sum_{1 \leq k < \|t\|} k\varphi^{k-1} A\left(\frac{\|t\|}{k}\right) + \frac{\varphi^{\|t\|}}{1-\varphi} \right) \|a\|.$$

3.2 Coupling

First note that the variable $\tilde{X}_t$ which approximates X_t does not follow the same distribution. For dealing with this issue, it is sufficient to construct a sequence of iid random variables ξ_i' which follow the same distribution as the one of the ξ_i, each term of the sequence being independent of all the ξ_i. We then set

$$\xi_t^* = \begin{cases} \xi_t \text{ if } t > 0, \\ \xi_t' \text{ if } t \leq 0, \end{cases}$$

$$X_t^* = \xi_t \left(a + \sum_{k=1}^{\infty} \sum_{j_1, \ldots, j_k} a_{j_1} \xi_{t-j_1}^* \cdots a_{j_k} \xi_{t-j_1-\cdots-j_k}^* a \right).$$

Coefficients τ_t for the τ–dependence introduced by Dedecker and Prieur [DP01] are easily computed. In this case, we find the upper bounds from above, up to a factor 2:

$$\tau_t = \mathbb{E}\|X_t - X_t^*\| \leq 2\mathbb{E}\|\xi_0\| \left(\mathbb{E}\|\xi_0\| \sum_{k=1}^{t-1} k\varphi^{k-1} A\left(\frac{t}{k}\right) + \frac{\varphi^t}{1-\varphi} \right) \|a\|;$$

see also Rüschendorf [RüS], Prieur [Pri05]. These coefficients τ_k are defined as $\tau_k = \tau(\sigma(X_i, i \leq 0), X_k)$ where for each random variable X and each σ-algebra $\mathcal{M}$ one sets

$$\tau(\mathcal{M}, X) = \mathbb{E}\left\{ \sup_{\mathrm{Lip}f \leq 1} \left| \int f(x) \mathbb{P}_{X|\mathcal{M}}(dx) - \int f(x) \mathbb{P}_X(dx) \right| \right\}$$

where $\mathbb{P}_X$ and $\mathbb{P}_{X|\mathcal{M}}$ denotes the distribution and the conditional distribution of X on the σ-algebra $\mathcal{M}$ and $\mathrm{Lip}f = \sup_{x \neq y} |f(x) - f(y)|/\|x - y\|$.

3.3 Markovian Approximation

We consider equation (1) truncated at the order N: $Y_t = \xi_t(a + \sum_{j=1}^{N} a_j Y_{t-j})$. The solution considered above can be rewritten as

$$X_t^N = \xi_t \left(a + \sum_{k=1}^{\infty} \sum_{N \geq j_1, \ldots, j_k \geq 1} a_{j_1} \xi_{t-j_1} \cdots a_{j_k} \xi_{t-j_1-\cdots-j_k} a \right) .$$

We can easily find an upper bound of the error: $\mathbb{E}\|X_t - X_t^N\| \leq \sum_{k=1}^{\infty} A(N)^k$. As in proposition 1, in the Riemannian case, this bound of the error writes as $C \sum_{k=1}^{\infty} N^{-bk} \leq C/(N^b - 1)$ with $b > 1$, while in the geometric case, this writes as $Cq^N/(1 - q^N) \leq Cq^N/(1 - q)$, $0 < q < 1$.

4 Weak Dependence

Consider integers $u, v \geq 1$. Let $i_1 < \cdots < i_u$, $j_1 < \cdots < j_v$ be integers with $j_1 - i_u \geq r$, we set U and V for the two random vectors $U = (X_{i_1}, X_{i_2}, \ldots, X_{i_u})$ and $V = (X_{j_1}, X_{j_2}, \ldots, X_{j_v})$. We fix a norm $\| \cdot \|$ on $\mathbb{R}^d$. For a function $h : (\mathbb{R}^d)^w \to \mathbb{R}$ we set

$$\mathrm{Lip}(h) = \sup_{x_1, y_1, \ldots, x_w, y_w \in \mathbb{R}^d} \frac{|h(x_1, \ldots, x_w) - h(y_1, \ldots, y_w)|}{\sum_{i=1}^{w} \|x_i - y_i\|} .$$

Theorem 3. *Assume that the coefficient defined by (2) satisfies $\varphi < 1$. The solution (3) to the equation (1) is $\theta-$weakly dependent, see [DD03]. This means that:*

$$|\mathrm{cov}(f(U), g(V))| \leq 2v\|f\|_\infty \mathrm{Lip}(g)\theta_r ,$$

for any integers $u, v \geq 1$, $i_1 < \cdots < i_u$, $j_1 < \cdots < j_v$ such that $j_1 - i_u \geq r$; with

$$\theta_r = \mathbb{E}\|\xi_0\| \left(\mathbb{E}\|\xi_0\| \sum_{k=1}^{r-1} k\varphi^{k-1} A\left(\frac{r}{k}\right) + \frac{\varphi^r}{1 - \varphi} \right) \|a\| .$$

Proof. For calculating a weak dependence bound, we approximate the vector V by the vector $\hat{V} = (\hat{X}_{j_1}, \hat{X}_{j_2}, \ldots, \hat{X}_{j_v})$, where we set

$$\hat{X}_t = \xi_t \left(a + \sum_{k=1}^{\infty} \sum_{j_1+\cdots+j_k < s} a_{j_1} \xi_{t-j_1} \cdots a_{j_k} \xi_{t-j_1-\cdots-j_k} a \right) .$$

Note that for each index $j \in \mathbb{Z}$, $\hat{X}_j$ is independent of $(X_{j-s})_{s \geq r}$. Note that for $1 \leq k \leq v$, $\mathbb{E}\|X_{j_k} - \hat{X}_{j_k}\| \leq \theta_r$ defined in theorem 3. Then

$$|\mathrm{cov}(f(U), g(V))| \leq \left| \mathbb{E}\left(f(U)(g(V) - g(\hat{V}))\right) - \mathbb{E}(f(U))\mathbb{E}(g(V) - g(\hat{V}))\right|$$

$$\leq 2\|f\|_\infty \mathbb{E}\left|g(V) - g(\hat{V})\right|$$

$$\leq 2\|f\|_\infty \mathrm{Lip}(g) \sum_{k=1}^{v} \mathbb{E}\|X_{j_k} - \hat{X}_{j_k}\|$$

$$\leq 2v\|f\|_\infty \mathrm{Lip}(g)\theta_r .$$

Remark 5. We obtain explicit expressions for this bound in Proposition 1 for the Riemannian and geometric decay rates.

Remark 6. In the case of random fields the η-weak dependence condition in [DL99] or [DL02] holds in a similar way with

$$\eta_r = 2\mathbb{E}\|\xi_0\| \left(\|\xi_0\|_\infty \sum_{k < r/2} k\varphi^{k-1} A\left(\frac{r}{k}\right) + \frac{\varphi^{[r/2]}}{1 - \varphi} \right) \|a\| ,$$

which means that the previous bound now reads

$$|\mathrm{cov}(f(U), g(V))| \leq \left(u\|g\|_\infty \mathrm{Lip}(f) + v\|f\|_\infty \mathrm{Lip}(g)\right)\eta_r .$$

The argument is the same except for the fact that now $\hat{U}$ and $\hat{V}$ are independent vectors with truncations at a level $s = [r/2]$ but $\hat{V}$ and U are not necessarily independent (recall that independence of U and $\hat{V}$ follows from $s \geq r$ in the proof for the causal case). This point makes the previous bound a bit more complicated than the one in theorem 3 and it explains the appearance of the factor 2 in the expression of η_r.

Remark 7. These weak dependence conditions imply various limit theorems both for partial sums processes and for the empirical process (see [DL99], [DD03] and [DL02]).

5 L^2 Properties

For the univariate case, the uniqueness of a stationary solution to (1) has been proved by [GKL00]. We present a criterion for existence and uniqueness of a solution in L^2. This solution is no longer necessarily weakly dependent.

Theorem 4. *Assume that the iid sequence $\{\xi_t\}$ is centered and the spectral radius $\rho(S)$ of the matrix $S = \sum_{k=1}^{\infty} a'_k \mathbb{E}(\xi'_k \xi_k) a_k$ satisfies $\rho(S) < 1$. Then there exists a unique stationary solution in L^2 to equation (1) given by (3).*

Remark 8.

- The assumption $\rho(S) < 1$ implies $\xi_t \in L^2$ for $t \in \mathbb{Z}$.

- The bilinear model of Example 2 is shown in [GS02] to display the double long memory property when the series $\{\alpha_j\}$ and $\{\beta_j\}$ are not summable but satisfy the condition

$$\sum_{j=1}^{\infty} \left(\alpha_j^2 \mathbb{E}\zeta_0^2 + \beta_j^2\right) < 1 .$$

  As a particular case, the squares of the LARCH(∞) process in Example 1 display long–range dependence. [GS02] prove that the corresponding partial sums process converges to the fractional Brownian Motion with normalization $\gg \sqrt{n}$.
- The GARCH(p, q) models in example 3, are always weakly dependent, in the sense of [DL99].
- Note that [GKL00] and [GS02] prove that the stationary ARCH(∞) model (Example 4), is not long range dependent in the previous sense; more precisely the partial sums process, normalized with $\sqrt{n}$, converges to the Brownian Motion.

Proof. Step 1: existence. Define $T = \mathbb{E}(\xi_k' \xi_k)$. Consider the chaotic solution (3) and set
$$C_t(k_2, \ldots, k_\ell) = \xi_t a_{k_2} \xi_{t-k_2} \cdots a_{k_\ell} \xi_{t-k_2-\cdots-k_\ell} a .$$
Write $\mathbb{E}(X_t' X_t) = a' \mathbb{E}\xi_t' \xi_t a + B = a'Ta + B$, where

$$
\begin{aligned}
B &= \sum_{\ell, k_1, \ldots, k_\ell \geq 1} \mathbb{E} C_{t-k_1}'(k_2, \ldots, k_\ell) a_{k_1}' T a_{k_1} C_{t-k_1}(k_2, \ldots, k_\ell) \\
&= \sum_{\ell, k_1, \ldots, k_\ell} \mathbb{E} C_{t-k_1}'(k_2, \ldots, k_\ell) a_{k_1}' \mathbb{E}\xi_{t-k_1}' \xi_{t-k_1} a_{k_1} C_{t-k_1}(k_2, \ldots, k_\ell) \\
&= \sum_{\ell, k_1, \ldots, k_\ell} \mathbb{E} C_{t-k_1}'(k_2, \ldots, k_\ell) \left(\mathbb{E} a_{k_1}' \xi_{t-k_1}' \xi_{t-k_1} a_{k_1}\right) C_{t-k_1}(k_2, \ldots, k_\ell) \\
&= \sum_{\ell, k_1, \ldots, k_\ell} \mathbb{E} C_t'(k_2, \ldots, k_\ell) \left(\mathbb{E} a_{k_1}' \xi_{t-k_1}' \xi_{t-k_1} a_{k_1}\right) C_t(k_2, \ldots, k_\ell) \\
&= \sum_{\ell, k_2, \ldots, k_\ell} \mathbb{E} C_t'(k_2, \ldots, k_\ell) S C_t(k_2, \ldots, k_\ell) \\
&\leq \rho(S) \sum_{\ell, k_2, \ldots, k_\ell} \mathbb{E} C_t'(k_2, \ldots, k_\ell) C_t(k_2, \ldots, k_\ell) \\
&\leq \mathbb{E}(\xi_0 a)'(\xi_0 a) \sum_{\ell=1}^{\infty} \rho(S)^\ell \qquad \text{(recursively)} \\
&\leq a'a\rho(T) \sum_{\ell=1}^{\infty} \rho(S)^\ell ,
\end{aligned}
$$

hence,

$$\mathbb{E}(X_t'X_t) \leq a'Ta + a'a\frac{\rho(T)}{1-\rho(S)} < \infty .\tag{6}$$

In the previous relations we used both the fact that the ξ_t are centered and iid and the relation $v'Av \leq v'v\rho(A)$ which holds if A denotes a non-negative $d \times d$ matrix and $v \in \mathbb{R}^d$. This conclude the proof of the existence of a solution in L^2.

Step 2: L^2 uniqueness. Let now X_t^1 and X_t^2 be two solutions to equation (1) in L^2. Define $\tilde{X}_t = X_t^1 - X_t^2$, then $\tilde{X}_t$ is solution to

$$\tilde{X}_t = \xi_t \tilde{A}_t , \quad \tilde{A}_t = \sum_{k=1}^{\infty} a_k \tilde{X}_{t-k} .\tag{7}$$

Now we use (7) and the fact that $\tilde{X}_t$ is centered and thus $\mathbb{E}\tilde{X}_s\tilde{X}_t = 0$ for $s \neq t$ to derive

$$\mathbb{E}\left((\tilde{X}_t g)'(\tilde{X}_t g)\right) = \sum_{k=1}^{\infty} g'\mathbb{E}\left(\tilde{X}_{t-k}' a_{t-k}' T a_{t-k} \tilde{X}_{t-k}\right) g$$

$$= \sum_{k=1}^{\infty} g'\mathbb{E}\left(\tilde{X}_t' a_{t-k}' T a_{t-k} \tilde{X}_t\right) g = g'\mathbb{E}\left(\tilde{X}_t' S \tilde{X}_t\right) g$$

$$= \mathbb{E}\left((\tilde{X}_t g)' S(\tilde{X}_t g)\right) \leq \rho(S)\mathbb{E}\left((\tilde{X}_t g)'(\tilde{X}_t g)\right) .$$

From equation (6), this expression is finite and thus the assumption $\rho(S) < 1$ concludes the proof.

Remark 9. The proof does not extend to the case of random fields because in this case the previous arguments of independence cannot be used. In that case we cannot address the question of uniqueness.

The previous L^2 existence and uniqueness assumptions do not imply that $\sum_{j\geq 1} \|a_j\| < \infty$, thus this situation is perhaps not a weakly dependent one. Giraitis and Surgailis [GS02], prove results both for the partial sums processes of X_t and $X_t^2 - \mathbb{E}X_t^2$. In our vector case the second problem is difficult and will be addressed in a forthcoming work. However X_t is now the increment of a (vector valued-)martingale and thus we partially extend Theorem 6.2 in [GS02], providing a version of Donsker's theorem for partial sums processes of $\{X_t\}$.

Proposition 3. *Let the assumptions of Theorem 4 hold. Define $S_n(t) = \sum_{1 \leq i \leq nt} X_i$ for $0 \leq t \leq 1$. Then $S_n(t)/\sqrt{\mathrm{var}S_n(t)}$ converges to $\Sigma W(t)$, for $0 \leq t \leq 1$ and where $W(t)$ is a $\mathbb{R}^d$ valued Brownian motion and Σ is a symmetric non negative matrix such that Σ^2 is the covariance matrix of X_0. The convergence holds for finite dimensional distributions.*

Remark 10.

- The convergence only holds for any k-tuples $(t_1, \ldots, t_k) \in [0, 1]^k$ and since the section is related to L^2 properties we cannot use the tightness arguments in [GS02] to obtain the Donsker theorem; indeed tightness is obtained through moment inequalities of order $p > 2$. L^p existence conditions are obtained in [GS02] for the bilinear case if $p = 4$; the method is based on the diagram formula and does not extend simply to this vector valued case. A bound for the moments of order $p > 2$ of the partial sum process $S_n(t)$ can be obtained using Rosenthal inequality, Theorem 2.11 in [HH80], if $\mathbb{E}\|X_t\|^p < \infty$. This inequality would imply the functional convergence in the Skohorod space $D[0, 1]$ if $p > 2$.

- If $\mathbb{E}\xi_0 \neq 0$ (as for the case of the bilinear model in [GS02]), we may write $X_t = \Delta M_t + \mathbb{E}\xi_0 \left(a + \sum_{j=1}^{\infty} a_j X_{t-j}\right)$ where

$$\Delta M_t = (\xi_t - \mathbb{E}\xi_t) \left(a + \sum_{j=1}^{\infty} a_j X_{t-j}\right)$$

  is a martingale increment. This martingale also obeys a central limit theorem. then,

$$n^{-1/2} S_n(t) \to \bar{\Sigma} W(t) ,$$

  where $W(t)$ is a vector Brownian motion, where $\bar{\Sigma}' \bar{\Sigma} = \Sigma$. If $\mathbb{E}\xi_0 = 0$ this is a way to prove proposition 3, which is a multi-dimensional extension of the proof in [GS02].

  For the case of the bilinear model, Giraitis and Surgailis also prove the (functional) convergence of the previous sequence of process to a Fractional Brownian Motion in [GS02]. For this, Riemannian decays of the coefficients are assumed. The covariance function of the process is also completely determined to prove such results; this is a quite difficult point to extend to our vector valued frame.

- A final comment concerns the analogue for powers of X_t which, if suitably normalized, are proved to converge to some higher order Rosenblatt process in [GS02] for the bilinear case. We have a structural difficulty to extend it; the only case which may reasonably be addressed is the real valued one ($d = 1$), but it also presents very heavy combinatorial difficulties. Computations for the covariances of the processes $(X_t^k)_{t\in\mathbb{Z}}$ will be addressed in a forthcoming work in order to extend those results.

Acknowledgements. The authors are grateful to the referees for their valuable comments.

References

[BP92] Bougerol, P., Picard, N. (1992). Stationarity of GARCH processes and of some nonnegative time series. *J. of Econometrics*, 52, 115–127.

[Bou98] Boussama, F. (1998). Ergodicité, mélange, et estimation dans les modèles GARCH. PhD Thesis, University Paris 7.

[Bou00] Boussama, F. (2000). Normalité asymptotique de l'estimateur du pseudo ma-ximum de vraisemblance d'un modèle GARCH. *Comptes Rendus de l'Académie des Sciences, Série I*, Volume 331-1, 81–84.

[DD03] Dedecker, J., Doukhan, P. (2003). A new covariance inequality and applications, *Stoch. Proc. Appl.*, 106, 63–80.

[DP01] Dedecker, J., Prieur, C. (2003). Coupling for τ-dependent sequences and applications. *J. of Theoret. Prob.*, forthcoming.

[DP04] Dedecker, J., Prieur, C. (2004). New dependence coefficients. Examples and applications to statistics. *Prob. Theory Relat. Fields*, In press.

[DL02] Doukhan, P., Lang, G. (2002). Rates of convergence in the weak invariance principle for the empirical repartition process of weakly dependent sequences. *Stat. Inf. for Stoch. Proc.*, 5, 199–228.

[DL99] Doukhan, P., Louhichi, S. (1999). A new weak dependence condition and applications to moment inequalities. *Stoch. Proc. Appl.*, 84, 313–342.

[EK96] Engle, R.F. Kroner, K.F. (1995). Multivariate simultaneous generalized ARCH. *Econometric Theory*, 11, 122–150.

[GLS05] Giraitis, L., Leipus, R., Surgailis, D. (2005). Recent advances in ARCH modelling, in: Teyssière, G. and Kirman, A., (Eds.), *Long–Memory in Economics*. Springer Verlag, Berlin. pp 3–38.

[GS02] Giraitis, L., Surgailis, D. (2002). ARCH-type bilinear models with double long memory, *Stoch. Proc. Appl.*, 100, 275–300

[GKL00] Giraitis, L., Kokoszka, P., Leipus, R. (2000). Stationary ARCH models: dependence structure and central limit theorems. *Econometric Theory*, 16, 3–22.

[HH80] Hall, P., Heyde, C.C. (1980). *Martingale Limit Theory and Its Applications*. Academic Press.

[MH04] Mandelbrot, B.B., Hudson R.L. (2004). *The Misbehavior of Markets: A Fractal View of Risk, Ruin, and Reward.* Basic Books, New-York.

[Man63] Mandelbrot, B.B. (1963). The variation of certain speculative prices. *Journal of Business*, 36, 394-419.

[Pri05] Prieur, C. (2004). Recent results on weak dependence. Statistical applications. Dynamical system's example. To appear in P. Bertail, P. Doukhan and Ph. Soulier (eds), Dependence in Probability and Statistics. Springer.

[RüS] Rüschendorf, L. (1985). The Wasserstein distance and approximation theorems. *Z. Wahrscheinlichkeitstheorie verw. Gebiete*, 70, 117–129.

[Tey97] Teyssière, G. (1997). Modelling Exchange rates volatility with multivariate long–memory ARCH processes. *Preprint*.

On a Szegö type limit theorem and the asymptotic theory of random sums, integrals and quadratic forms

Florin Avram[1] and Murad S. Taqqu[2]

[1] Dept. Mathematiques, Université de Pau, 64000 Pau, France
Florin.Avram@univ-pau.fr
[2] Dept. of Mathematics and Statistics, Boston University, Boston, MA 02215, USA
murad@math.bu.edu

1 Introduction

1.1 Time series motivation

One-dimensional discrete time series with long memory have been extensively studied [Hur51], [GJ80], [Hos81], [Ber94], [WTT99]. The FARIMA family, for example, models the series X_t, $t \in \mathbb{Z}$ as the solution of an equation

$$\phi(B)(1 - B)^d X_t = \theta(B)\epsilon_t \tag{1}$$

where B is the operator of backward translation in time, $\phi(B), \theta(B)$ are polynomials, d is a real number and ϵ_t is white noise. Using this family of models, it is usually possible via an extension of the classical Box-Jenkins methodology, to chose the parameter d and the coefficients of the polynomials $\phi(B), \theta(B)$ such that the residuals ϵ_t display white noise behaviour and hence may safely be discarded for prediction purposes.

In view of the prevalence of spatial statistics applications, it is important to develop models and methods which replace the assumption of a one-dimensional discrete time index with that of a multidimensional continuous one.

This paper was motivated by the attempt to extend to the case of multidimensional continuous indices certain central limit theorems of Giraitis and Surgailis, Fox and Taqqu and Giraitis and Taqqu, which were reorganised and generalized in [AB89], [AT89], [Avr92]. The approach of these papers reduced the central limit theorems considered to an application of three analytical tools:

(A1). The well-known diagram formula for computing moments/cumulants of Wick products – see Appendix B, section B.

(A2). A generalisation of the Hölder-Young inequality, which has become more recently known as the Hölder-Brascamp-Lieb-Barthe inequality – see Appendix A, section A.

(A3). Some generalizations of a Grenander-Szegö theorem on the trace of products of Toeplitz matrices –see Subsection 1.9.

We will review throughout the paper the one-dimensional discrete results in [AB89], [AT89], [Avr92], [AF92]; however, despite not having achieved yet the generalisation to continuous indices, we will formulate the results using a unifying measure theoretic notation, which includes the discrete one-dimensional setup and the continuous multi-dimensional one. This will allow us to discuss possible extensions.

1.2 The model

Let ξ_A, $A \subset \mathbb{R}^d$ denote a set indexed Lévy process with mean zero, finite second moments, independent values over disjoint sets and stationary intensity given by the Lebesgue measure. Let X_t, $t \in \mathbb{R}^d$ denote a linear random field

$$X_t = \int_{u \in \mathbb{R}^d} \hat{a}(t - u)\xi(\mathrm{d}u) , \quad t \in \mathbb{R}^d , \tag{2}$$

with a square-integrable kernel $\hat{a}(t)$, $t \in \mathbb{R}^d$. For various other conditions which ensure that (2) is well-defined, see for example Anh, Heyde and Leonenko [AHL02], pg. 733.

By choosing an appropriate "Green function" $\hat{a}(t)$, this wide class of processes includes the solutions of many differential equations with random noise $\xi(\mathrm{d}u)$.

The random field X_t is observed on a sequence I_T of increasing finite domains. In the discrete-time case, the cases $I_T = [1, T]^d$, $T \in \mathbb{Z}_+$ (in keeping with tradition) or $I_T = [-T/2, T/2]^d$, $T \in 2\mathbb{Z}_+$ will be assumed. In the continuous case, rectangles $I_T = \{t \in \mathbb{R}^d : -T_i/2 \leq t_i \leq T_i/2, i = 1, ..., d\}$ will be taken. For simplicity, we will assume always $T_1 = ... = T_d = T \to \infty$, but the extension to the case when all coordinates converge to ∞ at the same order of magnitude is immediate.

1.3 Spectral representations

We will assume throughout that all the existing cumulants of our stationary process X_t are expressed through Fourier transforms $c_k(t_1, t_2, ..., t_k)$ of "spectral densities" $f_k(\lambda_1, ..., \lambda_{k-1}) \in L_1$, i.e:

$$c_k(t_1, t_2, ..., t_k)$$

$$= \int_{\lambda_1,\dots,\lambda_{k-1}\in S} e^{i\sum_{j=1}^{k-1}\lambda_j(t_j-t_k)} f_k(\lambda_1,\dots,\lambda_{k-1})\,\mu(d\lambda_1)\dots\mu(d\lambda_{k-1})$$

$$= \int_{\lambda_1,\dots,\lambda_k\in S} e^{i\sum_{j=1}^{k}\lambda_j t_j} f_k(\lambda_1,\dots,\lambda_{k-1})\,\delta\Big(\sum_{j=1}^{k}\lambda_j\Big)\,\mu(d\lambda_1)\dots\mu(d\lambda_k)$$

where, throughout the paper, integrals involving delta functions will simply be used as a convenient notation for the corresponding integrals over lower dimensional subspaces. Throughout, S will denote the "spectral" space of discrete and continuous processes, i.e. $[-\pi,\pi]^d$ with Lebesgue measure normalized to unity, and $\mathbb{R}^d$ with Lebesgue measure, respectively.

For $k = 2$, we will denote the spectral density by $f(\lambda) = f_2(\lambda)$ and the second order cumulants/covariances by $r(t-s) = c_2(s,t)$.

1.4 The problem

Limit theory for quadratic forms is a subset of the more general task of providing limit theorems for sums/integrals and bilinear forms

$$S_T = \int_{t\in I_T} h(X_t)dt\,, \quad Q_T = \int_{t_1,t_2\in I_T} \hat{b}(t_1-t_2)\,h(X_{t_1},X_{t_2})dt_1dt_2 \qquad (3)$$

where X_t is a stationary sequence. In the discrete time case, S_T and Q_T become respectively

$$S_T = S_T(h) = \sum_{i=1}^{T} h(X_i)\,, \quad Q_T = Q_T(h) = \sum_{i=1}^{T}\sum_{j=1}^{T} \hat{b}(i-j)\,h(X_i,X_j)\,. \qquad (4)$$

These topics, first studied by Dobrushin and Major [DM79] and Taqqu [Taq79] in the Gaussian case, gave rise to very interesting non-Gaussian generalisations –see [GS86], [AT87], and are still far from fully understood in the case of the continuous, "spatial" multidimensional indices arising in spatial statistics.

It is well-known in the context of discrete time series that the expansion in univariate/bivariate Appell polynomials determines the type of central limit theorem (CLT) or non-central limit theorem (NCLT) satisfied by the sums/quadratic forms (3). Hence, this paper considers the problem (3) with h being an Appell polynomial.

1.5 Appell polynomials

In this paper we consider central limit theorems for quadratic forms

$$Q_T = Q_T^{(m,n)} = \int_{t,s\in I_T} P_{m,n}(X_t,X_s)\hat{b}(t-s)dsdt \qquad (5)$$

involving the bivariate Appell polynomials

$$P_{m,n}(X_t, X_s) =: \underbrace{X_t, \ldots, X_t}_{m}, \underbrace{X_s, \ldots, X_s}_{n} : \quad m, n \geq 0, m + n \geq 1 ,$$

which are defined via the Wick product $: X_1, \ldots, X_m :$ (see Appendix B). We will assume that $E|\xi_{I_1}|^{2(m+n)} < \infty$ in order to ensure that Q_T has a finite variance and suppose that t and s are discrete, that is, we consider in fact (4).

For a warm-up, we consider also sums

$$S_T = S_T^{(l)} = \int_{t \in I_T} P_l(X_t) \mathrm{d}t \tag{6}$$

involving the univariate Appell polynomials

$$P_l(X_t) =: \underbrace{X_t, \ldots, X_t}_{l} :$$

Note that when $\hat{b}(t-s)\mathrm{d}s\mathrm{d}t = \delta(t-s)\mathrm{d}t$, where δ is the Kronecker function, the quadratic forms (5) reduce to the sums (6) with $l = m + n$.

The variables X_t will be allowed to have short-range or long-range dependence (that is, with summable or non-summable sum of correlations), but the special short-range dependent case where the sum of correlations equals 0 will not be considered, since the tools described here are not sufficient in that case.

1.6 Asymptotic normality

We focus here on the asymptotic normality of $Q_T^{(1,1)}$ via the method of cumulants. One of the classical approaches to establish asymptotic normality for processes having all moments consists in computing all the scaled cumulants $\chi_{k,T}$ of the variables of interest, and in showing that they converge to those of a Gaussian distribution, that is, to 0 for $k \geq 3$.

We review now this method in the simplest case of symmetric bilinear forms $Q_T = Q_T^{(1,1)}$ in stationary Gaussian discrete-time series X_t, with covariances $r_{i-j}, i, j \in \mathbb{Z}$ and spectral density $f(\lambda)$ (note that $S_T^{(1)} = \int_{t \in I_T} X_t \mathrm{d}t$ is "too simple" for our purpose, since it is Gaussian and $\chi_k(S_T^{(1)}) = 0, \forall k \neq 2$). For $Q_T^{(1,1)}$, a direct computation yields the formula:

$$\chi_k = \chi(Q_T, \ldots, Q_T) = 2^{k-1}(k-1)! \, \mathrm{Tr}[(T_T(b)T_T(f))^k] \tag{7}$$

where

$$T_T(b) = (\hat{b}_{i-j}, \ i,j = 1, \ldots, T) , \qquad T_T(f) = (r(i-j), \ i,j = 1, \ldots, T)$$

denote Toeplitz matrices of dimension $T \times T$ and Tr denotes the trace. A limit theorem for traces of products of Toeplitz matrices

$$\mathrm{Tr}\left[\prod_{e=1}^{k} T_T(f_e)\right] = \sum_{j_1,\ldots,j_k=1}^{T} r_1(j_1 - j_2)r_2(j_2 - j_3)\ldots r_k(j_k - j_1)\,, \qquad (8)$$

obtained by Grenander and Szegö [GS58] and strengthened by Avram [Avr88], yields then the asymptotic normality of $Q_T^{(1,1)}$, under the condition that $b(\lambda)f(\lambda) \in L_2$. Note that replacing the sequences $r_e(j)$ by their Fourier representations $r_e(j) = \int_S f_e(\lambda)e^{ij\lambda}d\lambda$ in (8) and putting

$$\Delta_T(\lambda) = \sum_{t=1}^{T} e^{it\lambda}$$

yields the following alternative spectral integral representation for traces of products of Toeplitz matrices:

$$\mathrm{Tr}\left[\prod_{e=1}^{k} T_T(f_e)\right] \qquad (9)$$

$$= \int_{\lambda_1,\ldots,\lambda_k \in S} f_1(\lambda_1)f_2(\lambda_2)\ldots f_k(\lambda_k) \prod_{e=1}^{k} \Delta_T(\lambda_{e+1} - \lambda_e) \prod_{e=1}^{k} \mu(d\lambda_e)$$

$$= \int_{u_e \in S, e=1,\ldots,k-1} \Delta_T\!\left(-\sum_{1}^{k-1} u_e\right) \prod_{e=1}^{k-1} (\Delta_T(u_e)\mu(du_e)$$

$$\times \left(\int_{\lambda \in S} f_1(\lambda)f_2(\lambda + u_1)\ldots f_k\!\left(\lambda + \sum_{1}^{k-1} u_e\right)d\lambda\right)$$

where the index $k+1$ is defined to equal 1, and where we changed variables to $\lambda = \lambda_1$ and $u_e = \lambda_{e+1} - \lambda_e, e = 1, \ldots, k-1$. The first expression in the RHS of (9) is our first example of a "delta graph integral", to be introduced in the next subsection. These are integrals involving trigonometric kernels like Δ_T, applied to linear combinations which may be associated to the vertex-edge incidence structure of a directed graph: in this occurrence, the cycle graph on the vertices $\{1,\ldots,k\}$. The next expression in the RHS of (9) is a change of variables which will reveal the asymptotic behavior of delta graph integrals.

To obtain the central limit theorem by the method of cumulants for $T^{-1}Q_T^{(1,1)}$, one wants to show that:

(A1).

$$\lim_{T \to \infty} T^{-1}\chi_2\left(\frac{Q_T^{(1,1)}}{T}\right) \text{ is finite}$$

and that

(A2).

$$\lim_{T\to\infty} T^{-k/2}\chi_k\left(\frac{Q_T^{(1,1)}}{T}\right) = 0 , \quad k \geq 3 .$$

The asymptotic normality follows roughly from three facts:

(A1). Under appropriate integrability conditions, the function

$$J(u_1, ..., u_{k-1}) := \int_{\lambda \in S} f_1(\lambda)f_2(\lambda + u_1)...f_k(\lambda + \sum_1^{k-1} u_e)d\lambda$$

in the RHS of (9) is continuous in $u_e, e = 1, ..., k - 1$.

(A2). When $T \to \infty$, the measures

$$T^{-1}\Delta_T(-\sum_1^{k-1} u_e)\prod_{e=1}^{k-1}(\Delta_T(u_e)\mu(du_e)$$

converge weakly to the measure $\delta_0(u_1, ...u_{k-1})$. Together, the first two facts yield the convergence of the normalized variance:

$$\lim_{T\to\infty} \frac{\chi_{2,T}}{T} = J(0, ..., 0) = \int_{\lambda \in S} f_1(\lambda)f_2(\lambda)...f_k(\lambda)d\lambda$$

(A3). For $k \geq 3$, the cumulants satisfy

$$\lim_{T\to\infty} \frac{\chi_{k,T}}{T} = 0 .$$

In the case of bilinear forms in Hermite/Appell polynomials $P_{m,n}(X, Y)$ of Gaussian/linear processes, more complicated cumulant formulas arising from the so-called *diagram expansion* lead to spectral integral representations with a combinatorial structure similar to (9), but associated to graphs displaying a more complicated cycle structure – see Figure 1 and end of Appendix B.

This motivates introducing a class of spectral representations generalizing (9), in which the cycle graph is replaced by a general directed graph or matroid structure. The generalization of the limiting statements above will be referred to as "Szegö type limit theorems".

1.7 Delta graph integrals

Let $G = (\mathcal{V}, \mathcal{E})$ denote a directed graph with V vertices, E edges and $co(G)$ components. Let M denote the $V \times E$ incidence matrix of the graph.

Definition 1. *The incidence matrix M of a graph has entries $M_{v,e} = \pm 1$ if the vertex v is the end/start point of the edge e, and 0 otherwise.*

Definition 2. (Delta graph integrals). *Suppose that associated to the edges of a directed graph $G = (\mathcal{V}, \mathcal{E})$ there is a set $f_e(\lambda)$, $e = 1, \ldots, E$ of functions satisfying integrability conditions*

$$f_e \in L_{p_e}(\mu(\mathrm{d}\lambda)), \ 1 \leq p_e \leq \infty \,,$$

where μ is the normalized Lebesgue mesure on the torus $[-\pi, \pi]$ or the Lebesgue mesure on $\mathbb{R}$.

A Delta graph integral $J_T = J_T(G, f_e, e = 1, \ldots, E)$ is an integral of the form:

$$J_T = \int_{\lambda_1, \ldots, \lambda_E \in S} f_1(\lambda_1) f_2(\lambda_2) \ldots f_E(\lambda_E) \prod_{v=1}^{V} \Delta_T(u_v) \prod_{e=1}^{E} \mu(\mathrm{d}\lambda_e) \qquad (10)$$

where E, V and M denote respectively the number of edges, vertices, and the incidence matrix of the graph G, where

$$(u_1, \ldots, u_V)' = M(\lambda_1, \ldots, \lambda_E)'$$

and where a prime denotes a transpose. Finally,

$$\Delta_T(x) = \sum_{-T/2}^{T/2} \mathrm{e}^{\mathrm{i}tx} \mathrm{d}t = \frac{\sin((T+1)x/2)}{\sin(x/2)} \qquad (11)$$

or

$$\Delta_T(x) = \int_{-T/2}^{T/2} \mathrm{e}^{\mathrm{i}tx} \mathrm{d}t = \frac{\sin(Tx/2)}{x/2} \qquad (12)$$

is the Fejer kernel in discrete or continuous time, respectively.

The concept of Delta graph integrals arose from the study of cumulants of sums/quadratic forms in Appell polynomials of stationary Gaussian processes. A simple computation based on the diagram formula [Avr92] – see also Proposition 2 – shows that the cumulants $\chi_k(S_T^{(l)})$ and $\chi_k(Q_T^{(m,n)})$ are sums of Delta graph integrals.

The rest of the paper is devoted to reviewing "Szegö-type" limit theorems for Delta graph integrals in the one-dimensional discrete time case, following [AB89], [AT89], [Avr92].

1.8 Delta matroid integrals

Quoting Tutte [Tut59], it is probably true that "any theorem about graphs expressible in terms of edges and circuits exemplifies a more general result about matroids", a concept which formalizes the properties of the "rank function" $r(A)$ obtained by considering the rank of an arbitrary set of columns A

in a given arbitrary matrix M (thus, all matrices with the same rank function yield the same matroid). Tutte's "conjecture" holds in the case under consideration: a matroid Szegö-type limit theorem, in which the graph dependence structure is replaced with that of an arbitrary matroid, was given in Avram [Avr92].

A matroid is a pair $\mathcal{E}, r : 2^{\mathcal{E}} \to \mathbb{N}$ of a set $\mathcal{E}$ and a "rank like function" $r(A)$ defined on the subsets of $\mathcal{E}$. The most familiar matroids, associated to the set $\mathcal{E}$ of columns of a matrix and called *vector matroids*, may be specified uniquely by the rank function $r(A)$ which gives the rank of any set of columns A (matrices with the same rank function yield the same matroid). Matroids may also be defined in equivalent ways via their independent sets, via their bases (maximal independent sets) via their circuits (minimal dependent sets), via their spanning sets (sets containing a basis) or via their flats (sets which may not be augmented without increasing the rank). For excellent expositions on graphs and matroids, see [Oxl92], [Oxl04] and [Wel76]. We ask the reader not familiar with this concept to consider only the particular case of **graphic matroids**, which are associated to the incidence matrix of an oriented graph. It turns out that the algebraic dependence structure translates in this case into graph-theoretic concepts, with circuits corresponding to cycles.

Here, we will only need to use the fact that to each matroid one may associate a **dual matroid** with rank function

$$r^*(A) = |A| - r(M) + r(M - A) \, ,$$

and that in the case of graphic matroids, the dual matroid may be represented by the $C \times E$ matrix M^* whose rows $c = 1, ..., C$ are obtained by assigning arbitrary orientations to the circuits (cycles) c of the graph, and by writing each edge as a sum of $\pm$ the circuits it is included in, with the $\pm$ sign indicating a coincidence or opposition to the orientation of the cycle [3].

Definition 3. *Let $f_e(\lambda)$, $e = 1, \dots, E$ denote "base functions" associated with the columns of M and satisfying integrability conditions*

$$f_e \in L_{p_e}(d\mu) \, , \quad 1 \le p_e \le \infty \, , \tag{13}$$

where μ is Lebesgue measure on $\mathbb{R}$ or normalized Lebesgue measure on the torus $[-\pi, \pi]$. Let M denote an arbitrary matrix in the first case, and with integer coefficients in the second case. Let $\hat{f}_e(k), k \in I$ denote the Fourier transform of $f_e(\lambda)$, i.e.

$$\hat{f}_e(k) = \int_S e^{ik\lambda} f_e(\lambda)\mu(d\lambda)$$

A Delta matroid integral is defined by either one of two equivalent expressions, which correspond to the time and spectral domain, respectively:

---

[3] It is enough to include in M^* a basis of cycles, thus excluding cycles which may be obtained via addition modulo 2 of other cycles, after ignoring the orientation.

$$J_T = J_T(M, f_e, e = 1, ..., E)$$

$$= \int_{j_1,...,j_V \in I_T} \hat{f}_1(i_1)\hat{f}_2(i_2)...\hat{f}_E(i_E) \prod_{v=1}^{V} dj_v \qquad (14)$$

$$= \int_{\lambda_1,...,\lambda_E \in S} f_1(\lambda_1)f_2(\lambda_2)...f_E(\lambda_E) \prod_{v=1}^{V} \Delta_T(u_v) \prod_{e=1}^{E} \mu(d\lambda_e) \qquad (15)$$

where

$$(i_1, ..., i_E) = (j_1, ..., j_V)M$$

and

$$(u_1, ..., u_V)' = M(\lambda_1, ..., \lambda_E)'$$

and where, in the torus case, the linear combinations are computed modulo $[-\pi, \pi]$.

See [Avr92] for more information, in particular relations (1.3) and (3.1) of that paper. A "Delta matroid integral" is also called a "Delta graph integral" when the matroid is associated to the incidence matrix M of a graph (graphic matroid). Observe that (15), just as its graph precursor (10)), does not depend on the matrix M representing the matroid.

1.9 The Szegö-type limit theorem for Delta matroid integrals

Let

$$z_j = \frac{1}{p_j} \in [0, 1] , \quad j = 1, \cdots, E , \qquad (16)$$

where p_j is defined in (13). Theorem 1 below, which is a summary of Theorems 1, 2 and Corollary 1 of [Avr92], yields an upper bound and sometimes also the limit for Delta matroid integrals, in the case of discrete one-dimensional time series. The order of magnitude of the rate of convergence is

$$\boxed{\alpha_M(z) = V - r(M) + \max_{A \subset 1,...,E} [\sum_{j \in A} z_j - r^*(A)]} \qquad (17)$$

or equivalently,

$$\boxed{\alpha_M(z) = \max_{A \subset 1,...,E} [co(M - A) - \sum_{j \in A} (1 - z_j)]} \qquad (18)$$

where we define

$$co(M - A) = V - r(M - A) . \qquad (19)$$

Note: In the case of graph integrals, the function $co(M - A)$ represents the number of components, after the edges in A have been removed.

The function $\alpha_M(z)$ is thus found in the case of graph integrals by the following optimization problem:

$$\boxed{\text{The "graph breaking" problem:}}$$

Find a set of edges whose removal maximizes the number of remaining components, with

$$\sum_{j \in A}(1 - z_j) = \sum_{j \in A}(1 - p_j^{-1})$$

as small as possible.

The function $\alpha_M(z)$ is then used in the following theorem which is a Szegö-type limit theorem for Delta matroid integrals. The theorem, which is formulated in a general way, has been established for functions defined on $[-\pi, \pi]$ and extended periodically outside that interval. The following definition will be used.

Definition 4. *Let $\mathcal{L}_p(d\mu)$ be the Banach space of functions which may be approximated arbitrarily close by trigonometric polynomials, in the $L_p(d\mu)$ norm. The space $\mathcal{L}_p(d\mu)$ is endowed with the L_p norm.*

The space $\mathcal{L}_p(d\mu)$ is used in the following theorem because the proof of Part 2 involves considering first functions f that are linear combinations of trigonometric polynomials and then passing to the limit.

Theorem 1. *Let $J_T = J_T(M, f_1, ..., f_E)$ denote a Delta matroid integral and let $r(A), r^*(A)$ denote respectively the ranks of a set of columns in M and in the dual matroid M^*.*

Suppose that for every row l of the matrix M, one has $r(M) = r(M_l)$, where M_l is the matrix with the row l removed. Then:

(A1).

$$\boxed{J_T(M, f_1, ..., f_E) \leq c_M T^{\alpha_M(z)}} \tag{20}$$

where c_M is a constant and $\alpha_M(z)$ is given by (17).
(A2). Suppose now that $f_e \in \mathcal{L}_{p_e}(d\mu)$ and set $z = (p_1^{-1}, ... p_E^{-1})$.
a) If

$$\sum_{j \in A} z_j \leq r^*(A), \quad \forall A \tag{21}$$

(or, equivalently, $\alpha_M(z) = V - r(M) = co(M)$), then

$$\lim_{T \to \infty} \frac{J_T(M)}{T^{co(M)}} = J(M^*, f_1, ..., f_E) \tag{22}$$

where

$$\boxed{J(M^*, f_1, ..., f_E) = c_M \int_{S^C} f_1(\lambda_1) f_2(\lambda_2) ... f_E(\lambda_E) \prod_{c=1}^{C} \mu(dy_c)}$$

and where $(\lambda_1, ... \lambda_E) = (y_1, ..., y_C)M^*$ *(with every* λ_e *reduced modulo* $[-\pi, \pi]$ *in the discrete case), and* C *denotes the rank of the dual matroid* M^*.

b) *If a strict inequality* $\alpha_M(z) > co(M)$ *holds, then the inequality of part 1) may be strengthened to:*

$$J_T(M) = o(T^{\alpha(M)})$$

Remarks. 1) The conditions (21) applied to integrands $F_e(\lambda)$ with power type behavior are the famous **power counting conditions**, which were used to ensure the convergence of integrals with dependent variables. These conditions, in fact, also yield a Hölder type inequality, as shown in [BLL74], [AB89] and [AT89].

2) The results of the theorem, that is, the expression of $\alpha_M(z)$ and the limit integral $J(M^*) = J(M^*, f_1, ..., f_E)$ depend on the matrix M only via the two equivalent rank functions $r(A), r^*(A)$, i.e. only via the matroid dependence structure between the columns.

We refer to [Avr92] for the proof of parts 1 and 2 b) of this theorem.

1.10 Sketch of the proof

We sketch now the proof of Theorem 1, part 2 (a), for Delta graph integrals given in [AB89], in the discrete one-dimensional setup, and for a connected graph (w.l.o.g.). Note that in a connected graph there are only $V - 1$ independent rows of the incidence matrix M (or independent variables u_j), since the sum of all the rows is 0 (equivalently, $u_V = -\sum_{v=1}^{V-1} u_v$). Thus, $r(M) = V - 1$, $co(M) = 1$, and the order of magnitude appearing in the LHS of (22) is just T.

(A1). **Change of variables.** The main idea behind the proof of Theorem 1 is to identify a basis $y_1, ..., y_C$ in the complement of the space generated by the u_v's, $v = 1, ..., V$, switch to the variables $u_1, ..., u_{V-1}, y_1, ..., y_C$ and integrate in (15) first over the variables y_c's, $c = 1, ..., C$. This is easier in the graph case, when, fixing some spanning tree $\mathcal{T}$ in the graph, we have a one to one correspondence between a set of independent cycles (with cardinality C) and the complementary set of edges $\mathcal{T}^c$. Assume w.l.o.g. that in the list $(\lambda_1, ..., \lambda_E)$, the edges in $\mathcal{T}^c$ are listed first, namely $(\lambda_e, e \in \mathcal{T}^c) = (\lambda_1, ..., \lambda_C)$. We make the change of variables $y_1 = \lambda_1, ..., y_C = \lambda_C$, and $(u_1, ..., u_{V-1})' = \tilde{M}(\lambda_1, ..., \lambda_E)'$, where $\tilde{M}$ is the first $V - 1$ rows of M. Thus,

$$(y_1, ..., y_C, u_1, ..., u_{V-1})' = \begin{pmatrix} I_C & 0 \\ \tilde{M} & \end{pmatrix} (\lambda_1, ..., \lambda_E)'$$

where the first rows are given by an identity matrix I_C completed by zeroes. Inverting this yields

$$(\lambda_1, ..., \lambda_E) = (y_1, ..., y_C, u_1, ..., u_{V-1}) \ (M^* \mid N), \qquad (23)$$

that is, it turns out that the first columns of the inverse matrix are precisely the transpose of the dual matroid M^*.

Definition 5. *The function*

$$h_{M^*,N}(u_1, ..., u_{r(M)}) = \int_{y_1,...,y_C \in S} f_1(\lambda_1) f_2(\lambda_2)...f_E(\lambda_E) \prod_{c=1}^{C} d\mu(y_c) \quad (24)$$

*where λ_e are represented as linear combinations of $y_1, ..., y_C, u_1, ..., u_{V-1}$ via the linear transformation (23) will be called a **graph convolution**.*

The change to the variables $y_1, ..., y_C, u_1, ..., u_{V-1}$ and integration over $y_1, ..., y_C$ transforms the Delta graph integral into the following integral of the product of a " graph convolution" and a certain trigononetric kernel:

$$\int_{u_1,...,u_{V-1} \in S} h_{M^*,N}(u_1, ..., u_{V-1}) \prod_{v=1}^{V} \Delta_T(u_v) \prod_{v=1}^{V-1} d\mu(u_v) .$$

It turns out that the trigonometric kernel converges under appropriate conditions to Lebesgue measure on the graph $u_1 = ... = u_{V-1} = 0$. Since $(\lambda_1, ..., \lambda_E)$ are linear functions of $(y_1, ..., y_C, u_1, ..., u_{V-1})$ such that when $u_1 = ... = u_{V-1} = 0$ the relation $(\lambda_1, ..., \lambda_E) = (y_1, ..., y_C)M^*$ is satisfied, $h_{M^*,N}(0, ..., 0) = J(M^*)$, as defined in (22).

Thus, part 2 of Theorem 1 will be established once the convergence of the kernels and the continuity of the graph convolutions $h(u_1, ..., u_{r(M)})$ in the variables $(u_1, ..., u_{r(M)})$ is established.

(A2). **The continuity of graph convolutions**. Note that the function $h : \mathbb{R}^{V-1} \to \mathbb{R}$ is a composition

$$h_{M^*,N}(u_1, ..., u_{V-1}) = J(M^*, T_1(f_1), ..., T_E(f_E))$$

of the continuous functionals

$$T_e(u_1, ..., u_{V-1}) : \mathbb{R}^{V-1} \to \mathcal{L}_{p_e}$$

and of the functional

$$J(M^*, f_1, ..., f_E) : \prod_{e=1}^{E} \mathcal{L}_{p_e} \to \mathbb{R} .$$

The functional T_e is defined by $T_e(u_1, ..., u_{V-1}) = f_e(. + \sum_v u_v N_{v,e})$, where the $N_{v,e}$ are the components of the matrix N in (23). The functionals T_e are clearly continuous when f_e is a continuous function, and this continues to be true for functions $f_e \in \mathcal{L}_{p_e}$, since these can be approximated in the $\mathcal{L}_{p_e}$ sense by continuous functions. Thus, under our assumptions, the continuity of the functional $h_{M^*,N}(u_1, ..., u_{V-1})$ follows automatically from that of $J(M^*, f_1, ..., f_E)$.

Finally, under the "power counting conditions" (21), the continuity of the function $J(M^*, f_1, ..., f_E)$ follows from the Hölder-Brascamp-Lieb-Barthe inequality:

$$|J(M^*, f_1, ..., f_E)| \le \prod_{e=1}^{E} \|f_e\|_{p_e}$$

(see (c1), Theorem 2 below).

Remark. In the spatial statistics context ([ALS04], [ALS03]), the continuity of the graph convolutions $h_{M^*,N}(u_1, ..., u_{V-1})$ was usually assumed to hold, and indeed checking whether this assumption may be relaxed to L_p integrability conditions in the spectral domain is one of the outstanding difficulties for the spatial extension.

(A3). **Weak convergence of the measures.** The final step in proving Theorem 1 part 2 (a) is to establish that the measures

$$T^{-1}\Delta_T(-\sum_{v=1}^{V-1} u_v) \prod_{v=1}^{V-1} \Delta_T(u_v) \prod_{v=1}^{V-1} d\mu(u_v)$$

converge weakly to the measure $\delta_0(u_1, ..., u_{V-1})$ This convergence of measures holds since their Fourier coefficients converge – see [FT87], Lemma 7.1 – and since the absolute variations of these measures are uniformly bounded, as may be seen by applying the corresponding Hölder-Brascamp-Lieb-Barthe inequality (see Theorem 2 below) to the Delta graph integral, using estimates of the form

$$\|\Delta_T\|_{s_v^{-1}} \le k(s_v)T^{1-s_v}$$

with optimally chosen s_v, $v = 1, \cdots, V$.

In conclusion, the convergence of the product of Fejer kernels to a δ function implies the convergence of the scaled Delta graph integral

$$J_T(M, f_e, e = 1..., E)$$

to

$$J(0, ..., 0) = J(M^*, f_e, e = 1..., E),$$

establishing Part 2 (a) of the theorem.

Remark. It is not difficult to extend this change of variables to the case of several components and then to the matroid setup. In the first case, one would need to choose independent cycle and vertex variables $y_1, ..., y_{r(M^*)}$ and $u_1, ..., u_{r(M)}$, note the block structure of the matrices, with each block corresponding to a graph component, use the fact that for graphs with several components, the rank of the graphic matroid is $r(M) = V - co(G)$ and finally Euler's relation $E - V = C - co(G)$, which ensures that

$$E = (V - co(G)) + C = r(M) + r(M^*)$$

1.11 Central limit theorems for variables whose cumulants are Delta matroid integrals

We draw now the attention to the convenient simplifications offered by these tools for establishing central limit theorems, cf. [AB89], [AT89], [Avr92]. They arise from the fact that the cumulants are expressed as sums of integrals of the form (10) and their order of magnitude may be computed via the graph-optimization problem (17). Then a Szegö-type limit theorem (Theorem 1) is used to conclude the proof. This shows that the central limit theorem can sometimes be reduced to a simple optimization problem.

We quote now Corollary 2 of [Avr92].

Corollary 1. *Let Z_T be a sequence of zero mean random variables, whose cumulants of all orders are sums of Delta matroid integrals:*

$$\chi_k(Z_T) = \sum_{G \in \mathcal{G}_k} J_T(G)$$

Suppose that the "base functions" (see Definition 3) intervening in these integrals satisfy integrability conditions which imply that

$$\alpha_G(z) \leq k/2, \forall G \in \mathcal{G}_k$$

where z is the vector of reciprocals of the integrability indices of the base functions intervening in $J_T(G)$.

Then, a central limit theorem

$$\frac{Z_T}{\sigma T^{1/2}} \to N(0,1)$$

holds, with

$$\sigma^2 = \sum_{G \in \mathcal{G}_2} J(G^*),$$

where $J(G^*)$ is defined in the Theorem 1.2 (a).

These results reduce complicated central limit theorems for Gaussian processes to simple "graph breaking problems".

In all the examples currently studied, it turned out that the "Szegö theorems" ensure that the CLT holds whenever the integral defining the limiting variance converges.

For example, by Theorem 1.2 (a), the integral defining the limiting variance of sums S_T^m , of Hermite polynomials $P_m(X_t)$ of a Gaussian stationary sequence, $m \geq 2$, is:

$$\int_{y_1,\ldots,y_{m-1} \in S} f(y_1)f(y_2 - y_1)\ldots f(y_{m-1} - y_{m-2})f(-y_{m-1}) \prod_{c=1}^{m-1} dy_c = f^{(*,m)}(0)$$

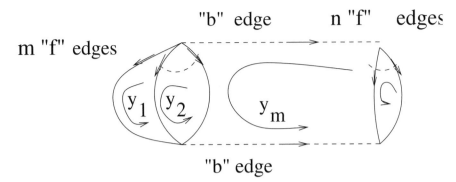

Fig. 1. Typical graph appearing in the expansion of the limiting variance of quadratic forms in bivariate Hermite polynomials $P^{(m,n)}$, m=3, n=2. Ignoring the right half of the picture yields the graph appearing in the limiting variance of sums of Hermite polynomials P_m. The arrows show the orientation of cycles and edges: an "edge variable" λ_e is the combination of all the "cycle variables" $\pm y_c$ containing that edge, with a minus sign when the direction of y_c is different from the direction of the edge.

where $f^{(*,m)}(0)$ denotes the m'th convolution, and the condition ensuring its convergence is that $f \in L_p$, where the integrability index $z = p^{-1}$ satisfies $z \le 1/m$.

In the case of Gaussian quadratic forms $Q_T^{(m,n)}$, the limiting variance is a sum of terms of the type:

$$\int_{y_1,\ldots,y_{m-1}\in S} f(y_1)f(y_2 - y_1)\ldots f(y_{m-1} - y_{m-2})f(-y_{m-1} + y_m)$$

$$\times\, b(y_m)b(-y_m)f(y_{m+1} - y_m)\ldots f(-y_{m+n-1}) \prod_{c=1}^{m+n-1} dy_c$$

$$= \int_{y_m\in S} f^{(*,m)}(y_m)b(y_m)b(-y_m)f^{(*,n)}(-y_m)dy_m$$

where one special cycle variable – y_m in our notation – corresponds to a four-cycle containing the two "b" edges (see Figure 1) and the rest correspond to cycles containing only two "f" edges (see Proposition 2, Appendix B). The fact that the convergence of this expression is enough to ensure the CLT was first shown in Giraitis and Taqqu [GT97] – see (2.5), Theorem 2.1 of that paper.

The convergence of the above integral is assured whenever $f \in L_{p_1}, b \in L_{p_2}$ and $z = (z_1, z_2)$ is within the polytope in Figure 2 by the Hölder-Brascamp-Lieb-Barthe inequality (see Theorem 2, c1)).

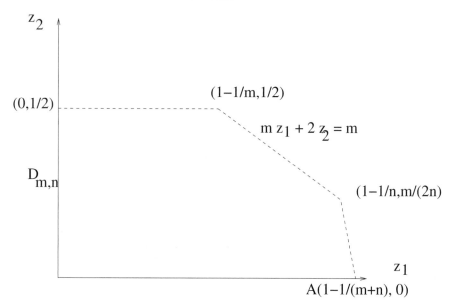

Fig. 2. The domain of the central limit theorem

Remarks

(A1). Giraitis and Taqqu expressed the upper boundary of this polytope in a convenient fashion:

$$d_m(z_1) + d_n(z_1) + 2z_2 = 1 \qquad (25)$$

where

$$d_m(z) = 1 - m(1 - z)^+.$$

(A2). The extremal points are solutions of equations $\alpha_G(z) = 1$, obtained for certain specific graphs $G \in \mathcal{G}_2$. The fact that at these points one has the inequalities $\alpha_G(z) \leq k/2$, $\forall G \in \mathcal{G}_k, \forall k \geq 2$ may be checked (see [Avr92]) by solving some graph-breaking problem, described after equation (17). An example is presented at the end of the paper.

Hence the conditions of Corollary 1 hold at the extremal points of the polytope in Figure 2, and hence throughout the polytope, yielding thus the asymptotic normality result of Avram [Avr92], Theorem 4, and of Giraitis and Taqqu [GT97], Theorem 2.3.

1.12 Conclusion

The problem of establishing the central limit theorem by the method of moments is related to some beautiful mathematics: the Hölder-Brascamp-Lieb-

Barthe inequality, the continuity of matroid convolutions and the matroid weak Szegö theorem.

This leads to fascinating mathematical questions like extending the matroid weak Szegö theorem to a strong one (i.e. providing correction terms).

The analytic methodology presented above suggests also the following conjecture:

Conjecture. A central limit theorem holds in the **continuous one dimensional index case**, with the same normalization and limiting variance as in the discrete one-dimensional index case, if

$$f \in L_{z_1^{-1}}, \quad \hat{b} \in L_{z_2^{-1}},$$

and the exponents z_1, z_2 lie on the upper boundary of the polytope in the Figure 2.

The tools described in this paper are also expected to be useful for studying processes with continuous multidimensional indices. Let us mention for example the versatile class of **isotropic** spatio-temporal models, of a form similar to (1) (with the Laplacian operator Δ replacing the operator B), recently introduced by Anh, Leonenko, Kelbert, McVinnish, Ruiz-Medina, Sakhno and coauthors [ALM01], [AHL02], [ALS04],[ALS03], [KLRM05]. These authors use the spectral approach as well and the tools described above hold the potential of simplifying their methods.

Even for unidimensional discrete processes, these tools might be useful for strengthening the central limit theorem to sharp large deviations statements, as in the work on one-dimensional Gaussian quadratic forms of Bercu, Gamboa, Lavielle and Rouault [BGL00], [BGR97].

Finally, these tools might improve asymptotic results concerning Whittle's estimator of spectral densities – see [Han70], [FT86], [GT99], [HG89], [HG93], [Hey97], [Guy95].

APPENDIX

A The Hölder-Brascamp-Lieb-Barthe inequality

Set here $V = m$ and $E = n$. Let M be a $m \times n$ matrix, $\mathbf{x} = (x_1, \ldots, x_m)$ and let $l_1(\mathbf{x}), \ldots l_n(\mathbf{x})$ be n linear transformations such that

$$(l_1(\mathbf{x}), \ldots, l_n(\mathbf{x})) = (x_1, \ldots, x_m)M .$$

Let f_j, $j = 1, \ldots, n$ be functions belonging respectively to

$$L_{p_j}(\mathrm{d}\mu), \ 1 \leq p_j \leq \infty, \ j = 1, \ldots, n .$$

We consider simultaneously three cases:

(C1) $\mu(dx)$ is the Lebesgue measure on the torus $[-\pi, \pi]^{n_j}$, and M has all its coefficients integers.

(C2) $\mu(dx)$ is the counting mesure on $\mathbb{Z}^{n_j}$, M has all its coefficients integers, and all its non-singular minors of dimension $m \times m$ have determinant ± 1.

(C3) $\mu(dx)$ is Lebesgue measure on $(-\infty, +\infty)^{n_j}$.

The following theorem, due when $n_j = 1, \forall j$ in the first case to [AB89], in the second to [AT89] and in the last to [Bar05], with arbitrary n_j yields conditions on

$$z_j = \frac{1}{p_j} , \; j = 1, \ldots, n ,$$

so that a generalized Hölder inequality holds. For a recent exposition and extensions, see [BCCT05], [BCCT].

The key idea of the proof in [AB89], [AT89], is that it is enough to find those points $z = (z_1, \ldots, z_n)$ with coordinates z_i equal to 0 or 1, for which the inequality (GH) below holds, then by the Riesz-Thorin interpolation theorem, (GH) will hold for the smallest convex set generated by these points. This yields:

Theorem 2 (Hölder-Brascamp-Lieb-Barthe inequality). *Suppose, respectively, that the conditions (C1), (C2) and (C3) hold and let f_j, $j = 1, \ldots, n$ be functions $f_j \in L_{p_j}(\mu(dx))$, $1 \leq p_j \leq \infty$, where the integration space is either $[-\pi, \pi]^{n_j}$, $\mathbb{Z}^{n_j}$, or $\mathbb{R}^{n_j}$, and $\mu(dx)$ is respectively normalized Lebesgue measure, counting measure and Lebesgue measure. Let $z_j = (p_j)^{-1}$.*

For every subset A of columns of M (including the empty set $\emptyset$), denote by $r(A)$ the rank of the matrix formed by these columns, and suppose respectively:

(c1) $\displaystyle\sum_{j \in A} z_j \leq r(A), \; \forall A$

(c2) $\displaystyle\sum_{j \in A} z_j + r(A^c) \geq r(M), \; \forall A$

(c3) $\displaystyle\sum_{j=1}^{T} z_j = m$, *and one of the conditions (c1) or (c2) is satisfied.*

Then, the following inequality holds:

(GH) $$\left| \int \prod_{j=1}^{T} f_j(l_j(\mathbf{x})) \prod_{i=1}^{m} d\mu(x_i) \right| \leq K \prod_{j=1}^{T} \|f_j\|_{p_j}$$

where the constant K in (GH) is equal to 1 in the cases (C1) and (C2) and is finite in the case (C3) (and given by the supremum over centered Gaussian functions – see [Bar05]).

Alternatively, the conditions (c1-c3) in the theorem are respectively equivalent to:

(A1). $z = (z_1, \ldots, z_n)$ lies in the convex hull of the indicators of the sets of independent columns of M, including the void set.

(A2). $z = (z_1, \ldots, z_n)$ lies **above** the convex hull of the indicators of the sets of columns of M which span its range.

(A3). $z = (z_1, \ldots, z_n)$ lies in the convex hull of the indicators of the sets of columns of M which form a basis.

Examples.

1. As an example of $(c1)$, consider the integral

$$J = \int_T \int_T f_1(x_1) f_2(x_2) f_3(x_1 + x_2) f_4(x_1 - x_2) dx_1 dx_2$$

where T denotes the torus $[0, 1]$, so $f_j(x \pm 1) = f_j(x)$, $j = 1, \ldots, 4$. Here $m = 2$, $n = 4$ and the matrix

$$M = \begin{pmatrix} 1 & 0 & 1 & 1 \\ 0 & 1 & 1 & -1 \end{pmatrix}$$

has rank $r(M) = 2$. The flats consist of $\emptyset$, the single columns and M. Only $\emptyset$ and M are flat, irreducible, and not singleton. Since $(a1')$ always holds for $\emptyset$, it is sufficient to apply it to M. The theorem yields

$$|J| \le \|f_1\|_{1/z_1} \|f_2\|_{1/z_2} \|f_3\|_{1/z_3} \|f_4\|_{1/z_4}$$

for any $z = (z_1, z_2, z_3, z_4) \in [0, 1]^4$ satisfying $z_1 + z_2 + z_3 + z_4 \le 2$, e.g. if $z = (0, 1, 1/4, 1/2)$, then

$$|J| \le \left(\sup_{0 \le x \le 1} |f_1(x)| \right) \left(\int_0^1 |f_2(x)| dx \right) \left(\int_0^1 f_3^4(x) dx \right)^{1/4} \left(\int_0^1 f_4^2(x) dx \right)^{1/2} .$$

2. To illustrate $(c2)$, consider

$$S = \sum_{x_1 = -\infty}^{\infty} \sum_{x_2 = -\infty}^{\infty} f_1(x_1) f_2(x_2) f_3(x_1 + x_2) f_4(x_1 - x_2) .$$

Since m, n and M are as in Example 1, we have $r(M) = m$ and the only flat and irreducible sets which are not singleton are $\emptyset$ and M. Since it is sufficient to apply $(c2)$ to $\emptyset$, the theorem yields $|S| \le \|f_1\|_{1/z_1} \|f_2\|_{1/z_2} \|f_3\|_{1/z_3} \|f_4\|_{1/z_4}$ for any $z = (z_1, z_2, z_3, z_4) \in [0, 1]^4$ satisfying $z_1 + z_2 + z_3 + z_4 \ge 2$, e.g.

$$|S| \le \prod_{j=1}^4 \left(\sum_{x=-\infty}^{+\infty} f_j^2(x) \right)^{1/2} .$$

B The application of the diagram formula for computing moments/cumulants of Wick products of linear processes

B.1 Wick products

We start with some properties of the Wick products (cf. [GS86], [Sur83]) and their application in our problem.

Definition 6. *The Wick products (also called Wick powers) are multivariate polynomials:*

$$: y_1, \ldots, y_n :^{(\nu)} =$$

$$\frac{\partial^n}{\partial z_1 \ldots \partial z_n} \left[\exp\left(\sum_1^n z_j y_j\right) \Big/ \int_{\mathbb{R}^n} \exp\left(\sum_1^n z_j y_j\right) d\nu(y) \right] \Bigg|_{z_1 = \cdots = z_n = 0}$$

corresponding to a probability measure ν on $\mathbb{R}^n$. Interpret this as a formal expression if ν does not have a moment generating function, the Wick products being then obtained by formal differentiation.

A sufficient condition for the Wick products $: y_1, \ldots, y_n :^{(\nu)}$ to exist is $E|Y_i|^n < \infty$, $i = 1, \ldots, n$.

When some variables appear repeatedly, it is convenient to use the notation

$$: \underbrace{Y_{t_1}, \ldots, Y_{t_1}}_{n_1}, \ldots, \underbrace{Y_{t_k}, \ldots, Y_{t_k}}_{n_k} := P_{n_1, \ldots, n_k}(Y_{t_1}, \ldots, Y_{t_k})$$

(the indices in P correspond to the number of times that the variables in ": :" are repeated). The resulting polynomials $P_{n_1, \ldots, n_k}$ are known as Appell polynomials. These polynomials are a generalization of the Hermite polynomials, which are obtained if Y_t are Gaussian; like them, they play an important role in the limit theory of quadratic forms of dependent variables (cf. [Sur83], [GS86], [AT87]).

For example, when $m = n = 1$, $P_{1,1}(X_t, X_s) = X_t X_s - \mathbb{E} X_t X_s$, and the bilinear form (5) is a weighted periodogram with its expectation removed.

Let W be a finite set and Y_i, $i \in W$ be a system of random variables. Let

$$Y^W = \prod_{i \in W} Y_i$$

be the ordinary product,

$$: Y^W :$$

the Wick product, and

$$\chi(Y^W) = \chi(Y_i, i \in W)$$

be the (mixed) cumulant of the variables $Y_i, i \in W$, respectively, which is defined as follows:

$$\chi(Y_1, \ldots, Y_n) = \frac{\partial^T}{\partial z_1 \ldots \partial z_n} \log \mathbb{E} \exp\left(\sum_{i=1}^T z_j Y_j\right) \Bigg|_{z_1 = \cdots = z_n = 0}.$$

The following relations hold ([Sur83], Prop. 1):

$$: Y^W := \sum_{U \subset W} Y^U \sum_{\{V\}} (-1)^r \chi(Y^{V_1}) \cdots \chi(Y^{V_r}),$$

$$Y^W = \sum_{U \subset W} : Y^U : \sum_{\{V\}} \chi(Y^{V_1}) \cdots \chi(Y^{V_r}) = \sum_{U \subset W} : Y^U : \mathbb{E}(Y^{W \setminus U})$$

where the sum $\sum_{U \subset W}$ is taken over all subsets $U \subset W$, including $U = \emptyset$, and the sum $\sum_{\{V\}}$ is over all partitions $\{V\} = (V_1, \ldots, V_r)$, $r \geq 1$ of the set $W \setminus U$. We define $Y^{\emptyset} =: Y^{\emptyset} := \chi(Y^{\emptyset}) = 1$.

B.2 The cumulants diagram representation

An important property of the Wick products is the existence of simple combinatorial rules for calculation of the (mixed) cumulants, analogous to the familiar diagrammatic formalism for the mixed cumulants of the Hermite polynomials with respect to a Gaussian measure [Mal80]. Let us assume that W is a union of (disjoint) subsets $W_1, \ldots, W_k$. If $(i, 1), (i, 2), \ldots, (i, n_i)$ represent the elements of the subset W_i, $i = 1, \ldots, k$, then we can represent W as a table consisting of rows $W_1, \ldots, W_k$, as follows:

$$\begin{pmatrix} (1, 1), \ldots, (1, n_1) \\ \cdots\cdots\cdots \\ (k, 1), \ldots, (k, n_k) \end{pmatrix} = W \ . \tag{26}$$

By a *diagram* γ we mean a partition $\gamma = (V_1, \ldots, V_r)$, $r = 1, 2, \ldots$ of the table W into nonempty sets V_i (the "edges" of the diagram) such that $|V_i| \geq 1$. We shall call the edge V_i of the diagram γ *flat*, if it is contained in one row of the table W; and *free*, if it consists of one element, i.e. $|V_i| = 1$. We shall call the diagram *connected*, if it does not split the rows of the table W into two or more disjoint subsets. We shall call the diagram $\gamma = (V_1, \ldots, V_r)$ *Gaussian*, if $|V_1| = \cdots = |V_r| = 2$. Suppose given a system of random variables $Y_{i,j}$ indexed by $(i, j) \in W$. Set for $V \subset W$,

$$Y^V = \prod_{(i,j) \in V} Y_{i,j}, \quad \text{and} \quad : Y^V := : (Y_{i,j}, (i, j) \in V) : \ .$$

For each diagram $\gamma = (V_1, \ldots, V_r)$ we define the number

$$I_\gamma = \prod_{j=1}^{r} \chi(Y^{V_j}) \ . \tag{27}$$

Proposition 1. (cf. [GS86], [Sur83]) *Each of the numbers*

$$(i) \quad \mathbb{E}Y^W = E(Y^{W_1} \ldots Y^{W_k}) \ ,$$
$$(ii) \quad \mathbb{E}(: Y^{W_1} : \cdots : Y^{W_k} :) \ ,$$
$$(iii) \quad \chi(Y^{W_1}, \ldots, Y^{W_k}) \ ,$$
$$(iv) \quad \chi(: Y^{W_1} :, \ldots, : Y^{W_k} :)$$

is equal to

$$\sum I_\gamma$$

where the sum is taken, respectively, over

(i) *all diagrams,*

(ii) *all diagrams without flat edges,*

(iii) *all connected diagrams,*

(iv) *all connected diagrams without flat edges.*

If $\mathbb{E} Y_{i,j} = 0$ *for all* $(i,j) \in W$, *then the diagrams in (i)-(iv) have no singletons.*

It follows, for exampgle, that $\mathbb{E} : Y^W := 0$ (take $W = W_1$, then W has only 1 row and all diagrams have flat edges).

B.3 Multilinearity

An important property of Wick products and of cumulants is their multilinearity. This implies that for Q_T defined in (5) that

$$\chi_k(Q_T, ..., Q_T) =$$

$$\int_{t_i, s_i \in I_T} \chi(: X_{t_{1,1}}, \ldots, X_{t_{1,m}}, X_{s_{1,1}}, \ldots, X_{s_{1,n}} :,$$

$$\ldots, : X_{t_{k,1}}, \ldots, X_{t_{k,m}}, X_{s_{k,1}}, \ldots, X_{s_{k,n}} :) \prod_{i=1}^{k} \hat{b}_{t_i - s_i} dt_i ds_i$$

where the cumulant in the integral needs to be taken for a table W of k rows $R_1, ..., R_k$, each containing the Wick product of m variables identically equal to X_{t_k} and of n variables identically equal to X_{s_k}.

A further application of part (iv) of Proposition 1 will decompose this as a sum of the form

$$\sum_{\gamma \in \Gamma(n_1, \ldots, n_k)} \int_{t_i, s_i \in I_T} R_\gamma(t_i, s_i) \prod_{i=1}^{k} \hat{b}_{t_i - s_i} dt_i ds_i$$

where $\Gamma(n_1, \ldots, n_k)$ denotes the set of all connected diagrams $\gamma = (V_1, \ldots, V_r)$ without flat edges of the table W and $R_\gamma(t_i, s_i)$ denotes the product of the cumulants corresponding to the partition sets of γ.

Example 1: When $m = n = 1$, the Gaussian diagrams are all products of correlations and the symmetry of $\hat{b}$ implies that all these $2^{k-1}(k-1)!$ terms are equal. We get thus the well-known formula (7) for the cumulants of discrete Gaussian bilinear forms.

Example 2: When $m = n = 1$ and $k = 2$, besides the Gaussian diagrams we have also one diagram including all the four terms, which makes intervene the fourth order cumulant of X_t.

B.4 The cumulants of sums and quadratic forms of linear processes

Consider

$$S_T^m = \sum_{i=1}^{T} P_m(X_{t_i}), \quad Q_T^{m,n} = \sum_{i=1}^{T}\sum_{j=1}^{T} \hat{b}(i-j) \, P_{m,n}(X_{t_i}, X_{t_j}) . \tag{28}$$

By part (iv) of proposition 1, applied to a table W of k rows $R_1..., R_k$, with $K = n_1 + ...n_k$ variables, and by the definition (27) and of $I\gamma$, we find the following formula for the cumulants of the Wick products of linear variables (2):

$$\chi(: X_{t_{1,1}}, \ldots, X_{t_{1,n_1}} :, \ldots, : X_{t_{k,1}}, \ldots, X_{t_{k,n_k}} :) = \sum_{\gamma \in \Gamma(n_1,\ldots,n_k)} \kappa_\gamma J_\gamma(\mathbf{t}) \tag{29}$$

where $\Gamma(n_1, \ldots, n_k)$ denotes the set of all connected diagrams $\gamma = (V_1, \ldots, V_r)$ without flat edges of the table W, $\kappa_\gamma = \chi_{|V_1|}(\xi_{I_1}) \cdots \chi_{|V_r|}(\xi_{I_1})$ and

$$J_\gamma(t_1, ..., t_K) = \prod_{j=1}^{r} J_{V_j}(t_{V_j}) \tag{30}$$

$$= \int_{s_1,\ldots,s_r \in I} \prod_{j=1}^{k} \Big[\hat{a}(t_{j,1} - s_{j,1}) \hat{a}(t_{j,n_1} - s_{j,n_1}) \cdots$$

$$\cdots \hat{a}(t_{k,1} - s_{k,1}) \cdots \hat{a}(t_{k,n_k} - s_{k,n_k}) \Big] ds_1 \ldots ds_r$$

$$= \int_{\lambda_1,\ldots,\lambda_K} e^{i \sum_{j=1}^{K} t_j \lambda_j} \prod_{i=1}^{K} a(\lambda_i) \prod_{j=1}^{r} \delta(\sum_{i \in V_j} \lambda_i) \prod_{i=1}^{K} d\lambda_i$$

where $s_{i,j} \equiv s_l$ if $(i,j) \in V_l$, $l = 1, \ldots, r$.

We will apply now this formula to compute the cumulants of (28), in which case each row j contains just one, respectively two random variables. We will see below that this yields decompositions as sums of Delta graph integrals with a specific graph structure.

For example, it is easy to check that the variance of $S_T^{(2)}$ is:

$$\chi_2(S_T^{(2)}) = 2 \int_{\lambda_1,\lambda_2 \in S} f(\lambda_1) f(\lambda_2) \Delta_T(\lambda_1 - \lambda_2) \Delta_T(\lambda_2 - \lambda_1) \prod_{e=1}^{2} \mu(d\lambda_e) .$$

Note that there are two possible diagrams of two rows of size 2, and that they yield both a graph on two vertices (corresponding to the rows), connected one to the other via two edges.

For another example, the third cumulant $\chi_3(S_T^{(2)})$ is a sum of terms similar to:

$$2^2 \int_{\lambda_1,\lambda_2,\lambda_3 \in S} f(\lambda_1)f(\lambda_2)f(\lambda_3)\Delta_T(\lambda_1-\lambda_2)\Delta_T(\lambda_2-\lambda_3)\Delta_T(\lambda_3-\lambda_1) \prod_{e=1}^{3} \mu(d\lambda_e).$$

This term comes from the 2^2 diagrams in which the row 1 is connected to row 2, 2 to 3 and 3 to 1.

The general structure of the intervening graphs is as follows (see [Avr92]):

(A1). In the case of cumulants of sums, we get graphs belonging to the set $\Gamma(m,k)$ of all connected graphs with no loops over k vertices, each of degree m.

(A2). In the case of cumulants of quadratic forms, we get – see Figure 3 – graphs belonging to the set $\Gamma(m,n,k)$ of all connected bipartite graphs with no loops whose vertex set consists of k pairs of vertices. The "left" vertex of each pair arises out of the first m terms : $X_{t_1}, ..., X_{t_m}$: in the diagram formula, and the "right" vertex of each pair arises out of the last n terms : $X_{s_1}, ..., X_{s_n}$: The edge set consists of:

a) k "kernel edges" pairing each left vertex with a right vertex. The kernel edges will contribute below terms involving the function $b(\lambda)$.

b) A set of "correlation edges", always connecting vertices in different rows, and contributing below terms involving the function $f(\lambda)$). They are arranged such that each left vertex connects to m and each right vertex connects to n such edges, yielding a total of $k(m+n)/2$ correlation edges.

Thus, the k "left vertices" are of degree $m+1$, and the other k vertices are of degree $n+1$. (The "costs" mentionned in Figure 3 refer to (17)).

The following proposition is easy to check.

Proposition 2. *Let $X_t, t \in I_T$ denote a stationary linear process given by (2) with $d = 1$. Then, the cumulants of the sums and quadratic forms defined in (28) are given respectively by:*

$$\chi_{k,l} = \chi_k(S_T^{(l)}, ..., S_T^{(l)}) = \sum_{\gamma \in \Gamma(m,k)} \kappa_\gamma \, \sigma_\gamma(T)$$

and

$$\chi_{k,m,n} = \chi_k(Q_T^{(m,n)}, ..., Q_T^{(m,n)}) = \sum_{\gamma \in \Gamma(m,n,k)} \kappa_\gamma \, \tau_\gamma(T)$$

where $\Delta_T(x)$ is the Fejer kernel, $\Gamma(l,k)$, $\Gamma(m,n,k)$ were defined above, and

$$\sigma_\gamma(T) = \int_{\mathbf{t} \in I_T^k} J_\gamma(\mathbf{t})d\mathbf{t}$$

$$= \int_{\lambda_1,...,\lambda_K} \prod_{j=1}^{k} \Delta_T\Big(\sum_{i=m(j-1)+1}^{mj} \lambda_i \Big) \prod_{i=1}^{K} a(\lambda_i) \prod_{j=1}^{r} \delta\Big(\sum_{i \in V_j} \lambda_i\Big) \prod_{i=1}^{K} d\lambda_i \,,$$

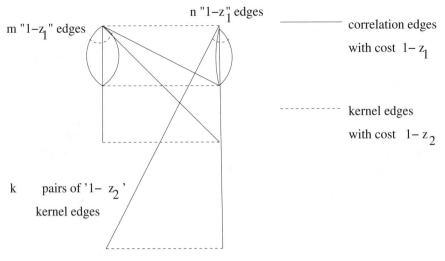

Fig. 3. The graphs appearing in the expansion of cumulants of quadratic forms. Here k=4, m=5, n=4. The figure displays only some of the k(m+n)/2=18 correlation edges.

$$\tau_\gamma(T) = \int_{\mu_1,\dots,\mu_k,\lambda_1,\dots,\lambda_K,\lambda'_1,\dots,\lambda'_{K'}}$$

$$\prod_{j=1}^{k}\left[\Delta_T\left(\mu_j + \sum_{i=m(j-1)+1}^{mj}\lambda_i\right)\Delta_T\left(-\mu_j+\sum_{i=n(j-1)+1}^{nj}\lambda'_i\right)b(\mu_j)\right]$$

$$\times\prod_{i=1}^{K}a(\lambda_i)\prod_{i=1}^{K'}a(\lambda'_i)\prod_{j=1}^{r}\delta(\sum_{i\in V_j}\lambda_i+\sum_{i\in V_j}\lambda'_i)\prod_{i=1}^{K}d\lambda_i\prod_{i=1}^{K'}d\lambda'_i\prod_{i=1}^{k}d\mu_i\;.$$

These graph structures are simple enough to allow a quick evaluation of the orders of magnitude $\alpha_M(z)$, via the corresponding graph-breaking problems; for the case of bilinear forms we refer to Lemma 1 in [Avr92].

For the case of sums, the domain of applicability of the CLT is $1 - z_1 \geq 1/m$. We check now that at the extremal point $1 - z_1 = 1/m$ we have

$$\alpha_G(z_1) = \max_A p(A)$$

$$= \max_A[co(G - A) - \sum_{e\in A}(1 - z_e)]$$

$$= \max_A[co(G - A) - |A|(1 - z_1)]$$

$$\leq k/2, \;\forall G \in \mathcal{G}_k$$

where we interpret $p(A)$ as a "profit," equal to the "gain" $co(G - A)$ minus the "cost" $\sum_{e\in A}(1 - z_e)$. We thus need to show that at the extremal point

$1 - z_1 = 1/m,$

$$co(G - A)] \leq |A|/m + k/2, \ \forall G \in \mathcal{G}_k \ .$$

Indeed, this inequality holds with equality for the "total breaking" $A = \mathcal{E}$ (which contains $(km)/2$ edges). It is also clear that no other set of edges A can achieve a bigger "profit" $p(A)$ (defined in (31)) than the total breaking, since for any other set A which leaves some vertex still attached to the others, the vertex could be detached from the others with an increase of the number of components by 1 and a cost no more than $m\frac{1}{m}$; thus the profit is nondecreasing with respect to the number of vertices left unattached and thus the total breaking achieves the maximum of $p(A)$.

Acknowledgments. Murad S. Taqqu was partially supported by the NSF Grants DMS-0102410 and DMS-0505747 at Boston University.

References

[AB89]   F. Avram and L. Brown. A generalized Hölder inequality and a generalized szegö theorem. *Proceedings of the American Mathematical Society*, 107(3):687–695, 1989.

[AF92]   F. Avram and R. Fox. Central limit theorems for sums of Wick products of stationary sequences. *Transactions of the American Mathematical Society*, 330(2):651–663, 1992.

[AHL02]   V.V. Anh, C.C. Heyde, and N.N. Leonenko. Dynamic models of long-memory processes driven by Lévy noise. *Journal of Applied Probability*, 39:730–747, 2002.

[ALM01]   V.V. Anh, N.N. Leonenko, and R. McVinish. Models for fractional riesz-bessel motion and related processes. *Fractals*, 9(3):329–346, 2001.

[ALS03]   V.V. Anh, N.N. Leonenko, and L.M. Sakhno. Higher-order spectral densities of fractional random fields. *Journal of Statistical Physics*, 111:789 – 814, 2003.

[ALS04]   V.V. Anh, N.N. Leonenko, and L.M. Sakhno. Quasi-likelihood-based higher-order spectral estimation of random fields with possible long-range dependence. *Journal of Applied Probability*, 41A:35–53, 2004.

[AT87]   F. Avram and M. S. Taqqu. Noncentral limit theorems and Appell polynomials. *The Annals of Probability*, 15:767–775, 1987.

[AT89]   F. Avram and M. S. Taqqu. Hölder's inequality for functions of linearly dependent arguments. *SIAM Journal of Mathematical Analysis*, 20(6):1484–1489, 1989.

[Avr88]   F. Avram. On bilinear forms in Gaussian random variables and Toeplitz matrices. *Probability Theory and Related Fields*, 79(1):37–46, 1988.

[Avr92]   F. Avram. Generalized Szegö theorems and asymptotics of cumulants by graphical methods. *Transactions of the American Mathematical Society*, 330(2):637–649, 1992.

[Bar05]   F. Barthe. On a reverse form of the Brascamp-Lieb inequality. *Inventiones Mathematicae*, 134:335–361, 2005.

[BCCT]   J. Bennett, A. Carbery, M. Christ, and T. Tao. The Brascamp-Lieb inequalities: finiteness, structure and extremals. Preprint: math.MG/0505065.

[BCCT05] J. Bennett, A. Carbery, M. Christ, and T. Tao. Finite bounds for Hölder-Brascamp-Lieb multilinear inequalities. Preprint, 2005.

[Ber94]   J. Beran. *Statistics for Long-Memory Processes*. Chapman & Hall, New York, 1994.

[BGL00]   B. Bercu, F. Gamboa, and M. Lavielle. Sharp large deviations for gaussian quadratic forms with applications. *ESAIM Probability and Statistics*, 4:1–24, 2000.

[BGR97]   B. Bercu, F. Gamboa, and A. Rouault. Large deviations for quadratic forms of stationary gaussian processes. *Stochastic Processes and their Applications*, 71:75–90, 1997.

[BLL74]   H.J Brascamp, E. H. Lieb, and J.M. Lutinger. A general rearrangement inequality for multiple integrals. *Journal of Functional Analysis*, 17:227–237, 1974.

[DM79]   R. L. Dobrushin and P. Major. Non-central limit theorems for non-linear functions of Gaussian fields. *Zeitschrift für Wahrscheinlichkeitstheorie und verwandte Gebiete*, 50:27–52, 1979.

[FT86]   R. Fox and M. S. Taqqu. Large-sample properties of parameter estimates for strongly dependent stationary Gaussian time series. *The Annals of Statistics*, 14:517–532, 1986.

[FT87]   R. Fox and M. S. Taqqu. Central limit theorems for quadratic forms in random variables having long-range dependence. *Probability Theory and Related Fields*, 74:213–240, 1987.

[GJ80]   C. W. J. Granger and R. Joyeux. An introduction to long-memory time series and fractional differencing. *Journal of Time Series Analysis*, 1:15–30, 1980.

[GS58]   U. Grenander and G. Szego. *Toeplitz forms and their applications*. Chelsea, New York, 1958.

[GS86]   L. Giraitis and D. Surgailis. Multivariate Appell polynomials and the central limit theorem. In E. Eberlein and M. S. Taqqu, editors, *Dependence in Probability and Statistics*, pages 21–71. Birkhäuser, New York, 1986.

[GT97]   L. Giraitis and M. S. Taqqu. Limit theorems for bivariate Appell polynomials: Part I. Central limit theorems. *Probability Theory and Related Fields*, 107:359–381, 1997.

[GT99]   L. Giraitis and M. S. Taqqu. Whittle estimator for non-Gaussian long-memory time series. *The Annals of Statistics*, 27:178–203, 1999.

[Guy95]   X. Guyon. *Random Fields on a Network: Modeling, Statistics and Applications*. Springer-Verlag, New York, 1995.

[Han70]   E.J. Hannan. *Multiple Time Series*. Springer-Verlag, New York, 1970.

[Hey97]   C.C. Heyde. *Quasi-Likelihood And Its Applications: A General Approach to Optimal Parameter Estimation*. Springer-Verlag, New York, 1997.

[HG89]   C. C. Heyde and R. Gay. On asymptotic quasi-likelihood. *Stochastic Processes and their Applications*, 31:223–236, 1989.

[HG93]   C. C. Heyde and R. Gay. Smoothed periodogram asymptotics and estimation for processes and fields with possible long-range dependence. *Stochastic Processes and their Applications*, 45:169–182, 1993.

[Hos81]   J. R. M. Hosking. Fractional differencing. *Biometrika*, 68(1):165–176, 1981.

[Hur51]   H. E. Hurst. Long-term storage capacity of reservoirs. *Transactions of the American Society of Civil Engineers*, 116:770–808, 1951.

[KLRM05] M. Kelbert, N.N. Leonenko, and M.D. Ruiz-Medina. Fractional random fields associated with stochastic fractional heat equations. *Advances of Applied Probability*, 37:108–133, 2005.

[Mal80]   V. A. Malyshev. Cluster expansions in lattice models of statistical physics and the quantum theory of fields. *Russian Mathematical Surveys*, 35(2):1–62, 1980.

[Oxl92]   J.G. Oxley. *Matroid Theory*. Oxford University Press, New York, 1992.

[Oxl04]   J.G. Oxley. What is a matroid? Preprint: www.math.lsu.edu/ oxley/survey4.pdf, 2004.

[Sur83]   D. Surgailis. On Poisson multiple stochastic integral and associated equilibrium Markov process. In *Theory and Applications of Random Fields*, pages 233–238. Springer-Verlag, Berlin, 1983. In: *Lecture Notes in Control and Information Science*, Vol. **49**.

[Taq79]   M. S. Taqqu. Convergence of integrated processes of arbitrary Hermite rank. *Zeitschrift für Wahrscheinlichkeitstheorie und verwandte Gebiete*, 50:53–83, 1979.

[Tut59]   W.T. Tutte. Matroids and graphs. *Transactions of the American Mathematical Society*, 90:527–552, 1959.

[Wel76]   D. Welsh. *Matroid Theory*. Academic Press, London, 1976.

[WTT99]   W. Willinger, M. S. Taqqu, and V. Teverovsky. Stock market prices and long-range dependence. *Finance and Stochastics*, 3:1–13, 1999.

Aggregation of Doubly Stochastic Interactive Gaussian Processes and Toeplitz forms of U-Statistics.

Didier Dacunha-Castelle[1] and Lisandro Fermín[1,2]

[1] Université Paris-Sud, Orsay-France. `didier.dacunha-castelle@math.u-psud.fr`
[2] Universidad Central de Venezuela. `fermin.lisandro@math.u-psud.fr`

1 Introduction

Granger [Gra80] has shown that by aggregating random parameter $AR(1)$ processes one may obtain long memory (LM) processes with spectral density equivalent to $1/\lambda^\alpha$ near $\lambda = 0$ for some α, $0 < \alpha < 1$. His study was the breakthrough to an enhanced technique of analyses and modeling of LM processes through elementary short memory (SM) processes. Following this discovery, several authors, e.g. [GG88, Lin99, LZ98], studied the aggregation of $AR(1)$ and $AR(2)$ processes. In particular, Lippi and Zaffaroni give a general presentation of Granger's results for $AR(1)$ processes and more generally for $ARMA$ processes.

In this work we develop an aggregation procedure by considering some doubly random, zero mean, second order and stationary elementary processes $Z^i = \{Z_t^i(Y^i), \, t \in T\}$ with spectral density $g(\lambda, Y^i)$. Here $Y = \{Y^i, i \in \mathbb{N}^d\}$ is a sequence of i.i.d. random variables with common distribution μ on $\mathbb{R}^s$. For every fixed path $Y(\omega)$, we define the sequence of partial aggregation processes $X^N = \{X_t^N(Y), t \in T\}$ of the elementary processes $\{Z^i\}$, by

$$X_t^N(Y) = \frac{1}{B_N} \sum_{i \in \{1,\dots,N\}^d} Z_t^i(Y^i), \tag{1}$$

where B_N is a sequence of positive numbers that will be presented in detail further in the article. Under some general conditions, for almost every path $Y(\omega)$, X^N converges in distribution to the same Gaussian process X, which is called the aggregation of the elementary processes $\{Z^i\}$. A different approach of aggregation using renewal switching processes is given in [TWS97].

First, we give a general framework for the aggregation procedure in the discrete time case as well as in the continuous case. Then, we extend the aggregation procedure in order to introduce dependence between elementary processes. It could be thought of as interactions between elementary processes, with the formalism for instance of statistical mechanics or as interactions between economical or sociological agents belonging to some sub-populations separated by a distance $|i - j|$.

Lippi and Zaffaroni [LZ98], introduce an innovation process of the form $\varepsilon_t + \eta_t^i$, where $\{\varepsilon_t, t \in \mathbb{Z}\}$ is the common innovation to all elementary process and $\eta^i = \{\eta_t^i, t \in \mathbb{Z}\}$ are independent sequences of white noises. In economical vocabulary η^i is the idiosyncratic component. But for aggregation they always consider the normalization $B_N = \sqrt{N}$ which forces to split the aggregation in two parts, a convergent part associated to η_t^i and a non convergent part associated to ε_t. Because of this inherent impediment, we approached the problem in a different way.

In our paper, we address the above issue by considering $Y = \{Y^i, i \in \mathbb{N}^d\}$ as the random environment model. Then we introduce interaction between elementary processes "living at i" starting from interaction between noises as $\mathbb{E}[\varepsilon_t^i \varepsilon_t^j] = \chi(i - j)$, where χ is a given covariance on $\mathbb{Z}^d$. Thus, the common innovation case is given by $\chi(j) = 1$, for all j, and the orthogonal innovation by $\chi(j) = 0$, for $j \neq 0$. This procedure induces a random stationary covariance between the elementary processes, given by $\mathbb{E}[Z_t^i Z_{t+\tau}^j] = \Psi_\tau(Y^i, Y^j)\chi(i - j)$, where $\{\chi(j), j \in \{0, ..., N\}^d\}$ is a Toeplitz operator sequence and $\Psi_\tau(Y^i, Y^j)$ is, for τ fixed, a second order U-statistic. The main tool used to prove the existence and the form of the aggregated process X is focused on second order U-statistic Toeplitz forms.

We obtain interesting qualitative behavior of our processes. First, we can show that the existence of aggregation is only linked to the asymptotic oscillation of the interaction and not, as believed previously, to the random environment, i.e. to the probability defined on the dynamics of elementary processes. In the case $d = 1$, the aggregation exists iff $s_k = \sum_{j=1}^{k-1} \chi(j)$ has a limit in the Cesaro sense, finite or infinite (in the last case necessarily positive).

If the aggregation exists in L^1 or $a.s.$ (where the existence in L^1 or $a.s$ means that the random covariance of X^N converges in L^1 or respectively $a.s.$) then its spectral density is always of the form $aF(\lambda) + bH^2(\lambda)$. The function $F(\lambda)$ is the mixture of the spectral densities $g(\lambda, Y^i)$ with respect to the common distribution μ of Y^i and the function $H(\lambda)$ is the mixture, with respect to μ, of the transfer functions $h(\lambda, Y^i)$ defined by $h^2(\lambda, y) = g(\lambda, y)$.

When there is no interaction, that is $\chi(j) = 0$ for $j \neq 0$, or when $s_k \to 0$, as for the increments of the fractional Brownian motion with index less than $1/2$, then $b = 0$ and $B_N \sim \sqrt{N}$.

When $s_k \to \infty$, which is here considered as a qualitatively strong interaction with large range, then $a = 0$ and $N = o(B_N^2)$.

If $s_k \to s < \infty$, for instance for short interaction, $\sum_{j=0}^{\infty} |\chi(j)| < \infty$, or for large range moderate oscillation when $\sum_{j=0}^{\infty} \chi(j) < \infty$ and $\sum_{j=0}^{\infty} |\chi(j)| = \infty$, then $a > 0$, $b > 0$.

So the limit is always a convex combination of the two extreme cases: independent innovations $(F(\lambda), B_N \sim \sqrt{N})$ and common innovations $(H^2(\lambda),$

$B_N \sim N$). The limit is reached for the normalization B_N which depends on the behavior of s_N.

Reaching LM for the aggregation has been one initial application of the procedure. It can be reached, for instance, aggregating $AR(1)$ elementary processes with random correlation parameter Y^i on $D = (-1,1)$. Then, if the distribution of Y^i is concentrated sufficiently near the boundary $\delta D = \{-1,1\}$ we can obtain LM. In our approach, it is seen clearly that F as well as H^2 can give the LM property. For modeling purposes, the simplest is to use H^2, since we only need to simulate one white noise. Therefore, LM is obtained by aggregating random parameter $AR(1)$ processes with the same innovation. Furthermore, in this case the concentration near the boundary δD is strong. The simplest way to illustrate this point is to choose $1/(1 - Y^i)$ as random variables with distribution μ'. If we want to simulate a LM processes with spectral density $G(\lambda)$ such that $G(\lambda) \sim 1/\lambda^\alpha$, when $\lambda \to 0$, using the classical simulation with independent innovations we have to take μ' as a p-stable positive distribution with $1 < p < 2$, and for the common innovation we have to choose $1/2 < p \leq 1$.

In Sect. 2 we introduce doubly stochastic processes. In Sect. 3 we define the aggregation procedure of doubly stochastic Gaussian processes considering dependence between elementary processes. We provide a Strong Law of Large Numbers (SLLN) for a random Toeplitz form of a second order U-statistic. Then we apply it to prove the convergence of the partial aggregations, see [Avr88, FT86, Gri76]. Finally in Sect. 4 we show the influence of this interaction on the LM character of the processes, obtained by means of aggregation of random parameter $AR(1)$ processes.

2 Doubly Stochastic Processes

We define the sequence of the elementary doubly stochastic processes for discrete and continuous time in the following way.

Let $Y = \{Y^i, i \in \mathbb{N}^d\}$ be a sequence of i.i.d. random variables with common distribution μ on $\mathbb{R}^s$ and define a family $\mathbb{W} = \{W^i(\lambda), \lambda \in \Lambda, i \in \mathbb{N}^d\}$ of complex Brownian random fields in the following way: let m denote Lebesgue's measure on Λ, where $\Lambda = (-\pi, \pi]^d$ in the case of discrete processes and $\Lambda = \mathbb{R}^d$ in the continuous case. Let χ be a correlation function on $\mathbb{Z}^d$. For simplicity we consider here only the one dimensional case, leaving for a follow-up paper the case $d > 1$ which is not very different from the case $d = 1$ except for very specific problems linked to anisotropy. We consider a sequence $\{W^i, i \in \mathbb{N}\}$ of complex Brownian random fields defined on $(\Omega, \mathcal{F}, \mathbb{P})$ by

$$\mathbb{E}[W^i(A)W^j(B)] = \chi(i - j)m(A \cap B) , \qquad (2)$$

where $i, j \in \mathbb{N}$ and $A, B \in \mathcal{A} = \{A \in \mathcal{B}(\Lambda) : m(A) < \infty\}$. The set of finite dimensional distributions of $\{W^i(A_i) : A_i \in \mathcal{A}, i = 1, ..., n\}$ is a coherent family of distributions, thus Kolmogorov's Theorem implies the existence of the family of random fields $\{W^i, i \in \mathbb{N}\}$. From equation (2) we obtain for all $f, g \in L^2(m)$

$$\mathbb{E}\left[\int_\Lambda f(\lambda)\mathrm{d}W^i(\lambda) \overline{\int_\Lambda g(\varphi)\mathrm{d}W^j(\varphi)}\right] = \chi(|i-j|) \int_\Lambda f(\lambda)\overline{g(\lambda)}\mathrm{d}\lambda . \quad (3)$$

Then the function χ is considered as the interaction correlation between the individual innovations W^i, with $\chi(0) = 1$.

Let $g(\lambda, y), y \in \mathbb{R}^s$, be a family of spectral densities, measurable on $\Lambda \bigotimes \mathbb{R}^s$ and such that $g(\lambda, y) \in L^1(m)$, μ−a.s. We will denote by h a particular square root of g, for simplicity we consider the real one.

We define the elementary processes $Z^i = \{Z_t^i(Y^i, \omega)\}$ by

$$Z_t^i(y, \omega) = \int_\Lambda \mathrm{e}^{\mathrm{i}t\lambda} h(\lambda, y) W^i(\omega, \mathrm{d}\lambda) . \quad (4)$$

For μ-almost all y, given $Y^i = y$, $Z^i = \{Z_t^i(y)\}$ is a stationary Gaussian process with spectral density $g(\lambda, y)$ and

$$\mathbb{E}^Y\left[Z_t^i \overline{Z_{t+\tau}^j}\right] = \chi(|i-j|) \int_\Lambda \mathrm{e}^{-\mathrm{i}\tau\lambda} h(\lambda, Y^i) h(\lambda, Y^j)\mathrm{d}\lambda ,$$

where $\mathbb{E}^Y[\cdot]$ is the conditional expectation given Y.

Remark 1. We can also define the elementary processes as follows.

- Discrete time case: Let $\{\varepsilon_n^i, n \in \mathbb{Z}\}$ be the discrete Fourier transform of W^i, then $\{\varepsilon_n^i, i \in \mathbb{N}, n \in \mathbb{Z}\}$ is an infinity array of normalized Gaussian random variables such that for a fixed i, $\{\varepsilon_n^i, n \in \mathbb{Z}\}$ is a white noise, where $\mathbb{E}[\varepsilon_n^i \varepsilon_n^j] = \chi(|i-j|)$ with $\chi(0) = 1$ and $\mathbb{E}[\varepsilon_n^i \varepsilon_m^j] = 0$ for $n \neq m$. Then we define the sequence $Z^i = \{Z_n^i(Y^i), n \in \mathbb{Z}\}$ of doubly random elementary processes by setting

$$Z_n^i(y) = \sum_{k \in \mathbb{Z}} c_k(y)\varepsilon_{n-k}^i , \quad (5)$$

where for μ-almost all y, $\{c_k(y)\}$ is the sequence of Fourier coefficients of μ.

- Continuous time case: Let B^i be the Fourier transform of W^i, which is a Brownian motion with the same properties as W^i. In this case Z_t^i is defined by

$$Z_t^i(y) = \int_{-\infty}^t c(t-s, y)\mathrm{d}B_s^i , \quad (6)$$

where for μ-almost all y, $c(\cdot, y)$ is the Fourier transform of $h(\cdot, y)$.

3 Aggregations and Mixtures

We will study the procedure of aggregation in the context of mixtures of spectral densities. The μ-mixture of the spectral densities $g(\lambda, y)$ is the spectral density F defined by

$$F(\lambda) = \int_{\mathbb{R}^s} g(\lambda, y) \, d\mu(y) \, . \tag{7}$$

The function $F(\lambda)$ is a well defined spectral density iff

$$V(F) = \int_\Lambda F(\lambda) \, d\lambda < \infty \, . \tag{8}$$

The mixture function given by

$$H(\lambda) = \int_{\mathbb{R}^s} h(\lambda, y) \, d\mu(y) \, , \tag{9}$$

is well defined and will be called a transfer function iff H^2 is a spectral density, i.e. if H^2 is is integrable with respect to Lebesgue's measure. This is a consequence of condition (8) and Jensen's inequality.

We will call $\{X_t^N, N \in \mathbb{N}\}$, defined in equation (1), the partial aggregation sequence of the elementary processes $\{Z^i\}$ associated to the mixture $F(\lambda)$.

We now study convergence in distribution of X^N to some Gaussian process X, called the aggregation of the processes $\{Z^i\}$, for almost all paths of Y. Let ν be the probability measure of the process Y, i.e. $\nu = \mu^{\otimes \mathbb{N}}$.

Definition 1. *Let $X^N(Y)$ be the partial aggregation of the sequence of elementary processes $\{Z^i(Y^i), i \in \mathbb{N}\}$. We say that the aggregation X exists $\nu - a.s.$ iff for ν-almost every sequence $Y = \{Y^i\}$, the partial aggregation process $X^N(Y)$ converges weakly to X.*

As the elementary processes Z^i are $\nu - a.s.$ Gaussian and zero-mean, then Definition 1 is equivalent to the following definition.

Definition 2. *The aggregation X exists $\nu - a.s.$ iff, for every $\tau \in T$, the covariance functions $\Gamma^N(\tau, Y)$ of $\{X_t^N(Y)\}$ converge $\nu - a.s.$ to the covariance $\Gamma(\tau)$ of X.*

We will need the following definition.

Definition 3. *Let $X^N(Y)$ be the partial aggregation of the sequence of elementary processes $\{Z^i(Y^i), i \in \mathbb{N}\}$. We say that the aggregation X exists in $L^p(\nu)$ (quadratic means q.m, for $p = 2$) iff $\Gamma^N(\tau, Y)$ converges in $L^p(\nu)$ to the covariance $\Gamma(\tau)$ of X.*

Let $\Gamma^N(\tau, B)$ be the covariance functions of the partial aggregation process $X^N(Y)$,

$$\Gamma^N(\tau, Y) = \frac{1}{B_N^2} \sum_{i=1}^N \Psi_\tau(Y^i, Y^i) + \frac{1}{B_N^2} \sum_{1 \le i \ne j \le N} \Psi_\tau(Y^i, Y^j) \chi(i-j) , \quad (10)$$

with

$$\Psi_\tau(Y^i, Y^j) = \int_\Lambda e^{-i\tau\lambda} h(\lambda, Y^i) h(\lambda, Y^j) d\lambda . \quad (11)$$

Let $[\chi]_{N,1} = \sum_{1 \le i \ne j \le N} \chi(i-j)$ and $[\chi]_{N,2} = \sum_{1 \le i \ne j \le N} \chi^2(i-j)$, where the function χ denotes the interaction correlation between innovations, and $\chi(0) = 1$. If $s_k = \sum_{j=1}^{k-1} \chi(j)$ then $[\chi]_{N,1} = 2 \sum_{k=1}^N s_k$.

As χ is a positive definite sequence, we have that $N\chi(0) + [\chi]_{N,1} \ge 0$. Hence $\frac{[\chi]_{N,1}}{N} = \frac{2}{N} \sum_{k=1}^N s_k \ge -1$, therefore $s_k \ge -\frac{1}{2}$. This implies that if s_k converges in the Cesaro sense to s, denoted by $s_k \overset{c}{\to} s$, then $-\frac{1}{2} \le s \le \infty$. So that there are only three cases to consider.

i. Weak interaction: when $s_k \overset{c}{\to} s < \infty$.
ii. Strong interaction: when $s_k \overset{c}{\to} \infty$.
iii. Oscillating interaction : when the sequence s_k does not converge in the Cesaro sense.

Let γ and ϕ be the respective covariances of F and H^2. By Jensen's inequality we have that $F \ge H^2$ and $\gamma_0 \ge \phi_0$ so that $\gamma_0 N + \phi_0 [\chi]_{N,1} \ge 0$. Since $2s \ge -1$ then $F(\lambda) + 2sH^2(\lambda) \ge 0$ and $\gamma_0 + 2s\phi_0 > 0$ for $\gamma_0 - \phi_0 > 0$.

First we will prove that $\Gamma^N(\tau, Y)$ converges $\nu - a.s.$ and in $L^1(\nu)$ iff condition (8) holds and the interaction correlation χ satisfies some additional conditions.

Theorem 1 (SLLN for $\Gamma^N(\tau, Y)$). *Let $B_N^2 = \gamma_0 N + \phi_0 [\chi]_{N,1}$. The covariance function $\Gamma^N(\tau, Y)$ converges in $L^1(\nu)$ iff condition (8) holds and s_k converges in the Cesaro sense to s, $-\frac{1}{2} \le s \le \infty$. Then its limit is given by*

$$\Gamma(\tau) = \frac{\gamma(\tau) + 2s\phi(\tau)}{\gamma_0 + 2s\phi_0} ,$$

with $\Gamma(\tau) = \phi(\tau)/\phi_0$ for $s = \infty$.

If condition (8) is satisfied then we have the following results:

(A1). $L^1(\nu)$ convergence is a necessary condition for $\nu - a.s.$ convergence, and in the case both limits exist, then they are the same.
(A2). In the case of weak interaction, let $r_k = \sum_{j=1}^k |\chi(j)|$. Then $\sum_{k=1}^\infty k^{-2} r_{k^2}^2 < \infty$ is a sufficient condition for $\nu - a.s.$ convergence.
(A3). In the case of strong interaction, the conditions $\sum_{N=1}^\infty [\chi]_{N,1}^{-2} [\chi]_{N,2} < \infty$ and $\sum_{k=1}^N s_k^2 \Theta \left(\sum_{k=1}^N s_k^2 \right) = O \left(\left\{ \sum_{k=1}^N s_k \right\}^2 \right)$, for some function Θ such that $\sum_{n=1}^\infty n^{-1} \Theta(n)^{-1} < \infty$, are jointly sufficient conditions for $\nu - a.s.$ convergence.

Remark 2. In Theorem 1:

(A1). The condition of item 2 holds if $\chi \in l^1$ or if $\chi(j) = (-1)^j L(j) j^{-\alpha}$ with $\alpha > 0$ and $\{L(j), j \in \mathbb{N}\}$ any sequence of slow variation.

(A2). Let $\zeta(N) = N^{-1} \sum_{k=1}^{N} s_k$, $s_k^+ = \sum_{j=1}^{k} (\chi(j) \vee 0)$ and $s_k^- = \sum_{j=1}^{k} (-\chi(j) \vee 0)$, then $s_k = s_k^+ - s_k^-$. Then the conditions of item 3 are satisfied if $s_k^- = o(s_k^+)$ and $\sum_{k=1}^{\infty} k^{-1} \zeta(k)^{-1} < \infty$.

In this case

$$[\chi]_{N,2} = O \left(\sum_{K=1}^{N} s_k^+ + \sum_{K=1}^{N} s_k^- \right) = O \left(\sum_{K=1}^{N} s_k^+ \right),$$

and $[\chi]_{N,1} \sim \sum_{K=1}^{N} s_k^+$. Then $[\chi]_{N,2}/[\chi]_{N,1}^2 = O \left(N^{-1} \zeta(N)^{-1} \right)$ and so the condition $\sum_{k=1}^{\infty} \{ [\chi]_{k,2}/[\chi]_{k,1}^2 \} < \infty$ holds.

On the other hand, $\sup_{k \leq N} \{ s_k^+ \} \sim \zeta(N)$, since s_k^+ is increasing, and $\zeta(N) \leq N$. Then, choosing $\Theta(n) = n^{\epsilon/3}$, for $0 < \epsilon < 1$, we have that

$$\frac{\sum_{k=1}^{N} s_k^2 \Theta \left(\sum_{k=1}^{N} s_k^2 \right)}{\left(\sum_{k=1}^{N} s_k \right)^2} = O \left(\frac{\Theta(N \zeta^2(N))}{N} \right) = O \left(\frac{1}{N^{1-\epsilon}} \right).$$

Two particular cases are: when $\chi(j) = j^{-\alpha}$ for $0 < \alpha < 1$ and when $\chi(j)$ has a fixed sign for j large enough and $\sum_{N=1}^{\infty} N^{-1} s_N^{-1} < \infty$.

Proof. The proof of theorem will be presented in two parts.

Part 1: (Convergence in L^1).

Since $\{h(\lambda, Y^i), i \in \mathbb{N}\}$ is an i.i.d sequence such that

$$\mathbb{E}[h(\lambda, Y^i)] = \int h(\lambda, y) d\mu(y) = H(\lambda),$$

$$\mathbb{E}[h(\lambda, Y^i)^2] = \int h^2(\lambda, y) d\mu(y) = \int g(\lambda, y) d\mu(y) = F(\lambda),$$

then $\mathbb{E}[\Psi_\tau(Y^i, Y^j)] = \phi(\tau)$ for $i \neq j$ and $\mathbb{E}[\Psi_\tau(Y^i, Y^i)] = \gamma(\tau)$. So condition (8) is equivalent to $\mathbb{E}[\Psi_\tau(Y^i, Y^j)] < \infty$ and by equation (10) we have

$$R_N(\tau, B) = \mathbb{E}[\Gamma^N(\tau, Y)] = \frac{N \gamma(\tau) + [\chi]_{N,1} \phi(\tau)}{N \gamma_0 + [\chi]_{N,1} \phi_0}.$$

Then $R_N(\tau, B)$ has a non-zero limit iff s_k converges in the Cesaro sense. In this case we obtain:

i. Weak interaction: if $s_k \xrightarrow{c} s < \infty$, then $R_N(\tau, B) \to \frac{\gamma(\tau) + 2s\phi(\tau)}{\gamma(0) + 2s\phi(0)}$.

ii. Strong interaction: if $s_k \xrightarrow{c} \infty$, then $R_N(\tau, B) \to \frac{\phi(\tau)}{\phi(0)}$.

iii. Oscillating interaction: if the sequence s_k does not converge in the Cesaro sense then $R_N(\tau, B)$ does not have a limit.

Part 2: (Convergence $\nu - a.s$).
Let $M_\tau(Y^i) = \mathbb{E}^i[\Psi_\tau(Y^i, Y^j)]$, for $i \neq j$. Then $\{M_\tau(Y^i), i \in \mathbb{N}\}$ is a sequence of i.i.d. random variables, such that $\mathbb{E}[M_\tau(Y^i)] = \mathbb{E}[\Psi_\tau(Y^i, Y^j)] = \phi(\tau)$ where $\mathbb{E}^i[\cdot]$ is the conditional expectation with respect to Y^i.

Let $\mathcal{H}$ be the Hilbert space generated by $\{\Psi_\tau(Y^i, Y^j); 1 \leq i \neq j \leq N\}$, $\mathcal{H}_1$ the linear space generated by $\{\Psi_\tau(Y^i, Y^j) - M_\tau(Y^i) - M_\tau(Y^j) + \phi(\tau); 1 \leq i < j \leq N\}$, $\mathcal{H}_2$ the linear space generated by $\{M_\tau(Y^i) + M_\tau(Y^j) - 2\phi(\tau); 1 \leq i < j \leq N\}$ and $\mathcal{C}$ the space of constants, then $\mathcal{H}_1$, $\mathcal{H}_2$, $\mathcal{C}$ form an orthogonal decomposition of $\mathcal{H}$; i.e. $\mathcal{H} = \mathcal{H}_1 \bigoplus \mathcal{H}_2 \bigoplus \mathcal{C}$. This can be checked by realizing that $\mathbb{E}^i[\Phi_\tau(Y^i, Y^j)] = 0$ $\mu - a.s.$ for $i \neq j$. Define

$$T_N(\tau, B) = \frac{1}{B_N^2} \sum_{i=1}^N \left(\Psi_\tau(Y^i, Y^i) - \gamma(\tau)\right),$$

$$Q_N(\tau, B) = \frac{1}{B_N^2} \sum_{1 \leq i \neq j \leq N} \left(M_\tau(Y^i) + M_\tau(Y^j) - 2\phi(\tau)\right) \chi(i - j),$$

$$U_N(\tau, B) = \frac{1}{B_N^2} \sum_{1 \leq i \neq j \leq N} \left(\Psi_\tau(Y^i, Y^j) - M_\tau(Y^i) - M_\tau(Y^j) + \phi(\tau)\right) \chi(i - j).$$

$\mathcal{H}$'s orthogonal decomposition applied to $\Gamma^N(\tau, Y) - T_N(\tau, B)$ gives the following orthogonal decomposition

$$\Gamma^N(\tau, Y) - T_N(\tau, B) = U_N(\tau, B) + Q_N(\tau, B) + R_N(\tau, B).$$

In what follows, we will show that T_N, Q_N and U_N converge to zero $\nu - a.s$ and in $L^1(\nu)$ under given conditions. If the limits $\nu - a.s$ and in $L^1(\nu)$ of Γ^N exist, then they are the same.

Step 1: (T_N's convergence to zero).
As $\{\Psi_\tau(Y^i, Y^i), i \in \mathbb{Z}\}$ is an i.i.d sequence and $\mathbb{E}[\Psi_\tau(Y^i, Y^i)] < \infty$ is equivalent to condition (8), if $N = O(B_N^2)$ then by applying the Strong Law of Large Numbers (SLLN) we obtain, under condition (8), for each τ that $T_N(\tau, B)$ converges in $L^1(\nu)$ and $\nu - a.s.$ to zero.

  i. Weak interaction: $s_k \xrightarrow{c} s$ and $-\frac{1}{2} \leq s < \infty$, then clearly $N = O(B_N^2)$.
  ii. Strong interaction: if $s_k \xrightarrow{c} \infty$ then $\frac{N}{B_N^2} = o(1)$ so the result holds.

Step 2: (Q_N's convergence to zero).
We can write

$$Q_N(\tau, B) = \frac{2}{B_N^2} \sum_{i=1}^N \xi^i,$$

where $\xi^i = \left(M_\tau(Y^i) - \phi(\tau)\right)(s_{N-i} + s_{i-1})$ is an i.i.d sequence such that $\mathbb{E}[\xi^i] = 0$, so that $\mathbb{E}[Q_N(\tau, B)] = 0$ and

$$\mathbb{E}[Q_N(\tau, B)]^2 = \frac{A_\tau}{B_N^4} \sum_{i=1}^N (s_{N-i} + s_{i-1})^2,$$

with

$$A_\tau = \mathbb{E}[M_\tau(Y^i) - \phi(\tau)]^2 = \int\int A(\lambda,\theta)H(\lambda)H(\theta)e^{i(\lambda+\theta)\tau}d\lambda d\theta - \phi^2(\tau) ,$$

where $A(\lambda,\theta) = \mathbb{E}[h(\lambda,Y^i)h(\theta,Y^i)]$. Condition (8) implies that $A(\tau) < \infty$.
As $\sum_{i=1}^N (s_{N-i} + s_{i-1})^2 = O(\sum_{i=1}^N s_i^2) \to \infty$ when $N \to \infty$, then it is
sufficient to show that $\frac{1}{B_N^4}\sum_{k=1}^N s_k^2 \to 0$ when $N \to \infty$.

i. Weak interaction: $s_k \xrightarrow{c} s < \infty$, $\sum_{k=1}^N s_k^2 = O(N)$ and $N = O(B_N^2)$,
therefore $\frac{1}{B_N^4}\sum_{k=1}^N s_k^2 = O(\frac{1}{N})$.

ii. Strong interaction: $s_k \xrightarrow{c} \infty$. As $\sup s_k = \infty$ and for j large enough
$|\chi(j)| \leq 1$, then $1 \leq |s_j| \leq j \leq N$, and $B_N^2 = O(\sum_{k=1}^N s_k)$, whence,
for some constant $C > 0$,

$$\frac{\sum_{k=1}^N s_k^2}{B_N^4} \leq C\frac{(\sup_{1\leq k\leq N} s_k)\sum_{k=1}^N s_k}{\left(\sum_{k=1}^N s_k\right)^2} = O\left(\frac{N}{\sum_{k=1}^N s_k}\right) = o(1) .$$

This implies that $Q_N(\tau,B)$ converges to zero in q.m. and therefore in $L^1(\nu)$.

In order to prove $\lim_{N\to\infty} Q_N(\tau,B) = 0$ $\nu - a.s.$, we can apply Petrov's
Theorem ([Pet75],6.17 p 222): let $\Delta_N = \sum_{i=1}^N \xi^i$ be a sum of independent
random variables such that its variance $V_N = A_\tau \sum_{i=1}^N (s_{N-i} + s_{i-1})^2$ diverge
to infinity. Then $\Delta_N - \mathbb{E}[\Delta_N] = o(\sqrt{V_N\Theta(V_N)})$ for some function Θ such that
$\sum_{n=}^\infty n^{-1}\Theta(n)^{-1} < \infty$.

So it is sufficient to find Θ such that

$$\sqrt{\sum_{k=1}^N s_k^2\,\Theta\left(\sum_{k=1}^N s_k^2\right)} = O(B_N^2) .$$

i. Weak interaction: $s_k \xrightarrow{c} s < \infty$, $\sum_{k=1}^N s_k^2 = O(N)$ and $N = O(B_N^2)$ so we
can apply Petrov's Theorem, taking $\Theta(n) = n^\epsilon$ for $0 < \epsilon < 1$.

ii. Strong interaction: $s_k \xrightarrow{c} \infty$. In this case, by hypothesis, we have

$$\frac{\sum_{k=1}^N s_k^2\,\Theta\left(\sum_{k=1}^N s_k^2\right)}{\left(\sum_{k=1}^N s_k\right)^2} = O(1) ,$$

for some function Θ such that $\sum_{n=1}^\infty n^{-1}\Theta(n)^{-1} < \infty$. Since $\sum_{k=1}^N s_k = O(B_N^2)$, we have

$$\frac{\sqrt{\sum_{k=1}^N s_k^2\Theta(\sum_{k=1}^N s_k^2)}}{B_N^2} \sim \frac{\sqrt{\sum_{k=1}^N s_k^2\Theta(\sum_{k=1}^N s_k^2)}}{\sum_{k=1}^N s_k} = O(1) .$$

Step 3: (U_N's convergence to zero).
We consider the kernel $\Phi(Y^i, Y^j)$ defined, for $i \neq j$, by

$$\Phi_\tau(Y^i, Y^j) = \Psi_\tau(Y^i, Y^j) - M_\tau(Y^i) - M_\tau(Y^j) + \phi(\tau) .$$

This kernel is symmetric and degenerated, i.e., $\mathbb{E}^j[\Phi_\tau(Y^i, Y^j)] = 0 \ \mu - a.s.$
Therefore we have that $\mathbb{E}[\Phi_\tau(Y^i, Y^j)] = 0$, $\mathbb{E}[\Phi_\tau(Y^i, Y^j)\Phi_\tau(Y^k, Y^l)] = 0$ for $(i, j) \neq (k, l)$, and

$$\sigma_\tau = \mathbb{E}[\Phi_\tau^2(Y^i, Y^j)] = \int \int (A(\lambda, \theta) - H(\lambda)H(\theta))^2 \, e^{i(\lambda+\theta)\tau} \, d\lambda d\theta .$$

Condition (8) implies $\mathbb{E}[\Phi_\tau^2(Y^i, Y^j)] < \infty$, for $i \neq j$. Then $U_N(\tau, B)$ is a random Toeplitz form of the second order U-statistic. The orthogonality of the random variables $\Phi_\tau(Y^i, Y^j)$, for $i \neq j$, implies that

$$\alpha_N = \mathbb{E}[|U_N(\tau, B)|^2] = \frac{\sigma_\tau [\chi]_{N,2}}{B_N^4} .$$

Then $U_N(\tau, B)$ converges to zero in q.m. iff $B_N^{-4}[\chi]_{N,2} \to 0$ when $N \to \infty$.
Let us see that this is always true when s_k has a limit $s \leq \infty$.

i. Weak interaction: $s_k \xrightarrow{c} s < \infty$, $\chi(j) \to 0$, then $B_N^{-4}[\chi]_{N,2} \sim N^{-2}[\chi]_{N,2}$ and

$$\frac{[\chi]_{N,2}}{N^2} \leq \frac{1}{N} \sum_{k=1}^{K_0} \left(1 - \frac{k}{N}\right) \chi^2(k)$$

$$+ \sup_{K_0 \leq k \leq N} |\chi(k)| \frac{1}{N} \sum_{k=K_0+1}^{N} \left(1 - \frac{k}{N}\right) |\chi(k)| ,$$

from where $B_N^{-4}[\chi]_{N,2} \to 0$ when $N \to \infty$. In particular, when $\chi \in l^2$ then $\alpha_N = O(N^{-1})$.

ii. Strong interaction: $s_k \xrightarrow{c} \infty$, then $N^{-1}[\chi]_{N,1} \to \infty$ and $[\chi]_{N,1} = O(B_N^2)$, from where

$$\frac{[\chi]_{N,2}}{B_N^4} = O\left(\frac{[\chi]_{N,2}}{[\chi]_{N,1}^2}\right) = o(1).$$

Let us now prove the $\nu - a.s.$ convergence for this random Toeplitz form, following the scheme of the classical proof for the SLLN in the case of i.i.d. random variables as in Petrov, [Pet75], See [Sur03] for the Central Limit Theorem.

First we consider the weak interaction case, $N = O(B_N^2)$. Let $a > 0$,

$$\mathbb{P}(\max_{n \geq N} |U_n(\tau, B)| \geq 2a) \leq \sum_{k=\lfloor \sqrt{N} \rfloor}^{\infty} \mathbb{P}(|U_{k^2}(\tau, B)| \geq a)$$

$$+ \sum_{k=\lfloor \sqrt{N} \rfloor}^{\infty} \mathbb{P}\left(\max_{k^2 \leq n \leq (k+1)^2} |U_n(\tau, B) - U_{k^2}(\tau, B)| \geq a\right).$$

From the estimation of α_N and applying Tchebychev's inequality, we have

$$\mathbb{P}(|U_{k^2}(\tau, B)| \geq a) \leq \frac{\sigma_\tau r_{k^2}}{a^2 k^2} \,,$$

where $r_k = \sum_{j=1}^k |\chi(j)|$. Then if $\sum_{k=1}^\infty k^{-2} r_{k^2}^2 < \infty$ we prove that the series $\sum_{k=\lfloor \sqrt{N} \rfloor}^\infty \mathbb{P}(|U_{k^2}(\tau, B)| \geq a)$ converges.

On the other hand,

$$U_n(\tau, B) - U_{k^2}(\tau, B) = \sum_{i,j \in A(n,k^2)} \Phi_\tau(Y^i, Y^j) \chi(i-j) \,,$$

where $A(n, k^2) = \{i,j : 1 \leq i < j, k^2 < j \leq n\} \cup \{i,j : 1 \leq j < i, k^2 < i \leq n\}$, and so

$$\max_{k^2 \leq n \leq (k+1)^2} |U_n(\tau, B) - U_{k^2}(\tau, B)| \leq a_k + b_k \,.$$

with

$$a_k = \max_{k^2 \leq n \leq (k+1)^2} \left| \frac{1}{n} \sum_{A(n,k^2)} \Phi_\tau(Y^i, Y^j) \chi(i-j) \right| \,.$$

$$b_k = \max_{k^2 \leq n \leq (k+1)^2} \left| \frac{1}{n} - \frac{1}{k^2} \right| \left| \sum_{1 \leq i < j \leq k^2} \Phi_\tau(Y^i, Y^j) \chi(i-j) \right| \,.$$

Since $a_k \leq \frac{1}{k^2} \sum_{A(k^2,(k+1)^2)} |\Phi_\tau(Y^i, Y^j)| |\chi(i-j)|$, then

$$\mathbb{P}(a_k \geq a) \leq \frac{1}{a^2 k^4} \mathbb{E}\left[\left(\sum_{A(k^2,(k+1)^2)} |\Phi_\tau(Y^i, Y^j) \chi(i-j)| \right)^2 \right]$$

$$\leq \frac{\sigma_\tau k^2}{a^2 k^4} \left(\sum_{l=1}^{(k+1)^2} |\chi(l)| \right)^2 = O\left(\frac{r_{k^2}^2}{k^2} \right) \,.$$

Furthermore, $b_k \leq \frac{2}{k^3} \left| \sum_{1 \leq i < j \leq k^2} \Phi_\tau(Y^i, Y^j) \chi(i-j) \right|$ so

$$\mathbb{P}(b_k \geq a) \leq \frac{4\sigma_\tau k^2 [\chi]_{k^2,2}}{a^2 k^6} = O\left(\frac{r_{k^2}}{k^2} \right) \,.$$

As $\mathbb{P}(\max_{k^2 \leq n \leq (k+1)^2} |U_n(\tau, B) - U_{k^2}(\tau, B)| \geq a) \leq \mathbb{P}(a_k \geq a) + \mathbb{P}(b_k \geq a)$ then

$$\sum_{k=\lfloor \sqrt{N} \rfloor}^\infty \mathbb{P}(\max_{k^2 \leq n \leq (k+1)^2} |U_n(\tau, B) - U_{k^2}(\tau, B)| \geq a) < \infty \,.$$

Finally, from Borel-Cantelli's lemma we derive the $\nu - a.s.$ convergence to zero of $U_N(\tau, B)$. This proves the convergence $\nu - a.s$ in the weak interaction case.

Note that, taking $k^{1+\beta}$ instead of k^2 and $N^{\frac{1}{1+\beta}}$ instead of $N^{\frac{1}{2}}$ in the previous proof, we can see that the result remains true for $\chi(j) = \frac{(-1)^j}{j^\alpha}$, with $\frac{1}{2} < \alpha < 1$.

Let us now prove the $\nu - a.s.$ convergence of U_N in the strong interaction case. In general we have

$$\mathbb{P}\left(\max_{N \geq K} |U_N(\tau, B)| \geq a\right) \leq \sum_{N=K}^{\infty} \mathbb{P}\left(|U_N(\tau, B)| \geq a\right)$$

$$\leq \frac{\sigma_\tau}{a^2} \sum_{N=K}^{\infty} \frac{[\chi]_{N,2}}{B_N^4}$$

$$\sim \frac{\sigma_\tau}{a^2} \sum_{N=K}^{\infty} \frac{[\chi]_{N,2}}{[\chi]_{N,1}^2} .$$

If $\sum_{N=1}^{\infty}\{[\chi]_{N,2}/[\chi]_{N,1}^2\} < \infty$, then Borel-Cantelli's lemma implies that $U_N(\tau, B)$ converges $\nu - a.s.$ to zero.

Part 3: (Convergence of Γ^N).
We have proved, in Part 2, that T_N, Q_N and U_N converge to zero in $L_1(\nu)$, then from decomposition

$$\Gamma^N(\tau, Y) - R_N(\tau, B) = T_N(\tau, B) + Q_N(\tau, B) + U_N(\tau, B) ,$$

we have that $\Gamma^N(\tau, Y) - R_N(\tau, B)$ converges to zero in $L_1(\nu)$. Additionally, in Part 1, we have proved that $R_N(\tau, B)$ converges to $\Gamma(\tau)$ iff $s_N \xrightarrow{c} s \leq \infty$. Then

$$\mathbb{E}[|\Gamma^N(\tau, Y) - \Gamma(\tau)|] \leq \mathbb{E}[|\Gamma^N(\tau, Y) - R_N(\tau, B)|] + |R_N(\tau, B) - \Gamma(\tau)| \to 0 ,$$

when $N \to \infty$. Therefore, $\Gamma^N(\tau, Y)$ is convergent in $L^1(\nu)$ iff $s_N \xrightarrow{c} s \leq \infty$.

The $\nu - a.s$ convergence of $\Gamma^N(\tau, Y)$ is implied by the $\nu - a.s$ convergence of T_N, Q_N and U_N, and the convergence of R_N. In this case, we obtain that $\Gamma^N(\tau, Y)$ also converges to $\Gamma(\tau)$. Notice that the convergence of s_k in the Cesaro sense is a necessary condition for the convergence $\nu - a.s.$ of Γ^N. This can be checked by considering a $\nu - a.s.$ convergent subsequence of $\{\Gamma^N\}$, and the earlier conclusion follows from the uniqueness of the limit.   □

Remark 3. It is easy to see that $\Gamma^N(\tau, Y)$ is convergent in *q.m.* if we suppose that

$$V_2(F) = \int_\Lambda F^2(\lambda)d\lambda < \infty , \tag{12}$$

and $B_N^{-4}[\chi]_{N,2} \to 0$, when $N \to \infty$, since condition (12) is equivalent to $\mathbb{E}[|\Psi_\tau(Y^i, Y^i)|^2] < \infty$ which implies that $T_N(\tau, B)$ converges to zero in L^2.

This explain why condition (12) is necessary to obtain $q.m.$ convergence of $\Gamma^N(\tau, Y)$.

Remark 4. In the strong interaction case, condition $\sum_{k=1}^{\infty} \{[\chi]_{N,2}/[\chi]_{N,1}^2\} < \infty$ is not necessary for the ν-a.s. convergence. For instance, if we consider an innovation process given by $\varepsilon_t = \varepsilon_{1,t} + \varepsilon_{2,t}$, where $\varepsilon_{1,t}$, $\varepsilon_{2,t}$ are independent Gaussian innovations with respective interaction functions $\chi_1(j) = (-1)^{j+1}$ and $\chi_2(j) = \frac{1}{j^\alpha}$. Then for $\frac{1}{2} < \alpha < 1$, taking $B_N^2 = N^{2-\alpha}$ and splitting X^N into the independent partial aggregation processes X_1^N and X_2^N, which corresponds respectively to $\varepsilon_{1,t}$ and $\varepsilon_{2,t}$, we obtain that the aggregated processes exist $\nu - a.s.$ However, $[\chi]_{N,2} \sim N^2$ and $[\chi]_{N,1} \sim N^{2-\alpha}$, therefore $\sum_{k=1}^{N} \{[\chi]_{k,2}/[\chi]_{k,1}^2\} \sim \sum_{k=1}^{N} k^{2\alpha-2} = \infty$.

Remark 5. As $\Phi(Y^i, Y^j)$, for $i \neq j$, is a symmetric and degenerated kernel, then using classical Freedholm theory we can write Φ as

$$\Phi(x,y) = \sum_{k=1}^{\infty} \lambda_k u_k(x) u_k(y) \,,$$

with $\sum_{k=1}^{\infty} \lambda_k^2 < \infty$, $\mathbb{E}[u_k(Y)] = 0$, $\mathbb{E}[u_k^2(Y)] = 1$ and $\mathbb{E}[u_k(Y)u_j(Y)] = 0$, for $k \neq j$. Then the proof of the convergence ($\nu - a.s$ and $q.m.$) for the random Toeplitz form $U_N(\tau, B)$ is equivalent to the proof of the convergence for the product kernel defined by $\tilde{\Phi}(x,y) = u_k(x)u_k(y)$. If $\hat{\chi}$ is the Fourier transform of χ, this is also equivalent to the proof of the convergence for the integrated periodogram

$$\frac{1}{B_N^2} \int_{-\pi}^{\pi} \left| \sum_{j=1}^{N} u(Y^j) e^{ij\lambda} \right| d\lambda \,,$$

for centered and second order i.i.d. random variables $\{u(Y^j), j \in \mathbb{N}\}$, see [DD83]. For a different approach see [LS01].

We will now prove that the aggregation always exists in $L^1(\nu)$ iff condition (8) is satisfied. Then under the conditions on the interaction correlation χ, given in Theorem 1, we deduce the $\nu - a.s.$ convergence. In these cases the spectral density of the aggregated process is a convex combination of F and H^2; F disappears for strong interactions, and H disappears for orthogonal innovations. The result is summarized in the following theorem.

Theorem 2 (Aggregation Convergence). *Under conditions in Theorem 1, the aggregated process X exists $\nu - a.s.$ or in $L^1(\nu)$. In this case its spectral density is given by $G(\lambda) = \{F(\lambda) + 2sH^2(\lambda)\}/\{\gamma_0 + 2s\phi_0\}$ with $G(\lambda) = H^2(\lambda)/\phi_0$ if $s = \infty$.*

Proof. Since the elementary processes Z^i are Gaussians, the convergence, for each τ, of $\Gamma^N(\tau, Y)$ to the covariance function

$$\Gamma(\tau) = \frac{\gamma(\tau) + 2s\phi(\tau)}{\gamma_0 + 2s\phi_0} \,,$$

implies the convergence of the finite dimensional distributions of the process $\{X_t^N(Y)\}$ to the Gaussian distribution with covariance function $\Gamma(\tau)$, in some sense ($L^1(\nu)$, q.m. or $\nu - a.s.$). Hence, in the discrete case we obtain that $\{X_t^N(Y)\}$ converges to a Gaussian process with covariance function $\Gamma(\tau)$ and spectral density $G(\lambda)$.

In the continuous time case it is necessary that $\Gamma^N(\tau, Y)$ converges $\nu - a.s.$ to a covariance function $\Gamma(\tau)$, for all $\tau \in \mathbb{R}$.

To see this, take a sequence $\{\psi_n\}$ such that ψ_n is a positive continuous function with compact support, and $\psi \to 1$. If $F_N(\lambda)$ denotes the Fourier transform of $\Gamma^N(\tau, Y)$, equivalently as for the discrete time case

$$\int \psi_j(\lambda)\mathrm{d}F_N(\lambda) \to \int \psi_j(\lambda)\mathrm{d}F(\lambda) \,,$$

for every j, $\nu - a.s.$. This convergence and $\int \mathrm{d}F_N(\lambda) \to \int \mathrm{d}F(\lambda)$ together imply that $\{F_n, n \in \mathbb{N}\}$ is a tight sequence and $\int \psi \mathrm{d}F_n \to \int \psi \mathrm{d}F$ for ψ in a denumerable dense set. Then every convergent subsequence has F as limit; i.e. $Fn \to F$ strongly $\nu - a.s.$, and so $\Gamma^N(\tau, Y)$ converges to $\Gamma(\tau)$ for all $\tau \in \mathbb{R}$, $\nu - a.s.$

Reciprocally, if the aggregated process X exists, in some sense, then X is a zero mean second order Gaussian process with spectral density given by a convex combination of F and H^2, in which case F satisfies condition (8). □

Proposition 1. *For independent innovations and $Y = \{Y^i, i \in \mathbb{N}\}$ any stationary ergodic process with invariant measure ν, the aggregated process X exists $\nu - a.s.$ and in $L^1(\nu)$. In this case its spectral density is given by $G(\lambda) = F(\lambda)/\gamma_0$.*

In the case of independent innovation, $\Gamma^N(\tau, Y)$ converges in $L^1(\nu)$ and $\nu - a.s.$ to $\gamma(\tau)$, for every τ. In this case, we only need the SLLN for $\{Y^i, I \in \mathbb{N}\}$ μ-integrable variables that can be chosen as any stationary ergodic process with invariant measure ν.

4 Aggregation of $AR(1)$ Processes and Long Memory

In this section we study the aggregation of random parameter $AR(1)$ processes considering dependence between individual innovations in order to show the influence of interactive innovations on the construction of LM processes.

We consider the elementary processes $Z_n^i(Y^i)$ as random parameter $AR(1)$ processes; i.e. Z_n^i has the $MA(\infty)$ expansion

$$Z_n^i(Y^i) = \sum_{k \in \mathbb{Z}} c_k(Y^i)\varepsilon_{n-k}^i \,,$$

where $\{\varepsilon_n^i, i \in \mathbb{N}, n \in \mathbb{Z}\}$ are interactive individual innovations such that $\mathbb{E}[\varepsilon_n^i \varepsilon_n^j] = \chi(|i - j|)$ with $\chi(0) = 1$ and $\mathbb{E}[\varepsilon_n^i \varepsilon_m^j] = 0$ for $n \neq m$, $\{Y^i\}$ is a sequence of i.i.d. random variables with common distribution μ on $D = (-1, 1)$ and $\{c_k(y), k \in \mathbb{Z}\}$ are the Fourier coefficients of $h(\lambda, y)$, where $h(\lambda, y)$ is the real root of the spectral density $g(\lambda, y) = \sigma^2\{1 - 2y \cos \lambda + y^2\}^{-1}$. In this case we have that the spectral densities mixture and the transfer function are given by

$$F(\lambda) = \int_{-1}^{1} \frac{\sigma^2}{1 - 2y \cos \lambda + y^2} \, d\mu(y) \ .$$

$$H(\lambda) = \int_{-1}^{1} \frac{\sigma^2}{(1 - 2y \cos \lambda + y^2)^{1/2}} \, d\mu(y) \ .$$

When μ is concentrated sufficiently near the boundary $\delta D = \{-1, 1\}$ of D, we can produce a singularity on F and on H at the frequencies 0 or π.

We take $d\mu(y) = |1 - y|^d \psi(y) dy$, where ψ is a bounded positive function supported in $[0, 1]$, continuous at $y = 1$ with $\psi(1) > 0$. Then we can verify that

i. If $-1 < d < 1$, then near $\lambda = 0$

$$F(\lambda) \sim \frac{\psi(1)}{|\lambda|^{1-d}} \int_0^{\infty} \frac{u^d}{1 + u^2} \, du \ .$$

ii. If $-1 < d < 0$, then near $\lambda = 0$

$$H(\lambda) \sim \frac{\psi(1)}{|\lambda|^{-d}} \int_0^{\infty} \frac{u^d}{(1 + u^2)^{1/2}} \, du \ .$$

From the above results follow two qualitative ways of obtaining α-LM processes, for $0 < \alpha < 1$; i.e. LM processes with spectral density $G(\lambda)$ such that $G(\lambda) \sim |\lambda|^{-\alpha}$ near $\lambda = 0$:

1. If $0 < d < 1$, we can obtain by aggregation α-LM processes with $0 < \alpha < 1$. In this case H^2 does not produce LM.
2. If $-\frac{1}{2} < d < 0$ then considering strong long interaction between innovations, we can also obtain by aggregation α-LM processes with $0 < \alpha < 1$, from H^2 contribution but for a much stronger concentration of the mixture measure near δD.

Acknowledgements

The research of L. Fermín was supported in part by a grant from the FONACIT and Proyecto Agenda Petróleo (Venezuela). We thank the referees for their valuable suggestions, which have substantially improved the presentation of the paper.

References

[Avr88] Avram, F. On bilinear forms in Gaussian random variable and Toeplitz matrices. *Probability Theory and Related Fields*, **79**, 37–45, 1988.

[DD83] Dacunha-Castelle, D., Duflo, M. *Probabilités et Statistiques* Tome 2. Masson, Paris, 1983.

[FT86] Fox, R., Taqqu, M.S. Central Limit Theorem for quadratic forms in random variables having long range dependence. *Probability Theory and Related Fields*, **74**, 213–240; 1987.

[GG88] Gonçalvez, E., Gourieroux, C. Agrégation de processus autorégressifs d'ordre 1. *Annales d'Economie et de Statistique*, **12**, 127–149, 1988.

[Gra80] Granger, C.W.J. Long Memory relationships and the aggregate of dynamic models. *Journal of Econometrics*, **14**, 227-238, 1980.

[Gri76] Grinblat, L.S. A limit theorem for measurable random processes and its application. *Proceeding American Mathematical Society*, **61**, 371–376, 1976.

[LS01] Lang, G., Soulier, Ph. Convergence de mesures spectrales aléatoires et applications à des principes d'invariance. *Statistical Inference for Stochastic Processes*, **3** (1-2), 41–51, 2000.

[Lin99] Linden, M. Time series properties of aggregated AR(1) processes with uniformly distributed coefficients. *Economics Letters*, **64**, 31–36, 1999.

[LZ98] Lippi, M., Zafferoni, P. Aggegation of simple linear dynamics: exact asymptotic results. *Econometrics Discussion Paper* 350, STICERD-LSE, 1998.

[Pet75] Petrov, V. *Sum of Independent Variables*. Springer Verlag, 1975.

[Sur03] Surgailis, D. Non CLT's: U-statistics multinomial formula and approximations of multiple Itô-Winer integrals. In *Doukhan, P., Oppenheim, G., Taqqu, M.S., (eds)* Long-Range Dependence. Birkhäuser, Boston, 130–142, 2003.

[TWS97] Taqqu, M.S., Willinger, W., Sherman, R. Proof of a fundamental results in self-similar trafic modeling. *Computer Communication Review*, **27** (2), 5–23, 1997.

Statistical Estimation and Applications

On Efficient Inference in GARCH Processes

Christian Francq[1] and Jean-Michel Zakoïan[2]

[1] Université Lille 3, GREMARS, BP 149, 59653 Villeneuve d'Ascq cedex, France
christian.francq@univ-lille3.fr
[2] Université Lille 3, GREMARS and CREST, 3 Avenue Pierre Larousse, 92245
Malakoff Cedex, France zakoian@ensae.fr

1 Introduction

The asymptotic properties of the QMLE of GARCH processes have attracted much attention in the recent years. It turned out that, contrary to what seemed to emerge from pioneering works, such estimators are consistent and asymptotically normal under fairly weak conditions on the parameter space and the true parameter value. See [1], [2], [3], [14], [12], for recent references on the QML estimation of general GARCH(p, q) models. See [20] for a recent comprehensive monograph on the estimation of GARCH models.

A serious limitation of the QMLE, however, is that some of its asymptotic properties require the existence of fourth-order moments for the underlying iid (independent and identically distributed) process (η_t). In [1] and [14] it is shown that the QMLE is not $\sqrt{n}$-consistent if $\mathbb{E}|\eta_0|^s = \infty$ with $0 < s < 4$. Another limitation of the QMLE is obviously its possible lack of efficiency when the underlying error distribution is not standard Gaussian.

This paper is concerned with efficiency properties in the framework of GARCH models. In a recent paper, [2] established the asymptotic properties of a class of estimators including the QMLE and the MLE. In the monograph [20], a chapter is devoted to ML estimation in a very general setting. In the present paper we limit ourselves to the MLE and, in Section 2, we establish its consistency and asymptotic normality, under weaker and/or more explicit conditions than those of the aforementioned references. Detailed comments on the assumptions are provided below. From a technical point of view, working with a general likelihood instead of the gaussian quasi-likelihood is far from being a trivial extension. Many problems arise from the need to put mild assumptions on the errors distribution, avoiding high-order moments. Next, we propose illustrations showing that (i) the gaussian QMLE is in general inefficient, (ii) it may reach the ML efficiency when the underlying distribution is not gaussian, (iii) the QML estimator can be improved by the so-called one-step method. In Section 3 we consider the problem of hypothesis testing based on the MLE and QMLE. The relative efficiency of Wald tests based on these two estimators is analyzed via a sequence of local alternatives. Section 4 concludes. All proofs are collected in an appendix.

For a matrix A of generic term $A(i,j)$ we use the norm $\|A\| = \sum |A(i,j)|$. The spectral radius of a square matrix A is denoted by $\rho(A)$. The symbol $\Rightarrow$ denotes the convergence in distribution. Let $x^- = \max(0,-x)$ and $x^+ = \max(0,x)$. Let λ denote the Lebesgue measure on $\mathbb{R}$.

2 Framework and main results

Let $(\epsilon_1, \ldots, \epsilon_n)$ be a realization of length n of a nonanticipative strictly stationary solution (ϵ_t) to the GARCH(p, q) model introduced by [10] and [6]:

$$\begin{cases} \epsilon_t = \sqrt{h_t}\,\eta_t \\ h_t = \omega_0 + \sum_{i=1}^{q} \alpha_{0i}\epsilon_{t-i}^2 + \sum_{j=1}^{p} \beta_{0j}h_{t-j}\,, \quad \forall t \in \mathbb{Z} \end{cases} \tag{1}$$

where $\omega_0 > 0$, $\alpha_{0i} \geq 0$ $(i = 1, \ldots, q)$, $\beta_{0j} \geq 0$ $(j = 1, \ldots, p)$, and

A1: (η_t) is a sequence of independent and identically distributed (i.i.d) random variables with a positive density f, such that for some $\delta_1 > 0$, $\sup_{y \in \mathbb{R}} |y|^{1-\delta_1} f(y) < \infty$ and $\sup_{y \in \mathbb{R}} |y|^{1+\delta_1} f(y) < \infty$.

In this paper we investigate the properties of the classical MLE, which means that f is assumed to be known. On the other hand, we do not need to assume $\mathbb{E}\eta_t = 0$ and $\mathbb{E}\eta_t^2 = 1$ (such moments may even not exist in our framework). This may seem surprising since for QML estimation, this kind of assumptions is crucial for identifiability. In the ML framework, the innovation density is supposed to be known so that every existing moment of η_t is also known. By avoiding the condition $\mathbb{E}\eta_t^2 = 1$ we allow for distributions with infinite variance (for instance the Cauchy distribution, see Comment 4 below for other examples). We then interpret the variable h_t as a conditional scale variable, instead of a conditional variance as is usually the case within the GARCH framework.

The vector of parameters is

$$\theta = (\theta_1, \ldots, \theta_{p+q+1})' = (\omega, \alpha_1, \ldots, \alpha_q, \beta_1, \ldots, \beta_p)'$$

and it belongs to a compact parameter space $\Theta \subset]0, +\infty[\times [0, \infty[^{p+q}$. The true parameter value is denoted by

$$\theta_0 = (\omega_0, \alpha_{01}, \ldots, \alpha_{0q}, \beta_{01}, \ldots, \beta_{0p})' = (\omega_0, \alpha_{0[1:q-1]}, \alpha_{0q}, \beta_{0[1:p-1]}, \beta_{0p})'\,.$$

We assume $\theta_0 \in \Theta$. Of course model (1) admits an infinity of representations parameterized by (θ_0, f), but once f has been fixed, θ_0 is uniquely determined (under Assumption **A4** below)

From [7] it is known that, assuming $\mathbb{E}\log^+ \|A_{0t}\| < \infty$, there exists a unique nonanticipative strictly stationary solution (ϵ_t) to Model (1) if and only if the sequence of matrices $\mathbf{A_0} = (A_{0t})$ has a strictly negative top Lyapunov exponent, $\gamma(\mathbf{A_0}) < 0$, where

$$
A_{0t} = \begin{pmatrix}
\alpha_{0[1:q-1]}\eta_t^2 & \alpha_{0q}\eta_t^2 & \beta_{0[1:p-1]}\eta_t^2 & \beta_{0p}\eta_t^2 \\
I_{q-1} & 0_{q-1\times1} & 0_{q-1\times p-1} & 0_{q-1\times1} \\
\alpha_{0[1:q-1]} & \alpha_{0q} & \beta_{0[1:p-1]} & \beta_{0p} \\
0_{p-1\times q-1} & 0_{p-1\times1} & I_{p-1} & 0_{p-1\times1}
\end{pmatrix},
$$

with I_k being the $k \times k$ identity matrix and $0_{k\times k'}$ being the $k \times k'$ null matrix. Note that **A1** entails $\mathbb{E}\log^+ \eta_0^2 < \infty$ and thus $\mathbb{E}\log^+ \|A_{0t}\| < \infty$, as required in [7] for the existence of $\gamma(\mathbf{A_0})$ in $\mathbb{R}\cup\{-\infty\}$. We do not make the stationarity assumption for all θ. Instead we will assume that

A2: $\gamma(\mathbf{A_0}) < 0$ and $\forall \theta \in \Theta, \ \sum_{j=1}^p \beta_j < 1$.

Note that $\sum_{j=1}^p \beta_{0j} < 1$ is implied by $\gamma(\mathbf{A_0}) < 0$. An important consequence of Assumptions **A1-A2** is that

$$
\mathbb{E}\eta_t^{2s} < \infty , \qquad \mathbb{E}h_t^s < \infty , \qquad \mathbb{E}\epsilon_t^{2s} < \infty , \qquad 0 \le s < \delta_1/2 \qquad (2)
$$

where (ϵ_t) stands for the strictly stationary solution to Model (1). The existence of a moment of order $2s$ for η_t follows directly from **A1**. The moment existence for h_t, and hence for ϵ_t, can be shown by a straightforward extension of the proofs given in [18] and [3] (Lemma 2.3) which assume $\mathbb{E}\eta_t^2 < \infty$.

Conditionally on initial values $\epsilon_0, \dots, \epsilon_{1-q}, \tilde{\sigma}_0^2, \dots, \tilde{\sigma}_{1-p}^2$, (see [12] for appropriate choices of the initial values), the likelihood is given by

$$
L_{n,f}(\theta) = L_{n,f}(\theta; \epsilon_1, \dots, \epsilon_n) = \prod_{t=1}^n \frac{1}{\tilde{\sigma}_t} f\left(\frac{\epsilon_t}{\tilde{\sigma}_t}\right),
$$

where the $\tilde{\sigma}_t^2$ are defined recursively, for $t \ge 1$, by

$$
\tilde{\sigma}_t^2 = \tilde{\sigma}_t^2(\theta) = \omega + \sum_{i=1}^q \alpha_i \epsilon_{t-i}^2 + \sum_{j=1}^p \beta_j \tilde{\sigma}_{t-j}^2 .
$$

The parameter space Θ is a compact subset of $[0, \infty[^{p+q+1}$ that bounds the first component away from zero. Namely, $\omega \ge \underline{\omega}$ for all $\theta \in \Theta$, for some positive constant $\underline{\omega}$. A MLE of θ is defined as any measurable solution $\hat{\theta}_n$ of

$$
\hat{\theta}_n = \arg\max_{\theta\in\Theta} L_{n,f}(\theta) = \arg\max_{\theta\in\Theta} \frac{L_{n,f}(\theta)}{L_{n,f}(\theta_0)} = \arg\max_{\theta\in\Theta} \tilde{Q}_{n,f}(\theta) \qquad (3)
$$

where

$$
\tilde{Q}_{n,f}(\theta) = \tilde{Q}_{n,f}(\theta; \epsilon_n, \dots, \epsilon_1) = n^{-1} \sum_{t=1}^n \tilde{\ell}_t(\theta) - \tilde{\ell}_t(\theta_0) ,
$$

and

$$
\tilde{\ell}_t(\theta) = \tilde{\ell}_t = \log f\left(\frac{\epsilon_t}{\tilde{\sigma}_t}\right) - \log \tilde{\sigma}_t .
$$

Write $h = \log f$ and $g(y) = yh'(y)$. The next conditions concern the smoothness of g.

A3: There is a $0 < C_0 < \infty$ and a $0 \leq \delta_2 < \infty$ such that $|g(y)| \leq C_0(|y|^{\delta_2} + 1)$ for all $y \in (-\infty, \infty)$.

Note that this assumption holds for the standard normal distribution (with $C_0 = 1$ and $\delta_2 = 2$). Let $\mathcal{A}_\theta(z) = \sum_{i=1}^q \alpha_i z^i$ and $\mathcal{B}_\theta(z) = 1 - \sum_{j=1}^p \beta_j z^j$. By convention, $\mathcal{A}_\theta(z) = 0$ if $q = 0$ and $\mathcal{B}_\theta(z) = 1$ if $p = 0$. Under the condition **A2**, which entails invertibility of the polynomial $\mathcal{B}_\theta(z)$, we can define $\mathcal{B}_\theta^{-1}(z)\mathcal{A}_\theta(z) = \sum_{i=1}^\infty \gamma_i z^i$ for $|z| \leq 1$. It will be convenient to consider ergodic and stationary approximations of $(\tilde{\sigma}_t^2)$ and $(\tilde{\ell}_t)$, defined by

$$\sigma_t^2 = \sigma_t^2(\theta) = \mathcal{B}_\theta^{-1}(1)\omega + \sum_{i=1}^\infty \gamma_i \epsilon_{t-i}^2 , \qquad \ell_t = \ell_t(\theta) = h\left(\frac{\epsilon_t}{\sigma_t}\right) - \log \sigma_t ,$$

and to introduce

$$Q_{n,f}(\theta) = Q_{n,f}(\theta; \epsilon_n, \ldots, \epsilon_1) = n^{-1} \sum_{t=1}^n \ell_t(\theta) - \ell_t(\theta_0) .$$

The following assumption is made for identifiability reasons.

A4: if $p > 0$, $\mathcal{A}_{\theta_0}(z)$ and $\mathcal{B}_{\theta_0}(z)$ have no common root, $\mathcal{A}_{\theta_0}(1) \neq 0$, and $\alpha_{0q} + \beta_{0p} \neq 0$.

2.1 Consistency and asymptotic normality

We are now in a position to state our first result.

Theorem 1. *Let* $(\hat{\theta}_n)$ *be a sequence of MLE satisfying (3). If* **A1-A4** *hold then*

$$\lim_{n \to \infty} \hat{\theta}_n = \theta_0 , \qquad a.s.$$

To obtain asymptotic normality of the MLE it is also necessary to replace the nonnegativity assumptions on the true parameter value by the stronger assumption

A5: $\theta_0 \in \overset{\circ}{\Theta}$, where $\overset{\circ}{\Theta}$ denotes the interior of Θ.

For the function $g^{(0)}(y) = g(y)$ and its derivatives $g^{(1)}(y) = g'(y)$ and $g^{(2)}(y) = g''(y)$, we strengthen the smoothness assumptions in **A3**.

A6: There is $0 < C_0 < \infty$ and $0 \leq \kappa < \infty$ such that $|y^k g^{(k)}(y)| \leq C_0(|y|^\kappa + 1)$ for all $y \in (-\infty, \infty)$ and such that $\mathbb{E}|\eta_t|^\kappa < \infty$ for $k = 0, 1, 2$.

The first condition of the next assumption ensures the existence of the information matrix of the scale parameter $\sigma > 0$ in the density family $\sigma^{-1}f(\cdot/\sigma)$. The second condition is a mild smoothness condition which is satisfied by all the standard densities (see Comment 4 below).

A7: $\tilde{I}_f = \int \{1 + g(y)\}^2 f(y)dy < \infty$, and $\lim_{y \to \pm\infty} y^2 f'(y) = 0$.

Theorem 2. *Let* $(\hat{\theta}_n)$ *be a sequence of MLE satisfying (3). If* **A1-A7** *hold*

$$\sqrt{n}\left(\hat{\theta}_n - \theta_0\right) \Rightarrow \mathcal{N}\left\{0, I_f^{-1}(\theta_0)\right\}, \qquad as \qquad n \to \infty,$$

where

$$I_f(\theta_0) = \frac{\tilde{I}_f}{4}\mathbb{E}\frac{1}{\sigma_t^4}\frac{\partial\sigma_t^2}{\partial\theta}\frac{\partial\sigma_t^2}{\partial\theta'}(\theta_0) .$$

In connection with Theorems 1 and 2, the following comments should be noted.

Comments: 1. The inverse of the asymptotic variance of the MLE is the product of the Fisher information $\tilde{I}_f$, evaluated at the value $\sigma = 1$, in the model $\frac{1}{\sigma}f(y/\sigma), \sigma > 0$, and a matrix depending on the sole GARCH coefficients. This means, for instance, that if a vector of GARCH parameters is difficult to estimate for a given density f, in the sense that the asymptotic variance of the MLE is large, it will also be difficult to estimate with any other distribution of the iid process. Note also that the asymptotic variance of the MLE is proportional to that of the QMLE relying on an ad hoc normal likelihood whatever the true underlying distribution of η_t.

2. Our results are closely connected with those of [2] and [20]. Our conditions are generally simpler then theirs because we limit ourselves to the maximum likelihood estimator and to standard GARCH. The results of [2] are useful because they can handle general criteria, assuming f is not necessarily the errors density in (3). On the other hand, contrary to [2], for consistency we allow θ_0 to be on the boundary of the parameter space. This point is of importance because it allows to handle cases where some components of θ_0 are null (provided **A4** is satisfied), for instance when one of the orders, p or q, is overidentified. The results of [20] are also useful because he considers a general setting, including the standard GARCH and allowing the unknown error distribution to depend on a nuisance parameter. In the general case, however, his conditions for consistency and asymptotic normality are not in closed form (see for instance his assumption (6.7)). In the particular case of student innovations, more explicit conditions are given in his Lemma 6.1.6., but it is easily seen that the density given in (7) below does not fulfill the conditions of this Lemma.

3. Our Assumptions **A1** and **A3**, required for consistency, do not impose any link between the constants δ_1 and δ_2. For instance consider a density of the form

$$f(y) = K(1 + |y|)^{-(1+\alpha)}(1 + \epsilon + \cos y^{2\gamma}) , \qquad where \qquad \alpha, \epsilon, \gamma > 0 . \qquad (4)$$

Then **A1** holds for any $\delta_1 \leq \alpha$, and **A3** holds for any $\delta_2 \geq 2\gamma$. It follows that the MLE is strongly consistent by Theorem 1. By contrast, condition (1.14) in [2] is, with our notations, in particular (2):

$$\left|\frac{\partial}{\partial t}\log\{tf(yt)\}\right| \leq C_1(y)(t^{\nu_1}+1)/t , \quad t > 0 , y \in \mathbb{R}, \qquad for\ some\ \ 0 \leq \nu_1 \leq 2s .$$

It is easily seen that this assumption does not hold true for the density in (4) with $\gamma \geq \alpha$.

4. It should be emphasized that the assumptions on the density f are mild and that they hold for all standard distributions. Easy computations show that Assumptions **A1**, **A3**, **A6** and **A7** hold for : (i) the standard Gaussian distribution, for any $\delta_1 \in (0, 1]$, $\delta_2 \geq 2$ and $\kappa \geq 2$; (ii) the two-sided exponential (Laplace) distribution, $f(y) = (1/2)e^{-|y|}$, for any $\delta_1 \in (0, 1)$, $\delta_2 \geq 1$ and $\kappa \geq 1$; (iii) the Cauchy distributions for any $\delta_1 \in (0, 1)$, $\delta_2 \geq 0$ and $\kappa < 1$; (iv) the Student distribution with parameter $\nu > 0$, for $\delta_1 = \nu$, $\delta_2 \geq 0$ and $\kappa < \nu$; (v) the density displayed in (7) below with $\delta_1 \leq 2a$, $\delta_2 \geq 2$ and $\kappa \geq 2$. For the distribution in (4), Assumption **A6** is satisfied whence $3\gamma < \alpha$.

2.2 Comparison with the QMLE

In this section we quantify the efficiency loss due to the use of the QMLE, compared with the MLE. It will be seen, in particular, that the QMLE can be efficient even for non normal densities.

Recall that the QMLE of θ is defined as any measurable solution $\hat{\theta}_n^{QML}$ of

$$\hat{\theta}_n^{QML} = \arg\min_{\theta \in \Theta} \sum_{t=1}^{n} \left\{ \frac{\epsilon_t^2}{\tilde{\sigma}_t^2(\theta)} + \log \tilde{\sigma}_t^2(\theta) \right\}, \tag{5}$$

and that, under **A2**, **A5** and the additional assumption that η_t^2 has a non-degenerate distribution with $\mathbb{E}\eta_t^2 = 1$ and $\mathbb{E}\eta_t^4 < \infty$, the QMLE is consistent and asymptotically normal with mean 0 and variance

$$(\mathbb{E}\eta_t^4 - 1) \left\{ \mathbb{E}\frac{1}{\sigma_t^4} \frac{\partial \sigma_t^2}{\partial \theta} \frac{\partial \sigma_t^2}{\partial \theta'} (\theta_0) \right\}^{-1} = \frac{(\mathbb{E}\eta_t^4 - 1)\tilde{I}_f}{4} I_f^{-1}(\theta_0). \tag{6}$$

The QMLE is not efficient since, in general, $(\mathbb{E}\eta_t^4 - 1)\tilde{I}_f > 4$. More precisely, we have the following result.

Corollary 1. *Suppose that the assumptions of Theorem 2 hold and assume that $\mathbb{E}\eta_t^2 = 1$ and $\mathbb{E}\eta_t^4 < \infty$. Then the QMLE has the same asymptotic variance as the MLE when the density of η_t is of the form*

$$f(y) = \frac{a^a}{\Gamma(a)} e^{-ay^2} |y|^{2a-1}, \quad a > 0, \quad \Gamma(a) = \int_0^\infty t^{a-1} e^{-t} dt. \tag{7}$$

When the density f of the noise η_t does not belong to the family defined by (7), the QMLE is asymptotically inefficient in the sense that

$$\text{var}_{as} \sqrt{n} \left\{ \hat{\theta}_n^{QML} - \theta_0 \right\} - \text{var}_{as} \sqrt{n} \left\{ \hat{\theta}_n - \theta_0 \right\} = \left(\frac{(\mathbb{E}\eta_t^4 - 1)\tilde{I}_f}{4} - 1 \right) I_f^{-1}(\theta_0).$$

is positive definite, when $\hat{\theta}_n$ is the MLE defined by (3).

Comments. 1. Obviously, the QMLE is efficient in the gaussian case, which corresponds to $a = 1/2$ in (7). The QMLE is also efficient when η_t follows some non gaussian distributions, which is less intuitive. As an illustration consider the density $f(y) = 4e^{-2y^2}|y|^3$, which corresponds to $a = 2$ in (7). For this distribution we have $\tilde{I}_f = 8$ and $\mathbb{E}\eta^4 = 3/2$. So we check that $(\mathbb{E}\eta_t^4 - 1)\tilde{I}_f = 4$, which entails that the QMLE is efficient.

Even if the QMLE is inefficient compared to the MLE when η_t follows a density which is not of the form (7), it should be noted that the QMLE is more robust than the MLE.

2. It is easily seen that the density in (7) is that of the variable

$$Z = \sqrt{\frac{\chi_{2a}^2}{2a}}\, u$$

for $2a$ integer, where χ_{2a}^2 has a χ^2 distribution with $2a$ degrees of freedom, and is independent of u, taking the values ± 1 with equal probabilities. For general a, the χ_{2a}^2 distribution can be replaced by a $\gamma(\frac{1}{2}, a)$.

2.3 Illustration

The loss of efficiency of the QMLE can be calculated explicitly for most standard densities f. For instance in the case of the Student distribution with ν degrees of freedom, rescaled such that they have the required zero mean and unit variance, the Asymptotic Relative Efficiency (ARE) of the MLE with respect to the QMLE is $\{1 - \frac{12}{\nu(\nu-1)}\}^{-1}$. Table 1 shows that the efficiency loss can be important.

Table 1. ARE of the MLE with respect to the QMLE: $\mathrm{var}_{as}\hat{\theta}_n^{QML} := ARE \times \mathrm{var}_{as}\hat{\theta}_n$, when $f(y) = \sqrt{\nu/\nu - 2}f_\nu(y\sqrt{\nu/\nu - 2})$, and f_ν denotes the Student density with ν degrees of freedom.

ν	5	6	7	8	9	10	20	30	∞
ARE	5/2	5/3	7/5	14/11	6/5	15/13	95/92	145/143	1

To compare the finite sample relative efficiencies of the QMLE and MLE with their ARE given in Table 1, we simulated $N = 1000$ independent samples of sizes $n = 100$ and $n = 1000$ of an ARCH(1) with true parameters values $\omega = 0.2$ and $\alpha = 0.9$. For the distribution of the noise η_t, we used the same densities as in Table 1. Table 2 summarizes the estimation results obtained for the QMLE $\hat{\theta}_n^{QML}$ and the one-step efficient estimator $\bar{\theta}_n$ defined in Theorem 3 (with $\tilde{\theta}_n = \hat{\theta}_n^{QML}$ as preliminary estimator). This table shows that the one-step estimator $\bar{\theta}_n$ may indeed be more accurate than the QMLE even in small samples. The observed efficiency gain is close to the asymptotic efficiency gain.

Table 2. Comparison of the QMLE and the efficient one-step estimator $\bar{\theta}_n$ for $N = 1000$ replications of the ARCH(1) model $\epsilon_t = \sigma_t \eta_t$, $\sigma_t^2 = \omega + \alpha \epsilon_{t-1}^2$, $\omega = 0.2$, $\alpha = 0.9$, $\eta_t \sim f(y) = \sqrt{\nu/\nu - 2} f_\nu(y\sqrt{\nu/\nu - 2})$. The AREs are estimated by taking the ratios of the RMSEs.

			$\hat{\theta}_n^{QML}$		$\bar{\theta}_n$		
ν	n	θ_0	Mean	RMSE	Mean	RMSE	$\widehat{ARE}$
5	100	$\omega = 0.2$	0.202	0.0794	0.211	0.0646	1.51
		$\alpha = 0.9$	0.861	0.5045	0.857	0.3645	1.92
	1000	$\omega = 0.2$	0.201	0.0263	0.201	0.0190	1.91
		$\alpha = 0.9$	0.897	0.1894	0.886	0.1160	2.67
	∞						2.5
6	100	$\omega = 0.2$	0.212	0.0816	0.215	0.0670	1.48
		$\alpha = 0.9$	0.837	0.3852	0.845	0.3389	1.29
	1000	$\omega = 0.2$	0.202	0.0235	0.202	0.0186	1.61
		$\alpha = 0.9$	0.889	0.1384	0.888	0.1060	1.70
	∞						1.67
20	100	$\omega = 0.2$	0.207	0.0620	0.209	0.0619	1.00
		$\alpha = 0.9$	0.847	0.2899	0.845	0.2798	1.07
	1000	$\omega = 0.2$	0.199	0.0170	0.199	0.0165	1.06
		$\alpha = 0.9$	0.899	0.0905	0.898	0.0885	1.05
	∞						1.03

2.4 One-step efficient estimator

It is possible to improve on the efficiency of the QMLE estimator without having to use optimization procedures. The technique is standard and consists in running one Newton-Raphson iteration with the QMLE, or any other $\sqrt{n}$-consistent preliminary estimator, as starting point. More precisely, we obtain the following result.

Theorem 3. *Suppose that the assumptions of Theorem 2 hold. Let $\tilde{\theta}_n$ be a preliminary estimator of θ_0 such that $\sqrt{n}(\tilde{\theta}_n - \theta_0) = O_{\mathbb{P}}(1)$, and let $\hat{I}_{n,f}$ be any weakly consistent estimator of $I_f(\theta_0)$. The so-called one-step estimator*

$$\bar{\theta}_n = \bar{\theta}_{n,f} = \tilde{\theta}_n + \hat{I}_{n,f}^{-1} \frac{1}{n} \frac{\partial}{\partial \theta} \log L_{n,f}(\tilde{\theta}_n)$$

has the same asymptotic distribution as the MLE :

$$\sqrt{n}\left(\bar{\theta}_n - \theta_0\right) \Rightarrow \mathcal{N}\left\{0, I_f^{-1}(\theta_0)\right\} \qquad as \qquad n \to \infty .$$

In practice the density f of the noise η_t is unknown so $\bar{\theta}_{n,f}$ and ML estimation are not feasible. However, one can estimate f from the standardized residuals $\hat{\eta}_t = \epsilon_t/\sigma_t(\hat{\theta}_n^{QML})$, $t = 1, \ldots, n$ (using for instance a non parametric kernel density estimator $\hat{f}$). One could wonder whether $\bar{\theta}_{n,\hat{f}}$ is an adaptive estimator, i.e. if it inherits the asymptotic optimality properties of $\bar{\theta}_{n,f}$. Adaptive estimation goes back to [19] and has been applied to GARCH models by several authors (see e.g [11], [17], [8], [16]. In particular, [8] showed that adaptive estimation of all GARCH coefficients is not possible, due to the presence of the scale parameter ω. An appropriate reparameterization of the model allows to estimate the volatility parameters (up to a scale parameter) with the same asymptotic precision as if the error distribution were known. In this sense adaptivity holds. In [13] the efficiency losses, with respect to the MLE, of the QMLE and Semi-Parametric estimators, are quantified. The simulation experiments of [11], and [8] confirm that the Semi-Parametric method is between the QMLE and MLE in terms of efficiency.

3 Application to efficient testing

Efficient tests can be derived from the efficient estimator $\hat{\theta}_n$ defined in (3). As an illustration consider the test of m linear restrictions about θ_0, defined by the null hypothesis

$$H_0 : R\theta_0 = r ,$$

where R is a known $m \times (p + q + 1)$ matrix, and r is a known $m \times 1$ vector.

3.1 Wald test based on the MLE and Wald test based on the QMLE

The Wald test based on the efficient estimator $\hat{\theta}_n$ has the rejection region $\{\mathbf{W}_{n,f} > \chi_m^2(1 - \alpha)\}$ where

$$\mathbf{W}_{n,f} = n(R\hat{\theta}_n - r)' \left(R\hat{I}_{n,f}^{-1}R'\right)^{-1} (R\hat{\theta}_n - r) ,$$

and $\chi_m^2(1 - \alpha)$ denotes the $(1 - \alpha)$-quantile of a chi-square distribution with m degree of freedom. Natural estimators of the information matrix $I_f(\theta_0)$ are

$$\hat{I}_{n,f} = -\frac{1}{n} \sum_{t=1}^{n} \frac{\partial^2}{\partial\theta\partial\theta'} \tilde{\ell}_t(\hat{\theta}_n) \quad \text{or} \quad \hat{I}_{n,f} = \frac{1}{n} \sum_{t=1}^{n} \frac{\partial}{\partial\theta} \tilde{\ell}_t(\hat{\theta}_n) \frac{\partial}{\partial\theta'} \tilde{\ell}_t(\hat{\theta}_n) .$$

One can also consider a Wald test based on the QMLE. This test has rejection region $\{\mathbf{W}_n^{QML} > \chi_m^2(1 - \alpha)\}$ where

$$\mathbf{W}_n^{QML} = n(R\hat{\theta}_n^{QML} - r)' \left(R\hat{I}_{\mathcal{N}}^{-1}R'\right)^{-1} (R\hat{\theta}_n^{QML} - r) ,$$

with

$$\hat{I}_N^{-1} = \frac{1}{n} \sum_{t=1}^{n} \left\{ \frac{\epsilon_t^2}{\tilde{\sigma}_t^2(\hat{\theta}_n)} - 1 \right\}^2 \left\{ \frac{1}{n} \sum_{t=1}^{n} \frac{1}{\tilde{\sigma}_t^4} \frac{\partial \tilde{\sigma}_t^2}{\partial \theta} \frac{\partial \tilde{\sigma}_t^2}{\partial \theta'}(\hat{\theta}_n) \right\}^{-1}$$

or any other asymptotically equivalent estimator.

If, under the null, the assumptions of Theorem 2 and those of Theorem 2.2 of [12] are satisfied (in particular $\theta_0 \in \overset{\circ}{\Theta}$), then both tests have the same asymptotic level α.

3.2 Local asymptotic powers of the two Wald tests

To compare the asymptotic powers of the two tests, let us consider local alternatives. In these sequences of local alternatives, the true value of the parameter is supposed to vary with n, and is therefore denoted by $\theta_{0,n}$:

$$H_n : \theta_{0,n} = \theta_0 + h/\sqrt{n}, \quad \text{with} \quad R\theta_0 = r \quad \text{and} \quad Rh \neq 0,$$

where $h \in \mathbb{R}^{p+q+1}$.

Consider the function

$$h \to \Lambda_{n,f}(\theta_0 + h/\sqrt{n}, \theta_0) := \log \frac{L_{n,f}(\theta_0 + h/\sqrt{n})}{L_{n,f}(\theta_0)}.$$

From [8], and [9] it is known that, under mild regularity conditions, ARCH and GARCH (1,1) processes are locally asymptotically normal (LAN) in the sense that, as $n \to \infty$

$$\Lambda_{n,f}(\theta_0 + h/\sqrt{n}, \theta_0) = h' \frac{1}{\sqrt{n}} \sum_{t=1}^{n} \frac{\partial}{\partial \theta} \ell_t(\theta_0) - \frac{1}{2} h' I_f(\theta_0) h + o_{\mathbb{P}_{\theta_0}}(1), \qquad (8)$$

with

$$\frac{1}{\sqrt{n}} \sum_{t=1}^{n} \frac{\partial}{\partial \theta} \ell_t(\theta_0) = \frac{-1}{\sqrt{n}} \sum_{t=1}^{n} \frac{1}{2\sigma_t^2} \left\{ 1 + \eta_t \frac{f'(\eta_t)}{f(\eta_t)} \right\} \frac{\partial \sigma_t^2}{\partial \theta}(\theta_0)$$
$$\Rightarrow \mathcal{N} \{0, I_f(\theta_0)\} \quad \text{under } \mathbb{P}_{\theta_0}. \qquad (9)$$

When (8) holds true, we deduce

$$\Lambda_{n,f}(\theta_0 + h/\sqrt{n}, \theta_0) = \mathcal{N} \left(-\frac{1}{2}\tau, \tau \right) + o_{\mathbb{P}_{\theta_0}}(1), \quad \tau = h' I_f(\theta_0) h.$$

The LAN property (8) entails that the MLE is locally asymptotically optimal (in the minimax sense and in various other senses, see [21] for details about LAN). As we will see in the sequel of this section, the LAN property also facilitates the computation of local asymptotic powers of tests.

Note that in the proof of Theorem 2 we have seen that

$$\sqrt{n}\left(\hat{\theta}_n - \theta_0\right) = I_f(\theta_0)^{-1} \frac{1}{\sqrt{n}} \sum_{t=1}^{n} \frac{\partial}{\partial\theta}\ell_t(\theta_0) + o_{\mathbb{P}_{\theta_0}}(1) \ . \tag{10}$$

In view of (8), (9) and (10), we have under H_0

$$\begin{pmatrix} \sqrt{n}(R\hat{\theta}_n - r) \\ \Lambda_{n,f}(\theta_0 + h/\sqrt{n}, \theta_0) \end{pmatrix} = \begin{pmatrix} RI_f^{-1/2}(\theta_0)X \\ h'I_f^{1/2}(\theta_0)X - \frac{h'I_f(\theta_0)h}{2} \end{pmatrix} + o_{\mathbb{P}_{\theta_0}}(1) \ , \tag{11}$$

where X follows a standard gaussian distribution in $\mathbb{R}^{1+p+q}$. Thus the asymptotic distribution of the vector defined in (11) is

$$\mathcal{N}\left\{ \begin{pmatrix} 0 \\ -\frac{h'I_f(\theta_0)h}{2} \end{pmatrix}, \begin{pmatrix} RI_f^{-1}(\theta_0)R' & Rh \\ h'R' & h'I_f(\theta_0)h \end{pmatrix} \right\} \quad \text{under } H_0 \ . \tag{12}$$

Le Cam's third lemma (see [21], p. 90) entails that

$$\sqrt{n}(R\hat{\theta}_n - r) \Rightarrow \mathcal{N}\left\{ Rh, RI_f^{-1}(\theta_0)R' \right\} \quad \text{under } H_n \ .$$

This result shows that the optimal Wald test of rejection region $\{\mathbf{W}_{n,f} > \chi_m^2(1-\alpha)\}$ has the asymptotic level α and the local asymptotic power

$$h \mapsto 1 - \Phi_{m,c_h}\left\{ \chi_m^2(1-\alpha) \right\} \ , \qquad c_h = h'R'\left\{ RI_f^{-1}(\theta_0)R' \right\}^{-1} Rh$$

where $\Phi_{m,c}(\cdot)$ denotes the distribution function of the noncentral chi-square distribution with m degrees of freedom and noncentrality parameter c. This test enjoys asymptotic optimalities (see [21]).

Recall that, under some regularity conditions, the QMLE satisfies

$$\sqrt{n}(\hat{\theta}_n^{QML} - \theta_0)$$

$$= 2\left\{ \mathbb{E}\frac{1}{\sigma_t^4}\frac{\partial\sigma_t^2}{\partial\theta}\frac{\partial\sigma_t^2}{\partial\theta'}(\theta_0) \right\}^{-1} \frac{-1}{\sqrt{n}}\sum_{t=1}^{n}\frac{1}{2\sigma_t^2}\left\{ 1 - \frac{\epsilon_t^2}{\sigma_t^2} \right\}\frac{\partial\sigma_t^2}{\partial\theta}(\theta_0) + o_{\mathbb{P}_{\theta_0}}(1)$$

$$= \frac{\tilde{I}_f}{2}I_f^{-1}(\theta_0)\frac{-1}{\sqrt{n}}\sum_{t=1}^{n}\frac{1}{2\sigma_t^2}\left\{ 1 - \eta_t^2 \right\}\frac{\partial\sigma_t^2}{\partial\theta}(\theta_0) + o_{\mathbb{P}_{\theta_0}}(1) \ .$$

Thus, using (8), (9), we obtain

$$\text{cov}_{as}\left\{ \sqrt{n}(\hat{\theta}_n^{QML} - \theta_0), \ \Lambda_{n,f}(\theta_0 + h/\sqrt{n}, \theta_0) \right\}$$

$$= \frac{\tilde{I}_f}{2}I_f^{-1}(\theta_0)\mathbb{E}\left\{ (1 - \eta_t^2)(1 + \eta_t\frac{f'(\eta_t)}{f(\eta_t)}) \right\}\mathbb{E}_{\theta_0}\left\{ \frac{1}{4\sigma_t^4}\frac{\partial\sigma_t^2}{\partial\theta}\frac{\partial\sigma_t^2}{\partial\theta'} \right\}h$$

$$= h + o_{\mathbb{P}_{\theta_0}}(1)$$

under H_0. So the previous arguments and Le Cam's third lemma show that

$$\sqrt{n}(R\hat{\theta}_n^{QML} - r) \Rightarrow \mathcal{N}\left\{ Rh, RI_N^{-1}(\theta_0)R' \right\} \quad \text{under } H_n.$$

Consequently the QMLE Wald test of rejection region $\{\mathbf{W}_n^{QML} > \chi_m^2(1-\alpha)\}$ has the asymptotic level α and the local asymptotic power

$$h \mapsto 1 - \Phi_{m,\tilde{c}_h}\left\{\chi_m^2(1-\alpha)\right\},$$

where

$$\tilde{c}_h = h'R'\left\{RI_N^{-1}(\theta_0)R'\right\}^{-1}Rh = \frac{4}{(\int x^4 f(x)dx - 1)\tilde{I}_f}c_h.$$

For simplicity, consider the case where $m = 1$, and f is the rescaled Student with 5 degrees of freedom, and θ_0 is such that $RI_f^{-1}(\theta_0)R' = 1$. Setting $c = Rh$, the local asymptotic powers of the optimal and QMLE Wald tests are respectively

$$c \mapsto 1 - \Phi_{1,c^2}\left\{\chi_1^2(0.95)\right\} \text{ and } c \mapsto 1 - \Phi_{1,\tilde{c}}\left\{\chi_1^2(0.95)\right\},$$

with

$$\tilde{c} = \frac{4}{(\int x^4 f(x)dx - 1)\tilde{I}_f}c^2.$$

Figure 1 compares these two local asymptotic powers. It is seen that, for this density, the Wald test based on the MLE has a considerably better power function than that based on the QMLE for c not too small and not too large.

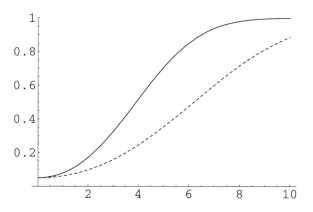

Fig. 1. Local asymptotic power of the optimal Wald test $\{\mathbf{W}_{n,f} > \chi_1^2(0.95)\}$ (full line) and local asymptotic power of the QMLE Wald test $\{\mathbf{W}_n^{QML} > \chi_1^2(0.95)\}$ (dotted line), when $m = 1$ and $f(y) = \sqrt{\nu/\nu - 2}f_\nu(y\sqrt{\nu/\nu - 2})$ with $\nu = 5$.

4 Conclusion

In this paper we established the consistency and asymptotic normality of the MLE, for the general GARCH model, under weak assumptions on the error distribution and the parameter space. In particular, contrary to the QMLE (see [1], [14]), the MLE remains $\sqrt{n}$-consistent under heavy-tailed errors. The MLE requires knowing the distribution of the errors, which is obviously a strong assumption. However, [15] showed that it is possible to fit this error distribution. Of course, caution is needed in practical uses of the MLE. Even if the efficiency loss of the QMLE compared to the MLE can be substantial, as was illustrated in this paper, the MLE fails in general to be consistent when the error distribution is misspecified.

A Appendix: Proofs

Let K and ρ be generic constants, whose values will be modified along the proofs, such that $K > 0$ and $0 < \rho < 1$.

A.1 Proof of Theorem 1

Following the scheme of proof of Theorem 2.1 in [12], this theorem will be a consequence of the following intermediate results:

$i)$ $\lim\limits_{n\to\infty} \sup\limits_{\theta\in\Theta} |Q_{n,f}(\theta) - \tilde{Q}_{n,f}(\theta)| = 0 ,$ $a.s.$

$ii)$ $\left(\exists t \in \mathbb{Z} \text{ such that } \sigma_t^2(\theta) = \sigma_t^2(\theta_0) \ \mathbb{P}_{\theta_0} \text{ a.s. }\right) \implies \theta = \theta_0 ,$

$iii)$ if $\theta \neq \theta_0$, $\mathbb{E}_{\theta_0}\{\ell_t(\theta) - \ell_t(\theta_0)\} < 0 ,$

$iv)$ any $\theta \neq \theta_0$ has a neighborhood $V(\theta)$ such that

$$\lim\sup_{n\to\infty} \ \sup_{\theta^*\in V(\theta)} \tilde{Q}_{n,f}(\theta^*) < 0 , \quad a.s.$$

We will use the following notations for σ_t^2 and its derivatives, introduced in [12],

$$\sigma_t^2 = \sum_{k=0}^{\infty} B^k(1,1)\left(\omega + \sum_{i=1}^{q} \alpha_i \epsilon_{t-k-i}^2\right) , \tag{13}$$

$$\frac{\partial \sigma_t^2}{\partial \omega} = \sum_{k=0}^{\infty} B^k(1,1), \qquad \frac{\partial \sigma_t^2}{\partial \alpha_i} = \sum_{k=0}^{\infty} B^k(1,1)\epsilon_{t-k-i}^2 , \tag{14}$$

$$\frac{\partial \sigma_t^2}{\partial \beta_j} = \sum_{k=1}^{\infty} B_{k,j}(1,1)\left(\omega + \sum_{i=1}^{q} \alpha_i \epsilon_{t-k-i}^2\right) \tag{15}$$

where

$$B_{k,j} = \frac{\partial B^k}{\partial \beta_j} = \sum_{m=1}^{k} B^{m-1} B^{(j)} B^{k-m} , \qquad B = \begin{pmatrix} \beta_1 & \beta_2 & \cdots & \beta_p \\ 1 & 0 & \cdots & 0 \\ \vdots & & & \\ 0 & \cdots & 1 & 0 \end{pmatrix} , \qquad (16)$$

and $B^{(j)}$ is a $p \times p$ matrix with $(1,j)$th element 1, and all other elements equal to zero. By **A2** and the compactness of Θ, we have

$$\sup_{\theta \in \Theta} \rho(B) < 1 . \qquad (17)$$

In the aforementioned paper, we showed that, almost surely, $\sup_{\theta \in \Theta} |\sigma_t^2 - \tilde{\sigma}_t^2| \leq K\rho^t$, $\forall t$. In view of $|\sigma_t - \tilde{\sigma}_t| = |\sigma_t^2 - \tilde{\sigma}_t^2|/(\sigma_t + \tilde{\sigma}_t)$ it follows that, almost surely,

$$\sup_{\theta \in \Theta} |\sigma_t - \tilde{\sigma}_t| \leq K\rho^t, \quad \forall t. \qquad (18)$$

Thus, using $\log x \leq x - 1$, almost surely

$$\sup_{\theta \in \Theta} n^{-1} \sum_{t=1}^{n} |\ell_t(\theta) - \tilde{\ell}_t(\theta)|$$

$$\leq n^{-1} \sum_{t=1}^{n} \sup_{\theta \in \Theta} \left\{ \left| h\left(\frac{\epsilon_t}{\sigma_t}\right) - h\left(\frac{\epsilon_t}{\tilde{\sigma}_t}\right) \right| + \left| \log\left(1 + \frac{\sigma_t - \tilde{\sigma}_t}{\tilde{\sigma}_t}\right) \right| \right\}$$

$$\leq \frac{K}{\omega} n^{-1} \sum_{t=1}^{n} \rho^t \sup_{\theta \in \Theta} \left| \frac{\epsilon_t}{\sigma_t^*} h'\left(\frac{\epsilon_t}{\sigma_t^*}\right) \right| + \frac{K}{\omega^{1/2}} n^{-1} \sum_{t=1}^{n} \rho^t$$

$$\leq K n^{-1} \sum_{t=1}^{n} \rho^t (|\epsilon_t|^{\delta_2} + 1) + K n^{-1} \qquad (19)$$

where $\sigma_t^* = \sigma_t^*(\theta)$ is between $\tilde{\sigma}_t$ and σ_t. The last inequality rests on Assumption **A3**. By the Markov inequality and (2), we deduce

$$\sum_{t=1}^{\infty} \mathbb{P}(\rho^t |\epsilon_t|^{\delta_2} > \varepsilon) \leq \sum_{t=1}^{\infty} \frac{\mathbb{E}(\rho^{2t/\delta_2} \epsilon_t^2)^s}{\varepsilon^{2s/\delta_2}} < \infty \qquad (20)$$

and thus $\rho^t |\epsilon_t|^{\delta_2} \to 0$ a.s. By the Cesaro lemma, the right-hand side of (19) tends to 0, and *i)* follows straightforwardly.

To prove *ii)*, we proceed as in [12]. It turns out that it is sufficient to prove that no exact linear combination of the $(\epsilon_{t-i}^2)_{i \geq 0}$ exists. Assume that ϵ_t^2 belongs to the σ-field generated by $\{\epsilon_{t-i}^2, i > 0\}$. It follows that

$$0 = \epsilon_t^{2s} - \mathbb{E}_{\theta_0}(\epsilon_t^{2s} | \epsilon_{t-i}, i > 0) = \sigma_t^{2s}(\theta_0)(\eta_t^{2s} - \mathbb{E}\eta_t^{2s}) \text{ with probability 1,}$$

which entails $\mathbb{P}(\eta_t^{2s} = \mathbb{E}\eta_t^{2s}) = 1$, in contradiction with the assumption that the law of η_t has a density.

Note that

$$\ell_t(\theta) - \ell_t(\theta_0) = \log \frac{\frac{1}{\sigma_t(\theta)} f\left(\frac{\epsilon_t}{\sigma_t(\theta)}\right)}{\frac{1}{\sigma_t(\theta_0)} f\left(\frac{\epsilon_t}{\sigma_t(\theta_0)}\right)}$$

is defined $\mathbb{P}_{\theta_0}$ (the distribution of the ϵ_t process)-almost surely. By Jensen we get

$$\mathbb{E}_{\theta_0}(\ell_t(\theta) - \ell_t(\theta_0) \mid \epsilon_{t-i}, i > 0) \leq \log \int \frac{1}{\sigma_t(\theta)} f\left(\frac{x}{\sigma_t(\theta)}\right) dx = 0 . \qquad (21)$$

The inequality is strict, unless if

$$\frac{1}{\sigma_t(\theta)} f\left(\frac{\epsilon_t}{\sigma_t(\theta)}\right) = \frac{1}{\sigma_t(\theta_0)} f\left(\frac{\epsilon_t}{\sigma_t(\theta_0)}\right), \qquad \mathbb{P}_{\theta_0}\text{-a.s.}$$

The latter equality would imply $\sigma_t(\theta_0)\eta_t = \sigma_t(\theta)\eta_t$, a.s, and thus, by integration $\sigma_t^{2s}(\theta_0)\mathbb{E}\eta_t^{2s} = \sigma_t^{2s}(\theta)\mathbb{E}\eta_t^{2s}$, and finally $\sigma_t(\theta_0) = \sigma_t(\theta)$. This is excluded for $\theta \neq \theta_0$ in view of $ii)$. Therefore the inequality in (21) is strict whenever $\theta \neq \theta_0$. By integration, $iii)$ follows.

Now we will show $iv)$. For any $\theta \in \Theta$ and any positive integer k, let $V_k(\theta)$ be the open ball with center θ and radius $1/k$. We have

$$\limsup_{n\to\infty} \sup_{\theta^* \in V_k(\theta)\cap\Theta} \tilde{Q}_{n,f}(\theta^*)$$

$$\leq \limsup_{n\to\infty} \sup_{\theta^* \in V_k(\theta)\cap\Theta} Q_{n,f}(\theta^*) + \limsup_{n\to\infty} \sup_{\theta\in\Theta} |Q_{n,f}(\theta) - \tilde{Q}_{n,f}(\theta)|$$

$$\leq \limsup_{n\to\infty} n^{-1} \sum_{t=1}^{n} \sup_{\theta^* \in V_k(\theta)\cap\Theta} \ell_t(\theta^*) - \ell_t(\theta_0)$$

$$:= \limsup_{n\to\infty} n^{-1} \sum_{t=1}^{n} X_{t,k}(\theta) \qquad (22)$$

where the second inequality comes from $i)$.

First we will show that, for any $\theta \in \Theta$ and for k sufficiently large, $\mathbb{E}_{\theta_0} X_{t,k}^+(\theta) < +\infty$. Let

$$\overline{f}(\epsilon) = \sup_{\sigma > \underline{\omega}} \left\{\frac{1}{\sigma} f\left(\frac{\epsilon}{\sigma}\right)\right\} .$$

Note that $\overline{f}(\epsilon) < \infty$ by Assumption **A1**. We have

$$X_{t,k}^+(\theta) = \log^+ \frac{\sup_{\theta^* \in V_k(\theta)\cap\Theta} \left\{\frac{1}{\sigma_t(\theta^*)} f\left(\frac{\epsilon_t}{\sigma_t(\theta^*)}\right)\right\}}{\frac{1}{\sigma_t(\theta_0)} f\left(\frac{\epsilon_t}{\sigma_t(\theta_0)}\right)}$$

$$\leq \log^+ \frac{\overline{f}(\epsilon_t)}{\frac{1}{\sigma_t(\theta_0)} f\left(\frac{\epsilon_t}{\sigma_t(\theta_0)}\right)} \leq K \left[\frac{\overline{f}(\epsilon_t)}{\frac{1}{\sigma_t(\theta_0)} f\left(\frac{\epsilon_t}{\sigma_t(\theta_0)}\right)}\right]^{\alpha}$$

for any $\alpha > 0$. It follows that, for any $m > 1$

$$\mathbb{E}_{\theta_0}\left(X_{t,k}^+(\theta) \mid \epsilon_{t-i}, i > 0\right)$$

$$\leq K \int \left\{\overline{f}(\epsilon)\right\}^{\alpha} \left\{\frac{1}{\sigma_t(\theta_0)} f\left(\frac{\epsilon}{\sigma_t(\theta_0)}\right)\right\}^{1-\alpha} d\lambda(\epsilon)$$

$$\leq K \left[\int \left\{\overline{f}(\epsilon)\right\}^{m\alpha} d\lambda(\epsilon)\right]^{\frac{1}{m}} \left[\int \left\{\frac{1}{\sigma_t(\theta_0)} f\left(\frac{\epsilon}{\sigma_t(\theta_0)}\right)\right\}^{\frac{m(1-\alpha)}{m-1}} d\lambda(\epsilon)\right]^{\frac{m-1}{m}}$$

$$= K \left[\int \left\{\overline{f}(\epsilon)\right\}^{m\alpha} d\lambda(\epsilon)\right]^{\frac{1}{m}} \left[\int \left\{f(\epsilon)\right\}^{\frac{m(1-\alpha)}{m-1}} d\lambda(\epsilon)\right]^{\frac{m-1}{m}} \left\{\sigma_t(\theta_0)\right\}^{\frac{m\alpha-1}{m}}. (23)$$

Choosing $\frac{1}{m} < \alpha < \frac{\delta_1 + \frac{1}{m}}{\delta_1 + 1}$, and in addition $\alpha < \frac{1}{m(1-\delta_1)}$ when $\delta_1 < 1$, we get

$$\int \left\{f(\epsilon)\right\}^{\frac{m(1-\alpha)}{m-1}} d\lambda(\epsilon)$$

$$\leq \int_{|\epsilon|\leq 1} \left\{|\epsilon|^{\delta_1-1} \sup_y |y|^{1-\delta_1} f(y)\right\}^{\frac{m(1-\alpha)}{m-1}} d\lambda(\epsilon) + \int_{|\epsilon|>1} \left\{f(\epsilon)\right\}^{\frac{m(1-\alpha)}{m-1}} d\lambda(\epsilon)$$

$$\leq K + K \int_{|\epsilon|>1} |\epsilon|^{\frac{-m(1-\alpha)(1+\delta_1)}{m-1}} d\lambda(\epsilon) < \infty, \tag{24}$$

and

$$\int \left\{\overline{f}(\epsilon)\right\}^{m\alpha} d\lambda(\epsilon) \leq \int_{|\epsilon|\leq 1} \left\{\overline{f}(\epsilon)\right\}^{m\alpha} d\lambda(\epsilon) + \int_{|\epsilon|>1} \left\{\overline{f}(\epsilon)\right\}^{m\alpha} d\lambda(\epsilon)$$

$$\leq \int_{|\epsilon|\leq 1} \left\{\frac{1}{\underline{\omega}^{\delta_1}|\epsilon|^{1-\delta_1}} \sup_{y\in\mathbb{R}} |y|^{1-\delta_1} f(y)\right\}^{m\alpha} d\lambda(\epsilon)$$

$$+ \int_{|\epsilon|>1} \left\{\frac{1}{|\epsilon|} \sup_{y\in\mathbb{R}} |y| f(y)\right\}^{m\alpha} d\lambda(\epsilon)$$

$$< \infty . \tag{25}$$

In the previous inequalities we used the expression, for all $\epsilon \neq 0$

$$\overline{f}(\epsilon) \leq \frac{1}{|\epsilon|} \sup_{|\tau|<|\epsilon|/\underline{\omega}} |\tau| f(\tau) \leq \frac{1}{\underline{\omega}^{\delta_1}|\epsilon|^{1-\delta_1}} \sup_{y\in\mathbb{R}} |y|^{1-\delta_1} f(y)$$

and

$$\overline{f}(\epsilon) \leq \frac{1}{|\epsilon|} \sup_{|\tau|<|\epsilon|/\underline{\omega}} |\tau| f(\tau) \leq \frac{1}{|\epsilon|} \sup_{y\in\mathbb{R}} |y| f(y) .$$

From (23), (24) and (25) we deduce that

$$\mathbb{E}_{\theta_0}\left(X_{t,k}^+(\theta) \mid \epsilon_{t-i}, i > 0\right) \leq K \left\{\sigma_t(\theta_0)\right\}^{\frac{m\alpha-1}{m}} .$$

Thus, because $\frac{m\alpha-1}{m} < \delta_1$, and in view of (2)

$$\mathbb{E}_{\theta_0}\{X_{t,k}^+(\theta)\} \leq K\mathbb{E}\{\sigma_t(\theta_0)\}^{\frac{m\alpha-1}{m}} < \infty . \tag{26}$$

Now we will use an ergodic theorem for stationary and ergodic processes (X_t) such that $\mathbb{E}(X_t)$ exists in $\mathbb{R} \cup \{-\infty, +\infty\}$ (see [5] p. 284 and 495). In view of (26), $\mathbb{E}_{\theta_0}\{X_{t,k}(\theta)\}$ exists and belongs to $\mathbb{R} \cup \{-\infty\}$. It follows that

$$\limsup_{n\to\infty} n^{-1} \sum_{t=1}^{n} X_{t,k}(\theta) = \mathbb{E}_{\theta_0}\{X_{t,k}(\theta)\} .$$

When k tends to infinity, the sequence $\{X_{t,k}(\theta)\}_k$ decreases to $X_t(\theta) = \ell_t(\theta) - \ell_t(\theta_0)$. Thus $\{X_{t,k}^-(\theta)\}_k$ increases to $X_t^-(\theta)$. By the Beppo-Levi theorem, $\mathbb{E}_{\theta_0} X_{t,k}^-(\theta) \uparrow \mathbb{E}_{\theta_0} X_t^-(\theta)$ when $k \uparrow +\infty$. By (26), the fact that the sequence $\{X_{t,k}^+(\theta)\}_k$ is decreasing, and the Lebesgue theorem, $\mathbb{E}_{\theta_0} X_{t,k}^+(\theta) \downarrow \mathbb{E}_{\theta_0} X_t^+(\theta)$ when $k \uparrow +\infty$. Thus we have shown that the right-hand side of (22) converges to $\mathbb{E}_{\theta_0}\{X(\theta)\}$ when $k \to \infty$. By iii) this limit is negative and iv) is proved.

Using a standard compactness argument we complete the proof of Theorem 1.

A.2 Proof of Theorem 2

Assumption **A5** and Theorem 1 entail that almost surely $\frac{\partial}{\partial\theta} \log L_{n,f}(\hat{\theta}_n) = 0$ for n large enough. In this case, a standard Taylor expansion gives

$$0 = \frac{1}{\sqrt{n}} \frac{\partial}{\partial\theta} \log L_{n,f}(\theta_0) + \frac{1}{n} \frac{\partial^2}{\partial\theta\partial\theta'} \log L_{n,f}(\theta^*)\sqrt{n}\left(\hat{\theta}_n - \theta_0\right) \tag{27}$$

where the elements of the matrix $\frac{\partial^2}{\partial\theta\partial\theta'} \log L_{n,f}(\theta^*)$ are $\frac{\partial^2}{\partial\theta_i\partial\theta_j} \log L_{n,f}(\theta_{ij}^*)$ with θ_{ij}^* between $\hat{\theta}_n$ and θ_0.

Recall that $\frac{\partial}{\partial\theta} \log L_{n,f}(\theta_0) = \sum_{t=1}^{n} \frac{\partial}{\partial\theta}\tilde{\ell}_t(\theta_0)$. We will show the following intermediate results.

a) $\lim_{n\to\infty} \frac{1}{\sqrt{n}} \left\{ \sum_{t=1}^{n} \frac{\partial}{\partial\theta}\tilde{\ell}_t(\theta_0) - \sum_{t=1}^{n} \frac{\partial}{\partial\theta}\ell_t(\theta_0) \right\} = 0 ,$ a.s.

b) $\frac{1}{\sqrt{n}} \sum_{t=1}^{n} \frac{\partial}{\partial\theta}\ell_t(\theta_0) \Rightarrow \mathcal{N}\{0, I_f(\theta_0)\} ,$ and $I_f(\theta_0)$ is invertible,

c) $\frac{1}{n} \frac{\partial^2}{\partial\theta\partial\theta'} \log L_{n,f}(\theta^*) \to -I_f(\theta_0) ,$ a.s.

The announced result will straightforwardly follow.

First mention that in [12] (see equation (4.14) and the proof of (iii)) we have shown that, under **A1**, **A2** and **A5**, for any $k \in \{1, 2, 3\}$, any $i_1, \ldots, i_k \in$

$\{1, \ldots, p+q+1\}$, and any $d \geq 1$, there exists a neighborhood $\mathcal{V}(\theta_0)$ of θ_0 such that

$$\mathbb{E}_{\theta_0} \sup_{\theta \in \mathcal{V}(\theta_0)} \frac{1}{\sigma_t^{2d}} \left| \frac{\partial^k \sigma_t^2}{\partial \theta_{i_1} \cdots \partial \theta_{i_k}} (\theta) \right|^d < \infty . \tag{28}$$

Similarly to (18), it can be shown that

$$\left\| \frac{1}{\sigma_t^2} \frac{\partial \sigma_t^2}{\partial \theta} (\theta) - \frac{1}{\tilde{\sigma}_t^2} \frac{\partial \tilde{\sigma}_t^2}{\partial \theta} (\theta) \right\| \leq K \rho^t + K \rho^t \left\| \frac{1}{\sigma_t^2} \frac{\partial \sigma_t^2}{\partial \theta} (\theta) \right\| . \tag{29}$$

Using

$$\sum_{t=1}^{n} \frac{\partial}{\partial \theta} \ell_t(\theta_0) = - \sum_{t=1}^{n} \frac{1}{2\sigma_t^2} \{1 + g(\eta_t)\} \frac{\partial \sigma_t^2}{\partial \theta} (\theta_0) , \tag{30}$$

a similar expression for $\sum_{t=1}^{n} \frac{\partial}{\partial \theta} \tilde{\ell}_t(\theta_0)$, Assumption **A6**, Inequalities (18) and (29), and a Taylor expansion similar to that used in (19), we obtain

$$\sum_{t=1}^{n} \left\| \frac{\partial}{\partial \theta} \tilde{\ell}_t(\theta_0) - \frac{\partial}{\partial \theta} \ell_t(\theta_0) \right\| \leq K \sum_{t=1}^{n} \left\| \frac{1}{\sigma_t^2} \frac{\partial \sigma_t^2}{\partial \theta} (\theta_0) \right\| |\epsilon_t|^\kappa \rho^t$$

$$+ K \sum_{t=1}^{n} \left| 1 + g \left(\frac{\epsilon_t}{\tilde{\sigma}_t} \right) \right| \rho^t \left\| \frac{1}{\sigma_t^2} \frac{\partial \sigma_t^2}{\partial \theta} (\theta_0) \right\|$$

$$= O(1) \quad a.s. \tag{31}$$

The last equality is obtained from (28) with $k = d = 1$, (20) and the expansion

$$\left| 1 + g \left(\frac{\epsilon_t}{\tilde{\sigma}_t} \right) \right| = \left| 1 + g(0) + \frac{\epsilon_t}{\tilde{\sigma}_t} g(x^*) \right| \leq 1 + |g(0)| + \frac{|\epsilon_t|^{1+\kappa}}{\omega^{1+\kappa}} ,$$

where x^* stands for some point between 0 and $\epsilon_t / \tilde{\sigma}_t$. Clearly a) is deduced from (31).

Let $\nu_t = - \frac{1}{2\sigma_t^2(\theta_0)} \{1 + g(\eta_t)\} \frac{\partial \sigma_t^2}{\partial \theta} (\theta_0)$, and let $\mathcal{F}_t$ be the σ-algebra generated by the random variables η_{t-i}, $i \geq 0$. Using **A1**, we obtain

$$\mathbb{E} g(\eta_t) = \int x f'(x) dx = \lim_{a,b \to \infty} [x f(x)]_{-b}^{a} - \int f(x) dx = -1 . \tag{32}$$

Thus

$$\mathbb{E}(\nu_t | \mathcal{F}_{t-1}) = [\mathbb{E} \{1 + g(\eta_t)\}] \frac{-1}{2\sigma_t^2(\theta_0)} \frac{\partial \sigma_t^2}{\partial \theta} (\theta_0) = 0 \quad a.s.,$$

and $(\nu_t, \mathcal{F}_t)$ is a martingale difference. Moreover (ν_t) is stationary and square integrable, with covariance matrix $\mathbb{E} \nu_1 \nu_1' = I_f(\theta_0)$. The existence of $I_f(\theta_0)$ is ensured by (28) with $k = 1$ and $d = 2$, and the first condition in **A7**. The invertibility of $\mathbb{E} \frac{1}{\sigma_t^4} \frac{\partial \sigma_t^2}{\partial \theta} \frac{\partial \sigma_t^2}{\partial \theta'} (\theta_0) = \frac{4}{\tilde{I}_f} I_f(\theta_0)$ is shown in [12] under the assumptions **A2** and **A4**. The convergence in law in b) is then obtained from the central limit theorem of [4].

A Taylor expansion yields

$$\frac{1}{n}\frac{\partial^2}{\partial\theta_i\partial\theta_j}\log L_{n,f}(\theta_{ij}^*) = \frac{1}{n}\frac{\partial^2}{\partial\theta_i\partial\theta_j}\log L_{n,f}(\theta_0)$$

$$+\frac{1}{n}\frac{\partial}{\partial\theta'}\left\{\frac{\partial^2}{\partial\theta_i\partial\theta_j}\log L_{n,f}(\tilde\theta_{ij})\right\}(\theta_{ij}^*-\theta_0)\ ,$$

where $\tilde\theta_{ij}$ is between θ_{ij}^* and θ_0 (and thus between $\hat\theta_n$ and θ_0). In view of the consistency of $\hat\theta_n$, to establish c) it suffices to show that

$$\text{d) } \lim_{n\to\infty}\frac{1}{n}\sum_{t=1}^{n}\frac{\partial^2}{\partial\theta\partial\theta'}\ell_t(\theta_0) = -I_f(\theta_0),\quad a.s.$$

$$\text{e) } \lim_{n\to\infty}\frac{1}{n}\sum_{t=1}^{n}\left|\frac{\partial^2}{\partial\theta_i\partial\theta_j}\tilde\ell_t(\theta_0) - \frac{\partial^2}{\partial\theta_i\partial\theta_j}\ell_t(\theta_0)\right| = 0\ ,\quad a.s.$$

and, for some neighborhood $\mathcal{V}(\theta_0)$ of θ_0,

$$\text{f) } \lim_{n\to\infty}\frac{1}{n}\sum_{t=1}^{n}\sup_{\theta\in\mathcal{V}(\theta_0)}\left|\frac{\partial^3}{\partial\theta_i\partial\theta_j\partial\theta_k}\tilde\ell_t(\theta) - \frac{\partial^3}{\partial\theta_i\partial\theta_j\partial\theta_k}\ell_t(\theta)\right| = 0\ ,\quad a.s.$$

$$\text{g) } \limsup_{n\to\infty}\frac{1}{n}\sum_{t=1}^{n}\sup_{\theta\in\mathcal{V}(\theta_0)}\left|\frac{\partial^3}{\partial\theta_i\partial\theta_j\partial\theta_k}\ell_t(\theta)\right| < \infty\ ,\quad a.s.$$

In view of (30) we obtain

$$\frac{\partial^2}{\partial\theta_i\partial\theta_j}\ell_t(\theta_0) = -\frac{1}{2}\{1+g(\eta_t)\}\left\{\frac{1}{\sigma_t^2}\frac{\partial^2\sigma_t^2}{\partial\theta_i\partial\theta_j}(\theta_0) - \frac{1}{\sigma_t^4}\frac{\partial\sigma_t^2}{\partial\theta_i}\frac{\partial\sigma_t^2}{\partial\theta_j}(\theta_0)\right\}$$

$$+\frac{1}{4}g'(\eta_t)\eta_t\frac{1}{\sigma_t^4}\frac{\partial\sigma_t^2}{\partial\theta_i}\frac{\partial\sigma_t^2}{\partial\theta_j}(\theta_0)\ . \tag{33}$$

Using (28) and (32), the expectation of the first summand in the right-hand side of (33) is equal to 0. In view of (32) and the last condition in **A7**,

$$\int x^2 f''(x)dx = \lim_{a,b\to\infty}[x^2 f'(x)]_{-b}^{a} - 2\int xf'(x)dx = 2\ .$$

It follows that

$$\mathbb{E}\{g'(\eta_t)\eta_t\} + \mathbb{E}\{1+g(\eta_t)\}^2 = 1 + \int\{3xf'(x)+x^2f''(x)\}\,dx = 0\ .$$

Thus $\mathbb{E}\{g'(\eta_t)\eta_t\} = -\tilde{I}_f$ and $\mathbb{E}\frac{\partial^2}{\partial\theta\partial\theta}\ell_t(\theta_0) = -I_f(\theta_0)$ in view of (33). The ergodic theorem entails d).

Using (29) and noting that the partial derivatives of $\sigma_t^2(\cdot)$ are all a.s. positive, it is easy to show that

$$\left| \frac{1}{\sigma_t^4} \frac{\partial \sigma_t^2}{\partial \theta_i}(\theta) \frac{\partial \sigma_t^2}{\partial \theta_j}(\theta) - \frac{1}{\tilde{\sigma}_t^4} \frac{\partial \tilde{\sigma}_t^2}{\partial \theta_i}(\theta) \frac{\partial \tilde{\sigma}_t^2}{\partial \theta_j}(\theta) \right|$$
$$\leq K \rho^t \left\{ 1 + \frac{1}{\sigma_t^2} \frac{\partial \sigma_t^2}{\partial \theta_i}(\theta) \right\} \left\{ 1 + \frac{1}{\sigma_t^2} \frac{\partial \sigma_t^2}{\partial \theta_j}(\theta) \right\} . \tag{34}$$

We also have

$$\left| \frac{1}{\sigma_t^2} \frac{\partial^2 \sigma_t^2}{\partial \theta_i \partial \theta_j}(\theta) - \frac{1}{\tilde{\sigma}_t^2} \frac{\partial^2 \tilde{\sigma}_t^2}{\partial \theta_i \partial \theta_j}(\theta) \right| \leq K \rho^t \left\{ 1 + \frac{1}{\sigma_t^2} \frac{\partial^2 \sigma_t^2}{\partial \theta_i \partial \theta_j}(\theta) \right\} . \tag{35}$$

Note that (33) continues to hold when $\ell_t(\theta_0)$, σ_t and η_t are replaced by $\tilde{\ell}_t(\theta_0)$, $\tilde{\sigma}_t$ and $\epsilon_t/\tilde{\sigma}_t$. Then, using (29), (34) and (35), we obtain

$$\left| \frac{\partial^2}{\partial \theta_i \partial \theta_j} \ell_t(\theta_0) - \frac{\partial^2}{\partial \theta_i \partial \theta_j} \tilde{\ell}_t(\theta_0) \right| \leq a_{1t} + a_{2t} + a_{3t} + a_{4t}$$

with

$$a_{1t} = K \rho^t \left| 1 + g(\eta_t) \right| \left[\left\{ 1 + \frac{1}{\sigma_t^2} \frac{\partial^2 \sigma_t^2}{\partial \theta_i \partial \theta_j}(\theta_0) \right\} \right.$$
$$\left. + \left\{ 1 + \frac{1}{\sigma_t^2} \frac{\partial \sigma_t^2}{\partial \theta_i}(\theta_0) \right\} \left\{ 1 + \frac{1}{\sigma_t^2} \frac{\partial \sigma_t^2}{\partial \theta_j}(\theta_0) \right\} \right] ,$$

$$a_{2t} = K \rho^t \left| g'(\eta_t) \eta_t \right| \left\{ 1 + \frac{1}{\sigma_t^2} \frac{\partial^2 \sigma_t^2}{\partial \theta_i \partial \theta_j}(\theta_0) \right\} ,$$

and, using **A6** to show

$$\left| g(\epsilon_t/\tilde{\sigma}_t) - g(\eta_t) \right| \leq K(|\epsilon_t|^\kappa + 1) \rho^t$$

and

$$\left| g'(\epsilon_t/\tilde{\sigma}_t) \epsilon_t/\tilde{\sigma}_t - g'(\eta_t) \eta_t \right| \leq K(|\epsilon_t|^\kappa + 1) \rho^t,$$

$$a_{3t} = K \rho^t (1 + |\eta_t|^\kappa) \left\{ \frac{1}{\sigma_t^2} \frac{\partial^2 \sigma_t^2}{\partial \theta_i \partial \theta_j}(\theta_0) + \frac{1}{\sigma_t^4} \frac{\partial \sigma_t^2}{\partial \theta_i}(\theta) \frac{\partial \sigma_t^2}{\partial \theta_j}(\theta) \right\} ,$$

$$a_{4t} = K \rho^t (|\epsilon_t|^\kappa + 1) \frac{1}{\sigma_t^2} \frac{\partial^2 \sigma_t^2}{\partial \theta_i \partial \theta_j}(\theta_0) .$$

Using the Borel-Cantelli lemma and the Markov inequality, as in (20), we show that $a_{kt} \to 0$ a.s. as $t \to \infty$, for $k = 1, \ldots, 4$, and e) follows. Now note that (29), (34) and (35) are valid uniformly over any neighborhood $\mathcal{V}(\theta_0) \subset \Theta$. For instance we have

$$\sup_{\theta \in \mathcal{V}(\theta_0)} \left\| \frac{1}{\sigma_t^2} \frac{\partial \sigma_t^2}{\partial \theta}(\theta) - \frac{1}{\tilde{\sigma}_t^2} \frac{\partial \tilde{\sigma}_t^2}{\partial \theta}(\theta) \right\| \leq K \rho^t + K \rho^t \sup_{\theta \in \mathcal{V}(\theta_0)} \left\| \frac{1}{\sigma_t^2} \frac{\partial \sigma_t^2}{\partial \theta}(\theta) \right\| . \tag{36}$$

where $\mathcal{V}(\theta_0)$ satisfies (28). Consequently e) can be strengthened to obtain

$$\lim_{n\to\infty} \frac{1}{n} \sum_{t=1}^{n} \sup_{\theta \in \mathcal{V}(\theta_0)} \left| \frac{\partial^2}{\partial \theta_i \partial \theta_j} \ell_t(\theta) - \frac{\partial^2}{\partial \theta_i \partial \theta_j} \tilde{\ell}_t(\theta) \right| = 0 , \quad a.s.$$

for some neighborhood $\mathcal{V}(\theta_0)$ of θ_0. A direct extension to the third order derivatives gives f).

In view of (30) and (33), the third derivatives $\frac{\partial^3}{\partial \theta_i \partial \theta_j \partial \theta_k} \ell_t(\theta)$ can be written as a sum of products of several terms involving

$$\frac{1}{\sigma_t^2} \frac{\partial \sigma_t^2}{\partial \theta_i} , \quad \frac{1}{\sigma_t^2} \frac{\partial^2 \sigma_t^2}{\partial \theta_i \partial \theta_j} , \quad \frac{1}{\sigma_t^2} \frac{\partial^3 \sigma_t^2}{\partial \theta_i \partial \theta_j \partial \theta_k} ,$$

$$g\left(\frac{\epsilon_t}{\sigma_t}\right) , \quad g'\left(\frac{\epsilon_t}{\sigma_t}\right) \frac{\epsilon_t}{\sigma_t} , \quad g''\left(\frac{\epsilon_t}{\sigma_t}\right) \frac{\epsilon_t^2}{\sigma_t^2} . \tag{37}$$

All these products contain one and only one of the last 3 terms. In view of (28), for each of the first 3 terms the sup over $\theta \in \mathcal{V}(\theta_0)$ has finite moments of any order. In [12] it is shown (see (4.25)) that for a sufficiently small $\mathcal{V}(\theta_0)$, $\sup_{\theta \in \mathcal{V}(\theta_0)} \sigma_t^2(\theta_0)/\sigma_t^2$ admits moments of all orders. Thus, using **A6** and writing $\epsilon_t/\sigma_t = \eta_t \sigma_t(\theta_0)/\sigma_t$, we have

$$\sup_{\theta \in \mathcal{V}(\theta_0)} \mathbb{E} \left| g\left(\frac{\epsilon_t}{\sigma_t}\right) \right| \le K + K \mathbb{E} |\eta_t|^\kappa \, \mathbb{E} \sup_{\theta \in \mathcal{V}(\theta_0)} \left| \frac{\sigma_t^2(\theta_0)}{\sigma_t^2} \right|^\kappa < \infty .$$

We have the same inequality when $g(\epsilon_t/\sigma_t)$ is replaced by the last two terms of (37). We deduce that

$$\mathbb{E} \sup_{\theta \in \mathcal{V}(\theta_0)} \left| \frac{\partial^3}{\partial \theta_i \partial \theta_j \partial \theta_k} \ell_t(\theta) \right| < \infty$$

and g) follows from the ergodic theorem.

A.3 Proof of Corollary 1

In view of the expression (6) of the asymptotic variance of the QMLE, it suffices to show that

$$(\mathbb{E}\eta_t^4 - 1)\tilde{I}_f \ge 4 , \tag{38}$$

with equality iff f satisfies (7). Using $\mathbb{E}\eta_t^2 = 1$ and (32), we have

$$\int (y^2 - 1) \left(1 + \frac{f'(y)}{f(y)} y\right) f(y) dy = \int y^3 f'(y) dy - \int y f'(y) dy$$

$$= \lim_{a,b\to\infty} [y^3 f(y)]_{-b}^{a} - \int 3y^2 f(y) dy + 1 = -2 .$$

Thus, the Cauchy-Schwarz inequality yields

$$4 \le \int (y^2 - 1)^2 f(y) dy \int \left(1 + \frac{f'(y)}{f(y)} y\right)^2 f(y) dy = (\mathbb{E}\eta_t^4 - 1)\tilde{I}_f$$

with equality iff there exists $a \neq 0$ such that $1+\eta_t f'(\eta_t)/f(\eta_t) = -2a\left(\eta_t^2 - 1\right)$ a.s. The latter equality holds iff $f'(y)/f(y) = -2ay + (2a-1)/y$ almost everywhere. The solution of this differential equation, under the constraint $f \geq 0$ and $\int f(y)dy = 1$, is given by (7).

A.4 Proof of Theorem 3

Using d)-g) in the proof of Theorem 2, a Taylor expansion around θ_0 gives

$$\frac{1}{\sqrt{n}}\frac{\partial}{\partial\theta}\log L_{n,f}(\tilde{\theta}_n) = \frac{1}{\sqrt{n}}\frac{\partial}{\partial\theta}\log L_{n,f}(\theta_0) - I_f(\theta_0)\sqrt{n}\left(\tilde{\theta}_n - \theta_0\right) + o_{\mathbb{P}_{\theta_0}}(1) \ .$$

Moreover

$$\sqrt{n}\left(\bar{\theta}_n - \tilde{\theta}_n\right) = \hat{I}_{n,f}^{-1}\frac{1}{\sqrt{n}}\frac{\partial}{\partial\theta}\log L_{n,f}(\tilde{\theta}_n)$$

$$= I_f^{-1}(\theta_0)\frac{1}{\sqrt{n}}\frac{\partial}{\partial\theta}\log L_{n,f}(\tilde{\theta}_n) + o_{\mathbb{P}_{\theta_0}}(1) \ .$$

Thus we have

$$\sqrt{n}\left(\bar{\theta}_n - \theta_0\right) = \sqrt{n}\left(\bar{\theta}_n - \tilde{\theta}_n\right) + \sqrt{n}\left(\tilde{\theta}_n - \theta_0\right)$$

$$= I_f^{-1}(\theta_0)\frac{1}{\sqrt{n}}\frac{\partial}{\partial\theta}\log L_{n,f}(\tilde{\theta}_n)$$

$$+ I_f^{-1}(\theta_0)\left\{\frac{1}{\sqrt{n}}\frac{\partial}{\partial\theta}\log L_{n,f}(\theta_0) - \frac{1}{\sqrt{n}}\frac{\partial}{\partial\theta}\log L_{n,f}(\tilde{\theta}_n)\right\} + o_p(1)$$

$$= I_f^{-1}(\theta_0)\frac{1}{\sqrt{n}}\frac{\partial}{\partial\theta}\log L_{n,f}(\theta_0) + o_p(1) \Rightarrow \mathcal{N}\left\{0, I_f^{-1}(\theta_0)\right\} \ ,$$

using b) in the proof of Theorem 2.

References

1. Berkes, I., Horváth, L.: The rate of consistency of the quasi-maximum likelihood estimator. Statistics and Probability Letters, **61**, 133–143 (2003)
2. Berkes, I., Horváth, L.: The efficiency of the estimators of the parameters in GARCH processes. Annals of Statistics, **32**, 633–655 (2004)
3. Berkes, I., Horváth, L., Kokoszka, P.S.: GARCH processes: structure and estimation. Bernoulli, **9**, 201–227 (2003)
4. Billingsley, P.: The Lindeberg-Levy theorem for martingales. Proceedings of the American Mathematical Society, **12**, 788–792 (1961)
5. Billingsley, P.: Probability and Measure. John Wiley, New York (1995)
6. Bollerslev, T.: Generalized autoregressive conditional heteroskedasticity. Journal of Econometrics, **31**, 307–327 (1986)
7. Bougerol, P., Picard, N.: Stationarity of GARCH processes and of some non-negative time series. Journal of Econometrics, **52**, 115–127 (1992)

8. Drost, F.C., Klaassen, C.A.J.: Efficient estimation in semiparametric GARCH models. Journal of Econometrics, **81**, 193–221 (1997)
9. Drost, F.C., Klaassen, C.A.J., Werker, B.J.M.: Adaptive estimation in time-series models. Annals of Statistics, **25**, 786–817 (1997)
10. Engle, R.F.: Autoregressive conditional heteroskedasticity with estimates of the variance of the United Kingdom inflation. Econometrica, **50**, 987–1007 (1982)
11. Engle, R.F., González-Rivera, G.: Semiparametric ARCH models. Journal of Business and Econometric Statistics, **9**, 345–359 (1991)
12. Francq, C., Zakoïan, J.M.: Maximum Likelihood Estimation of Pure GARCH and ARMA-GARCH Processes. Bernoulli, **10**, 605–637 (2004)
13. González-Rivera, G., Drost, F.C.: Efficiency comparisons of maximum-likelihood based estimators in GARCH models. Journal of Econometrics, **93**, 93–111 (1999)
14. Hall, P., Yao, Q.: Inference in ARCH and GARCH models with heavy-tailed errors. Econometrica, **71**, 285–317 (2003)
15. Koul H.K., Ling, S.: Fitting an error distribution in some heteroscedastic time series models. Forthcoming in Annals of Statistics (2005)
16. Ling, S., McAleer, M.: Adaptive estimation in nonstationry ARMA models with GARCH noises. Annals of Statistics, **31**, 642–674 (2003)
17. Linton, O.: Adaptive estimation in ARCH models. Econometric Theory, **9**, 539–564 (1993)
18. Nelson, D.B.: Stationarity and persistence in the GARCH(1,1) model. Econometric Theory, **6**, 318–334 (1990)
19. Stein, C.: Efficient nonparametric testing and estimation. Proceedings of the third Berkeley Symposium on Mathematical Statistics and Probability, **1**, 187–195 (1956)
20. Straumann, D.: Estimation in conditionally heteroscedastic time series models. Lecture Notes in Statistics, Springer Berlin Heidelberg (2005)
21. van der Vaart, A.W.: Asymptotic statistics. Cambridge University Press, Cambridge (1998)

Acknowledgement: We are very grateful to anonymous referees for their comments.

Almost sure rate of convergence of maximum likelihood estimators for multidimensional diffusions

Dasha Loukianova[1] and Oleg Loukianov[2]

[1] Département de Mathématiques, Université d'Evry-Val d'Essonne, France
`dasha@maths.univ-evry.fr`
[2] Département d'Informatique, IUT de Fontainebleau, France
`oleg@iut-fbleau.fr`

1 Introduction

Though the maximum likelihood estimation (MLE) of the drift for continuously observed diffusions is now quite a classical topic, many recent articles and monographs contribute to complete known results (see e.g. the works of Basawa and Prakasa Rao [BP80], Feigin [F76], Jankunas and Khasminskii [JKh97], Küchler and Sørensen [KS99], Dietz, Höpfner, Kutoyants [Ku03], [DKu03], [HKu03], van Zanten [Z03a, Z03b], Yoshida [Y90]). In particular, one question of interest is the rate of convergence of normalized MLE in concrete models. The large majority of papers treats the case of linear dependence on the parameter (as in [DKu03, JKh97, HKu03, KS99]) or that of one-dimensional ergodic diffusions ([L03, Z03a, Z03b]). In these frameworks many important results concerning the rate of convergence *in probability* have been obtained. This rate depends on the entropy of the parameter set and the regularity of diffusion coefficients, and is deterministic. An extensive survey of the state-of-art in this domains can be found in the book [Ku03] by Kutoyants.

In the present article we contribute to two less known aspects of this problem: firstly, we are interested in *almost sure* rates of convergence; secondly, we provide a unified treatment for multidimensional recurrent diffusions, including the ergodic but also less studied null-recurrent case.

Let us give a rough formulation of the problem. Suppose that Θ is a compact metric space with a distance $d(\theta, \theta')$, such that the entropy of Θ grows at most exponentially. This is the case for all compacts in $\mathbb{R}^d$, but also for some interesting functional spaces (see [VW96] for a recent treatment of the topic). Consider a diffusion X_t in $\mathbb{R}^n$ given by

$$dX_t = \sigma(X_t)\,dB_t + b(\theta_0, X_t)\,dt,$$

where B_t is a Brownian motion, and put $a = \sigma\sigma^*$. Suppose that $\theta_0 \in \Theta$ is an unknown parameter to be found. It is well-known that any maximum likelihood estimator $\hat\theta_t$ of θ_0 can be written as

$$\hat\theta_t = \arg\sup_{\theta\in\Theta}\left(M_t(\theta) - \frac{1}{2}A_t(\theta)\right),$$

where

$$M_t(\theta) = \int_0^t (b_\theta - b_{\theta_0})^* a^{-1}\sigma(X_s^{\theta_0})\,\mathrm{d}W_s$$

are martingales under the law $\mathbb{P}$ of X_t, with $M_t(\theta_0) = 0$, and $A_t(\theta) = \langle M(\theta)\rangle_t$ are their quadratic variations.

Recall that an upper rate of convergence of $\hat\theta_t$ is a "maximal" process r_t such that $r_t d(\hat\theta_t, \theta_0) < \infty$ in some sense. Namely, two important notions of boundedness can be considered:

$$\mathbb{P}\left(\limsup_{t\to\infty} r_t d(\hat\theta_t, \theta) = \infty\right) = 0 \qquad \text{(a.s. rate)}$$

and

$$\lim_{M\to\infty}\limsup_{t\to\infty}\mathbb{P}\left(r_t d(\hat\theta_t,\theta) > M\right) = 0 \qquad \text{(rate in probability)}$$

A standard way (that we also follow) to obtain a lower bound for r_t is exposed in [VW96]. Its main idea is to controll the properly normalized families $M_t(\theta)$ and $A_t(\theta)$ by some functions of $d(\theta,\theta_0)$ *uniformly* in t. Indeed, since $M_t(\theta_0) = 0$, any MLE $\hat\theta_t$ must satisfy $2M_t(\hat\theta_t) \geq A_t(\hat\theta_t)$. Bounding $M_t(\hat\theta_t)$ from above and $A_t(\hat\theta_t)$ from below in terms of $d(\hat\theta_t,\theta_0)$ will then provide an estimate for $d(\hat\theta_t,\theta_0)$.

To gain such control, one firstly imposes some kind of regularity (almost sure or in the mean) on the normed family $\langle M(\theta) - M(\theta')\rangle_t$. In the present article we restrict ourselves to the Hölder framework, since all the applications we have in mind belong to it. Namely, we suppose that there exist two continuous processes U_t and $V_t \nearrow \infty$ such that $\liminf_{t\to\infty} U_t/V_t > 0$ and for t great enough

$$\forall\theta\in\Theta,\ d^{2\kappa}(\theta,\theta_0)U_t \leq \langle M(\theta) - M(\theta_0)\rangle_t$$
$$\text{and}\quad \forall(\theta,\theta')\in\Theta^2,\ \langle M(\theta) - M(\theta')\rangle_t \leq d^{2\delta}(\theta,\theta')V_t \quad \text{(H)}$$

for some constants $\delta \leq 1$ and $\kappa \geq \delta$. As we shall see below, these properties are rather natural and easy to check in many situations, and the first of them provides the required lower bound on $A_t(\theta)$.

The second (and more difficult) step is to deduce from (H) corresponding normalizer and upper bound (or modulus of continuity) for $M_t(\theta)$. The main difficulty comes from the fact that this bound has to be uniform in t, which typically implies its dependence on the entropy of Θ.

When studying the convergence in probability, this task can be achieved by Nishiyama maximal inequality [N99], further developed in van Zanten [Z05] (see also the forthcoming paper [VZ05] by van der Vaart and van Zanten). Without entering into the details, let us just note that for $\Theta \subset \mathbb{R}^d$ and Hölder framework as above, one obtains that $|M_t(\theta) - M_t(\theta')|/\sqrt{V_t} \sim d^\delta(\theta, \theta')$ in the mean.

To apply this general scheme in our almost sure setting, we need to derive an a.s. modulus of continuity of normalized martingales $M_t(\theta)$. Since we have not found a suitable result in the literature[3], this is done in section 2, theorem 1. Namely, we show that under the hypothesis (H), $\mathbb{P}$-a.s. for t great enough

$$\forall (\theta, \theta') \in \Theta^2, \ \frac{|M_t(\theta) - M_t(\theta')|}{\sqrt{V_t \ln \ln V_t}} \leq d^\delta(\theta, \theta') l(d(\theta, \theta'))$$

for some decreasing function $l(x)$ depending on the entropy of Θ.

Two points can be noted by comparison with the corresponding modulus in the mean. The first one is the presence of $\ln \ln V_t$ term, which is natural since we consider the almost sure convergence and the best normalizer is there given by the law of the iterated logarithm.

The second point is that even for $\Theta \subset \mathbb{R}^d$, our Kolmogorov-type proof does not allow to get rid of the additional factor $l(x)$. In fact, one can take $l(x) = \sqrt{\ln(1/x)}$ in this case, which recalls the Lévy modulus of continuity for Brownian motion (see [RY94, p. 30]). On the other hand, $l(x)$ is unnecessary in some simple situations (for example, if $M_t(\theta) = \theta \cdot M_t$), so it would be interesting to know whether in general theorem 1 holds with $l(x) = \text{const}$ for $\Theta \subset \mathbb{R}^d$.

Once the necessary moduli of continuity are obtained, we use them to deduce an a.s. rate of convergence r_t in the way outlined above. Assuming the hypothesis (H), we show in section 3 (theorem 2) that r_t can be found from

$$r_t^{2\kappa - \delta} l\left(\frac{1}{r_t}\right) = \sqrt{\frac{V_t}{\ln \ln V_t}} \,,$$

where $l(x)$ is the function from Theorem 1. This kind of equation (without $\ln \ln V_t$ factor) is known for the rate of convergence in probability (see [VW96, Theorem 3.2.5]), so theorem 2 can be considered as its almost sure counterpart.

Finally, in section 4 we apply theorem 2 to find an upper rate of convergence of MLE in some concrete examples. From what precedes, given a diffusion

$$dX_t = \sigma(X_t) \, dB_t + b(\theta_0, X_t) \, dt, \quad \theta_0 \in \Theta \,,$$

---

[3] A uniform law of the iterated logarithm was already used by van de Geer and Stougie in [GS91] to find an a.s. rate of convergence, but in some rather different context.

we only have to check that the hypothesis (H) holds. The verification is based on the fact that $\langle M(\theta) - M(\theta') \rangle_t$ can be expressed in terms of b and σ. To simplify the formulations, suppose that $X_t \in \mathbb{R}$ and $\sigma(x) = 1$, then

$$\langle M(\theta) - M(\theta') \rangle_t = \int_0^t (b(\theta, X_s) - b(\theta', X_s))^2 \, \mathrm{d}s \,.$$

An obvious way to fulfill the necessary conditions is to suppose that the function $b(\theta, x)$ satisfies for all $(\theta, \theta', x) \in \Theta^2 \times \mathbb{R}$,

$$K(x) d^\kappa(\theta, \theta') \le |b(\theta, x) - b(\theta', x)| \le C(x) d^\delta(\theta, \theta') \qquad (*)$$

for some non-negative $C(x)$ and $K(x)$. Then we can take $U_t = \int_0^t K^2(X_s) \, \mathrm{d}s$ and $V_t = \int_0^t C^2(X_s) \, \mathrm{d}s$ and, as soon as

$$\lim_{t \to \infty} \int_0^t C^2(X_s) \, \mathrm{d}s = \infty \quad \text{and} \quad \liminf_{t \to \infty} \frac{\int_0^t K^2(X_s) \, \mathrm{d}s}{\int_0^t C^2(X_s) \, \mathrm{d}s} > 0 \,, \qquad (**)$$

theorem 2 applies.

An important framework where $(**)$ is easy to check is that of Harris recurrent diffusions. Recall that $X_t \in \mathbb{R}^n$ is called recurrent if it admits an invariant measure μ such that $\mu(f) = \int f(x) \mu(\mathrm{d}x) > 0$ implies $\int_0^\infty f(X_s) \, \mathrm{d}s = \infty$ a.s. for every μ-integrable $f(x) \ge 0$. According to the case $\mu(\mathbb{R}^n) < \infty$ or $\mu(\mathbb{R}^n) = \infty$ the diffusion is called ergodic or null-recurrent.

Assuming that X_t is recurrent, suppose that $C^2(x)$ and $K^2(x)$ are both integrable and positive with respect to μ. This μ-integrability condition is generally satisfied for ergodic diffusions (since μ is finite), and is not really restrictive for null-recurrent ones. Then $V_t \nearrow \infty$, and the Chacon-Ornstein theorem (see e.g. [RY94]) insures that the ratio of two integrals in $(**)$ converges to $\mu(K^2)/\mu(C^2) > 0$, as required. Thus for recurrent diffusions with a Hölder drift we obtain r_t as a function of an integrable additive functional (IAF) V_t. All IAF of a recurrent diffusion have the same order of growth and express thereby a natural random scale for the rate of convergence of MLE. In particular, for ergodic diffusions any IAF is equivalent to t, so the rate r_t becomes deterministic, as usual. For example, one obtains $r_t \sqrt{\ln r_t} \sim \sqrt{t / \ln \ln t}$ for $\Theta \subset \mathbb{R}^d$ and a Lipschitz drift. For null-recurrent diffusions r_t preserves a random form, but depends only on trajectory, so it can be calculated from observations.

The scope of application of theorem 2 is not limited by the assumption $(*)$, which is often too restrictive. Using some kind of "uniform Chacon-Ornstein theorem", we can find the rate of convergence for diffusions with discontinuous drift (example 3). As noticed before, Θ can also be a functional space, provided that its entropy is not too great. In particular, we are able to treat an equation

$$\mathrm{d}X_t = \theta(X_t) \, \mathrm{d}t + \mathrm{d}B_t$$

if X_t is recurrent (exemple 5).

The rate of convergence of MLE for null-recurrent diffusions is the main statistical contribution of our paper, since there are only few known results on this topic (see [DKu03, HKu03, LL05]). Nevertheless, theorem 2 applies also in some transient cases (example 1).

2 Regularity of martingale families

Let Θ be a bounded metric space with a distance d. Consider a filtered probability space $\boldsymbol{\Omega} = (\boldsymbol{\Omega}, \mathcal{F_t}, \mathbb{P})$ and a family of real processes $\{M_t(\theta);\ t \geq 0,\ \theta \in \Theta\}$ on Ω such that for all θ the process $\{M_t(\theta)\}$ is a continuous local martingale starting at zero. For every martingale $M_t(\theta)$ denote by $A_t(\theta)$ its quadratic variation. Denote also by $M_t(\theta, \theta')$ and $A_t(\theta, \theta')$ respectively the martingale $M_t(\theta) - M_t(\theta')$ and its quadratic variation.

Suppose there exist a constant $\delta \in\ ]0, 1]$, a random variable $\tau(\omega) \geq 0$ and an adapted continuous increasing process $V_t \nearrow \infty$ such that almost surely

$$\forall (\theta, \theta') \in \Theta^2,\ \forall t > \tau, \quad A_t(\theta, \theta') \leq d^{2\delta}(\theta, \theta')V_t .$$

In this section we prove a uniform law of the iterated logarithm for $M_t(\theta)$ and find a modulus of continuity of the family $\left\{ \frac{M_t(\theta)}{\sqrt{V_t \ln \ln V_t}} \right\}$.

Let us start with some auxiliary results. Without loss of generality we suppose $\mathrm{diam}(\Theta) \leq 1$. Let $m \in \mathbb{N}$ and cover Θ by open balls of radius 2^{-m}. Amongst all such coverings choose a minimal one and define Θ_m as the set of its centers and $N(2^{-m})$ as the number of points in Θ_m. We suppose that Θ has a finite entropy, i.e. $N(2^{-m}) < \infty$ for all m. Note that Θ_0 is reduced to a single point. Let also

$$\Psi_m = \bigcup_{l=0}^{m} \Theta_l \quad \text{and} \quad \Psi = \bigcup_{l=0}^{\infty} \Theta_l .$$

For $m > 0$ define a projection $\pi : \Theta_m \mapsto \Theta_{m-1}$ such that $\pi(\theta)$ is the center of a ball of radius $2^{-(m-1)}$ covering θ. Actually, we consider here a family of projections (one for each m) all denoted by π, for notational simplicity. Clearly, $d(\theta, \pi(\theta)) < 2^{-(m-1)}$ for $\theta \in \Theta_m$.

Now construct the projection $\theta \to \theta_p$ of Ψ to Ψ_p in the following way. Let m be the smallest integer such that $\theta \in \Theta_m$. If $p \geq m$ then put $\theta_p = \theta$, else define θ_p by the well-known chaining rule:

$$\theta \to \pi(\theta) \to \ldots \to \pi^{m-p}(\theta) = \theta_p .$$

By the triangle inequality we have

$$d(\theta, \theta_p) \leq d(\theta, \pi(\theta)) + \ldots + d(\pi^{m-p-1}(\theta), \pi^{m-p}(\theta))$$

$$< \frac{1}{2^{m-1}} + \ldots + \frac{1}{2^p} < \frac{2}{2^p} . \tag{1}$$

Note that this bound is obviously true also for $p \geq m$.

Let $\varphi : [0, 10] \mapsto \mathbb{R}^+$ be a continuous function such that for all $0 < t \leq 10$ and some constants $c(t)$

$$\varphi(x) \text{ is increasing, and } \forall x \in [0,1], \ \varphi(tx) \leq c(t)\varphi(x), \text{ with } c\left(\frac{1}{2}\right) < 1 . \quad (\varPhi)$$

Such $\varphi(x)$ will play the role of modulus of continuity of normalized martingale families in the sequel. Namely, we will be interested in the functions of the form $\varphi(x) = x^\delta \cdot l(x)$, where $l(x)$ is decreasing. Note that this form implies that $\varphi(tx) \leq t^\delta \varphi(x)$ and $l(tx) \geq t^{-\delta} l(x)$ for $t \geq 1$. On the other hand, if $\varphi(x)/x^\gamma$ is increasing for some $\gamma > 0$, then $\varphi(x)$ is increasing too and $\varphi(tx) \leq t^\gamma \varphi(x)$ for $t \leq 1$. This provides a simple way to construct a function satisfying the condition $(\varPhi)$. Some examples are $\varphi(x) = x^\gamma$ or $\varphi(x) = x^\gamma \ln(c/x)$ for $\gamma > 0$ and c large enough.

The following classical type lemma will be used to extend the φ-continuity from "near" points of $\varPsi_m$ on all $\varPsi$.

Lemma 1. *Let $\varphi(x)$ satisfy $(\varPhi)$ and let $f : \Theta \mapsto \mathbb{R}$ be a function such that for all $m \geq 1$*

$$\forall (\theta, \theta') \in \varPsi_m^2, \quad d(\theta, \theta') < 5 \cdot 2^{-m} \Rightarrow |f(\theta) - f(\theta')| \leq \varphi(d(\theta, \theta')) .$$

Then there exists a constant C such that

$$\forall (\theta, \theta') \in \varPsi^2, \quad |f(\theta) - f(\theta')| \leq C\varphi(d(\theta, \theta')) .$$

Proof. We only have to consider the case $\theta \neq \theta'$. Take $(\theta, \theta') \in \varPsi^2$, $\theta \neq \theta'$ and let p be such that $2^{-(p+1)} < d(\theta, \theta') \leq 2^{-p}$. Then (1) implies

$$d(\theta_p, \theta'_p) \leq d(\theta_p, \theta) + d(\theta, \theta') + d(\theta', \theta'_p) < 5 \cdot 2^{-p} ,$$

hence

$$|f(\theta_p) - f(\theta'_p)| \leq \varphi\left(5 \cdot 2^{-p}\right) \leq \varphi(10 \cdot d(\theta, \theta')) \leq c(10) \cdot \varphi(d(\theta, \theta')) . \quad (2)$$

Now write

$$|f(\theta) - f(\theta')| \leq |f(\theta) - f(\theta_p)| + |f(\theta_p) - f(\theta'_p)| + |f(\theta'_p) - f(\theta')| . \quad (3)$$

Suppose that $\theta \in \Theta_m$ where m is minimal. The first term of the last sum equals zero if $m \leq p$, otherwise note that $\pi^l(\theta)$ and $\pi^{l+1}(\theta)$ belong to $\varPsi_{m-l}$ and that $d(\pi^l(\theta), \pi^{l+1}(\theta)) < 2 \cdot 2^{-(m-l)}$ for $0 \leq l < m - p$, so

$$\begin{aligned}
|f(\theta) - f(\theta_p)| &\leq |f(\theta) - f(\pi(\theta))| + \ldots + |f(\pi^{m-p-1}(\theta)) - f(\pi^{m-p}(\theta))| \\
&\leq \varphi(2^{-(m-1)}) + \ldots + \varphi(2^{-p}) \\
&\leq \left(c^{m-p-1}(1/2) + \ldots + c(1/2) + 1\right) \varphi(2^{-p}) \\
&\leq \frac{c(2)}{1 - c(1/2)} \varphi(d(\theta, \theta')) . \quad (4)
\end{aligned}$$

The same holds for $|f(\theta') - f(\theta_p')|$, so we get from (3), (2) and (4)

$$|f(\theta) - f(\theta')| \le C\varphi(d(\theta, \theta')) ,$$

where $C = c(10) + 2c(2)/(1 - c(1/2))$.   □

Our first theorem concerns the uniform φ-continuity of martingale families. The Kolmogorov-like proof is typical for the situation when we want to obtain a uniform continuity property of the paths using similar property of the moments (see e.g. Senoussi [Se00]). A notable exception is that we use a random normalization and prove an equicontinuity property on the infinite time interval.

Theorem 1. *Suppose there exist a constant $\delta \in\ ]0,1]$, a random variable $\tau(\omega) \ge 0$ and an adapted continuous increasing process $V_t \nearrow \infty$ such that almost surely*

$$\forall(\theta, \theta') \in \Theta^2, \ \forall t > \tau, \quad A_t(\theta, \theta') \le d^{2\delta}(\theta, \theta')V_t .$$

Let $\varphi(x) = x^\delta \cdot l(x)$ satisfy the condition (Φ), where $l(x)$ is decreasing and such that

$$\sum_{m \ge 1} m^2 N^2(2^{-m}) \exp\left(-l^2(2^{-m})\right) < \infty . \tag{5}$$

Then there exists a version $\tilde{M}_t(\theta)$ of $M_t(\theta)$ such that for almost all $\omega \in \Omega$ there is a $t(\omega)$ such that the family $\left\{\frac{\tilde{M}_t(\theta)}{\sqrt{V_t \ln \ln V_t}}\right\}_{t > t(\omega)}$ of functions of θ is $C\varphi$-continuous on Θ for some constant C depending only on φ. In particular, $\tilde{M}_t(\theta)$ is continuous in θ for all $t > t(\omega)$.

Proof. Put $c = 4e \cdot 5^{2\delta}/l^2(1/2)$. We start by showing the "local" $\sqrt{c}\varphi$-regularity of $M_t/\sqrt{V_t \ln \ln V_t}$. Denote by $\tilde{\Psi}_m^2$ the set of all couples $(\theta, \theta') \in \Psi_m^2$ such that $0 < d(\theta, \theta') < 5 \cdot 2^{-m}$. Let $\tau_n = \inf\{t : V_t = e^n\}$ and

$$\mathbb{A}_n = \left\{\forall(\theta, \theta'), \ A_{\tau_n}(\theta, \theta') \le d^{2\delta}(\theta, \theta')e^n\right\} ,$$

so $\{\tau < \tau_n\} \subseteq \mathbb{A}_n$. For $n \ge 1$ denote

$$\mathbb{B}_n = \left\{\exists m \ge 1, \ \exists(\theta, \theta') \in \tilde{\Psi}_m^2, \quad \sup_{[\tau_{n-1}, \tau_n[} \frac{|M_t(\theta, \theta')|}{\sqrt{V_t \ln \ln V_t}} > \sqrt{c}\varphi(d(\theta, \theta'))\right\} \cap \mathbb{A}_n$$

and let $\mathbb{P}(A; B; \ldots) = \mathbb{P}(A \cap B \cap \ldots)$. Recall the Bernstein inequality for continuous local martingales (see e.g. [RY94])

$$\mathbb{P}\left(\sup_{t \le \tau} |M_t| \ge z; \langle M \rangle_\tau \le L\right) \le 2 \exp\left(-\frac{z^2}{2L}\right) .$$

We have for n great enough

$$\mathbb{P}(\mathbb{B}_n) \leq \sum_{m \geq 1} \sum_{\tilde{\Psi}_m^2} \mathbb{P} \left(\sup_{[\tau_{n-1}, \tau_n[} |M_t(\theta, \theta')| > \varphi(d(\theta, \theta')) \sqrt{ce^{n-1} \ln \ln e^{n-1}} \; ; \; \mathbb{A}_n \right)$$

$$\leq \sum_{m \geq 1} \sum_{\tilde{\Psi}_m^2} 2 \exp \left(-\frac{ce^{n-1} \ln(n-1)}{2e^n d^{2\delta}(\theta, \theta')} \cdot d^{2\delta}(\theta, \theta') l^2(d(\theta, \theta')) \right)$$

$$\leq 2 \sum_{m \geq 1} \sum_{\tilde{\Psi}_m^2} \exp \left(-\frac{4e \cdot 5^{2\delta}}{2e \cdot l^2(1/2)} \ln(n-1) \cdot 5^{-2\delta} l^2(2^{-m}) \right)$$

$$\leq 2 \sum_{m \geq 1} m^2 N^2(2^{-m}) \exp \left(-l^2(2^{-m}) \right) \left(\frac{e}{(n-1)^{2/l^2(1/2)}} \right)^{l^2(2^{-m})}$$

$$\leq \frac{2 \exp \left(l^2(1/2) \right)}{(n-1)^2} \sum_{m \geq 1} m^2 N^2(2^{-m}) \exp \left(-l^2(2^{-m}) \right) .$$

The bound on the second line is due to the Bernstein inequality and the third one to the fact that $l(d(\theta, \theta')) \geq l(5 \cdot 2^{-m}) \geq 5^{-\delta} l(2^{-m})$. On the fourth line we use the bound Card $\tilde{\Psi}_m^2 \leq m^2 N^2(2^{-m})$ (see the remark 2 below). Finally, the fifth one is valid as soon as $(n-1)^{2/l^2(1/2)} > e$.

By the assumption (5), the m-series converges and we get $\mathbb{P}(\mathbb{B}_n) \leq C(n-1)^{-2}$. So the n-series converges too, and by the Borel-Cantelli lemma, $\mathbb{P}(\limsup \mathbb{B}_n) = 0$.

Since $\tau_n \to +\infty$, for almost all $\omega \in \Omega$ and all t greater than some $t(\omega)$ we have:

$$\forall m \geq 1, \; \forall (\theta, \theta') \in \tilde{\Psi}_m^2, \quad \frac{|M_t(\theta) - M_t(\theta')|}{\sqrt{V_t \ln \ln V_t}} \leq \sqrt{c} \varphi(d(\theta, \theta')) .$$

Here ω is fixed and for a given t the ratio $M_t(\theta)/\sqrt{V_t \ln \ln V_t}$ is a real function of $\theta \in \Theta$. By the lemma 1 the family $\left\{ \frac{M_t(\theta)}{\sqrt{V_t \ln \ln V_t}} \right\}_{t > t(\omega)}$ is then $C\varphi$-continuous on Ψ with some constant C not depending on t.

Now, since Ψ is dense in Θ, the existence of a version claimed in the theorem follows by classical technique, taking

$$\tilde{M}_t(\theta) = \lim_{\substack{\theta' \to \theta \\ \theta' \in \Psi}} M_t(\theta')$$

for t large enough. $\square$

Remark 1. Using $a\varphi(x)$ (with $a > 0$) instead of $\varphi(x)$, we can replace the entropy condition (5) by

$$\sum_{m \geq 1} m^2 N^2(2^{-m}) \exp \left(-a^2 l^2(2^{-m}) \right) < \infty .$$

Remark 2. Since $l^2(\varepsilon) \leq c\varepsilon^{-2\delta}$, the entropy condition (5) can be satisfied only by compacts with $N(\varepsilon)$ bounded by $\exp(\text{const} \cdot \varepsilon^{-\alpha})$, $\alpha \leq 2\delta$. This is the case for any compact of $\mathbb{R}^d$, since $N(\varepsilon) \sim \varepsilon^{-d}$. There are also interesting functional spaces satisfying this condition for some $\alpha < 2$ (see examples in section 4).

In fact, the bound $\text{Card}\,\tilde{\Psi}_m^2 \leq \text{Card}\Psi_m^2 \leq (mN(2^{-m}))^2$ used in the proof may seem too rough; for example, if $\Theta \subset \mathbb{R}^d$, it could be replaced by $\text{const} \cdot N(2^{-m})$. But for the cases we are interested in, namely when $N(\varepsilon) \leq \exp(\text{const} \cdot \varepsilon^{-\alpha})$, the terms $(mN(2^{-m}))^2$ and $N(2^{-m})$ are of the same order compared with $\exp\left(-l^2(2^{m\beta})\right)$, so we do not need such a refinement.

Let us formulate a result that will be referred to in the sequel. It concerns the modulus of continuity $\varphi(x)$ in theorem 1.

Corollary 1. *Let Θ be a metric compact as above.*

- *If $\Theta \subset \mathbb{R}^d$ then we can take $\varphi(x) = x^\delta \sqrt{\ln(c/x)}$ with c large enough;*
- *If $N(\varepsilon) \sim \exp(c\varepsilon^{-\alpha})$ with $0 < \alpha < 2\delta$ then $\varphi(x) = x^{\delta-\alpha/2}$.*

The proof is a direct application of remark 1 with an appropriate choice of $a\varphi(x)$.

3 Upper rate of convergence of MLE

Consider a metric compact Θ and let X_t^θ be a family of n-dimensional diffusions given by

$$\mathrm{d}X_t^\theta = \sigma(X_t^\theta)\,\mathrm{d}B_t + b(X_t^\theta, \theta)\,\mathrm{d}t\ , \quad X_0^\theta = x \in \mathbb{R}^n\ , \quad \theta \in \Theta\ , \tag{6}$$

where B_t is a k-dimensional Brownian motion. We suppose that for each $\theta \in \Theta$ the functions b and σ satisfy the usual assumptions for the existence of a weak solution on $[0, +\infty[$ (see e.g. [RY94]); we put $a = \sigma\sigma^*$ supposed positive definite.

If the true value of θ is unknown and one observes a trajectory $(X_s^\theta, s \leq t)$ of the solution of (6), one can estimate θ by the maximum likelihood method. In this section we show the existence of such estimator and give its rate of convergence in terms of some stochastic process written explicitly as a function of the coefficients of (6).

Take $\Omega = C([0, \infty[\to \mathbb{R}^n)$ and let $\mathcal{F}$ be its borelian σ-field, and $(\mathcal{F}_t)$ its natural filtration. Denote $\mathbb{P}_\theta^x$ the law of the solution of (6) issued from $x \in \mathbb{R}^n$. The measures $\mathbb{P}_\theta^x$ are locally absolutely continuous w.r.t. the law $\mathbb{P}^x$ of the solution of

$$\mathrm{d}X_t = \sigma(X_t)\,\mathrm{d}B_t\ , \quad X_0 = x\ .$$

Let $L_t^{\theta,x}$ be the local density of $\mathbb{P}_\theta^x$ w.r.t. $\mathbb{P}^x$:

$$L_t^{\theta,x} = \exp\left[\int_0^t b_\theta^* a^{-1}\,\mathrm{d}X_s - \frac{1}{2}\int_0^t b_\theta^* a^{-1} b_\theta\,\mathrm{d}s\right]\ ,$$

where $a = a(X_s)$ and $b_\theta = b(\theta, X_s)$. We suppose that $\forall (\theta, x)$ the process $L_t^{\theta, x}$ is given and for fixed (x, t) consider $L_t^{\theta, x}(\omega)$ as a function of θ and of the continuous trajectory $\omega \to X_s(\omega)$, $s \leq t$.

Denote by θ_0 the (unknown) true value of the parameter. The maximum likelihood estimator $\hat{\theta}_t$ of θ_0 is defined as a maximizer of the random map $\theta \to L_t^{\theta, x}$ provided it exists. ¿From now on we omit the superscript x, all the results being true for all $x \in \mathbb{R}^n$.

Notice that the maximizer of $\theta \to L_t^\theta$ coincides with that of $\theta \to L_t^\theta / L_t^{\theta_0}$. The last function is more convenient for showing the existence and studying the properties of MLE. Indeed, we have under $\mathbb{P}_{\theta_0}$

$$L_t^\theta / L_t^{\theta_0} = \exp \left(\int_0^t (b_\theta - b_{\theta_0})^* a^{-1} \sigma(X_s^{\theta_0}) \, dW_s \right.$$
$$\left. - \frac{1}{2} \int_0^t (b_\theta - b_{\theta_0})^* a^{-1} (b_\theta - b_{\theta_0})(X_s^{\theta_0}) \, ds \right) = \exp \left(M_t(\theta) - \frac{1}{2} A_t(\theta) \right),$$

where W_t is a Brownian motion under $\mathbb{P}_{\theta_0}$,

$$M_t(\theta) = \int_0^t (b_\theta - b_{\theta_0})^* a^{-1} \sigma(X_s^{\theta_0}) \, dW_s$$

and

$$A_t(\theta) = \langle M(\theta) \rangle_t = \int_0^t (b_\theta - b_{\theta_0})^* a^{-1} (b_\theta - b_{\theta_0})(X_s^{\theta_0}) \, ds .$$

In particular,

$$\hat{\theta}_t = \arg \sup_{\theta \in \Theta} \left(M_t(\theta) - \frac{1}{2} A_t(\theta) \right),$$

and since $M(\theta_0) = 0$, we see that $M_t(\hat{\theta}_t) - \frac{1}{2} A_t(\hat{\theta}_t) \geq 0$.

Recall that $A_t(\theta, \theta')$ denotes the quadratic variation of $M_t(\theta) - M_t(\theta')$:

$$A_t(\theta, \theta') = \langle M(\theta) - M(\theta') \rangle_t = \int_0^t (b_\theta - b_{\theta'})^* a^{-1} (b_\theta - b_{\theta'})(X_s^{\theta_0}) \, ds . \quad (7)$$

The foregoing theorem establishes an upper rate of convergence of MLE provided that $A_t(\theta, \theta')$ can be controlled by $d(\theta, \theta')$.

Theorem 2. *Consider a family of diffusions given by (6). Suppose that there are two adapted continuous positive processes U_t and V_t, two constants $\delta \in \,]0, 1]$ and $\kappa \geq \delta$, a function $\varphi(x) = x^\delta \cdot l(x)$ satisfying the condition (Φ), where $l(x)$ is decreasing and a random variable $\tau = \tau(\omega)$ such that the following points hold $\mathbb{P}_{\theta_0}$-almost surely:*

- *The families $M_t(\theta)$ and $A_t(\theta)$ are continuous in θ for $t > \tau$, and*

$$\frac{|M_t(\theta)|}{\sqrt{V_t \ln \ln V_t}} \leq \varphi(d(\theta, \theta_0)) . \quad (8)$$

- *For all $t > \tau$ and $\theta \in \Theta$,*

$$d^{2\kappa}(\theta, \theta_0)U_t \leq A_t(\theta) . \tag{9}$$

- $V_t \nearrow \infty$ *and* $\liminf_{t \to \infty} U_t/V_t > 0.$

  Let r_t be a positive increasing process such that for t large enough

$$r_t^{2\kappa}\varphi\left(\frac{1}{r_t}\right) = r_t^{2\kappa-\delta}l\left(\frac{1}{r_t}\right) \leq \sqrt{\frac{V_t}{\ln\ln V_t}} . \tag{10}$$

Then, $\mathbb{P}_{\theta_0}$-a.s. a MLE exists for t large enough and, if $\mathbb{P}_{\theta_0}$ is complete

$$\mathbb{P}_{\theta_0}\left(\limsup_{t\to\infty} r_t(\omega)d(\hat{\theta}_t, \theta_0) = \infty\right) = 0 .$$

Proof. By our assumptions $L_t^\theta/L_t^{\theta_0}$ is continuous on Θ and since Θ is compact, a MLE $\hat{\theta}_t$ exists for t great enough.

Let us estimate the distance $d(\hat{\theta}_t, \theta_0)$. Denote

$$S_{t,j}(\omega) = \left\{\theta : 2^j < r_t d(\theta, \theta_0) \leq 2^{j+1}\right\} \subset \Theta .$$

Let $J = J(\omega) \in \mathbb{N}$ be a random variable. We have for fixed t

$$\left\{r_t d(\hat{\theta}_t, \theta_0) > 2^J\right\} = \left\{\exists j \geq J, \ 2^j < r_t d(\hat{\theta}_t, \theta_0) \leq 2^{j+1}\right\} \subseteq$$

$$\left\{\exists j \geq J, \ \sup_{S_{t,j}}\left(\frac{M_t(\theta)}{V_t} - \frac{1}{2}\frac{A_t(\theta)}{V_t}\right) \geq 0\right\} \subseteq$$

$$\left\{\exists j \geq J, \ \sup_{S_{t,j}}\frac{M_t(\theta)}{V_t} \geq \inf_{S_{t,j}}\frac{1}{2}\frac{A_t(\theta)}{V_t}\right\} .$$

By the first two assumptions for all $t > \tau$,

$$\frac{|M_t(\theta)|}{\sqrt{V_t \ln\ln V_t}} \leq \varphi(d(\theta, \theta_0)) \quad \text{and} \quad A_t(\theta) \geq d^{2\kappa}(\theta, \theta_0)U_t . \tag{11}$$

Let $T_1 = T_1(\omega)$ be such that

$$\forall t > T_1, \quad \frac{U_t}{V_t} \geq l = l(\omega) > 0 . \tag{12}$$

Finally, let T_2 be such that (10) holds for $t > T_2$.

Put $T = \max(\tau, T_1, T_2)$ and denote B a set of the full probability such that (12) and (11) hold for $t > T$. Let also

$$B_T = \left\{\omega \in B : \ \exists t > T, \ r_t d(\hat{\theta}_t, \theta_0) > 2^J\right\} .$$

Then

$$
B_T \subseteq \left\{ \omega \in B : \exists t > T, \ \exists j \geq J, \quad \sup_{S_{t,j}} \frac{M_t(\theta)}{V_t} \geq \frac{1}{2} \inf_{S_{t,j}} \frac{A_t(\theta)}{V_t} \right\}
$$

$$
\subseteq \left\{ \exists t > T, \ \exists j \geq J, \quad \sup_{S_{t,j}} \varphi(d(\theta, \theta_0)) \sqrt{\frac{\ln \ln V_t}{V_t}} \geq \inf_{S_{t,j}} d^{2\kappa}(\theta, \theta_0) \frac{U_t}{V_t} \right\}
$$

$$
\subseteq \left\{ \exists t > T, \ \exists j \geq J, \quad \varphi(2^{j+1}/r_t) \sqrt{\frac{\ln \ln V_t}{V_t}} \geq (2^j/r_t)^{2\kappa} \cdot l \right\}
$$

$$
\subseteq \left\{ \exists t > T, \ \exists j \geq J, \quad C(\omega) r_t^{2\kappa} \varphi(1/r_t) \sqrt{\frac{\ln \ln V_t}{V_t}} \geq 2^{(2\kappa - \delta)j} \right\}
$$

$$
\subseteq \left\{ \exists j \geq J, \quad C(\omega) \geq 2^{(2\kappa - \delta)j} \right\} = \left\{ C(\omega) \geq 2^{(2\kappa - \delta)J} \right\}.
$$

For an appropriate choice of $J(\omega)$, the last set is empty. Hence, if $\mathbb{P}_{\theta_0}$ is complete,

$$
\mathbb{P}_{\theta_0} \left(\exists t > T, \ r_t d(\hat{\theta}_t, \theta_0) > 2^J \right) = 0
$$

and the theorem follows.   □

Remark 3. The assumption (8) of theorem 2 can be replaced by

$$
\forall t > \tau, \ \forall (\theta, \theta') \in \Theta^2, \quad A_t(\theta, \theta') \leq V_t d^{2\delta}(\theta, \theta'), \tag{8'}
$$

and then deduced from it using theorem 1 for an appropriate choice of $\varphi(x)$ (see corollary 1). Generally, (8') is much easier to check than (8), but in some cases the function $\varphi(x)$ obtained from theorem 1 is not optimal (see the linear example below).

4 Examples

Before proceeding to examples, let us present two classical frameworks where (some of) the assumptions of theorem 2 are easy to check. For the sake of simplicity, denote by X_t the diffusion $X_t^{\theta_0}$ corresponding to the true parameter θ_0. Recall that we consider a multidimensional diffusion

$$
dX_t = b(\theta, X_t) \, dt + \sigma(X_t) \, dW_t, \quad \theta \in \Theta,
$$

such that $a(x) = \sigma(x)\sigma^*(x)$ is invertible.

Hölder drift

Suppose that

$$
\forall (\theta, \theta', x) \in \Theta^2 \times \mathbb{R}^n, \quad |b(\theta, x) - b(\theta', x)| \leq C(x) d^\delta(\theta, \theta') \tag{13}
$$

for some borelian $C(x) > 0$. Further, suppose that for some $K(x) \geq 0$ and $\kappa \geq \delta$ it satisfies the following distinguishability condition:

$$\forall (\theta, x) \in \Theta \times \mathbb{R}^n, \quad |b(\theta, x) - b(\theta_0, x)| \geq K(x) d^\kappa (\theta, \theta_0) . \tag{14}$$

Typically, one can take

$$C(x) = \sup_{\Theta^2} \frac{|b(\theta, x) - b(\theta', x)|}{d^\delta (\theta, \theta')} \quad \text{and} \quad K(x) = \inf_{\Theta^2} \frac{|b(\theta, x) - b(\theta', x)|}{d^\kappa (\theta, \theta')} .$$

As easily seen from (7), the assumptions (8') and (9) then hold with

$$V_t = \int_0^t C^2 |a^{-1}|(X_s) \, ds \quad \text{and} \quad U_t = \int_0^t K^2 |a|^{-1} (X_s) \, ds ,$$

where $|\cdot|$ denotes the usual matrix norm: $|a| = \sup_{|u|=1} u^* a u$. Note that U_t and V_t are here additive functionals of X_t.

Recurrent diffusion

Suppose that $X_t \in \mathbb{R}^n$ is Harris recurrent, with invariant measure μ, so $\mu(f) > 0$ implies $\int_0^\infty f(X_s) \, ds = \infty$ for every measurable $f(x) \geq 0$.

Suppose that U_t and V_t are *integrable* additive functionals of X_t, say $U_t = \int_0^t f(X_s) \, ds$ and $V_t = \int_0^t g(X_s) \, ds$, with $0 < \mu(f), \mu(g) < \infty$. For example, if the drift is Hölder in θ as above, it means that $K^2 |a|^{-1}$ and $C^2 |a^{-1}|$ have to be μ-integrable. Then it follows immediately that $V_t \nearrow \infty$, and $\lim_{t \to \infty} U_t / V_t = \mu(f) / \mu(g)$ by Chacon-Ornstein limit ratio theorem, hence the third assumption of theorem 2 holds in this case.

In fact, the pointwise inequalities (13) and (14) are too restrictive in some recurrent examples (a drift with a single cusp or jump). Since $A_t(\theta, \theta') / V_t$ converges to some limit $f(\theta, \theta')$ by Chacon-Ornstein theorem, we can replace it by the Hölder assumption on $f(\theta, \theta')$, provided the convergence is uniform in (θ, θ').

Concrete examples

The first group concerns the conventional parametric estimation, when $\Theta \subset \mathbb{R}^d$. Note that in this case the assumptions (8') and $V_t \nearrow \infty$ imply by theorem 1 that (8) holds for $\varphi(x) = x^\delta \sqrt{\ln(c/x)}$ (see corollary 1). In particular, the continuity in θ of (some version of) $M_t(\theta)$ follows from the continuity of φ at zero.

The second group of examples belongs to the framework of semi-parametric estimation, since the compact Θ therein is infinite-dimensional. Its covering numbers $N(\varepsilon)$ will satisfy $N(\varepsilon) \leq \exp (a \cdot \varepsilon^{-\alpha})$ with $\alpha < 2\delta$, so we will take $\varphi(x) = cx^{\delta - \alpha/2}$ in theorem 2.

Example 1 (Linear model). Consider a one-dimensional diffusion

$$dX_t = \theta \cdot b(X_t) + dW_t, \quad \theta \in \Theta \subset \mathbb{R}.$$

We have in this case

$$M_t(\theta) = (\theta - \theta_0) \int_0^t b(X_s)\,dW_s \quad \text{and} \quad A_t(\theta) = (\theta - \theta_0)^2 \int_0^t b^2(X_s)\,ds.$$

According to the remark above, the assumptions of theorem 2 are satisfied with $\delta = k = 1$, $U_t = V_t = \int_0^t b^2(X_s)\,ds$ and $\varphi(x) = x\sqrt{\ln(c/x)}$ as soon as $V_t \to \infty$. But the obtained modulus $\varphi(x)$ is not optimal. In fact, if we put $M_t = \int_0^t b(X_s)\,dW_s$ then $V_t = \langle M \rangle_t$, and

$$\frac{M_t(\theta) - M_t(\theta')}{\sqrt{V_t \ln \ln V_t}} = (\theta - \theta')\frac{M_t}{\sqrt{V_t \ln \ln V_t}} \sim (\theta - \theta')$$

by the law of the iterated logarithm for t great enough. Hence one can take $\varphi(x) = x$ and we get $r_t \sim \sqrt{V_t / \ln \ln V_t}$.

Actually, this is a trivial result, since one finds easily that $\hat{\theta}_t - \theta_0 = M_t/V_t$, so $d(\hat{\theta}_t, \theta_0) \preceq \sqrt{\ln \ln V_t/V_t}$, which shows that the rate r_t above is optimal.

In some cases one can calculate V_t explicitly. For example, consider the one-dimensional Ornstein-Uhlenbeck process

$$dX_t = \theta X_t\,dt + dW_t.$$

It is easy to see that X_t is ergodic for $\theta < 0$, null-recurrent for $\theta = 0$ and transient for $\theta > 0$. Also, it is well known that $X_t \sim e^{\theta t} Z_\theta$ when $\theta > 0$, where $Z_\theta = \int_0^\infty e^{-\theta t}\,dW_t$. So we get

$$V_t \sim \begin{cases} t & \text{if } \theta < 0 \\ \int_0^t W_s^2\,ds & \text{if } \theta = 0 \\ e^{2\theta t} & \text{if } \theta > 0 \end{cases}$$

which shows that theorem 2 applies for any Θ and gives the upper rate of convergence of MLE.

Another example is given by a one-dimensional diffusion

$$dX_t = \theta|X_t|^\alpha dt + dW_t$$

with $0 < \alpha < 1$. In this case X_t is transient for any $\theta \neq 0$, and $|X_t| \sim t^{\frac{1}{1-\alpha}}$ (see [DKu03]). So we find

$$V_t \sim \begin{cases} t^{\frac{1+\alpha}{1-\alpha}} & \text{if } \theta \neq 0 \\ \int_0^t W_s^{2\alpha}\,ds & \text{if } \theta = 0 \end{cases}$$

which gives (up to $\ln \ln V_t$, due to the almost sure convergence) the rate of [DKu03]. Note that the consistency of MLE in these examples does not depend on the nature (recurrent or transient) of X_t, but the rate does. As we see, it is the best in the transient case and the worst in the ergodic one.

Example 2 (Recurrent diffusions with Hölder drift). Suppose that $\Theta \subset \mathbb{R}^d$ and that $X_t \in \mathbb{R}^n$ is Harris recurrent, with invariant measure μ. Recall that according to the case $\mu(\mathbb{R}^n) < \infty$ or $\mu(\mathbb{R}^n) = \infty$ the diffusion is said ergodic or null-recurrent.

Suppose that the coefficient $b(\theta, x)$ satisfies (13) and (14), where the functions $C^2(x)|a^{-1}(x)|$ and $K^2(x)|a(x)|^{-1}$ are μ-integrable and $\mu(K^2|a|^{-1}) > 0$. Then it follows from the discussion above that theorem 2 applies, and the rate we obtain satisfies

$$r_t^{2\kappa - \delta} \sqrt{\ln(cr_t)} = \sqrt{\frac{V_t}{\ln \ln V_t}} \, .$$

It is easy to see that V_t in this case can be replaced by any integrable positive additive functional. In particular, if X_t is ergodic then every additive functional is equivalent to t. If $b(\theta, x)$ is Lipschitz ($\delta = \kappa = 1$), we obtain $r_t \sqrt{\ln r_t} \sim \sqrt{t/\ln \ln t}$, which is slightly worse then the expected rate $r_t = \sqrt{t/\ln \ln t}$ known for the case of linear dependence, but we do not know if one can get rid of $\ln r_t$ factor in general.

The assumptions of μ-integrability can seem restrictive for null-recurrent diffusions, since μ is infinite. But consider a one-dimensional diffusion $dX_t = b(X_t)\,dt + dW_t$ and put $B(x) = \int_0^x b(y)\,dy$. It is recurrent if and only if the function

$$S(x) = \int_0^x \exp(-2B(y))\,dy$$

is a space transformation of $\mathbb{R}$: $\lim_{x \to \pm \infty} S(x) = \pm \infty$. If $b(x)$ depends on θ, typically it means that $\exp(-2B(x))$ grows at least as $1/x^{1-\varepsilon}$ for the values of θ such that X_t is recurrent, which gives $b(x) \preceq c/x$. Since in this case $d\mu/dx = 2\exp(2B(x)) \preceq x^{1-\varepsilon}$, we see that $b^2(x)$ is μ-integrable. Now if $b(x)$ is Hölder on Θ, the order of its Hölder coefficient $C(x)$ is in general the same as $b(x)$, so $C^2(x)$ is μ-integrable too.

To illustrate this informal reasoning, consider the following SDE:

$$dX_t = \frac{\theta X_t}{1 + \theta^2 + X_t^2}\,dt + dW_t \, .$$

It is easy to see that its solution is recurrent if and only if $\theta \le 1/2$ and null-recurrent for $|\theta| \le 1/2$. A simple calculation shows that $\delta = \kappa = 1$ and $C(x) \sim K(x) \sim (1 + |x|)^{-1}$ for $|x|$ great enough. If $\theta_0 \in]-\infty, 1/2[$, then $\mu(dx) \sim (1 + |x|)^{2\theta_0}\,dx$, so the integrability assumptions are fulfilled and theorem 2 applies. But it is not clear what happens in the "border" case $\theta_0 = 1/2$ as well as for $\theta_0 > 1/2$ when X_t becomes transient.

Example 3 (Recurrent diffusion with discontinuous drift). We borrow this example from [Ku03, p. 270]. Consider a diffusion with a switching drift

$$dX_t = -\operatorname{sgn}(X_t - \theta)\,dt + dW_t, \quad \theta \in [a, b] \subset \mathbb{R} \, .$$

It is easy to see that X_t is ergodic with invariant measure $\mu(dx) = \exp(-2|x - \theta|)\,dx$. By the ergodic theorem

$$\frac{1}{t} A_t(\theta, \theta') = \frac{4}{t} \int_0^t \mathbb{1}_{[\theta, \theta']}(X_s) \, ds \to 4 \int \mathbb{1}_{[\theta, \theta']}(x) \mu(dx) = f(\theta, \theta') \ .$$

The function $f(\theta, \theta')$ satisfies $K|\theta-\theta'| \le f(\theta, \theta') \le C|\theta-\theta'|$ for some constants $0 < K < C$. Hence, if we prove that the convergence above is uniform in (θ, θ'), then the assumptions of theorem 2 will hold with $\kappa = \delta = 1/2$, $U_t = Kt/2$ and $V_t = 2t$. The proof follows immediately from the uniform ergodic lemmas 4.1 and 4.2 in [Z99], which claim that

$$\sup_\theta \left| \frac{1}{t} \int_0^t \mathbb{1}_{]-\infty, \theta]}(X_s) \, ds - \int \mathbb{1}_{]-\infty, \theta]}(x) \mu(dx) \right| \to 0 \quad \text{as } t \to \infty \ .$$

So, an upper rate of convergence in this model can be found from $r_t \ln r_t \sim t / \ln \ln t$.

Example 4 (Signal transform). Suppose that Θ is a set of real functions on $[0, 1]$ such that

$$\forall \theta \in \Theta, \ \sup_{0 \le s \le 1} |\theta(s)| + \sup_{0 \le s < t \le 1} \frac{|\theta(t) - \theta(s)|}{|t - s|^\alpha} \le 1 \tag{15}$$

for some $\alpha \in]1/2, 1]$, and

$$\forall (\theta_1, \theta_2) \in \Theta^2, \ (\exists s \in [0, 1], \ \theta_1(s) < \theta_2(s)) \Rightarrow (\forall s \in [0, 1], \ \theta_1(s) \le \theta_2(s))$$

Then the covering numbers $N_\infty(\varepsilon)$ of Θ in the uniform norm satisfy

$$N_\infty(\varepsilon) \le \exp\left(\mathbf{const} \cdot \varepsilon^{-\frac{1}{\alpha}}\right)$$

(see [VW96]). Since $\| \cdot \|_1 \le \| \cdot \|_\infty$ on Θ, we have $N_1(\varepsilon) \le N_\infty(\varepsilon)$. Hence the assumption (5) is satisfied for $l(x) = cx^{-1/(2\alpha)}$ (see corollary 1).

Now consider the following family of SDE:

$$dX_t = dW_t + b(\theta, X_t)dt \ ,$$

where $b(\theta, x) = \int_0^1 \theta(s)\Psi(s, x) \, ds$ with $0 < m \le \Psi(s, x) \le M$. This equation can be viewed as describing a dynamical system, where $\theta(s)$ is an unknown input signal transformed by a known filter Ψ. Then

$$|b(\theta, x) - b(\theta_1, x)| \le \int_0^1 |\theta(s) - \theta_1(s)| \Psi(s, x) \, ds \le M \|\theta - \theta_1\|_1$$

and

$$|b(\theta, x) - b(\theta_0, x)| \ge m \|\theta - \theta_0\|_1 \ ,$$

so it is easy to check that all assumptions of theorem 2 are satisfied for $l(x)$ above, with $V_t = M^2 t$, $U_t = m^2 t$, $d(\theta, \theta_1) = \|\theta - \theta_1\|_1$ and $\delta = \kappa = 1$.

Hence the MLE for this model is consistent and its a.s. upper rate of convergence is found from $cr_t \cdot r_t^{1/(2\alpha)} = \sqrt{\frac{t}{\ln \ln t}}$, that is $r_t \sim \left(\frac{t}{\ln \ln t}\right)^{\frac{1}{2+1/\alpha}}$.

Example 5 (Infinite-dimensional parametric space). Consider a diffusion

$$\mathrm{d}X_t = \theta(X_t)\,\mathrm{d}t + \mathrm{d}W_t\ ,$$

where $\theta(x)$ is an unknown function to be estimated. Suppose that any candidate $\theta(x)$ is supported by $[0,1]$ and satisfies (15) for some $\alpha > 1/2$, and let Θ be the set of all such θ. Then X_t is null-recurrent for any $\theta \in \Theta$ and its invariant measure μ satisfies

$$0 < m_\theta \le \mathrm{d}\mu/\mathrm{d}x \le M_\theta$$

for some constants.

Denote by μ_0 the invariant measure corresponding to the unknown true parameter θ_0. The metrics on Θ we will consider is $d(\theta, \theta_1) = \|\theta - \theta_1\|_{L^2(\mu_0)}$. We have

$$m_{\theta_0}\|\theta\|^2_{L^2(\mathrm{d}x)} \le \|\theta\|^2_{L^2(\mu_0)} \le M_{\theta_0}\|\theta\|^2_{L^2(\mathrm{d}x)} \le \text{const}$$

and

$$\int_0^1 |\theta(x+h) - \theta(x)|^2\,\mathrm{d}x \le \int_0^1 h^{2\alpha}\,\mathrm{d}t = h^{2\alpha},$$

which implies by M. Riesz theorem that Θ is compact in $L^2(\mu_0)$. Moreover, since $\|\cdot\|_{L^2(\mu_0)} \le c\|\cdot\|_\infty$, we have $N(\Theta, \varepsilon, d) \le N(\Theta, c\varepsilon, \|\cdot\|_\infty) \sim \exp\left(A\varepsilon^{-1/\alpha}\right)$, so the entropy condition (5) is satisfied for $l(x) = cx^{-1/(2\alpha)}$.

Now let us find two processes U_t and V_t such that (4) holds. In our case

$$A_t(\theta, \theta_1) = \int_0^t (\theta - \theta_1)^2(X_s)\,\mathrm{d}s\ .$$

Put

$$Z_t = \int_0^t \mathbb{1}_{[0,1]}(X_s)\,\mathrm{d}s\ ,$$

then $A_t(\theta, \theta_1)$ and Z_t are integrable additive functionals of X_t, hence by Chacon-Ornstein theorem

$$\forall(\theta, \theta_1),\quad \frac{A_t(\theta, \theta_1)}{Z_t} \to \frac{d^2(\theta, \theta_1)}{\mu_0([0,1])}\quad \text{a.s.}$$

We will show that this convergence is uniform on Θ^2. Choose a $\|\cdot\|_\infty$-dense countable subset Θ' of Θ such that the convergence above almost surely holds for all $(\theta, \theta_1) \in \Theta' \times \Theta'$. We have

$$\frac{|A_t(\theta, \theta_1) - A(\psi, \psi_1)|}{Z_t} = \frac{1}{Z_t}\left|\int_0^t ((\theta - \theta_1)^2 - (\psi - \psi_1)^2)(X_s)\,\mathrm{d}s\right|$$

$$\le \frac{1}{Z_t}\int_0^t |\theta - \theta_1 - \psi + \psi_1| \cdot |\theta - \theta_1 + \psi - \psi_1|(X_s)\,\mathrm{d}s$$

$$\le (\|\theta - \psi\|_\infty + \|\theta_1 - \psi_1\|_\infty)\frac{\int_0^t 4 \cdot \mathbb{1}_{[0,1]}(X_s)\,\mathrm{d}s}{Z_t}$$

$$= 4(\|\theta - \psi\|_\infty + \|\theta_1 - \psi_1\|_\infty)\ .$$

A.s. the family $A_t(\theta, \theta_1)/Z_t$ (indexed by t) being equicontinuous and convergent on dense subset $\Theta' \times \Theta'$ to a continuous limit $d^2(\theta, \theta_1)/\mu_0([0,1])$, it converges uniformly on whole Θ^2. Hence a.s. there exists some $\tau(\omega)$ such that

$$\forall t > \tau, \ \forall (\theta, \theta_1) \in \Theta^2, \quad \frac{d^2(\theta, \theta_1)}{2\mu_0([0,1])} \leq \frac{A_t(\theta, \theta_1)}{Z_t} \leq \frac{2d^2(\theta, \theta_1)}{\mu_0([0,1])} \ .$$

Taking $U_t = Z_t/2\mu_0([0,1])$ and $V_t = 2Z_t/\mu_0([0,1])$ we satisfy all the assumptions of theorem 2 and obtain

$$r_t \sim \left(\frac{Z_t}{\ln \ln Z_t} \right)^{\frac{1}{2+1/\alpha}} \ .$$

In the same manner, if we take for Θ the space of decreasing functions $\theta : \mathbb{R} \to [-1, 1]$ such that $\lim_{x \to \pm\infty} \theta(x) = \mp 1$, then $N(\Theta, \varepsilon, d) \sim \exp(1/\varepsilon)$ (see [VW96]). In this case X_t is ergodic, and the calculations above hold with $Z_t = t$, hence we get $r_t \sim \left(\frac{t}{\ln \ln t} \right)^{\frac{1}{3}}$.

Acknowledgment

The authors gratefully acknowledge the careful reading of the manuscript by the referees. Their remarks have helped us to correct a number of misprints and to improve the overall readability of the paper.

References

[BP80]   I.V. Basawa, B.L.S. Prakasa Rao, *Statistical Inference for Stochastic Processes*, Academic Press, London (1980)

[DKu03] H.M. Dietz, Yu.A. Kutoyants, Parameter estimation for some non-recurrent solutions of SDE, *Statistics and Decisions* **21**, 29–45 (2003)

[F76]   P.D. Feigin, Maximum likelihood estimation for continuous-time stochastic processes, *Advances in Applied Probability* **8**, 712–736 (1976)

[JKh97] A. Jankunas, R.Z. Khasminskii, Estimation of parameters of linear homogeneous stochastic differential equations, *Stochastic Processes and their Applications* **72** , 205–219 (1997)

[HKu03] R. Höpfner, Yu.A. Kutoyants, On a problem of statistical inference in null recurrent diffusion, *Statistical Inference for Stochastic Processes* **6**(1), 25–42 (2003)

[KS99]   U. Küchler, M. Sørensen, A note on limit theorems for multivariate martingales, *Bernoulli* **5**(3), 483–493 (1999)

[Ku03]   Yu.A. Kutoyants, *Statistical Inference for Ergodic Diffusion Processes*, Springer Series in Statistics, New York (2003)

[L03]   D. Loukianova, Remark on semigroup techniques and the maximum likelihood estimation, *Statistics and Probability Letters* **62**, 111–115 (2003)

[LL05]   D. Loukianova, O. Loukianov, Uniform law of large numbers and consistency of estimators for Harris diffusions, to appear in *Statistics and Probability Letters*

[N99]   Y. Nishiyama, A maximal inequality for continuous martingales and M-estimation in gaussian white noise model, *The Annals of Statistics* **27**(2), 675–696 (1999)

[RY94]   D. Revuz, M. Yor, *Continuous Martingales and Brownian Motion*, Second Edition, Springer-Verlag, Berlin, Heidelberg (1994)

[Se00]   R. Senoussi, Uniform iterated logarithm laws for martingales and their application to functional estimation in controlled Markov chains, *Stochastic Processes and their Applications* **89**, 193–211 (2000)

[GS91]   S.A. van de Geer, L. Stougie, On rates of convergence and asymptotic normality in the multiknapsack problem, *Mathematical Programming* **51**, 349-358 (1991)

[G00]   S.A. van de Geer, *Empirical Processes in M-estimation*, Cambridge University Press, Cambridge (2000)

[VW96]   A.W. van der Vaart, J.A. Wellner, *Weak Convergence and Empirical Processes*, Springer-Verlag (1996)

[VZ05]   A. van der Vaart, H. van Zanten, Donsker theorems for diffusions: necessary and sufficient conditions, to appear in *The Annals of Probability* (2005)

[Z99]   H. van Zanten, On the Uniform Convergence of Local Time and the Uniform Consistency of Density Estimators for Ergodic Diffusions, *Probability, Networks and Algorithms* PNA-R9909 (1999)

[Z03a]   H. van Zanten, On empirical processes for ergodic diffusions and rates of convergence of *M*-estimators, *Scandinavian Journal of Statistics* **30**(3), 443-458 (2003)

[Z03b]   H. van Zanten, On uniform laws of large numbers for ergodic diffusions and consistency of estimators, *Statistical Inference for Stochastic Processes* **6**(2), 199-213 (2003)

[Z05]   H. van Zanten, On the rate of convergence of the maximum likelihood estimator in Brownian semimartingale models, *Bernoulli* **11**(4), 643–664 (2005).

[Y90]   N. Yoshida, Asymptotic behavior of *M*-estimators and related random field for diffusion process, *Ann. Inst. Statist. Math.* **42**(2), 221–251 (1990)

Convergence rates for density estimators of weakly dependent time series

Nicolas Ragache[1] and Olivier Wintenberger[2]

[1] MAP5, Université René Descartes 45 rue des Saints-Pères, 75270 Paris, France
`nicolas.ragache@ensae.fr`
[2] SAMOS, Statistique Appliquée et MOdélisation Stochastique, Université Paris 1,
Centre Pierre Mendès France, 90 rue de Tolbiac, F-75634 Paris Cedex 13,
France. `olivier.wintenberger@univ-paris1.fr`

1 Introduction

Assume that $(X_n)_{n \in \mathbb{Z}}$ is a sequence of $\mathbb{R}^d$ valued random variables with common distribution which is absolutely continuous with respect to Lebesgue's measure, with density f. Stationarity is not assumed so that the case of a sampled process $\{X_{i,n} = x_{h_n(i)}\}_{1 \leq i \leq n}$ for any sequence of monotonic functions $(h_n(.))_{n \in \mathbb{Z}}$ and any stationary process $(x_n)_{n \in \mathbb{Z}}$ that admits a marginal density is included. This paper investigates convergence rates for density estimation in different cases. First, we consider two concepts of weak dependence:

- Non-causal η-dependence introduced in [DL99] by Doukhan & Louhichi,
- Dedecker & Prieur's $\tilde{\phi}$-dependence (see [DP04]).

These two notions of dependence cover a large number of examples of time series (see section § 3). Next, following Doukhan (see [Dou90]) we propose a unified study of linear density estimators $\hat{f}_n$ of the form

$$\hat{f}_n(x) = \frac{1}{n} \sum_{i=1}^{n} K_{m_n}(x, X_i) , \tag{1}$$

where $\{K_{m_n}\}$ is a sequence of kernels. Under classical assumptions on $\{K_{m_n}\}$ (see section § 2.2), the results in the case of independent and identically distributed (i.i.d. in short) observations X_i are well known (see for instance [Tsy04]). At a fixed point $x \in \mathbb{R}^d$, the sequence m_n can be chosen such that

$$\|\hat{f}_n(x) - f(x)\|_q = O\left(n^{-\rho/(2\rho+d)}\right) . \tag{2}$$

The coefficient $\rho > 0$ measures the regularity of f (see Section 2.2 for the definition of the notion of regularity). The same rate of convergence also holds for the Mean Integrated Square Error (MISE), defined as $\int \|\hat{f}_n(x) - f(x)\|_2^2 p(x) \, dx$

for some nonnegative and integrable function p. The rate of uniform convergence on a compact set incurs a logarithmic loss appears. For all $M > 0$ and for a suitable choice of the sequence m_n,

$$\mathbb{E} \sup_{\|x\| \leq M} |\hat{f}_n(x) - f(x)|^q = \mathcal{O}\left(\frac{\log n}{n}\right)^{q\rho/(d+2\rho)}, \tag{3}$$

and

$$\sup_{\|x\| \leq M} |\hat{f}_n(x) - f(x)| =_{a.s.} \mathcal{O}\left(\frac{\log n}{n}\right)^{\rho/(d+2\rho)}. \tag{4}$$

These rates are optimal in the minimax sense. We thus have no hope to improve on them in the dependent setting. A wide literature deals with density estimation for absolutely regular or β-mixing processes (for a definition of mixing coefficients, see [Dou94]). For instance, under the assumption $\beta_r = o\left(r^{-3-2d/\rho}\right)$, Ango Nze & Doukhan prove in [AD98] that (2), (3) and (4) still hold. The sharper condition $\sum_r |\beta_r| < \infty$ entails the optimal rate of convergence for the MISE (see [Vie97]). Results for the MISE have been extended to the more general $\tilde{\phi}$- and η-dependence contexts by Dedecker & Prieur ([DP04]) and Doukhan & Louhichi in [DL01]. In this paper, our aim is to extend the bounds (2), (3) and (4) in the η- and $\tilde{\phi}$-weak dependence contexts.

We use the same method as in [DL99] based on the following moment inequality for weakly dependent and centered sequences $(Z_n)_{n\in\mathbb{Z}}$. For each even integer q and for each integer $n \geq 2$:

$$\left\|\sum_{i=1}^n Z_i\right\|_q^q \leq \frac{(2q-2)!}{(q-1)!} \left\{V_{2,n}^{q/2} \vee V_{q,n}\right\}, \tag{5}$$

where $\|X\|_q^q = \mathbb{E}|X|^q$ and for $k = 2, \ldots, q$,

$$V_{k,n} = n \sum_{r=0}^{n-1} (r+1)^{k-2} C_k(r), $$

with

$$C_k(r) := \sup\{|\mathrm{cov}(Z_{t_1} \cdots Z_{t_p}, Z_{t_{p+1}} \cdots Z_{t_k})|\}, \tag{6}$$

where the supremum is over all the ordered k-tuples $t_1 \leq \cdots \leq t_k$ such that $\sup_{1 \leq i \leq k-1} t_{i+1} - t_i = r$.

We will apply this bound when the Z_is are defined in such a way that $\sum_{i=1}^n Z_i$ is proportional to the fluctuation term $\hat{f}_n(x) - \mathbb{E}\hat{f}_n(x)$. The inequality (5) gives a bound for this part of the deviation of the estimator which depends on the covariance bounds $C_k(r)$. The other part of the deviation is the bias, which is treated by deterministic methods. In order to obtain suitable controls of the fluctuation term, we need two different type of bounds for $C_k(r)$. Conditions on the decay of the weak dependence coefficients give a first bound. Another type of condition is also required to bound $C_k(r)$ for the smaller values of r; this is classically achieved with a regularity condition on the joint law of the pairs (X_j, X_k) for all $j \neq k$. In Doukhan & Louhichi (see [DL01]), rates of convergence are obtained when the coefficient η decays geometrically fast and the joint densities are bounded. We relax these conditions to cover the case when the joint distributions are not absolutely continuous and when the η- and $\tilde{\phi}$-dependence coefficients decrease slowly (sub-geometric and Riemannian decays are considered).

Under our assumptions, we prove that (2) still holds (see Theorem 1). Unfortunately, additional losses appear for the uniform bounds. When η_r or $\tilde{\phi}_r = O(e^{-ar^b})$ with $a > 0$ and $b > 0$, we prove in Theorem 2 that (3) and (4) hold with $\log(n)$ replaced by $\log^{2(b+1)/b}(n)$. If η_r or $\tilde{\phi} = O(r^{-a})$ with $a > 1$, Theorem 3 gives bounds similar to (3) and (4) with the right hand side replaced by $O(n^{-q\rho q_0/\{2\rho q_0 + d(q_0+2)\}})$ and $O(\{\log^{2+4/(q_0-2)}(n)/n\}^{\rho(q_0-2)/\{2\rho q_0 + d(q_0+2)\}})$, respectively, and with $q_0 = 2\lceil (a-1)/2 \rceil$ (by definition $\lceil x \rceil$ is the smallest integer larger than or equal to the real number x). As already noticed in [DL01], the loss w.r.t the i.i.d. case highly depends on the decay of the dependence coefficients. In the case of geometric decay, the loss is logarithmic while it is polynomial in the case of polynomial decays.

The paper is organized as follows. In Section 2.1, we introduce the notions of η and $\tilde{\phi}$ dependence. We give the notation and hypothesis in Section 2.2. The main results are presented in Section 2.3. We then apply these results to particular cases of weak dependence processes, and we provide examples of kernel K_m in Section 3. Section 4 contains the proof of the Theorems and three important lemmas.

2 Main results

We first describe the notions of dependence considered in this paper, then we introduce assumptions and formulate the main results of the paper (convergence rates).

2.1 Weak dependence

We consider a sequence $(X_i)_{i \in \mathbb{Z}}$ of $\mathbb{R}^d$ valued random variables, and we fix a norm $\| \cdot \|$ on $\mathbb{R}^d$. Moreover, if $h : \mathbb{R}^{du} \to \mathbb{R}$ for some $u \geq 1$, we define

$$\text{Lip } (h) = \sup_{(a_1,\ldots,a_u) \neq (b_1,\ldots,b_u)} \frac{|h(a_1,\ldots,a_u) - h(b_1,\ldots,b_u)|}{\|a_1 - b_1\| + \cdots + \|a_u - b_u\|}.$$

Definition 1 (η-dependence, Doukhan & Louhichi (1999)). *The process $(X_i)_{i \in \mathbb{Z}}$ is η-weakly dependent if there exists a sequence of non-negative real numbers $(\eta_r)_{r \geq 0}$ satisfying $\eta_r \to 0$ when $r \to \infty$ and*

$$\left| \operatorname{cov} \left(h \left(X_{i_1}, \ldots X_{i_u} \right), k \left(X_{i_{u+1}}, \ldots, X_{i_{u+v}} \right) \right) \right| \leq \left(u \operatorname{Lip}(h) + v \operatorname{Lip}(k) \right) \eta_r \ ,$$

for all $(u+v)$-tuples, $(i_1, \ldots, i_{u+v})$ with $i_1 \leq \cdots \leq i_u \leq i_u + r \leq i_{u+1} \leq \cdots \leq i_{u+v}$, and $h, k \in \Lambda^{(1)}$ where

$$\Lambda^{(1)} = \left\{ h : \exists u \geq 0, h : \mathbb{R}^{du} \to \mathbb{R}, \operatorname{Lip}(h) < \infty, \|h\|_\infty = \sup_{x \in \mathbb{R}^{du}} |h(x)| \leq 1 \right\} \ .$$

Remark The η-dependence condition can be applied to non-causal sequences because information "from the future" (i.e. on the right of the covariance) contributes to the dependence coefficient in the same way as information "from the past" (i.e. on the left). It is the non-causal alternative to the θ condition in [DD03] and [DL99].

Definition 2 ($\tilde{\phi}$-dependence, Dedecker & Prieur (2004)). *Let $(\Omega, \mathcal{A}, \mathbb{P})$ be a probability space and $\mathcal{M}$ a σ-algebra of $\mathcal{A}$. For any $l \in \mathbb{N}^*$, any random variable $X \in \mathbb{R}^{dl}$ we define:*

$$\tilde{\phi}(\mathcal{M}, X) = \sup\{\|\mathbb{E}(g(X)|\mathcal{M}) - \mathbb{E}(g(X))\|_\infty, g \in \Lambda_{1,l}\} \ ,$$

where $\Lambda_{1,l} = \{h : \mathbb{R}^{dl} \mapsto \mathbb{R}/\operatorname{Lip}(h) < 1\}$. The sequence of coefficients $\tilde{\phi}_k(r)$ is then defined by

$$\tilde{\phi}_k(r) = \max_{l \leq k} \frac{1}{l} \sup_{i+r \leq j_1 < j_2 < \cdots < j_l} \tilde{\phi}(\sigma(\{X_j; j \leq i\}), (X_{j_1}, \ldots, X_{j_l})) \ .$$

The process is $\tilde{\phi}$-dependent if $\tilde{\phi}(r) = \sup_{k>0} \tilde{\phi}_k(r)$ tends to 0 with r.

Remark The $\tilde{\phi}$ dependence coefficients provide covariance bounds. For a Lipschitz function k and a bounded function h,

$$\left| \operatorname{cov} \left(h \left(X_{i_1}, \ldots, X_{i_u} \right), k \left(X_{i_{u+1}}, \ldots, X_{i_{u+v}} \right) \right) \right|$$
$$\leq v \mathbb{E} \left| h \left(X_{i_1}, \ldots, X_{i_u} \right) \right| \operatorname{Lip}(k) \tilde{\phi}(r) \ . \quad (7)$$

2.2 Notations and definitions

Assume that $(X_n)_{n \in \mathbb{Z}}$ is an η or $\tilde{\phi}$ dependent sequence of $\mathbb{R}^d$ valued random variables. We consider two types of decays for the coefficients. The geometric case is the case when Assumption [H1] or [H1'] holds.

[H1]: $\eta_r = O\left(e^{-ar^b}\right)$ with $a > 0$ and $b > 0$,

[H1']: $\tilde{\phi}(r) = O\left(e^{-ar^b}\right)$ with $a > 0$ and $b > 0$.

The Riemannian case is the case when Assumption [H2] or [H2'] holds.

[H2]: $\eta_r = O(r^{-a})$ with $a > 1$,
[H2']: $\tilde{\phi}(r) = \mathcal{O}(r^{-a})$ with $a > 1$.

As usual in density estimation, we shall assume:

[H3]: The common marginal distribution of the random variables X_n, $n \in \mathbb{Z}$ is absolutely continuous with respect to Lebesgue's measure, with common bounded density f.

The next assumption is on the density with respect to Lebesgue's measure (if it exists) of the joint distribution of the pairs (X_j, X_k), $j \neq k$.

[H4] The density $f_{j,k}$ of the joint distribution of the pair (X_j, X_k) is uniformly bounded with respect to $j \neq k$.

Unfortunately, for some processes, these densities may not even exist. For example, the joint distributions of Markov chains $X_n = G(X_{n-1}, \epsilon_n)$ may not be absolutely continous. One of the simplest example is

$$X_k = \frac{1}{2}(X_{k-1} + \epsilon_k) , \tag{8}$$

where $\{\epsilon_k\}$ is an i.i.d. sequence of Bernoulli random variables and X_0 is uniformly distributed on $[0, 1]$. The process $\{X_n\}$ is strictly stationary but the joint distributions of the pairs (X_0, X_k) are degenerated for any k. This Markov chain can also be represented (through an inversion of the time) as a dynamical system $(T_{-n}, \ldots, T_{-1}, T_0)$ which has the same law as $(X_0, X_1, \ldots, X_n)$ (T_0 and X_0 are random variables distributed according to the invariant measure, see [BGR00] for more details). Let us recall the definition of a dynamical system.

Definition 3 (dynamical system). *A one-dimensional dynamical system is defined by*

$$\forall k \in \mathbb{N}, \ T_k := F^k(T_0) , \tag{9}$$

where $F : I \to I$, I is a compact subset of $\mathbb{R}$ and in this context, F^k denotes the k-th iterate of the appplication F: $F^1 = F$, $F^{k+1} = F \circ F^k$, $k \geq 1$. We assume that there exists an invariant probability measure μ_0, i.e. $F(\mu_0) = \mu_0$, absolutely continuous with respect to Lebesgue's measure, and that T_0 is a random variable with distribution μ_0.

We restrict our study to one-dimensional dynamical systems T in the class $\mathcal{F}$ of dynamical systems defined by a transformation F that satisfies the following assumptions (see [Pri01]).

- $\forall k \in \mathbb{N}$, $\forall x \in \text{int}(I)$, $\lim_{t \to 0^+} F^k(x+t) = F^k(x^+)$ and $\lim_{t \to 0^-} F^k(x+t) = F^k(x^-)$ exist;
- $\forall k \in \mathbb{N}^*$, denoting $D_+^k = \{x \in \text{int}(I), F^k(x^+) = x\}$ and $D_-^k = \{x \in \text{int}(I), F^k(x^-) = x\}$, we assume $\lambda\left(\bigcup_{k \in \mathbb{N}^*}\left(D_+^k \bigcup D_-^k\right)\right) = 0$, where λ is the Lebesgue measure.

When the joint distributions of the pairs (X_j, X_k) are not assumed absolutely continuous (and then [H4] is not satisfied), we shall instead assume:

[H5] The dynamical system $(X_n)_{n \in \mathbb{Z}}$ belongs to $\mathcal{F}$.

We consider in this paper linear estimators as in (1). The sequence of kernels K_m is assumed to satisfy the following assumptions.

(a) The support of K_m is a compact set with diameter $O(1/m_n)$;
(b) The functions $x \mapsto K_m(x, y)$ and $x \mapsto K_m(y, x)$ for all y are Lipschitz functions with Lipschitz constant $O\left(m^{1/d}\right)$;
(c) For all x, $\int K_m(x, y)\, dy = 1$;
(d) The sequence K_m is uniformly bounded.
(e) The bias of the estimator $\hat{f}_n$ defined in (1) is of order $m_n^{-\rho/d}$, uniformly on compact sets.

$$\sup_{\|x\| \leq M} \left| \mathbb{E}[\hat{f}_n(x)] - f(x) \right| = O(m_n^{-\rho/d}) . \tag{10}$$

2.3 Results

In all our results we consider kernels K_m and a density estimator of the form (1) such that assumptions (a), (b), (c), (d), (e) hold.

Theorem 1 ($\mathbb{L}^q$-convergence).

Geometric case. *Under Assumptions [H4] or [H5] and [H1] or [H1'], the sequence m_n can be chosen such that inequality (2) holds for all $0 < q < +\infty$.*

Riemannian case. *Under the assumptions [H4] or [H5], if additionally*
- *[H2] holds with $a > \max\left(1 + 2/d + (d+1)/\rho, 2 + 1/d\right)$ (η-dependence),*
- *or [H2'] holds with $a > 1 + 2/d + 1/\rho$ (ϕ-dependence),*

then the sequence m_n can be chosen such that inequality (2) holds for all $0 < q \leq q_0 = 2\lceil (a-1)/2 \rceil$.

Theorem 2 (Uniform rates, geometric decays). *For any $M > 0$, under Assumptions [H4] or [H5] and [H1] or [H1'] we have, for all $0 < q < +\infty$, and for a suitable choice of the sequence m_n,*

$$\mathbb{E} \sup_{\|x\| \leq M} |\hat{f}_n(x) - f(x)|^q = O\left(\left(\frac{\log^{2(b+1)/b}(n)}{n} \right)^{q\rho/(d+2\rho)} \right) ,$$

$$\sup_{\|x\| \leq M} |\hat{f}_n(x) - f(x)| =_{a.s.} O\left(\left(\frac{\log^{2(b+1)/b}(n)}{n} \right)^{\rho/(d+2\rho)} \right) .$$

Theorem 3 (Uniform rates, Riemannian decays). *For any $M > 0$, under Assumptions [H4] or [H5], [H2] or [H2'] with $a \geq 4$ and $\rho > 2d$, for $q_0 = 2\lceil (a-1)/2 \rceil$ and $q \leq q_0$, the sequence m_n can be chosen such that*

$$\mathbb{E} \sup_{\|x\| \leq M} |\hat{f}_n(x) - f(x)|^q = O\left(n^{-\frac{q\rho q_0}{2\rho q_0 + d(q_0+2)}} \right),$$

or such that

$$\sup_{\|x\| \leq M} |\hat{f}_n(x) - f(x)| =_{a.s.} O\left(\left(\frac{\log^{2+4/(q_0-2)} n}{n} \right)^{\frac{\rho(q_0-2)}{2\rho q_0 + d(q_0+2)}} \right).$$

Remarks.

- Theorem 1 shows that the optimal convergence rate of (2) still holds in the weak dependence context. In the Riemannian case, when $a \geq 4$, the conditions are satisfied if the density function f is sufficient regular, namely, if $\rho > d + 1$.
- The loss with respect to the i.i.d. case in the uniform convergence rates (Theorems 2 and 3) is due to the fact that the probability inequalities for dependent observations are not as good as Bernstein's inequality for i.i.d. random variables (Bernstein inequalities in weak dependence context are proved in [KN05]). The convergence rates depend on the decay of the weak dependence coefficients. This is in contrast to the case of independent observations.
- In Theorem 2 the loss is a power of the logarithm of the number of observations. Let us remark that this loss is reduced when b tends to infinity. In the case of η-dependence and geometric decreasing, the same result is in [DL99] for the special case $b = 1$. In the framework of $\tilde{\phi}$-dependence, Theorem 2 seems to provide the first result on uniform rates of convergence for density estimators.
- In Theorem 3, the rate of convergence in the mean is better than the almost sure rate for technical reasons. Contrary to the geometric case, the loss is no longer logarithmic but is a power of n. The rate gets closer to the optimal rate as $q_0 \to \infty$, or equivalently $a \to \infty$.
- These results are new under the assumption of Riemannian decay of the weak dependence coefficients. The condition on a is similar to the condition on β in [AD03]. Even if the rates are better than in [DL01], there is a huge loss with respect to the mixing case. It would be interesting to know the minimax rates of convergence in this framework.

3 Models, applications and extensions

The class of weak dependent processes is very large. We apply our results to three examples: **two-sided moving averages**, **bilinear models** and **ex-**

panding maps. The first two will be handled with the help of the coefficients η, the third one with the coefficients $\tilde{\phi}$.

3.1 Examples of η-dependent time series.

It is of course possible to define η-dependent random fields (see [DDLLLP04] for further details); for simplicity, we only consider processes indexed by $\mathbb{Z}$.

Definition 4 (Bernoulli shifts). *Let $H : \mathbb{R}^{\mathbb{Z}} \to \mathbb{R}$ be a measurable function. A Bernoulli shift is defined as $X_n = H(\xi_{n-i}, i \in \mathbb{Z})$ where $(\xi_i)_{i \in \mathbb{Z}}$ is a sequence of i.i.d random variables called the innovation process.*

In order to obtain a bound for the coefficients $\{\eta_r\}$, we introduce the following regularity condition on H. There exists a sequence $\{\delta_r\}$ such that

$$\sup_{i \in \mathbb{Z}} \mathbb{E} \left| H\left(\xi_{i-j}, j \in \mathbb{Z}\right) - H\left(\xi_{i-j} \mathbb{1}_{|j|<r}, j \in \mathbb{Z}\right)\right| \leq \delta_r ,$$

Bernoulli shifts are η-dependent with $\eta_r = 2\delta_{r/2}$ (see [DL99]). In the following, we consider two special cases of **Bernoulli shifts**.

(A1). **Non causal linear processes.** A real valued sequence $(a_i)_{i \in \mathbb{Z}}$ such that $\sum_{j \in \mathbb{Z}} a_j^2 < \infty$ and the innovation process $\{\xi_n\}$ define a **non-causal linear process** $X_n = \sum_{-\infty}^{+\infty} a_i \xi_{n-i}$. If we control a moment of the innovations, the **linear process** (X_n) is η-dependent. The sequence $\{\eta_r\}_{r \in \mathbb{N}}$ is directly linked to the coefficients $\{a_i\}_{i \in \mathbb{Z}}$ and various types of decay may occur. We consider only Riemannian decays $a_i = \mathcal{O}\left(i^{-A}\right)$ with $A \geq 5$ since results for geometric decays are already known. Here $\eta_r = \mathcal{O}\left(\sum_{|i|>r/2} a_i\right) = O(r^{1-A})$ and [H2] holds. Furthermore, we assume that the sequence $(\xi_i)_{i \in \mathbb{Z}}$ is i.i.d. and satisfies the condition $|\mathbb{E}e^{iu\xi_0}| \leq C(1 + |u|)^{-\delta}$, for all $u \in \mathbb{R}$ and for some $\delta > 0$ and $C < \infty$. Then, the densities f and $f_{j,k}$ exist for all $j \neq k$ and they are uniformly bounded (see the proof in the causal case in Lemma 1 and Lemma 2 in [GKS96]); hence [H4] holds. If the density f of X_0 is ρ-regular with $\rho > 2$, our estimators converge to the density with the rates:

  - $n^{-\rho/(2\rho+1)}$ in $\mathbb{L}^q$-norm ($q \leq 4$) at each point x,
  - $n^{-\rho/(2\rho+3/2)}$ in $\mathbb{L}^q$-norm ($q \leq 4$) uniformly on an interval,
  - $\left(\log^4(n)/n\right)^{\rho/(4\rho+3)}$ almost surely on an interval.

  In the first case, the rate we obtain is the same as in the i.i.d. case. For such linear models, the density estimator also satisfies the Central Limit Theorem (see [HLT01] and [Ded98]).

(A2). **Bilinear model.** The process $\{X_t\}$ is a **bilinear model** if there exist two sequences $(a_i)_{i \in \mathbb{N}^*}$ and $(b_i)_{i \in \mathbb{N}^*}$ of real numbers and real numbers a and b such that:

$$X_t = \xi_t \left(a + \sum_{j=1}^{\infty} a_j X_{t-j} \right) + b + \sum_{j=1}^{\infty} b_j X_{t-j} . \qquad (11)$$

Squared **ARCH**(∞) or **GARCH**(p, q) processes satisfy such an equation, with $b = b_j = 0$ for all $j \geq 1$. Define

$$\lambda = \|\xi_0\|_p \sum_{j=1}^{\infty} a_j + \sum_{j=1}^{\infty} b_j .$$

If $\lambda < 1$, then the equation (11) has a strictly stationary solution in L^p (see [DMR05]). This solution is a **Bernoulli shift** for which we have the behavior of the coefficient η:

- $\eta_r = O\left(e^{-\lambda r}\right)$ for some $\lambda > 0$ if there exists an integer N such that $a_i = b_i = 0$ for $i \geq N$.
- $\eta_r = O(e^{-\lambda\sqrt{r}})$ for some $\lambda > 0$ if $a_i = O(e^{-Ai})$ and $b_i = O(e^{-Bi})$ with $A > 0$ and $B > 0$.
- $\eta_r = O(\{r/\log(r)\}^{-\lambda})$ for some $\lambda > 0$ if $a_i = O(i^{-A})$ and $b_i = O(i^{-B})$ with $A > 1$ and $B > 1$.

Let us assume that the i.i.d. sequence $\{\xi_t\}$ has a marginal density $f_\xi \in C_\rho$, for some $\rho > 2$. The density of X_t conditionally to the past can be written as a function of f_ξ. We then check recursively that the common density of X_t for all t, say f, also belongs to C_ρ. Furthermore, the regularity of f_ξ ensures that f and the joint densities $f_{j,k}$ for all $j \neq k$ are bounded (see [DMR05]) and [H4] holds. The assumptions of Theorem 1 are satisfied, and the estimator $\hat{f}_n$ achieves the minimax bound (2) if either:

- There exists an integer N such that $a_i = b_i = 0$ for $i \geq N$;
- There exist $A > 0$ and $B > 0$ such that $a_i = O(e^{-Ai})$ and $b_i = O(e^{-Bi})$;
- There exist $A \geq 4$ and $B \geq 5$ such that $a_i = O(i^{-A})$ and $b_i = O(i^{-B})$. Then, this optimal bound holds only for $2 \leq q < q(A, B)$ where $q(A, B) = 2[((B-1) \wedge A)/2]$.

Note finally that the rates of uniform convergence provided by Theorems 2 and 3 are sub-optimal.

3.2 Examples of $\tilde{\phi}$-dependent time series.

Let us introduce an important class of **dynamical systems**:

Example 1. $(T_i = F^i(T_0))_{i \in \mathbb{N}}$ is an **expanding map** or equivalently F is a Lasota-Yorke function if it satisfies the three following criteria.

- (Regularity) There exists a grid $0 = a_0 \leq a_1 \cdots \leq a_n = 1$ such as $F \in C_1$ and $|F'(x)| > 0$ on $]a_{i-1}, a_i[$ for each $i = 1, \ldots, n$.
- (Expansivity) Let I_n be the set on which $(F^n)'$ is defined. There exists $A > 0$ and $s > 1$ such that $\inf_{x \in I_n} |(F^n)'| > As^n$.

- (Topological mixing) For any nonempty open sets U, V, there exists $n_0 \geq 1$ such as $F^{-n}(U) \cap V \neq \varnothing$ for all $n \geq n_0$.

Examples of **Markov chains** $X_n = G(X_{n+1}, \epsilon_n)$ associated to an **expanding map** $\{T_n\}$ belonging to $\mathcal{F}$ are given in [BGR00] and [DP04]. The simplest one is $X_k = (X_{k-1} + \epsilon_k)/2$ where the ϵ_k follows a binomial law and X_0 is uniformly distributed on $[0,1]$. We easily check that $F(x) = 2x \mod 1$, the transformation of the associated **dynamical system** T_n, satisfies all the assumptions such as T_n is an **expanding map** belonging to $\mathcal{F}$.

The coefficients of $\tilde{\phi}$-dependence of such a **Markov chain** satisfy $\tilde{\phi}(r) = O(e^{-ar})$ for some $a > 0$ (see [DP04]). Theorems 1 and 2 give the $\mathbb{L}^q$ rate $n^{-\rho/(2\rho+1)}$, the uniform $\mathbb{L}^q$ rate and the almost sure rate $\left(\log^4(n)/n\right)^{\rho/(2\rho+1)}$ of the estimators of the density of μ_0.

3.3 Sampled process

Since we do not assume stationarity of the observed process, the following observation scheme is covered by our results. Let $(x_n)_{n \in \mathbb{Z}}$ be a stationary process whose marginal distribution is absolutely continuous, let $(h_n)_{n \in \mathbb{Z}}$ be a sequence of monotone functions and consider the sampled process $\{X_{i,n}\}_{1 \leq i \leq n}$ defined by $X_{i,n} = x_{h_n(i)}$. The dependence coefficients of the sampled process may decay to zero faster than the underlying unobserved process. For instance, if the dependence coefficients of the process $(x_n)_{n \in \mathbb{Z}}$ have a Riemannian decay, those of the sampled process $\{x_{h_n(i)}\}$ with $h_n(i) = i2^n$ decay geometrically fast. The observation scheme is thus a crucial factor that determines the rate of convergence of density estimators.

3.4 Density estimators and bias

In this section, we provide examples of kernels K_m and smoothness assumptions on the density f such that assumptions (a), (b), (c), (d) and (e) of subsection 2.2 are satisfied.

Kernel estimators The kernel estimator associated to the bandwidth parameter m_n is defined by:

$$\hat{f}_n(x) = \frac{m_n}{n} \sum_{i=1}^{n} K\left(m_n^{1/d}(x - X_i)\right).$$

We briefly recall the classical analysis for the deterministic part R_n in this case (see [Tsy04]). Since the sequence $\{X_n\}$ has a constant marginal distribution, we have $\mathbb{E}[\hat{f}_n(x)] = f_n(x)$ with $f_n(x) = \int_D K(s) f\left(x - s/m_n^{1/d}\right) ds$. Let us assume that K is a Lipschitz function compactly supported in $D \subset \mathbb{R}^d$. For $\rho > 0$, let K satisfy, for all $j = j_1 + \cdots + j_d$ with $(j_1, \ldots, j_d) \in \mathbb{N}^d$:

$$\int x_1^{j_1} \cdots x_d^{j_d} K(x_1, \ldots, x_d) \mathrm{d}x_1 \cdots \mathrm{d}x_d = \begin{cases} 1 & \text{if } j = 0, \\ 0 & \text{for } j \in \{1, \ldots, \lceil \rho - 1 \rceil - 1\}, \\ \neq 0 & \text{if } j = \lceil \rho - 1 \rceil. \end{cases}$$

Then the kernels $K_{m_n}(x, y) = m_n K\left(m_n^{1/d}(x - y)\right)$ satisfy (a), (b), (c) and (d). Assumption (e) holds and if $f \in C_\rho$, where C_ρ is the class of function f such that for $\rho = \lceil \rho - 1 \rceil + c$ with $0 < c \le 1$, f is $\lceil \rho - 1 \rceil$-times continuously differentiable and there exists $A > 0$ such that $\forall (x, y) \in \mathbb{R}^d \times \mathbb{R}^d, |f^{(\lceil \rho - 1 \rceil)}(x) - f^{(\lceil \rho - 1 \rceil)}(y)| \le A|x - y|^c$.

Projection estimators We only consider in this section the case $d = 1$. Under the assumption that the family $\{1, x, x^2, \ldots\}$ belongs to $L^2(I, \mu)$, where I is a bounded interval of $\mathbb{R}$ and μ is a measure on I, an orthonormal basis of $L^2(I, \mu)$ can be defined which consists of polynomials $\{P_0, P_1, P_2, \ldots\}$. The fact that I is compact and the Christoffel-Darboux formula and its corollary (see [Sze33]) ensure properties (a), (b) and (d) for the elements of the basis. We assume that f belongs to a class C'_ρ which is slightly more restrictive than the class C_ρ (see Theorem 6.23 p.218 in [DS01] for details). Then for any $f \in L^2(I, \mu) \cap C'_\rho$, there exists a function $\pi_{f,m_n} \in V_{m_n}$ such that $\sup_{x \in I} |f(x) - \pi_{f,m_n}(x)| = O(m_n^{-\rho})$. Consider then the projection $\pi_{m_n} f$ of f on the subspace $V_{m_n} = \mathrm{Vect}\{P_0, P_1, \ldots, P_{m_n}\}$. It can be expressed as

$$\pi_{m_n} f(x) = \sum_{j=0}^{m_n} \left\{ \int_I P_j(s) f(s) \mathrm{d}\mu(s) \right\} P_j(x).$$

The projection estimator of the density f of the real valued random variables $\{X_i\}_{1 \le i \le n}$ is naturally defined as

$$\hat{f}_n(x) = \frac{1}{n} \sum_{i=1}^n K_{m_n}(x, X_i) = \frac{1}{n} \sum_{i=1}^n \sum_{j=0}^{m_n} P_j(X_i) P_j(x).$$

Then $\mathbb{E}\hat{f}_n(x) = \pi_{m_n} f(x)$ is an approximation of $f(x)$ in V_{m_n}. We easily check that properties (a), (b), (c) and (d) hold for the kernels K_{m_n}. Unfortunately, the optimal rate $(m_n^{-\rho})$ does not necessarily hold. We then have to consider the weighted kernels $K_m^a(x, y)$ defined by:

$$K_m^a(x, y) = \sum_{j=0}^m a_{m,j} \sum_{k=0}^j P_k(x) P_k(y),$$

where $\{a_{m,j}; m \in \mathbb{N}, 0 \le j \le m\}$ is a weight sequence satisfying $\sum_{j=0}^m a_{m,j} = 1$ and for all j: $\lim_{m \to \infty} a_{m,j} = 0$. If the sequence $\{a_{m,j}\}$ is such that K_m^a is a nonnegative kernel then $\|K_m^a\|_1 = \int_I K_m^a(x, s) \mathrm{d}\mu(s) = 1$ and the kernel K_m^a satisfies (a), (b), (c), (d). Moreover, the uniform norm of the operator $f \mapsto K_m^a * f(x)$ is $\sup_{\|f\|_\infty = 1} \|K_m^a * f\|_\infty = \|K_m^a\|_1 = 1$. The linear estimator built with this kernel is

$$\hat{f}_n^a(x) = \frac{1}{n} \sum_{i=1}^{n} \sum_{j=0}^{m_n} a_{m_n,j} \sum_{k=0}^{j} P_k(X_i) P_k(x) ,$$

and its bias has the optimal rate:

$$|\mathbb{E}\hat{f}_n^a(x) - f(x)| = |K_{m_n}^a * f(x) - \pi_{f,m_n} f(x) + \pi_{f,m_n} f(x) - f(x)| ,$$
$$\leq |K_{m_n}^a * (f(x) - \pi_{f,m_n} f(x)) + \pi_{f,m_n} f(x) - f(x)| ,$$
$$\leq (\|K_{m_n}^a\|_1 + 1) m_n^{-\rho} = \mathcal{O}(m_n^{-\rho}) .$$

Such an array $\{a_{m,j}\}$ cannot always be defined. We give an example where it is possible.

Example 2 (Fejer kernel). For the trigonometric basis $\{\cos(nx), \sin(nx)\}_{n \in \mathbb{N}}$, we can find a 2π-periodic function $f \in \mathcal{C}_1'$ such that $\sup_{x \in [-\pi;\pi]} |f(x) - \pi_m f(x)| = O(m^{-1} \log m)$. The associated estimator reads:

$$\hat{f}_n(x) = \frac{1}{2\pi} + \frac{1}{n\pi} \sum_{i=1}^{n} \sum_{k=1}^{m_n} \cos(kX_i)\cos(kx) + \sin(kX_i)\sin(kx) .$$

We remark that $\mathbb{E}\hat{f}_n$ is the Fourier series of f truncated at order m_n:

$$D_{m_n} f(x) = \frac{1}{2\pi} \int_0^{2\pi} f(t) D_{m_n}(x - t) dt .$$

where

$$D_m(x) = \sum_{k=-m}^{m} e^{ikx} = \frac{\sin(\{2m+1\}x/2)}{\sin(x/2)}$$

is (the symmetric) Dirichlet's kernel. Recall that Fejer's kernel is defined as

$$F_m(x) = \frac{1}{m} \sum_{k=0}^{m-1} D_k(x) = \sum_{k=-(m-1)}^{m-1} \left(1 - \frac{|k|}{m}\right) e^{ikx} = \frac{\sin^2(mx/2)}{m\sin^2(x/2)} .$$

The kernel F_m is a nonnegative weighted kernel corresponding to Dirichlet's kernel and the sequence of weights $a_{m,j} = 1/m$ and satisfies (a), (b), (c) and (d). The estimator associated to the Fejer's kernels is defined by

$$\tilde{f}_n(x) = \frac{1}{2\pi} + \frac{1}{n\pi} \sum_{i=1}^{n} \sum_{j=1}^{m_n} \frac{1}{m_n} \sum_{k=1}^{j} \cos kX_i \cos kx + \sin kX_i \sin kx ,$$

If the common density f is 2π-periodic and belongs to $\mathcal{C}_1'$, then assumption (e) holds.

Using general Jackson's kernels (see [DS01]), we can find an estimator such that $R_n = O(m_n^{-\rho/d})$ for other values of ρ, but the weight sequence $a_{m,j}$ highly depends of the value of ρ.

Wavelet estimation Wavelet estimation is a particular case of projection estimation. For the sake of simplicity, we restrict hte study to $d = 1$.

Definition 5 (Scaling function [Dou88]). *A function $\phi \in L^2(\mathbb{R})$ is called a scaling function if the family $\{\phi(\cdot - k)\,;\, k \in \mathbb{Z}\}$ is orthonormal.*

We choose the bandwidth parameter $m_n = 2^{j(n)}$ and define $V_j = \text{Vect}\{\phi_{j,k}, k \in \mathbb{Z}\}$, where $\phi_{j,k} = 2^{j/2}\phi(2^j(x-k))$. Under the assumption that ϕ is compactly supported, we define (the sum over the index k is in fact finite):

$$\hat{f}_n(x) = \frac{1}{n}\sum_{k=-\infty}^{\infty}\sum_{i=1}^{n}\phi_{j(n),k}(X_i)\phi_{j(n),k}(x)\,.$$

The wavelets estimator is of the form (1) with $K(x,y) = \sum_{k=-\infty}^{\infty}\phi(y-k)\phi(x-k)$ and $K_m(x,y) = mK(mx,my)$. Under the additionnal assumption that $\sum_{k\in\mathbb{Z}}\phi(x-k) = 1$ for almost all x, we can write:

$$\left|\mathbb{E}(\hat{f}_n(x)) - f(x))\right| \le \left|\int K_{m_n}(y,x)f(y)dy - f(x)\right|\,,$$

$$= \left|\int m_n K(m_n y, m_n x)(f(y) - f(x))dy\right|\,,$$

$$= \left|\int m_n K(m_n x + t, m_n x)(f(x + t/m_n) - f(x))dt\right|\,.$$

If ϕ is a Lipschitz function such that $\int \phi(x)x^j dx = 0$ if $0 < j < \lceil \rho - 1 \rceil$ and $\int \phi(x)x^{\lceil\rho-1\rceil}dx \ne 0$, then the kernel K_m satisfy properties (a), (b), (c) and (d). If $f \in C_\rho$, then Assumption (e) holds.

4 Proof of the Theorems

The proof of our results is based on the decomposition:

$$\hat{f}_n(x) - f(x) = \underbrace{\hat{f}_n(x) - \mathbb{E}\left(\hat{f}_n(x)\right)}_{FL_n(x)=\text{fluctuation}} + \underbrace{\mathbb{E}\left(\hat{f}_n(x)\right) - f(x)}_{\text{bias}}\,. \tag{12}$$

The bias term is of order $m_n^{-\rho/d}$ by Assumption (e). We now present three lemmas useful to derive the rate of the fluctuation term.

Lemma 1 (Moment inequalities). *For each even integer q, under the assumption [H4] or [H5] and if moreover one of the following assumption holds:*

- *[H1] or [H1'] holds (geometric case);*
- *[H2] holds, $m_n = n^\delta \log(n)^\gamma$ with $\delta > 0$, $\gamma \in \mathbb{R}$ and*

$$a > \max\left(q - 1, \frac{(q-1)\delta(4 + 2/d)}{q - 2 + \delta(4 - q)}, 2 + \frac{1}{d}\right)\,,$$

- *[H2'] holds, $m_n = n^\delta \log(n)^\gamma$ with $\delta > 0$ and $\gamma \in \mathbb{R}$ and*

$$a > \max\left(q - 1, \frac{(q-1)\delta(2 + 2/d)}{q - 2 + \delta(4 - q)}, 1 + \frac{1}{d}\right) .$$

Then, for each $x \in \mathbb{R}^d$,

$$\limsup_{n \to \infty} (n/m_n)^{q/2} \|FL_n(x)\|_q^q < +\infty .$$

Lemma 2 (Probability inequalities).

- *Geometric case. Under Assumptions [H4] or [H5] and [H1] or [H1'] there exist positive constants C_1, C_2 such that*

$$\mathbb{P}\left(|FL_n(x)| \geq \epsilon \sqrt{m_n/n}\right) C_1 \leq \exp\{-C_2 \epsilon^{b/(b+1)}\} .$$

- *Riemannian case. Under Assumptions [H4] or [H5], if $m_n = n^\delta \log(n)^\gamma$ and if one of the following assumtions holds:*
  - *[H2] with $a > \max\{1 + 2(\delta + 1/d)/(1 - \delta), 2 + 1/d\}$,*
  - *[H2'] with $a > \max\left(1 + 2\{1/d(1 - \delta)\}, 1 + 1/d\right)$,*

  then, there exists $C > 0$ such that

$$\mathbb{P}\left(|FL_n(x)| \geq \epsilon \sqrt{m_n/n}\right) \leq C\epsilon^{-q_0} ,$$

with $q_0 = 2\lceil(a-1)/2\rceil$.

Lemma 3 (Fluctuation rates). *Under the assumptions of Lemma 2, we have for any $M > 0$,*

- *Geometric case.*

$$\sup_{\|x\| \leq M} |FL_n(x)| =_{a.s.} O\left(\sqrt{\frac{m_n}{n}} \log^{(b+1)/b}(n)\right) ;$$

- *Riemannian case.*

$$\sup_{\|x\| \leq M} |FL_n(x)| =_{a.s.} O\left(\sqrt{\frac{m_n^{1+2/q_0}}{n^{1-2/q_0}}} \log n\right) ,$$

with $q_0 = 2\lceil(a-1)/2\rceil$.

Remarks.

- In Lemma 1, we improve the moment inequality of [DL01], where the condition in the case of coefficient η is $a > 3(q - 1)$, which is always stronger than our condition.

- In the i.i.d. case a Bernstein type inequality is available:

$$\mathbb{P}\left(|FL_n(x)| \geq \epsilon\sqrt{\frac{m}{n}}\right) \leq C_1 \exp\left(-C_2\epsilon^2\right) ,$$

  Lemma 2 provides a weaker inequality for dependent sequences. Other probability inequalities for dependent sequences are presented in [DP04] and [KN05].
- Lemma 3 gives the almost sure bounds for the fluctuation. It is derived directly from the two previous lemmas.

Proof of the lemmas

Proof (Proof of Lemma 1). Let x be a fixed point in $\mathbb{R}^d$. Denote $Z_i = u_n(X_i) - \mathbb{E}u_n(X_i)$ where $u_n(.) = K_{m_n}(.,x)/\sqrt{m_n}$. Then

$$\sum_{i=1}^{n} Z_i = \sum_{i=1}^{n} u_n(X_i) - \mathbb{E}u_n(X_i) = \frac{n}{\sqrt{m_n}}(\hat{f}_n(x) - \mathbb{E}\hat{f}_n(x)) = \frac{n}{\sqrt{m_n}}FL_n(x) .$$

$$\tag{13}$$

The order of magnitude of the fluctuation $FL_n(x)$ is obtained by applying the inequality (5) to the centered sequence $\{Z_i\}_{1\leq i\leq n}$ defined above. We then control the normalized fluctuation of (13) with the covariance terms $C_k(r)$ defined in equation (6). Firstly, we bound the covariance terms:

- **Case $r = 0$.** Here $t_1 = \cdots = t_k = i$. Then we get:

$$C_k(r) = \left|\text{cov}\left(Z_{t_1}\cdots Z_{t_p}, Z_{t_{p+1}}\cdots Z_{t_k}\right)\right| \leq 2\mathbb{E}|Z_i|^k .$$

  By definition of Z_i:

$$\mathbb{E}|Z_i|^k \leq 2^k\mathbb{E}|u_n(X_i)|^k \leq 2^k\|u_n\|_\infty^{k-1}\mathbb{E}|u_n(X_0)| . \tag{14}$$

- **Case $r > 0$.** $C_k(r) = \left|\text{cov}\left(Z_{t_1}\cdots Z_{t_p}, Z_{t_{p+1}}\cdots Z_{t_k}\right)\right|$ is bounded in different ways, either using weak-dependence property or by direct bound.
    - **Weak-dependence bounds:**
        · η-*dependence:* Consider the following application:

$$\phi_p : (x_1,\ldots,x_p) \mapsto (u_n(x_1)\cdots u_n(x_p)) .$$

      Then $\|\phi_p\|_\infty \leq 2^p\|u_n\|_\infty^p$ and $\text{Lip}\,\phi_p \leq 2^p\|u_n\|_\infty^{p-1}\text{Lip}\,u_n$. Thus by η-dependence, for all $k \geq 2$ we have:

$$C_k(r) \leq \left(p2^p\|u_n\|_\infty^{p-1} + (k-p)2^{p-k}\|u_n\|_\infty^{p-k-1}\right)\text{Lip}\,u_n\eta_r ,$$
$$\leq k2^k\|u_n\|_\infty^{k-1}\text{Lip}\,u_n\eta_r . \tag{15}$$

· $\tilde{\phi}$-*dependence:* We use the inequality (7). Using the bound

$$\mathbb{E}|\phi_p(X_1,\ldots,X_p)| \le \|u_n\|_\infty^{p-1}\mathbb{E}|u_n(X_0)| \, ,$$

we derive a bound for the covariance terms:

$$C_k(r) \le k2^k\|u_n\|_\infty^{k-2}\mathbb{E}|u_n(X_0)|\mathrm{Lip}\,u_n\tilde{\phi}(r) \, . \tag{16}$$

— **Direct bound:** Triangular inequality implies for $C_k(r)$:

$$\left|\mathrm{cov}\left(Z_{t_1}\cdots Z_{t_p}, Z_{t_{p+1}}\cdots Z_{t_k}\right)\right| \le \underbrace{\left|\mathbb{E}\prod_{i=1}^{k} Z_{t_i}\right|}_{A} + \underbrace{\left|\mathbb{E}\prod_{i=1}^{p} Z_{t_i}\right|}_{B_p}\underbrace{\left|\mathbb{E}\prod_{i=p+1}^{k} Z_{t_i}\right|}_{B_{k-p}} \, ,$$

$$A = |\mathbb{E}\left(u_n(X_{t_1}) - \mathbb{E}u_n(X_{t_1})\right)\cdots(u_n(X_{t_k}) - \mathbb{E}u_n(X_{t_k}))| \, ,$$
$$= |\mathbb{E}u_n(X_0)|^k + |\mathbb{E}\left(u_n(X_{t_1})\cdots u_n(X_{t_k})\right)|$$
$$+ \sum_{s=1}^{k-1}|\mathbb{E}u_n(X_0)|^{k-s}\sum_{t_{i_1}\le\cdots\le t_{i_s}}\left|\mathbb{E}\left(u_n(X_{t_{i_1}})\cdots u_n(X_{t_{i_s}})\right)\right| \, .$$

Firstly, with $k \ge 2$:

$$|\mathbb{E}u_n(X_0)|^k \le \|u_n\|_\infty^{k-2}(\mathbb{E}|u_n(X_0)|)^2 \, .$$

Secondly, if $1 \le s \le k-1$:

$$\left|\mathbb{E}\left(u_n(X_{t_{i_1}})\cdots u_n(X_{t_{i_s}})\right)\right| \le \mathbb{E}|u_n(X_{t_{i_1}})\cdots u_n(X_{t_{i_s}})| \, ,$$
$$\le \|u_n\|_\infty^{s-1}\mathbb{E}|u_n(X_0)| \, ,$$
$$|\mathbb{E}u_n(X_0)|^{k-s} \le \|u_n\|_\infty^{k-s-1}\mathbb{E}|u_n(X_0)| \, .$$

Thirdly there is at least two different observations with a gap of $r > 0$ among $X_{t_1},\ldots,X_{t_k}$ so for any integer $k \ge 2$:

$$|\mathbb{E}\left(u_n(X_{t_1})\cdots u_n(X_{t_k})\right)| \le \|u_n\|_\infty^{k-2}\mathbb{E}|u_n(X_0)u_n(X_r)| \, .$$

Then, collecting the last four inequations yields:

$$A \le \|u_n\|_\infty^{k-2}(\mathbb{E}|u_n(X_0)|)^2$$
$$+(\mathbb{E}|u_n(X_0)|)^2\sum_{s=1}^{k-1}C_s^k\|u_n(X_0)\|_\infty^{k-2} + \|u_n\|_\infty^{k-2}\mathbb{E}|u_n(X_0)u_n(X_r)| \, .$$

So:

$$A \le \|u_n\|_\infty^{k-2}\left((2^k - 1)(\mathbb{E}|u_n(X_0)|)^2 + \mathbb{E}|u_n(X_0)u_n(X_r)|\right) \, . \tag{17}$$

Now, we bound B_i with $i < k$. As before:

$$B_i = |\mathbb{E}\left(u_n(X_{t_1}) - \mathbb{E}u_n(X_{t_1})\right) \cdots (u_n(X_{t_i}) - \mathbb{E}u_n(X_{t_i}))| \ ,$$

$$= \sum_{s=0}^{i} |\mathbb{E}(u_n(X_0)|^{i-s} \sum_{t_{j_1} \leq \cdots \leq t_{j_s}} |\mathbb{E}\left(u_n(X_{t_{j_1}}) \cdots u_n(X_{t_{j_s}})\right)| \ ,$$

$$\leq 2^i \|u_n\|_\infty^{i-2} (\mathbb{E}|u_n(X_0)|)^2 \ .$$

Then:

$$B_p \times B_{k-p} \leq 2^k \|u_n\|_\infty^{k-4} (\mathbb{E}|u_n(X_0)|)^4 \leq 2^k \|u_n\|_\infty^{k-2} (\mathbb{E}|u_n(X_0)|)^2 \ . \tag{18}$$

Another interesting bound for $r > 0$ follows, because according to inequalities (17) and (18) we have:

$$C_k(r) \leq \|u_n\|_\infty^{k-2} \left((2^{k+1} - 1)(\mathbb{E}|u_n(X_0)|)^2 + \mathbb{E}|u_n(X_0)u_n(X_r)|\right) \ .$$

Noting $\gamma_n(r) = \mathbb{E}|u_n(X_0)u_n(X_r)| \vee (\mathbb{E}|u_n(X_0)|)^2$, we have:

$$C_k(r) \leq 2^{k+1} \|u_n\|_\infty^{k-2} \gamma_n(r) \ . \tag{19}$$

We now use the different values of the bounds in inequalities (14), (15), (16) and (19). If we define the sequence $(w_r)_{0 \leq r \leq n-1}$ as:

- $w_0 = 1$,
- $w_r = \gamma_n(r) \wedge \|u_n\|_\infty \mathrm{Lip}\, u_n \eta_r \wedge \mathbb{E}|u_n(X_0)| \mathrm{Lip}\, u_n \tilde{\phi}(r)$,

then, for all r such that $0 \leq r \leq n - 1$ and for all $k \geq 2$:

$$C_k(r) \leq k 2^k \|u_n\|_\infty^{k-2} w_r \ .$$

We derive from this inequality and from (5):

$$\left\| \sum_{i=1}^n Z_i \right\|_q^q \leq \frac{(2q-2)!}{(q-1)!} \left\{ \left(n \sum_{r=0}^{n-1} C_2(r) \right)^{q/2} \vee n \sum_{r=0}^{n-1} (r+1)^{q-2} C_q(r) \right\} \ ,$$

$$\preceq (q\sqrt{n})^q \left\{ \left(\sum_{r=0}^{n-1} w_r \right)^{q/2} \vee \left(\frac{\|u_n\|_\infty}{\sqrt{n}} \right)^{q-2} \sum_{r=0}^{n-1} (r+1)^{q-2} w_r \right\} \ .$$

The symbol $\preceq$ means $\leq$ up to an universal constant. In order to control w_r, we give bounds for the terms $\gamma_n(r) = \mathbb{E}|u_n(X_0)u_n(X_r)| \vee (\mathbb{E}|u_n(X_0)|)^2$:

- In the case of [H4], we have:

$$\mathbb{E}|u_n(X_0)u_n(X_r)| \leq \sup_{j,k} \|f_{j,k}\|_\infty \|u_n\|_1^2 \ ,$$
$$(\mathbb{E}|u_n(X_0)|)^2 \leq \|f\|_\infty^2 \|u_n\|_1^2 \ .$$

- In the case of [H5], Lemma 2.3 of [Pri01] proves that $\mathbb{E}|u_n(X_0)u_n(X_r)| \leq (\mathbb{E}|u_n(X_0)|)^2$ for n sufficiently large and the same bound as above remains true for the last term.

In both cases, we conclude that $\gamma_n(r) \preceq \|u_n\|_1^2$. The properties (a), (b), (c) and (d) of section 2.2 ensures that $\|u_n\|_1^2 \preceq \dfrac{1}{m_n}$, $\|u_n\|_\infty \mathrm{Lip}\, u_n \preceq m_n^{1+1/d}$ and $\mathbb{E}|u_n(X_0)|\mathrm{Lip}\, u_n \preceq m_n^{1/d}$. We then have for $r \geq 1$:

$$w_r \preceq \frac{1}{m_n} \wedge m_n^{1+1/d}\eta_r \wedge m_n^{1/d}\tilde{\phi}_r \ . \tag{20}$$

In order to prove Lemma 1, it remains to control the sums

$$\left(\frac{\|u_n\|_\infty}{\sqrt{n}}\right)^{k-2} \sum_{r=0}^{n-1}(r+1)^{k-2}w_r \ , \tag{21}$$

for $k = 2$ and $k = q$ in both Riemannian and geometric cases.

- **Geometric case.**
  Under [H1] or [H1']: We remark that $a \wedge b \leq a^\alpha b^{1-\alpha}$ for all $\alpha \in [0;1]$. Using (20), we obtain first that $w_r \preceq (\eta_r \wedge \tilde{\phi}_r)^\alpha m_n^{\alpha(1+1/d)-(1-\alpha)}$ for n sufficiently large. Then for $0 < \alpha \leq d/(2d+1)$ we bound w_r independently of m_n: $w_r \preceq (\eta_r \wedge \tilde{\phi}_r)^\alpha$. For all even integer $k \geq 2$ we derive from the form of $\eta_r \wedge \tilde{\phi}_r$ that (in the third inequality $u = ar^b$):

$$\sum_{r=1}^{n-1}(r+1)^{k-2}w_r \preceq \sum_{r=0}^{n-1}(r+1)^{k-2}\exp(-\alpha a r^b) \ ,$$

$$\preceq \int_0^\infty r^{k-2}\exp(-\alpha a r^b)dr \ ,$$

$$\preceq \frac{1}{ba^{\frac{k-1}{b}}}\int_1^\infty u^{\frac{k-1}{b}-1}\exp(-u)du \ ,$$

$$\preceq \frac{1}{ba^{\frac{k-1}{b}}}\Gamma\left(\frac{k-1}{b}\right) \ .$$

Using the Stirling formula, we can find a constant B such that, for the special cases $k = 2$ and $k = q$:

$$\sum_{r=1}^{n-1}(r+1)^{k-2}w_r \preceq \frac{1}{ba^{\frac{k-1}{b}}}\Gamma\left(\frac{k-1}{b}\right) \preceq (Bk)^{\frac{k}{b}} \ .$$

- **Riemannian case.**
  Under [H2]: we have $m_n = n^\delta \log(n)^\beta$ for some $0 < \delta < 1$ and $\gamma > 0$ and the assumption of Lemma 1 implies that:

$$a > \max\left(q-1, \frac{\delta(q-1)(4+2/d)}{q-2+\delta(4-q)}, 2+\frac{1}{d}\right) \ .$$

Then, we have $a > \max\left(k - 1, \dfrac{\delta(k-1)(4+2/d)}{k-2+\delta(4-k)}\right)$ for both cases $k = q$ or $k = 2$. This assumption on a implies that:

$$\frac{(k+2/d)\delta + 2 - k}{2(a - k + 1)} < \frac{(4-k)\delta + k - 2}{2(k-1)}.$$

Furthermore, reminding that $0 < \delta < 1$:

$$0 < \frac{(4-k)\delta + k - 2}{2(k-1)} = 1 - \frac{k(1+\delta) - 4\delta}{2(k-1)} \leq 1.$$

We derive from the two previous inequalities that there exists $\zeta_k \in]0, 1[$ verifying $\dfrac{(k+2/d)\delta + 2 - k}{2(a - k + 1)} < \zeta_k < \dfrac{(4-k)\delta + k - 2}{2(k-1)}$.

For $k = q$ or $k = 2$, we now use Tran's technique as in [ABD02]. We divide the sum (21) in two parts in order to bound it by sequences tending to 0, due to the choice of ζ_k:

$$\left(\sqrt{\frac{m_n}{n}}\right)^{k-2} \sum_{r=0}^{[n^{\zeta_k}]-1} (r+1)^{k-2} w_r \preceq \left(\sqrt{\frac{m_n}{n}}\right)^{k-2} \frac{[n^{\zeta_k}]^{k-1}}{m_n},$$

$$\preceq n^{(2\zeta_k(k-1)-((4-k)\delta+k-2))/2},$$

$$= O(1),$$

$$\left(\sqrt{\frac{m_n}{n}}\right)^{k-2} \sum_{r=[n^{\zeta_k}]}^{n-1} (r+1)^{k-2} w_r \leq \left(\sqrt{\frac{m_n}{n}}\right)^{k-2} m_n^{1+1/d} [n^{\zeta_k}]^{k-1-a},$$

$$\leq n^{(-2\zeta_k(a-k-1)+((k+2/d)\delta+2-k))/2},$$

$$= O(1).$$

Under [H2']: Under the assumption of Lemma 1:

$$a > \max\left(q - 1, \frac{\delta(q-1)(2+2/d)}{q-2+\delta(4-q)}, 1 + \frac{1}{d}\right),$$

we derive exactly as in the previous case that there exists $\zeta_k \in]0; 1[$ for $k = q$ or $k = 2$ such that

$$\frac{(k-2+2/d)\delta + 2 - k}{2(a - k + 1)} < \zeta_k < \frac{(4-k)\delta + k - 2}{2(k-1)}.$$

We then apply again the Tran's technique that bound the sum (21) in that case.

Lemma 1 directly follow from (13).   □

Remarks. We have in fact proved the following sharper result. There exists a universally constant C such that

$$\left(\frac{n}{m_n}\right)^{q/2} \|FL_n(x)\|_q^q \leq \begin{cases} (Cq)^q & \text{in the Riemaniann case,} \\ (Cq^{1+1/b}\sqrt{n})^q & \text{in the geometric case.} \end{cases} \quad (22)$$

Proof (Proof of Lemma 2). The cases of Riemannian or geometric decay of the dependence coefficients are considered separately.

- **Geometric decay** We present a technical lemma useful to deduce exponential probabilities from moment inequalities at any even order.

    Lemma 4. *If the variables* $\{V_n\}_{n\in\mathbb{Z}}$ *satisfies, for all* $k \in \mathbb{N}^*$

$$\|V_n\|_{2k} \leq \phi(2k) , \quad (23)$$

    where ϕ *is an increasing function with* $\phi(0) = 0$. *Then:*

$$\mathbb{P}(|V_n| \geq \epsilon) \leq e^2 \exp\left(-\phi^{-1}(\epsilon/e)\right) .$$

    Proof. By Markov's inequality and Assumption (23), we obtain

$$\mathbb{P}\left(|V_n| \geq \epsilon\right) \leq \left(\frac{\phi(2k)}{\epsilon}\right)^{2k} .$$

    With the convention $0^0 = 1$, the inequality is true for all $k \in \mathbb{N}$. Reminding that $\phi(0) = 0$, there exists an integer k_0 such that $\phi(2k_0) \leq \epsilon/e < \phi(2(k_0 + 1))$. Noting ϕ^{-1} the generalized inverse of ϕ, we have:

$$\mathbb{P}\left(|V_n| \geq \epsilon\right) \leq \left(\frac{\phi(2k_0)}{\epsilon}\right)^{2k_0} \leq e^{-2k_0} = e^2 e^{-2(k_0+1)} ,$$
$$\leq e^2 \exp\left(-\phi^{-1}(\epsilon/e)\right) .$$

    $\square$

    We rewrite the inequality (22): $\left\|\sqrt{\frac{n}{m_n}}FL_n\right\|_{2k} \leq \phi(2k)$ with $\phi(x) = Cx^{\frac{b+1}{b}}$ for a convenient constant C. Applying Lemma 4 to $V_n = \sqrt{\frac{n}{m_n}}FL_n$ we obtain:

$$\mathbb{P}\left(|FL_n| \geq \epsilon\sqrt{\frac{m_n}{n}}\right) \leq e^2 \exp\left(-\phi^{-1}(\epsilon/e)\right) ,$$

    and we obtain the result of the Lemma 2.
- **Riemannian decay** In this case, the result of Lemma 1 is obtained only for some values of q depending of the value of the parameter a:
    - In the case of η-dependence:

$$a > \max\left(q - 1, \frac{1 + \delta + 2/d}{1 - \delta}, 2 + \frac{1}{d}\right) .$$

– In the case of $\tilde{\phi}$-dependence:

$$a > \max\left(q - 1, 1 + \frac{2}{d(1-\delta)}, 1 + \frac{1}{d}\right).$$

We consider that the assumptions of the Lemma 2 on a are satisfied in both cases of dependence. Then $q_0 = 2\lceil(a-1)/2\rceil$ is the even integer such that $a - 1 \leq q_0 < a + 1$. It is the largest order such that the assumptions of Lemma 1 (recalled above) are verified and then the Lemma 1 gives us directly the rate of the moment: $\limsup_{n\to\infty}(n/m_n)^{q_0/2}\|FL_n(x)\|_{q_0}^{q_0} < +\infty$. We apply Markov's inequality to obtain the result of Lemma 2:

$$\mathbb{P}\left(|FL_n(x)| \geq \epsilon\sqrt{m_n/n}\right) \leq \frac{\left(\sqrt{n/m_n}\,\|FL_n(x)\|_{q_0}\right)^{q_0}}{\epsilon^{q_0}}.$$

$\square$

Proof (Proof of Lemma 3). We follow here Liebscher's strategy as in [AD03]. We recover $B := B(0, M)$, the ball of center 0 and radius M, by at least $(4M\mu+1)^d$ balls $B_j = B(x_j, 1/\mu)$. Then, under the assumption that $K_m(., y)$ is supported on a compact of diameter proportional to $1/m$, we have, for all j:

$$\sup_{x\in B_j}|FL_n(x)| \leq |\hat{f}(x_j) - \mathbb{E}\hat{f}(x_j)|$$

$$+ C\frac{m_n^{1/d}}{\mu}(|\tilde{f}(x_j) - \mathbb{E}\tilde{f}(x_j)| + 2|\mathbb{E}\tilde{f}(x_j)|), \quad (24)$$

with C a constant and $\tilde{f} = n^{-1}m_n\sum_{i=1}^n \tilde{K}_{m_n}(x, X_i)$ where $\tilde{K}_{m_n}$ is a kernel of type $\tilde{K}_{m_n}(x, y) = K_{m_n}(x_j, y)\mathbb{1}_{|x-x_j|\leq m_n^{-1}}$ satisfying properties (a), (b), (c) and (d) of section 2.2. Then using (24) and with obvious short notation:

$$\mathbb{P}\left(\sup_{\|x\|\leq M}|FL_n(x)| > \epsilon\sqrt{\frac{m_n}{n}}\right) \leq \sum_{j=1}^{(4M\mu+1)^d}\mathbb{P}\left(\sup_{x\in B_j}|FL_n(x)| > \epsilon\sqrt{\frac{m_n}{n}}\right),$$

$$\leq (4M\mu+1)^d\left[\sup_j\mathbb{P}\left(|FL_n(x_j)| > \epsilon\sqrt{\frac{m_n}{n}}\right)\right.$$

$$+\mathbb{P}\left(C\frac{m_n^{1/d}}{\mu}|\tilde{FL}_n(x)| > \epsilon\sqrt{\frac{m_n}{n}}\right)$$

$$\left.+\mathbb{P}\left(2C\frac{m_n^{1/d}}{\mu}|\mathbb{E}\tilde{f}_n(x)| > \epsilon\sqrt{\frac{m_n}{n}}\right)\right].$$

Using the fact that f is bounded, $\mathbb{E}\tilde{f}_n = \int \tilde{K}_{m_n}(s)f(x-hs)ds$ is bounded independently of n. We deduce that the last probability term of the sum tends to 0 with n.

With the choice $\mu = m_n^{1/d}$, we apply Lemma 2 on f and $\tilde{f}$. We have then the same rate (uniform in x) for both terms $\mathbb{P}\left(C\mu^{-1}m_n^{1/d}|\tilde{FL}_n(x)| > \epsilon\sqrt{m_n/n}\right)$ and $\mathbb{P}\left(|FL_n(x)| > \epsilon\sqrt{m_n/n}\right)$. Then replacing μ^d by m_n, we obtain uniform probability inequalities in both cases of geometric or Riemannian decays:

$$\mathbb{P}\left(\sup_{\|x\|\leq M} |FL_n(x)| \geq \epsilon_n\sqrt{m_n/n}\right) \preceq m_n \exp\left(-C\epsilon_n^{b/b+1}\right), \qquad (25)$$

$$\mathbb{P}\left(\sup_{\|x\|\leq M} |FL_n(x)| \geq \epsilon_n\sqrt{m_n/n}\right) \preceq m_n\epsilon_n^{-q_0}. \qquad (26)$$

In the geometric case, we take ϵ_n with the form $G(\log n)^{(b+1)/b}$ such that the bound in the inequality (25) becomes $m_n n^{-GC}$. Reminding that $m_n \leq n$, the sequence $m_n n^{-GC}$, bounded by n^{1-GC}, is summable for a conveniently chosen constant G. Borel-Cantelli's Lemma then concludes the proof in this case.

In the Riemannian case, we take $\epsilon_n = (m_n n)^{q_0^{-1}} \log n$ such that the bound in the inequality (26) becomes $n^{-1}\log^{-q_0} n$. Reminding that $q_0 \geq 2$, this sequence is summable and here again we conclude by applying Borel-Cantelli's Lemma. $\square$

Proof of the theorems

The order of magnitude of the bias is given by Assumption (e) and the Lemmas provide bounds for fluctuation term. There only remain to determine the optimal bandwidth m_n in each case.

Proof (Proof of Theorem 1). Applying Lemma 1 yields Theorem 1 when q is an even integer. For any real q, Lemma 1 with $2(\lceil q/2\rceil + 1) \geq 2$ and Jensen's inequalities yields:

$$\left(\frac{n}{m_n}\right)^{q/2} \mathbb{E}|FL_n(x)|^q = \left(\frac{n}{m_n}\right)^{q/2} \mathbb{E}\left(FL_n(x)^{2(\lceil q/2\rceil+1)}\right)^{q/\{2(\lceil q/2\rceil+1)\}},$$

$$\leq \left(\left(\frac{n}{m_n}\right)^{\lceil q/2\rceil+1} \mathbb{E}FL_n(x)^{2(\lceil q/2\rceil+1)}\right)^{q/\{2(\lceil q/2\rceil+1)\}}.$$

Plugging this bound and the bound for the bias in (12), we obtain a bound for the $\mathbb{L}^q$-error of estimation:

$$\|\hat{f}_n(x) - f(x)\|_q \leq \|FL_n(x)\|_q + |R_n(x)| = O\left(\sqrt{\frac{m_n}{n}} + m_n^{-\rho/d}\right).$$

The optimal bandwidth $m_n^* = n^{d/(2\rho+d)}$ is the same as in the i.i.d. case. Set $\delta = d/(2\rho + d)$. For this value of δ, the conditions on the parameter a of Lemma 2 are equivalent to those of Theorem 1. $\square$

Proof (Proof of Theorem 2). Applying the probability inequality (25) in the proof of Lemma 3 and the identity $\mathbb{E}|Y|^q = \int_0^{+\infty} \mathbb{P}\left(|Y| \geq t^{1/q}\right) dt$, we obtain

$$\mathbb{E} \sup_{\|x\| \leq M} |\hat{f}_n(x) - f(x)|^q = O\left(\left\{\sqrt{\frac{m_n}{n}} \log^{(b+1)/b}(n)\right\}^q + m_n^{-q\rho/d}\right).$$

Lemma 3 gives the rate of almost sure convergence:

$$\sup_{\|x\| \leq M} |\hat{f}_n(x) - f(x)| =_{a.s.} O\left(\sqrt{\frac{m_n}{n}} \log^{\frac{b+1}{b}} n + m_n^{-\rho/d}\right).$$

In both cases, the optimal bandwidth is $m_n^* = (n/\log^{2(b+1)/b}(n))^{d/(2\rho+d)}$, which yields the rates claimed in Theorem 2. □

Proof (Proof of Theorem 3). Applying the probability inequality (26) and the same line of reasoning as in the previous proof, we obtain

$$\mathbb{E} \sup_{\|x\| \leq M} |\hat{f}_n(x) - f(x)|^q = O\left(\left(\sqrt{\frac{m_n}{n}} m_n^{1/q_0}\right)^q + m_n^{-q\rho/d}\right),$$

where $q_0 = 2\lceil(a-1)/2\rceil$. The optimal bandwidth is $m_n^* = n^{dq_0/(d(q_0+2)+2\rho q_0)}$. Set $\delta = dq_0/(2\rho q_0 + d(q_0+2))$. For this value of δ, the conditions on a of Lemma 2 are satisfied as soon as $a \geq 4$ and $\rho > 2d$.

Lemma 3 gives the rate for the fluctuation in the almost sure case. This leads the optimal bandwidth

$$m_n^* = \left(n/\log^{2+4/(q_0-2)}(n)\right)^{d(q_0-2)/(2\rho q_0 + d\{q_0+2\})}.$$

We deduce the two rates of Theorem 3 in the almost sure and $\mathbb{L}^q$ framework. □

References

[AD98] P. ANGO NZE and P. DOUKHAN (1998), *Functional estimation for time series: uniform convergence properties*, Journal of Statistical Planning and Inference, vol. 68, pp. 5-29.

[ABD02] P. ANGO NZE, P. BÜHLMANN and P. DOUKHAN (2002), *Weak dependence beyond mixing and asymptotics for nonparametric regression*, Annals of Statistics, vol. 30, n. 2, pp. 397-430.

[AD03] P. ANGO NZE and P. DOUKHAN (2003), *Weak Dependence: Models and Applications to econometrics*, Econometric Theory, vol. 20, n. 6, pp. 995-1045.

[BGR00] A.D. BARBOUR, R.M. GERRARD and G. REINERT (2000), *Iterates of expanding maps*, Probability Theory and Related Fields, vol. 116, pp. 151-180.

[Ded98]   J. DEDECKER (1998), *A central limit theorem for random fields*, Probability Theory and Related Fields, vol. 110, pp. 397-426.

[DD03]   J. DEDECKER and P. DOUKHAN (2003), *A new covariance inequality and applications*, Stochastic Processes and their Applications, vol. 106, n. 1, pp. 63-80.

[DP04]   J. DEDECKER and C. PRIEUR (2004), *New dependence coefficients. Examples and applications to statistics*, To appear in Probability Theory and Related Fields.

[DDLLLP04]   J. DEDECKER, P. DOUKHAN, G. LANG, J.R. LEON, S. LOUHICHI and C. PRIEUR (2004), *Weak dependence: models, theory and applications*, Merida, XVII escuela venezolana de matematicas.

[Dou88]   P. DOUKHAN (1988), *Formes de Toeplitz associées à une analyse multi-échelle*, Compte rendus des Séances de l'Académie des Sciences, Série I. Mathématique. 306, vol. 84, n. 15, pp. 663-666.

[Dou90]   P. DOUKHAN (1991), *Consistency of delta-sequence estimates of a density or of a regression function for a weakly dependent stationary sequence*, Séminaire de statistique d'Orsay, Estimation Fonctionnelle 91-55.

[Dou94]   P. DOUKHAN (1994) *Mixing: properties and examples*, Lecture Notes in Statistics, vol. 85, Springer-Verlag.

[DL99]   P. DOUKHAN and S. LOUHICHI (1999), *A new weak dependence condition and applications to moment inequalities*, Stochastic Process and their Applications, vol. 84, pp. 313-342.

[DL01]   P. DOUKHAN and S. LOUHICHI (2001), *Functional estimation for weakly dependent stationary time series*, Scandinavian Journal of Statistics, vol. 28, n. 2, pp. 325-342.

[DMR05]   P. DOUKHAN H. MADRE and M. ROSENBAUM (2005). *ARCH type bilinear weakly dependent models*, submitted.

[DS01]   P. DOUKHAN and J.C. SIFRE (2001), *Cours d'analyse - Analyse réelle et intégration*, Dunod.

[GKS96]   L. GIRAITIS, H.L. KOUL and D. SURGAILIS (1996), *Asymptotic normality of regression estimators with long memory errors*, Statistics & Probability Letters, vol. 29, pp. 317-335.

[HLT01]   M. HALLIN, Z. LU, L.T. TRAN (2001), *Density estimation for spatial linear processes*, Bernoulli, pp. 657-668.

[KN05]   R. S. KALLABIS and M. H. NEUMANN (2005), *A Bernstein inequality under weak dependence*, prepublication.

[Pri01]   C. PRIEUR (2001), *Density Estimation For One-Dimensional Dynamical Systems*, ESAIM , Probability & Statististics, pp. 51-76.

[Sze33]   G. SZEGÖ (1933), *Orthogonal polynomials*, American Mathematical Society Colloquium Publication, vol. 23.

[Tsy04]   A.B. TSYBAKOV (2004), *Introduction à l'estimation non-paramétrique*, Springer.

[Vie97]   G. VIENNET (1997), *Inequalities for absolutely regular sequences : application to density estimation*, Probability Theory and Related Fields, vol. 107, pp. 467-492.

Variograms for spatial max-stable random fields

Dan Cooley[1], Philippe Naveau[1,2], and Paul Poncet[3]

[1] Department of Applied Mathematics, University of Colorado at Boulder,USA
[2] Laboratoire des Sciences du Climat et de l'Environnement, IPSL-CNRS
   philippe.naveau@colorado.edu
[3] Ecole des Mines de Paris, France

1 Introduction

For the univariate case in extreme value theory, it is well known (e.g. [EKM97]) that parametric models characterize the asymptotic behavior of extremes (e.g. sample maxima, exceedances above a high threshold) as the sample size increases. More precisely, a Generalized Extreme Value (GEV) distribution approximates the distribution of sample maxima and a Generalized Pareto Distribution fits asymptotically exceedances. These results can be extended to stationary observations under some general conditions, e.g. [LLR83]. In a multivariate setting and therefore, for spatial extremes, a general structure of the limiting behavior of component-wise maxima has also been proposed in terms of max-stable processes, e.g. [HR77], see Equation (3) for a definition of such processes. However, an important distinction between the univariate and the multivariate case is that no parametrized model can entirely represent max-stable processes. Hence, statistical inference has mostly focused on special bivariate cases, e.g. logistic or bilogistic model (see Chapter 8 in [Col01]). When dealing with spatial data, there are as many variables as locations and consequently, additional assumptions have to be made in order to work with manageable models. The multivariate framework has rarely been treated from a statistical point of view (e.g. [HT04]), but there has been a growing interest in the analysis of spatial extremes in recent years. For example, [HP05] proposed two specific stationary models for extreme values. These models depend on one parameter that varies as a function of the distance between two points. [DM04] proposed space-time processes for heavy tail distributions by linearly filtering i.i.d. sequences of random fields at each location. [Sch02, Sch03] simulated stationary max-stable random fields and studied the extremal coefficients for such fields. Bayesian or latent processes for modeling extremes in space has been also investigated by several authors. In this case, the spatial structure was modeled by assuming that the extreme value parameters were realizations of a smoothly varying process, typically a Gaussian process with a spatial dependence [CC99]. In geostatistics, a classical approach called "Gaussian anamorphosis" consists of transforming a spatial field into a Gaussian one (e.g. Chapter 33 in [Wac03]). Indicator functions are then used to obtain information about the tails. Although the approach performs well for some specific cases, it is not based on extreme value theory and therefore, modeling extremes with a Gaussian anamorphosis does not take advantage of the theoretical foundation provided by extreme value theory.

In comparison with all these past developments, our research can be seen as a further step in the direction taken by [Sch03] and [Sch02]. We work with stationary max-stable fields and focus on capturing the spatial structure with extremal coefficients. The novelty is that we propose different estimators of extremal coefficients that are more clearly linked to the field of geostatistics. In contrast with the work of [HP05] in which a special structure was imposed (i.e. a parameter measured the dependence as a function of the distance between locations), our estimator can be used for any max-stable field. It is also worthwhile to emphasize the following two points. First, we will show that our estimators are closely related to a distance introduced in [DR93]. Second, although we do not pursue the Gaussian anomorphosis approach used in geostatistics, one should not forget that the field of geostatistics has a long history of modeling spatial data sets, and its development has been tremendous in terms of environmental, climatological and ecological studies. Hence, finding connections between geostatistics and extreme value theory is of primary interest for improving the analysis of complex fields of extreme values.

In geostatistics, it is classical to define the following second-order statistic

$$\gamma(h) = \frac{1}{2}\mathbb{E}|Z(x+h) - Z(x)|^2 \, , \qquad (1)$$

where $\{Z(x)\}$ represents a spatial and stationary process with a well-defined covariance function and $x \in \mathbb{R}^2$. The function $\gamma(.)$ is called the (non-centered) (semi) variogram and it has been extensively used by the geostatistic community, see [Wac03], [CD99], [Ste99] and [Cre93]. With respect to extremes, this definition is not well adapted, because a second order statistic is difficult to interpret inside the framework of extreme value theory. Instead of taking the squared difference in $\gamma(.)$, we will show that working with the absolute difference $|Z(x+h) - Z(x)|$ is more appropriate when dealing with extreme values. The first-order moment of this difference leads to the definition of the madogram

$$\nu(h) = \frac{1}{2}\mathbb{E}|Z(x+h) - Z(x)| \, , \qquad (2)$$

where the stationary process $\{Z(x)\}$ with an assumed finite mean represents extreme values at different locations x, say annual maximum precipitation at given weather stations. The basic properties of this first-order variogram have been studied by [Mat87] but he did not explore the connection between madograms and extreme value theory. Besides the basic properties $\nu(0) = 0$, $\nu(h) \geq 0$, $\nu(h) = \nu(-h)$ and $\nu(h+k) \leq \nu(h) + \nu(k)$, he showed that

$$\nu(h) = \int\limits_{-\infty}^{+\infty} \gamma(h; u)\, du \text{ with } \gamma(h; u) = \frac{1}{2}\mathbb{E}|\mathbb{1}(Z(x + h) > u) - \mathbb{1}(Z(x) > u)| ,$$

and

$$\sum_i \sum_j \lambda_i\, \nu(h_i - h_j)\, \lambda_j \le 0, \text{ for all } \lambda_1, \dots, \lambda_k \text{ such that } \sum_{i=1}^{k} \lambda_i = 0 .$$

Here $\mathbb{1}(A)$ represents the indicator function that is equal to 1 if A true and equal to 0 otherwise. In addition, the madogram satisfies $\sqrt{2}\nu(h) \le \sqrt{\gamma(h)}$ and it is differentiable at the origin, if and only if, it has a linear behavior near the origin. A detailed discussion of the properties of the madogram can be found in Chapter 6 of [Pon04].

Our research can be seen as an extension of the work by [NPC05] who studied the bivariate case but did not treat the spatial aspect of the madogram for extreme value theory. In addition, the estimation of the madogram for simulated max-stable random fields is novel.

This paper is organized as follows. In Section 2.1, we recall the basic principles of max-stable processes and two specific classes of max-stables processes that are used in our simulations. Section 2.2 links the extremal coefficient and the madogram in the context of max-stable processes. This part provides a bridge between geostatistics and extreme value theory. To illustrate how this relationship works, simulations of the estimated madograms are presented in Section 2.3. Conclusions and future research directions are described in Section 3.

2 Spatial extremes

2.1 Max-stable processes

In this work, we assume that $\{Z(x)\}$ is a max-stable process [Res87, Smi04]. To define such a process, we consider that the marginal distribution has been transformed to unit Fréchet

$$\mathbb{P}(Z(x) \le u) = \exp(-1/u), \text{ for any } u > 0 .$$

Then, we impose that all the finite-dimensional distributions are max-stable, i.e.

$$\mathbb{P}\left(Z(x_1) \le tu_1, ..., Z(x_r) \le tu_r\right)^t = \mathbb{P}\left(Z(x_1) \le u_1, ..., Z(x_r) \le u_r\right) ,$$

for any $t \ge 0$, $r \ge 1$, $x_i \in \mathbb{R}^2$, $u_i > 0$ with $i = 1, .., r$. Such processes can be rewritten [Sch02, Sch03, DR93] as

$$\mathbb{P}\left(Z(x) \leq u(x), \text{ for all } x \in \mathbb{R}^2\right) = \exp\left[-\int \max_{x \in \mathbb{R}^2}\left\{\frac{g(s,x)}{u(x)}\right\}\delta(ds)\right] , \quad (3)$$

where the function $g(.)$ is a nonnegative function which is measurable in the first argument, upper semi-continuous in the second and has to satisfy $\int g(s,x)\delta(ds) = 1$.

To display the characteristics of such random fields, we plot two types of max-stable fields in figures 1 and 2. The first class was developed by [Smi90], where points in $(0,\infty) \times \mathbb{R}^2$ are simulated according to a point process:

$$Z(x) = \sup_{(y,s) \in \Pi} [s\, f(x-y)], \text{ with } x \in \mathbb{R} , \quad (4)$$

where Π is a Poisson point process on $\mathbb{R}^2 \times (0,\infty)$ with intensity $s^{-2}dyds$ and f is a non-negative function such that $\int f(x)dx = 1$. In this paper f is taken to be a two-dimensional Gaussian density function with the identity covariance matrix. Figure 1 displays a field realization with dimensions of 40×40.

The second type proposed by [Sch02] stems from a Gaussian random field which is then scaled by the realization of a point process on $(0,\infty)$:

$$Z(x) = \max_{s \in \Pi} sY_s(x) , \quad (5)$$

where Y_s are i.i.d. stationary Gaussian processes and Π is a Poisson point process on $(0,\infty)$ with intensity $\sqrt{2\pi}r^{-2}dr$. Figure 2 shows one realization of such a process. Both figures of max-stable random fields have Gumbel margins; it is more convenient to visualize the spatial structure with Gumbel margins than with Fréchet ones. They were simulated using a code developed by Schlather, see http://cran.r-project.org contributed package RandomFields. The spatial structure of these two examples will be detailed in the next section.

2.2 Extremal coefficient and madogram

Concerning the spatial dependence, we note that Equation (3) gives

$$\mathbb{P}\left(Z(x) \leq u \text{ and } Z(x+h) \leq u\right) = \exp\left[-\theta(h)/u\right] ,$$

where the quantity $\theta(h)$ equals

$$\theta(h) = \int \max\{g(s,x), g(s,x+h)\}\delta(ds) ,$$

and is called the *extremal coefficient*. It has been used in many dependence studies for bivariate vectors [Fou04, LT97]. The coefficient belongs to the interval $[1,2]$ and gives partial, but not complete, information about the dependence structure of the bivariate couple $(Z(x), Z(x+h))$. If the two variables of this couple are independent, then $\theta(h)$ is 2. At the other end of the spectrum, if the variables are equal in probability, then $\theta(h) = 1$. Hence, the extremal coefficient value $\theta(h)$ provides some dependence information as a function of the distance between two locations.

The bivariate distribution for the Schlather model $\mathbb{P}(Z(x) \leq s, Z(x+h) \leq t)$ corresponds to

$$\exp\left\{-\frac{1}{2}\left(\frac{1}{t}+\frac{1}{s}\right)\left(1+\sqrt{1-2\left(\rho(h)+1\right)\frac{st}{(s+t)^2}}\right)\right\}, \qquad (6)$$

where $\rho(h)$ is the covariance function of the underlying Gaussian process. This yields an extremal coefficient of

$$\theta(h) = 1 + \sqrt{1-\frac{1}{2}\left(\rho(h)+1\right)}. \qquad (7)$$

In Figure 2, the covariance function $\rho(h)$ is chosen to be equal to $\rho(h) = \exp(-h/40)$.

For the Smith model the bivariate distribution $\mathbb{P}(Z(x) \leq s, Z(x+h) \leq t)$ equals

$$\exp\left\{-\left[\frac{1}{s}\Phi\left(\frac{a}{2}+\frac{1}{a}\log\frac{t}{s}\right)+\frac{1}{t}\Phi\left(\frac{a}{2}+\frac{1}{a}\log\frac{s}{t}\right)\right]\right\}, \qquad (8)$$

where $\Phi(.)$ corresponds to the cumulative distribution function of the standardized Normal distribution, see Figure 2. In equality (8), the extremal coefficient equals

$$\theta(h) = 2\Phi\left(\sqrt{h^T\Sigma^{-1}h}/2\right), \qquad (9)$$

where $a^2 = h^T\Sigma^{-1}h$.

The immediate statistical question is: how can we estimate $\theta(h)$ when observations from a max-stable field are available? Proposition 1 and Section 2.3 will provide elements to answer this question.

In this paragraph, we assume that the marginals of $\{Z(x)\}$ follows a unit Fréchet distribution. This implies that the first-order moment of $\{Z(x)\}$ is not finite and the madogram defined by (2) can not be computed. To bypass this hurdle, we introduce the following rescaled madogram

$$\eta_t(h) = \frac{1}{2}\mathbb{E}\left|t\mathbb{1}(Z(x+h)\leq t)-t\mathbb{1}(Z(x)\leq t)\right|, \qquad (10)$$

where $t \in (0,\infty)$. By construction, the madogram $\eta_t(h)$ is always defined.

Proposition 1. *For any stationary max-stable spatial random field with unit Fréchet marginals and extremal coefficient $\theta(h)$, the madogram $\eta_t(h)$ defined by (10) is equal to $\eta_t(h) = t\left(e^{-1/t}-e^{-\theta(h)/t}\right)$, and consequently, the limiting madogram, $\eta(h) = \lim \eta_t(h)$ as $t \uparrow \infty$, is $\eta(h) = \theta(h) - 1$.*

Hence, this proposition indicates that a madogram completely characterizes the extremal coefficient and vice-versa. The extremal coefficient enables us to make the link between max-stable processes and madograms. The proof of Proposition 1 can be found in the Appendix. Note that $\eta(h)$ is a madogram, as [Mat87] stated that the limit of madograms is a madogram. The following argument justifies this claim.

[DR93] defined a L_1 distance for max-stable bivariate random variables (X, Y), i.e.

$$\mathbb{P}(X \leq x, Y \leq y) = \exp \left(- \int_{[0,1]} \max \left[\frac{f(s)}{x}, \frac{g(s)}{y} \right] ds \right) ,$$

where $f, g \in L_1([0, 1])$ are called *spectral functions*. This definition can be compared to (3) and it provides another way to characterize max-stable processes. The distance $d(X, Y)$ was defined as the distance between the spectral functions

$$d(X, Y) = \int_{[0,1]} |f(s) - g(s)| ds .$$

If we denote $\theta(X, Y) = \int \max[f(s), g(s)] ds$, the equality $|a - b| = 2 \max(a, b) - (a + b)$ gives us

$$d(X, Y) = 2\theta(X, Y) - \left(\int f(s)ds + \int g(s)ds \right) ,$$
$$= 2(\theta(X, Y) - 1) , \tag{11}$$

because assuming unit Fréchet margins implies that the spectral functions have unit L_1 norms, i.e. $\int f(s)ds = \int g(s)ds = 1$. Equality (11) has to be compared with the one stated in Proposition 1, $\eta(h) = \theta(h) - 1$. Both equalities contain the same message, a L_1 norm (either expressed in terms of madograms or spectral functions) fully characterizes the extremal coefficient. From the madogram, it is straightforward to define an asymptotically unbiased estimator from Equation (10)

$$\hat{\eta}_t(h) = \frac{1}{2|\mathcal{N}_h|} \sum_{(x_i, x_j) \in \mathcal{N}_h} |t\mathbb{1}(Z(x_i) \leq t) - t\mathbb{1}(Z(x_j) \leq t)| ,$$

where $\mathcal{N}_h$ is the set of sample pairs lagged by the distance h and $|\mathcal{N}_h|$ is the cardinal of this set.

Numerous simulations (not shown here but available from the authors upon request) indicate that the choice of t in $\hat{\eta}_t(h)$ is difficult and that $\hat{\eta}_t(h)$ only provides reasonable estimates when the dependence is strong, i.e. when the extremal coefficient is close to 1. One drawback is that $\hat{\eta}_t(h)$ converges very slowly if the extremal coefficient is close to 2, i.e. weak dependence between two sites. The reason for this stems for the archetypical assumption of Fréchet margins used in multivariate extremes. As an alternative, we will present other madograms in the next paragraph that have the advantage of giving much better estimates of $\theta(h)$. The fundamental difference is that we will not assume having Fréchet margins anymore but instead Generalized Extreme Value (GEV) margins with finite second moments. This can be done without loss of generality. To implement this strategy, we recall that the GEV distribution is defined as

$$\mathbb{P}(Z(x) \leq u) = \exp(-1/u_\beta(u)), \text{ with } u_\beta(u) = \left(1 + \xi \frac{u - \mu}{\sigma}\right)_+^{1/\xi}, \quad (12)$$

where a_+ is equal to a if $a > 0$ and 0 otherwise. The parameter vector β is defined as $\beta = (\mu, \sigma, \xi)$. In Equation (12), μ, $\sigma > 0$ and ξ are the location, scale and shape parameters, respectively. The shape parameter drives the tail behavior of F. For example, the classical Gumbel distribution corresponds to $\xi = 0$. If $\xi > 0$ then the distribution is heavy tailed (e.g. the variance is infinite if $\xi \geq 0.5$). If $\xi < 0$, it has a bounded upper tail. With these new notations, the relationship between the madogram and the extremal coefficient for the GEV case can be elegantly expressed by the following proposition.

Proposition 2. *Let $\{Z(x)\}$ be any stationary max-stable spatial random field with GEV marginals and the extremal coefficient $\theta(h)$. If the GEV shape parameter ξ satisfies $\xi < 1$, then the madogram $\nu(h)$ and the extremal coefficient $\theta(h)$ verify*

$$\theta(h) = \begin{cases} u_\beta \left(\mu + \frac{\nu(h)}{\Gamma(1-\xi)}\right), & \text{if } \xi < 1 \text{ and } \xi \neq 0, \\ \exp\left(\frac{\nu(h)}{\sigma}\right), & \text{if } \xi = 0, \end{cases} \quad (13)$$

where $u_\beta(.)$ is defined by (12) and $\Gamma(.)$ is the classical Gamma function.

The proof is a direct extension of the result obtained by [NPC05] for the bivariate case. For completeness, the proof is included in the Appendix. A direct application of Proposition 2 gives $\nu(h) = 1 - \frac{1}{\theta(h)}$, if $\{Z(x)\}$ has unit Weibull margins ($\mu = 0$, $\sigma = 1, \xi = -1$) and $\nu(h) = \log \theta(h)$, if $\{Z(x)\}$ has unit Gumbel margins ($\mu = 0$, $\sigma = 1, \xi = 0$).

It is important to note that there is not a unique way to make the connection between madograms and the extremal coefficient. For example, we can introduce the F-madogram

$$\nu^F(h) = \frac{1}{2}\mathbb{E}|F(Z(x + h)) - F(Z(x))|,$$

where the margin of $\{Z(x)\}$ equals $F(u) = \exp(-1/u_\beta(u))$. Implementing the same strategy to derive Proposition 2, it is easy to show the equality

$$\nu^F(h) = \theta\mathbb{E}[F(Z(x))F(Z(x))^{\theta(h)-1}] - \mathbb{E}[F(Z(x))].$$

We also have $\mathbb{E}[F^r(Z(x))] = 1/(1 + r)$ because $F(Z(x))$ follows a uniform distribution. Hence, we can write $2\nu^F(h) = \frac{\theta(h)-1}{\theta(h)+1}$, or conversely

$$\theta(h) = \frac{1 + 2\nu^F(h)}{1 - 2\nu^F(h)} \ . \tag{14}$$

This last equality shows a different way of linking $\theta(h)$ to a madogram, compared to Equality (13).

Another important goal when working with spatial extremes is to be able to construct valid extremal coefficients $\theta(h)$. [Sch03] investigated this issue. For example, these authors deduced that if $\theta(h)$ is a real function such that $1 - 2(\theta(h) - 1)^2$ is positive semi-definite, then $\theta(h)$ is a valid extremal coefficient. To prove this assertion, they simply defined the function $\rho(h) = 1 - 2(\theta(h) - 1)^2$ as a covariance. From equations (6) and (7), it followed there exists a max-stable random field with such a $\theta(h)$.

Using the madogram, we can show, in a different way, some of propositions obtained by [Sch03], but our proofs are much shorter and simpler. For example, taking advantage of the link between $\theta(h)$ and madograms, it is possible to derive properties for $\theta(h)$.

Proposition 3. *Any extremal coefficient $\theta(h)$ is such that $2 - \theta(h)$ is positive semi-definite.*

In addition, new conditions for the extremal coefficient function $\theta(h)$ are found via the madogram.

Proposition 4. *Any extremal coefficient $\theta(h)$ satisfies the following inequalities*

$$\theta(h + k) \leq \theta(h)\theta(k),$$
$$\theta(h + k)^\tau \leq \theta(h)^\tau + \theta(k)^\tau - 1, \text{ for all } 0 \leq \tau \leq 1 \ ,$$
$$\theta(h + k)^\tau \geq \theta(h)^\tau + \theta(k)^\tau - 1, \text{ for all } \tau \leq 0 \ .$$

The proofs of propositions 3 and 4 can also be found in the Appendix.

2.3 Extremal coefficient estimators

Estimating the extremal coefficient is of primary interest in spatial extreme value theory. In practice, it is assumed that the margins have first been transformed to unit Fréchet, i.e. the GEV parameters (μ, σ, ξ) were estimated and then used to transform the data. In a second step, the extremal coefficient was estimated. Similarly, if we suppose that (μ, σ, ξ) are known, we can assume without loss of generality that, instead of having unit Fréchet margins, the margins follow any GEV with $\xi < 0.5$. This restriction on ξ allows us to work with a finite variogram and to take advantage of the relationship between the madogram and the extremal coefficient stated in (13). Hence, a natural estimator of $\theta(h)$ is

$$\widehat{\theta}(h) = u_\beta \left(\mu + \frac{\widehat{\nu}(h)}{\Gamma(1 - \xi)} \right) , \tag{15}$$

where

$$\widehat{\nu}(h) = \frac{1}{2|\mathcal{N}_h|} \sum_{(x_i, x_j) \in \mathcal{N}_h} |Z(x_j) - Z(x_i)| \,,$$

and $\mathcal{N}_h$ is the set of sample pairs lagged by the distance h. The main advantages of working with the estimator defined by (15) are: (i) it is based on a simple concept (madogram and Equation (13)), (ii) it is straightforward to implement (see the definition of $\widehat{\nu}$), (iii) it works well in practice (see figures 5 and 4) and (iv) its main theoretical properties can be easily derived if the properties of $\widehat{\nu}(h)$ are known.

To determine the quality of the fit between the theoretical value of $\theta(h)$ and its estimate $\hat{\theta}(h)$, a sequence of 300 random max-stable fields from the Smith and Schlather models was simulated. Figure 3 shows the sample points used to estimate the empirical variogram $\hat{\nu}(h)$. In figures 4 and 5, the solid blue lines represent the theoretical values (either for the madogram or the extremal coefficient) and the red blue points correspond to the estimated values (empirical mean of the 300 fields). In addition, the box plots display the variability of the estimator. As for the madogram, the adequation between the theorical extremal coefficient $\theta(h)$ and $\hat{\theta}(h)$ is good overall. The best fit is obtained for short distances and the worst one for points that are far away.

The asymptotic properties of the estimator $\hat{\theta}(h)$ depend on the asymptotic behavior of the empirical madogram $\hat{\nu}(h)$. By construction, this madogram is simply a sum of stationary absolute increments. Depending on the type of asymptotic behavior under study, e.g. a growing number of observations in a fixed and bounded region, one can derive the asymptotic distribution of $\hat{\nu}(h)$. The δ-method (see Proposition 6.4.1 of Brockwell and Davis' book ([BD91])) can be then applied to get the asymptotic properties of the estimator $\hat{\theta}(h)$.

3 Conclusion

In this paper, we apply a classical concept in geostatistics, variograms, to the analysis of max-stable random fields. Working with the first-order variogram called the madogram has a few advantages: it is simple to define and easy to compute, and it has a clear link with extreme value theory throughout the extremal coefficient. While the unit Fréchet distribution plays a central role in multivariate extreme value theory and deserves special attention, our results indicate that working with GEV margins with $\xi < 0.5$ brings more flexibility in statistical modeling. To illustrate this point, two different types of max-stable fields were studied. Simulations show that the madogram can capture relatively well the prescribed extremal coefficient. In addition, the proof of Proposition 2 shows the relationship between the madogram and probability weighted moments. These moments have been used to estimate the parameters of GEV and GP distributions in the i.i.d. case (see [HW87], [HWW85]). Despite being appreciated by hydrologists ([KPN02]) for its simplicity and having good properties for small samples, this method-of-moment has recently fallen out of fashion because of its lack of flexibility for non-stationary time series (in comparison to the maximum likelihood estimation method, see [Col99]). It is interesting that these moments appear again in the analysis of spatial extremes through the madogram.

Still, much more research, theoretical as well as practical, has to be undertaken to determine the full potential of such a statistic. In particular, the madogram only captures a partial spatial dependence structure for bivariate vectors such as $(Z(x + h), Z(x))$. Two important questions remain. First, how to extend the madogram in order to capture the full bivariate structure? Second, how to completely characterize the full spatial dependence with a madogram-base statistic? Currently, we are working on these two issues and preliminary results indicats that the first question could be answered in the near future. Investigating the second issue provides a bigger challenge and more research is needed in this area.

Acknowledgments: This work was supported by the National Science Foundation (grant: NSF-GMC (ATM-0327936)), the Weather and Climate Impact Assessment Science Initiative at the National Center for Atmospheric Research (NCAR) and the European project E2C2. Some of the figures were obtained with the freely available R package developed by Martin Schlather.

References

[BD91] Brockwell, P.J. and Davis, R.A. (1991). *Times Series: Theory and Methods.* Springer Verlag.

[CD99] Chilès, J.-P. and Delfiner, P. (1999). *Geostatistics: Modeling Spatial Uncertainty.* John Wiley & Sons Inc., New York. A Wiley-Interscience Publication.

[CC99] Coles, S. and Casson, E. A. (1999). Spatial regression for extremes. *Extremes*, 1:339–365.

[Col99] Coles, S. and Dixon, M. (1999). Likelihood-based inference for extreme value models. *Extremes*, 2:523.

[Col01] Coles, S. G. (2001). *An Introduction to Statistical Modeling of Extreme Values.* Springer Series in Statistics. Springer-Verlag London Ltd., London.

[Cre93] Cressie, N. A. C. (1993). *Statistics for Spatial Data.* John Wiley & Sons Inc., New York.

[DM04] Davis, R. and Mikosch, T. (2004). Extreme value theory for space-time processes with heavy-tailed distributions. *Lecture notes from the MaPhySto Workshop on "Nonlinear Time Series Modeling", Copenhagen, Denmark.*

[DR93] Davis, R. and Resnick, S. (1993). Prediction of stationary max-stable processes. *Ann. of Applied Prob*, 3:497–525.

[HP05] de Haan, L. and Pereira, T. (2005). Spatial extremes: the stationary case. *submitted.*

[HR77] de Haan, L. and Resnick, S. (1977). Limit theory for multivariate sample extremes. *Z. Wahrscheinlichkeitstheorie*, 4:317–337.

[EKM97] Embrechts, P., Klüppelberg, C., and Mikosch, T. (1997). *Modelling Extremal Events for Insurance and Finance*, volume 33 of *Applications of Mathematics*. Springer-Verlag, Berlin. For insurance and finance.

[Fou04] Fougères, A. (2004). Multivariate extremes. In B. Finkenstadt and H. Rootzen, editor, *Extreme Values in Finance, Telecommunications and the Environment*, pages 373–388. Chapman and Hall CRC Press, London.

[GLMW79] Greenwood, J., Landwehr, J., Matalas, N., and Wallis, J. (1979). Probability weighted moments: Definition and relation to parameters of several distributions expressable in inverse form. *Water Resources Research*, 15:1049–1054.

[HT04] Heffernan, J. E. and Tawn, J. A. (2004). A conditional approach for multivariate extreme values. *Journal of the Royal Statistical Society, Series B*, 66.

[HW87] Hoskings, J. and Wallis, J. (1987). Parameter and quantile estimation for the Generalized Pareto Distribution. *Technometrics*, 29:339–349.

[HWW85] Hoskings, J., Wallis, J., and Wood, E. (1985). Estimation of the generalized extreme-value distribution by the method of probability-weighted-moments. *Technometrics*, 27:251–261.

[KPN02] Katz, R., Parlange, M., and Naveau, P. (2002). Extremes in hydrology. *Advances in Water Resources*, 25:1287–1304.

[LMW79] Landwher, J., Matalas, N., and Wallis, J. (1979). Probability weighted moments compared with some traditionnal techniques in estimating Gumbel's parameters and quantiles. *Water Resources Research*, 15:1055–1064.

[LLR83] Leadbetter, M., Lindgren, G., and Rootzén, H. (1983). *Extremes and related properties of random sequences and processes*. Springer Verlag, New York.

[LT97] Ledford, A. and Tawn, J. (1997). Modelling dependence within joint tail regions. *J. R. Statist. Soc.*, B:475–499.

[Mat87] Matheron, G. (1987). Suffit-il, pour une covariance, d'être de type positif? *Sciences de la Terre, série informatique géologique*, 26:51–66.

[NPC05] Naveau, P., Poncet, P., and Cooley, D. (2005). First-order variograms for extreme bivariate random vectors. *Submitted*.

[Pon04] Poncet, P. (2004). Théorie des valeurs extrêmes: vers le krigeage et le downscaling par l'introduction du madogramme. Ecole nationale supérieure des Mines de Paris, Paris, France.

[Res87] Resnick, S. (1987). *Extreme Values, Regular Variation, and Point Processes*. Springer-Verlag, New York.

[Sch02] Schlather, M. (2002). Models for stationary max-stable random fields. *Extremes*, 5(1):33–44.

[Sch03] Schlather, M. and Tawn, J. (2003). A dependence measure for multivariate and spatial extreme values: Properties and inference. *Biometrika*, 90:139–156.

[Smi90] Smith, R. (1990). Max-stable processes and spatial extremes. *Unpublished manusript*.

[Smi04] Smith, R. (2004). Statistics of extremes, with applications in environment, insurance and finance. In B. Finkenstadt and H. Rootzen, editor, *Extreme Values in Finance, Telecommunications and the Environment*, pages 1–78. Chapman and Hall CRC Press, London.

[Ste99] Stein, M. L. (1999). *Interpolation of Spatial Data*. Springer-Verlag, New York. Some theory for Kriging.

[Wac03] Wackernagel, H. (2003). *Multivariate Geostatistics. An Introduction with Applications*. Springer, Heidelberg, third edition.

4 APPENDIX

Proof (Proof of Proposition 1). By definition of $\eta_t(h)$ and the equality $|a-b| = (a+b) - 2\min(a,b)$, we can write the following equalities

$$
\begin{aligned}
\frac{\eta_t(h)}{t} &= \frac{1}{2}\mathbb{E}\left|\mathbb{1}(Z(x+h) \le t) - \mathbb{1}(Z(x) \le t)\right|, \\
&= \mathbb{E}[\mathbb{1}(Z(x) \le t)] - \mathbb{E}[\mathbb{1}(Z(x+h) \le t)\mathbb{1}(Z(x) \le t)], \\
&= \mathbb{P}[Z(x) \le t] - \mathbb{P}[Z(x+h) \le t, Z(x) \le t], \\
&= \exp(-1/t) - \exp(-\theta(h)/t).
\end{aligned}
$$

The required result follows immediately. □

Proof (Proof of Proposition 2). The equality $|a - b| = 2\max(a,b) - (a+b)$ allows us to rewrite the madogram $\nu(h) = \mathbb{E}|Z(x+h) - Z(x)|/2$ as

$$
\nu(h) = \mathbb{E}[M(h)] - \mathbb{E}[Z(x)], \quad \text{where } M(h) = \max(Z(x+h), Z(x)) .
$$

In addition, we have $\mathbb{P}(M(h) \le u) = F^{\theta(h)}(u)$ with $F(u) = \mathbb{P}(Z(x) \le u)$. This is a direct consequence from the original definition of θ, see [Sch03], eq. (1) for example. It follows that the mean of $M(h)$ is

$$
\mathbb{E}[M(h)] = \theta(h) \int z F^{\theta(h)-1}(z) \, dF(z) .
$$

It is straightforward to recognize the link between $\mathbb{E}[M(h)]$ and probability weighted moments (PWM) $\mathbb{E}[Z(x)F^r(Z(x))]$. From the work of Hoskings and Wallis [HW87, HWW85], we know the analytical expressions of the PWM for the GEV distribution, also see [GLMW79] and [LMW79]. More precisely, we have

$$
\mathbb{E}[Z(x)F^r(Z(x))] = \frac{1}{1+r}\left(\mu - \frac{\sigma}{\xi} + \frac{\sigma}{\xi}\Gamma(1-\xi)(1+r)^\xi\right) ,
$$

for $r \ge 0$ and $\xi < 1$. We deduce $\mathbb{E}[Z(x)] = \mu - \frac{\sigma}{\xi} + \frac{\sigma}{\xi}\Gamma(1-\xi)$ and

$$
\mathbb{E}[Z(x)F^{\theta(h)-1}(Z(x))] = \frac{1}{\theta(h)}\left(\mu - \frac{\sigma}{\xi} + \frac{\sigma}{\xi}\Gamma(1-\xi)\theta^\xi(h)\right) .
$$

Equality (13) for the GEV is then obtained. For the special case $\xi = 0$, we use an asymptotic argument by letting $\xi \to 0$. □

Proof (Proof of Proposition 3). Let $Z(x)$ be any spatial max-stable random field with extremal coefficient $\theta(h)$ and unit Fréchet margins. For all $z \ge 0$, it is possible to define the function C_z as

$$
\begin{aligned}
C_z(h) = {} & \mathbb{E}[\sqrt{z}\mathbb{1}(Z(x) \le z)\sqrt{z}\mathbb{1}(Z(x+h) \le z)] - \\
& \mathbb{E}[\sqrt{z}\mathbb{1}(Z(x) \le z)]\mathbb{E}[\sqrt{z}\mathbb{1}(Z(x+h) \le z)] .
\end{aligned}
$$

By construction $C_z(h)$ is a covariance function and it equals

$$C_z(h) = z\mathbb{P}(Z(x) \le z, Z(x+h) \le z)) - z\mathbb{P}(Z(x) \le z)\mathbb{P}(Z(x+h) \le z)),$$
$$= z\left(\exp(-\theta(h)/z) - \exp(2/z)\right), \text{ because } \mathbb{P}(Z(x) \le z) = \exp(-1/z) .$$

It follows that

$$\lim_{z \to +\infty} C_z(h) = 2 - \theta(h) .$$

Hence, $2 - \theta(h)$, as a limit of covariance functions, is a covariance function.   □

Proof (Proof of Proposition 4). For the first inequality, we assume that the max-stable field has Gumbel margins. This implies $\nu(h) = \log\theta(h)$. Applying the triangular inequality and the definition of the madogram gives

$$\log\theta(h+k) \le \log\theta(h) + \log\theta(k) .$$

The second and third inequalities are derived in a similar way. We now choose the margins such that the madogram is equal to $\frac{\Gamma(1-\tau)}{\tau}(\theta(h)^\tau - 1)$ with $0 < \tau < 1$. Again, this madogram satisfies the triangular inequality, i.e.

$$\frac{\Gamma(1-\tau)}{\tau}(\theta(h+k)^\tau - 1) \le \frac{\Gamma(1-\tau)}{\tau}(\theta(h)^\tau - 1) + \frac{\Gamma(1-\tau)}{\tau}(\theta(k)^\tau - 1) .$$

The required inequalities follow. The special case $\tau = 1$ is obtained by letting $\tau \uparrow 1$.   □

5 Figures

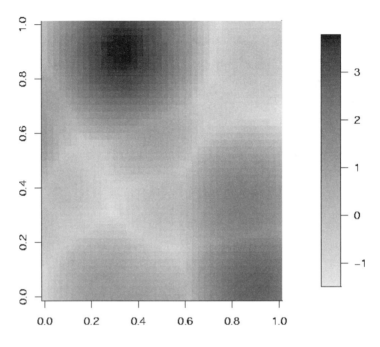

Fig. 1. One realization of a max-stable random field with Gumbel margins. The spatial structure is based on the Smith model, see equations (4) and (9).

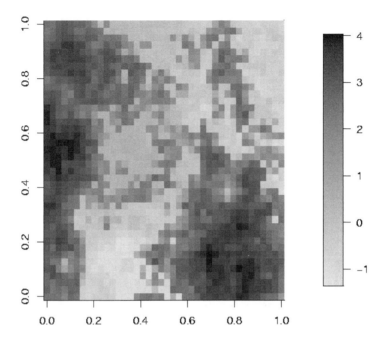

Fig. 2. One realization of a max-stable random field with Gumbel margins. The spatial structure is based on the Schlather model, see equations (5) and (7).

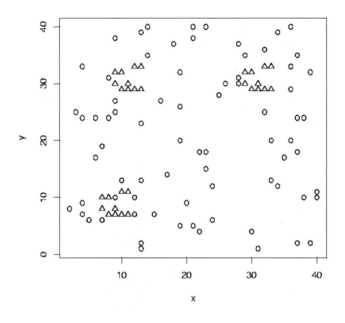

Fig. 3. Locations of the sampling used to estimate the madogram (see Equation (15)). The triangles represent points that were used to get the estimated madogram at short distances, i.e. near the origin (h close to zero).

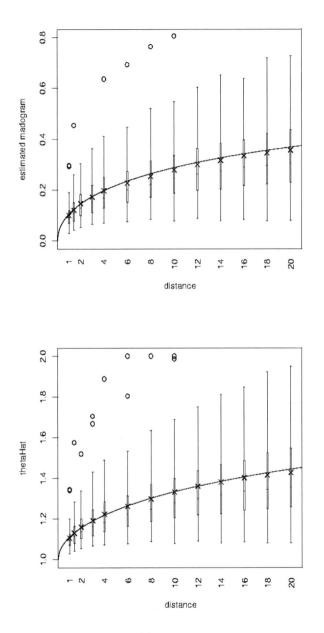

Fig. 4. Schlather model defined by (6): the upper panel shows the theoretical mado-gram and the boxplots for the estimates of 300 max-stable realizations with Gumbel margins; the bottom panel corresponds to the extremal coefficients, see (9) obtained from Equality (15). For both panels, the solid line gives the theoretical values as a function of the distance h between two locations. The crosses represent the empirical mean.

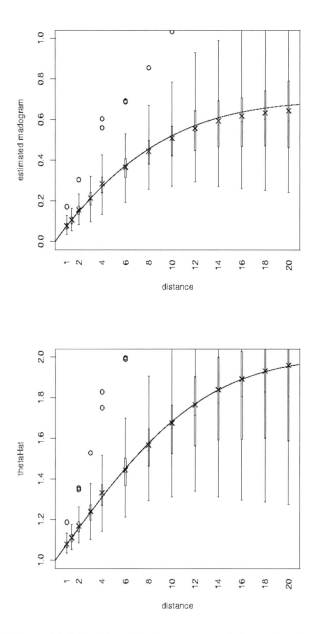

Fig. 5. Smith model defined by (8): the upper panel shows the theoretical mado-gram and the boxplots for the estimates of 300 max-stable realizations with Gumbel margins; the bottom panel corresponds to the extremal coefficients, see (9) obtained from Equality (15). For both panels, the solid line gives the theoretical values as a function of the distance h between two locations. The crosses represent the empirical mean.

A non-stationary paradigm for the dynamics of multivariate financial returns

Stefano Herzel[1], Cătălin Stărică[2] and Reha Tütüncü[3]

[1] Department of Economics, University of Perugia, Italy. herzel@unipg.it,
[2] Department of Mathematical Statistics, Chalmers University of Technology and Department of Economics, Gothenburg University, Sweden. starica@math.chalmers.se
[3] Department of Mathematical Sciences, Carnegie Mellon University, Pittsburgh. reha@andrew.cmu.edu

1 Introduction

This paper discusses a non-stationary, unconditional approach to understanding the dynamic of multivariate financial returns. Non-stationary modeling has a long tradition in financial econometric literature predating the currently prevalent *stationary, conditional* paradigm of which the autoregressive conditionally heteroscedastic (ARCH) -type processes and stochastic volatility models are outstanding examples (see for example, Officer [Off76] or Hsu, Miller and Wichern [HMW74]). Our work is motivated by growing evidence of instability in the stochastic features of stock returns. More concretely, a growing body of econometric literature (Diebold [Die86], Lamoureux and Lastrapes [LL90], Simonato [Sim92], Cai [Cai94], Lobato and Savin [LS98], Mikosch and Stărică [MS02], [MS04] among others) argues that most of the features of return series that puzzle through their omni-presence, the so called "stylized facts", including the ARCH effects, the slowly decaying sample ACF for absolute returns and the IGARCH effect (for definitions and details see Mikosch and Stărică [MS04]) could be manifestations of non-stationary changes in the second moment dynamic of returns (see also Stock and Watson [SW96]). We illustrate our methodology through a detailed analysis of a tri-variate sample of daily log-returns consisting of the foreign exchange rate Euro/Dollar (EU), the FTSE 100 index, and the 10 year US T-bond. The three series are common examples of risk factors[4].

---

[4] For a definition and examples of the importance of modeling the joint dynamic of risk factors, see for example the RiskMetrics document [Risk95]. Briefly, a common current approach to modeling the joint dynamic of large portfolios of financial instruments consists in reducing the size of the model by relating the movements of a large number of the instruments in the portfolio to a relatively

The paper concentrates on answering the following methodological question: How can one analyze the multivariate dynamic of returns in the non-stationary conceptual framework? We argue that a possible adequate set-up could be that of classical non-parametric regression with fixed equidistant design points (see Campbell *et al.* [CLM96] or Wand and Jones [WJ95]). More concretely, the vectors of financial returns are assumed have a time-varying *unconditional* covariance matrix that evolves smoothly. Its dynamics is estimated by a local weighted average or local smoothing. The vectors of standardized innovations are assumed to have asymmetric heavy tails and are modeled parametrically. The careful description of the extremal behavior of the standardized innovations yields a model suited for precise VaR calculations and for generation of stress-testing scenarios.

A closely related issue to the methodological question discussed is: What type of non-stationarities might affect the multivariate dynamic of financial returns? The in-depth analysis in Section 5 as well as the forecasting results in Section 6 indicate the time-varying second unconditional moment as a possible main source of non-stationarity of returns on the three financial instruments we use to exemplify our approach[5].

An important aspect of the methodology we propose is related to answering the following: How should we interpret the slow decay of the sample autocorrelation function (SACF) of absolute returns (see Figures 1 and 2)? Should we take it at face value, supposing that events that happened a number of months (or years) ago bear a strong impact on the present dynamics of returns? Or are the non-stationarities in the returns responsible for its presence as a number of authors have argued lately (the list of related relevant references includes Hidalgo and Robinson [HR96], Lobato and Savin [LS98], Granger and Hyung [GH99], Granger and Teräsvirta [GT99], Diebold and Inoue [DI01], Mikosch and Stărică [MS02], [MS04])? In a recent paper, Stărică and Granger [SG05] have documented the superiority of the paradigm of time-varying unconditional variance over some specifications of stationary long memory and stationary conditional autoregressive heteroscedastic methodology in longer horizon volatility forecasting. Our approach is based

---

small number of so called risk factors (market indices, foreign exchange rates, interest rates). The modeling then concentrates on describing the dynamics of the risk factors.

[5] Our findings and the modeling methodology that they motivate extend to the multivariate framework the work of Officer [Off76] and Hsu, Miller and Wichern [HMW74]. The former, using a non-parametric approach to volatility estimation, reports evidence of time-varying second moment for the time series of returns on the *S&P 500* index and industrial production. The later modeled the returns as a non-stationary process with discrete shifts in the unconditional variance. Note also that, although the paper only reports the detailed results of an analysis of three risk factors, qualitatively similar results are obtained for a large number of other risk factors.

on interpreting the slow decay of the SACF/SCCF of absolute returns as a sign of the presence of non-stationarities in the second moment structure.

Our primary goal is to propose an approach that, while capable of explaining the multivariate dynamics of financial data adequately, is simple and easy to implement. For this reason, at each step of our modeling and estimation approach, we deliberately choose simple and well known methodologies rather than complex estimation techniques. Our empirical study, to which a substantial portion of this article is devoted, demonstrates that the non-stationary paradigm is capable of fitting multivariate data accurately and that it outperforms the plain-vanilla specification of the industry standard Riskmetrics in a simulation study of distributional forecasts.

Non-parametric techniques have been extensively used in the econometric literature on financial and macro-economic time series. For example, Rodríguez-Poo and Linton [RL01] use kernel-based inference technique to estimate the time-dependent volatility structure of residuals of an VAR process. They apply their methodology to macro-economic time series. Fan et al. [FJZZ03] also use kernel regression to estimate time-dependent parametric models for means and covariances in a Gaussian setting. These models are time-dependent generalizations of the time-homogeneous, stationary models discussed in Fan and Yao [FY98]. Unlike these studies, we focus on the the dynamic modeling of the *full distribution* of multivariate returns and not only on particular features of it (like mean or second moment structure). We emphasize a non-Gaussian, heavy-tailed modeling of the standardized innovations for an accurate description of the extremal behavior of the multivariate distribution of returns.

The rest of the paper is organized as follows. Section 2 introduces our non-stationary paradigm, Section 3 collects the relevant results from the statistical literature on non-parametric curve estimation. Section 4 discusses a heavy-tail parametric model for the innovation series. In Section 5, the non-stationary paradigm described in Section 2 is used to analyse the dynamics on a tri-variate sample of returns on the foreign exchange rate Euro/Dollar (EU), the FTSE 100 index, and the 10 year US T-bond (the dimension of the multivariate vector of returns has been intentionally kept low to facilitate an in-depths statistical analysis). Section 6 evaluates the performance of our model in forecasting the distribution of multivariate returns. In Section 7 we comment on the relationship between our modeling approach and the RiskMetrics methodology while Section 8 concludes.

2 A simple non-stationary paradigm for multivariate return modeling

Denote by $\mathbf{r}_k$ the $d \times 1$-dimensional vector of returns $k = 1, 2, \ldots, n$. ARCH-type models assume that $(\mathbf{r}_k)$ is a *stationary*, *dependent*, white noise sequence with a certain conditional second moment structure. More specifically, the

$d \times d$ conditional variance-covariance matrix $\mathbf{H}_k := \mathbb{E}(\mathbf{r}_k \mathbf{r}_k' \mid \mathbf{r}_{k-1}, \mathbf{r}_{k-2}, \ldots)$ is assumed to follow a stationary stochastic process defined in terms of past $\mathbf{r}$'s and past $\mathbf{H}$'s. Often, it is assumed that $P(\mathbf{r}_k \in \cdot \mid \mathbf{r}_{k-1}, \mathbf{r}_{k-2}, \ldots) = P(N(0, \mathbf{H}_k) \in \cdot)$. The common assumptions of the ARCH-type models imply that $(\mathbf{r}_k)$ is a strongly stationary sequence. In particular, the unconditional covariance does not change in time (see Stărică [Stă03] for a discussion on the implications of this assumption on modeling and forecasting univariate index returns).

Our alternative approach assumes $(\mathbf{r}_k)$ to be a *non-stationary* sequence of *independent* random vectors. More concretely, the distribution of $\mathbf{r}_k$ is characterized by a changing unconditional covariance structure that is a manifestation of complex market conditions. The covariance dynamics is hence driven by exogenous factors. We emphasize that, in our approach, the presence of autocorrelation structure in absolute (square) returns is explained by a non-stationary covariance structure[6]. To acknowledge the slow nature of the changes, i.e. the persistence in the second moment structure, the covariance is modeled as a *smooth* function of time. This approach leads to the following regression-type model[7]:

$$\mathbf{r}_k = \mathbf{S}(t_{k,n})\,\boldsymbol{\varepsilon}_{k,n}, \qquad k = 1, 2, \ldots, n, \quad \text{where} \quad t_{k,n} := k/n, \quad t_k \in [0,1]$$

$\mathbf{S}(t) : [0,1] \to \mathbf{R}^{d \times d}$ is an invertible matrix and a smooth function of time, $(\boldsymbol{\varepsilon}_{k,n})$ is an iid sequence of random vectors with mutually independent coordinates, such that $\mathbb{E}\,\boldsymbol{\varepsilon}_{k,n} = 0$, $Var\,\boldsymbol{\varepsilon}_{k,n} = I_d$. $\hspace{2cm}$ (1)

(The notation is that of the classical non-parametric regression set-up and is motivated by the specific nature of the asymptotic results[8]. We will omit indices n whenever feasible.)

The precise smoothness assumptions on $\mathbf{S}(t)$ are discussed in the sequel. The elements of the sequence $(\boldsymbol{\varepsilon}_{k,n})$ are called the *standardized innovations*. From (1), it follows that

---

[6] Sequences of independent observations will display spurious autocorrelation structure if there is a break in the unconditional variance. In other words, the presence of autocorrelation structure is not incompatible with the assumption of independence. See Diebold and Inoue [DI01] and Mikosch and Stărică [MS04].

[7] A mean term could be included in model (1). Denoting by $\mathbf{u}_k := \mathbf{r}_k - \mathbb{E}\mathbf{r}$, $k = 1, \ldots, n$, the model would then assume $\mathbf{u}_t$ to be independent with covariance matrix $\mathbf{S}(t)\mathbf{S}'(t)$, a smooth function of t. We have implemented both procedures, i.e. with and without removing of a mean estimate in a preliminary step, and obtained qualitatively equal results. Hence, in the sequel, we work under the simplifying assumption of a negligeable mean of the return series.

[8] Unlike in other fields of statistics, the asymptotic results involve not only an increasing number of observations but also an increase in the frequency with which the unknown function is observed. To attain this goal, the observations are indexed between 0 and 1. In this way an increase in the sample size implies also an increase in the frequency with which we observe the regression function.

$$\mathbb{E}(\mathbf{r}_k\mathbf{r}_k' \mid \mathbf{r}_{k-1}, \mathbf{r}_{k-2}, \ldots) = \mathbb{E}(\mathbf{r}_k\mathbf{r}_k') = \mathbf{S}(t_{k,n})\mathbf{S}'(t_{k,n}) := \mathbf{\Sigma}(t_{k,n}), \quad \text{and}$$

$$\mathbb{P}(\mathbf{r}_k \in \cdot \mid \mathbf{r}_{k-1}, \mathbf{r}_{k-2}, \ldots) = P(\mathbf{r}_k \in \cdot), \quad k = 1, 2, \ldots, n.$$

This modeling approach reflects the belief that the distribution of the vector of future returns incorporates a changing pool of information which is partly expressed in the recent past of the time series and the fact that we are not aware of exogenous variables capable of reliably explaining the dynamics of the volatility. In other words, our uncertainty about the form of the model is manifestly expressed in the choice of the non-parametric regression approach.

Furthermore, we will assume the existence of a smooth[9] function $\mathbf{V}(t)$: $[0, 1] \rightarrow \mathbf{R}^{d \times d}$ such that var $r_{i,k}r_{j,k} = v_{ij}(t_k)$ (in short, var $\mathbf{r}_k\mathbf{r}_k' = \mathbf{V}(t_k)$) where $r_{i,k}$ is the i-th coordinate of $\mathbf{r}_k$ and $t_k = k/n$[10]. With this notation

$$r_{i,k}r_{j,k} = \Sigma_{ij}(t_k) + v_{ij}^{1/2}(t_k)\tilde{\varepsilon}_k^{ij}, \quad k = 1, 2, \ldots, n, \quad i, j = 1, 2, \ldots d, \quad (2)$$

where the errors $\tilde{\varepsilon}_k^{ij}$ are iid vectors with independent coordinates, such that $\mathbb{E}\,\tilde{\varepsilon}_k^{ij} = 0$, var $\tilde{\varepsilon}_k^{ij} = 1$. Hence the function $\mathbf{\Sigma}(t)$ can be estimated by standard non-parametric heteroscedastic regression methods for non-random, equidistant design points using the series $\mathbf{r}_k\mathbf{r}_k'$, $k = 1, \ldots, n$.

The non-stationary paradigm that we have introduced above can be used both for describing the dynamics of multivariate data as well as for short horizon forecasting. The methodological difference between applying it for data description or for forecasting will become clear in the next section.

3 Non-parametric smoothing

Non-parametric regression develops a special type of asymptotics based on the so-called "infill" assumption. Under this paradigm, increasing the sample size entails having more observations covering the same time span. It is important to understand that in the framework of the current paper the reading we give to the "infill" assumption is quite different from the common working hypothesis of financial time series analysis where, commonly, more observations in the same time interval amounts to a higher sampling frequency. Bluntly, for us the increase of the sample size *does not mean* observing the data on a finer grid[11]. In the framework of the present work, an increase of the sample is understood as an increase in the number of daily returns available for estimation. The following simple-minded example might clarify the use of non-parametric asymptotics at work in the sequel. If initially the sample available

---

[9] The precise smoothness assumptions on $\mathbf{V}(t)$ are discussed in the sequel.

[10] In words, we assume that the covariance structure and the variance of the covariance are evolving smoothly through time.

[11] Changing the sampling frequency entails changing the deterministic function that needs to be estimated. This situation is not covered by the classical theory of non-parametric regression and will be dealt with somewhere else.

for estimation was the daily returns on the first Monday of every year for, say, eleven consecutive years, i.e. eleven observations, the sample size increases to twenty two observations by adding the daily returns on the first Monday in the month of June of the eleven years, to thirty three by adding the daily returns on all the first Mondays in every trimester, to forty four observations adding those of every quarter, and so on. Note that the sample covers always eleven years of financial history and that the sample increases without any change in the sampling frequency. All the returns in the sample are daily returns. Our assumption is that, at the point when the sample includes all the daily returns in the eleven year interval, the condition are met for the non-parametric regression asymptotics to work.

Having said all that, let us now turn to present the main statistical results that form the basis of the empirical analysis to follow. Our main reference in the context of non-parametric regression is Müller and Stadtmüller [MS87] on kernel curve estimation in the heteroscedastic regression model

$$y_{k,n} = \mu(t_{k,n}) + \sigma(t_{k,n})\,\epsilon_{k,n}, \qquad k = 1, 2, \ldots, n \ . \tag{3}$$

The random variables y_k are observations of the unknown regression function $\mu(t) : [0,1] \to \mathbf{R}$, perturbed by heteroscedastic errors $\sigma(t_k)\epsilon_k$. The standardized errors ϵ_k are iid with mean zero and unit variance not necessarily Gaussian. The functions $\mu : [0,1] \to \mathbf{R}$ and $\sigma : [0,1] \to \mathbf{R}_+$ are assumed smooth (the smoothness requirements will be made precise in the sequel).

Our analysis uses kernel regression smoothing. For an introduction on smoothing estimator and in particular, on kernel estimators, see Section 12.3 of Campbell et al. [CLM96] or Wand and Jones [WJ95]. The following kernel estimator will be used in the various steps of mean and variance estimation in the heteroscedastic regression model (3)

$$\hat{f}(t; h) = \sum_{k=1}^{n} W_k(t; h)\, U_k \ , \tag{4}$$

where U_k stand for $\tilde{\sigma}^2(t_k)$, preliminary variance estimates, in the estimation of $f := \sigma^2(t)$ and for y_k in the estimation of $f := \mu(t)$. The weights $W_k(t)$ satisfy

$$W_k(t; h) = W_{k,n}(t; h) = \frac{1}{h} \int_{s_{k-1}}^{s_k} K\left(\frac{t-u}{h}\right) du\,, \quad s_k = \frac{t_{k-1} + t_k}{2} \ . \tag{5}$$

The quantity $h > 0$ is the bandwidth of the estimator and the kernel function K on $[-1,1]$ satisfies the basic condition $\int K(u)du = 1$ and some other assumptions. These are satisfied by the Gaussian kernel density function when it is first truncated at $[-3,3]$, then rescaled to $[-1,1]$ and finally made Lipschitz continuous such that $K(-1) = K(1) = 0$ by changing the kernel appropriately in $[-1,1]\backslash[-1+\delta, 1-\delta]$ for $\delta = 0.01$. This is the kernel used in Section 5. We note that such estimates use past and future information. A modified kernel, only based on the past, will be introduced later.

3.1 Estimation of the variance

Let us summarize now some of the necessary theory for the estimation of $\hat{\sigma}$ in the heteroscedastic model (3). The kernel estimator of $\sigma(t)$ in the heteroscedastic regression model (3) is defined in two steps.

(1) First, a preliminary smoothing removes the mean function μ in (3) in some neighborhood of t_k. The preliminary estimator of the variance at an inner point t_k in $[0,1]$ is given by

$$\tilde{\sigma}^2(t_k) = \left(\sum_{j=-m_1}^{m_2} w_j \, y_{j+k} \right)^2 , \qquad (6)$$

with the weights w_j satisfying $\sum_{j=-m_1}^{m_2} w_j = 0$ and $\sum_{j=-m_1}^{m_2} w_j^2 = 1$ for some fixed $m_1, m_2 \geq 0$.

(2) Second, we view the preliminary estimates of the variance, $\tilde{\sigma}^2(t_k)$ as measurements from the following regression model:

$$\tilde{\sigma}^2(t_k) = \sigma^2(t_k) + \tilde{\epsilon}_k, \quad 1 \leq k \leq n , \qquad (7)$$

where the errors $\tilde{\epsilon}_k$ form an $m_1 + m_2$-dependent sequence, $\mathbb{E}\tilde{\epsilon}_k = 0$. The estimator of the variance is then given by

$$\hat{\sigma}^2(t) := \hat{\sigma}^2(t; h_{\sigma^2}) = \sum_{k=1}^{n} W_k(t; h_{\sigma^2}) \, \tilde{\sigma}^2(t_k) , \qquad (8)$$

where the weights $W_k(t; h)$ are defined in (5).

In the sequel we assume that σ^2 is twice differentiable with a continuous second derivative, μ is Lipschitz continuous of order $\alpha \geq 0.25$ and $\mathbb{E}|\epsilon_i|^{5+\delta} < \infty$ for some $\delta > 0$. Then the following statements can be derived from Theorem 3.1 and Remark at the bottom of p. 622 in Müller and Stadtmüller [MS87]:

(1) The estimated variance $\hat{\sigma}^2(t)$ satisfies

$$\left| \hat{\sigma}^2(t) - \sigma^2(t) \right| \leq c \left(h_{\sigma^2}^2 + (\log n / n h_{\sigma^2})^{1/2} \right) ,$$

almost surely, for some unspecified positive constant c, uniformly on any compact of the interval (0,1), if the bandwidth h_{σ^2} satisfies

$$\liminf n^{1/5+\delta'} h_{\sigma^2} / \log n > 0, \quad \liminf n h_{\sigma^2}^2 > 0 ,$$

where $0 < \delta' < \delta$.

(2) The expected value $\mathbb{E}\hat{\sigma}^2(t)$ satisfies

$$|\mathbb{E}\hat{\sigma}^2(t) - \sigma^2(t)| \leq c \left(h_{\sigma^2}^2 + n^{-1} \right)$$

for some unspecified positive constant c, uniformly on any compact of the interval (0,1) .

3.2 Estimation of the mean in the heteroscedastic regression model

If moreover, μ is twice differentiable with continuous second derivative, Lemma 5.3 of Müller and Stadtmüller [MS87] gives the following results for $\hat{\mu}_{\text{He}}(t; h_\mu)$, the estimator given by (4) with $f := \mu$:

(1) The expected value $\mathbb{E}\hat{\mu}_{\text{He}}(t)$ satisfies, as $n \to \infty$ and $h_\mu := h_{\mu,n} \to 0$, $nh_\mu \to \infty$

   (i)

$$\mathbb{E}\hat{\mu}_{\text{He}}(t) - \mu(t) = \mu^{''}(t)h_\mu^2\, B + o(h_\mu^2) + O(n^{-1}) \,, \tag{9}$$

   where $B = \int K(u)u^2 du/2$,

   (ii)

$$|\mathbb{E}\hat{\mu}_{\text{He}}(t) - \mu(t)| \le c\,(h_\mu^2 + n^{-1}) \,,$$

   for some unspecified positive constant c, uniformly for $t \in [\delta, 1 - \delta]$, any fixed $\delta \in (0, 1)$.

(2) The variance of $\hat{\mu}_{\text{He}}(t)$ satisfies for every t, as $n \to \infty$ and $h_\mu := h_{\mu,n} \to 0$, $nh_\mu \to \infty$

$$\text{var}(\hat{\mu}_{\text{He}}(t)) = \frac{\sigma^2(t)}{nh_\mu}\, U\,(1 + o(1)) \,, \tag{10}$$

where $U = \int K^2(u)du = 0.84$ for the normal kernel used in our analysis. Note that the bandwidths h_μ in the estimation of μ and h_{σ^2} in that of σ^2 are in general very different.

These results apply to the concrete heteroscedastic regressions of interest (2) as follows. The estimator of $\mathbf{\Sigma}(t)$ as given by the heteroscedatic approach described in subsection 3.2 is

$$\hat{\mathbf{\Sigma}}(t;\, h) := \sum_{k=1}^{n} W_{k,\, n}(t; h)\, \mathbf{r}_k \mathbf{r}_k^{'} \,, \tag{11}$$

where the weights $W_{k,\, n}$ are defined in (5). Note that the matrix $\hat{\mathbf{\Sigma}}(t;\, h)$ is positive definite by construction.

The estimator of $\mathbf{V}(t)$ in (2) given by the methodology described in subsection 3.1 is

$$\hat{\mathbf{V}}(t;\, \tilde{h}) := \sum_{k=1}^{n} W_{k,\, n}(t; \tilde{h}) \left(\sum_{l=-m_1}^{m_2} w_l\, \mathbf{r}_{l+k} \mathbf{r}_{l+k}^{'} \right)^2 \,, \tag{12}$$

where the square operation has to be intended component-wise. The weights w_l satisfying $\sum_{l=-m_1}^{m_2} w_l = 0$ and $\sum_{l=-m_1}^{m_2} w_l^2 = 1$ for some fixed $m_1, m_2 \ge 0$.

Note that the matrix V is not a covariance matrix and that we have simply used the convenience of the matrix notation in order to write concisely $d \times d$ (coordinate-wise) equations. In the analysis that follows, we have used $m_1 = 1$, $m_2 = 0$ and $w_1 = w_2 = 1/\sqrt{(2)}$.

If the coordinates of $\mathbf{S}$ are twice differentiable with continuous second derivatives and $\mathbb{E}|\varepsilon_1|^{5+\delta} < \infty$ for some $\delta > 0$ then, as $n \to \infty$

(i)

$$\left|\mathbb{E}\, \hat{\sigma}_{ij}(t) - \sigma_{ij}(t)\right| \le c\left(h^2 + n^{-1}\right),$$

as $h \to 0$, $nh \to \infty$, for some unspecified positive constant c, uniformly on any compact in $(0,1)$;

(ii) as $h \to 0$, $nh \to \infty$, $\hat{\sigma}_{ij}(t) - \sigma_{ij}(t)$ is approximately distributed as

$$N\left(\sigma_{ij}''(t)\, h^2\, B, \frac{v_{ij}(t)}{nh}\, U\right);$$

(iii) if the bandwidth $\tilde{h}$ is chosen as $\tilde{h} \sim (\log n)/n^{1/5+\delta'}$, where $0 < \delta' < \delta$, then as $n \to \infty$

$$\hat{v}_{ij}(t) \to v_{ij}(t),$$

almost surely, uniformly on any compact in $(0,1)$. Moreover

$$\left[\hat{\sigma}_{ij}(t) - z_{\alpha/2}\sqrt{\frac{\hat{v}_{ij}(t)U}{nh}},\ \hat{\sigma}_{ij}(t) + z_{\alpha/2}\sqrt{\frac{\hat{v}_{ij}(t)U}{nh}}\right] \qquad (13)$$

are approximate $(100 - \alpha)\%$ point-wise confidence intervals for $\sigma_{ij}(t)$, where $z_{\alpha/2}$ are the $(100 - \alpha/2)\%$ normal quantile.

In the analysis of the multivariate return time series in Section 5, a Gaussian kernel is used. We note that, according to our experience, an exponential kernel or the LOESS procedure produce very close results. This is in accordance with the established fact that for the equidistant design set-up, the shape of the kernel function makes little difference; see the monographes by Müller [Mul88] and Wand and Jones [WJ95].

As we have already emphasized, the non-stationary paradigm under discussion can be used both for understanding the nature of past changes in the dynamics of multivariate data as well as for short horizon forecasting. The methodological difference between the use of the paradigm for data description and that for forecasting consists in the type of kernel used in estimation of the regression function. A symmetric kernel will be used when interested in describing the dynamic of the changes in the historical sample while an asymmetric one, giving weights only to the past and current observations will be applied in the forecasting exercises. See Sections 5 and 6 for detailed applications of the paradigm in the two set-ups.

3.3 Bandwidth selection

The equations (9) and (10) yield the asymptotic integrated square error (MISE) of $\hat{\mu}_{He}(t)$, the estimator of μ in (3) given by (4) :

$$MISE = h_\mu^4 B^2 \int \mu''(u)^2 du + \frac{\int \sigma^2(u) du}{nh_\mu} U .$$

Minimizing the MISE with respect to the bandwidth h_μ yields the globally optimal bandwidth

$$h_\mu^{(g)} = \left(\frac{\int \sigma^2(u) du U}{4nB^2 \int \mu''(u)^2 du} \right)^{1/5} . \tag{14}$$

The choice of smoothing parameter or bandwidth is crucial when applying non-parametric regression estimators, such as kernel estimators. For this reason we applied a set of different methods of bandwidth selection. Cross-validation is a method based on minimizing residual mean squared error criteria frequently used to infer the optimal smoothing parameter. Another method builds on estimating the asymptotically optimal global bandwidth (14) from the data. Since estimators for the residual variance and for an asymptotic expression for the bias (9) are plugged into the asymptotic formula (14), such selection rules are called 'plug-in' estimators. The functional that quantifies bias is approximated by the integrated squared second derivative of the regression function. This functional is determined by an iterative procedure introduced in Gasser *et al.* [GKK91] based on a kernel estimator $\hat{\mu}''(t; h_{\mu''})$ for the derivative. Such an estimator has the form (4) with the kernel K tailored to estimate second derivatives (see Gasser *et al.* [GMM85]; for our application we used the optimal (2,4) kernel).

4 A heavy-tailed model for the distribution of the innovations

The final step is modeling the distribution of the *estimated standardized innovations* defined as

$$\hat{\underline{\varepsilon}}_k := \hat{\mathbf{S}}^{-1}(t_k)\,\hat{\mathbf{r}}_k, \qquad k = 1, 2, \ldots, n \tag{15}$$

with $\hat{\mathbf{S}}(t)$, the square root of the estimate $\hat{\mathbf{\Sigma}}(t)$ of $\mathbf{S}(t)\mathbf{S}'(t)$ in (11). One possibility is to use the empirical cumulative distribution function (cdf) of $\hat{\underline{\varepsilon}}$ as a model for the standardized innovations as done in Barone-Adesi *et al.* [BGV99]. However, since the estimated standardized innovations are usually heavy tailed (see Section 5.3 for evidence supporting this claim), the use of the empirical cdf [12] will underestimate the probability of extreme standardized innovations and, hence, the risk of extreme returns, with potentially serious consequences for risk managing.

Since we assume the estimated standardized innovations to have d independent coordinates, it is sufficient to specify the distributions of $\hat{\varepsilon}_i$, $i = 1, \ldots, d$. A flexible and parcimonious family of distributions that allow for asymmetry between the distributions of positive and negative standardized innovations and, in addition, for arbitrary tail indices can be defined starting from the Pearson type VII distribution with shape parameter m and scale parameter c; see Drees and Stărică [DS02]. The density of this distribution is

$$f(x; m, c) = \frac{2\Gamma(m)}{c\Gamma(m - 1/2)\pi^{1/2}} \left(1 + \left(\frac{x}{c}\right)^2\right)^{-m}, \qquad x > 0. \qquad (16)$$

Note that f is the density of a t-distributed random variable with $\nu = 2m - 1$ degrees of freedom multiplied by the scale parameter $c\nu^{-1/2}$. This family was also used to model the distribution of financial returns in an univariate stochastic volatility framework by Nagahara and Kitagawa [NK99].

According to our experience, this distribution (concentrated on the positive axis) fits well the positive standardized innovations and the absolute value of the negative ones. Because usually there are about as many positive standardized innovations as there are negative ones, it may be assumed that the cdf of the standardized innovations has median 0. Hence, denoting the densities of the negative and positive standardized innovations by $f_-(\cdot; m_-, c_-)$ and $f_+(\cdot; m_+, c_+)$, respectively, the density of the distribution of the coordinates of the standardized innovations is

$$f^{VII}(x; m_-, c_-, m_+, c_+) = \qquad (17)$$
$$\tfrac{1}{2}\left(f_-(x; m_-, c_-)1_{(-\infty,0)}(x) + f_+(x; m_+, c_+)1_{[0,\infty)}(x)\right).$$

We refer to the distribution with density (17) (that covers the whole real axis) as the asymmetric Pearson type VII and denote its cdf by F^{VII}.

To summarize, for a given coordinate, f^{VII} is determined by four parameters m_-, c_-, m_+, and c_+, with (m_-, c_-) and (m_+, c_+) being estimated separately by fitting a one-sided Pearson type VII distribution to the absolute values of the negative and positive standardized innovations, respectively, e.g. by maximum likelihood. These parameters, together with the covariance estimates $\hat{\mathbf{S}}(t)$ fully specify the distribution of the time series of returns in the model (1).

---

[12] Using the empirical cdf is tantamount to assuming that the worse case scenarios cannot be any worse than what we have in the sample. Using extreme value techniques for modeling the tails of the innovations allows to extrapolate outside the range of the observed data producing events that are more extreme than the limited history available and that are in line with the distributional features of the observed sample.

5 Understanding the dynamics of multivariate returns. An example.

In this section, we apply the methods described in the previous section to the 2927 observations (from January 2, 1990 until September 12, 2001) of the time series of daily returns of three qualitatively different financial instruments: one foreign exchange rate, the Euro/Dollar (EU), an index, the FTSE 100 and an interest rate, the 10 year US T-bond. The EU and the US T-bond series are available on the site of the US Federal Reserve Board: *http://www.federalreserve.gov/releases/*. To facilitate a graphical display of the empirical analysis, we conduct our study in a tri-variate setup. Note that similar modeling results have been achieved with higher dimensional vectors of returns.

The goal of the discussion in this section is to provide a picture of the changes in the dynamic of the multivariate vector of returns and to check the quality of the non-parametric paradigm applied in the set-up of modeling. The next section will consider the performance of the paradigm in the forecasting set-up.

Figures 1 and 2 display the SACF-SCCF of the data and that of the absolute values of the data.

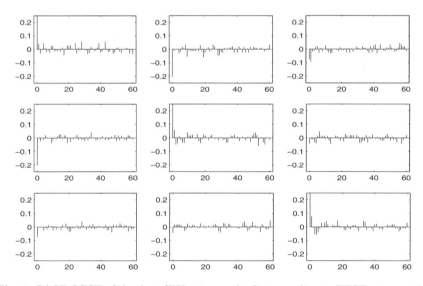

Fig. 1. SACF-SCCF of the data (EU returns, the first coordinate, FTSE returns, the second coordinate, the 10 year T-bond returns, the third coordinate respectively). On the diagonal the SACF of the 3 series. Off the diagonal the SCCF of pairs. Since the dependency structure in the data is unknown, no confidence intervals for the correlations are displayed.

The SACF/SCCF of the returns (Figure 1) show extremely small auto- or cross-correlations at lags greater then 4 between the EU, the FTSE or the 10 year T-bond returns. In contrast to this, the SACF/SCCF of absolute returns in Figure 2 show larger correlations in the absolute values.

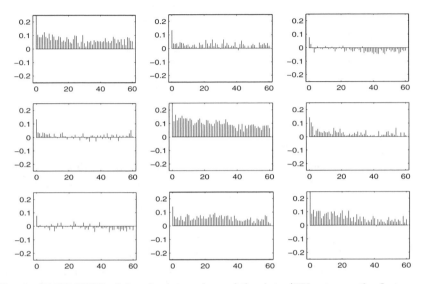

Fig. 2. SACF-SCCF of the absolute values of the data (EU returns, the first coordinate, FTSE returns, the second coordinate, the 10 year T-bond returns, the third coordinate respectively). On the diagonal the SACF of the 3 series. Off the diagonal the SCCF of pairs. Since the dependency structure in the data is unknown, no confidence intervals for the correlations are displayed.

Note that an SACF/SCCF that displays positive correlations at large lags (like that in Figure 2) is not evidence of dependent data. Independent and *non-stationary* observations with a time-varying unconditional variance can produce SACF/SCCF like the ones in Figure 2. Positive correlations at large lags could be a sign of non-stationarities in the second moment structure of the time series as well as a proof of stationary, non-linear, long-range dependence; see Mikosch and Stărică [MS04]. As emphasized earlier, our working paradigm is consistent with the non-stationary interpretation of the SACF/SCCF.

5.1 The evolution of the unconditional covariance structure.

We estimated the optimal bandwidth in the set-up of the model (3) with $y_k = |\mathbf{r}_k \mathbf{r}'_k|$, $k = 1, 2, \ldots, n$, using cross-validation and the method of Gasser *et al.* [GKK91][13]. Figure 3 displays the cross-validation graph. Based on this

---

[13] Using $y_k = |\mathbf{r}_k|$ yields qualitatively egual results.

graph, the choice for the bandwidth is $h_\mu^{(c)} \in [0.005, 0.008]$ with a minimum at 0.006. The procedure of Gasser *et al.* [GKK91] produced $h_\mu^{(g)} = 0.0076$. This is the bandwidth that we use in defining $\hat{\Sigma}(t)$.

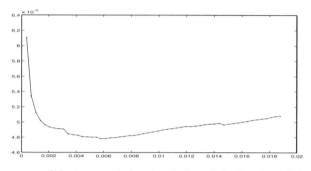

Fig. 3. The cross validation graph for the choice of the bandwidth h_μ for $\hat{\mu}$ in (4) and $y_k = |\mathbf{r}_k \mathbf{r}_k'|$. The bandwidths $h_\mu^{(c)}$ that minimize the cross validation function belongs to the interval [0.005, 0.008].

The graphs in Figure 4 display two estimates of the time-varying standard deviations (sd's) of the three time series. Those in Figure 5 show two estimates of the time-dependent correlation between the three pairs of univariate time series (in the top graph, EU and FTSE, in the middle FTSE and T-bill, in the lower one, EU and T-bill). In all the pictures, the solid line is the estimate obtained using $\hat{\Sigma}$, defined in (11) with bandwidth $h = 0.0076$. The dotted line is the estimate obtained using the estimator

$$\hat{\Sigma}_1(t) := \sum_{k=1}^{n} \tilde{W}_{k,\,n}(t)\, \mathbf{r}_k \mathbf{r}_k' \,, \tag{18}$$

where the weights $\tilde{W}_{k,\,n}$ are defined as in (5) with the symmetric kernel K replaced by $\tilde{K}(u) = K(u) 1_{u \le 0}$. The bandwidth used in (18) was $h = 0.007$ [14]. Note that $\hat{\Sigma}_1(t)$, estimate of $\mathbf{S}(t)\mathbf{S}'(t)$, uses only the information available at day t. This estimator will be used to produce the forecasting results presented in Section 6 [15].

The 95% confidence intervals given by (13) are also plotted. Note that the estimated volatilities and correlations that use only the past information belong almost always to the 95% confidence intervals. Hence using only past information seems to yield a rather precise estimates.

The graphs in Figure 4 and 5 show rather large variations in the estimated standard deviations as well as in the estimated correlation structure

---

[14] For the choice of this value, see Section 6.
[15] The boundary modification proposed in Rice [Ric84] has been used to take care of the boundary effect.

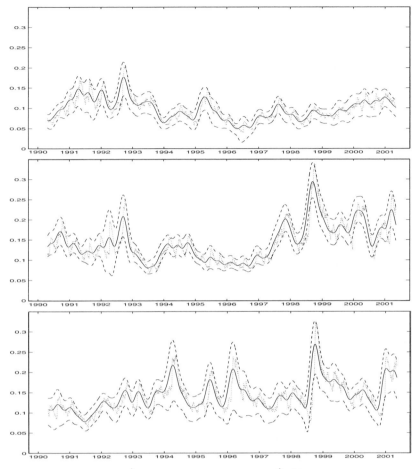

Fig. 4. Local estimates ($\hat{\boldsymbol{\Sigma}}(t)$, $h = 0.0076$, solid line, $\hat{\boldsymbol{\Sigma}}_1(t)$, $h = 0.007$, dotted line) of the (annualized) standard deviation (sd) of the data: EU, (*Top*), FTSE (*Middle*) and the 10-year T-bond (*Bottom*). The annualized sd is obtained by multiplying the daily sd by a factor of $\sqrt{250}$. The 95% confidence intervals given by (13) are also displayed.

of the series. They individuate the existence of periods with unconditional volatilities and unconditional correlations that are statistically significant different. In particular, the estimated correlation between the EU and the T-bond switched from negative values in the interval (-0.3, -0.2) in the beginning of the 90's, to positive ones around 0.2 in the beginning of the second half of the decade. The largest fluctuations in the estimated sd are displayed by the FTSE with increases from values around 10% in the middle of the decade to a peak of roughly 25% towards the end of the 90's. The two figures support

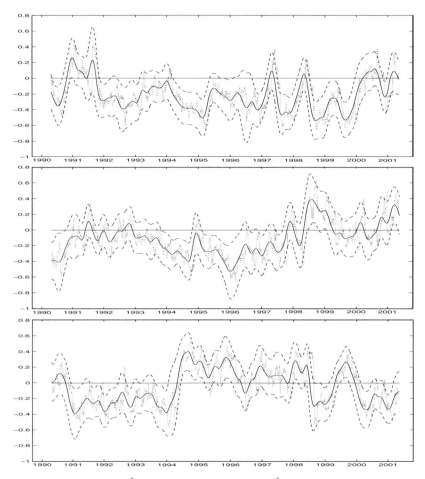

Fig. 5. Local estimates ($\hat{\boldsymbol{\Sigma}}(t)$, $h = 0.0076$, solid line, $\hat{\boldsymbol{\Sigma}}_1(t)$, $h = 0.0076$, dotted line) of the correlations between the data: EU and FTSE, (*Top*), FTSE and 10-year T-bond (*Middle*) and 10-year T-bond and EU (*Bottom*). The 95% confidence intervals given by (13) are also displayed.

the assumption of time-varying unconditional covariance structure that is the basis of our non-stationary paradigm.

A discussion of the behavior of the proposed methodology on return data generated by state-of-the-art models from the stationary, conditional volatility paradigm is included in Section 9[16]. Since the comparison with the stationary conditional paradigm is not the focus of this paper, the discussion is reduced in size. However, the simulation results presented suggest that the proposed procedure can distinguish between a stationary dependent series with conditional heteroscedasticity from a non-stationary independent series.

---

[16] This discussion was suggested by an anonimous referee.

5.2 The dependence structure of the standardized innovations

In this section we analyze the dependency structure of the estimated standardized innovations $\hat{\underline{\varepsilon}}_t$ defined in (15). A battery of three tests is used to achieve this goal. In the sequel we are ignoring the fact that the innovations come from a kernel regression and we treat them as if they were directly observed. In doing this we neglect the possible effect of the estimation error on the asymptotic properties of the statistics we present. As a consequence, the p-values of the tests should be interpreted more as upper limits than as precise values.

With this caveat in mind, we begin by verifying that the marginal distributions of the coordinates of the estimated standardized innovations $\hat{\varepsilon}_i$, $i = 1, 2, 3$, do not change through time. Towards this goal, for a given coordinate i, we split the sample $(\hat{\varepsilon}_{i,t})$ in three subsamples of equal length, $(\hat{\varepsilon}_{i,t}^{(1)})$, $(\hat{\varepsilon}_{i,t}^{(2)})$, $(\hat{\varepsilon}_{i,t}^{(3)})$ respectively. Then, we perform a pairwise comparison of the three resulting empirical cumulative distribution functions using a 2-sample Kolmogorov-Smirnov test, producing three p-values.

For the pair $(\hat{\varepsilon}_i^{(1)}, \hat{\varepsilon}_i^{(2)})$, the working assumptions are that $\hat{\varepsilon}_i^{(1)}$'s and $\hat{\varepsilon}_i^{(2)}$'s are mutually independent (see the independence tests (20) in the sequel for evidence supporting this assumption) and that all the observations in the sample $(\hat{\varepsilon}_i^{(1)})$ come from the same continuous population $F_i^{(1)}$, while all the observations in the sample $(\hat{\varepsilon}_i^{(2)})$ come from the same continuous population $F_i^{(2)}$. The null hypothesis is

$$H_0 : \quad F_i^{(1)} \text{ and } F_i^{(2)} \text{ are identical.} \tag{19}$$

Table 1 reports the nine p-values (3 for each coordinate) for the estimated standardized innovations $(\hat{\underline{\varepsilon}}_t)$ (left) together with the nine values corresponding to their absolute values $(|\hat{\underline{\varepsilon}}_t|)$ (right).

Table 1. The p-values corresponding to the 2-sample Kolmogorov-Smirnov tests on subsamples of estimated standardized innovations $(\hat{\underline{\varepsilon}}_t)$ (left) and their absolute values $(|\hat{\underline{\varepsilon}}_t|)$ (right). The column labels code the pairs of subsamples.

	1 and 2	1 and 3	2 and 3		1 and 2	1 and 3	2 and 3
$\hat{\varepsilon}_1$	0.93	0.23	0.05	$|\hat{\varepsilon}_1|$	0.89	0.39	0.21
$\hat{\varepsilon}_2$	0.10	0.46	0.59	$|\hat{\varepsilon}_2|$	0.20	0.43	0.75
$\hat{\varepsilon}_3$	0.23	0.63	0.65	$|\hat{\varepsilon}_3|$	0.09	0.48	0.64

Table 1 supports the hypothesis of stationarity of the coordinates of the sequence of estimated standardized innovations $(\hat{\underline{\varepsilon}}_t)$.

In the sequel we use the covariance/autocovariance structure of the estimated standardized innovations $(\hat{\underline{\varepsilon}}_t)$ and their absolute values $(|\hat{\underline{\varepsilon}}_t|)$ (see Figures 6 and 7) to test the hypothesis

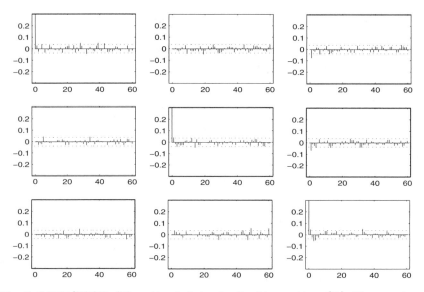

Fig. 6. SACF/SCCF of the estimated standardized innovations $(\hat{\underline{\varepsilon}}_t)$. The covariance structure was estimated using $\hat{\Sigma}$.

$$H_0 : \hat{\underline{\varepsilon}}_t \text{ are iid vectors with independent coordinates.} \qquad (20)$$

The confidence intervals in Figures 6 and 7 correspond to the null hypothesis (20). These figures show that accounting for the changing covariance produces standardized innovations that are practically uncorrelated, removing the long memory look of the SACF of absolute returns in Figure 2. They support the choice of modeling the standardized innovations as a sequence of iid vectors with independent coordinates.

The visual test of the hypothesis (20) is complemented by a Ljung-Box test for the first 25 lags. Table 2 gives the p-values for the estimated standardized innovations $(\hat{\underline{\varepsilon}}_t)$ (the left half) and their absolute values (the right half). The value reported at the intersection of the i-th line with the j-th column is the p-value of the Ljung-Box statistic obtained by summing the first 25 values of the SCCF between the coordinate i and past lags of the coordinate j. Besides the pair (1,3), all other p-values do not reject the hypothesis (20) at 5% significance levels.

Finally, the hypothesis that the coordinates of the estimated standardized innovations, $\hat{\varepsilon}_1$, $\hat{\varepsilon}_2$, $\hat{\varepsilon}_3$ are pair-wise independent is tested using Kendall's τ distribution-free statistic. Kendall's τ takes values between -1 and 1 (independent variables have $\tau=0$) and provides an alternative measure of dependence between two variables to the usual correlation. While the easy-to-compute correlation is the natural scalar measure of linear dependence, Kendall's τ is a valuable measure of dependency also in the case of non-normality and non-linearity. In large samples, as the sample size n goes to ∞,

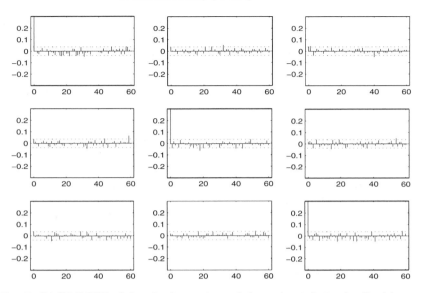

Fig. 7. SACF/SCCF of the absolute values of the estimated standardized innovations ($|\hat{\varepsilon}_t|$). The covariance structure was estimated using $\hat{\Sigma}$.

Table 2. The p-values for the Ljung-Box test at lag 25 of the estimated standardized innovations ($\hat{\varepsilon}_t$) (left) and their absolute values ($|\hat{\varepsilon}_t|$) (right). The row and column numbers represent the coordinates.

	1	2	3	1	2	3
1	0.15	0.17	0.03	0.20	0.89	0.12
2	0.81	0.12	0.11	0.62	0.21	0.16
3	0.70	0.88	0.07	0.25	0.50	0.22

$$3\tau\sqrt{\frac{n(n-1)}{2(2n+5)}} \xrightarrow{d} N(0,1).$$

Therefore Kendall's τ can be used as a test statistic for testing the null hypothesis of independent variables. (For more details on Kendall's τ we refer to Kendall and Stuart [KS79].)

The test is applied to all pairs of coordinates ($\hat{\varepsilon}_i$, $\hat{\varepsilon}_j$) ($i, j = 1, 2, 3$, $i < j$) and all pairs of their absolute values. observations ($\hat{\varepsilon}_{i,t}$, $\hat{\varepsilon}_{j,t}$), $t = 1, \ldots, n$, are mutually independent (see the independence tests (20) for evidence supporting this assumption) and come from the same continuous bivariate population. The null hypothesis is

$$H_0: \quad \text{the random variables } \hat{\varepsilon}_i \text{ and } \hat{\varepsilon}_j \text{ are independent.} \qquad (21)$$

The resulting p-values are given in Table 3. For all pairs the hypothesis of independence (21) is not rejected at usual statistical levels of significance..

At this point, we conclude that the battery of test described above do not reject the hypothesis that the estimated standardized innovations $(\hat{\underline{\varepsilon}}_t)$ is a stationary sequence of iid vectors with independent coordinates.

5.3 The multivariate distribution of the standardized innovations

In this section we concentrate on modeling the marginal distribution of the estimated standardized innovations $(\hat{\underline{\varepsilon}}_t)$. We begin by presenting some evidence that supports our claim that the marginal distributions of the three coordinate series $(\hat{\varepsilon}_i)$, $i = 1, 2, 3$, are heavy tailed. Figure 8 displays the standard normal plots of three coordinate series of the estimated standardized innovations $(\hat{\underline{\varepsilon}}_t)$. The graphs seem to show departures from normality for at least two of the three coordinates (the first and the third) with the right tail apparently heavier than the left one.

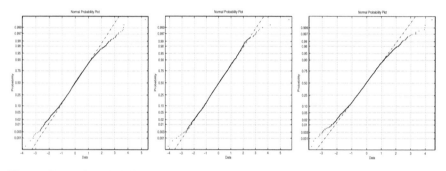

Fig. 8. Normal probability plots of the three series of coordinates the of estimated standardized innovations $(\hat{\underline{\varepsilon}}_t)$.

The impression given by Figure 8 is confirmed by the p-values of the Kolmogorov-Smirnov (K-S) and Andersen-Darling (A-D) tests (for details on these tests see [SW86]) applied to the coordinate series $(\hat{\varepsilon}_i)$, $i = 1, 2, 3$ reported in the left half of Table 4. The null hypothesis is

$$H_0 : \quad F_i \text{ is the standard normal distribution.} \tag{22}$$

The K-S and A-D tests are chosen for their complementary nature. It is well known that the Kolmogorov-Smirnov test is sensitive to departures from the

Table 3. The p-values for the Kendall's τ distribution-free test of independence applied to the estimated standardized innovations sequence $(\hat{\underline{\varepsilon}}_t)$ (left) and to the absolute values $(|\hat{\underline{\varepsilon}}_t|)$ (right). The pairs on the top are pairs of coordinates.

	(1,2)	(1,3)	(2,3)	(1,2)	(1,3)	(2,3)
Kendall	0.34	0.60	0.89	0.98	0.99	0.71

hypothesized law affecting the middle of the distribution while the Andersen-Darling test has been proved to be effective in identifying departures that affect the tails. The normality assumption is rejected at the 5% level by the A-D test for all three coordinates, while the K-S rejects it for the first and third coordinate.

Table 4. The p-values for the Andersen-Darling and Kolmogorov-Smirnov tests of normality (left) and of asymmetric VII Pearson (right) applied to the 3 coordinate series of the estimated standardized innovations ($\hat{\underline{\varepsilon}}_t$).

H_0: Normal	A-D	K-S	H_0: Pearson VII	A-D	K-S
$\hat{\varepsilon}_1$	0.007	0.038	$\hat{\varepsilon}_1$	0.21	0.20
$\hat{\varepsilon}_2$	0.037	0.108	$\hat{\varepsilon}_2$	0.17	0.48
$\hat{\varepsilon}_3$	¡0.001	0.003	$\hat{\varepsilon}_3$	0.10	0.25

Figure 8 and the values on the left side of Table 4 show that the estimated standardized innovations have tails that are heavier than normal tails.

We continue with the parametric modeling of the marginals of the estimated standardized innovations ($\hat{\underline{\varepsilon}}_t$) as asymmetric Pearson type VII heavy tailed distributions. Table 5 contains the estimated parameters obtained by fitting an asymmetric Pearson VII distribution (17) to the three coordinates of the estimated standardized innovations ($\hat{\underline{\varepsilon}}_t$).

Table 5. The parameters of the asymmetric Pearson distribution corresponding to the 3 series of estimated standardized innovations $\hat{\underline{\varepsilon}}_t$ (the standard deviations are provided in parentheses). The tail indices are given by $\nu = 2m - 1$.

	m_-	c_-	m_+	c_+	Left tail	Right tail
$\hat{\varepsilon}_1$	5.94 (1.48)	2.92 (0.47)	3.88 (0.60)	2.24 (0.25)	10.87	6.75
$\hat{\varepsilon}_2$	9.24 (3.71)	3.87 (0.91)	9.84 (4.22)	4.14 (1.03)	17.48	18.67
$\hat{\varepsilon}_3$	6.62 (1.86)	3.16 (0.55)	4.30 (0.75)	2.40 (0.29)	12.23	7.59

The estimated parameters in Table 5 confirm the results of the previous tail analysis: the first and the third coordinates have heavier tails then the second, with the right tail being heavier then the left one.

The estimated tail indexes reported in Table 5 are all sufficiently large for the regularity conditions needed for nonparametric regression to hold. More concretely, recall that the results in Section 3 that guarantee the asymptotic normality of the estimators of time-varying variance-covariance matrix assume, at least, a fifth finite moment[17]. The consistency of the estimator

---

[17] If this condition is violated, the confidence intervals around our estimated time-varying variance-covariance structure would be quite different.

continues to hold as long as a finite moment of order larger than two exists[18]. However, as observed by an anonimous referee, the estimated tail indexes in Table 5 could be affected by a positive bias induced by estimation of the tail parameters based on the residuals $\hat{\underline{\varepsilon}}_t$ and not on the true standardized innovations $\underline{\varepsilon}_t$. As residuals tend to look more normal than innovations, this could lead to an underestimation of the tails. The true tail indexes of the standardized innovations $\underline{\varepsilon}_t$ could be in fact heavier than they appear to be according to the estimates in Table 5.

To measure the bias induced by using $\hat{\underline{\varepsilon}}_t$ instead of $\underline{\varepsilon}_t$ we conducted the following simulation study. Series with the variance-covariance structure displayed in Figures 4 and 5 structure and standardized innovations from the asymmetric Pearson distribution with tail indexes ranging from $\nu = 4$ to $\nu = 6$ were simulated. Note that the interval of tail indexes covered by the simulations is centered in five, the minimal value needed for the asymptotic normality results in Section 3 to hold. We then smoothed the simulated series as described in Section 3 and produced the estimated standardized innovations. These residuals were then used to estimate the tail parameters m. Recall that the tail index $\nu = 2m - 1$. The exercise was repeated 1000 times.

Table 6. Estimation results for the parameter m on 1000 simulations of series with the variance-covariance structure displayed in Figures 4 and 5 and standardized innovations from the asymmetric Pearson distribution with given parameters m_+ and m_-. The second and fourth columns represent the true parameters, the third and fifth columns are the mean of the estimates with the standard deviation (in parentheses). The simulated series were first smoothed as described in Section 3 to produce the estimated standardized innovations $\hat{\underline{\varepsilon}}$ on which the parameters m_+ and m_- were then estimated. For tail indexes around the theoretically critical value of five, the estimation of the tail index based on estimated standardized innovations $\hat{\underline{\varepsilon}}_t$ is reasonably reliable.

	m_-	$\hat{m}_-$	m_+	$\hat{m}_+$
$\hat{\varepsilon}_1$	2.5	2.8065 (0.37)	3.5	3.5743 (0.65)
$\hat{\varepsilon}_2$	3	3.3550 (0.55)	3	3.4018 (0.61)
$\hat{\varepsilon}_3$	3.5	3.9515 (0.80)	2.5	2.8017 (0.38)

Table 6 reports the results of the simulation exercise. They show that, at least for tail indexes around the theoretically critical value of five, the estimation of the tail index based on estimated standardized innovations $\hat{\underline{\varepsilon}}_t$ is reasonably reliable. Although we confirm the presence of a positive bias, this is relatively small and the confidence intervals cover the true parameter. According to our simulations, for the value $m = 3$ which corresponds to a tail index $\nu = 5$ (the hypothesis for the asymptotic normality of the estimator holds), the limit of the 90% one sided upper confidence interval rests at 4.18.

---

[18] If the tail is so heavy that the variance is infinite, it is not clear that smoothing heavy-tailed data with linear smoothers would be a fruitful endeavor.

Based on this value and on the values in Table 5, only the right tail of the first coordinate might not have a finite fifth moment (although the probability of this happening is rather small). All the other standardized innovations seem to fulfill, with high probability, the moment condition needed for the asymptotic normality of the variance-covariance estimator. Moreover, for a tail index $\nu = 4$ (corresponding to a value of $m = 2.5$), the limit of the 99% one sided upper confidence interval is 3.68. Together with Table 5, this result confirms that all the innovation sequences have, with high probability, a finite forth moment. Hence the consistency of the variance-covariance estimator comfortably holds.

To test the hypothesis

$$H_0: \quad Var(\hat{\underline{\varepsilon}}_t) = I_d , \tag{23}$$

two estimates of the variances of the coordinates of the estimated standardized innovations $(\hat{\underline{\varepsilon}}_t)$ together with the corresponding standard deviations are produced. The first estimate is the sample variance with the standard deviation given by $\sqrt{m_i^4/n}$, $i = 1, 2, 3$, where m_i^4 is the sample fourth moment of $(\hat{\varepsilon}_i)$. The second estimate is the variance of the estimated asymmetric Pearson type VII given by (17). Since the variance of any coordinate is a function of the parameters reported in Table 5, the standard deviation for this variance estimate is obtained from the covariance matrix of the MLE estimates using the delta method. The three pairs of point estimates together with the standard deviations are reported on the left half of Table 7. The right half of the same table reports the sample covariance together with the corresponding standard deviation. According to the values in Table 7 the hypothesis that $Var(\hat{\underline{\varepsilon}}_t) = I_d$ is not rejected at the 5% significance level.

Table 7. The estimated variances of the coordinates of the estimated standardized innovations $(\hat{\underline{\varepsilon}}_t)$. The first column reports the sample variance while the second one is the variance of the estimated asymmetric Pearson type VII. The last column reports the sample covariance. The standard deviations are reported in parentheses.

	Empirical	Pearson VII		Covariance
$\hat{\varepsilon}_1$	0.971 (0.043)	1.007 (0.10)	$\hat{\varepsilon}_1, \hat{\varepsilon}_2$	0.0080 (0.020)
$\hat{\varepsilon}_2$	0.957 (0.037)	0.997 (0.17)	$\hat{\varepsilon}_1, \hat{\varepsilon}_3$	0.0026 (0.020)
$\hat{\varepsilon}_3$	0.944 (0.041)	1.002 (0.10)	$\hat{\varepsilon}_2, \hat{\varepsilon}_3$	0.0004 (0.019)

To verify the goodness of fit of the asymmetric Pearson type VII distribution, the probability plot of the coordinates of the estimated standardized innovations $(\hat{\underline{\varepsilon}}_t)$ using the estimated asymmetric Pearson VII distributions are displayed in Figure 9. A good fit of the asymmetric Pearson VII distributions should translate in linear graphs close to the first diagonal. The null hypothesis is

H_0 : F_i is the asymmetric Pearson VII distribution with parameters given in Table 4.

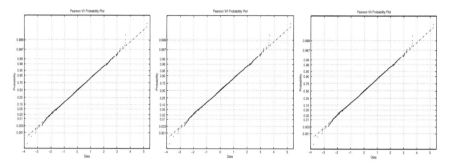

Fig. 9. The asymmetric Pearson VII probability plots of the three coordinate series of the estimated standardized innovations $(\hat{\varepsilon}_{i,t})$, $i = 1, 2, 3$.

The straight plots in Figure 9 are a confirmation of the good fit of the asymmetric Pearson VII distribution.

The hypothesis (24) is formally tested using the Kolmogorov-Smirnov and Andersen-Darling tests. The p-values of these tests are reported on the right in Table 4. The hypothesis is not rejected at usual levels of significance.

The plots in Figure 9 and the results in Table 4 provide evidence that the parametric family described by (17) is indeed an appropriate model for the estimated standardized innovations $(\hat{\underline{\varepsilon}}_t)$.

This concludes the evaluation of the goodness of fit of the model (1). The statistical analysis seems to show that the model provides an overall good description of the multivariate data set considered. We now direct our attention towards evaluating the forecasting performance of the non-stationary paradigm.

6 Forecasting multivariate returns

In this section we discuss aspects related to forecasting the multivariate returns using the non-stationary paradigm described in Section 2. We emphasize that we are interested in forecasting the whole distribution of the vector of future returns and not only the second moment structure.

We begin by specifying the m-day ahead forecasting methodology. Then we check the quality of our 1-day multivariate distributional forecasts. We end the section with a comparison (in the univariate framework) between the forecasting behavior of the industry standard Riskmetrics and that of our methodology on randomly generated portfolios containing the three instruments EU, the FTSE, and the US T-bond at one-day, ten-day and twenty-day horizons.

The main reason for the focus on relatively short-horizon forecasts is the fact that Riskmetrics, the benchmark methodology, is mainly used for this type of forecasting. A second reason is the non-stationary nature of the data as well as the frequency we chose to analyze. In a non-stationary set-up, the dynamics of the data is unknown and forecasting is possible only due to the gradual nature of the changes. An estimation based on data of a given frequency, say days, will provide information for forecasts of the order of a few time units (in our case, days) ahead (one, ten or twenty). If the goal of the forecast is a longer horizon, the frequency of the data used for estimation would change as to match the desired forecasting horizon. (For month- (year-) ahead forecasts, monthly (yearly) data would be used, etc). We have not investigated the forecasting behavior of our methodology at longer horizon[19].

6.1 The m-day ahead forecasting methodology

Given $\hat{\Sigma}_1(\cdot)$, an estimate of the unconditional covariance matrix $\Sigma(\cdot) = S(\cdot)S'(\cdot)$ based only on past information, denote by $\hat{F}_{i,t}^{VII}$, $i = 1, 2, 3$, the asymmetric Pearson type VII distributions (17) with parameters estimated on the coordinates of the series $(\hat{S}_1^{-1}(1)\,r_1,\ \hat{S}_1^{-1}(2)\,r_2, \ldots,\ \hat{S}_1^{-1}(t)\,r_t)$, where $\hat{S}_1(\cdot)$ is the square root of $\hat{\Sigma}_1(\cdot)$.

Based on the model (1), the distributional forecast at time t of the m-day ahead return $r_{t+1,m} := r_{t+1} + \ldots + r_{t+m}$ is given by

$$r_{t,m}^{REG} = \sum_{l=1}^{m} \hat{S}_1(t)\,\underline{\epsilon}_{lt}\ , \tag{25}$$

$\underline{\epsilon}_{lt}$, $l = 1, \ldots, m$, are iid d-dimensional random vectors,

$\epsilon_{i,lt}$ are mutually independent with distributions $\hat{F}_{i,t}^{VII}$, $(i = 1, 2, 3)$.

In other words, since the covariance matrix evolves slowly through time, to produce the m-day ahead forecast, the next m multivariate returns are assumed iid with a covariance matrix and parameters of the distribution of the standardized innovations estimated on recent past data.

For our forecasting exercise we use the one-sided-kernel estimate of the unconditional covariance matrix $S(\cdot)S'(\cdot)$ defined in (18). While the theoretical discussion in Section 2 focused on symmetric kernels, similar results are available for estimators of the type (18) (see Gijbels, Pope and Wand [GPW99] for the homoscedastic case). In particular, the bias and the variance of these

---

[19] We have no reason to believe that stationary models from the conditional heteroscedastic variance class would automatically produce better longer horizon forecasts due to their mean-reversion as an anonimous referee suggested. By contrary, at least in the univariate case, the results in Stărică ([Stă03]) and Stărică et al. ([SHN05]) show that for extensive periods, the Garch(1,1) model produces significantly poorer forecasts than a non-stationary set-up closely related to the methodology in Section 3.

estimators are also given by (9) and (10). Moreover, for forecasting, cross-validation can be safely employed as a method of bandwidth selection even when the errors are serially correlated [20].

6.2 One-day ahead multivariate density forecast evaluation

Evaluating the multivariate distributional forecast (see Diebold *et al.* [DHT99]) is particularly simple in the case of the model (1), due to the assumption of independence of the sequence $(\mathbf{r}_t)$. Verifying that the distribution of $\mathbf{r}_{t,1}^{REG}$ defined by (25) coincides with that of $\mathbf{r}_{t+1,1} = \mathbf{r}_{t+1}$ is equivalent to checking that the m-dimensional vectors $(\mathbf{z}_t)$

$$z_{i,\,t} := F_{i,t}^{VII}(v_{i,t}), \quad i = 1, 2, 3, \quad \text{where} \quad \mathbf{v}_t = \hat{\mathbf{S}}_1^{-1}(t)\mathbf{r}_t\,, \tag{26}$$

are iid, with independent, uniformly (0,1) distributed coordinates.

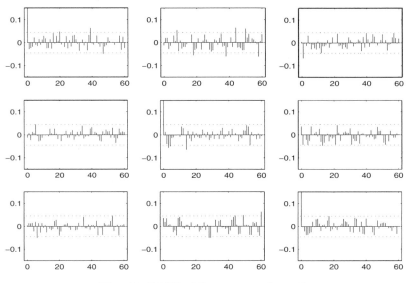

Fig. 10. SACF of the sequence $(\mathbf{z}_t - \bar{\mathbf{z}})$.

For evaluating the forecasting performance the sample is split in two: the first 1000 observations are used to produce the initial parameter estimates

---

[20] Cross-validation mistakes the smoothness of the series caused by positive correlation for low variability, yielding bandwidth choices usually smaller then than the optimal one. While this can be disastrous in mean estimation, it is the correct type of behavior in the forecasting context since averaging over a small number of past observations is more likely to be close to the next value in the series when there are positive correlations.

while the remaining 1926 observations are used to check the goodness of fit of the distribution forecast.

For an informed decision on the bandwidth to be used in the estimation of the unconditional covariance matrix (18), the cross-validation was run (using only the first 1000 observations) in the set-up of the model (3) with $y_k = |\mathbf{r}_k\mathbf{r}'_k|$, $k = 1, 2, \ldots, n$ both for K, the symmetric Gaussian kernel and for the asymmetric $\tilde{K}$, $\tilde{K}(u) = K(u)1_{u \leq 0}$. The results are displayed in Figure 11: on the left, the graph for the Gaussian kernel K, on the right, the one for $\tilde{K}$. The cross-validation optimal bandwidth seem to belong to the interval [0.0025, 0.008] for the Gaussian kernel and to the interval [0.004, 0.007] for $\tilde{K}$. The empirical relationship between the two intervals of optimal bandwidth is in accordance with the equivalent kernel theory according to which the optimal bandwidths corresponding to the two kernels are related by: $h_{(\tilde{K})} = 2^{1/5}h_{(K)}$. Although in the forecasting exercise a fixed band-width, $h_\mu = 0.007$ was used, an adaptative choice is also available. A time-depending bandwidth can be obtained by running the cross-validation on the sample up to the moment when the forecast is made.

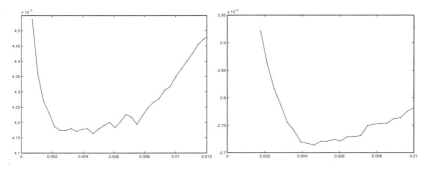

Fig. 11. The cross validation graph for the choice of the bandwidth h_μ for $\hat{\mu}$ in (4) and $y_k = |\mathbf{r}_k\mathbf{r}'_k|$ using only the first 1000 observations for the Gaussian kernel (*Left:* $h_\mu^{(c)} \in [0.0025, 0.008]$) and the kernel $\tilde{K}$ (*Right:* $h_\mu^{(c)} \in [0.004, 0.007]$).

A battery of tests similar to the one in Section 5 is employed to verify the hypotheses of iid-ness of the sequence $(\mathbf{z}_t)$ and those of uniformity and mutual independence of the coordinate sequences $(z_{i,t})$, $i = 1, 2, 3$. The precise working assumptions are those of the corresponding tests in Section 5. The same caveat on the impact of the uncertainty with respect to the estimated covariance matrix as in Section 5 applies. Figures 10 and 12 display the SACF/SCCF of the sequence $(\mathbf{z}_t - \bar{\mathbf{z}})$ and that of its absolute values ($\bar{\mathbf{z}}$ is the sample mean). Overall, they seem to support the hypothesis of iid vectors with independent coordinates for the sequence $(\mathbf{z}_t)$, although small violations of the confidence intervals are observed in the absolute values at the first lag of pairs (1,3) and (2,3).

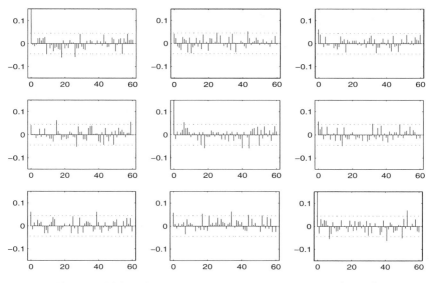

Fig. 12. SACF of the absolute values of the sequence $(\mathbf{z}_t - \bar{\mathbf{z}})$.

The visual test of the SACF/SCCF is complemented by the Ljung-Box test for the first 25 lags (the p-values are reported in Table 8). The value at the intersection of the row i with column j corresponds to the p-value of the Ljung-Box statistic associated with the SACF/SCCF of the coordinate i and past lags of the coordinate j. The p-values confirm the validity of the assumption of iid vectors with independent coordinates for the sequence $(\mathbf{z}_t)$.

Table 8. The p-values for the Ljung-Box test at lag 25 of the sequence $(\mathbf{z}_t - \bar{\mathbf{z}})$ (left) and the absolute values $(|\mathbf{z}_t - \bar{\mathbf{z}}|)$(right). The row and column numbers represent the coordinates.

	1	2	3	1	2	3
1	0.53	0.20	0.38	0.50	0.34	0.62
2	0.96	0.20	0.09	0.71	0.51	0.32
3	0.80	0.75	0.67	0.52	0.47	0.49

The hypothesis of pair-wise, mutual independence of the coordinates of the vector $\mathbf{z}$ is tested using the already familiar distribution-free test of Kendall's τ. The p values corresponding to the pairs of coordinates are given in Table 9. For all pairs the hypothesis of independent coordinates is not rejected at usual levels of statistical significance.

Figure 13 displays the uniform probability plots for the three coordinates z_i, $i = 1, 2, 3$. The straight plots in this figure together with the p-values of the Andersen-Darling and Kolmogorov-Smirnov tests of uniformity given in

Table 9. The p values for Kendall's τ distribution-free test of independence applied to the sequence $(\mathbf{z}_t - \bar{\mathbf{z}})$ (left) and to that of absolute values $(|\mathbf{z}_t - \bar{\mathbf{z}}|)$ (right).

	(1,2)	(1,3)	(2,3)	(1,2)	(1,3)	(2,3)
Kendall	0.31	0.55	0.95	0.97	0.99	0.98

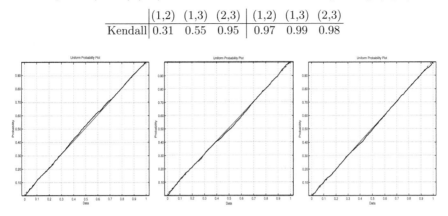

Fig. 13. The uniform probability plot of the three series of coordinates z_i, $i = 1, 2, 3$, $t = 1, \ldots, 1926$.

Table 10 support the conclusion that the marginal distributions of the three sequences $(z_{i,t})$, $i = 1, 2, 3$ are uniform $(0,1)$.

Table 10. p-values for the Andersen-Darling and Kolmogorov-Smirnov tests of uniformity applied to the coordinates of the sequence $(\mathbf{z}_t)$.

	1	2	3		1	2	3
A-D	0.14	0.19	0.77	K-S	0.10	0.24	0.58

6.3 Univariate density forecast evaluation

We conclude this section with a distributional forecast comparison in a univariate framework. The comparison is done between the industry standard RiskMetrics and the approach described in Section 6.1 for forecasting horizons of one, ten and twenty days. Both methodologies are used to produce daily distributional forecasts for the returns of randomly generated portfolios containing the (by now familiar) three financial instruments. More specifically, for a given day t, the two approaches are first used to produce two multivariate distributional forecasts for the next day vector of returns. For RiskMetrics, the distributional forecast is

$$\mathbf{r}_{t,m}^{RM} \stackrel{d}{=} N(0, m\hat{\Sigma}_t^2) , \tag{27}$$

where

$$\hat{\Sigma}_t^2 := \sum_{i=0}^{l-1} \lambda^{t-i} \mathbf{r}_{t-i} \mathbf{r}'_{t-i} / \sum_{i=0}^{l-1} \lambda^{t-i} , \tag{28}$$

is the exponential moving average estimate of the conditional covariance matrix Σ_t^2. The parameters used were $\lambda = 0.94$ and $l = 120$ for one-day ahead forecasts and $\lambda = 0.97$ and $l = 200$ for ten- and twenty-day ahead forecasts (as stipulated in the RiskMetrics documents [Risk95]). For the regression-type model (1), the m-day forecast $\mathbf{r}_{t,m}^{REG}$ is given by (25).

Note that our comparison focuses on the most common specification of the distributional forecast of RiskMetrics, i.e. that where the future returns are jointly normal. We chose this specification due to the fact that it is widely used in practice. Comparisons with other specifications (normal mixture models, GED models) are currently under investigation and the results will be reported elsewhere.

The return of a given portfolio $\mathbf{w}$ with weights $\mathbf{w} = (w_1, w_2, w_3)$ over the period $[t + 1, t + m]$ is denoted by $r_{t+1,m}^{(\mathbf{w})}$. The distribution of $r_{t+1,m}^{(\mathbf{w})}$ forecasted by the RiskMetrics methodology, which we denote by $F_{t,m}^{RM}$, is the distribution of $\mathbf{wr}_{t,m}^{RM}$ (a normal with mean 0 and variance $m\mathbf{w}\hat{\Sigma}_t^2 \mathbf{w}'$). The distribution forecasted by the regression-type model (1), denoted by $F_{t,m}^{REG}$, is that of $\mathbf{wr}_{t,m}^{REG}$.

As explained in Diebold *et al.* [DGT98], evaluating the correct distributional forecast $F_{mt,m}$ at the realized portfolio returns $r_{mt+1,m}^{(\mathbf{w})}$, $t = 1, \ldots, [n/m] - 1$ yields an iid sequence $(F_{mt,m}(r_{mt+1,m}^{(\mathbf{w})}))$ of uniform (0,1) random variables. Hence the quality of a distributional forecast $G_{mt,m}$ can then be assessed by testing the hypothesis

H_0: $(G_{mt,m}(r_{mt+1,m}^{(\mathbf{w})}))$ is an iid sequence with uniform (0,1)

marginal distribution. (29)

In the sequel we test hypothesis (29) for $G_{t,m} = F_{t,m}^{RM}$ and $G_{t,m} = F_{t,m}^{REG}$. More concretely the sequences $(F_{mt,m}^{RM}(r_{mt+1,m}^{(\mathbf{w})}))$ and $(F_{mt,m}^{REG}(r_{mt+1,m}^{(\mathbf{w})}))$, $t = 1, \ldots, [n/m] - 1$ are tested for variance $1/12$ (the variance of a uniform (0,1)), using a test based on the Central Limit Theorem, for uniform (0,1) marginal distribution, employing the Kolmogorov-Smirnov and Andersen-Darling test and for independence, using the Ljung-Box statistic at lag 10. We used the following simulation set-up.

For every horizon ($m = 1$, $m = 10$, $m = 20$) three thousand portfolios were randomly generated. The weights of each portfolio $\mathbf{w}$ were sampled from a uniform (0,1) distribution then normalized such that they added up to 1. As in Section 6.2, the sample is split into two parts: the first 1000 observations serve to produce the initial parameter estimates for the regression-type model while the remaining 1926 observations are used to compute the sequences

[21] $(F_{mt,m}^{RM}(r_{mt+1,m}^{(\mathbf{w})}))$ and $(F_{mt,m}^{REG}(r_{mt+1,m}^{(\mathbf{w})}))$, $t = 1, \ldots, [n/m] - 1$, for each portfolio $\mathbf{w}$. (We kept the weights of the portfolios constant during the testing period.) For every sequence we produced the p-values corresponding to the four mentioned statistics.

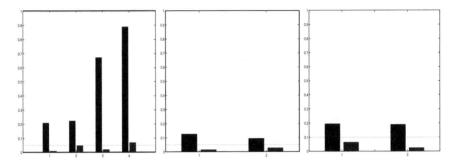

Fig. 14. The percentage of the p-values for the K-S (1), A-D (2), the L-B at lag 10 (3) and the variance test (4) that are smaller than 5% (Left and Center) and 10% (Right). For a given test, the first bar concerns the RiskMetrics methodology while the second one refers to the forecasting methodology described in Section 6.1. *Left:* One-day ahead, *Center:* Ten-day ahead, *Right:* Twenty-day ahead.

The results of these simulations are summarized in Figure 14 where the percentage of p-values smaller than 5% for $m = 1$, $m = 10$ and than 10% for $m = 20$ is reported [22]. For a given test, the first bar concerns RiskMetrics while the second one refers to the forecasting methodology described in Section 6.1. It is interesting to notice that, for one-day ahead forecasting, for almost 90% of the portfolios, plain-vanilla RiskMetrics fails (at the 5% level) the variance test. This should be compared to the 94% acceptance rate for our methodology. Moreover, 25% of the sequences $(F_{t,1}^{RM}(r_{t+1}^{(\mathbf{w})}))$ fail at least one of the uniformity tests (either K-S or A-D) compared to only 5% of the $(F_{t,1}^{REG}(r_{t+1}^{(\mathbf{w})}))$ sequences. Finally, RiskMetrics fails at least one of the four tests in 94% of the cases compared to only 9% for our methodology.

For $m = 10$ and $m = 20$, the empirical percentage of p-values for the last two tests were, for both methods, below the fixed theoretical level of 5% for ten-day forecasts and 10% for the twenty-day forecasts[23]. For the variance test, this is not surprising, since, due to averaging, for ten- and twenty-day

---

[21] The sequence of the m-days ahead forecast would exhibit an intertemporal $m - 1$ dependence. To keep the independence between observations, necessary for the statistical tests, we restricted the sample to the sub-sequence of m-days apart forecasts.

[22] For $m = 20$, a higher percentage of 10 has been used due to the small number of observations in the sequences $(G_{mt,m}(r_{mt+1,m}^{(\mathbf{w})}))$, $t = 1, \ldots, [n/m] - 1$.

[23] For this reason, they are not reported in the graphs in the center and right of Figure 14.

returns, the multivariate normality assumption of RiskMetrics is more adequate than for daily returns. However, the normality tests show once again the superiority of our methodology over the plain-vanilla RiskMetrics[24]. As mentioned before, comparisons involving other RiskMetrics specifications of the conditional distribution are under study and the results will be reported elsewhere.

7 RiskMetrics vs. non-parametric regression

We conclude with a few remarks on the relationship between our approach and RiskMetrics. Univariately, the probabilistic model that forms the basis of RiskMetrics forecasting methodology outlined in (27) and (28) is the following conditional, multiplicative process

$$r_t = \sigma_t \, \epsilon_t, \qquad \epsilon_t \sim \mathcal{N}(0,1) \,, \tag{30}$$

(see page 73 of [Risk95]) where

$$\sigma_t^2 = \lambda \sigma_{t-1}^2 + (1-\lambda) r_{t-1}^2 \,, \tag{31}$$

according to Section B.2.1 of the Appendix B of [Risk95]. This specification is, up to a constant term, that of a IGARCH process explaining why in the literature the RiskMetrics model is often thought of as being an IGARCH model.

From a probabilistic point of view, the model (30) and (31) is faulty. Results by Kesten [Kes73] and Nelson [Nel90] imply that a time series evolving according to the dynamics (30) and (31) will tend to 0 almost surely.

The claimed close relationship between the RiskMetrics methodology and GARCH-type models, prompted by the deceiving formal analogy between the GARCH(1,1) specification

$$\sigma_t^2 = \alpha_0 + \alpha_1 X_{t-1}^2 + \beta_1 \sigma_{t-1}^2 \,,$$

and (31) and emphasized by the comparisons in Section 5.2.3 of ([Risk95]), is hence misleading. Instead, the RiskMetrics approach can be motivated by the non-stationary model (1).

Note that the forecast (28) is just a kernel smoother of the type (18) with an one-sided exponential kernel $K_{exp}(x) = a^x 1_{[-\infty,0]}(x)$, $a = \lambda^m$ and $h = 1/m$. Our experience shows that replacing the normal kernel with the exponential leads to results very similar to the ones reported in Section 6.

---

[24] This forecasting methodology has been thoroughly investigated in the univariate case in a companion paper by Drees and Stărică [DS02]. There the authors show by the example of the S&P 500 time series of returns that this apparently structureless forecasting methodology outperforms conventional GARCH-type models both over one day and over time horizons of up to forty days.

This finding is in line with the well-known fact that the choice of the band-width h affects the performance of a kernel regression estimator much more strongly than the choice of the kernel. In fact, in the Sections 5 and 6 we have deliberately chosen the normal kernel instead of the exponential filter (more common in time series analysis) to demonstrate that the choice of the kernel does not matter much.

Besides providing a solid statistical framework, the set-up of the non-stationary paradigm introduced in Section 2 allows for a optimal choice of the bandwidth, motivated by results from the statistical theory of curve estimation. By contrast, the choice of the parameters λ and l is empirical.

While the volatility forecasts by the RiskMetrics methodology are similar to ours, the assumption of normal innovations is too restrictive to yield accurate forecasts of the distribution of future returns. This has also been observed in [Risk95]. In Appendix B of the RiskMetrics document normal mixture models or GED models for the innovations are proposed. However, these alternative models lack two features that are essential for a successful fit of many real data sets: they do not allow for asymmetry of the distribution of innovations and they assume densities with exponentially decaying tails, thus excluding heavy tails.

8 Conclusions

In this paper a simple multivariate non-stationary paradigm for modeling and forecasting the distribution of returns on financial instruments is discussed.

Unlike most of the multivariate econometric models for financial returns, our approach supposes the volatility to be exogenous. The vectors of returns are assumed to be independent and to have a changing unconditional covariance structure. The methodological frame is that of non-parametric regression with fixed equidistant design points where the regression function is the evolving unconditional covariance. The vectors of standardized innovations have independent coordinates and asymmetric heavy tails and are modeled parametrically. The use of the non-stationary paradigm is exemplified on a tri-variate sample of risk factors consisting of a foreign exchange rate Euro/Dollar (EU), an index, FTSE 100 index, and an interest rate, the 10 year US T-bond. The paradigm provides both a good description of the changes in the dynamic of the three risk factors and good multivariate distributional forecasts.

We believe that the careful parametric modeling of the extremal behavior of the standardized innovations makes our approach amenable for precise VaR calculations. Evaluating its behavior in these settings is, however, subject of further research.

Acknowledgement

This research has been supported by The Bank of Sweden Tercentenary Foundation and the Jan Wallanders and Tom Hedelius Stiftelse (Handelsbanken

Forsknings Stiftelser), by the NSF through grant CCR-9875559 and by the Consiglio Nazionale delle Ricerche. The comments of an anonymous referee helped improve the paper.

References

[BGV99] Barone-Adesi, G., Giannopoulos, K., Vosper, L.: VaR without correlations for portfolio of derivative securities. *J. of Futures Markets*, **19**, 583–602, 1999.

[CLM96] CAMPBELL, J., LO A., AND MACKINLAY, A. The Econometrics of Financial Markets. Princeton University Press, 1996.

[Cai94] Cai, J. A Markov model of unconditional variance in ARCH. *J. Business and Economic Statist.* **12**, 309–316, 1994.

[Die86] Diebold, F.X. Modeling the persistence of the conditional variances: a comment. *Econometric Reviews* **5**, 51–56, 1986.

[DI01] Diebold, F.X. and Inoue, A. Long memory and regime switching. *J. Econometrics* **105**, 131–159, 2001.

[DGT98] Diebold, F.X., Gunther, T. and Tay, A. Evaluating Density Forecasts with Applications to Financial Risk Management. *International Economic Review*, **39**, 863–883, 1998.

[DHT99] Diebold, F.X., Hahn, J. and Tay, A. Multivariate Density Forecast Evaluation and Calibration in Financial Risk Management: High-Frequency Returns on Foreign Exchange. *Review of Economics and Statistics*, **81**, 661–673, 1999.

[DS02] Drees, H. and Stărică , C. A simple non-stationary model for stock returns. Preprint, 2002. Available at http://www.math.chalmers.se/~starica/

[ES02] Engle, R.F. and Sheppard, K. Theoretical and Empirical Properties of Dynamic Conditional Correlation Multivariate GARCH. Preprint, 2002. Available at http://weber.ucsd.edu/~mbacci/engle/index_recent.html

[FJZZ03] Fan, J., Jiang, J., Zhang C. and Zhou Z. Time-dependent Diffusion Models for Term Structure Dynamics and the Stock Price Volatility. *Statistica Sinica*, **13**, 965-992, 2003.

[FY98] Fan, J. and Yao, Q. Efficient Estimation of Conditional Variance Functions in Stochastic Regression. *Biometrica*, **85**, 645–660, 1998.

[GKK91] Gasser, T., Kneip, A., K'ohler, W. A flexible and fast method for automatic smoothing. *J. Amer. Statist. Assoc.*, **86**, 643–652, 1991.

[GMM85] Gasser, T., Müller, H.-G., Mammitzsch, V. Kernels for nonparametric curve estimation. *J. Roy. Statist. Soc. Ser. B*, **47**, 238–252, 1985.

[GPW99] Gijbels, I., Pope, A. and Wand, M. P. Understanding exponential smoothing via kernel regression. *J. R. Stat. Soc. Ser. B*, **61**, 39–50, 1999.

[GH99] Granger, C.W.J., Hyung, N. Occasional structural breaks and long-memory. Discussion paper 99-14, University of California, San Diego, 1999.

[GT99] Granger, C., W., Teräsvirta, T. A simple non-linear time series model with misleading linear properties. *Economics Letters* **62**, 161–165, 1999.

[HR96] Hidalgo, J. and Robinson, P.M. Testing for structural change in a long–memory environment. *J. Econometrics* **70**, 159 – 174, 1996.

[HMW74] Hsu, D.A., Miller, R., and Wichern, D. On the stable Paretian behavior of stock-market prices. *J. Amer. Statist. Assoc.* **69**, 108–113, 1974.

[KS79] Kendall, M. and Stuart, A. The advanced theory of statistics. Charles Griffin, London, 1979.

[Kes73] Kesten, H. Random difference equations and renewal theory for products of random matrices. *Acta Math.* **131**, 207–248, 1973.

[LS98] Lobato, I. N. and Savin, N. E. Real and Spurious Long-Memory Properties of Stock-Market Data. *J. of Business & Economic Statist.*, 261–268, 1998.

[LL90] Lamoureux, C.G. and Lastrapes, W.D. Persistence in variance, structural change and the GARCH model. *J. Business and Economic Statist.* **8**, 225–234, 1990.

[MS02] Mikosch, T. and Stărică, C. Is it really long memory we see in financial returns? In: Extremes and Integrated Risk Management. Ed. P. Embrechts, Risk Books, 2002.

[MS04] Mikosch, T. and Stărică, C. Non-stationarities in financial time series, the long range dependence and the IGARCH effects. *Rev. of Economics and Statist.*, **86**, 378–390, 2004.

[Mul88] Müller, H., G. Nonparametric Regression Analysis of Longitudinal Data. Springer, Berlin, 1988.

[MS87] Müller, H.-G. and Stadtmüller, U. Estimation of heteroscedasticity in regression analysis. *Annals of Statistics* **15**, 610–625, 1987.

[NK99] Nagahara, Y., and Kitagawa, G. A Non-Gaussian Stochastic Volatility Model. *J. of Computational Finance*, **2**, 33–47, 1999.

[Nel90] Nelson, D.B. Stationarity and persistence in the GARCH(1, 1) model. *Econometric Theory* **6**, 318–334, 1990.

[Off76] Officer, R. A time series examination of the market factor of the New York Stock Exchange. Ph.D. Dissertation. University of Chicago, 1971.

[RL01] Rodríguez-Poo, J. M. and Linton, O. Nonparametric factor analysis of residual time series. *Test*, **10**, 161–182, 2001.

[Ric84] Rice, J. Boundary modification for kernel regression. *Comm. Statist. A— Theory Methods* **13**, 893–900, 1984.

[Risk95] RiskMetrics. Technical Document, 1995. Available at http://www.riskmetrics.com/techdoc.html.

[SW86] Shorack, G. R. and Wellner J. A. Empirical processes with applications to statistics. John Wiley & Sons, New York, 1996.

[Sim92] Simonato, J., G. Estimation of GARCH processes in the presence of structural change. *Economic Letters* **40**, 155–158, 1992.

[Stă03] Stărică, C. Is Garch(1,1) as good a model as the accolades of the Noble prize would imply?, Preprint 2003. Available at http://www.math.chalmers.se/~starica

[SG05] Stărică, C. and Granger, C. Non-stationarities in stock returns. *Rev. of Economics and Statist.* **87**, 2005. Available at http://www.math.chalmers.se/~starica

[SHN05] Stărică, C., Herzel, S., and Nord, T. Why does the Garch(1,1) model fail to produce reasonable longer-horizon forecasts? Preprint, 2005. Available at http://www.math.chalmers.se/~starica

[SW96] Stock, J. and Watson, M. Evidence on structural instability in macroeconomic time series relations. *J. Business and Economic Statist.*, **14**, 11–30, 1996.

[WJ95] Wand, M.P. and Jones, M.C. *Kernel Smoothing.* Chapman and Hall, London, 1995.

9 Appendix

As the most commonly used paradigm for modeling of financial returns is that of stationary conditional heteroscedastic volatility, it is of interest to evaluate the behavior of our methodology were the returns produced by a data generating processes of this type.

More concretely, we simulated returns both from a stationary conditional heteroscedastic volatility model, using parameters as fitted to the real data and from our model, using the variance-covariance structure and the innovation's distribution estimated on the real data. We then ran the smoothing procedure in Section 3 on the simulated data and evaluated the null hypothesis that the estimate innovation vectors are independent[25]. If the hypothesis is not rejected on the simulated stationary conditional heteroscedastic volatility data, then the proposed procedure may be overly flexible, in the sense that it can make a stationary dependent series with conditional heteroscedasticity look indistinguishable from a non-stationary independent series. Otherwise, the simulation exercise brings additional evidence supporting the modeling of returns as non-stationary, independent vectors.

Given the large number of possible specifications of a multivariate conditional heteroscedastic volatility set-up in the econometric literature, the choice of the model to simulate from is, of course, subjective and possibly open to criticism. Since an exhaustive comparison with the stationary conditional volatility paradigm is not the aim of the paper, we chose to have a closer look at one specification that is both parsimonious and widely used in the literature: the dynamic conditional correlation model of Engle and Sheppard [ES02].

The model was first estimated on the data. The estimated parameters were used to simulate 1000 samples of the length of the return data, i.e. 3000 observations. For every sample the smoothing procedure in Section 3 is applied. First the cross-validation procedure ran on the norm series of the tri-dimensional vector data to select the bandwidth to be used for the smoothing of the sample. Then the kernel smoothing procedure with the bandwidth chosen in the previous step was applied and a sample of residuals was produced. The same steps were applied to simulated independent vectors with the time-varying variance-covariance structure displayed in Figures 4 and 5. Both steps, i.e. the bandwidth estimation as well as the estimation of the innovations, produce qualitatively different results depending on the type of data used in simulations.

---

[25] As seen previously, this hypothesis was not rejected on the actual data

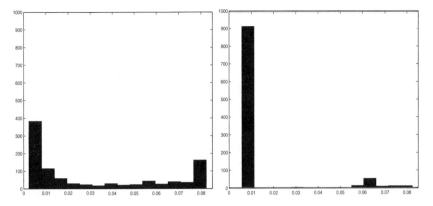

Fig. 15. The estimated optimal bandwidth for the DCC samples (left) and samples from our model (right). The bandwidth were obtained for each sample by cross validation.

The graphs in Figure 15 show strong qualitative differences between the estimation of the smoothing bandwidth h. While for the independent, non-stationary simulations, the values of the bandwidth are strongly concentrated around the value obtained when using the real data, for the simulations of the DCC model, the smoothing parameter shows a wide dispersion with more than 15% of the samples yielding values of the bandwidth larger or equal to 0.08 or more than ten times the desired value (to keep the scale of the graph within a reasonable range all the values larger than 0.08 were set to this value).

The graphs in Figure 16 display the histogram of the sample autocorrelation at lag one for the absolute values of the estimated standardized innovations $\hat{\varepsilon}_i^{(1)}$, $i = 1, 2, 3$. The left hand side histograms correspond to the residuals of the DCC model while the ones on the right hand side to those from the non-stationary model. The residuals were obtained by smoothing the simulated samples using the methodology in Section 3. The bandwidth used in smoothing were obtained for each sample by cross validation. The vertical lines are the 95% confidence intervals corresponding to a sample of length 3000 of iid data. While the autocorrelation of the residuals obtained from samples generated by the non-stationary model do not reject the hypothesis of independent residuals, those corresponding to the residuals from samples generated by the DCC model indicate more often than the statistical error would grant that the estimated innovations are not iid.

To conclude, the evidence presented seems to support the fact that the smoothing methodology can distinguish between a stationary dependent model with conditional heteroscedasticity and a non-stationary independent model. However, since the differences are measured based on the behavior of statistical instruments, it may take more than one sample to be able to detect the difference.

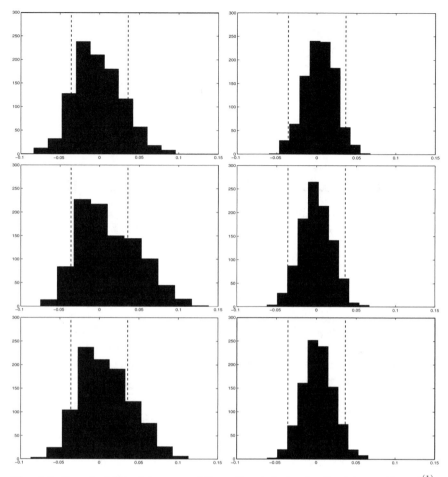

Fig. 16. Sample ACF at lag one of the estimated standardized innovations $\hat{\varepsilon}_i^{(1)}$, $i = 1, 2, 3$ of the DCC model (left) and our model (right). The lines correspond to the three sequences of innovations. The bandwidth used in smoothing were obtained for each sample by cross validation. The distribution of the SACF at lag 1 of the residuals of the non-stationary model is consistent with the hypothesis of independence while that of the residuals of the DCC model is not.

To end the section, we show that the estimated DCC model cannot reproduce the covariance structure of the real data. The samples simulated in the previous study were used to determine the range of possible covariance structures that the DCC model and our model can produce. The graphs in Figure 17 display the mean (stem) and the 95% one-sided point-wise simulation-based confidence interval (dotted) for the sample cross-correlation between the absolute values of the second and the third coordinates of samples generated from the DCC model with the parameters estimated on our real data (left)

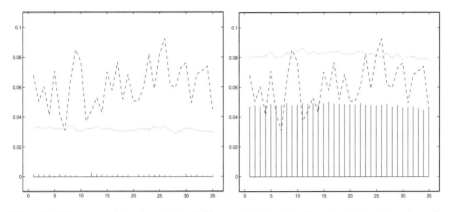

Fig. 17. The mean (stem) and the 95% one-sided point-wise simulation-based confidence interval (dotted) for the sample cross-correlation between the absolute values of the second and the third return series. The samples generated from the DCC model with the parameters estimated on our real data were used to produce the graph on the left while the simulated independent vectors with the time-varying variance-covariance structure displayed in Figure 5 were used to produce the graph on the right. 1000 samples of each model were generated.

and that of the simulated independent vectors with the time-varying variance-covariance structure displayed in Figures 4 and 5 (right). The dashed line is the sample cross-correlation of the data. The graphs show clearly that, while our model could produce a covariance structure for this pair of coordinates that comes close to that of the real data, it is highly unlikely that the DCC model could match it.

Multivariate Non-Linear Regression with Applications

Tata Subba Rao[1] and Gyorgy Terdik[2]

[1] School of Mathematics, University of Manchester, POB 88, Manchester M60
1QD, United Kingdom `tata.subbarao@umist.ac.uk`
[2] Department of Information Technology, Faculty of Informatics, University of
Debrecen, 4010 Debrecen, Pf.12, Hungary `terdik@delfin.unideb.hu`

1 Introduction

Observations, especially those of geophysics, are rarely stationary; in most of
the cases they are of the form

$$\underline{Y}_t = \underline{X}_t(\vartheta) + \underline{Z}_t,$$

where $\underline{X}_t(\vartheta)$ is some deterministic "trend" and $\underline{Z}_t$ is some stationary "noise".
We discuss the problem of estimating the parameter ϑ included in some pa-
rameter space $\Theta \subset \mathbb{R}^p$ and $\underline{X}_t(\vartheta)$ some uniformly continuous known function of ϑ with vector values. Robinson [Rob72] established some asymptotic
properties of generalized least squares estimates of ϑ in the non-linear multiple model. Earlier Hannan [Han71], [Han73] obtained fundamental results
on weighted least squares estimates of ϑ in the frequency domain, occurring
in the scalar model. The frequency domain approach proves to be very natural because the classical Grenander's conditions [Gre54] are given in terms of
the regression spectrum, and most of the trends have nice Fourier transforms
that can be computed easily and fast. Constrained non-linear regression was
studied by [Wan96] and [Wan04] and we refer to [RH97], [Hos97] and [CT01]
for stochastic regression models with long-range dependence.

In this chapter we consider the multivariate non-linear regression model
with multiple stationary residuals. The minimum contrast estimate [Guy95],
[Hey97] of the unknown parameter is constructed in the frequency domain
using Hannan's idea [Han71]. Robinson's result [Rob72] on the strong consistency of the estimate is quoted. We do not give very general conditions although we point out some possibilities for weakening some assumptions. Our
main result is the exact form of the asymptotic variance of the estimator in
terms of the weight function and the regression spectrum. We introduce a scaling technique for the application of the general CLT to several cases where the
original assumptions of the asymptotic results do not hold. The mixed model

containing linear regression and linear combinations of non-linear regression is considered in detail. One of the most important applications is linear regression with harmonic trend when the harmonic frequencies are unknown. It is applied to the real data set referred to as the Chandler wobble. Based on high-resolution GPS data it is shown that beside the well-known Chandler period (410 days in our case) the period of 12 hours is also present and the residual series is long-range dependent.

2 Non-Linear Time Series Regression

2.1 Model

Consider a multiple model of dimension d; that is, the observation $\underline{Y}_t$, the random disturbances $\underline{Z}_t$ and the function $\underline{X}_t(\underline{\vartheta}_0)$ containing the regressors are d-dimensional vectors and

$$\underline{Y}_t = \underline{X}_t(\underline{\vartheta}_0) + \underline{Z}_t . \tag{1}$$

The function $\underline{X}_t(\underline{\vartheta})$ is a possibly non-linear function of both the regressors and the multiple parameter $\underline{\vartheta}$ of dimension p, and $\underline{Z}_t$ is a stationary time series. The parameter $\underline{\vartheta} \in \Theta \subset \mathbb{R}^p$. The set Θ of admissible parameters $\underline{\vartheta}$ is defined by a number of possibly nonlinear equations (see [Rob72] for more details). We assume that the set Θ is chosen suitably in each case. There should be no confusion when we call $\underline{X}_t(\underline{\vartheta})$ the regressor, although in particular cases the picture is more colorful. The regressor, more strictly, is a function of t depending non-linearly on some parameters. A particular model that we keep in mind, has $\underline{X}_t(\underline{\vartheta})$ in the form

$$\begin{aligned}
\underline{X}_t(\underline{\vartheta}) &= \mathbf{B}_1\underline{X}_{1,t} + \mathbf{B}_2\underline{X}_{2,t}(\underline{\lambda}) \\
&= [\mathbf{B}_1, \mathbf{B}_2]\begin{bmatrix} \underline{X}_{1,t} \\ \underline{X}_{2,t}(\underline{\lambda}) \end{bmatrix} \\
&= \mathbf{B}\underline{X}_{3,t}(\underline{\lambda}) ,
\end{aligned}$$

see Section 5.2 for details. Here $\underline{X}_{2,t}(\underline{\lambda})$ is some non-linear function of $\underline{\lambda}$. The multiple parameter $\underline{\vartheta}$ contains the entries of the matrices $\mathbf{B}_1$ and $\mathbf{B}_2$ and moreover the vector $\underline{\lambda}$. The admissible set Θ is the union of three subsets. There is no restriction on the entries of matrix $\mathbf{B}_1$ with size $d \times p_1$, say. The matrix $\mathbf{B}_2$ and the vector $\underline{\lambda}$ have some delicate connection, because $\underline{\lambda}$ must be entirely identifiable. The parameter $\underline{\lambda}$ is identifiable unless some particular entries of $\mathbf{B}_2$ annihilate an entry, say λ_k, from the model. Finally $\underline{\lambda}$ is constrained to lie within a compact set; for instance, for harmonic regressors $\underline{\lambda} \in [-1/2, 1/2]^2$, see Section 5.3.

We assume that $\underline{Z}_t$ is linear; that is, it has moving average representation

$$\underline{Z}_t = \sum_{k=-\infty}^{\infty} \mathbf{A}_k \underline{W}_{t-k} \ , \quad \sum_{k=-\infty}^{\infty} \mathrm{Tr}\left(\mathbf{A}_k \mathbf{C}_W \mathbf{A}_k^*\right) < \infty \ ,$$

where $\underline{W}_t$ is an i.i.d. series and $\mathbf{C}_W = \mathrm{Var}\,\underline{W}_t$, is non-singular. Moreover $\underline{Z}_t$ has a piecewise continuous spectral density $\mathbf{S}_Z(\omega)$ [Han70], [Bri01]. The model is feedback free; that is, $\underline{Z}_t$ does not depend on $\underline{X}_t$.

2.2 The regression spectrum

We start with a regressor $\underline{X}_t(\vartheta)$ which is a function of t depending non-linearly on some parameters and with a compact set Θ. It is known that the so-called Grenander's conditions ([Gre54], [GR57]) for the regressor $\underline{X}_t(\vartheta)$ are sufficient and in some situations ([Wu81]) are necessary as well for the consistency of the LS estimators. We state them below. Denote

$$\|X_{k,t}(\vartheta)\|_T^2 = \sum_{t=1}^{T} X_{k,t}^2(\vartheta) \ .$$

Condition 1 (G1) *For all $k = 1, 2, \ldots, d$,*

$$\lim_{T \to \infty} \|X_{k,t}(\vartheta)\|_T^2 = \infty \ .$$

Condition 2 (G2) *For all $k = 1, 2, \ldots, d$,*

$$\lim_{T \to \infty} \frac{X_{k,T+1}^2(\vartheta)}{\|X_{k,t}(\vartheta)\|_T^2} = 0 \ .$$

Without any restriction of generality we assume that the regressor $\underline{X}_t(\vartheta)$ is properly scaled: $\|X_{k,t}(\vartheta)\|_T^2 \simeq T$, for all k; see Definition 1 and a note therein. Define the following matrices for any integer $h \in [0, T)$,

$$\widehat{\mathbf{C}}_{\underline{X},T}(h, \vartheta_1, \vartheta_2) = \frac{1}{T} \sum_{t=1}^{T-h} \underline{X}_{t+h}(\vartheta_1)\, \underline{X}_t^{\mathsf{T}}(\vartheta_2) \ , \tag{2}$$

$$\widehat{\mathbf{C}}_{\underline{X},T}(-h, \vartheta_1, \vartheta_2) = \widehat{\mathbf{C}}_{\underline{X},T}^{\mathsf{T}}(h, \vartheta_2, \vartheta_1) \ .$$

If $\vartheta_1 = \vartheta_2 = \vartheta$, here and everywhere else, we use the shorter notation $\widehat{\mathbf{C}}_{\underline{X},T}(h, \vartheta_1, \vartheta_2)\Big|_{\vartheta_1=\vartheta_2=\vartheta} = \widehat{\mathbf{C}}_{\underline{X},T}(h, \vartheta)$. The next condition is essentially saying that the regressor $\underline{X}_t(\vartheta)$ is changing "slowly" in the following sense: for each integer h, $\|X_{k,t}(\vartheta)\|_{T+h}^2 \simeq T$.

Condition 3 (G3) *For each integer h,*

$$\lim_{T \to \infty} \widehat{\mathbf{C}}_{\underline{X},T}(h, \vartheta) = \mathbf{C}_{\underline{X}}(h, \vartheta) \ .$$

Condition 4 (G4) $\mathbf{C}_{\underline{X}}(0, \vartheta)$ *is non-singular.*

One can apply Bochner's theorem to the limit $\mathbf{C}_{\underline{X}}$:

$$\mathbf{C}_{\underline{X}}(h, \vartheta) = \int_{-1/2}^{1/2} e^{2i\pi\lambda h}\, d\mathbf{F}(\lambda, \vartheta) \ ,$$

where $\mathbf{F}$ is a *spectral distribution matrix function of the regressors* whose entries are of bounded variations, SDFR for short. The SDFR can be reached as the limit of the periodogram in the following sense; see [Bri01] for details. Let us introduce the discrete Fourier transform

$$\underline{d}_{\underline{X},T}(\omega, \vartheta) = \sum_{t=0}^{T-1} \underline{X}_t(\vartheta)\, z^{-t} \ , \quad z = e^{2i\pi\omega} \ , \quad -\frac{1}{2} \le \omega < \frac{1}{2} \ ,$$

and the periodogram of the nonrandom series $\underline{X}_t(\vartheta)$,

$$\mathbf{I}_{\underline{X},T}(\omega, \vartheta_1, \vartheta_2) = \frac{1}{T}\underline{d}_{\underline{X},T}(\omega, \vartheta_1)\, \underline{d}_{\underline{X},T}^*(\omega, \vartheta_2) \ ,$$

where * denotes the transpose and complex conjugate. Both $\underline{d}_{\underline{X},T}$ and $\mathbf{I}_{\underline{X},T}$ depend on some parameters. We have the well-known connections

$$\widehat{\mathbf{C}}_{\underline{X},T}(h, \vartheta_1, \vartheta_2) = \int_{-1/2}^{1/2} e^{2i\pi\lambda h}\, \mathbf{I}_{\underline{X},T}(\lambda, \vartheta_1, \vartheta_2)\, d\lambda \ ,$$

$$\mathbf{I}_{\underline{X},T}(\omega, \vartheta_1, \vartheta_2) = \sum_{|h|<T} \widehat{\mathbf{C}}_{\underline{X},T}(h, \vartheta_1, \vartheta_2)\, e^{-2i\pi\omega h} \ ,$$

$$\mathbf{I}_{\underline{X},T}^{\mathsf{T}}(\omega, \vartheta_1, \vartheta_2) = \overline{\mathbf{I}_{\underline{X},T}(\omega, \vartheta_2, \vartheta_1)} \ ,$$

between $\widehat{\mathbf{C}}_{\underline{X},T}$ and the periodogram. The definition (2), which Jennrich [Jen69] calls the tail product, reminds us of the empirical cross-covariance matrix of a stationary time series. It is scaled by $1/T$ (which might not work in some particular cases of the regressors without some additional scaling), which implies that the series $\underline{X}_t$ does not belong to L_2; that is,

$$\lim_{T\to\infty} \left\| \underline{d}_{\underline{X},T}(\omega, \vartheta) \right\|^2 = \infty \ ,$$

and the rate of divergence is T. For the univariate case, we refer to the classical books of Grenander and Rosenblatt [GR57] and Anderson [And71], and for the vector-valued case, of Hannan [Han70] and Brillinger [Bri01].

Now, define the empirical SDFR $\mathbf{F}_T$ as

$$\mathbf{F}_T(\omega, \vartheta_1, \vartheta_2) = \int_0^\omega \mathbf{I}_{\underline{X},T}(\lambda, \vartheta_1, \vartheta_2)\, d\lambda \ .$$

It follows from the Grenander's conditions above that $\mathbf{F}$ is the weak limit of $\mathbf{F}_T$. This is the condition what we need later. This was noticed by Ibragimov and Rozanov [IR78, Chapter 7].

Condition 5 (I-R) *The matrix function* $\mathbf{F}_T$ *converges to* $\mathbf{F}$ *weakly; more precisely for each continuous bounded function* $\varphi(\omega)$ *the limit*

$$\lim_{T\to\infty} \int_{-1/2}^{1/2} \varphi(\omega)\, \mathrm{d}\mathbf{F}_T(\omega, \underline{\vartheta}_1, \underline{\vartheta}_2) = \int_{-1/2}^{1/2} \varphi(\omega)\, \mathrm{d}\mathbf{F}(\omega, \underline{\vartheta}_1, \underline{\vartheta}_2) \qquad (3)$$

holds.

If $\mathbf{F}_T$ converges to $\mathbf{F}$ weakly then (3) is valid not only for continuous bounded functions but for some wider class of functions as well, in particular for piecewise continuous functions having discontinuity in finitely many ω points with $\mathbf{F}$-measure zero. This is very important, in particular for disturbances with long memory. In fact, if the random disturbances $\underline{Z}_t$ have some long memory components, then the corresponding entries of the spectral density matrix of $\underline{Z}_t$ have discontinuities at zero. Yajima [Yaj91] has proved that the standard results of Grenander are valid as far as $\mathrm{d}\mathbf{F}(0, \underline{\vartheta}_1, \underline{\vartheta}_2) = 0$. Hence special attention is necessary only for those long memory components of $\underline{Z}_t$ for which the corresponding $\mathrm{d}\mathbf{F}_{k,k}(0, \underline{\vartheta}_1, \underline{\vartheta}_2) > 0$.

The matrix function $\mathbf{F}$ is Hermite symmetric because $\mathbf{F}_T$ fulfils the following equations

$$\mathbf{F}_T^\mathsf{T}(\omega, \underline{\vartheta}_1, \underline{\vartheta}_2) = \overline{\mathbf{F}_T(\omega, \underline{\vartheta}_2, \underline{\vartheta}_1)} = \mathbf{F}_T(-\omega, \underline{\vartheta}_2, \underline{\vartheta}_1) \ .$$

The regressor $\underline{X}_t(\underline{\vartheta})$ depends on the parameter $\underline{\vartheta} \in \Theta$, therefore we require all Grenander's conditions uniformly in $\underline{\vartheta}$.

2.3 The Objective Function

Frequency domain analysis has a number of advantages. First of all, the Fourier transform of a large stationary sample behaves like i.i.d. complex Gaussian under some broad assumptions; see [Bri01]. The FFT, a technically simple and easy procedure, turns the data $\underline{Y}_t$, $t = 1, 2, \ldots, T$, from the time domain into frequency domain $\underline{d}_{Y,T}(\omega_k)$. We deal with Fourier frequencies $\omega_k = k/T \in [-1/2, 1/2]$, $k = -T_1, \ldots, -1, 0, 1, \ldots, T_1$, where $T_1 = \mathsf{Int}\,[(T-1)/2]$, only. From (1) we have the equation

$$\underline{d}_{Y,T}(\omega) = \underline{d}_{X,T}(\omega, \underline{\vartheta}_0) + \underline{d}_{Z,T}(\omega) \ ,$$

for the Fourier transforms, with obvious notation. The parameter $\underline{\vartheta}_0$ denotes the true unknown value and we would like to adjust the regressor $\underline{X}_t(\underline{\vartheta})$ to the model, finding a parameter $\underline{\vartheta}$ such that the distance

$$\underline{d}_{Y,T}(\omega) - \underline{d}_{X,T}(\omega, \underline{\vartheta}) = \underline{d}_{X,T}(\omega, \underline{\vartheta}_0) - \underline{d}_{X,T}(\omega, \underline{\vartheta}) + \underline{d}_{Z,T}(\omega) \ , \qquad (4)$$

is minimal, in some sense. The Euclidean distance, for instance, is

$$\sum_{k=-T_1}^{T_1} \left\| \underline{d}_{Y,T}(\omega_k) - \underline{d}_{X,T}(\omega_k, \vartheta) \right\|^2$$

$$= \sum_{k=-T_1}^{T_1} \left\| \underline{d}_{X,T}(\omega_k, \vartheta_0) - \underline{d}_{X,T}(\omega_k, \vartheta) + \underline{d}_{Z,T}(\omega_k) \right\|^2 ,$$

which, by the Parseval theorem, actually corresponds to the sum of squares in time domain

$$\sum_{t=0}^{T-1} \left\| \underline{Y}_t - \underline{X}_t(\vartheta) \right\|^2 = \sum_{t=0}^{T-1} \left\| \underline{X}_t(\vartheta_0) - \underline{X}_t(\vartheta) + \underline{Z}_t \right\|^2 .$$

Therefore minimizing either expression leads to the same result. The sequence $\{\underline{Z}_t\}$ itself is not necessarily i.i.d. hence we are facing a generalized non-linear regression problem with stationary residuals. The quadratic function, suggested by Hannan [Han71] for the scalar-valued case, that we are going to minimize is

$$Q_T(\vartheta)$$

$$= \frac{1}{T^2} \sum_{k=-T_1}^{T_1} \left(\underline{d}_{Y,T}(\omega_k) - \underline{d}_{X,T}(\omega_k, \vartheta) \right)^* \boldsymbol{\Phi}(\omega_k) \left(\underline{d}_{Y,T}(\omega_k) - \underline{d}_{X,T}(\omega_k, \vartheta) \right)$$

$$= \frac{1}{T} \sum_{k=-T_1}^{T_1} \operatorname{Tr} \left(\mathbf{I}_{Y,T}(\omega_k) \boldsymbol{\Phi}(\omega_k) \right) + \operatorname{Tr} \left(\mathbf{I}_{X,T}(\omega_k, \vartheta) \boldsymbol{\Phi}(\omega_k) \right)$$

$$- 2 \operatorname{Re} \operatorname{Tr} \left(\mathbf{I}_{Y,X,T}(\omega_k, \vartheta) \boldsymbol{\Phi}(\omega_k) \right) , \tag{5}$$

where $\boldsymbol{\Phi}(\omega_k)$ is a series of matrix weights, originated from a continuous, Hermitian matrix function $\boldsymbol{\Phi}$, satisfying $\boldsymbol{\Phi}(\omega) \geq 0$. Equation (4) provides the more informative form for the above

$$Q_T(\vartheta) = \frac{1}{T} \sum_{k=-T_1}^{T_1} \operatorname{Tr} \left(\mathbf{I}_{X,T}(\omega_k, \vartheta_0) \boldsymbol{\Phi}(\omega_k) \right)$$

$$+ \operatorname{Tr} \left(\mathbf{I}_{X,T}(\omega_k, \vartheta) \boldsymbol{\Phi}(\omega_k) \right) + \operatorname{Tr} \left(\mathbf{I}_{Z,T}(\omega_k) \boldsymbol{\Phi}(\omega_k) \right)$$

$$+ 2 \operatorname{Tr} \left(\left[\mathbf{I}_{X,Z,T}(\omega_k, \vartheta_0) - \mathbf{I}_{X,Z,T}(\omega_k, \vartheta) \right] \boldsymbol{\Phi}(\omega_k) \right)$$

$$- 2 \operatorname{Tr} \left(\mathbf{I}_{X,T}(\omega_k, \vartheta, \vartheta_0) \boldsymbol{\Phi}(\omega_k) \right) .$$

The proof of $\mathbf{I}_{X,Z,T}(\omega_k, \vartheta) \to 0$, a.s. and uniformly in ϑ is given by Robinson [Rob72, Lemma 1]. Now, suppose Conditions I-R, (or G1-G4) and take the limit

$$Q(\vartheta) = \lim_{T\to\infty} Q_T(\vartheta)$$

$$= \int_{-1/2}^{1/2} \mathrm{Tr}\left(\boldsymbol{\Phi}(\omega)\, d\left[\mathbf{F}(\omega,\vartheta_0) + \mathbf{F}(\omega,\vartheta) - \mathbf{F}(\omega,\vartheta_0,\vartheta) - \mathbf{F}(\omega,\vartheta,\vartheta_0)\right]\right)$$

$$+ \int_{-1/2}^{1/2} \mathrm{Tr}\left[\mathbf{S}_{\underline{Z}}(\omega)\,\boldsymbol{\Phi}(\omega)\right]\,d\omega$$

$$= R(\vartheta,\vartheta_0) + \int_{-1/2}^{1/2} \mathrm{Tr}\left[\boldsymbol{\Phi}(\omega)\,\mathbf{S}_{\underline{Z}}(\omega)\right]\,d\omega\ . \tag{6}$$

The function

$$R(\vartheta,\vartheta_0)$$
$$= \int_{-1/2}^{1/2} \mathrm{Tr}\left(\boldsymbol{\Phi}(\omega)\,\mathrm{d}\left[\mathbf{F}(\omega,\vartheta_0) + \mathbf{F}(\omega,\vartheta) - \mathbf{F}(\omega,\vartheta_0,\vartheta) - \mathbf{F}(\omega,\vartheta,\vartheta_0)\right]\right)\ ,$$

is the only part of $Q(\vartheta)$ depending on ϑ. We require the following condition for the existence of the minimum, see [Rob72].

Condition 6 (R)

$$R(\vartheta,\vartheta_0) > 0, \quad \vartheta \in \Theta, \quad \vartheta \neq \vartheta_0\ .$$

Obviously

$$\lim_{T\to\infty} \left[Q_T(\vartheta) - Q_T(\vartheta_0)\right] = R(\vartheta_0,\vartheta)\ .$$

The *minimum contrast estimator* $\widehat{\vartheta}_T$ is the value which realizes that minimum value of $Q_T(\vartheta)$

$$\widehat{\vartheta}_T = \arg\min_{\vartheta\in\Theta} Q_T(\vartheta)\ .$$

One can easily see (using [MN99, Theorem 7, Chapter 7]) under some additional assumptions given below, that $Q_T(\vartheta)$ is convex because the Hessian $HQ(\vartheta_0)$ is non-negative definite. Therefore the next theorem, due to Robinson [Rob72], is valid not only for a compact Θ but for a more general case such as a convex parameter set Θ as well. The minimum contrast method is also called quasi-likelihood and it is very efficient in several cases, even in non-Gaussian situations; for instance, see [ALS04].

Theorem 1. *Under assumptions I-R (or G1-4), and R, the minimum contrast estimator $\widehat{\vartheta}_T$ converges a.s. to ϑ_0.*

3 Asymptotic Normality

For the asymptotic normality it is necessary to consider the second order derivatives of the SDFR and their limits for the objective function as usual, see [Rob72]. The matrix of the second derivatives of $\widehat{\mathbf{C}}_{X,T}(h, \underline{\vartheta}_1, \underline{\vartheta}_2)$ can be calculated, by the matrix differential calculus [MN99];

$$\frac{\partial^2 \widehat{\mathbf{C}}_{X,T}(h, \underline{\vartheta}_1, \underline{\vartheta}_2)}{\partial \underline{\vartheta}_2^\mathsf{T} \partial \underline{\vartheta}_1^\mathsf{T}} = \frac{\partial}{\partial \underline{\vartheta}_1^\mathsf{T}} \operatorname{Vec}\left(\frac{\partial \operatorname{Vec} \widehat{\mathbf{C}}_{X,T}(h, \underline{\vartheta}_1, \underline{\vartheta}_2)}{\partial \underline{\vartheta}_2^\mathsf{T}}\right).$$

Here the differentiating of the right-hand side can be carried out directly; see Section 7.2. Notice, the order of the variables $\underline{\vartheta}_1, \underline{\vartheta}_2$ in $\widehat{\mathbf{C}}_{X,T}$ is opposite to the order of the partial derivatives: $\partial \underline{\vartheta}_2^\mathsf{T} \partial \underline{\vartheta}_1^\mathsf{T}$. The latter means that one differentiates first by $\underline{\vartheta}_2$ then by $\underline{\vartheta}_1$; that is, the operator acting on the right hand side. Starting the differentiating by $\underline{\vartheta}_1$, then followed by $\underline{\vartheta}_2$ is "indirect". It can be expressed by the help of the "direct" one

$$\frac{\partial^2 \widehat{\mathbf{C}}_{X,T}(h, \underline{\vartheta}_1, \underline{\vartheta}_2)}{\partial \underline{\vartheta}_1^\mathsf{T} \partial \underline{\vartheta}_2^\mathsf{T}} = (\mathbf{K}_{p \cdot d} \otimes \mathbf{U}_d) \mathbf{K}_{d \cdot dp} \frac{\partial^2 \widehat{\mathbf{C}}_{X,T}(-h, \underline{\vartheta}_2, \underline{\vartheta}_1)}{\partial \underline{\vartheta}_1^\mathsf{T} \partial \underline{\vartheta}_2^\mathsf{T}};$$

here we apply the commutation matrix $\mathbf{K}_{p \cdot d}$ (see (20), $\otimes$ denotes the Kronecker product, and $\mathbf{U}_d$ is the $d \times d$ identity matrix. Following Hannan [Han71] we assume

Condition 7 (H) *The second partial derivatives of the regressor $\underline{X}_t(\underline{\vartheta})$ exist and $(\partial^2 \widehat{\mathbf{C}}_{X,T}(h, \underline{\vartheta}_1, \underline{\vartheta}_2))/\underline{\vartheta}_2^\mathsf{T} \partial \underline{\vartheta}_1^\mathsf{T}$ converges to some limit, denoted by*

$$\frac{\partial^2 \mathbf{C}_X(h, \underline{\vartheta}_1, \underline{\vartheta}_2)}{\partial \underline{\vartheta}_2^\mathsf{T} \partial \underline{\vartheta}_1^\mathsf{T}}.$$

It is necessary to emphasize that Condition H is

$$\frac{\partial^2 \mathbf{C}_X(h, \underline{\vartheta}_1, \underline{\vartheta}_2)}{\partial \underline{\vartheta}_2^\mathsf{T} \partial \underline{\vartheta}_1^\mathsf{T}} \overset{\bullet}{=} \lim_{T \to \infty} \frac{\partial^2 \widehat{\mathbf{C}}_{X,T}(h, \underline{\vartheta}_1, \underline{\vartheta}_2)}{\partial \underline{\vartheta}_2^\mathsf{T} \partial \underline{\vartheta}_1^\mathsf{T}},$$

where the left-hand side is defined by the limit but is necessarily the derivative of $\mathbf{C}_X$. From now on we use the symbol $\overset{\bullet}{=}$ for the definition of the left side of an expression.

The above notation is used for the regression spectrum as well.

Condition 8 (I-R-H) *The derivative $(\partial^2 \mathbf{F}_T(\omega, \underline{\vartheta}_1, \underline{\vartheta}_2))/\partial \underline{\vartheta}_1^\mathsf{T} \partial \underline{\vartheta}_2^\mathsf{T}$ of the matrix function $\mathbf{F}_T$ converges weakly to some function denoted by*

$$\frac{\partial^2 \mathbf{F}(\omega, \underline{\vartheta}_1, \underline{\vartheta}_2)}{\partial \underline{\vartheta}_2^\mathsf{T} \partial \underline{\vartheta}_1^\mathsf{T}}.$$

Again

$$\frac{\partial^2 \mathbf{F}(h, \underline{\vartheta}_1, \underline{\vartheta}_2)}{\partial \underline{\vartheta}_2^\mathsf{T} \partial \underline{\vartheta}_1^\mathsf{T}} \overset{\bullet}{=} \lim_{T \to \infty} \frac{\partial^2 \mathbf{F}_T(\omega, \underline{\vartheta}_1, \underline{\vartheta}_2)}{\partial \underline{\vartheta}_1^\mathsf{T} \partial \underline{\vartheta}_2^\mathsf{T}},$$

by definition. According to the above derivatives we calculate the Hessian $\mathbf{HF}$ for the SDFR $\mathbf{F}$ as well; see Section 7.2 for the proof.

Lemma 1. *Assume Condition I-R-H, then*

$$\mathsf{HF}(\omega, \underline{\vartheta}) = \left[\mathsf{H}_{\underline{\vartheta}_1} \mathbf{F}(\omega, \underline{\vartheta}_1, \underline{\vartheta}_2) + \frac{\partial^2 \mathbf{F}(\omega, \underline{\vartheta}_1, \underline{\vartheta}_2)}{\partial \underline{\vartheta}_1^\mathsf{T} \partial \underline{\vartheta}_2^\mathsf{T}} \right.$$
$$\left. + \mathsf{H}_{\underline{\vartheta}_2} \mathbf{F}(\omega, \underline{\vartheta}_1, \underline{\vartheta}_2) + \frac{\partial^2 \mathbf{F}(\omega, \underline{\vartheta}_1, \underline{\vartheta}_2)}{\partial \underline{\vartheta}_2^\mathsf{T} \partial \underline{\vartheta}_1^\mathsf{T}} \right] \Bigg|_{\underline{\vartheta}_1 = \underline{\vartheta}_2 = \underline{\vartheta}} , \tag{7}$$

where the indirect derivative fulfills

$$\frac{\partial^2 \mathbf{F}(\omega, \underline{\vartheta}_1, \underline{\vartheta}_2)}{\partial \underline{\vartheta}_1^\mathsf{T} \partial \underline{\vartheta}_2^\mathsf{T}} = (\mathbf{K}_{p \cdot d} \otimes \mathbf{U}_d) \mathbf{K}_{d \cdot dp} \frac{\partial^2 \mathbf{F}(-\omega, \underline{\vartheta}_2, \underline{\vartheta}_1)}{\partial \underline{\vartheta}_1^\mathsf{T} \partial \underline{\vartheta}_2^\mathsf{T}} .$$

3.1 Asymptotic Variance

For the variance of $\mathrm{Vec}(\partial Q_T(\underline{\vartheta}_0))/\partial \underline{\vartheta}^\mathsf{T}$ consider the expression

$$T \, \mathrm{Vec} \, \frac{\partial Q_T(\underline{\vartheta})}{\partial \underline{\vartheta}^\mathsf{T}} \tag{8}$$
$$= \sum_{k=-T_1}^{T_1} \left[\frac{\partial \, \mathrm{Vec} \, \mathbf{I}_{\underline{X},T}(\omega_k, \underline{\vartheta})}{\partial \underline{\vartheta}^\mathsf{T}} - \frac{\partial \left(\mathrm{Vec} \, \mathbf{I}_{\underline{Y},\underline{X},T}(\omega_k, \underline{\vartheta}) + \mathrm{Vec} \, \mathbf{I}_{\underline{X},\underline{Y},T}(\omega_k, \underline{\vartheta}) \right)}{\partial \underline{\vartheta}^\mathsf{T}} \right]^\mathsf{T}$$
$$\times \left[\mathrm{Vec} \, \boldsymbol{\Phi}^\mathsf{T}(\omega_k) \right] .$$

Let $\boldsymbol{\Psi}$ be some matrix function of appropriate dimension, and introduce the following expression, which is frequently used below,

$$\mathbf{J}(\boldsymbol{\Psi}, \mathbf{F}) = \int_{-1/2}^{1/2} (\mathbf{U}_p \otimes [\mathrm{Vec}(\boldsymbol{\Psi}^\mathsf{T}(\omega_k))]^\mathsf{T}) \, \mathrm{d} \left(\frac{\partial^2 \mathbf{F}(\omega, \underline{\vartheta}_1, \underline{\vartheta}_2)}{\partial \underline{\vartheta}_2^\mathsf{T} \partial \underline{\vartheta}_1^\mathsf{T}} \Bigg|_{\underline{\vartheta}_1 = \underline{\vartheta}_2 = \underline{\vartheta}_0} \right) ,$$

where $\mathbf{U}_p$ denotes the identity matrix of order p.

Lemma 2.

$$\lim_{T \to \infty} \mathrm{Var} \left[\sqrt{T} \, \mathrm{Vec} \, \frac{\partial Q_T(\underline{\vartheta}_0)}{\partial \underline{\vartheta}^\mathsf{T}} \right] = 4 \mathbf{J}(\boldsymbol{\Phi} \mathbf{S}_{\underline{Z}} \boldsymbol{\Phi}, \mathbf{F}) .$$

See Section 7.3 for the proof.

The limit of the Hessian is calculated from (8). The Hessians according to $\mathsf{H}_{\underline{\vartheta}_1} \mathbf{I}_{\underline{X},T}(\omega_k, \underline{\vartheta}_1, \underline{\vartheta}_2)$ and $\mathsf{H}_{\underline{\vartheta}_2} \mathbf{I}_{\underline{X},T}(\omega_k, \underline{\vartheta}_1, \underline{\vartheta}_2)$ of the terms in (8) at $\underline{\vartheta}_1 = \underline{\vartheta}_2 = \underline{\vartheta}_0$ is canceled with $\mathsf{H}_{\underline{\vartheta}} \mathbf{I}_{\underline{Y},\underline{X},T}(\omega_k, \underline{\vartheta})$ and $\mathsf{H}_{\underline{\vartheta}} \mathbf{I}_{\underline{X},\underline{Y},T}(\omega_k, \underline{\vartheta})$, respectively. So we have to deal only with the mixed derivatives of $\mathbf{I}_{\underline{X},T}(\omega_k, \underline{\vartheta})$. See Section 7.4. Hence the Hessian of $Q(\underline{\vartheta})$ at $\underline{\vartheta} = \underline{\vartheta}_0$ follows.

Lemma 3.

$$\mathsf{H}Q(\underline{\vartheta}_0) = \lim_{T \to \infty} [\mathsf{H}Q_T(\underline{\vartheta}_0)] = 2 \mathbf{J}(\boldsymbol{\Phi}, \mathbf{F}) .$$

Notice that the matrix $\mathbf{J} = \mathbf{J}\left(\mathbf{\Phi}\mathbf{S}_{\underline{Z}}\mathbf{\Phi}, \mathbf{F}\right)$ and the Hessian $\mathsf{H}Q\left(\underline{\vartheta}_0\right)$ are the same except that the latter depends only on $\mathbf{\Phi}$; that is, $\mathsf{H}Q(\underline{\vartheta}_0) = \mathbf{J}(\mathbf{\Phi}, \mathbf{F})$.
Put

$$\mathbf{J}_T = \mathrm{Var}\left[\sqrt{T}\,\mathrm{Vec}\,\frac{\partial Q_T\left(\underline{\vartheta}_0\right)}{\partial \underline{\vartheta}^{\mathsf{T}}}\right]\,,$$

and suppose

Condition 9 (R) *The limit variance matrix* $\mathbf{J}\left(\mathbf{\Phi}\mathbf{S}_{\underline{Z}}\mathbf{\Phi}, \mathbf{F}\right)$ *of* $\mathbf{J}_T$ *is positive definite, for all admissible spectral density* $\mathbf{S}_{\underline{Z}}$ *and SDFR* $\mathbf{F}$, *and moreover* $\mathbf{J}\left(\mathbf{\Phi}, \mathbf{F}\right) > 0$.

Theorem 2. *Under assumptions I-R, I-R-H, and R,*

$$\sqrt{T}\mathbf{J}_T^{-1/2}\mathsf{H}Q_T\left(\widehat{\widehat{\underline{\vartheta}}}\right)\left(\widehat{\underline{\vartheta}}_T - \underline{\vartheta}_0\right) \overset{\mathcal{D}}{\to} N\left(0, \mathbf{U}_p\right)\,,$$

where $\widehat{\widehat{\underline{\vartheta}}}$ *is closer to* $\underline{\vartheta}_0$ *than* $\widehat{\underline{\vartheta}}_T$. *In other words*

$$\lim_{T\to\infty}\mathrm{Var}\left[\sqrt{T}\left(\widehat{\underline{\vartheta}}_T - \underline{\vartheta}_0\right)\right] = \mathbf{J}^{-1}\left(\mathbf{\Phi}, \mathbf{F}\right)\mathbf{J}\left(\mathbf{\Phi}\mathbf{S}_{\underline{Z}}\mathbf{\Phi}, \mathbf{F}\right)\mathbf{J}^{-1}\left(\mathbf{\Phi}, \mathbf{F}\right)\Big|_{\underline{\vartheta}=\underline{\vartheta}_0}\,. \quad (9)$$

The optimal choice of $\mathbf{\Phi}\left(\omega\right)$ is $\mathbf{S}_{\underline{Z}}^{-1}\left(\omega\right)$ assuming $\mathbf{S}_{\underline{Z}}\left(\omega\right) > 0$. The choice $\mathbf{S}_{\underline{Z}}^{-1}\left(\omega\right)$ is appropriate because the "residual" series $\underline{d}_{\underline{Z},T}(\omega_k)$ is asymptotically independent Gaussian with variance $T\mathbf{S}_{\underline{Z}}\left(\omega_k\right)$. The variance in this case $\left(\mathbf{\Phi} = \mathbf{S}_{\underline{Z}}^{-1}\right)$ follows from (9)

$$\lim_{T\to\infty}\mathrm{Var}\left[\sqrt{T}\left(\widehat{\underline{\vartheta}}_T - \underline{\vartheta}_0\right)\right] = \mathbf{J}^{-1}\left(\mathbf{S}_{\underline{Z}}^{-1}, \mathbf{F}\right)\,, \quad (10)$$

where

$$\mathbf{J}^{-1}\left(\mathbf{S}_{\underline{Z}}^{-1}, \mathbf{F}\right)$$
$$= \left[\int_{-1/2}^{1/2}\left[\mathbf{U}_p \otimes \left(\mathrm{Vec}\left[\mathbf{S}_{\underline{Z}}^{-1}\left(\omega\right)\right]^{\mathsf{T}}\right)^{\mathsf{T}}\right]\mathrm{d}\frac{\partial^2\mathbf{F}\left(\omega, \underline{\vartheta}_1, \underline{\vartheta}_2\right)}{\partial\underline{\vartheta}_2^{\mathsf{T}}\partial\underline{\vartheta}_1^{\mathsf{T}}}\Big|_{\underline{\vartheta}_1=\underline{\vartheta}_2=\underline{\vartheta}_0}\right]^{-1}\,.$$

4 Scaling

To assess the generality of scaling consider the linear case

$$\underline{Y}_t = \mathbf{B}\underline{X}_t + \underline{Z}_t\,,$$

first. In this case $\underline{\vartheta} = \mathrm{Vec}\,\mathbf{B}$, so the regressor $\underline{X}_t$ depends on the parameter $\underline{\vartheta}$ linearly ($\underline{X}_t$ depends on t but not on $\underline{\vartheta}$). Here $\mathbf{B}$ is $d \times p$ and $\underline{X}_t$ is $p \times 1$. If $\|X_{k,t}\|_T \simeq D_k\left(T\right)$ which tends to infinity by the Grenander's Condition G1, then the matrix

$$\widehat{\mathbf{C}}_{\underline{X},T}\left(h,\underline{\vartheta}_1,\underline{\vartheta}_2\right) = \frac{1}{T}\sum_{t=1}^{T-h}\underline{X}_{t+h}\left(\underline{\vartheta}_1\right)\underline{X}_t^{\mathsf{T}}\left(\underline{\vartheta}_2\right),$$

might not converge unless each $D_k\left(T\right) \simeq \sqrt{T}$. This is not the case for the important problem of polynomial regression, say. Grenander's solution to this problem can be interpreted in the following way. Define the diagonal matrix $\mathbf{D}_T = \mathrm{diag}(D_1, D_2, \ldots, D_p)$, where $D_k = D_k(T) \simeq \|X_{k,t}\|_T$. Now, consider the linear regression problem

$$\underline{Y}_t = \widetilde{\mathbf{B}}\underline{V}_t + \underline{Z}_t,$$

where $\underline{V}_t = \sqrt{T}\mathbf{D}_T^{-1}\underline{X}_t$. One solves this linear regression problem and observes the connection

$$\widetilde{\mathbf{B}}\underline{V}_t = \left(\frac{1}{\sqrt{T}}\mathbf{B}\mathbf{D}_T\right)\left(\sqrt{T}\mathbf{D}_T^{-1}\underline{X}_t\right)$$

between the original and the scaled equation. Therefore the asymptotic variance of the estimate of the unknown matrix $\mathbf{B}$ is connected by

$$\lim_{T\to\infty}\mathrm{Var}\,\sqrt{T}(\widehat{\widetilde{\mathbf{B}}} - \widetilde{\mathbf{B}}_0) = \lim_{T\to\infty}\mathrm{Var}[(\widehat{\mathbf{B}} - \mathbf{B}_0)\mathbf{D}_T].$$

We call this type of transformation "primary" scaling and the result is the properly scaled regressor. Note here that the procedure of scaling opens the possibility of considering random regressors that are not necessarily weakly stationary, either because the second-order moment does not exist (see [KM96]), or because stationarity holds only asymptotically.

Definition 1. *The series $\underline{X}_t$ is properly scaled if*

$$\|X_{k,t}\|_T^2 \simeq T,$$

as $T \to \infty$, for each $k = 1, 2, \ldots, d$.

In general, let $D_k(T) \simeq \|X_{k,t}\|_T$, for each k and define

$$\mathbf{D}_T = \mathrm{diag}(D_1, D_2, \ldots, D_d).$$

Then it is easy to see that the new series $\sqrt{T}\mathbf{D}_T^{-1}\underline{X}_t$ is properly scaled. The primary scaling of the non-linear regressors $\underline{X}_t\left(\underline{\vartheta}\right)$ is possible if $D_k\left(T\right)$ does not depend on the unknown parameter $\underline{\vartheta}$. Even if the regressors $\underline{X}_t\left(\underline{\vartheta}\right)$ are properly scaled, some problem may arise when we take the limit of the derivatives because there is no guarantee for their convergence. Therefore we introduce some further scaling of the properly scaled regressors $\underline{X}_t\left(\underline{\vartheta}\right)$.

First, a diagonal matrix $\mathbf{D}_T = \mathrm{diag}(D_{X,k}(T), k = 1, 2, \ldots, d)$ applies; the result is $\sqrt{T}\mathbf{D}_T^{-1}\underline{X}_t\left(\underline{\vartheta}\right)$. Another type of scaling goes through the process of differentiating. We define the *scaled* partial derivative $\partial_{s,T}\left(\underline{\vartheta}\right)$ according to the diagonal matrix $\mathbf{D}_{1,T} = \mathrm{diag}(D_k^{(1)}(T), k = 1, 2, \ldots, p)$ by $\partial(\mathbf{D}_{1,T}^{-1}\underline{\vartheta})$; hence

$$\frac{\partial}{\partial_{s,T}\underline{\vartheta}^{\mathsf{T}}}\left[\mathbf{D}_T^{-1}\underline{X}_t\left(\vartheta\right)\right] = \left(\frac{\partial}{\partial\underline{\vartheta}^{\mathsf{T}}}\left[\mathbf{D}_T^{-1}\underline{X}_t\left(\vartheta\right)\right]\right)\mathbf{D}_{1,T}^{-1} . \tag{11}$$

The result of these scalings is

$$\frac{\partial}{\partial_{s,T}\underline{\vartheta}^{\mathsf{T}}}\left[\mathbf{D}_T^{-1}\underline{X}_t\left(\vartheta\right)\right] = \mathbf{D}_T^{-1}\left[\frac{\partial}{\partial\underline{\vartheta}^{\mathsf{T}}}\underline{X}_t\left(\vartheta\right)\right]\mathbf{D}_{1,T}^{-1} .$$

The entries of the scaled partial derivatives are

$$\left[D_{X,j}\left(T\right)D_k^{(1)}\left(T\right)\right]^{-1}\partial X_{j,t}\left(\vartheta\right)/\partial\vartheta_k.$$

The second scaled derivative of $\widehat{\mathbf{C}}_{\underline{X},T}\left(h,\underline{\vartheta}_1,\underline{\vartheta}_2\right)$ is of interest:

$$\frac{\partial_{s,T}^2\widehat{\mathbf{C}}_{\mathbf{D}_T\underline{X},T}(h,\underline{\vartheta}_1,\underline{\vartheta}_2)}{\partial_{s,T}\underline{\vartheta}_2^{\mathsf{T}}\partial_{s,T}\underline{\vartheta}_1^{\mathsf{T}}} = \left(\mathbf{D}_{1,T}^{-1}\otimes\mathbf{U}_{d^2}\right)\frac{\partial^2\widehat{\mathbf{C}}_{\sqrt{T}\mathbf{D}_T^{-1}\underline{X},T}(h,\underline{\vartheta}_1,\underline{\vartheta}_2)}{\partial\underline{\vartheta}_2^{\mathsf{T}}\partial\underline{\vartheta}_1^{\mathsf{T}}}\mathbf{D}_{1,T}^{-1}$$

$$= T\left(\mathbf{D}_{1,T}^{-1}\otimes\mathbf{D}_T^{-1}\otimes\mathbf{D}_T^{-1}\right)\frac{\partial^2\widehat{\mathbf{C}}_{\underline{X},T}\left(h,\underline{\vartheta}_1,\underline{\vartheta}_2\right)}{\partial\underline{\vartheta}_2^{\mathsf{T}}\partial\underline{\vartheta}_1^{\mathsf{T}}}\mathbf{D}_{1,T}^{-1} ;$$

see Section 7.5 for the proof. Notice that the $1/T$ in the expression of $\widehat{\mathbf{C}}_{\underline{X},T}$ is canceled and the role of scaling has been taken totally by the scaling matrices.

Condition 10 (H$'$) *All the second partial derivatives of the regressor $\underline{X}_t\left(\vartheta\right)$ exist. There exist diagonal matrices $\mathbf{D}_T$ and $\mathbf{D}_{1,T}$ such that uniformly in $\underline{\vartheta}$, the scaled derivative*

$$\frac{\partial_{s,T}^2\widehat{\mathbf{C}}_{\sqrt{T}\mathbf{D}_T^{-1}\underline{X},T}\left(h,\underline{\vartheta}_1,\underline{\vartheta}_2\right)}{\partial_{s,T}\underline{\vartheta}_2^{\mathsf{T}}\partial_{s,T}\underline{\vartheta}_1^{\mathsf{T}}}$$

converges to some limit, denoted by

$$\frac{\partial_s^2\mathbf{C}_{\underline{X}}(h,\underline{\vartheta}_1,\underline{\vartheta}_2)}{\partial_s\underline{\vartheta}_1\partial_s\underline{\vartheta}_2} .$$

Condition H means

$$\frac{\partial_s^2\mathbf{C}_{\underline{X}}\left(h,\underline{\vartheta}_1,\underline{\vartheta}_2\right)}{\partial_s\underline{\vartheta}_2^{\mathsf{T}}\partial_s\underline{\vartheta}_1^{\mathsf{T}}} \overset{\bullet}{=} \lim_{T\to\infty}\frac{\partial_{s,T}^2\widehat{\mathbf{C}}_{\mathbf{D}_T\underline{X},T}\left(h,\underline{\vartheta}_1,\underline{\vartheta}_2\right)}{\partial_{s,T}\underline{\vartheta}_2^{\mathsf{T}}\partial_{s,T}\underline{\vartheta}_1^{\mathsf{T}}} .$$

The diagonal matrices $\mathbf{D}_T$ and $\mathbf{D}_{1,T}$ can be chosen directly if the effects of the entries and the partial derivatives are separate; that is,

$$\left\|\partial X_{j,t}\left(\vartheta\right)/\partial\vartheta_k\right\|_T \simeq B_{X,j}\left(T\right)B_k^{(1)}\left(T\right) ,$$

then $\mathbf{D}_T = \mathrm{diag}(B_{X,j}(T), j = 1,\ldots,d)$, and $\mathbf{D}_{1,T} = \mathrm{diag}(B_k^{(1)}(T), k = 1,\ldots,p)$, say. Note here that the matrix $\mathbf{D}_T$ contains the factors of primary scaling. There are regressors $\underline{X}_t\left(\vartheta\right)$, of course, having more sophisticated derivatives and the above procedure does not apply.

The above notation is used for the regression spectrum as well.

Condition 11 (I-R-H′) *The scaled derivative of matrix function* $\mathbf{F}_T$,

$$\frac{\partial_{s,T}^2 \mathbf{F}_T(\omega, \underline{\vartheta}_1, \underline{\vartheta}_2)}{\partial_{s,T} \underline{\vartheta}_1^\mathsf{T} \partial_{s,T} \underline{\vartheta}_2^\mathsf{T}} ,$$

converges weakly to some function denoted by

$$\frac{\partial_s^2 \mathbf{F}(h, \underline{\vartheta}_1, \underline{\vartheta}_2)}{\partial_s \underline{\vartheta}_2^\mathsf{T} \partial_s \underline{\vartheta}_1^\mathsf{T}} .$$

Introduce the notation

$$\mathbf{J}_T(\mathbf{D}_T, \boldsymbol{\Psi}, \mathbf{F})$$
$$= \int_{-1/2}^{1/2} (\mathbf{U}_p \otimes [\mathrm{Vec}\,(\mathbf{D}_T \boldsymbol{\Psi}^\mathsf{T}(\omega_k) \mathbf{D}_T)]^\mathsf{T}) \, \mathrm{d} \left(\frac{\partial_s^2 \mathbf{F}(\omega, \underline{\vartheta}_1, \underline{\vartheta}_2)}{\partial_s \underline{\vartheta}_2^\mathsf{T} \partial_s \underline{\vartheta}_1^\mathsf{T}} \bigg|_{\underline{\vartheta}_1 = \underline{\vartheta}_2 = \underline{\vartheta}_0} \right) .$$

Theorem 3. *Under conditions I-R and I-R-H′ we have*

$$\sqrt{T} \mathbf{J}_T^{-1/2} \mathsf{H}_{s,T} Q_T(\underline{\vartheta}_0) \mathbf{D}_{1,T} \left(\widehat{\underline{\vartheta}}_T - \underline{\vartheta}_0 \right) \xrightarrow{\mathcal{D}} N(0, \mathbf{U}_p) .$$

In other words the variance of $\left(\widehat{\underline{\vartheta}}_T - \underline{\vartheta}_0 \right)$ *can be approximated by*

$$\mathbf{D}_{1,T}^{-1} \mathbf{J}_T^{-1} (\mathbf{D}_T, \boldsymbol{\Phi}, \mathbf{F}) \mathbf{J}_T (\mathbf{D}_T, \boldsymbol{\Phi} \mathbf{S}_{\underline{Z}} \boldsymbol{\Phi}, \mathbf{F}) \mathbf{J}_T^{-1} (\mathbf{D}_T, \boldsymbol{\Phi}, \mathbf{F}) \mathbf{D}_{1,T}^{-1} \bigg|_{\underline{\vartheta} = \underline{\vartheta}_0} .$$

Moreover if $\boldsymbol{\Phi} = \mathbf{S}_{\underline{Z}}^{-1}$, one has the asymptotic variance

$$\mathbf{D}_{1,T} \mathbf{J}_T^{-1} \left(\mathbf{D}_T, \mathbf{S}_{\underline{Z}}^{-1}, \mathbf{F} \right) \mathbf{D}_{1,T}$$

of $(\widehat{\underline{\vartheta}}_T - \underline{\vartheta}_0)$. We show in the next section that the linear regressors are scaled directly.

The spectrum $\mathbf{S}_{\underline{Z}}$ in general is not known. This leads to a semiparametric problem, therefore one uses recursion for the estimation of the parameters. Fortunately, the additional term to the function R in the objective function is the Whittle likelihood up to a constant. As far as we are concerned with a rational spectral density, the method of Hannan [Han71] applies and both the estimator of the unknown parameter $\underline{\vartheta}$ and the estimator for the parameters of the spectrum are consistent.

5 Some Particular Cases

We derive from the above general formulae some particular cases of interest.

5.1 Multiple Linear Regression with Stationary Errors

Consider the linear case

$$\underline{Y}_t = \mathbf{B}\underline{X}_t + \underline{Z}_t \ .$$

In this case $\vartheta = \text{Vec}\,\mathbf{B}$, so the regressors depend on the parameter ϑ linearly ($\underline{X}_t$ depends only on t but not on ϑ). Here $\mathbf{B}$ is $d \times p$ and $\underline{X}_t$ is $p \times 1$. The primary scaling, if it is necessary, is given by the diagonal matrix $\sqrt{T}\mathbf{D}_T^{-1}$ with $\mathbf{D}_T = \text{diag}(D_1, D_2, \ldots, D_p)$, where $D_k(T) \simeq \|X_{k,t}\|_T$. It is easy to see that

$$\widehat{\mathbf{C}}_{\underline{X},T}\,(h, \vartheta_1, \vartheta_2) = \frac{1}{T} \sum_{t=1}^{T-h} \underline{X}_{t+h}\,(\vartheta_1)\,\underline{X}_t^{\mathsf{T}}\,(\vartheta_2) \ ,$$

where $\underline{X}_t\,(\vartheta) = \mathbf{B}\underline{X}_t$, converges for all possible values of $\mathbf{B}$ if and only if

$$\widehat{\mathbf{C}}_{\underline{X},T}\,(h) = \frac{1}{T} \sum_{t=1}^{T-h} \underline{X}_{t+h}\underline{X}_t^{\mathsf{T}}$$

converges. Assume that $\underline{X}_t$ is properly scaled (otherwise scale it; that is, replace it by $\sqrt{T}\mathbf{D}_T^{-1}\underline{X}_t$). Observe that

$$\frac{\partial \underline{X}_t(\vartheta)}{\partial \vartheta_k} = \frac{\partial \mathbf{B}\underline{X}_t}{\partial \vartheta_k} = [0, \ldots, 0, \underline{X}_{j_k,t}, 0, \ldots, 0]^{\mathsf{T}} \ ,$$

therefore the secondary scaling is $\mathbf{D}_{1,T} = \mathbf{U}_{dp}$. The discrete Fourier transform simplifies to

$$\underline{d}_{\underline{Y},T}(\omega) = \mathbf{B}\underline{d}_{\underline{X},T}(\omega) + \underline{d}_{\underline{Z},T}(\omega) \ ,$$

together with the periodograms

$$\mathbf{I}_{\underline{X},T}\,(\omega_k, \vartheta) = \mathbf{B}\mathbf{I}_{\underline{X},T}(\omega_k)\mathbf{B}^{\mathsf{T}} \ ,$$
$$\mathbf{I}_{\underline{Y},\underline{X},T}\,(\omega_k, \vartheta) = \mathbf{I}_{\underline{Y},\underline{X},T}\,(\omega_k)\,\mathbf{B}^{\mathsf{T}} \ ,$$
$$\mathbf{I}_{\underline{X},\underline{Y},T}\,(\omega_k, \vartheta) = \mathbf{B}\mathbf{I}_{\underline{X},\underline{Y},T}\,(\omega_k) \ .$$

The normal equation

$$\frac{\partial Q_T\,(\mathbf{B})}{\partial \mathbf{B}} = \mathbf{0} \ ,$$

gives the estimate

$$\text{Vec}\,\left(\widehat{\mathbf{B}}\right) = \left(\sum_{k=-T_1}^{T_1} \mathbf{I}_{\underline{X},T}^{\mathsf{T}}\,(\omega_k) \otimes \mathbf{\Phi}\,(\omega_k)\right)^{-1} \text{Vec}\,\sum_{k=-T_1}^{T_1} \mathbf{\Phi}\,(\omega_k)\,\mathbf{I}_{\underline{Y},\underline{X},T}\,(\omega_k) \ .$$

The inverse here can be taken in the Moore-Penrose sense as well. This estimate is *linear and unbiased* because

$$\mathsf{E}\,\sum_{k=-T_1}^{T_1} \mathbf{\Phi}\,(\omega_k)\,\mathbf{I}_{\underline{Y},\underline{X},T}\,(\omega_k) = \mathbf{\Phi}\,(\omega_k)\,\mathbf{B}_0\mathbf{I}_{\underline{X},T}\,(\omega_k) \ .$$

The Hessian of $Q_T(\mathbf{B})$ is

$$
\begin{aligned}
\mathsf{H}Q_T(\mathbf{B}) &= \frac{1}{T} \sum_{k=-T_1}^{T_1} \left(\mathbf{I}_{\underline{X},T}^{\mathsf{T}}(\omega_k) \otimes \mathbf{\Phi}(\omega_k) + \mathbf{I}_{\underline{X},T}(\omega_k) \otimes \mathbf{\Phi}^{\mathsf{T}}(\omega_k) \right) \\
&= \int_{-1/2}^{1/2} \mathrm{d}\mathbf{F}^{\mathsf{T}}(\omega) \otimes \mathbf{\Phi}(\omega) + o(1) .
\end{aligned}
$$

The variance matrix of the estimate $\widehat{\mathbf{B}}$

$$
\lim_{T \to \infty} \operatorname{Var} \left(\frac{1}{\sqrt{T}} \operatorname{Vec} \sum_{k=-T_1}^{T_1} \mathbf{\Phi}(\omega_k) \mathbf{I}_{\underline{Y},\underline{X},T}(\omega_k) \right)
$$

$$
= 4 \operatorname{Vec} \int_{-1/2}^{1/2} \mathrm{d}\mathbf{F}^{\mathsf{T}}(\omega) \otimes \left[\mathbf{\Phi}(\omega) \mathbf{S}_{\underline{Z}}(\omega) \mathbf{\Phi}(\omega) \right] ;
$$

see (29). In particular we have the variance for two frequently used estimates: the linear least squares (LS) estimator if $\mathbf{\Phi}(\omega) = \mathbf{U}_d$, or the best linear unbiased estimator (BLUE) if $\mathbf{\Phi}(\omega) = \mathbf{S}_{\underline{Z}}^{-1}(\omega)$. Grenander shows that under some assumption the LS and BLUE are equivalent. Actually, both have the same limit variance

$$
\lim_{T \to \infty} \operatorname{Var} \left[\sqrt{T} \operatorname{Vec}(\widehat{\mathbf{B}}) \right] = \left[\int_{-1/2}^{1/2} \mathrm{d}\mathbf{F}^{\mathsf{T}}(\omega) \otimes \mathbf{S}_{\underline{Z}}^{-1}(\omega) \right]^{-1} .
$$

This limit does not depend on $\mathbf{B}_0$. This result can be reached from the general formula (10) for the variance.

Remark 1. The estimation of the transpose of matrix $\mathbf{B}$ is customary and more direct in the time domain; see [Han70]. The variance of $\widehat{\mathbf{B}}^{\mathsf{T}}$ follows from (12) easily

$$
\operatorname{Var}[\sqrt{T} \operatorname{Vec}(\widehat{\mathbf{B}}^{\mathsf{T}})] = \operatorname{Var}[\mathbf{K}_{p \cdot d} \operatorname{Vec}(\widehat{\mathbf{B}})] = \mathbf{K}_{p \cdot d} \operatorname{Var} \operatorname{Vec}(\widehat{\mathbf{B}}) \mathbf{K}_{d \cdot p} ;
$$

hence

$$
\lim_{T \to \infty} \operatorname{Var}[\sqrt{T} \operatorname{Vec}(\widehat{\mathbf{B}}^{\mathsf{T}})] = \left[\int_{-1/2}^{1/2} \mathbf{S}_{\underline{Z}}^{-1}(\omega) \otimes \mathrm{d}\mathbf{F}^{\mathsf{T}}(\omega) \right]^{-1} .
$$

In practice, we are interested in the asymptotic variance of $\widehat{\mathbf{B}}$ according to the original unscaled regressors. We have estimated the matrix $T^{-1/2}\mathbf{B}\mathbf{D}_T$, because

$$
\mathbf{B}\underline{X}_t = \left(\frac{1}{\sqrt{T}} \mathbf{B}\mathbf{D}_T \right) \sqrt{T} \mathbf{D}_T^{-1} \underline{X}_t .
$$

It implies that the asymptotic variance of $\operatorname{Vec} \widehat{\mathbf{B}}$ is

$$\left[\int_{-1/2}^{1/2}\mathbf{D}_T \mathrm{d}\mathbf{F}^{\mathsf{T}}(\omega)\mathbf{D}_T \otimes \mathbf{S}_{\underline{Z}}^{-1}(\omega)\right]^{-1} \; ; \tag{12}$$

see [Han70, Theorem 10, Chapter VII]. For instance, if we are concerned with polynomial regressor $\underline{X}_{j,t} = t^{j-1}, j = 1,,\ldots,p$, then the corresponding scales are $T_j(T) \simeq \sqrt{T^{2j-1}/(2j-1)}$ (this latter one applies for any fractional $j > 1/2$ as well), and $\mathbf{D}_T = \mathrm{diag}(T_1, T_2, \ldots T_d)$. The SDFR $\mathbf{F}$ is concentrated at zero with values $\mathrm{d}\mathbf{F}_{j,k}(0) = \sqrt{(2k-1)(2j-1)}/(k+j-1)$, so the asymptotic variance of Vec $\widehat{\mathbf{B}}$ is

$$\left[\mathbf{D}_T^{-1}\mathrm{d}\mathbf{F}^{-1}(0)\mathbf{D}_T^{-1}\right] \otimes \mathbf{S}_{\underline{Z}}(0) \; ;$$

see [GR57, p. 247], for the scalar-valued case.

5.2 Mixed Model

A very realistic model to consider is the following

$$\underline{Y}_t = \underline{X}_t(\underline{\vartheta}_0) + \underline{Z}_t \; ,$$

where the regressor has the form

$$\underline{X}_t(\underline{\vartheta}) = \mathbf{B}_1 \underline{X}_{1,t} + \mathbf{B}_2 \underline{X}_{2,t}(\underline{\lambda}) \tag{13}$$

$$= [\mathbf{B}_1, \mathbf{B}_2] \begin{bmatrix} \underline{X}_{1,t} \\ \underline{X}_{2,t}(\underline{\lambda}) \end{bmatrix}$$

$$= \mathbf{B}\underline{X}_{3,t}(\underline{\lambda}) \; .$$

Here the unknown parameter is $\underline{\vartheta} = \mathrm{Vec}(\mathrm{Vec}\,\mathbf{B}_1, \mathrm{Vec}\,\mathbf{B}_2, \underline{\lambda})$, where $\mathbf{B}_1$ is $d \times p$, $\mathbf{B}_2$ is $d \times q$, $\underline{\lambda}$ is $r \times 1$, $\underline{X}_{1,t}$ has dimension p, $\underline{X}_{2,t}(\underline{\lambda})$ has dimension q, $\mathbf{B} = [\mathbf{B}_1, \mathbf{B}_2]$, and $\underline{X}_{3,t}(\underline{\lambda}) = \begin{bmatrix} \underline{X}_{1,t} \\ \underline{X}_{2,t}(\underline{\lambda}) \end{bmatrix}$. First we turn to the problem of estimation. Minimize the objective function

$$Q_T(\mathbf{B}, \underline{\lambda})$$

$$= \frac{1}{T}\sum_{k=-T_1}^{T_1}\left[\mathrm{Tr}\left(\mathbf{I}_{\underline{Y},T}(\omega_k)\mathbf{\Phi}(\omega_k)\right) + \mathrm{Tr}\left(\mathbf{B}\mathbf{I}_{\underline{X}_3,T}(\omega_k,\underline{\lambda})\mathbf{B}^{\mathsf{T}}\mathbf{\Phi}(\omega_k)\right) \right.$$

$$\left. - \mathrm{Tr}\left(\mathbf{I}_{\underline{Y},\underline{X}_3,T}(\omega_k,\underline{\lambda})\mathbf{B}^{\mathsf{T}}\mathbf{\Phi}(\omega_k)\right) - \mathrm{Tr}\left(\mathbf{B}\mathbf{I}_{\underline{X}_3,\underline{Y},T}(\omega_k,\underline{\lambda})\mathbf{\Phi}(\omega_k)\right)\right] \; . \tag{14}$$

Then, take the derivative with respect to $\mathbf{B}_1, \mathbf{B}_2$, and $\underline{\lambda}$. Actually, we can apply the linear method for $\mathbf{B}$ in terms of $\underline{X}_{3,t}(\underline{\lambda})$. Suppose that $\widehat{\mathbf{B}} = \left[\widehat{\mathbf{B}}_1, \widehat{\mathbf{B}}_2\right]$ and $\widehat{\underline{\lambda}}$ fulfill the system of equations

$$\frac{\partial Q_T(\mathbf{B}, \underline{\lambda})}{\partial \mathbf{B}} = \mathbf{0} \; ,$$

$$\mathrm{Vec}\,\frac{\partial Q_T(\underline{\lambda})}{\partial \underline{\lambda}^{\mathsf{T}}} = \underline{0} \; .$$

The estimation of the linear parameters $\mathbf{B}_1$ and $\mathbf{B}_2$ can be carried out as linear regression when the parameter $\underline{\lambda}$ is fixed. It leads to a recursive procedure. When $\underline{\lambda} = \underline{\tilde{\lambda}}$ is a fixed initial value, the normal equation gives the estimates

$$\mathrm{Vec}(\widehat{\mathbf{B}})$$

$$= \left(\sum_{k=-T_1}^{T_1} \mathbf{I}_{\underline{X}_3,T}^{\mathsf{T}}(\omega_k, \underline{\tilde{\lambda}}) \otimes \mathbf{\Phi}(\omega_k) \right)^{-1} \mathrm{Vec} \sum_{k=-T_1}^{T_1} \mathbf{\Phi}(\omega_k) \mathbf{I}_{\underline{Y},\underline{X}_3,T}(\omega_k, \underline{\tilde{\lambda}}) .$$

Now, to get the estimate for $\underline{\lambda}$, we keep $\mathbf{B} = \widehat{\mathbf{B}}$ fixed and minimize (14), that is, find the solution to the equation

$$\sum_{k=-T_1}^{T_1} \left[\frac{\partial \mathbf{BI}_{\underline{X}_3,T}(\omega_k, \underline{\lambda})\mathbf{B}^{\mathsf{T}}}{\partial \underline{\lambda}^{\mathsf{T}}} - \frac{\partial \mathbf{I}_{\underline{Y},\underline{X}_3,T}(\omega_k, \underline{\lambda})\mathbf{B}^{\mathsf{T}}}{\partial \underline{\lambda}^{\mathsf{T}}} - \frac{\partial \mathbf{BI}_{\underline{X}_3,\underline{Y},T}(\omega_k, \underline{\lambda})}{\partial \underline{\lambda}^{\mathsf{T}}} \right]_{\underline{\lambda}=\widehat{\underline{\lambda}}}^{\mathsf{T}}$$

$$\times \mathrm{Vec}\, \mathbf{\Phi}^{\mathsf{T}}(\omega_k) = \underline{0} .$$

The primary scaling of $\underline{X}_{3,t}(\underline{\lambda}) = \begin{bmatrix} \underline{X}_{1,t} \\ \underline{X}_{2,t}(\underline{\lambda}) \end{bmatrix}$ is given by

$$\mathbf{D}_T = \mathrm{diag}\left(\mathbf{D}_{X_1,T}, \mathbf{D}_{X_2,T} \right) ,$$

where $\mathbf{D}_{X_1,T} = \mathrm{diag}(D_{X_1,k}(T), k = 1, \ldots, p)$ and $\mathbf{D}_{X_2,T} = \mathrm{diag}(D_{X_2,k}(T), k = 1, \ldots, q)$. The secondary scaling of regressors is $\mathbf{D}_{1,T} = \mathrm{diag}(\mathbf{U}_{dp+dq}, \mathbf{D}_{3,T})$. Let us denote the limit variance of the derivative

$$\begin{bmatrix} \left[\frac{\partial Q_T(\mathbf{B}_1,\mathbf{B}_2,\underline{\lambda})}{\partial \mathrm{Vec}\, \mathbf{B}_1^{\mathsf{T}}} \right]^{\mathsf{T}} \\ \left[\frac{\partial Q_T(\mathbf{B}_1,\mathbf{B}_2,\underline{\lambda})}{\partial \mathrm{Vec}\, \mathbf{B}_2^{\mathsf{T}}} \right]^{\mathsf{T}} \\ \left[\frac{\partial Q_T(\mathbf{B}_1,\mathbf{B}_2,\underline{\lambda})}{\partial \underline{\lambda}^{\mathsf{T}}} \right]^{\mathsf{T}} \end{bmatrix}$$

by

$$\Sigma = 2 \begin{bmatrix} \Sigma_{11} & \Sigma_{12} & \Sigma_{1\lambda}\mathbf{D}_{3,T} \\ \Sigma_{21} & \Sigma_{22} & \Sigma_{2\lambda}\mathbf{D}_{3,T} \\ \mathbf{D}_{3,T}\Sigma_{\lambda 1} & \mathbf{D}_{3,T}\Sigma_{\lambda 2} & \mathbf{D}_{3,T}\Sigma_{\lambda\lambda}\mathbf{D}_{3,T} \end{bmatrix} ,$$

where the blocks of Σ already contain the scaling $\mathbf{D}_T$ of the regressor. Here as well as later on $\mathbf{\Phi} = \mathbf{S}_{\underline{Z}}^{-1}$. The linear part

$$\Sigma_{11} = \int_{-1/2}^{1/2} \mathbf{D}_{X_1,T}\, d\mathbf{F}_{11}^{\mathsf{T}}(\omega)\, \mathbf{D}_{X_1,T} \otimes \mathbf{S}_{\underline{Z}}^{-1}(\omega) ,$$

$$\Sigma_{12} = \int_{-1/2}^{1/2} \mathbf{D}_{X_2,T}\, d\mathbf{F}_{12}^{\mathsf{T}}(\omega, \underline{\lambda}_0)\, \mathbf{D}_{X_1,T} \otimes \mathbf{S}_{\underline{Z}}^{-1}(\omega) ,$$

$$\Sigma_{22} = \int_{-1/2}^{1/2} \mathbf{D}_{X_2,T}\, d\mathbf{F}_{22}^{\mathsf{T}}(\omega, \underline{\lambda}_0)\, \mathbf{D}_{X_2,T} \otimes \mathbf{S}_{\underline{Z}}^{-1}(\omega) .$$

The mixed blocks are

$$\Sigma_{1\lambda} = \int_{-1/2}^{1/2} \left(\mathbf{D}_{X_1,T} \otimes \mathbf{S}_{\underline{Z}}^{-1}(\omega) \, \mathbf{B}_{2,0} \mathbf{D}_{X_2,T} \right) \mathrm{d} \frac{\partial \mathbf{F}_{1,2}(\omega,\underline{\lambda}_0)}{\partial \underline{\lambda}^{\mathsf{T}}} ,$$

$$\Sigma_{2\lambda} = \int_{-1/2}^{1/2} \left(\mathbf{D}_{X_2,T} \otimes \mathbf{S}_{\underline{Z}}^{-1}(\omega) \, \mathbf{B}_{2,0} \mathbf{D}_{X_2,T} \right) \mathrm{d} \frac{\partial \mathbf{F}_{2,2}(\omega,\underline{\lambda}_0)}{\partial \underline{\lambda}^{\mathsf{T}}} .$$

The nonlinear block $\Sigma_{\lambda\lambda}$ comes from the general result (10):

$$\Sigma_{\lambda\lambda} = 2 \int_{-1/2}^{1/2} \left(\mathbf{U}_r \otimes \mathrm{Vec} \left(\left[\mathbf{D}_{X_2,T} \mathbf{B}_{2,0}^{\mathsf{T}} \mathbf{S}_{\underline{Z}}^{-1}(\omega) \, \mathbf{B}_{2,0} \mathbf{D}_{X_2,T} \right]^{\mathsf{T}} \right)^{\mathsf{T}} \right)$$

$$\times \mathrm{d} \left. \frac{\partial^2 \mathbf{F}_{2,2}(\omega,\underline{\lambda}_1,\underline{\lambda}_2)}{\partial \underline{\lambda}_2^{\mathsf{T}} \partial \underline{\lambda}_1^{\mathsf{T}}} \right|_{\underline{\lambda}_1 = \underline{\lambda}_2 = \underline{\lambda}_0} .$$

Finally the variance matrix of the estimates $\mathrm{Vec}(\mathrm{Vec}\,\widehat{\mathbf{B}}_1, \mathrm{Vec}\,\widehat{\mathbf{B}}_2, \widehat{\underline{\lambda}})$ is

$$\mathrm{Var}[\mathrm{Vec}(\mathrm{Vec}\,\widehat{\mathbf{B}}_1, \mathrm{Vec}\,\widehat{\mathbf{B}}_2, \widehat{\underline{\lambda}})] \simeq \begin{bmatrix} \Sigma_{11} & \Sigma_{12} & \Sigma_{1\lambda}\mathbf{D}_{3,T} \\ \Sigma_{21} & \Sigma_{22} & \Sigma_{2\lambda}\mathbf{D}_{3,T} \\ \mathbf{D}_{3,T}\Sigma_{\lambda 1} & \mathbf{D}_{3,T}\Sigma_{\lambda 2} & \mathbf{D}_{3,T}\Sigma_{\lambda\lambda}\mathbf{D}_{3,T} \end{bmatrix}^{-1} .$$

5.3 Linear Trend with Harmonic Components

Here we consider a particular case of the mixed model above. Let

$$\underline{Y}_t = \underline{X}_t(\underline{\vartheta}_0) + \underline{Z}_t ,$$

where

$$\underline{X}_t(\underline{\vartheta}) = \mathbf{B} \begin{bmatrix} 1 \\ t \end{bmatrix} + \mathbf{A} \begin{bmatrix} \cos(2\pi t\lambda_1) \\ \sin(2\pi t\lambda_1) \\ \cos(2\pi t\lambda_2) \\ \sin(2\pi t\lambda_2) \end{bmatrix} ,$$

The parameter is $\underline{\vartheta}^{\mathsf{T}} = ([\mathrm{Vec}\,\mathbf{B}_1]^{\mathsf{T}}, [\mathrm{Vec}\,\mathbf{B}_2]^{\mathsf{T}}, [\lambda_1,\lambda_2])$, $|\lambda_i| \le \pi$, $\lambda_1 \ne \lambda_2$, $\lambda_i \ne 0, \pm 1/2$. It is readily seen that the estimation of the coefficient $\mathbf{B}$ of the linear regression

$$\mathbf{B} = \begin{bmatrix} b_{11} & b_{12} \\ b_{21} & b_{22} \end{bmatrix}$$

can be done separately as it has no influence on the estimation of the rest of the parameters.

$$\mathbf{A} = \begin{bmatrix} a_{11} & a_{12} & a_{13} & a_{14} \\ a_{21} & a_{22} & a_{23} & a_{24} \end{bmatrix} .$$

The primary scaling for $\underline{X}_{1,t}$ is $\mathbf{D}_{X_1,T} = \mathrm{diag}\left(T^{1/2}, T^{3/2}/\sqrt{3}\right)$, and for $\underline{X}_{2,t}(\underline{\lambda})$ is $\mathbf{D}_{X_2,T} = T^{1/2}/\sqrt{2}\,\mathbf{U}_4$, because $\underline{X}_{1,t} = [1,t]^{\mathsf{T}}$ and

$$\underline{X}_{2,t}(\underline{\lambda}) = [\cos(2\pi t\lambda_1), \sin(2\pi t\lambda_1), \cos(2\pi t\lambda_2), \sin(2\pi t\lambda_2)] .$$

The secondary scaling for the linear part $\underline{X}_{1,t}$, as we have already seen, is $\mathbf{U}_2$, and the secondary one for the nonlinear part is $\mathbf{U}_4$. The scaled partial derivative according to $\underline{\lambda}$ is $\mathbf{D}_{3,T} = 2\pi T/\sqrt{3}\mathbf{U}_2$ because the primary scaling $\sqrt{T/2}$ has already been applied. Therefore the scaling matrix $\mathbf{D}_T$ of the regressors $\left[\underline{X}_{1,t}^\mathsf{T}, \underline{X}_{2,t}^\mathsf{T}(\underline{\lambda})\right]^\mathsf{T}$ is $\mathbf{D}_T = \operatorname{diag}(\mathbf{D}_{\underline{X}_1,T}, \mathbf{D}_{\underline{X}_2,T})$ and $\mathbf{D}_{1,T} = \operatorname{diag}(\mathbf{U}_{12}, \mathbf{D}_{3,T})$. The asymptotic variance is

$$\mathbf{D}_{1,T}^{-1}\mathbf{J}^{-1}\left(\mathbf{D}_T\mathbf{S}_{\underline{Z}}^{-1}\mathbf{D}_T, \mathbf{F}\right)\mathbf{D}_{1,T}^{-1} .$$

The proper scaling for the term $X_{k,t}X_{m,t+h}$ in $\widehat{\mathbf{C}}_{\underline{X},T}$ is $\left(\|X_{k,t}\|_T \|X_{m,t}\|_T\right)^{-1}$, in general. Here it can be changed into an equivalent function of T; instead of (2) we have

$$\widehat{\mathbf{C}}_{\underline{X},T}(h, \underline{\vartheta}_1, \underline{\vartheta}_2) = \mathbf{D}_T^{-1}\sum_{t=1}^{T-h} \underline{X}_{t+h}(\underline{\vartheta}_1)\underline{X}_t^\mathsf{T}(\underline{\vartheta}_2)\mathbf{D}_T^{-1} .$$

Let us partition the second derivative of SDFR according to the parameters. With obvious notation, denote

$$\frac{\partial^2 \mathbf{F}(\omega, \underline{\lambda}_1, \underline{\lambda}_2)}{\partial\underline{\lambda}_2^\mathsf{T}\partial\underline{\lambda}_1^\mathsf{T}} = \begin{bmatrix} \mathbf{F}_{11} & \mathbf{F}_{12} & \mathbf{F}_{1\lambda} \\ \mathbf{F}_{21} & \mathbf{F}_{22} & \mathbf{F}_{2\lambda} \\ \mathbf{F}_{\lambda 1} & \mathbf{F}_{\lambda 2} & \mathbf{F}_{\lambda\lambda} \end{bmatrix} ,$$

and assume $\lambda_1 \neq \lambda_2, \lambda_i \neq 0, \pm 1/2$.

(A1). The regression spectrum of the linear part is

$$d\mathbf{F}_{11}(\omega) = \begin{bmatrix} 1 & \sqrt{3}/2 \\ \sqrt{3}/2 & 1 \end{bmatrix} d\delta_{\omega\geq 0}$$

where $\delta_{\omega\geq 0}$ denotes the Kronecker delta. Hence the block Σ_{11} follows

$$\Sigma_{11} = \mathbf{D}_{\underline{X}_1,T}d\mathbf{F}_{11}(0)\mathbf{D}_{\underline{X}_1,T} \otimes \mathbf{S}_{\underline{Z}}^{-1}(0) .$$

(A2). It is seen that there is no mixed effect: $\mathbf{F}_{12}(\omega, \underline{\lambda}_0) = \mathbf{0}$, $\Sigma_{12} = \mathbf{0}$, and $\mathbf{F}_{1\lambda}(\omega, \underline{\lambda}_0) = \mathbf{0}$, $\Sigma_{1\lambda} = \mathbf{0}$.

(A3). The $\mathbf{F}_{22}(\omega, \underline{\lambda}_0)$ corresponds to the coefficient $\mathbf{A}$. Let

$$\mathbf{H}_{1h}(\lambda) = \begin{bmatrix} \cos(2\pi\lambda h) & -\sin(2\pi\lambda h) \\ \sin(2\pi\lambda h) & \cos(2\pi\lambda h) \end{bmatrix} ,$$

Notice

$$\widehat{\mathbf{C}}_{\underline{X}_2,T}(h, \underline{\lambda}, \underline{\mu}) = \mathbf{D}_{\underline{X}_2,T}^{-1}\sum_{t=1}^{T-h} \underline{X}_{2,t+h}(\underline{\lambda})\underline{X}_{2,t}^\mathsf{T}(\underline{\mu})\mathbf{D}_{\underline{X}_2,T}^{-1}$$

$$\rightarrow \begin{bmatrix} \delta_{\lambda_1=\mu_1}\mathbf{H}_{1h}(\lambda_1) & \delta_{\lambda_1=\mu_2}\mathbf{H}_{1h}(\lambda_1) \\ \delta_{\lambda_2=\mu_1}\mathbf{H}_{1h}(\lambda_2) & \delta_{\lambda_2=\mu_2}\mathbf{H}_{1h}(\lambda_2) \end{bmatrix} ,$$

where $\delta_{\lambda=\omega}$ denotes the Kronecker delta. Define the step functions

$$
g_{c\lambda}(\omega) = \begin{cases} 0, & \omega < -\lambda, \\ 1/2, & -\lambda \le \omega < \lambda, \\ 1, & \lambda \le \omega, \end{cases} \quad g_{s\lambda}(\omega) = \begin{cases} 0, & \omega < -\lambda, \\ i/2, & -\lambda \le \omega < \lambda, \\ 0, & \lambda \le \omega, \end{cases}
$$

and

$$
\mathbf{G}_{1\lambda}(\omega) = \begin{bmatrix} g_{c\lambda}(\omega) & -g_{s\lambda}(\omega) \\ g_{s\lambda}(\omega) & g_{c\lambda}(\omega) \end{bmatrix}.
$$

Now we have

$$
\lim_{T\to\infty} \widehat{\mathbf{C}}_{\underline{X}_2,T}\left(h,\underline{\lambda},\underline{\mu}\right) = \int_{-1/2}^{1/2} e^{2i\pi\omega h}\, d\mathbf{F}_{22}(\omega,\underline{\lambda},\underline{\mu}),
$$

where

$$
\mathbf{F}_{22}(\omega,\underline{\lambda},\underline{\mu}) = \begin{bmatrix} \delta_{\lambda_1=\mu_1}\mathbf{G}_{1\lambda_1}(\omega) & \delta_{\lambda_1=\mu_2}\mathbf{G}_{1\lambda_1}(\omega) \\ \delta_{\lambda_2=\mu_1}\mathbf{G}_{1\lambda_2}(\omega) & \delta_{\lambda_2=\mu_2}\mathbf{G}_{1\lambda_2}(\omega) \end{bmatrix}.
$$

The scaled version of the block is

$$
\frac{2}{T}\Sigma_{22} = \int_{-1/2}^{1/2} \left(\mathbf{D}_{X_2,T} d\mathbf{F}_{22}^{\mathsf{T}}\left(\omega,\underline{\lambda}_0\right)\mathbf{D}_{X_2,T}\right) \otimes \mathbf{S}_{\underline{Z}}^{-1}(\omega)
$$

$$
= \begin{bmatrix} \begin{bmatrix} \operatorname{Re}\mathbf{S}_{\underline{Z}}^{-1}(\lambda_1) & \operatorname{Im}\mathbf{S}_{\underline{Z}}^{-1}(\lambda_1) \\ -\operatorname{Im}\mathbf{S}_{\underline{Z}}^{-1}(\lambda_1) & \operatorname{Re}\mathbf{S}_{\underline{Z}}^{-1}(\lambda_1) \end{bmatrix} & \mathbf{0} \\ \mathbf{0} & \begin{bmatrix} \operatorname{Re}\mathbf{S}_{\underline{Z}}^{-1}(\lambda_2) & \operatorname{Im}\mathbf{S}_{\underline{Z}}^{-1}(\lambda_2) \\ -\operatorname{Im}\mathbf{S}_{\underline{Z}}^{-1}(\lambda_2) & \operatorname{Re}\mathbf{S}_{\underline{Z}}^{-1}(\lambda_2) \end{bmatrix} \end{bmatrix}.
$$

(A4). For $\mathbf{F}_{2\lambda}(\omega,\underline{\lambda}_0)$, define the matrices

$$
\mathbf{U}_2(1) = \begin{bmatrix} 1 & 0 \\ 0 & 0 \end{bmatrix},
$$

$$
\mathbf{U}_2(2) = \begin{bmatrix} 0 & 0 \\ 0 & 1 \end{bmatrix};
$$

then we have

$$
\frac{1}{\pi T}\frac{\partial \operatorname{Vec}\widehat{\mathbf{C}}_{\underline{X}_2,\underline{X}_2,T}\left(h,\underline{\lambda},\underline{\mu}\right)}{\partial\underline{\mu}^{\mathsf{T}}} \to \begin{bmatrix} \delta_{\lambda_1=\mu_1}\mathbf{U}_2(1)\otimes\begin{bmatrix} -\sin(2\pi\lambda_1 h) \\ \cos(2\pi\lambda_1 h) \end{bmatrix} \\ \delta_{\lambda_1=\mu_2}\mathbf{U}_2(1)\otimes\begin{bmatrix} -\cos(2\pi\lambda_1 h) \\ -\sin(2\pi\lambda_1 h) \end{bmatrix} \\ \delta_{\lambda_2=\mu_1}\mathbf{U}_2(2)\otimes\begin{bmatrix} -\sin(2\pi\lambda_2 h) \\ \cos(2\pi\lambda_2 h) \end{bmatrix} \\ \delta_{\lambda_2=\mu_2}\mathbf{U}_2(2)\otimes\begin{bmatrix} -\cos(2\pi\lambda_2 h) \\ -\sin(2\pi\lambda_2 h) \end{bmatrix} \end{bmatrix}.
$$

Notice that if $\underline{\lambda} = \underline{\mu}$ and $\lambda_1 \ne \lambda_2$ then this latter matrix is written

$$[\text{Vec}[\mathbf{U}_2(1) \otimes \mathbf{H}_{2h}(\lambda_1)], \text{Vec}[\mathbf{U}_2(2) \otimes \mathbf{H}_{2h}(\lambda_2)]] \ ,$$

where

$$\mathbf{H}_{2h}(\lambda) = \begin{bmatrix} -\sin(2\pi\lambda h) & -\cos(2\pi\lambda h) \\ \cos(2\pi\lambda h) & -\sin(2\pi\lambda h) \end{bmatrix} \ .$$

Notice that for three frequencies $\underline{\lambda} = [\lambda_1, \lambda_2, \lambda_3]$ we would have,

$$[\text{Vec}[\mathbf{U}_3(k) \otimes \mathbf{H}_{2h}(\lambda_k)]]|_{k=1,2,3} \ ,$$

where $\mathbf{U}_3(j)$ is a 3×3 matrix with zero elements except for the jth entry in the diagonal which is 1.

$$\mathbf{F}_{2\lambda}(\omega, \underline{\lambda}) = \frac{\sqrt{3}}{2} [\text{Vec}[\mathbf{U}_2(1) \otimes \mathbf{G}_{2\lambda_1}(\omega)], \text{Vec}[\mathbf{U}_2(2) \otimes \mathbf{G}_{2\lambda_2}(\omega)]] \ ,$$

where

$$\mathbf{G}_{2\lambda}(\omega) = \begin{bmatrix} -g_{s\lambda}(\omega) & -g_{c\lambda}(\omega) \\ g_{c\lambda}(\omega) & -g_{s\lambda}(\omega) \end{bmatrix} \ .$$

Let us apply the general formula for $\Sigma_{2\lambda}$, so

$$\Sigma_{2\lambda} = \int_{-1/2}^{1/2} \left(\mathbf{U}_4 \otimes \mathbf{S}_{\underline{Z}}^{-1}(\omega) \mathbf{A}_0 \mathbf{D}_{\underline{X}_2, T} \right) \left(\mathbf{D}_{\underline{X}_2, T} \otimes \mathbf{U}_4 \right) d\mathbf{F}_{2\lambda}(\omega, \underline{\lambda}_0)$$

$$= \frac{T}{2} \int_{-1/2}^{1/2} \left(\mathbf{U}_4 \otimes \mathbf{S}_{\underline{Z}}^{-1}(\omega) \mathbf{A}_0 \right) d\mathbf{F}_{2\lambda}(\omega, \underline{\lambda}_0) \ .$$

Put

$$\Gamma_2 = \begin{bmatrix} i & -1 \\ 1 & i \end{bmatrix}$$

$$\Lambda_2(\omega) = \mathbf{U}_4 \otimes \mathbf{S}_{\underline{Z}}^{-1}(\omega) \mathbf{A}_0$$

$$\Sigma_{2\lambda} = \frac{\sqrt{3}T}{4} [\Lambda_2(\lambda_1) \text{Vec}[\mathbf{U}_2(1) \otimes \Gamma_2], \Lambda_2(\lambda_2) \text{Vec}[\mathbf{U}_2(2) \otimes \Gamma_2]] \ .$$

(A5). Finally, $\mathbf{F}_{\lambda\lambda}(\omega, \underline{\lambda}_0)$ is

$$\frac{3}{(2\pi)^2 T^2} \frac{\partial^2 \text{Vec}\, \widehat{\mathbf{C}}_{\underline{X}_2, T}(h, \underline{\lambda}, \underline{\mu})}{\partial \underline{\mu}^{\mathsf{T}} \partial \underline{\lambda}^{\mathsf{T}}} \rightarrow \begin{bmatrix} \delta_{\lambda_1=\mu_1} \mathbf{U}_2(1) \otimes \begin{bmatrix} \cos(2\pi\lambda_1 h) \\ \sin(2\pi\lambda_1 h) \end{bmatrix} \\ \delta_{\lambda_1=\mu_1} \mathbf{U}_2(1) \otimes \begin{bmatrix} -\sin(2\pi\lambda_1 h) \\ \cos(2\pi\lambda_1 h) \end{bmatrix} \\ \mathbf{0}_{16\times 2} \\ \delta_{\lambda_2=\mu_2} \mathbf{U}_2(2) \otimes \begin{bmatrix} \cos(2\pi\lambda_2 h) \\ \sin(2\pi\lambda_2 h) \end{bmatrix} \\ \delta_{\lambda_2=\mu_2} \mathbf{U}_2(2) \otimes \begin{bmatrix} -\sin(2\pi\lambda_2 h) \\ \cos(2\pi\lambda_2 h) \end{bmatrix} \end{bmatrix} \ .$$

Define now the matrix $\mathbf{U}_{2,4}(1,1)$ of 2×4 with all elements zero except the entry $(1,1)$ which is one; we have

$$\left[\mathrm{Vec}[\mathbf{U}_{2,4}(1,1) \otimes \mathbf{H}_{3h}(\lambda_1)], \ \mathrm{Vec}[\mathbf{U}_{2,4}(2,4) \otimes \mathbf{H}_{3h}(\lambda_2)] \right] \ ,$$

where

$$\mathbf{H}_{3h}(\lambda) = \begin{bmatrix} \cos(2\pi\lambda h) & -\sin(2\pi\lambda h) \\ \sin(\lambda 2\pi h) & \cos(2\pi\lambda h) \end{bmatrix} .$$

The SDFR

$$\mathbf{F}_{\lambda\lambda}(\omega, \underline{\lambda})$$
$$= (2\pi)^2 \left[\mathrm{Vec}[\mathbf{U}_{2,4}(1,1) \otimes \mathbf{G}_{3\lambda_1}(\omega)], \ \mathrm{Vec}[\mathbf{U}_{2,4}(2,4) \otimes \mathbf{G}_{3\lambda_2}(\omega)] \right] \ ,$$

where

$$\mathbf{G}_{3\lambda}(\omega) = \begin{bmatrix} g_{c\lambda}(\omega) & -g_{s\lambda}(\omega) \\ g_{s\lambda}(\omega) & g_{c\lambda}(\omega) \end{bmatrix} .$$

The corresponding variance matrix is

$$\frac{2}{T}\Sigma_{\lambda\lambda} = \frac{T}{2} \int_{-1/2}^{1/2} \left(\mathbf{U}_2 \otimes \mathrm{Vec}[\mathbf{A}^\mathsf{T}[\mathbf{S}_{\underline{Z}}^{-1}(\omega)]^\mathsf{T}\mathbf{A}]^\mathsf{T} \right) d\mathbf{F}_{\lambda\lambda}(\omega, \underline{\lambda})$$

$$= \int_{-1/2}^{1/2} \begin{bmatrix} \left[\mathrm{Vec}(\mathbf{A}^\mathsf{T}[\mathbf{S}_{\underline{Z}}^{-1}(\omega)]^\mathsf{T}\mathbf{A}) \right]^\mathsf{T} & \mathbf{0}_{1\times 16} \\ \mathbf{0}_{1\times 16} & \left[\mathrm{Vec}(\mathbf{A}^\mathsf{T}[\mathbf{S}_{\underline{Z}}^{-1}(\omega)]^\mathsf{T}\mathbf{A}) \right]^\mathsf{T} \end{bmatrix}$$
$$\times d\mathbf{F}_{\lambda\lambda}(\omega, \underline{\lambda}) \ .$$

Put, for computational purposes,

$$\Gamma_3 = \begin{bmatrix} 1 & i \\ -i & 1 \end{bmatrix} ,$$
$$\Lambda(\omega) = \mathbf{U}_2 \otimes \left[\mathrm{Vec}(\mathbf{A}^\mathsf{T}[\mathbf{S}_{\underline{Z}}^{-1}(\omega)]^\mathsf{T}\mathbf{A}) \right]^\mathsf{T} ;$$

then the variance matrix has the form

$$\Sigma_{\lambda\lambda} =$$
$$\frac{2\pi^2 T}{2} \mathrm{Re} \left[\Lambda(\lambda_1) \mathrm{Vec}[\mathbf{U}_{2,4}(1,1) \otimes \Gamma_3], \ \Lambda(\lambda_2) \mathrm{Vec}[\mathbf{U}_{2,4}(2,4) \otimes \Gamma_3] \right] \ .$$

It simplifies further:

$$\Sigma_{\lambda\lambda} = \frac{T}{2} \begin{bmatrix} \sigma_{11} & 0 \\ 0 & \sigma_{22} \end{bmatrix} ;$$

the entries are given in terms of the entries $A_{mn}(\omega) = \left[\mathbf{A}^\mathsf{T} \left[\mathbf{S}_{\underline{Z}}^{-1} \right]^\mathsf{T} \mathbf{A} \right]_{mn}$,

$$\sigma_{11} = \mathrm{Re}\, A_{11}(\lambda_1) + \mathrm{Im}\, A_{21}(\lambda_1) - \mathrm{Im}\, A_{12}(\lambda_1) + \mathrm{Re}\, A_{22}(\lambda_1) \ ,$$
$$\sigma_{22} = \mathrm{Re}\, A_{33}(\lambda_2) + \mathrm{Im}\, A_{43}(\lambda_2) - \mathrm{Im}\, A_{34}(\lambda_2) + \mathrm{Re}\, A_{44}(\lambda_2) \ .$$

Now we return to the asymptotic variance matrix of the parameters; let us collect the blocks of the variance matrix

$$\mathbf{D}_{1,T} \begin{bmatrix} \Sigma_{11} & 0 & 0 \\ 0 & \Sigma_{22} & \Sigma_{2\lambda} \\ 0 & \Sigma_{\lambda 2} & \Sigma_{\lambda\lambda} \end{bmatrix} \mathbf{D}_{1,T} \;,$$

where $\mathbf{D}_{1,T} = \mathrm{diag}(\mathbf{U}_{12}, \mathbf{D}_{2,T})$. The variance matrix of the coefficient $\widehat{\mathbf{A}}$ is

$$\frac{2}{T} \left(\Sigma'_{22} - \Sigma'_{2\lambda} \Sigma'^{-1}_{\lambda\lambda} \Sigma'_{\lambda 2} \right)^{-1} \;,$$

and of $\widehat{\lambda}$

$$\mathbf{D}^{-1}_{3,T} \left(\Sigma_{\lambda\lambda} - \Sigma_{\lambda 2} \Sigma^{-1}_{22} \Sigma_{2\lambda} \right)^{-1} \mathbf{D}^{-1}_{3,T} = \frac{6}{(2\pi)^2 T^3} \left(\Sigma'_{\lambda\lambda} - \Sigma'_{\lambda 2} \Sigma'^{-1}_{22} \Sigma'_{2\lambda} \right)^{-1} \;,$$

where $\mathbf{D}_{3,T} = 2\pi T/\sqrt{3}\mathbf{U}_{12}$, and Σ' denotes the covariance matrix without scaling. The speed of convergence of the variance matrix of the coefficient $\widehat{\mathbf{A}}$ is $T/2$ and that of the frequency $\widehat{\lambda}$ is $(2\pi)^2 T^3/6$.

6 Chandler Wobble

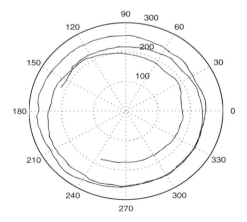

Fig. 1. Centralized wobbling motion in polar coordinates

Data Description: The Chandler wobble, named after its 1891 discoverer, Seth Carlo Chandler, Jr., is one of several wobbling motions exhibited by the Earth as it rotates on its axis, much as a top wobbles as it spins. The period of this wobbling is 430 to 435 days. It has been estimated by several workers, for instance, Brillinger [Bri73] and Arató et al. [AKS62]. Some properties of the monthly data have been shown in [IT97].

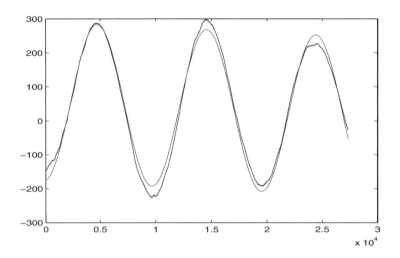

Fig. 2. x-values and the fitted ones in mas versus hours.

Since 1995, a combined solution to the various GPS (Global Positioning System) series has been performed. Here the hourly measurements between MJD 49719 (JAN 1, '95) and 50859 (FEB 15, '98) (MJD is for Modified Julian Day), are used in our current analyses. The values of the data are given in mas = milli-arcseconds, 1 arcsec $\sim$ 30 m. The number of data points is $T = 27,361$.

Rotational variations of polar motion are due to the superposition of the influences of six partial tides. Different techniques suggest that these are real oscillations of polar motion. Rapid oscillations with periods of 12 h have already been considered; see IVS 2004 General Meeting Proceedings [Gro00].

The aim of this investigation is to give some statistical evidence of the presence of 12 h oscillation, in other words to show that the frequency $2\pi/12$ has significantly non-zero weight. It is also a question whether there is any significant shift in the position of the center.

The model [SNS01] to be fitted is a linear trend with harmonic components,

$$\underline{Y}_t = \mathbf{B}_{2\times 2} \begin{bmatrix} 1 \\ t \end{bmatrix} + \mathbf{A}_{4\times 2} \begin{bmatrix} \cos(2\pi t\lambda_1) \\ \sin(2\pi t\lambda_1) \\ \cos(2\pi t\lambda_2) \\ \sin(2\pi t\lambda_2) \end{bmatrix} + \underline{Z}_t \,,$$

where $\underline{Y}_t$ is the measurement and the matrices $\mathbf{A}$ and $\mathbf{B}$ together with the frequencies λ_i, ($|\lambda_i| \leq \pi$) are unknown parameters. So we are faced with a non-linear regression problem.

6.1 Chandler Wobble, Results

We started the computation with the initial values $\lambda_1 = 2\pi/410/24$ and $\lambda_2 = 2\pi/12$, and the number of Fourier frequencies was 2^{13}. The estimated parameters are

$$\widehat{\mathbf{B}} = \begin{bmatrix} 41.6043 & 0.0003 \\ 323.4485 & -0.0007 \end{bmatrix} ,$$

$$\widehat{\mathbf{A}} = \begin{bmatrix} -244.8065 & 16.5279 & 0.1248 & -0.0521 \\ 25.3854 & 256.5682 & 0.0166 & 0.1064 \end{bmatrix} ,$$

and

$$\widehat{\underline{\lambda}} = \begin{bmatrix} 0.0001 \\ 0.0833 \end{bmatrix} .$$

The estimated frequencies correspond to the periods 410.5626 days and 11.9999 hours. Analyzing the residual series $\underline{Z}_t$ we found ourselves in the situation of long-range dependent data.

6.2 Disturbance with Long Memory

Let $\underline{Z}_t$ be a stationary time series with piecewise continuous spectral density

$$\mathbf{S}_Z(\omega) = \Lambda(\omega)\, \mathbf{S}_2^\sharp(\omega)\, \Lambda^*(\omega) ,$$

where

$$\Lambda(\omega) = \mathrm{diag}([1 - e^{2i\pi\omega}]^{-h_1}, [1 - e^{2i\pi\omega}]^{-h_2}, \ldots, [1 - e^{2i\pi\omega}]^{-h_d}) ,$$

$h_k \in [0, 1/2)$, $k = 1, 2, \ldots, d$, and the matrix $\mathbf{S}_2^\sharp(\omega)$ is a positive continuous spectral density matrix (we often have in mind a stationary, physically realizable, vector-ARIMA time series). The Hurst exponents $(h_1, h_2, \ldots, h_d)$ are not necessarily different; denote them $\underline{h} = (h_1, h_2, \ldots, h_d)$. Yajima [Yaj91] shows that for each fixed h_k the regressors are classified according to the discontinuity of their spectrum at zero. In our case it concerns the linear part only. We introduce the third-rate scaling according to the long memory. Let $\mathbf{D}_{L,T} = \mathrm{diag}(T^{h_k}, k = 1, 2, \ldots, d)$ be the diagonal matrix; then

$$\Sigma_{11} = \mathbf{D}_{\underline{X}_1,T}\, \mathrm{d}\mathbf{F}_{11}(0)\mathbf{D}_{\underline{X}_1,T} \otimes \mathbf{D}_{L,T}\mathbf{S}_{\underline{Z}}^{-1}(0)\mathbf{D}_{L,T} .$$

Robinson and Hidalgo [RH97, Theorem 5] establish that the weights $\mathbf{S}_{\underline{Z}}^{-1}$ are consistently estimated via recursion even if the data are long-range dependent.

The technique of estimation we follow is a multiple recursion. First we put a constant for $\mathbf{S}_Z$, and some initial value for the nonlinear parameter ϑ. We keep the initial value of $\underline{\vartheta}$ fixed and iterate for the linear parameters updating the estimated residual spectrum $\mathbf{S}_Z$ step by step. We then fix the estimated linear parameter and find the estimate of the non-linear parameter $\underline{\vartheta}$ through a weighted nonlinear least squares procedure; meanwhile the estimated residual spectrum $\mathbf{S}_Z$ is updated; then we iterate the whole process.

6.3 Conclusions

- The estimation of the Hurst parameter is calculated by the method of [IT03], based on the behavior of the cumulants up to order 5. Both are very close to $1/2$; that is, $h_1 = 0.4986$ and $h_2 = 0.4860$. This might be the reason that other procedures failed, including the quasi-maximum likelihood estimates of Lobato [Lob97], [Lob99], for vector-valued time series. Therefore we proceed with the marginal Hurst parameters.
- As is expected, there is no real information on the location parameter (the constant in the model) because the estimated variances of the parameters and b_{21} are so large. Some improvement can be reached by Dahlhaus's [Dah95] result. The diagonals of the variance matrix of Vec $\widehat{\mathbf{B}}$ are $[1.1733 * 10^6, 0.7725 * 10^6, 0.2097, 0.1381]$.
- The standard deviation for the parameters b_{12} and b_{22} are 0.4579 and 0.3716; hence there is no evidence of the shifting of either coordinate, at least with larger than 95% confidence.
- Actually, we have only "two observations" of the period ~ 410 days, therefore it is not surprising that the standard deviation of the parameters $a_{1:2,1:2}$ again, are large, showing no information on the values. More precisely, the standard deviations are $[397.1890, 481.8903, 436.7575, 442.9037]$.
- The quantity of main interest is the standard deviation of the parameters λ_2 and $a_{1:2,3:4}$. The standard deviation of the estimates $\widehat{a}_{1:2,3:4}$ is $[0.0154, 0.0218, 0.0233, 0.0146]$ so we conclude that all of them are non-zero except $a_{1,3}$, at least with probability .95. There is some empirical evidence for fitting a model with an additional frequency $\lambda_3 = 30$, $\lambda_4 = 2\lambda_2$. The latter one raises some special problems, known from biology; see [Bro90].

Acknowledgement. The Chandler wobble data with high resolution are through the courtesy of Professor József Závoti.
This research was partially supported by the Hungarian NSF, OTKA No. T047067.

7 Appendix

7.1 Some Matrix Relations

$$\text{Vec}(\underline{a}\underline{b}^{\mathsf{T}}) = \underline{b} \otimes \underline{a} , \tag{15}$$
$$\underline{a}^{\mathsf{T}} \otimes \underline{b} = \underline{b}\underline{a}^{\mathsf{T}} = \underline{b} \otimes \underline{a}^{\mathsf{T}}. \tag{16}$$

see [MN99, p. 28].

$$(\operatorname{Vec} \mathbf{A})^{\mathsf{T}} \operatorname{Vec} \mathbf{B} = \operatorname{Tr}(\mathbf{A}^{\mathsf{T}} \mathbf{B}) \,, \tag{17}$$

see [MN99, p. 30]. The vectors $\underline{a}$, $\underline{b}$, and $\underline{c}$ fulfill

$$(\underline{a} \otimes \underline{b}) \, \underline{c}^{\mathsf{T}} = \underline{a} \otimes (\underline{b}\underline{c}^{\mathsf{T}}) \tag{18}$$

$$(\underline{a} \otimes \underline{b}) \, \underline{c}^{\mathsf{T}} = (\underline{a}\underline{c}^{\mathsf{T}}) \otimes \underline{b} \,, \tag{19}$$

The commutation matrix $\mathbf{K}_{m \cdot n}$ is defined by the relation

$$\mathbf{K}_{m \cdot n} \operatorname{Vec} \mathbf{A} = \operatorname{Vec} \mathbf{A}^{\mathsf{T}} \,, \tag{20}$$

for any matrix A with dimension $m \times n$. The next identity is

$$(\underline{a} \otimes \underline{b}) \, \underline{c}^{\mathsf{T}} = \mathbf{K}_{d \cdot d} \, (\underline{b} \otimes \underline{a}) \, \underline{c}^{\mathsf{T}} \,. \tag{21}$$

We have (see [MN99, p. 47]): if A is $m \times n$ and B is $p \times q$ then

$$\operatorname{Vec} (\mathbf{A} \otimes \mathbf{B}) = (\mathbf{U}_n \otimes \mathbf{K}_{q \cdot m} \otimes \mathbf{U}_p) \, (\operatorname{Vec} \mathbf{A} \otimes \operatorname{Vec} \mathbf{B}) \tag{22}$$

$$\mathbf{K}_{p \cdot m} \, (\mathbf{A} \otimes \mathbf{B}) \, \mathbf{K}_{n \cdot q} = \mathbf{B} \otimes \mathbf{A} \,. \tag{23}$$

One can prove the following identity

$$\mathbf{A} \mathbf{B} \underline{b} \underline{a}^{\mathsf{T}} = [\mathbf{U} \otimes (\operatorname{Vec} \mathbf{B})^{\mathsf{T}}] \, (\mathbf{K}_{n \cdot d} \otimes \mathbf{U}_m) \, \mathbf{K}_{d \cdot mn} \, [\underline{b} \underline{a}^{\mathsf{T}} \otimes \operatorname{Vec} \mathbf{A}^{\mathsf{T}}] \,, \tag{24}$$

where the only assumption for the matrices $\mathbf{A}$ and $\mathbf{B}$, vectors $\underline{b}$ and $\underline{a}$, is that the matrix product on the left-hand side should be valid; and $\mathbf{U}$ is the identity matrix with appropriate order. We also have the following

$$\mathbf{K}_{dp \cdot d} \, (\mathbf{K}_{d \cdot p} \otimes \mathbf{U}_d) = \mathbf{U}_p \otimes \mathbf{K}_{d \cdot d} \,. \tag{25}$$

7.2 Jacobian and Hessian of SDFR, Proofs

Consider the Jacobian $\partial \underline{X}_t (\underline{\vartheta}) / \partial \underline{\vartheta}^{\mathsf{T}}$ of the regressor $\underline{X}_t (\underline{\vartheta})$. Then the Jacobian of $\widehat{\mathbf{C}}_{X,T} (h, \underline{\vartheta})$ is

$$\frac{\partial \widehat{\mathbf{C}}_{X,T} (h, \underline{\vartheta})}{\partial \underline{\vartheta}^{\mathsf{T}}} = \frac{1}{T} \sum_{t=1}^{T-h} \left[\underline{X}_t (\underline{\vartheta}) \otimes \frac{\partial \underline{X}_{t+h} (\underline{\vartheta})}{\partial \underline{\vartheta}^{\mathsf{T}}} + \frac{\partial \underline{X}_t (\underline{\vartheta})}{\partial \underline{\vartheta}^{\mathsf{T}}} \otimes \underline{X}_{t+h} (\underline{\vartheta}) \right]$$

$$= \frac{\partial \widehat{\mathbf{C}}_X (h, \underline{\vartheta}_1, \underline{\vartheta}_2)}{\partial \underline{\vartheta}_1^{\mathsf{T}}} + \left. \frac{\partial \widehat{\mathbf{C}}_X (h, \underline{\vartheta}_1, \underline{\vartheta}_2)}{\partial \underline{\vartheta}_2^{\mathsf{T}}} \right|_{\underline{\vartheta}_1 = \underline{\vartheta}_2 = \underline{\vartheta}} \,;$$

see (18), (19), and (16). Now take the limit of $\partial \widehat{\mathbf{C}}_{X,T} (h, \underline{\vartheta}) / \partial \underline{\vartheta}^{\mathsf{T}}$ and define the Jacobian $\partial \mathbf{F} (\lambda, \underline{\vartheta}) / \partial \underline{\vartheta}^{\mathsf{T}}$ for SDFR $\mathbf{F}$ by

$$\frac{\partial \mathbf{C}_X (h, \underline{\vartheta})}{\partial \underline{\vartheta}^{\mathsf{T}}} = \int_{-1/2}^{1/2} \exp (i 2\pi \lambda h) \, \mathrm{d} \frac{\partial \mathbf{F} (\lambda, \underline{\vartheta})}{\partial \underline{\vartheta}^{\mathsf{T}}} \,;$$

that is, the $\partial \mathbf{C}_X(h, \underline{\vartheta})/\partial \underline{\vartheta}^{\mathsf{T}}$ is the inverse Fourier transform of $\partial \mathbf{F}(\lambda, \underline{\vartheta})/\partial \underline{\vartheta}^{\mathsf{T}}$. If the limit of the Jacobian

$$\frac{\partial \mathbf{F}_T(\omega, \underline{\vartheta})}{\partial \underline{\vartheta}^{\mathsf{T}}} = \int_0^\omega \frac{\partial \mathbf{I}_{X,T}(\lambda, \underline{\vartheta})}{\partial \underline{\vartheta}^{\mathsf{T}}} \, d\lambda \,,$$

exists and the differential operator and the limit are exchangeable then we have

$$\lim_{T \to \infty} \frac{\partial \mathbf{F}_T(\omega, \underline{\vartheta})}{\partial \underline{\vartheta}^{\mathsf{T}}} = \frac{\partial}{\partial \underline{\vartheta}^{\mathsf{T}}} \lim_{T \to \infty} \mathbf{F}_T(\omega, \underline{\vartheta}) = \frac{\partial \mathbf{F}(\omega, \underline{\vartheta})}{\partial \underline{\vartheta}^{\mathsf{T}}} \,.$$

This is not always the case. Notice

$$\begin{aligned}
\mathbf{F}_T(\omega, \underline{\vartheta}) &= \int_0^\omega \mathbf{I}_{X,T}(\lambda, \underline{\vartheta}) \, d\lambda \\
&= \int_0^\omega \frac{1}{T} \underline{d}_{X,T}(\lambda, \underline{\vartheta}_1) \underline{d}_{X,T}^*(\lambda, \underline{\vartheta}_2) \, d\lambda \Big|_{\underline{\vartheta}_1 = \underline{\vartheta}_2 = \underline{\vartheta}} \\
&= \mathbf{F}_T(\omega, \underline{\vartheta}_1, \underline{\vartheta}_2)|_{\underline{\vartheta}_1 = \underline{\vartheta}_2 = \underline{\vartheta}} \,,
\end{aligned}$$

therefore

$$\frac{\partial \mathbf{F}(\omega, \underline{\vartheta})}{\partial \underline{\vartheta}^{\mathsf{T}}} = \frac{\partial \mathbf{F}(\omega, \underline{\vartheta}_1, \underline{\vartheta}_2)}{\partial \underline{\vartheta}_1^{\mathsf{T}}} + \frac{\partial \mathbf{F}(\omega, \underline{\vartheta}_1, \underline{\vartheta}_2)}{\partial \underline{\vartheta}_2^{\mathsf{T}}}\Big|_{\underline{\vartheta}_1 = \underline{\vartheta}_2 = \underline{\vartheta}} \,.$$

This corresponds to the Jacobian

$$\frac{\partial \mathbf{C}_X(h, \underline{\vartheta})}{\partial \underline{\vartheta}^{\mathsf{T}}} = \frac{\partial \mathbf{C}_X(h, \underline{\vartheta}_1, \underline{\vartheta}_2)}{\partial \underline{\vartheta}_1^{\mathsf{T}}} + \frac{\partial \mathbf{C}_X(h, \underline{\vartheta}_1, \underline{\vartheta}_2)}{\partial \underline{\vartheta}_2^{\mathsf{T}}}\Big|_{\underline{\vartheta}_1 = \underline{\vartheta}_2 = \underline{\vartheta}} \,.$$

The Hessian $\mathsf{H}\mathbf{F}$ of $\mathbf{F}(\lambda, \underline{\vartheta})$ is defined similarly; first the Hessian of $\widehat{\mathbf{C}}_{X,T}(h, \underline{\vartheta})$,

$$\mathsf{H}\widehat{\mathbf{C}}_{X,T}(h, \underline{\vartheta})$$

$$= \frac{1}{T} \sum_{t=1}^{T-h} \frac{\partial \operatorname{Vec}\left[\underline{X}_t(\underline{\vartheta}) \otimes \frac{\partial \underline{X}_{t+h}(\underline{\vartheta})}{\partial \underline{\vartheta}^{\mathsf{T}}} + \frac{\partial \underline{X}_t(\underline{\vartheta})}{\partial \underline{\vartheta}^{\mathsf{T}}} \otimes \underline{X}_{t+h}(\underline{\vartheta})\right]}{\partial \underline{\vartheta}^{\mathsf{T}}}$$

$$= \frac{1}{T} \sum_{t=1}^{T-h} \frac{\partial \left(\mathbf{K}_{p \cdot d} \otimes \mathbf{U}_d\right) \left(\underline{X}_t(\underline{\vartheta}) \otimes \operatorname{Vec} \frac{\partial \underline{X}_{t+h}(\underline{\vartheta})}{\partial \underline{\vartheta}^{\mathsf{T}}}\right)}{\partial \underline{\vartheta}^{\mathsf{T}}}$$

$$+ \frac{\partial \left(\operatorname{Vec} \frac{\partial \underline{X}_t(\underline{\vartheta})}{\partial \underline{\vartheta}^{\mathsf{T}}} \otimes \underline{X}_{t+h}(\underline{\vartheta})\right)}{\partial \underline{\vartheta}^{\mathsf{T}}}$$

$$= \frac{1}{T} \sum_{t=1}^{T-h} \left(\mathbf{K}_{p \cdot d} \otimes \mathbf{U}_d\right) \left[\frac{\partial \underline{X}_t(\underline{\vartheta})}{\partial \underline{\vartheta}^{\mathsf{T}}} \otimes \operatorname{Vec} \frac{\partial \underline{X}_{t+h}(\underline{\vartheta})}{\partial \underline{\vartheta}^{\mathsf{T}}} + \underline{X}_t(\underline{\vartheta}) \otimes \mathsf{H}\underline{X}_{t+h}(\underline{\vartheta})\right]$$

$$+ \left[\operatorname{Vec} \frac{\partial \underline{X}_t(\underline{\vartheta})}{\partial \underline{\vartheta}^{\mathsf{T}}} \otimes \frac{\partial \underline{X}_{t+h}(\underline{\vartheta})}{\partial \underline{\vartheta}^{\mathsf{T}}} + \mathsf{H}\underline{X}_t(\underline{\vartheta}) \otimes \underline{X}_{t+h}(\underline{\vartheta})\right] \,.$$

Notice

$$\left(\mathbf{K}_{p\cdot d}\otimes\mathbf{U}_d\right)\left[\frac{\partial \underline{X}_t\left(\vartheta\right)}{\partial\underline{\vartheta}^{\mathsf{T}}}\otimes\operatorname{Vec}\frac{\partial \underline{X}_{t+h}\left(\vartheta\right)}{\partial\underline{\vartheta}^{\mathsf{T}}}\right]$$

$$=\left(\mathbf{K}_{p\cdot d}\otimes\mathbf{U}_d\right)\mathbf{K}_{d\cdot dp}\left[\operatorname{Vec}\frac{\partial \underline{X}_{t+h}\left(\vartheta\right)}{\partial\underline{\vartheta}^{\mathsf{T}}}\otimes\frac{\partial \underline{X}_t\left(\vartheta\right)}{\partial\underline{\vartheta}^{\mathsf{T}}}\right],$$

$$\left(\mathbf{K}_{p\cdot d}\otimes\mathbf{U}_d\right)\left[\underline{X}_t\left(\vartheta\right)\otimes\mathsf{H}\underline{X}_{t+h}\left(\vartheta\right)\right]$$

$$=\left(\mathbf{K}_{p\cdot d}\otimes\mathbf{U}_d\right)\mathbf{K}_{d\cdot dp}\left[\mathsf{H}\underline{X}_{t+h}\left(\vartheta\right)\otimes\underline{X}_t\left(\vartheta\right)\right];$$

see (23). Let us denote the limit of $\mathsf{H}\widehat{\mathbf{C}}_{\underline{X},T}\left(h,\underline{\vartheta}\right)$ by $\mathsf{H}\mathbf{C}_{\underline{X}}\left(h,\underline{\vartheta}\right)$, and its inverse Fourier transform by $\mathsf{H}\mathbf{F}\left(\lambda,\underline{\vartheta}\right)$; that is,

$$\mathsf{H}\mathbf{C}_{\underline{X}}(h,\underline{\vartheta})=\int_{-1/2}^{1/2}\mathrm{e}^{2i\pi\lambda h}\,\mathrm{d}\mathsf{H}\mathbf{F}(\lambda,\underline{\vartheta})\,.$$

Similarly to the above

$$\mathsf{H}\mathbf{C}_{\underline{X}}(h,\underline{\vartheta})$$

$$=\left(\mathbf{K}_{p\cdot d}\otimes\mathbf{U}_d\right)\mathbf{K}_{d\cdot dp}\left[\mathsf{H}_{\underline{\vartheta}_1}\mathbf{C}_{\underline{X}}\left(-h,\underline{\vartheta}_2,\underline{\vartheta}_1\right)+\frac{\partial^2\mathbf{C}_{\underline{X}}\left(-h,\underline{\vartheta}_2,\underline{\vartheta}_1\right)}{\partial\underline{\vartheta}_1^{\mathsf{T}}\partial\underline{\vartheta}_2^{\mathsf{T}}}\right]\Bigg|_{\underline{\vartheta}_1=\underline{\vartheta}_2=\underline{\vartheta}}$$

$$+\,\mathsf{H}_{\underline{\vartheta}_2}\mathbf{C}_{\underline{X}}(h,\underline{\vartheta}_1,\underline{\vartheta}_2)+\frac{\partial^2\mathbf{C}_{\underline{X}}(h,\underline{\vartheta}_1,\underline{\vartheta}_2)}{\partial\underline{\vartheta}_2^{\mathsf{T}}\partial\underline{\vartheta}_1^{\mathsf{T}}}\Bigg|_{\underline{\vartheta}_1=\underline{\vartheta}_2=\underline{\vartheta}};$$

where we used the shorthand notation

$$\frac{\partial^2\mathbf{C}_{\underline{X}}\left(h,\underline{\vartheta}_1,\underline{\vartheta}_2\right)}{\partial\underline{\vartheta}_2^{\mathsf{T}}\partial\underline{\vartheta}_1^{\mathsf{T}}}=\frac{\partial}{\partial\underline{\vartheta}_1^{\mathsf{T}}}\operatorname{Vec}\left(\frac{\partial\operatorname{Vec}\mathbf{C}_{\underline{X}}\left(h,\underline{\vartheta}_1,\underline{\vartheta}_2\right)}{\partial\underline{\vartheta}_2^{\mathsf{T}}}\right);$$

here the partial derivative of the right side can be carried out directly. (The order of the variables $\underline{\vartheta}_1,\underline{\vartheta}_2$ is opposite to the order of the derivatives; $\partial\underline{\vartheta}_2^{\mathsf{T}}\partial\underline{\vartheta}_1^{\mathsf{T}}$ means differentiating first by $\underline{\vartheta}_2$ then by $\underline{\vartheta}_1$, that is, the operator acting on right-hand side.) Starting with $\underline{\vartheta}_1$ then followed by $\underline{\vartheta}_2$ is indirect, because

$$\frac{\partial^2\mathbf{C}_{\underline{X}}\left(h,\underline{\vartheta}_1,\underline{\vartheta}_2\right)}{\partial\underline{\vartheta}_1^{\mathsf{T}}\partial\underline{\vartheta}_2^{\mathsf{T}}}=\left(\mathbf{K}_{p\cdot d}\otimes\mathbf{U}_d\right)\mathbf{K}_{d\cdot dp}\frac{\partial^2\mathbf{C}_{\underline{X}}\left(-h,\underline{\vartheta}_2,\underline{\vartheta}_1\right)}{\partial\underline{\vartheta}_1^{\mathsf{T}}\partial\underline{\vartheta}_2^{\mathsf{T}}};$$

for the reason behind this see (22). Note that

$$\mathbf{C}_{\underline{X}}\left(-h,\underline{\vartheta}_2,\underline{\vartheta}_1\right)=\mathbf{C}_{\underline{X}}^{\mathsf{T}}\left(h,\underline{\vartheta}_1,\underline{\vartheta}_2\right)\,.$$

Similarly the Hessian with respect to $\underline{\vartheta}_2$ is direct and with respect to $\underline{\vartheta}_1$ is indirect; that is,

$$\mathsf{H}_{\underline{\vartheta}_1}\mathbf{C}_{\underline{X}}\left(h,\underline{\vartheta}_1,\underline{\vartheta}_2\right)=\left(\mathbf{K}_{p\cdot d}\otimes\mathbf{U}_d\right)\mathbf{K}_{d\cdot dp}\mathsf{H}_{\underline{\vartheta}_1}\mathbf{C}_{\underline{X}}\left(-h,\underline{\vartheta}_2,\underline{\vartheta}_1\right)\,.$$

According to the above notations we write

$$\mathsf{HF}\left(\omega,\underline{\vartheta}\right) = \left[\mathsf{H}_{\underline{\vartheta}_1}\,\mathbf{F}\left(\omega,\underline{\vartheta}_1,\underline{\vartheta}_2\right) + \frac{\partial^2\mathbf{F}\left(\omega,\underline{\vartheta}_1,\underline{\vartheta}_2\right)}{\partial\underline{\vartheta}_1^\mathsf{T}\partial\underline{\vartheta}_2^\mathsf{T}}\right.$$
$$\left.+\ \mathsf{H}_{\underline{\vartheta}_2}\,\mathbf{F}\left(\omega,\underline{\vartheta}_1,\underline{\vartheta}_2\right) + \frac{\partial^2\mathbf{F}\left(\omega,\underline{\vartheta}_1,\underline{\vartheta}_2\right)}{\partial\underline{\vartheta}_2^\mathsf{T}\partial\underline{\vartheta}_1^\mathsf{T}}\right]\Bigg|_{\underline{\vartheta}_1=\underline{\vartheta}_2=\underline{\vartheta}}.$$

Again here, for instance

$$\frac{\partial^2\mathbf{F}\left(\omega,\underline{\vartheta}_1,\underline{\vartheta}_2\right)}{\partial\underline{\vartheta}_1^\mathsf{T}\partial\underline{\vartheta}_2^\mathsf{T}} = \left(\mathbf{K}_{p\cdot d}\otimes\mathbf{U}_d\right)\mathbf{K}_{d\cdot dp}\frac{\partial^2\mathbf{F}\left(-\omega,\underline{\vartheta}_2,\underline{\vartheta}_1\right)}{\partial\underline{\vartheta}_1^\mathsf{T}\partial\underline{\vartheta}_2^\mathsf{T}}.$$

7.3 Variance of the Derivative

The summands in

$$T\,\mathrm{Vec}\,\frac{\partial Q_T\left(\underline{\vartheta}\right)}{\partial\underline{\vartheta}^\mathsf{T}}$$
$$= \sum_{k=-T_1}^{T_1}\left[\frac{\partial\,\mathrm{Vec}\,\mathbf{I}_{X,T}\left(\omega_k,\underline{\vartheta}\right)}{\partial\underline{\vartheta}^\mathsf{T}} - \frac{\partial\left(\mathrm{Vec}\,\mathbf{I}_{Y,X,T}\left(\omega_k,\underline{\vartheta}\right) + \mathrm{Vec}\,\mathbf{I}_{X,Y,T}\left(\omega_k,\underline{\vartheta}\right)\right)}{\partial\underline{\vartheta}^\mathsf{T}}\right]^\mathsf{T}$$
$$\times\left[\mathrm{Vec}\,\boldsymbol{\Phi}^\mathsf{T}(\omega_k)\right]$$

are asymptotically independent. Therefore first we are interested in the variance separately. Notice that

$$\left[\frac{\partial\left(\mathrm{Vec}\,\mathbf{I}_{\underline{Z},\underline{X},T}\left(\omega_k,\underline{\vartheta}\right) + \mathrm{Vec}\,\mathbf{I}_{\underline{X},\underline{Z},T}\left(\omega_k,\underline{\vartheta}\right)\right)}{\partial\underline{\vartheta}^\mathsf{T}}\right]^\mathsf{T}\mathrm{Vec}\,\boldsymbol{\Phi}^\mathsf{T}(\omega_k)$$
$$= 2\left[\frac{\partial\,\mathrm{Vec}\,\mathbf{I}_{\underline{Z},\underline{X},T}\left(\omega_k,\underline{\vartheta}\right)}{\partial\underline{\vartheta}^\mathsf{T}}\right]^\mathsf{T}\mathrm{Vec}\,\boldsymbol{\Phi}^\mathsf{T}(\omega_k).$$

Indeed

$$\left[\frac{\partial\,\mathrm{Vec}\,\mathbf{I}_{\underline{Z},\underline{X},T}\left(\omega_k,\underline{\vartheta}\right)}{\partial\underline{\vartheta}^\mathsf{T}}\right]^\mathsf{T}\left[\mathrm{Vec}\,\boldsymbol{\Phi}^\mathsf{T}(\omega_k)\right] = \mathrm{Vec}\,\frac{\partial\,\mathrm{Tr}\,\boldsymbol{\Phi}^\mathsf{T}(\omega_k)\,\mathbf{I}_{\underline{X},\underline{Z},T}^\mathsf{T}\left(\omega_k,\underline{\vartheta}\right)}{\partial\underline{\vartheta}^\mathsf{T}}$$
$$= \left[\frac{\partial\,\mathrm{Vec}\,\mathbf{I}_{\underline{X},\underline{Z},T}\left(\omega_k,\underline{\vartheta}\right)}{\partial\underline{\vartheta}^\mathsf{T}}\right]^\mathsf{T}\mathrm{Vec}\,\boldsymbol{\Phi}^\mathsf{T}(\omega_k),$$

and

$$T\sum_{k=-T_1}^{T_1}\frac{\partial\,\mathrm{Vec}\,\mathbf{I}_{\underline{Z},\underline{X},T}(\omega_k,\underline{\vartheta})}{\partial\underline{\vartheta}^\mathsf{T}}\,\mathrm{Vec}\,\boldsymbol{\Phi}^\mathsf{T}(\omega_k)$$
$$= \sum_{k=-T_1}^{T_1}\left[\left(\mathrm{Vec}\,\frac{\partial\overline{d_{\underline{X},T}(\omega_k,\underline{\vartheta})}}{\partial\underline{\vartheta}^\mathsf{T}}\right)\otimes\underline{d}_{\underline{Z},T}(\omega_k)\right]\mathrm{Vec}\,\boldsymbol{\Phi}^\mathsf{T}(\omega_k),$$

therefore we consider the variance matrix of the complex variable; see [Bri01, p. 89].

$$\text{Var}\left(\left[\frac{\partial\,\text{Vec}\,\mathbf{I}_{\underline{X},\underline{Z},T}\,(\omega_k,\underline{\vartheta})}{\partial\underline{\vartheta}^{\mathsf{T}}}\right]^{\mathsf{T}}[\text{Vec}\,\mathbf{\Phi}^{\mathsf{T}}\,(\omega_k)]\right)$$

$$=\frac{1}{T^2}\,\text{Var}\left(\left[\frac{\partial\underline{d}_{\underline{X},T}\,(\omega_k,\underline{\vartheta})}{\partial\underline{\vartheta}^{\mathsf{T}}}\right]^{\mathsf{T}}\mathbf{\Phi}^{\mathsf{T}}\,(\omega_k)\,\overline{\underline{d}_{\underline{Z},T}\,(\omega_k)}\right)$$

$$=\frac{1}{T}\left[\frac{\partial\underline{d}_{\underline{X},T}\,(\omega_k,\underline{\vartheta})}{\partial\underline{\vartheta}^{\mathsf{T}}}\right]^{\mathsf{T}}\mathbf{\Phi}^{\mathsf{T}}\,(\omega_k)\,\mathbf{S}_{\underline{Z}}^{\mathsf{T}}\,(\omega_k)\,\mathbf{\Phi}^{\mathsf{T}}\,(\omega_k)\left[\frac{\partial\underline{d}_{\underline{X},T}\,(\omega_k,\underline{\vartheta})}{\partial\underline{\vartheta}^{\mathsf{T}}}\right]+o\,(1)\ .$$

Because of (24), this limit is written

$$\left[\frac{\partial\underline{d}_{\underline{X},T}\,(\omega_k,\underline{\vartheta})}{\partial\underline{\vartheta}_1^{\mathsf{T}}}\right]^{\mathsf{T}}\mathbf{\Phi}^{\mathsf{T}}\,(\omega_k)\,\mathbf{S}_{\underline{Z}}^{\mathsf{T}}\,(\omega_k)\,\mathbf{\Phi}^{\mathsf{T}}\,(\omega_k)\left[\frac{\partial\overline{\underline{d}_{\underline{X},T}\,(\omega_k,\underline{\vartheta})}}{\partial\underline{\vartheta}_2^{\mathsf{T}}}\right]$$

$$=T\left(\mathbf{U}_p\otimes[\text{Vec}(\mathbf{\Phi}^{\mathsf{T}}(\omega_k)\mathbf{S}_{\underline{Z}}^{\mathsf{T}}(\omega_k)\mathbf{\Phi}^{\mathsf{T}}(\omega_k))]^{\mathsf{T}}\right)$$

$$\times\left(\frac{\partial\overline{\underline{d}_{\underline{X},T}\,(\omega_k,\underline{\vartheta}_2)}}{\partial\underline{\vartheta}_2^{\mathsf{T}}}\otimes\text{Vec}\,\frac{\partial\underline{d}_{\underline{X},T}\,(\omega_k,\underline{\vartheta}_1)}{\partial\underline{\vartheta}_1^{\mathsf{T}}}\right)\ .$$

The variance matrix of the derivative $(\partial Q_T(\underline{\vartheta}_0))/\partial\underline{\vartheta}^{\mathsf{T}}$ has the limit

$$\lim_{T\to\infty}\,\text{Var}\left[\sqrt{T}\,\text{Vec}\,\frac{\partial Q_T(\underline{\vartheta}_0)}{\partial\underline{\vartheta}^{\mathsf{T}}}\right]$$

$$=\int_{-1/2}^{1/2}\left(\mathbf{U}_p\otimes\left[\text{Vec}(\mathbf{\Phi}^{\mathsf{T}}(\omega)\mathbf{S}_{\underline{Z}}^{\mathsf{T}}(\omega)\mathbf{\Phi}^{\mathsf{T}}(\omega))\right]^{\mathsf{T}}\right)$$

$$\times\text{d}\left(\frac{\partial^2\mathbf{F}\,(\omega,\underline{\vartheta}_1,\underline{\vartheta}_2)}{\partial\underline{\vartheta}_2^{\mathsf{T}}\partial\underline{\vartheta}_1^{\mathsf{T}}}+\frac{\partial^2\mathbf{F}\,(\omega,\underline{\vartheta}_1,\underline{\vartheta}_2)}{\partial\underline{\vartheta}_1^{\mathsf{T}}\partial\underline{\vartheta}_2^{\mathsf{T}}}\bigg|_{\underline{\vartheta}_1=\underline{\vartheta}_2=\underline{\vartheta}_0}\right)\ .$$

It is worth noticing that

$$\left(\mathbf{U}_p\otimes[\text{Vec}(\mathbf{\Phi}^{\mathsf{T}}(\omega)\mathbf{S}_{\underline{Z}}^{\mathsf{T}}(\omega)\mathbf{\Phi}^{\mathsf{T}}(\omega))]^{\mathsf{T}}\right)(\mathbf{K}_{p\cdot d}\otimes\mathbf{U}_d)\mathbf{K}_{d\cdot dp}$$

$$=\left(\mathbf{U}_p\otimes[\text{Vec}(\mathbf{\Phi}(\omega)\mathbf{S}_{\underline{Z}}(\omega)\mathbf{\Phi}(\omega))]^{\mathsf{T}}\right)\ ,\quad(26)$$

and

$$\frac{\partial^2\mathbf{F}\,(\omega,\underline{\vartheta}_1,\underline{\vartheta}_2)}{\partial\underline{\vartheta}_1^{\mathsf{T}}\partial\underline{\vartheta}_2^{\mathsf{T}}}=(\mathbf{K}_{p\cdot d}\otimes\mathbf{U}_d)\,\mathbf{K}_{d\cdot dp}\,\frac{\partial^2\mathbf{F}\,(-\omega,\underline{\vartheta}_2,\underline{\vartheta}_1)}{\partial\underline{\vartheta}_1^{\mathsf{T}}\partial\underline{\vartheta}_2^{\mathsf{T}}}\ ;$$

hence the asymptotic variance is written

$$\int_{-1/2}^{1/2} \left(\mathbf{U}_p \otimes [\mathrm{Vec}(\mathbf{\Phi}^\mathsf{T}(\omega_k)\mathbf{S}_{\underline{Z}}^\mathsf{T}(\omega_k)\mathbf{\Phi}^\mathsf{T}(\omega_k))]^\mathsf{T} \right)$$

$$\times\, \mathrm{d} \left(\frac{\partial^2 \mathbf{F}(\omega, \underline{\vartheta}_1, \underline{\vartheta}_2)}{\partial \underline{\vartheta}_2^\mathsf{T} \partial \underline{\vartheta}_1^\mathsf{T}} + \frac{\partial^2 \mathbf{F}(\omega, \underline{\vartheta}_1, \underline{\vartheta}_2)}{\partial \underline{\vartheta}_1^\mathsf{T} \partial \underline{\vartheta}_2^\mathsf{T}} \bigg|_{\underline{\vartheta}_1 = \underline{\vartheta}_2 = \underline{\vartheta}_0} \right)$$

$$= 2 \int_{-1/2}^{1/2} \left(\mathbf{U}_p \otimes [\mathrm{Vec}(\mathbf{\Phi}^\mathsf{T}(\omega_k)\mathbf{S}_{\underline{Z}}^\mathsf{T}(\omega_k)\mathbf{\Phi}^\mathsf{T}(\omega_k))]^\mathsf{T} \right)$$

$$\times\, \mathrm{d} \left(\frac{\partial^2 \mathbf{F}(\omega, \underline{\vartheta}_1, \underline{\vartheta}_2)}{\partial \underline{\vartheta}_2^\mathsf{T} \partial \underline{\vartheta}_1^\mathsf{T}} \bigg|_{\underline{\vartheta}_1 = \underline{\vartheta}_2 = \underline{\vartheta}_0} \right) .$$

7.4 Hessian of Q

We are interested in, for instance,

$$\frac{\partial}{\partial \underline{\vartheta}_2^\mathsf{T}} \left(\left[\frac{\partial \mathrm{Vec}\, \mathbf{I}_{X,T}(\omega_k, \underline{\vartheta}_1, \underline{\vartheta}_2)}{\partial \underline{\vartheta}_1^\mathsf{T}} \right]^\mathsf{T} \mathrm{Vec}\, \mathbf{\Phi}^\mathsf{T}(\omega_k) \right) \bigg|_{\underline{\vartheta}_1 = \underline{\vartheta}_2 = \underline{\vartheta}} .$$

Now, using the chain rule for the derivatives, we have

$$\frac{\partial}{\partial \underline{\vartheta}_2^\mathsf{T}} \mathrm{Vec} \left([\mathrm{Vec}\, \mathbf{\Phi}^\mathsf{T}(\omega_k)]^\mathsf{T} \left[\underline{d}_{X,T}(\omega_k, \underline{\vartheta}_2) \otimes \frac{\partial \underline{d}_{X,T}(\omega_k, \underline{\vartheta}_1)}{\partial \underline{\vartheta}_1^\mathsf{T}} \right] \right)$$

$$= (\mathbf{U}_p \otimes [\mathrm{Vec}\, \mathbf{\Phi}^\mathsf{T}(\omega_k)]^\mathsf{T}) \frac{\partial}{\partial \underline{\vartheta}_2^\mathsf{T}} \mathrm{Vec} \left(\underline{d}_{X,T}(\omega_k, \underline{\vartheta}_2) \otimes \frac{\partial \underline{d}_{X,T}(\omega_k, \underline{\vartheta}_1)}{\partial \underline{\vartheta}_1^\mathsf{T}} \right)$$

$$= (\mathbf{U}_p \otimes [\mathrm{Vec}\, \mathbf{\Phi}^\mathsf{T}(\omega_k)]^\mathsf{T}) (\mathbf{K}_{p \cdot d} \otimes \mathbf{U}_d)$$

$$\mathbf{K}_{d \cdot dp} \left(\mathrm{Vec}\, \frac{\partial \underline{d}_{X,T}(\omega_k, \underline{\vartheta}_1)}{\partial \underline{\vartheta}_1^\mathsf{T}} \otimes \frac{\partial \overline{\underline{d}_{X,T}}(\omega_k, \underline{\vartheta}_2)}{\partial \underline{\vartheta}_2^\mathsf{T}} \right)$$

$$= (\mathbf{U}_p \otimes [\mathrm{Vec}\, \mathbf{\Phi}(\omega_k)]^\mathsf{T}) \left(\mathrm{Vec}\, \frac{\partial \underline{d}_{X,T}(\omega_k, \underline{\vartheta}_1)}{\partial \underline{\vartheta}_1^\mathsf{T}} \otimes \frac{\partial \overline{\underline{d}_{X,T}}(\omega_k, \underline{\vartheta}_2)}{\partial \underline{\vartheta}_2^\mathsf{T}} \right)$$

$$= T\, (\mathbf{U}_p \otimes [\mathrm{Vec}\, \mathbf{\Phi}(\omega_k)]^\mathsf{T}) \frac{\partial^2 \mathbf{I}_{X,T}(-\omega, \underline{\vartheta}_2, \underline{\vartheta}_1)}{\partial \underline{\vartheta}_1^\mathsf{T} \partial \underline{\vartheta}_2^\mathsf{T}} ; \tag{27}$$

see (24) and (25). Similar steps lead to the mixed derivative $\partial^2 / \partial \underline{\vartheta}_2^\mathsf{T} \partial \underline{\vartheta}_1^\mathsf{T}$,

$$\frac{\partial}{\partial \underline{\vartheta}_1^\mathsf{T}} \mathrm{Vec} \left([\mathrm{Vec}\, \mathbf{\Phi}^\mathsf{T}(\omega_k)]^\mathsf{T} \left[\frac{\partial \overline{\underline{d}_{X,T}}(\omega_k, \underline{\vartheta}_2)}{\partial \underline{\vartheta}_2^\mathsf{T}} \otimes \underline{d}_{X,T}(\omega_k, \underline{\vartheta}_1) \right] \right)$$

$$= (\mathbf{U}_p \otimes [\mathrm{Vec}\, \mathbf{\Phi}^\mathsf{T}(\omega_k)]^\mathsf{T}) \left[\mathrm{Vec}\, \frac{\partial \overline{\underline{d}_{X,T}}(\omega_k, \underline{\vartheta}_2)}{\partial \underline{\vartheta}_2^\mathsf{T}} \otimes \frac{\partial \underline{d}_{X,T}(\omega_k, \underline{\vartheta}_1)}{\partial \underline{\vartheta}_1^\mathsf{T}} \right]$$

$$= T\, (\mathbf{U}_p \otimes [\mathrm{Vec}\, \mathbf{\Phi}^\mathsf{T}(\omega_k)]^\mathsf{T}) \frac{\partial^2 \mathbf{I}_{X,T}(\omega, \underline{\vartheta}_1, \underline{\vartheta}_2)}{\partial \underline{\vartheta}_2^\mathsf{T} \partial \underline{\vartheta}_1^\mathsf{T}} ; \tag{28}$$

clearly at $\underline{\vartheta}_1 = \underline{\vartheta}_2 = \underline{\vartheta}$ the expressions (27) and (28) are complex conjugate.

7.5 Scaled Derivative

Consider the second scaled derivative of $\underline{X}_{t+h}(\underline{\vartheta}_1)\underline{X}_t^{\mathsf{T}}(\underline{\vartheta}_2)$,

$$\frac{\partial^2_{s,T}\underline{X}_{t+h}(\underline{\vartheta}_1)\underline{X}_t^{\mathsf{T}}(\underline{\vartheta}_2)}{\partial_{s,T}\underline{\vartheta}_2^{\mathsf{T}}\partial_{s,T}\underline{\vartheta}_1^{\mathsf{T}}} = \mathrm{Vec}\,\frac{\partial_{s,T}\underline{X}_t(\underline{\vartheta}_2)}{\partial_{s,T}\underline{\vartheta}_2^{\mathsf{T}}} \otimes \frac{\partial_{s,T}\underline{X}_{t+h}(\underline{\vartheta}_1)}{\partial_{s,T}\underline{\vartheta}_1^{\mathsf{T}}}$$
$$+ (\mathbf{K}_{p\cdot d} \otimes \mathbf{U}_d)\,\mathbf{K}_{d\cdot dp}\left[\mathrm{Vec}\,\frac{\partial_{s,T}\underline{X}_{t+h}(\underline{\vartheta}_1)}{\partial_{s,T}\underline{\vartheta}_1^{\mathsf{T}}} \otimes \frac{\partial_{s,T}\underline{X}_t(\underline{\vartheta}_2)}{\partial_{s,T}\underline{\vartheta}_2^{\mathsf{T}}}\right],$$

at $\mathbf{D}_T\underline{X}_t(\underline{\vartheta})$. The scaled derivative of each term is

$$\mathbf{D}_T\left[\frac{\partial}{\partial\underline{\vartheta}^{\mathsf{T}}}\underline{X}_t(\underline{\vartheta})\right]\mathbf{D}_{1,T}\,,$$

by definition; hence

$$\mathrm{Vec}\,\frac{\partial_{s,T}\mathbf{D}_T\underline{X}_t(\underline{\vartheta}_2)}{\partial_{s,T}\underline{\vartheta}_2^{\mathsf{T}}} \otimes \frac{\partial_{s,T}\mathbf{D}_T\underline{X}_{t+h}(\underline{\vartheta}_1)}{\partial_{s,T}\underline{\vartheta}_1^{\mathsf{T}}}$$
$$= (\mathbf{D}_{1,T}\otimes\mathbf{D}_T\otimes\mathbf{D}_T)\times\left[\mathrm{Vec}\,\frac{\partial\underline{X}_t(\underline{\vartheta}_2)}{\partial\underline{\vartheta}_2^{\mathsf{T}}}\otimes\frac{\partial\underline{X}_{t+h}(\underline{\vartheta}_1)}{\partial\underline{\vartheta}_1^{\mathsf{T}}}\right]\mathbf{D}_{1,T}\,.$$

The matrices $(\mathbf{D}_{1,T}\otimes\mathbf{D}_T\otimes\mathbf{D}_T)$ and $(\mathbf{K}_{p\cdot d}\otimes\mathbf{U}_d)\,\mathbf{K}_{d\cdot dp}$ commute. We conclude

$$\frac{\partial^2_{s,T}\mathbf{D}_T\underline{X}_{t+h}(\underline{\vartheta}_1)\,\mathbf{D}_T\underline{X}_t^{\mathsf{T}}(\underline{\vartheta}_2)}{\partial_{s,T}\underline{\vartheta}_2^{\mathsf{T}}\partial_{s,T}\underline{\vartheta}_1^{\mathsf{T}}}$$
$$= (\mathbf{D}_{1,T}\otimes\mathbf{D}_T\otimes\mathbf{D}_T)\left[\frac{\partial^2\underline{X}_{t+h}(\underline{\vartheta}_1)\underline{X}_t^{\mathsf{T}}(\underline{\vartheta}_2)}{\partial\underline{\vartheta}_2^{\mathsf{T}}\partial\underline{\vartheta}_1^{\mathsf{T}}}\right]\mathbf{D}_{1,T}\,.$$

7.6 Asymptotic Variance for the Linear Model

The variance matrix of the complex vector $\mathrm{Vec}\sum_{k=-T_1}^{T_1}\mathbf{\Phi}(\omega_k)\mathbf{I}_{\underline{Y},\underline{X},T}(\omega_k)$ is an easy calculation based on the calculus worked out in [Ter02]. It turns out that

$$\lim_{T\to\infty}\mathrm{Var}\left(\frac{1}{T}\mathrm{Vec}\sum_{k=-T_1}^{T_1}\mathbf{\Phi}(\omega_k)\mathbf{I}_{\underline{Y},\underline{X},T}(\omega_k)\right)$$
$$= \mathrm{Vec}\int_{-1/2}^{1/2}d\mathbf{F}^{\mathsf{T}}(\omega)\otimes\left[\mathbf{\Phi}(\omega)\mathbf{S}_{\underline{Z}}(\omega)\mathbf{\Phi}(\omega)\right]. \quad (29)$$

Taking $\mathbf{\Phi}(\omega) = \mathbf{S}_{\underline{Z}}^{-1}(\omega)$, we derive the variance (12) directly from (10). The mixed derivative

$$\frac{\partial^2\mathbf{F}(\omega,\underline{\vartheta}_1,\underline{\vartheta}_2)}{\partial\underline{\vartheta}_2^{\mathsf{T}}\partial\underline{\vartheta}_1^{\mathsf{T}}},$$

is the inverse Fourier transform of the same mixed derivative of $\mathbf{C}_X(h, \vartheta)$ which is the limit of the $(\partial^2 \mathbf{I}_{X,T}(\omega, \vartheta_1, \vartheta_2))/\partial \underline{\vartheta}_2^\mathsf{T} \partial \underline{\vartheta}_1^\mathsf{T}$ at $\underline{\vartheta}_1 = \underline{\vartheta}_2 = \underline{\vartheta}$; in our case,

$$
\begin{aligned}
\frac{\partial^2 \mathbf{I}_{X,T}(\omega, \vartheta_1, \vartheta_2)}{\partial \underline{\vartheta}_2^\mathsf{T} \partial \underline{\vartheta}_1^\mathsf{T}} &= \mathrm{Vec}\, \frac{\partial \overline{d_{X,T}(\omega_k, \vartheta_2)}}{\partial \underline{\vartheta}_2^\mathsf{T}} \otimes \frac{\partial \underline{d}_{X,T}(\omega_k, \vartheta_1)}{\partial \underline{\vartheta}_1^\mathsf{T}} \quad (30) \\
&= \mathrm{Vec}\, \frac{\partial \mathbf{B}_2 \overline{d_{X,T}(\omega_k)}}{\partial (\mathrm{Vec}\, \mathbf{B}_2)^\mathsf{T}} \otimes \frac{\partial \mathbf{B}_1 \underline{d}_{X,T}(\omega_k)}{\partial (\mathrm{Vec}\, \mathbf{B}_1)^\mathsf{T}} \\
&= \overline{\underline{d}_{X,T}}(\omega_k) \otimes \mathrm{Vec}\, \mathbf{U}_d \otimes \underline{d}_{X,T}^\mathsf{T}(\omega_k) \otimes \mathbf{U}_d \,.
\end{aligned}
$$

The product

$$
(\mathbf{U}_{pd} \otimes [\mathrm{Vec}\, \mathbf{\Phi}^\mathsf{T}(\omega_k)]^\mathsf{T}) \left(\overline{\underline{d}_{X,T}}(\omega_k) \otimes \mathrm{Vec}\, \mathbf{U}_d \otimes \underline{d}_{X,T}^\mathsf{T}(\omega_k) \otimes \mathbf{U}_d \right)
$$

equals

$$
\left[\underline{d}_{X,T}(\omega_k)\, \underline{d}_{X,T}^*(\omega_k) \right]^\mathsf{T} \otimes \mathbf{\Phi}(\omega_k) \,,
$$

and (12) follows.

7.7 Variance Matrix for the Mixed Model

Put the objective function in terms of the parameters

$$
\begin{aligned}
Q_T(\mathbf{B}_1, \mathbf{B}_2, \underline{\lambda}) &\qquad\qquad (31) \\
= \sum_{k=-T_1}^{T_1} &[\mathrm{Tr}(\mathbf{I}_{\underline{Y},T}(\omega_k)\mathbf{\Phi}(\omega_k)) \\
&+ \mathrm{Tr}(\mathbf{B}_1 \mathbf{I}_{\underline{X}_1,T}(\omega_k)\mathbf{B}_1^\mathsf{T} \mathbf{\Phi}(\omega_k)) + \mathrm{Tr}(\mathbf{B}_1 \mathbf{I}_{\underline{X}_1,\underline{X}_2,T}(\omega_k,\underline{\lambda})\mathbf{B}_2^\mathsf{T} \mathbf{\Phi}(\omega_k)) \\
&+ \mathrm{Tr}(\mathbf{B}_2 \mathbf{I}_{\underline{X}_2,\underline{X}_1,T}(\omega_k,\underline{\lambda})\mathbf{B}_1^\mathsf{T} \mathbf{\Phi}(\omega_k)) + \mathrm{Tr}(\mathbf{B}_2 \mathbf{I}_{\underline{X}_2,T}(\omega_k,\underline{\lambda})\mathbf{B}_2^\mathsf{T} \mathbf{\Phi}(\omega_k)) \\
&- \mathrm{Tr}(\mathbf{I}_{\underline{Y},\underline{X}_1,T}(\omega_k)\mathbf{B}_1^\mathsf{T} \mathbf{\Phi}(\omega_k)) - \mathrm{Tr}(\mathbf{B}_1 \mathbf{I}_{\underline{X}_1,\underline{Y},T}(\omega_k)\mathbf{\Phi}(\omega_k)) \\
&- \mathrm{Tr}(\mathbf{I}_{\underline{Y},\underline{X}_2,T}(\omega_k,\underline{\lambda})\mathbf{B}_2^\mathsf{T} \mathbf{\Phi}(\omega_k)) - \mathrm{Tr}(\mathbf{B}_2 \mathbf{I}_{\underline{X}_2,\underline{Y},T}(\omega_k,\underline{\lambda})\mathbf{\Phi}(\omega_k))] \,.
\end{aligned}
$$

Consider now the normal equations

$$
\frac{\partial Q_T(\mathbf{B}_1, \mathbf{B}_2, \underline{\lambda})}{\partial \mathbf{B}_1} = \mathbf{0} \,,
$$

$$
\frac{\partial Q_T(\mathbf{B}_1, \mathbf{B}_2, \underline{\lambda})}{\partial \mathbf{B}_2} = \mathbf{0} \,,
$$

$$
\mathrm{Vec}\, \frac{\partial Q_T(\mathbf{B}_1, \mathbf{B}_2, \underline{\lambda})}{\partial \underline{\lambda}^\mathsf{T}} = \underline{0} \,.
$$

They are written

$$\frac{\partial Q_T\left(\mathbf{B}_1,\mathbf{B}_2,\underline{\lambda}\right)}{\partial\mathbf{B}_1} = \sum_{k=-T_1}^{T_1} \boldsymbol{\Phi}^{\mathsf{T}}\left(\omega_k\right)\mathbf{B}_1\mathbf{I}_{\underline{X}_1,T}^{\mathsf{T}}\left(\omega_k\right) + \boldsymbol{\Phi}\left(\omega_k\right)\mathbf{B}_1\mathbf{I}_{\underline{X}_1,T}\left(\omega_k\right)$$
$$+ \boldsymbol{\Phi}^{\mathsf{T}}\left(\omega_k\right)\mathbf{B}_2\mathbf{I}_{\underline{X}_1,\underline{X}_2,T}^{\mathsf{T}}(\omega_k,\underline{\lambda}) + \boldsymbol{\Phi}\left(\omega_k\right)\mathbf{B}_2\mathbf{I}_{\underline{X}_2,\underline{X}_1,T}(\omega_k,\underline{\lambda})$$
$$- \boldsymbol{\Phi}\left(\omega_k\right)\mathbf{I}_{\underline{Y},\underline{X}_1,T}(\omega_k) - \boldsymbol{\Phi}^{\mathsf{T}}\left(\omega_k\right)\mathbf{I}_{\underline{X}_1,\underline{Y},T}^{\mathsf{T}}(\omega_k) .$$

Similarly

$$\frac{\partial Q_T\left(\mathbf{B}_1,\mathbf{B}_2,\underline{\lambda}\right)}{\partial\mathbf{B}_2} = \sum_{k=-T_1}^{T_1} \boldsymbol{\Phi}^{\mathsf{T}}\left(\omega_k\right)\mathbf{B}_2\mathbf{I}_{\underline{X}_2,T}^{\mathsf{T}}\left(\omega_k,\underline{\lambda}\right) + \boldsymbol{\Phi}\left(\omega_k\right)\mathbf{B}_2\mathbf{I}_{\underline{X}_2,T}\left(\omega_k,\underline{\lambda}\right)$$
$$+ \boldsymbol{\Phi}^{\mathsf{T}}\left(\omega_k\right)\mathbf{B}_1\mathbf{I}_{\underline{X}_1,\underline{X}_2,T}^{\mathsf{T}}(\omega_k,\underline{\lambda}) + \boldsymbol{\Phi}\left(\omega_k\right)\mathbf{B}_1\mathbf{I}_{\underline{X}_2,\underline{X}_1,T}(\omega_k,\underline{\lambda})$$
$$- \boldsymbol{\Phi}\left(\omega_k\right)\mathbf{I}_{\underline{Y},\underline{X}_2,T}(\omega_k,\underline{\lambda}) - \boldsymbol{\Phi}^{\mathsf{T}}\left(\omega_k\right)\mathbf{I}_{\underline{X}_2,\underline{Y},T}^{\mathsf{T}}(\omega_k,\underline{\lambda}) ,$$

and finally

$$\mathrm{Vec}\frac{\partial Q_T(\mathbf{B}_1,\mathbf{B}_2,\underline{\lambda})}{\partial\underline{\lambda}^{\mathsf{T}}}$$
$$= \sum_{k=-T_1}^{T_1}\left[\frac{\partial\mathbf{B}_1\mathbf{I}_{\underline{X}_1,\underline{X}_2,T}(\omega_k,\underline{\lambda})\mathbf{B}_2^{\mathsf{T}}}{\partial\underline{\lambda}^{\mathsf{T}}} + \frac{\partial\mathbf{B}_2\mathbf{I}_{\underline{X}_2,\underline{X}_1,T}(\omega_k,\underline{\lambda})\mathbf{B}_1^{\mathsf{T}}}{\partial\underline{\lambda}^{\mathsf{T}}}\right.$$
$$+ \frac{\partial\mathbf{B}_2\mathbf{I}_{\underline{X}_2,T}(\omega_k,\underline{\lambda})\mathbf{B}_2^{\mathsf{T}}}{\partial\underline{\lambda}^{\mathsf{T}}} - \frac{\partial\mathbf{I}_{\underline{Y},\underline{X}_2,T}(\omega_k,\underline{\lambda})\mathbf{B}_2^{\mathsf{T}}}{\partial\underline{\lambda}^{\mathsf{T}}}$$
$$\left. - \frac{\partial\mathbf{B}_2\mathbf{I}_{\underline{X}_2,\underline{Y},T}(\omega_k,\underline{\lambda})}{\partial\underline{\lambda}^{\mathsf{T}}}\right]\mathrm{Vec}[\boldsymbol{\Phi}^{\mathsf{T}}(\omega_k)] .$$

The variance of the derivative Let us denote the limit of the Hessian matrix of the estimates $\mathrm{Vec}(\mathrm{Vec}\,\widehat{\mathbf{B}}_1,\mathrm{Vec}\,\widehat{\mathbf{B}}_2,\widehat{\underline{\lambda}})$ by

$$\varSigma = 2\begin{bmatrix} \varSigma_{11} & \varSigma_{12} & \varSigma_{1\lambda} \\ \varSigma_{21} & \varSigma_{22} & \varSigma_{2\lambda} \\ \varSigma_{\lambda 1} & \varSigma_{\lambda 2} & \varSigma_{\lambda\lambda} \end{bmatrix} .$$

The second derivative of $\partial Q_T(\mathbf{B}_1,\mathbf{B}_2,\underline{\lambda})/\partial\mathbf{B}_1$ by $\mathbf{B}_1$ depends neither on $\mathbf{B}_2$ nor $\underline{\lambda}$. Therefore, according to (12),

$$\varSigma_{11} = \int_{-1/2}^{1/2} \mathrm{d}\mathbf{F}_{11}^{\mathsf{T}}(\omega) \otimes \mathbf{S}_{\underline{Z}}^{-1}(\omega) ;$$

here as well as later on $\boldsymbol{\Phi} = \mathbf{S}_{\underline{Z}}^{-1}$. The matrix $\varSigma_{12}$ between $\mathrm{Vec}\,\widehat{\mathbf{B}}_1$ and $\mathrm{Vec}\,\widehat{\mathbf{B}}_2$ follows from

$$\frac{\partial^2 Q_T(\mathbf{B}_1,\mathbf{B}_2,\underline{\lambda})}{\partial\mathbf{B}_2\partial\mathbf{B}_1}$$
$$= \sum_{k=-T_1}^{T_1} \mathbf{I}_{\underline{X}_1,\underline{X}_2,T}(\omega_k,\underline{\lambda}) \otimes \boldsymbol{\Phi}^{\mathsf{T}}(\omega_k) + \mathbf{I}_{\underline{X}_2,\underline{X}_1,T}^{\mathsf{T}}(\omega_k,\underline{\lambda}) \otimes \boldsymbol{\Phi}(\omega_k) ,$$

so it is

$$\Sigma_{12} = \int_{-1/2}^{1/2} \mathrm{d}\mathbf{F}_{1,2}^{\mathsf{T}}(\omega, \underline{\lambda}_0) \otimes \mathbf{S}_{\underline{Z}}^{-1}(\omega) \ .$$

The second derivative of $\partial Q_T(\mathbf{B}_1, \mathbf{B}_2, \underline{\lambda})/\partial \mathbf{B}_2$ by $\mathbf{B}_2$ is similar except the SDFR depends on $\underline{\lambda}$

$$\Sigma_{22} = \int_{-1/2}^{1/2} \mathrm{d}\mathbf{F}_2(\omega, \underline{\lambda}_0) \otimes [\mathbf{S}_{\underline{Z}}^{-1}(\omega)]^{\mathsf{T}} \ .$$

Now for the matrix $\Sigma_{1\lambda}$ consider

$$\frac{\partial^2 Q_T(\mathbf{B}_1, \mathbf{B}_2, \underline{\lambda})}{\partial \mathbf{B}_1 \partial \underline{\lambda}^{\mathsf{T}}} = \sum_{k=-T_1}^{T_1} \frac{\partial \mathrm{Vec}\,\mathbf{\Phi}^{\mathsf{T}}(\omega_k)\mathbf{B}_2\mathbf{I}_{\underline{X}_1,\underline{X}_2,T}^{\mathsf{T}}(\omega_k, \underline{\lambda})}{\partial \underline{\lambda}^{\mathsf{T}}}$$

$$+ \frac{\partial \mathrm{Vec}\,\mathbf{\Phi}(\omega_k)\mathbf{B}_2\mathbf{I}_{\underline{X}_2,\underline{X}_1,T}(\omega_k, \underline{\lambda})}{\partial \underline{\lambda}^{\mathsf{T}}}$$

$$= \frac{1}{T}\sum_{k=-T_1}^{T_1}\left[\frac{\partial \mathrm{Vec}\,\mathbf{\Phi}^{\mathsf{T}}(\omega_k)\mathbf{B}_2\overline{d_{\underline{X}_2,T}(\omega_k, \underline{\lambda})}d_{\underline{X}_1,T}^{\mathsf{T}}(\omega_k)}{\partial \underline{\lambda}^{\mathsf{T}}}\right.$$

$$\left.+ \frac{\partial \mathrm{Vec}\,\mathbf{\Phi}(\omega_k)\mathbf{B}_2\underline{d}_{\underline{X}_2,T}(\omega_k, \underline{\lambda})\,\underline{d}_{\underline{X}_1,T}^{*}(\omega_k)}{\partial \underline{\lambda}^{\mathsf{T}}}\right]$$

$$= \frac{1}{T}\sum_{k=-T_1}^{T_1}(\mathbf{U}_p \otimes \mathbf{\Phi}^{\mathsf{T}}(\omega_k)\mathbf{B}_2)\left[\underline{d}_{\underline{X}_1,T}(\omega_k) \otimes \frac{\partial \overline{d_{\underline{X}_2,T}(\omega_k, \underline{\lambda})}}{\partial \underline{\lambda}^{\mathsf{T}}}\right]$$

$$+ (\mathbf{U}_p \otimes \mathbf{\Phi}(\omega_k)\mathbf{B}_2)\left[\overline{\underline{d}_{\underline{X}_1,T}(\omega_k)} \otimes \frac{\partial \underline{d}_{\underline{X}_2,T}(\omega_k, \underline{\lambda})}{\partial \underline{\lambda}^{\mathsf{T}}}\right] \ ;$$

hence the limit

$$\Sigma_{1\lambda} = \int_{-1/2}^{1/2}(\mathbf{U}_p \otimes \mathbf{\Phi}(\omega)\mathbf{B}_{2,0})\,\mathrm{d}\frac{\partial \mathbf{F}_{1,2}(\omega, \underline{\lambda}_0)}{\partial \underline{\lambda}^{\mathsf{T}}}$$

$$= \int_{-1/2}^{1/2}(\mathbf{U}_p \otimes \mathbf{S}_{\underline{Z}}^{-1}(\omega)\mathbf{B}_{2,0})\,\mathrm{d}\frac{\partial \mathbf{F}_{1,2}(\omega, \underline{\lambda}_0)}{\partial \underline{\lambda}^{\mathsf{T}}} \ .$$

The matrix $\Sigma_{2\lambda}$ based on

$$\frac{\partial Q_T(\mathbf{B}_1, \mathbf{B}_2, \underline{\lambda})}{\partial \mathbf{B}_2} = \sum_{k=-T_1}^{T_1}\mathbf{\Phi}^{\mathsf{T}}(\omega_k)\mathbf{B}_2\mathbf{I}_{\underline{X}_2,T}^{\mathsf{T}}(\omega_k, \underline{\lambda}) + \mathbf{\Phi}(\omega_k)\mathbf{B}_2\mathbf{I}_{\underline{X}_2,T}(\omega_k, \underline{\lambda})$$

$$+ \mathbf{\Phi}^{\mathsf{T}}(\omega_k)\mathbf{B}_1\mathbf{I}_{\underline{X}_1,\underline{X}_2,T}^{\mathsf{T}}(\omega_k, \underline{\lambda}) + \mathbf{\Phi}(\omega_k)\mathbf{B}_1\mathbf{I}_{\underline{X}_2,\underline{X}_1,T}(\omega_k, \underline{\lambda})$$

$$- \mathbf{\Phi}(\omega_k)\mathbf{I}_{\underline{Y},\underline{X}_2,T}(\omega_k, \underline{\lambda}) - \mathbf{\Phi}^{\mathsf{T}}(\omega_k)\mathbf{I}_{\underline{X}_2,\underline{Y},T}^{\mathsf{T}}(\omega_k, \underline{\lambda}) \ ,$$

$$\frac{\partial^2 Q_T\left(\mathbf{B}_1,\mathbf{B}_2,\underline{\lambda}\right)}{\partial \mathbf{B}_2 \partial \underline{\lambda}^{\mathsf{T}}}$$

$$= \sum_{k=-T_1}^{T_1} \frac{\partial \operatorname{Vec} \boldsymbol{\Phi}^{\mathsf{T}}\left(\omega_k\right)\mathbf{B}_2 \mathbf{I}^{\mathsf{T}}_{\underline{X}_2,T}\left(\omega_k,\underline{\lambda}\right)}{\partial \underline{\lambda}^{\mathsf{T}}} + \frac{\partial \operatorname{Vec}\boldsymbol{\Phi}\left(\omega_k\right)\mathbf{B}_2 \mathbf{I}_{\underline{X}_2,T}\left(\omega_k,\underline{\lambda}\right)}{\partial \underline{\lambda}^{\mathsf{T}}}$$

$$+ \frac{\partial \operatorname{Vec} \boldsymbol{\Phi}^{\mathsf{T}}\left(\omega_k\right)\mathbf{B}_1 \mathbf{I}^{\mathsf{T}}_{\underline{X}_1,\underline{X}_2,T}\left(\omega_k,\underline{\lambda}\right)}{\partial \underline{\lambda}^{\mathsf{T}}} + \frac{\partial \operatorname{Vec} \boldsymbol{\Phi}\left(\omega_k\right)\mathbf{B}_1 \mathbf{I}_{\underline{X}_2,\underline{X}_1,T}\left(\omega_k,\underline{\lambda}\right)}{\partial \underline{\lambda}^{\mathsf{T}}}$$

$$- \frac{\partial \operatorname{Vec} \boldsymbol{\Phi}\left(\omega_k\right)\mathbf{I}_{\underline{Y},\underline{X}_2,T}\left(\omega_k,\underline{\lambda}\right)}{\partial \underline{\lambda}^{\mathsf{T}}} - \frac{\partial \operatorname{Vec} \boldsymbol{\Phi}^{\mathsf{T}}\left(\omega_k\right)\mathbf{I}^{\mathsf{T}}_{\underline{X}_2,\underline{Y},T}\left(\omega_k,\underline{\lambda}\right)}{\partial \underline{\lambda}^{\mathsf{T}}}$$

$$= 2\sum_{k=-T_1}^{T_1} \frac{\partial \operatorname{Vec}\boldsymbol{\Phi}^{\mathsf{T}}\left(\omega_k\right)\mathbf{B}_2 \mathbf{I}^{\mathsf{T}}_{\underline{X}_2,T}\left(\omega_k,\underline{\lambda}_1,\underline{\lambda}_2\right)}{\partial \underline{\lambda}_1^{\mathsf{T}}}\Bigg|_{\underline{\lambda}_1=\underline{\lambda}_2=\underline{\lambda}} .$$

By equation (13), the limit of the derivative at $\underline{\vartheta}=\underline{\vartheta}_0$ is zero, thus

$$\Sigma_{2\lambda} = \int_{-1/2}^{1/2} \left(\mathbf{U}_q \otimes \boldsymbol{\Phi}\left(\omega\right)\mathbf{B}_{2,0}\right)\frac{\partial \mathbf{F}_{2,2}\left(\omega,\underline{\lambda}_1,\underline{\lambda}_2\right)}{\partial \underline{\lambda}_1^{\mathsf{T}}}\Bigg|_{\underline{\lambda}_1=\underline{\lambda}_2=\underline{\lambda}_0} .$$

The matrix $\Sigma_{\lambda\lambda}$ comes from the general result (10):

$$\Sigma_{\lambda\lambda} = \int_{-1/2}^{1/2}\left(\mathbf{U}_r \otimes [\operatorname{Vec}([\mathbf{S}_{\underline{Z}}^{-1}(\omega)]^{\mathsf{T}})^{\mathsf{T}}(\mathbf{B}_{2,0}\otimes\mathbf{B}_{2,0})]\right)$$

$$\times\, \mathrm{d}\,\frac{\partial^2 \mathbf{F}_{2,2}\left(\omega,\underline{\lambda}_1,\underline{\lambda}_2\right)}{\partial \underline{\lambda}_2^{\mathsf{T}}\partial \underline{\lambda}_1^{\mathsf{T}}}\Bigg|_{\underline{\lambda}_1=\underline{\lambda}_2=\underline{\lambda}_0}$$

$$= \int_{-1/2}^{1/2}\left(\mathbf{U}_r \otimes \operatorname{Vec}([\mathbf{B}_{2,0}^{\mathsf{T}}\mathbf{S}_{\underline{Z}}^{-1}(\omega)\mathbf{B}_{2,0}]^{\mathsf{T}})^{\mathsf{T}}\right)$$

$$\times\, \mathrm{d}\,\frac{\partial^2 \mathbf{F}_{2,2}\left(\omega,\underline{\lambda}_1,\underline{\lambda}_2\right)}{\partial \underline{\lambda}_2^{\mathsf{T}}\partial \underline{\lambda}_1^{\mathsf{T}}}\Bigg|_{\underline{\lambda}_1=\underline{\lambda}_2=\underline{\lambda}_0} .$$

Finally

$$\lim_{T\to\infty}\operatorname{Var}[\operatorname{Vec}(\operatorname{Vec}\widehat{\mathbf{B}}_1,\operatorname{Vec}\widehat{\mathbf{B}}_2,\widehat{\underline{\lambda}})] = \begin{bmatrix} \Sigma_{11} & \Sigma_{12} & \Sigma_{1\lambda} \\ \Sigma_{21} & \Sigma_{22} & \Sigma_{2\lambda} \\ \Sigma_{\lambda 1} & \Sigma_{\lambda 2} & \Sigma_{\lambda\lambda} \end{bmatrix}^{-1} .$$

Use Theorem 2 of [MN99, p. 16] for the inverse.

We use the general formula for the variance of the estimate of $\underline{\vartheta} = \operatorname{Vec}(\operatorname{Vec}\mathbf{B}_1,\operatorname{Vec}\mathbf{B}_2,\underline{\lambda})$, where $\mathbf{B}_1$ is $d\times p$, $\mathbf{B}_2$ is $d\times q$, $\underline{\lambda}$ is $r\times 1$, $\underline{X}_{1,t}$ has dimension p, and $\underline{X}_{2,t}(\underline{\lambda})$ has dimension q. For the mixed derivative

$$\frac{\partial^2 \mathbf{F}\left(\omega,\underline{\vartheta}_1,\underline{\vartheta}_2\right)}{\partial \underline{\vartheta}_2^{\mathsf{T}}\partial \underline{\vartheta}_1^{\mathsf{T}}} \tag{32}$$

we use

$$\mathbf{I}_{\underline{X},T}(\omega_k, \underline{\vartheta}) = \mathbf{B}\mathbf{I}_{\underline{X}_3,T}(\omega_k, \underline{\lambda})\mathbf{B}^{\mathsf{T}}$$

$$= [\mathbf{B}_1 \underline{d}_{\underline{X}_1,T}(\omega_k) + \mathbf{B}_2 \underline{d}_{\underline{X}_2,T}(\omega_k, \underline{\lambda})]$$

$$\times [\mathbf{B}_1 \underline{d}_{\underline{X}_1,T}(\omega_k) + \mathbf{B}_2 \underline{d}_{\underline{X}_2,T}(\omega_k, \underline{\lambda})]^* \,|_{\mathbf{B}_1 = \mathbf{B}_2} \ .$$

To the parameters $\underline{\vartheta}_1, \underline{\vartheta}_2$ will correspond $\underline{\vartheta}_i = \mathrm{Vec}(\mathrm{Vec}\,\mathbf{B}_{1i}, \mathrm{Vec}\,\mathbf{B}_{2i}, \underline{\lambda}_i)$, $i = 1, 2$. Write

$$\mathbf{I}_{\underline{X},T}(\omega_k, \underline{\vartheta}_1, \underline{\vartheta}_2) = [\mathbf{B}_{11}\underline{d}_{\underline{X}_1,T}(\omega_k) + \mathbf{B}_{21}\underline{d}_{\underline{X}_2,T}(\omega_k, \underline{\lambda}_1)]$$

$$\times [\mathbf{B}_{12}\underline{d}_{\underline{X}_1,T}(\omega_k) + \mathbf{B}_{22}\underline{d}_{\underline{X}_2,T}(\omega_k, \underline{\lambda}_2)]^*$$

$$= \mathbf{B}_{11}\underline{d}_{\underline{X}_1,T}(\omega_k)\underline{d}^*_{\underline{X}_1,T}(\omega_k)\mathbf{B}^{\mathsf{T}}_{12}$$

$$+ \mathbf{B}_{21}\underline{d}_{\underline{X}_2,T}(\omega_k, \underline{\lambda}_1)\underline{d}^*_{\underline{X}_1,T}(\omega_k)\mathbf{B}^{\mathsf{T}}_{12}$$

$$+ \mathbf{B}_{11}\underline{d}_{\underline{X}_1,T}(\omega_k)\underline{d}^*_{\underline{X}_2,T}(\omega_k, \underline{\lambda}_2)\mathbf{B}^{\mathsf{T}}_{22}$$

$$+ \mathbf{B}_{21}\underline{d}_{\underline{X}_2,T}(\omega_k, \underline{\lambda}_1)\underline{d}^*_{\underline{X}_2,T}(\omega_k, \underline{\lambda}_2)\mathbf{B}^{\mathsf{T}}_{22} \ .$$

The variance of

$$\left[\begin{array}{c} \left[\frac{\partial Q_T(\mathbf{B}_1,\mathbf{B}_2,\underline{\lambda})}{\partial\,\mathrm{Vec}\,\mathbf{B}^{\mathsf{T}}_1} \right]^{\mathsf{T}} \\ \left[\frac{\partial Q_T(\mathbf{B}_1,\mathbf{B}_2,\underline{\lambda})}{\partial\,\mathrm{Vec}\,\mathbf{B}^{\mathsf{T}}_2} \right]^{\mathsf{T}} \\ \left[\frac{\partial Q_T(\mathbf{B}_1,\mathbf{B}_2,\underline{\lambda})}{\partial\underline{\lambda}^{\mathsf{T}}} \right]^{\mathsf{T}} \end{array} \right]$$

according to the mixed derivative (32) contains nine nonzero terms.

(A1). We have already seen the case

$$\frac{\partial^2 \mathbf{B}_{11}\underline{d}_{\underline{X}_1,T}(\omega_k)\underline{d}^*_{\underline{X}_1,T}(\omega_k)\mathbf{B}^{\mathsf{T}}_{12}}{\partial\mathbf{B}_{12}\partial\mathbf{B}_{11}}$$

$$= \overline{\underline{d}_{\underline{X}_1,T}}(\omega_k) \otimes \mathrm{Vec}\,\mathbf{U}_d \otimes \underline{d}^{\mathsf{T}}_{\underline{X}_1,T}(\omega_k) \otimes \mathbf{U}_d \ .$$

See (30). The linear model shows

$$(\mathbf{U}_{pd} \otimes [\mathrm{Vec}\,\mathbf{\Phi}^{\mathsf{T}}(\omega_k)]^{\mathsf{T}})(\overline{\underline{d}_{\underline{X}_1,T}}(\omega_k) \otimes \mathrm{Vec}\,\mathbf{U}_d \otimes \underline{d}^{\mathsf{T}}_{\underline{X}_1,T}(\omega_k) \otimes \mathbf{U}_d)$$

$$= [\underline{d}_{\underline{X}_1,T}(\omega_k)\underline{d}^*_{\underline{X}_1,T}(\omega_k)]^{\mathsf{T}} \otimes \mathbf{\Phi}(\omega_k) \ .$$

The cases

$$\frac{\partial^2 \mathbf{B}_{21}\underline{d}_{\underline{X}_2,T}(\omega_k, \underline{\lambda}_1)\,\underline{d}^*_{\underline{X}_1,T}(\omega_k)\,\mathbf{B}^{\mathsf{T}}_{12}}{\partial\mathbf{B}_{12}\partial\mathbf{B}_{21}}$$

$$= \overline{\underline{d}_{\underline{X}_1,T}}(\omega_k) \otimes \mathrm{Vec}\,\mathbf{U}_d \otimes \underline{d}^{\mathsf{T}}_{\underline{X}_2,T}(\omega_k, \underline{\lambda}_1) \otimes \mathbf{U}_d \ ,$$

$$\frac{\partial^2 \mathbf{B}_{11}\underline{d}_{\underline{X}_1,T}(\omega_k)\,\underline{d}^*_{\underline{X}_2,T}(\omega_k, \underline{\lambda}_2)\,\mathbf{B}^{\mathsf{T}}_{22}}{\partial\mathbf{B}_{22}\partial\mathbf{B}_{11}}$$

$$= \overline{\underline{d}_{\underline{X}_2,T}}(\omega_k, \underline{\lambda}_2) \otimes \mathrm{Vec}\,\mathbf{U}_d \otimes \underline{d}^{\mathsf{T}}_{\underline{X}_1,T}(\omega_k) \otimes \mathbf{U}_d \ ,$$

and

$$\frac{\partial^2 \mathbf{B}_{21}\underline{d}_{\underline{X}_2,T}(\omega_k,\underline{\lambda}_1)\,\underline{d}^*_{\underline{X}_2,T}(\omega_k,\underline{\lambda}_2)\,\mathbf{B}_{22}^{\mathsf{T}}}{\partial \mathbf{B}_{22}\partial \mathbf{B}_{21}}$$

$$= \overline{\underline{d}_{\underline{X}_2,T}}(\omega_k,\underline{\lambda}_2) \otimes \mathrm{Vec}\,\mathbf{U}_d \otimes \underline{d}^{\mathsf{T}}_{\underline{X}_2,T}(\omega_k,\underline{\lambda}_1) \otimes \mathbf{U}_d$$

are similar because the parameters $\underline{\lambda}_i$ are fixed here. Also

$$\left(\mathbf{U}_{pd} \otimes [\mathrm{Vec}\,\mathbf{\Phi}^{\mathsf{T}}(\omega_k)]^{\mathsf{T}}\right)\left(\overline{\underline{d}_{\underline{X}_1,T}}(\omega_k) \otimes \mathrm{Vec}\,\mathbf{U}_d \otimes \underline{d}^{\mathsf{T}}_{\underline{X}_2,T}(\omega_k,\underline{\lambda}_1) \otimes \mathbf{U}_d\right)$$

$$= \left[\underline{d}_{\underline{X}_2,T}(\omega_k,\underline{\lambda}_1)\,\underline{d}^*_{\underline{X}_1,T}(\omega_k)\right]^{\mathsf{T}} \otimes \mathbf{\Phi}(\omega_k)\ ,$$

$$\left(\mathbf{U}_{pd} \otimes [\mathrm{Vec}\,\mathbf{\Phi}^{\mathsf{T}}(\omega_k)]^{\mathsf{T}}\right)\left(\overline{\underline{d}_{\underline{X}_1,T}}(\omega_k) \otimes \mathrm{Vec}\,\mathbf{U}_d \otimes \underline{d}^{\mathsf{T}}_{\underline{X}_2,T}(\omega_k,\underline{\lambda}_2) \otimes \mathbf{U}_d\right)$$

$$= \left[\underline{d}_{\underline{X}_1,T}(\omega_k)\,\underline{d}^*_{\underline{X}_2,T}(\omega_k,\underline{\lambda}_2)\right]^{\mathsf{T}} \otimes \mathbf{\Phi}(\omega_k)\ ,$$

$$\left(\mathbf{U}_{qd} \otimes [\mathrm{Vec}\,\mathbf{\Phi}^{\mathsf{T}}(\omega_k)]^{\mathsf{T}}\right)\left(\overline{\underline{d}_{\underline{X}_2,T}}(\omega_k,\underline{\lambda}_2) \otimes \mathrm{Vec}\,\mathbf{U}_d \otimes \underline{d}^{\mathsf{T}}_{\underline{X}_2,T}(\omega_k,\underline{\lambda}_1) \otimes \mathbf{U}_d\right)$$

$$= \left[\underline{d}_{\underline{X}_2,T}(\omega_k,\underline{\lambda}_1)\,\underline{d}^*_{\underline{X}_2,T}(\omega_k,\underline{\lambda}_2)\right]^{\mathsf{T}} \otimes \mathbf{\Phi}(\omega_k)\ .$$

(A2). Taking the derivative with respect to $\underline{\lambda}_2$ and $\underline{\lambda}_1$,

$$\frac{\partial^2 \mathbf{B}_{21}\underline{d}_{\underline{X}_2,T}(\omega_k,\underline{\lambda}_1)\underline{d}^*_{\underline{X}_2,T}(\omega_k,\underline{\lambda}_2)\mathbf{B}_{22}^{\mathsf{T}}}{\partial\underline{\lambda}_2^{\mathsf{T}}\partial\underline{\lambda}_1^{\mathsf{T}}}$$

$$= \mathrm{Vec}\,\mathbf{B}_{22}\frac{\partial\overline{\underline{d}_{\underline{X}_2,T}(\omega_k,\underline{\lambda}_2)}}{\partial\underline{\lambda}_2^{\mathsf{T}}} \otimes \mathbf{B}_{21}\frac{\partial\underline{d}_{\underline{X},T}(\omega_k,\underline{\lambda}_1)}{\partial\underline{\lambda}_1^{\mathsf{T}}}$$

$$= \left[(\mathbf{U}_r \otimes \mathbf{B}_{22})\,\mathrm{Vec}\,\frac{\partial\overline{\underline{d}_{\underline{X}_2,T}(\omega_k,\underline{\lambda}_2)}}{\partial\underline{\lambda}_2^{\mathsf{T}}}\right] \otimes \mathbf{B}_{21}\frac{\partial\underline{d}_{\underline{X},T}(\omega_k,\underline{\lambda}_1)}{\partial\underline{\lambda}_1^{\mathsf{T}}}$$

$$= (\mathbf{U}_r \otimes \mathbf{B}_{22} \otimes \mathbf{B}_{21})\frac{\partial^2\underline{d}_{\underline{X}_2,T}(\omega_k,\underline{\lambda}_1)\underline{d}^*_{\underline{X}_2,T}(\omega_k,\underline{\lambda}_2)}{\partial\underline{\lambda}_2^{\mathsf{T}}\partial\underline{\lambda}_1^{\mathsf{T}}}\ ,$$

therefore we can apply the earlier result.

(A3). Consider now

$$(\mathbf{U}_{pd} \otimes [\mathrm{Vec}\, \boldsymbol{\Phi}^{\mathsf{T}}(\omega_k)]^{\mathsf{T}}) \frac{\partial^2 \mathbf{B}_{21} \underline{d}_{X_2,T}(\omega_k, \underline{\lambda}_1) \underline{d}^*_{X_1,T}(\omega_k) \mathbf{B}^{\mathsf{T}}_{12}}{\partial \mathbf{B}_{12} \partial \underline{\lambda}^{\mathsf{T}}_1}$$

$$= (\mathbf{U}_{pd} \otimes [\mathrm{Vec}\, \boldsymbol{\Phi}^{\mathsf{T}}(\omega_k)]^{\mathsf{T}})$$

$$\times \frac{\partial}{\partial \underline{\lambda}^{\mathsf{T}}_1} \mathrm{Vec} \frac{\partial\, \mathrm{Vec}[\mathbf{B}_{21} \underline{d}_{X_2,T}(\omega_k, \underline{\lambda}_1) \underline{d}^*_{X_1,T}(\omega_k) \mathbf{B}^{\mathsf{T}}_{12}]}{\partial (\mathrm{Vec}\, \mathbf{B}_{12})^{\mathsf{T}}}$$

$$= (\mathbf{U}_{pd} \otimes [\mathrm{Vec}\, \boldsymbol{\Phi}^{\mathsf{T}}(\omega_k)]^{\mathsf{T}}) \frac{\partial}{\partial \underline{\lambda}^{\mathsf{T}}_1} (\mathbf{U}_{dp} \otimes \mathbf{K}_{d \cdot d})$$

$$\times \mathrm{Vec}(\mathbf{B}_{21} \underline{d}_{X_2,T}(\omega_k, \underline{\lambda}_1) \underline{d}^*_{X_1,T}(\omega_k) \otimes \mathbf{U}_d)$$

$$= (\mathbf{U}_{pd} \otimes [\mathrm{Vec}\, \boldsymbol{\Phi}(\omega_k)]^{\mathsf{T}})$$

$$\times \frac{\partial}{\partial \underline{\lambda}^{\mathsf{T}}_1} \mathrm{Vec}(\mathbf{B}_{21} (\underline{d}_{X_2,T}(\omega_k, \underline{\lambda}_1) \underline{d}^*_{X_1,T}(\omega_k)) \otimes \mathbf{U}_d)$$

$$= \frac{\partial}{\partial \underline{\lambda}^{\mathsf{T}}_1} \mathrm{Vec}[[\mathrm{Vec}\, \boldsymbol{\Phi}(\omega_k)]^{\mathsf{T}} (\mathbf{B}_{21}(\underline{d}_{X_2,T}(\omega_k, \underline{\lambda}_1) \underline{d}^*_{X_1,T}(\omega_k)) \otimes \mathbf{U}_d)]$$

$$= \frac{\partial}{\partial \underline{\lambda}^{\mathsf{T}}_1} \mathrm{Vec}[(\mathbf{B}_{21}(\underline{d}_{X_2,T}(\omega_k, \underline{\lambda}_1) \underline{d}^*_{X_1,T}(\omega_k)) \otimes \mathbf{U}_d)\, \mathrm{Vec}\, \boldsymbol{\Phi}(\omega_k)]$$

$$= \frac{\partial}{\partial \underline{\lambda}^{\mathsf{T}}_1} \mathrm{Vec}[\boldsymbol{\Phi}(\omega_k) \mathbf{B}_{21}(\underline{d}_{X_2,T}(\omega_k, \underline{\lambda}_1) \underline{d}^*_{X_1,T}(\omega_k))]$$

$$= (\mathbf{U}_p \otimes \boldsymbol{\Phi}(\omega_k) \mathbf{B}_{21}) \left[\overline{\underline{d}_{X_1,T}(\omega_k)} \otimes \frac{\partial \underline{d}_{X_2,T}(\omega_k, \underline{\lambda})}{\partial \underline{\lambda}^{\mathsf{T}}} \right].$$

(A4). The case

$$(\mathbf{U}_{qd} \otimes [\mathrm{Vec}\, \boldsymbol{\Phi}(\omega_k)]^{\mathsf{T}}) \frac{\partial^2 \mathbf{B}_{21} \underline{d}_{X_2,T}(\omega_k, \underline{\lambda}_1) \underline{d}^*_{X_2,T}(\omega_k, \underline{\lambda}_2) \mathbf{B}^{\mathsf{T}}_{22}}{\partial \mathbf{B}_{22} \partial \underline{\lambda}^{\mathsf{T}}_1}$$

$$= (\mathbf{U}_q \otimes \boldsymbol{\Phi}(\omega_k) \mathbf{B}_{21}) \left[\overline{\underline{d}_{X_2,T}(\omega_k, \underline{\lambda}_2)} \otimes \frac{\partial \underline{d}_{X_2,T}(\omega_k, \underline{\lambda}_1)}{\partial \underline{\lambda}_1^{\mathsf{T}}} \right].$$

is as the previous one.
(A5).

$$\frac{\partial^2 \mathbf{B}_{21}\underline{d}_{\underline{X}_2,T}\left(\omega_k,\underline{\lambda}_1\right)\underline{d}^*_{\underline{X}_2,T}\left(\omega_k,\underline{\lambda}_2\right)\mathbf{B}_{22}^{\mathsf{T}}}{\partial\underline{\lambda}_2^{\mathsf{T}}\partial\mathbf{B}_{21}}$$

$$=\frac{\partial}{\partial\left(\operatorname{Vec}\mathbf{B}_{21}\right)^{\mathsf{T}}}\operatorname{Vec}\left[\left(\mathbf{B}_{22}\otimes\mathbf{B}_{21}\right)\left(\frac{\overline{\underline{d}_{\underline{X}_2,T}\left(\omega_k,\underline{\lambda}_2\right)}}{\partial\underline{\lambda}_2^{\mathsf{T}}}\otimes\underline{d}_{\underline{X}_2,T}\left(\omega_k,\underline{\lambda}_1\right)\right)\right]$$

$$=\frac{\partial}{\partial\left(\operatorname{Vec}\mathbf{B}_{21}\right)^{\mathsf{T}}}$$

$$\left(\left[\left(\frac{\overline{\underline{d}_{\underline{X}_2,T}\left(\omega_k,\underline{\lambda}_2\right)}}{\partial\underline{\lambda}_2^{\mathsf{T}}}\otimes\underline{d}_{\underline{X}_2,T}\left(\omega_k,\underline{\lambda}_1\right)\right)^{\mathsf{T}}\otimes\mathbf{U}_{d^2}\right]\operatorname{Vec}\left(\mathbf{B}_{22}\otimes\mathbf{B}_{21}\right)\right)$$

$$=\left[\left(\frac{\overline{\underline{d}_{\underline{X}_2,T}\left(\omega_k,\underline{\lambda}_2\right)}}{\partial\underline{\lambda}_2^{\mathsf{T}}}\otimes\underline{d}_{\underline{X}_2,T}\left(\omega_k,\underline{\lambda}_1\right)\right)^{\mathsf{T}}\otimes\mathbf{U}_{d^2}\right]$$

$$\times\left(\mathbf{U}_q\otimes\mathbf{K}_{q\cdot d}\otimes\mathbf{U}_d\right)\left(\operatorname{Vec}\mathbf{B}_{22}\otimes\mathbf{U}_{dq}\right)\ .$$

(A6).

$$\frac{\partial^2\mathbf{B}_{11}\underline{d}_{\underline{X}_1,T}(\omega_k)\underline{d}^*_{\underline{X}_2,T}(\omega_k,\underline{\lambda}_2)\mathbf{B}_{22}^{\mathsf{T}}}{\partial\underline{\lambda}_2^{\mathsf{T}}\partial\mathbf{B}_{11}}$$

$$=\left[\left(\frac{\overline{\underline{d}_{\underline{X}_2,T}\left(\omega_k,\underline{\lambda}_2\right)}}{\partial\underline{\lambda}_2^{\mathsf{T}}}\otimes\underline{d}_{\underline{X}_1,T}\left(\omega_k\right)\right)^{\mathsf{T}}\otimes\mathbf{U}_{d^2}\right]$$

$$\left(\mathbf{U}_q\otimes\mathbf{K}_{p\cdot d}\otimes\mathbf{U}_d\right)\left(\operatorname{Vec}\mathbf{B}_{22}\otimes\mathbf{U}_{dp}\right)\ .$$

The block follows

$$\int_{-1/2}^{1/2}\mathbf{U}_r\otimes\left(\left[\operatorname{Vec}\left(\mathbf{\Phi}^{\mathsf{T}}\left(\omega_k\right)\right)\right]^{\mathsf{T}}\left[\mathbf{B}_{22}\otimes\mathbf{B}_{21}\right]\right)\ \mathrm{d}\frac{\partial^2\mathbf{F}_2\left(\omega,\underline{\lambda}_1,\underline{\lambda}_2\right)}{\partial\underline{\lambda}_2^{\mathsf{T}}\partial\underline{\lambda}_1^{\mathsf{T}}}\ .$$

References

[AKS62] M. Arató, A. N. Kolmogorov, and J. G. Sinay. Evaluation of the parameters of a complex stationary Gauss-Markov process. *Dokl. Akad. Nauk SSSR*, 146:747–750, 1962.

[ALS04] V. V. Anh, N. N. Leonenko, and L. M. Sakhno. Quasilikelihood-based higher-order spectral estimation of random processes and fields with possible long-range dependence. *J. of Appl. Probab.*, 41A:35–54, 2004.

[And71] T. W. Anderson. *The statistical analysis of time series*. Wiley, New York, 1971.

[Bri73] D. R. Brillinger. An empirical investigation of the chandler wobble and two proposed excitation processes. *Bull. Int. Statist. Inst.*, 45(3):413–434, 1973.

[Bri01] D. R. Brillinger. *Time Series; Data Analysis and Theory*. Society for Industrial and Applied Mathematics (SIAM), Philadelphia, 2001. Reprint of the 1981 edition.

[Bro90] E. N. Brown. A note on the asymptotic distribution of the parameter estimates for the harmonic regression model. *Biometrika*, 77(3):653–656, 1990.

[CT01] K. Choy and M. Taniguchi. Stochastic regression model with dependent disturbances. *J. Time Ser. Anal.*, 22(2):175–196, 2001.

[Dah95] R. Dahlhaus. Efficient location and regression estimation for long range dependent regression models. *Ann. Statist.*, 23(3):1029–1047, 1995.

[GR57] U. Grenander and M. Rosenblatt. *Statistical Analysis of Stationary Time Series*. Wiley, New York, 1957.

[Gre54] U. Grenander. On the estimation of regression coefficients in the case of an autocorrelated disturbance. *Ann. Math. Statistics*, 25:252–272, 1954.

[Gro00] R. S. Gross. The excitation of the chandler wobble. *Geophys. Res. Lett.*, 27(15):2329, 2000.

[Guy95] X. Guyon. *Random Fields on a Network Modeling, Statistics, and Applications*. Springer-Verlag, New York, 1995.

[Han70] E. J. Hannan. *Multiple Time Series*. Springer-Verlag, New York, 1970.

[Han71] E. J. Hannan. Non-linear time series regression. *Applied Probability Trust*, pages 767–780, 1971.

[Han73] E. J. Hannan. The asymptotic theory of linear time series models. *J. Appl. Proab.*, 10:130–145, 1973.

[Hey97] C. C. Heyde. *Quasi-Likelihood and Its Application*. Springer Series in Statistics. Springer-Verlag, New York, 1997.

[Hos97] Y. Hosoya. A limit theory for long-range dependence and statistical inference on related models. *Ann. Statist.*, 25(1):105–137, 1997.

[IR78] I. A. Ibragimov and Yu. A. Rozanov. *Gaussian Random Processes*. Springer-Verlag, New York, 1978. Translated from the Russian by A. B. Aries.

[IT97] E. Iglói and Gy. Terdik. Bilinear modelling of Chandler wobble. *Theor. of Probab. Appl.*, 44(2):398–400, 1997.

[IT03] E. Iglói and Gy. Terdik. Superposition of diffusions with linear generator and its multifractal limit process. *ESAIM Probab. Stat.*, 7:23–88 (electronic), 2003.

[Jen69] R. I. Jennrich. Asymptotic properties of non-linear least squares estimators. *Ann. Math. Statist.*, 40:633–643, 1969.

[KM96] C. Klüppelberg and T. Mikosch. The integrated periodogram for stable processes. *Ann. Statist.*, 24(5):1855–1879, 1996.

[Lob97] I. Lobato. Consistency of the averaged cross-periodogram in long memory series. *J. Time Series Anal.*, 18(2):137–155, 1997.

[Lob99] I. N. Lobato. A semiparametric two-step estimator in a multivariate long memory model. *J. Econometrics*, 90(1):129–153, 1999.

[MN99] J. R. Magnus and H. Neudecker. *Matrix Differential Calculus with Applications in Statistics and Econometrics*. Wiley, Chichester, 1999. Revised reprint of the 1988 original.

[RH97] P. M. Robinson and F. J. Hidalgo. Time series regression with long-range dependence. *Ann. Statist.*, 25(1):77–104, 1997.

[Rob72] P. M. Robinson. Non-linear regression for multiple time-series. *J. Appl. Probab.*, 9:758–768, 1972.

[SNS01] H. Schuh, T. Nagel, and T. Seitz. Linear drift and periodic variations observed in long time series of polar motion. *J. Geodesy*, 74:701–710, 2001.

[Ter02] Gy. Terdik. Higher order statistics and multivariate vector Hermite polyno-
         mials for nonlinear analysis of multidimensional time series. *Teor. Ǐmovīr.
         Mat. Stat.*, 66:147–168, 2002.

[Wan96] J. Wang. Asymptotics of least-squares estimators for constrained nonlinear
         regression. *Ann. Statist.*, 24(3):1316–1326, 1996.

[Wan04] L. Wang. Asymptotics of estimates in constrained nonlinear regression with
         long-range dependent innovations. *Ann. Inst. Statist. Math.*, 56(2):251–264,
         2004.

[Wu81]   Chien-Fu Wu. Asymptotic theory of nonlinear least squares estimation.
         Ann. Statist., 9(3):501–513, 1981.

[Yaj88] Y. Yajima. On estimation of a regression model with long-memory station-
         ary errors. *Ann. Statist.*, 16(2):791–807, 1988.

[Yaj91] Y. Yajima. Asymptotic properties of the LSE in a regression model with
         long-memory stationary errors. *Ann. Statist.*, 19(1):158–177, 1991.

Nonparametric estimator of a quantile function for the probability of event with repeated data

Claire Pinçon[1] and Odile Pons[2]

[1] Laboratoire de Biomathématiques - Faculté de Pharmacie 3, rue du Professeur Laguesse - B.P. 83 - 59006 Lille cedex - France
[2] Mathématiques et Informatique Appliquée - INRA Domaine de Vilvert - 78352 Jouy-en-Josas cedex - France

1 Introduction

Consider the case where an event occurs when a process Y decreases under a specific value μ_Y. The knowledge of the threshold μ_Y is of particular interest, but there are situations where Y is not observable. Instead, suppose that we are able to observe a process X associated to Y by a function m such that $E\{X\} = E\{m(Y)\}$. When m is monotonic, it becomes possible to estimate the threshold $\mu_X = m(\mu_Y)$ for the process X related to the occurrence of event.

These issues are frequently encountered in biomedical fields, where longitudinal data, such as demographic or clinical characteristics, are recorded to reflect a more complex disease evolution process. Think for example to the diabetes mellitus that is caused by the destruction or the dysfunction of β-cells of the islets of Langerhans in pancreas; indirect measures of this β-cells dysfunction rely on biological parameters such as glycosolated hemoglobin increase. Another example, studied in this paper, is the Duchenne Muscular Dystrophy (DMD), a disease resulting from a deficit in a protein called dystrophin, and passed by a mother to her son. The diagnosis of being a DMD gene carrier for a mother is difficult to make since only slight symptoms are experienced. However, several enzymes levels have been shown to increase significantly for the carrier mothers. One can then develop diagnostic tests by estimating on a sample of mothers with known status (DMD carrier or not) the limiting values of the enzymes levels above which the diagnosis is considered as positive.

We suppose that for a sample of n individuals, a *random* number of observations, say J_i, is recorded for each individual. For subject i, $i = 1, ..., n$, for observation j, $j = 1, ..., J_i$, let Y_{ij} be the value of the unmeasurable process Y, such that $Y_{ij} \geq Y_{ij+1}$. We observe (δ_{ij}, X_{ij}) with $\delta_{ij} = I\{Y_{ij} \leq \mu_Y\}$ and $X_{ij} = m(Y_{ij}) + \varepsilon_{ij}$, where (i) the $\{\varepsilon_{ij}\}$ have zero mean, are independent from the $\{Y_{ij}\}$, with continuous distribution function F_ε and with density function f_ε continuously differentiable and symmetrical at 0; (ii) m is continuous and monotonic, assumed increasing in the following, so that $E\{X_{ij}\} \geq E\{X_{ij+1}\}$. Then, for any $i = 1, ..., n$, $Y_i = (Y_{i1}, ..., Y_{iJ_i})'$, $X_i = (X_{i1}, ..., X_{iJ_i})'$ and $\varepsilon_i = (\varepsilon_{i1}, ..., \varepsilon_{iJ_i})'$ are vectors of *dependent* variables.

Conditionally to X_i, $\{\delta_{ij}, j = 1, ..., J_i\}$ is a sequence of Bernoulli variables with parameter $p(x) = \Pr\{\delta_{ij} = 1 | X_{ij} = x\} = 1 - F_\varepsilon(x - \mu_X)$, and with $E\{\delta_{ij}\delta_{ij'} | X_{ij} = x, X_{ij'} = x'\} = p(x)$ for $j' > j$.

By symmetry of f_ε, $p(\mu_X) = 1/2$, and reciprocally $\mu_X = p^{-1}(1/2)$; since p is a decreasing continuous function, its inverse distribution p^{-1} is defined by

$$p^{-1}(u) = \sup\{x : p(x) \geq u\}$$

for $u \in (0, 1)$. Let $q(u)$ denote $p^{-1}(u)$ for u in $(0, 1)$, so that $\mu_X = q(1/2)$.

The nonparametric estimation of the quantile distribution of a cumulative density function has been extensively studied in the literature for independent data. Quantiles estimators may be obtained by smoothing the empirical quantile function, for example with kernel functions (see among others Parzen, 1979; Sheather and Marron, 1990; Cheng and Parzen, 1997, and the references therein), or by quantile regression (see e.g. Koenker and Bassett, 1978). Methods to handle dependent data differ with the structure of the dependence between observations. For stationary time series, mixing assumptions ensure the asymptotic independence of blocks of observations (Billingsley, 1968, Doukhan, 1994). For repeated data, when the function m is linear, quantile regression provides estimators that are conditional to the value of the process Y and whose properties have been explored as both the number of subjects in the sample n and the number of measures per subject $J_i = J$ tend to infinity, provided J grows faster than n (Koenker, 2004). In this paper, we let the function m unspecified, and we consider that asymptotics is achieved for large n whereas the number of observations per subject remains bounded, supposing an ergodic property for the unknown densities of the X_{ij}'s. We propose to estimate the quantile function q by inverting a smooth estimator of the probability curve p, leading to the following estimator:

$$\hat{q}_{N,h}(u) = \sup\{x : \hat{p}_{N,h}(x) \geq u\}$$

for u in $(0, 1)$, where $\hat{p}_{N,h}$ denotes a kernel estimator of p depending on the total number of observations N and of a smoothing parameter h.

In section 2, we begin to establish the asymptotics for the probability estimator $\hat{p}_{N,h}(x)$: by classical arguments (Hall, 1984; Shorack and Wellner, 1986; Nadaraya, 1989, Härdle, 1990), we derive its bias and variance, and their convergence to zero implies the L^2-convergence of $\hat{p}_{N,h}(x)$ to $p(x)$. We then prove the convergence in distribution of the process $(Nh)^{1/2}\{\hat{p}_{N,h} - p\}$ to a Gaussian variable of which expectation and variance are deduced from the bias and variance of $\hat{p}_{N,h}$. These results allow to state the main theorem of this article about the convergence in distribution of the process $(Nh)^{1/2}\{\hat{q}_{N,h} - q\}$. In section 3, we give an asymptotic expression of the Mean Squared Error of $\hat{q}_{N,h}(u)$ for u in $(0; 1)$, and its minimization in h is shown equivalent to the minimization in h of the MSE of $\widehat{p}_{N,h}(x)$ for $x = q(u)$, resulting in a unique optimal local bandwidth, an approximation of which is possible with a bootstrap procedure. The last section illustrates the method with an application to real data about the diagnosis of Duchenne Muscular Dystrophy.

2 Asymptotics

The conditional probability $p(x)$ is estimated by the smooth estimator

$$\widehat{p}_{N,h}(x) = \frac{N^{-1} \sum_{i=1}^{n} \sum_{j=1}^{J_i} K_h(x - X_{ij})\delta_{ij}}{N^{-1} \sum_{i=1}^{n} \sum_{j=1}^{J_i} K_h(x - X_{ij})}, \tag{1}$$

with $N = \sum_{i=1}^{n} J_i$ and with $K_h(x) = h^{-1}K(h^{-1}x)$, where K is a kernel function for which we suppose the following assumptions hold:

(K_1) K is a positive symmetrical function having compact support $[-1; 1]$;
(K_2) $\int K(v)dv = 1$, $\kappa = \int v^2 K(v)dv < \infty$, $\kappa_\gamma = \int K^\gamma(v)dv < \infty$ for $\gamma \geq 0$, $\int |K'(v)|^\gamma dv < \infty$ for $\gamma = 1, 2$;
(K_3) as $n \to \infty$, $h \to 0$ and Nh^5 converges to a finite N_0 a.s.;
(K_4) for $\sup_{i=1,\dots,n} J_i < \infty$, a.s.
    This condition implies that, as $n \to \infty$, $N \to \infty$ with $N/n < \infty$ a.s. and $J(n)/N < \infty$ a.s. where $J(n) = \sum_{i=1}^{n} J_i(J_i - 1)$.

Let $f_{X_{ij}}$ and $f_{(X_{ij}, X_{ij'})}$ be the unknown density functions of the X_{ij}'s and of the $(X_{ij}, X_{ij'})$'s for $j \neq j'$ respectively. We restrict the study to a finite sub-interval $I_X = [a, b]$ of the support of the process X, and, for $h > 0$, let $I_{X,h}$ denote $[a + h; b - h]$. In addition of the hypotheses formulated above, we suppose that

(D_1) the function p is twice continuously differentiable and strictly monotonic on I_X;
(D_2) $\mathbb{E}\{N^{-1} \sum_{i=1}^{n} \sum_{j=1}^{J_i} f_{X_{ij}}(x)\}$ converges as $N \to \infty$ to a function $s^{(0)}(x)$, where $s^{(0)}$ is strictly positive and twice continuously differentiable on I_X. We denote by $s^{(\gamma)}$ its γth derivative for $\gamma \leq 2$;
(D_3) $\mathbb{E}\{N^{-1} \sum_{i=1}^{n} \sum_{j=1}^{J_i} f_{X_{ij}}(x_1)f_{X_{ij}}(x_2)\}$ converges as $N \to \infty$ to a continuous function $v(x_1, x_2)$ on $I_X^{\otimes 2}$;
(D_4) The expectations $\mathbb{E}\{J(n)^{-1} \sum_{i=1}^{n} \sum_{j=1}^{J_i} \sum_{j' \neq j} f_{(X_{ij}, X_{ij'})}(x_1, x_2)\}$ and $\mathbb{E}\{J(n)^{-1} \sum_{i=1}^{n} \sum_{j=1}^{J_i} \sum_{j' \neq j} f_{X_{ij}}(x_1)f_{X_{ij'}}(x_2)\}$ converge as $N \to \infty$ to continuous functions $c_1(x_1, x_2)$ and $c_2(x_1, x_2)$ respectively on $I_X^{\otimes 2}$.

We introduce $\widehat{n}_{N,h}(x)$ and $\widehat{d}_{N,h}(x)$ as the numerator and denominator of (1) respectively; let also $d_{N,h}(x) = \mathbb{E}\{\widehat{d}_{N,h}(x)\}$, $n_{N,h}(x) = \mathbb{E}\{\widehat{n}_{N,h}(x)\}$ and

$p_{N,h}(x) = \mathbb{E}\{\widehat{p}_{N,h}(x)\}$. When N, and $J_1, \ldots, J_n$ are actually random, our results hold conditionally on N.

Proposition 1 states different results about the kernel estimator $\widehat{p}_{N,h}(x)$: part (a) concerns the convergence in probability of $\widehat{p}_{N,h}$; its bias, its variance and its higher order moments are given in part (b); the derivation of its variance allows an expansion of $(Nh)^{1/2}\{\widehat{p}_{N,h} - p_{N,h}\}(x)$ in part (c) that will be central to prove the convergence in distribution of the process $(Nh)^{1/2}\{\widehat{p}_{N,h} - p_{N,h}\}$ in Theorem 1.

Proposition 1. *For $h > 0$, on $I_{X,h}$,*

(a) $\widehat{p}_{N,h}$ *converges in probability to p uniformly in x.*

(b) $\widehat{p}_{N,h}(x)$ *converges in norm L^2 to $p(x)$, with bias*

$$b_{N,h}(x) = h^2 B(x) + o(h^2), \ \ B(x) = \kappa\{p^{(1)}(x)s^{(1)}(x)\{s^{(0)}(x)\}^{-1} + \frac{1}{2}p^{(2)}(x)\},$$

and with variance

$$v_{N,h}(x) = N^{-1}h^{-1}\sigma^2(x) + o(N^{-1}h^{-1}), \ \ \sigma^2(x) = p(x)\{1 - p(x)\}\kappa_2\{s^{(0)}(x)\}^{-1}.$$

(c) *The following expansion holds:*

$$(Nh)^{1/2}\{\widehat{p}_{N,h} - p_{N,h}\}(x) = (Nh)^{1/2}\left\{s^{(0)}(x)\right\}^{-1}$$
$$\times \left[\{\widehat{n}_{N,h} - n_{N,h}\}(x) - p(x)\{\widehat{d}_{N,h} - d_{N,h}\}(x)\right] + o_{L^2}(1) \ . \quad (2)$$

Theorem 1. *For $h > 0$, the process $U_{N,h} = (Nh)^{1/2}\{\widehat{p}_{N,h} - p\}I_{\{I_{X,h}\}}$ converges in distribution to $W + N_0^{1/2}B$ where W is a centered Gaussian variable on I_X with variance σ^2 and with null covariances.*

Proof (Proof of Theorem 1). Under assumptions $(K_1 - K_4)$ and $(D_1 - D_4)$, $(Nh)^{1/2}\left(\widehat{n}_{N,h}(x) - n_{N,h}(x), \widehat{d}_{N,h}(x) - d_{N,h}(x)\right)'$ converges in distribution to a centered Gaussian variable for any $x \in I_{X,h}$.

The convergence of $(Nh)^{1/2}\{\widehat{p}_{N,h}(x) - p_{N,h}(x)\}$ to a centered Gaussian variable follows then from formula (2), and its limiting variance is $\sigma^2(x)$.

Now, for any $x \in I_{X,h}$, write $U_{N,h}(x) = (Nh)^{1/2}\{\widehat{p}_{N,h}(x) - p_{N,h}(x)\} + (Nh)^{1/2}b_{N,h}(x)$. With Proposition 1,

$$U_{N,h}(x) = (Nh)^{1/2}\{\widehat{p}_{N,h}(x) - p_{N,h}(x)\} + (Nh^5)^{1/2}B(x) + o\left((Nh^5)^{1/2}\right).$$

Since, as $N \to \infty$, Nh^5 converges a.s. to a finite N_0, $U_{N,h}(x)$ converges to $\{W + N_0^{1/2}B\}(x)$.

To ensure the weak convergence of the process $U_{N,h}$ on $I_{X,h}$, we have to prove the weak convergence of its finite dimensional distributions and its tightness (Billingsley, 1968). Let $x_1, ..., x_m$ be elements of $I_{X,h}$ each. As shown above, for $k = 1, ..., m$, $U_{N,h}(x_k)$ converges to a Gaussian variable with variance $\sigma^2(x_k)$; the vector $(U_{N,h}(x_1), ..., U_{N,h}(x_m))'$ converges then to a multivariate Gaussian distribution with variance covariance matrix $\{\sigma^2(x_k, x_l)\}$ for $k = 1, ..., m$ and $l = 1, ..., m$, where $\sigma^2(x_k, x_l)$ remains to be expressed for $k \neq l$. By (2),

$$\text{cov}\{U_{N,h}(x_k), U_{N,h}(x_l)\} = \frac{Nh}{s^{(0)}(x_k)s^{(0)}(x_l)} \Big[\text{cov}\{\widehat{n}_{N,h}(x_k), \widehat{n}_{N,h}(x_l)\}$$
$$- p(x_k)\text{cov}\{\widehat{d}_{N,h}(x_k), \widehat{n}_{N,h}(x_l)\} - p(x_l)\text{cov}\{\widehat{n}_{N,h}(x_k), \widehat{d}_{N,h}(x_l)\}$$
$$+ p(x_k)p(x_l)\text{cov}\{\widehat{d}_{N,h}(x_k), \widehat{d}_{N,h}(x_l)\}\Big] + o(1).$$

With assumptions (D_3) and (D_4),

$$\text{cov}\{\widehat{n}_{N,h}(x_k), \widehat{n}_{N,h}(x_l)\} = N^{-1}h^{-1}p\left(\frac{x_k + x_l}{2}\right)s^{(0)}\left(\frac{x_k + x_l}{2}\right)$$
$$\times \mathbb{1}\{0 \leq \alpha < 1\}\int K(v - \alpha)K(v + \alpha)\,dv + o(N^{-1}h^{-1}),$$

where $\alpha = |x_l - x_k|/(2h)$. For $0 \leq \alpha < 1$,

$$p\left(\frac{x_k + x_l}{2}\right) = p(x_k) + o(1) = p(x_l) + o(1),$$
$$s^{(0)}\left(\frac{x_k + x_l}{2}\right) = s^{(0)}(x_k) + o(1) = s^{(0)}(x_l) + o(1),$$

and the first term of $\text{cov}\{U_{N,h}(x_k), U_{N,h}(x_l)\}$ becomes

$$\frac{1}{2}\left\{\frac{p(x_k)}{s^{(0)}(x_k)} + \frac{p(x_l)}{s^{(0)}(x_l)}\right\}\mathbb{1}\{0 \leq \alpha < 1\}\left(\int K(v - \alpha)K(v + \alpha)\,dv\right) + o(1).$$

By similar developments for the other covariance terms,

$$\text{cov}\{U_{N,h}(x_k), U_{N,h}(x_l)\} = \frac{1}{2}\left\{\frac{p(x_k)(1 - p(x_k))}{s^{(0)}(x_k)} + \frac{p(x_l)(1 - p(x_l))}{s^{(0)}(x_l)}\right\}$$
$$\mathbb{1}\{0 \leq \alpha < 1\}\left(\int K(v - \alpha)K(v + \alpha)\,dv\right) + o(1).$$

As $N \to \infty$, $h \to 0$ and $\mathbb{1}\{0 \leq \alpha < 1\} \to 0$ unless $x_k = x_l$: for $x_k \neq x_l$, $\text{cov}\{U_{N,h}(x_k), U_{N,h}(x_l)\} = o(1)$, so that the limiting variance covariance matrix of the Gaussian variable W is such that $\sigma^2(x_k, x_l) = 0$.

The tightness of the sequence $\{U_{N,h}\}$ on $I_{X,h}$ will follow from (i) the tightness of $\{U_{N,h}(a)\}$ and (ii) a bound of the increments $\mathbb{E}\,|U_{N,h}(x_2) - U_{N,h}(x_1)|^2$.

For condition (i), let $\varepsilon > 0$ and $c > N_0^{1/2}|B(a)| + \left(2\varepsilon^{-1}\sigma^2(a)\right)^{1/2}$. Then

$$\Pr\{|U_{N,h}(a)| > c\} \le \Pr\left\{(Nh)^{1/2}|(\widehat{p}_{N,h} - p_{N,h})(a)| + (Nh)^{1/2}|b_{N,h}(a)| > c\right\}$$

$$\le \frac{\operatorname{var}\{(Nh)^{1/2}(\widehat{p}_{N,h} - p_{N,h})(a)\}}{\{c - (Nh)^{1/2}|b_{N,h}(a)|\}^2};$$

for N sufficiently large,

$$\Pr\{|U_{N,h}(a)| > c\} \le \frac{\sigma^2(a)}{\{c - N_0^{1/2}|B(a)|\}^2} + o(1) < \varepsilon.$$

Consider condition (ii). $U_{N,h}$ is written $W_{N,h} + (Nh)^{1/2}b_{N,h}$ where

$$\{b_{N,h}(x_1) - b_{N,h}(x_2)\}^2 \le c_{b,N,h}(x_1 - x_2)^2$$

with $c_{b,N,h}$ a positive constant, and

$$W_{N,h} = (Nh)^{1/2}\{\widehat{p}_{N,h} - p_{N,h}\}$$
$$= \frac{(Nh)^{1/2}}{\widehat{d}_{N,h}}\left\{[\widehat{n}_{N,h} - n_{N,h}] - p\left[\widehat{d}_{N,h} - d_{N,h}\right]\right.$$
$$\left. - \left(\frac{n_{N,h}}{d_{N,h}} - p\right)\left[\widehat{d}_{N,h} - d_{N,h}\right]\right\} + (Nh)^{1/2}\left(\frac{n_{N,h}}{d_{N,h}} - p_{N,h}\right).$$

One can then compute

$$(Nh)^{1/2}\left(\frac{n_{N,h}}{d_{N,h}} - p_{N,h}\right) = o\left((Nh^5)^{1/2}\right) = o(1),$$

$$\frac{1}{d_{N,h}} = \frac{1}{f^{(0)}}\left\{1 - \frac{[d_{N,h} - f^{(0)}][1 + o(1)]}{f^{(0)}}\right\} = \frac{1}{f^{(0)}}\left[1 - O(h^2)\right],$$

$$\frac{1}{\widehat{d}_{N,h}} = \frac{1}{d_{N,h}}\sum_{k\ge 0}\left(-\frac{\widehat{d}_{N,h} - d_{N,h}}{d_{N,h}}\right)^k = \frac{[1 - O(h^2)]}{f^{(0)}}\sum_{k\ge 0}\left\{-\frac{\widehat{d}_{N,h} - d_{N,h}}{f^{(0)}}\right\}^k.$$

As for Proposition 1, we can prove that $\operatorname{var}\{\widehat{d}_{N,h}(x_1) - \widehat{d}_{N,h}(x_2)\} \le c_d(x_1 - x_2)^2(Nh)^{-1}$ for a constant c_d, and similar bounds hold for $\operatorname{var}\{\widehat{n}_{N,h}(x_1) - \widehat{n}_{N,h}(x_2)\}$ and $\operatorname{var}\{\widehat{p}_{N,h}(x_1) - \widehat{p}_{N,h}(x_2)\}$. Under conditions (D_1)-(D_4), this implies that there exists a constant $c_{N,h}$ such that

$$\mathbb{E}\left\{\mathbb{1}_{\Omega_{N,h}}[U_{N,h}(x_1) - U_{N,h}(x_2)]\right\}^2 \le c_{N,h}(x_1 - x_2)^2,$$

where $\Omega_{N,h}$ is a probability set where all the o converging to zero in the approximation of the process $U_{N,h}$ by the previous expansions are sufficiently small for large N. $\square$

Since the limiting distribution of the process $U_{N,h}$ does not depend on the bandwidth h, one can state the following corollary:

Corollary 1. $\sup_{h>0:Nh^5 \to N_0}$ *a.s.* $U_{N,h}$ *converges in distribution to* $W+N_0^{1/2}B$.

Consider now the quantile estimator

$$\widehat{q}_{N,h}(u) = \sup\{x \in I_{X,h} : \widehat{p}_{N,h}(x) \geq u\} \tag{3}$$

introduced in section 1; as asserted in the following lemma, the uniqueness of $\widehat{q}_{N,h}(u)$ for any u in $\widehat{I}_{U,h} = \widehat{p}_{N,h}(I_{X,h})$ is guaranteed by the asymptotic monotonicity of $\widehat{p}_{N,h}$, due to its convergence in probability (Proposition 1) and to the monotonicity of p.

Lemma 1. *As* $N \to \infty$ *and* $h \to 0$, *for any* $x_1 < x_2$ *in* $I_{X,h}$ *and for every* $\zeta > 0$, *there exists* $C > 0$ *such that* $\Pr\{\widehat{p}_{N,h}(x_1) - \widehat{p}_{N,h}(x_2) > C\} \geq 1 - \zeta$.

We can now state the main theorem that establishes the convergence in distribution of the process $(Nh)^{1/2}\{\widehat{q}_{N,h} - q\}$ to a non centered Gaussian variable, the necessary condition for this variable to be centered being that the limit N_0 of Nh^5 as $N \to \infty$ has to be null.

Theorem 2. *For* $h > 0$, *on* $\widehat{I}_{U,h}$,
(a) $\widehat{q}_{N,h}$ *converges in probability to* q *uniformly in* u.

(b) *The process* $(Nh)^{1/2}\{\widehat{q}_{N,h} - q\}\mathbb{1}_{\{\widehat{I}_{U,h}\}}$ *converges in distribution to the process*
$$cess \quad \frac{W - N_0^{1/2}B}{p^{(1)}} \circ q.$$

Proof. For $u \in \widehat{I}_{U,h}$, there exists a unique x in $I_{X,h}$ such that $u = \widehat{p}_{N,h}(x)$; then we have

$$\widehat{q}_{N,h}(u) - q(u) = q \circ p(x) - q \circ \widehat{p}_{N,h}(x) = -\frac{\widehat{p}_{N,h}(x) - p(x)}{p^{(1)}(x)} + o(\widehat{p}_{N,h}(x) - p(x));$$

by Proposition 1(a), and since $\inf_{x \in I_X} |p^{(1)}| > 0$ by assumption (D_1), $\widehat{q}_{N,h}(u)$ converges in probability to $q(u)$ uniformly on $\widehat{I}_{U,h}$. Now write

$$(Nh)^{1/2}\{\widehat{q}_{N,h} - q\} = (Nh)^{1/2}\frac{p \circ \widehat{q}_{N,h} - p \circ q}{\widehat{\Delta}_{N,h}}$$

with

$$\widehat{\Delta}_{N,h} = \frac{p \circ \widehat{q}_{N,h} - p \circ q}{\widehat{q}_{N,h} - q}.$$

Since $p \circ \widehat{q}_{N,h} - p \circ q = \{p - \widehat{p}_{N,h}\} \circ \widehat{q}_{N,h}$, the numerator becomes

$$p \circ \widehat{q}_{N,h} - p \circ q = \{p - \widehat{p}_{N,h}\} \circ \widehat{q}_{N,h},$$

hence

$$(Nh)^{1/2}\{\widehat{q}_{N,h} - q\} = \frac{-U_{N,h} \circ \widehat{q}_{N,h}}{\widehat{\Delta}_{N,h}}.$$

By Theorem 1 and by uniform convergence in probability of $\widehat{q}_{N,h}$ to $q(u)$, the numerator converges in distribution to $\{W - N_0^{1/2}B\} \circ q$; again by uniform convergence in probability of $\widehat{q}_{N,h}$, the denominator converges to $p^{(1)} \circ q$ uniformly on $\widehat{I}_{U,h}$.

3 Optimal bandwidth

Although we do not express explicitly the bias and the variance of the estimator $\widehat{q}_{N,h}(u)$ for a fixed u in $\widehat{I}_{U,h}$, it is possible to find out an asymptotic equivalent of its Mean Squared Error using the results of Proposition 1(b): for any x in $I_{X,h}$, an asymptotic equivalent of the MSE of $\widehat{p}_{N,h}(x)$ is

$$\text{AMSE}_p(x, h) = N^{-1}h^{-1}\sigma^2(x) + h^4 B^2(x).$$

As in proof of Theorem 2, write, for any u in $\widehat{I}_{U,h}$,

$$\widehat{q}_{N,h}(u) - q(u) = -\frac{\{\widehat{p}_{N,h} - p\} \circ \widehat{q}_{N,h}(u)}{p^{(1)} \circ \widehat{q}_{N,h}(u)} + o(\{\widehat{p}_{N,h} - p\} \circ \widehat{q}_{N,h}(u)),$$

so that

$$\widehat{q}_{N,h}(u) - q(u) = -\frac{\{\widehat{p}_{N,h} - p\} \circ q(u)}{p^{(1)} \circ q(u)} + o_{L^2}(1),$$

and an asymptotic equivalent of the MSE of $\widehat{q}_{N,h}(u)$ is

$$\text{AMSE}_q(u, h) = \frac{\text{AMSE}_p(q(u), h)}{\{p^{(1)} \circ q(u)\}^2}.$$

The minimization in h of $\text{AMSE}_q(u, h)$ leads to the following optimal bandwidth, varying with u, and called the *optimal local bandwidth*:

$$h_{opt}(u) = N^{-1/5}\left\{\frac{\sigma^2 \circ q(u)}{4B^2 \circ q(u)}\right\}^{1/5}.$$

Note that $h_{opt}(u)$ equals the optimal local bandwidth minimizing $\text{AMSE}_p(x, h)$ for the unique value of x such that $x = q(u)$.

Several remarks can be made at this point. First of all, Theorem 1 extends to a variable bandwidth h, that is, a function such that $N \sup_{x \in I_X} |h(x)|^5$ is finite, and the optimal bandwidth has a rate of convergence given by $N^{-1/5}$, so that the rate of convergence of $\widehat{q}_{N,h_{opt}}$ to q is $N^{-4/5}$.

Then, the limit N_0 of Nh^5 as $N \to \infty$ is not null, so that the process $(Nh)^{1/2}\{\widehat{q}_{N,h_{opt}} - q\}$ converges in distribution to a *non centered* Gaussian variable.

Last, the optimal bandwidth depends on the probability function p, on its first derivative $p^{(1)}$, on the limiting density $s^{(0)}$ and its first derivative $s^{(1)}$, each of them being unknown. To obtain the optimal bandwidth estimate in avoiding the estimation of these unknown distributions, we propose to use a bootstrap procedure. For a fixed bandwidth h, let $\widehat{\mu}_X = \widehat{q}_{N,h}(1/2)$ be the estimator (3) of μ_X in the whole sample of size N. B bootstrapped samples of size N^* are generated, including in a sample the entire block of observations of a subject to maintain the dependence structure; estimating μ_X in these B samples allow to estimate the bias and variance of $\widehat{\mu}_X$ and then $\text{AMSE}_q(1/2, h)$. Reitering this procedure for different values of h leads to the optimal local bandwidth corresponding to the minimal AMSE.

4 Application

The Duchenne Muscular Dystrophy (DMD) is an inherited disorder passed by a mother to her son. The disorder is caused by a mutation of the dystrophin gene, that codes for the dystrophin, a protein that is required inside muscle cells for structural support. A deficit in dystrophin has for consequence a progressive muscular atrophy. Since the first signs of the disorder appear between 3 and 5, sometimes earlier, the Duchenne dystrophy is rapidly diagnosed for children. But it is much more difficult to confirm that a mother is a carrier of the DMD gene, because the carriers experience only slight symptoms; however, their levels of several enzymes were shown to be highest than for the non-carrier mothers (Andrews and Herzberg, 1985). Our objective was to estimate the threshold values for these enzymes (several μ_X) corresponding to the threshold value for dystrophin (μ_Y) above which a mother is declared carrier of the DMD gene ($\delta = 1$); the function m linking X and Y is decreasing. There were 125 subjects, 38 (30.4%) of them carrying the DMD gene. For each woman, data consisted in the levels of creatine kinase, hemopexin, lactate dehydrogenase and pyruvate kinase measured on a single blood sample, up to seven times several days apart for some women.

Table 1 shows the sample mean, median and variance of each first measurement enzyme level depending on the mother status; means and medians were all tested significantly different for carrier and non carrier mothers.

The large values of sample variances in the carrier mother group are due to several outliers. Because of this heterogeneity, the monotonicity of the X_{ij}'s is not guaranteed; in this case, the value $1/2$ may be reached more than once by the function $\widehat{p}_{N,h}$. We then compare the quantile estimator $\widehat{\mu}_{X,\text{sup}}$ as defined by (3) to

$$\widehat{\mu}_{X,\text{inf}} = \inf\left\{x \in I_{X,h} : \widehat{p}_{N,h}(x) \leq 1/2\right\}$$

Enzyme		Non-Carrier	Carrier
Creatine kinase (UI/l)	mean (s.d)	39.82 (19.03)	175.87 (192.82)
	median	35.00	102.50
Hemopexin (mg/dl)	mean (s.d)	83.40 (13.01)	93.99 (10.84)
	median	82.50	93.05
Pyruvate kinase (UI/l)	mean (s.d)	12.96 (3.91)	23.13 (17.60)
	median	12.70	19.45
Lactate dehydrogenase (U/g)	mean (s.d)	170.16 (41.63)	243.08 (62.26)
	median	167.50	242.50

Table 1. Enzymes levels description according to the status of the mothers (carriers vs non-carriers of the DMD gene) for the first blood sample.

to observe if there are notable differences between these estimates. For convenience, we suppress the subscript X in the names of the above estimators.

Each estimate is computed with the Epanechnikov kernel $K(x) = 0.75(1 - x^2)I\{-1 < x < 1\}$ for $x \in I_{X,h}$. The bootstrap procedure introduced in section 3 is used to estimate their bias and variance, so that we can compute unbiased estimates equal to initial estimates minus estimated bias, and 95% confidence intervals. We compare our estimators with a naive one defined as $\inf\{X_{ij} : \delta_{ij} = 1\}$; again, a bootstrap procedure allows to compute naive unbiased threshold estimates, denoted by $\widehat{\mu}_{\text{naive}}$, and 95% confidence intervals.

Results are described in Table 2. For each enzyme, the threshold estimates allows to compute the sensitivity, the specificity, the positive and negative predicted values (denoted by PPV and NPV respectively) of the diagnostic. Recall that the PPV is the probability of being a carrier conditionally to a positive diagnosis, and the NPV is the probability of being a non-carrier conditionally to a negative diagnosis. The threshold estimators we propose produce a more accurate diagnosis than the naive estimator, since the PPV and NPV are more balanced. The differences between $\widehat{\mu}_{\text{inf}}$ and $\widehat{\mu}_{\text{sup}}$ are explained by the variances of the enzymes levels.

Appendix: Proof of Proposition 1

To prove Proposition 1, we need the following lemma about the numerator $\widehat{n}_{N,h}(x)$ and the denominator $\widehat{d}_{N,h}(x)$ of $\widehat{p}_{N,h}(x)$. These results are derived in using arguments that apply for kernel density estimators; expressions of the bias and the moments are obtained with Taylor expansions, most of the terms in these developments vanishing by assumptions $(D3)$ and $(D4)$.

Lemma 2. For $x \in I_{X,h}$,
(a) $\widehat{n}_{N,h}(x)$ converges in norm L^2 to $p(x)s^{(0)}(x)$, with bias

$$b\{\widehat{n}_{N,h}(x)\} = \frac{h^2}{2}\kappa\{p\,s^{(0)}\}^{(2)}(x) + o(h^2),$$

		$\widehat{\mu}$ $[CI_{95\%}]$	Se (%)	Sp (%)	PPV (%)	NPV (%)
Creatine kinase	$\widehat{\mu}_{\text{naive}}$	15.90 [11.08;26.92]	100	1.15	30.65	100
(UI/l)	$\widehat{\mu}_{\text{inf}}$	39.91 [34.45;45.37]	89.47	60.92	50.00	92.98
	$\widehat{\mu}_{\text{sup}}$	137.07 [110.22;163.91]	34.21	100	100	77.68
Hemopexin	$\widehat{\mu}_{\text{naive}}$	7.48 [0;19.18]	100	0	30.4	0
(mg/dl)	$\widehat{\mu}_{\text{inf}}$	78.17 [69.60;86.74]	94.74	29.89	37.11	92.86
	$\widehat{\mu}_{\text{sup}}$	90.56 [86.69;94.44]	60.53	71.26	47.92	80.52
Pyruvate kinase	$\widehat{\mu}_{\text{naive}}$	8.04 [7.52;9.08]	100	11.49	33.04	100
(UI/l)	$\widehat{\mu}_{\text{inf}}$	10.00 [9.99;10.01]	94.74	24.14	35.29	91.30
	$\widehat{\mu}_{\text{sup}}$	15.82 [12.73;18.92]	68.42	78.16	57.78	85.00
Lactate	$\widehat{\mu}_{\text{naive}}$	116.41 [105.89;138.11]	100	4.88	32.76	100
dehydrogenase	$\widehat{\mu}_{\text{inf}}$	202.03 [184.34;219.72]	68.42	80.49	61.90	84.62
(U/g)	$\widehat{\mu}_{\text{sup}}$	199.95 [187.70;212.20]	68.42	79.27	60.47	84.42

Table 2. Enzymes thresholds estimates and bootstrap 95% confidence intervals (from 1000 bootstrapped samples); sensitivity (Se), specificity (Sp), positive and predictive values (PPV and NPV resp.) are estimated on the sample of first blood measures ($n = 125$).

and variance

$$\text{var}\{\widehat{n}_{N,h}(x)\} = N^{-1}h^{-1}\kappa_2 \, p(x)s^{(0)}(x) + o(N^{-1}h^{-1}) \; ;$$

its higher order moment expansions are $o(N^{-1}h^{-1})$.
(b) $\widehat{d}_{N,h}(x)$ converges in norm L^2 to $s^{(0)}(x)$, with bias

$$b\{\widehat{d}_{N,h}(x)\} = \frac{h^2}{2}\kappa s^{(2)}(x) + o(h^2) \; ,$$

and variance

$$\text{var}\{\widehat{d}_{N,h}(x)\} = N^{-1}h^{-1}\kappa_2 \, s^{(0)}(x) + o(N^{-1}h^{-1}) \; ;$$

its higher order moments expansions are $o(N^{-1}h^{-1})$.
(c) Their joint moment equals $N^{-1}h^{-1}\kappa_2 \, p(x)s^{(0)}(x) + o(N^{-1}h^{-1})$.

Proof (Proof of Proposition 1). Let $x \in I_{X,h}$, and begin with the proof of (a). By assumption (K_2), $\int |K'(v)|dv < \infty$, and by (K_3), h may be written $h_n = c_n n^{-1/5}$ with a bounded sequence c_n, and for $a_n = (\exp\{-\gamma c_n^2\})$ it satisfies $\sum_{n=1}^{\infty} \exp\{-n\gamma h_n^2\} = \sum_{n=1}^{\infty}\{a_n\}^{n^{3/5}} < \infty$ for every $\gamma > 0$. Then (Rao, 1983), $\widehat{n}_{N,h} - n_{N,h}$ and $\widehat{d}_{N,h} - d_{N,h}$ converge both in probability to 0 uniformly on $I_{X,h}$. Hence, $\sup_{x \in I_{X,h}} |\widehat{p}_{N,h}(x) - p(x)|$ converges to 0 in probability.

Consider now the proof of (b). To compute the bias of $\widehat{p}_{N,h}(x)$, we begin to prove that

$$p_{N,h}(x) = \frac{n_{N,h}(x)}{d_{N,h}(x)} + O\left(N^{-1}h^{-1}\right) \; .$$

Write

$$p_{N,h}(x) = \frac{n_{N,h}(x)}{d_{N,h}(x)} - \frac{\mathbb{E}\left[\widehat{n}_{N,h}(x)\{\widehat{d}_{N,h}(x) - d_{N,h}(x)\}\right]}{d_{N,h}^2(x)}$$

$$+ \frac{\mathbb{E}\left[\widehat{p}_{N,h}(x)\{\widehat{d}_{N,h}(x) - d_{N,h}(x)\}^2\right]}{d_{N,h}^2(x)} \; ; \quad (4)$$

then

$$\left| p_{N,h}(x) - \frac{n_{N,h}(x)}{d_{N,h}(x)} \right| \leq |\varphi_1(x)| + |\varphi_2(x)| \, ,$$

with

$$\varphi_1(x) \leq \frac{\left[\mathrm{var}\{\widehat{n}_{N,h}(x)\} \mathrm{var}\{\widehat{d}_{N,h}(x)\} \right]^{1/2}}{d_{N,h}^2(x)} = O\left(N^{-1}h^{-1}\right),$$

$$\varphi_2(x) \leq \frac{\mathrm{var}\{\widehat{d}_{N,h}(x)\}}{d_{N,h}^2(x)} = O\left(N^{-1}h^{-1}\right) \, .$$

Therefore, for any $x \in I_{X,h}$,

$$p_{N,h}(x) = \frac{n_{N,h}(x)}{d_{N,h}(x)} + O\left(N^{-1}h^{-1}\right) \, . \quad (5)$$

The bias of $\widehat{p}_{N,h}(x)$ is

$$b_{N,h}(x) = \left(\frac{n_{N,h}(x)}{d_{N,h}(x)} - p(x) \right) + \left(p_{N,h}(x) - \frac{n_{N,h}(x)}{d_{N,h}(x)} \right) \, .$$

With a second order Taylor expansion of $d_{N,h}^{-1}(x)$, one obtains

$$\frac{n_{N,h}(x)}{d_{N,h}(x)} = p(x) + h^2 \{ p^{(1)}(x) s^{(1)}(x) \{ s^{(0)}(x) \}^{-1}$$

$$+ \frac{1}{2} p^{(2)}(x) \} \int v^2 K(v) dv + o(h^2) \, .$$

Furthermore, by (5),

$$p_{N,h}(x) - \frac{n_{N,h}(x)}{d_{N,h}(x)} = o(h^2) \, . \quad (6)$$

The bias of $\widehat{p}_{N,h}(x)$ follows immediately. The variance $v_{N,h}(x)$ of $\widehat{p}_{N,h}(x)$ is

$$v_{N,h}(x) = \mathbb{E}\left\{ \widehat{p}_{N,h}(x) - \frac{n_{N,h}(x)}{d_{N,h}(x)} \right\}^2 - \left\{ p_{N,h}(x) - \frac{n_{N,h}(x)}{d_{N,h}(x)} \right\}^2 \, .$$

By using twice relation (4),

$$d_{N,h}(x) \left\{ \widehat{p}_{N,h}(x) - \frac{n_{N,h}(x)}{d_{N,h}(x)} \right\}$$

$$= \widehat{n}_{N,h}(x) - n_{N,h}(x) - p_{N,h}(x)\left\{ \widehat{d}_{N,h}(x) - d_{N,h}(x) \right\} \qquad (7)$$

$$- \frac{\{\widehat{n}_{N,h}(x) - n_{N,h}(x)\}\left\{ \widehat{d}_{N,h}(x) - d_{N,h}(x) \right\}}{d_{N,h}(x)}$$

$$+ \frac{\widehat{p}_{N,h}(x)\left\{ \widehat{d}_{N,h}(x) - d_{N,h}(x) \right\}^2}{d_{N,h}(x)}$$

$$+ \left\{ p_{N,h}(x) - \frac{n_{N,h}(x)}{d_{N,h}(x)} \right\}\left\{ \widehat{d}_{N,h}(x) - d_{N,h}(x) \right\},$$

so that

$$d_{N,h}^2(x)\ \mathbb{E}\left\{ \widehat{p}_{N,h}(x) - \frac{n_{N,h}(x)}{d_{N,h}(x)} \right\}^2$$

$$= \mathrm{var}\{\widehat{n}_{N,h}(x)\} + p_{N,h}^2(x)\mathrm{var}\{\widehat{d}_{N,h}(x)\} - 2p_{N,h}(x)\pi_{0,1,1}(x)$$

$$+2\frac{p_{N,h}(x)}{d_{N,h}(x)}\pi_{0,1,2}(x) - 2\frac{\pi_{0,2,1}(x)}{d_{N,h}(x)} + \frac{\pi_{0,2,2}(x)}{d_{N,h}^2(x)} + \frac{\pi_{2,0,4}(x)}{d_{N,h}^2(x)} + 2\frac{\pi_{1,1,2}(x)}{d_{N,h}(x)}$$

$$-2p_{N,h}(x)\frac{\pi_{1,0,3}(x)}{d_{N,h}(x)} - 2\frac{\pi_{1,1,3}(x)}{d_{N,h}^2(x)} + \left\{ p_{N,h}(x) - \frac{n_{N,h}(x)}{d_{N,h}(x)} \right\}\varphi_3(x),$$

where

$$\pi_{k,k',k''}(x) = \mathbb{E}\left[\widehat{p}_{N,h}^k(x)\{\widehat{n}_{N,h}(x) - n_{N,h}(x)\}^{k'}\{\widehat{d}_{N,h}(x) - d_{N,h}(x)\}^{k''} \right]$$

for $k \geq 0$, $k' \geq 0$ and $k'' \geq 0$, and with

$$\varphi_3(x) = \left\{ p_{N,h}(x) - \frac{n_{N,h}(x)}{d_{N,h}(x)} \right\}\mathrm{var}\{\widehat{d}_{N,h}(x)\}$$

$$+ 2\left[p_{N,h}(x)\mathrm{var}\{\widehat{d}_{N,h}(x)\} + \pi_{0,1,1}(x) - \frac{\pi_{0,1,2}(x)}{d_h(x)} + \frac{\pi_{1,0,3}(x)}{d_h(x)} \right].$$

Since $\widehat{p}_{N,h}(x) \in]0,1[$, using Cauchy-Schwarz inequalities, and by Lemma 2, one can prove that the $\pi_{k,k',k''}(x)$'s in the above expression are $o(N^{-1}h^{-1})$ except $\pi_{0,1,1}(x)$. By (5),

$$\left\{ p_{N,h}(x) - \frac{n_{N,h}(x)}{d_{N,h}(x)} \right\}\varphi_3(x) = o(N^{-1}h^{-1}).$$

Since $d_{N,h}(x) = s^{(0)}(x) + O(h^2)$ and $p_{N,h}(x) = p(x) + O(h^2)$,

$$v_{N,h}(x) = \{s^{(0)}(x)\}^{-2}[\mathrm{var}\{\widehat{n}_{N,h}(x)\} + p^2(x)\,\mathrm{var}\{\widehat{d}_{N,h}(x)\}$$
$$- 2p(x)\,\mathrm{cov}\{\widehat{n}_{N,h}(x), \widehat{d}_{N,h}(x)\}] + o(N^{-1}h^{-1}) \ .$$

From the above expression we can deduce the following results. First,

$$(Nh)^{1/2}\{\widehat{p}_{N,h}(x) - p_{N,h}(x)\} = (Nh)^{1/2}\{s^{(0)}(x)\}^{-1}[\{\widehat{n}_{N,h}(x)$$
$$- n_{N,h}(x)\} - p(x)\{\widehat{d}_{N,h}(x) - d_{N,h}(x)\}] + o_{L_2}(1) \ ,$$

as stated in (2), and, by Lemma 2,

$$v_{N,h}(x) = N^{-1}h^{-1}\left\{s^{(0)}(x)\right\}^{-1} p(x)\{1 - p(x)\}\kappa_2 + o\left(N^{-1}h^{-1}\right) \ .$$

Consider now $\mathbb{E}\{\widehat{p}_{N,h}(x) - p_{N,h}(x)\}^l$ for $l \geq 3$; we have

$$\mathbb{E}\left|\widehat{p}_{N,h}(x) - p_{N,h}(x)\right|^l$$
$$\leq 2^l \left\{\mathbb{E}\left|\widehat{p}_{N,h}(x) - \frac{n_{N,h}(x)}{d_{N,h}(x)}\right|^l + \mathbb{E}\left|\frac{n_{N,h}(x)}{d_{N,h}(x)} - p_{N,h}(x)\right|^l\right\}$$

by relation (7), and since the moments of order ≥ 3 of $\widehat{n}_{N,h}(x)$ and of $\widehat{d}_{N,h}(x)$ are $o(N^{-1}h^{-1})$ by Lemma 2, the first term of this sum is $o(N^{-1}h^{-1})$. By formula (5), the second term of the sum is $O(N^{-l}h^{-l}) = o(N^{-1}h^{-1})$. Hence, $E\{\widehat{p}_{N,h}(x) - p_{N,h}(x)\}^l = o(N^{-1}h^{-1})$.

References

[AH85]   Andrews, D. F., and Herzberg, A. M. *Data : a collection of problems from many fields for the student and research worker.* Springer-Verlag, New York, 1985.

[Bil68]   Billingsley, P. (1968). *Convergence of probability measures.* Wiley, New York.

[CP97]   Cheng, C. and Parzen, E. (1997). Unified estimators of smooth quantile and quantile density functions. *Journal of Statistal Planning Inference,* **59**, 291-307.

[Dou94]   Doukhan, P. (1994). *Mixing: properties and examples.* Lecture Notes in Statist. 85, Springer Verlag, New York.

[Hal84]   Hall, P. (1984). Integrated square error properties of kernel estimators of regression functions. *Annals of Statistics.,* **12**, 241-260.

[Här90]   Härdle, W. (1990). *Applied nonparametric regression.* Cambridge University Press, Cambridge.

[Koe04]   Koenker, R. (2004). Quantile regression for longitudinal data. *Journal of Multivariate Analysis,* **91**, 74-89.

[KB78]   Koenker, R. and Bassett, G. (1978). Regression quantiles. *Econometrica,* **46**, 33-50.

[Nad89]   Nadaraya, E. A. (1989). *Nonparametric estimation of probability densities and regression curves.* Kluwer Academic Publisher, Boston.

[Par79]   Parzen, E. (1979). Nonparametric statistical data modeling (with comments). *Journal of the American Statistical Association,* **74**, 105-131.

[Pra83]   Prakasa Rao, B. L. S. (1983). *Nonparametric Functional Estimation.* Academic Press, New York.

[SM90]   Sheather, S. J. and Marron, J. S. (1990). Kernel quantile estimators, *Journal of the American Statistical Association,* **85**, 410-416.

[SW86]   Shorack, G.R. and Wellner, J.A. (1986). *Empirical processes with statistical application.* Wiley, New York.

Lecture Notes in Statistics
For information about Volumes 1 to 132, please contact Springer-Verlag

133: Dipak Dey, Peter Müller, and Debajyoti Sinha (Editors), Practical Nonparametric and Semiparametric Bayesian Statistics. xv, 408 pp., 1998.

134: Yu. A. Kutoyants, Statistical Inference For Spatial Poisson Processes. vii, 284 pp., 1998.

135: Christian P. Robert, Discretization and MCMC Convergence Assessment. x, 192 pp., 1998.

136: Gregory C. Reinsel, Raja P. Velu, Multivariate Reduced-Rank Regression. xiii, 272 pp., 1998.

137: V. Seshadri, The Inverse Gaussian Distribution: Statistical Theory and Applications. xii, 360 pp., 1998.

138: Peter Hellekalek and Gerhard Larcher (Editors), Random and Quasi-Random Point Sets. xi, 352 pp., 1998.

139: Roger B. Nelsen, An Introduction to Copulas. xi, 232 pp., 1999.

140: Constantine Gatsonis, Robert E. Kass, Bradley Carlin, Alicia Carriquiry, Andrew Gelman, Isabella Verdinelli, and Mike West (Editors), Case Studies in Bayesian Statistics, Volume IV. xvi, 456 pp., 1999.

141: Peter Müller and Brani Vidakovic (Editors), Bayesian Inference in Wavelet Based Models. xiii, 394 pp., 1999.

142: György Terdik, Bilinear Stochastic Models and Related Problems of Nonlinear Time Series Analysis: A Frequency Domain Approach. xi, 258 pp., 1999.

143: Russell Barton, Graphical Methods for the Design of Experiments. x, 208 pp., 1999.

144: L. Mark Berliner, Douglas Nychka, and Timothy Hoar (Editors), Case Studies in Statistics and the Atmospheric Sciences. x, 208 pp., 2000.

145: James H. Matis and Thomas R. Kiffe, Stochastic Population Models. viii, 220 pp., 2000.

146: Wim Schoutens, Stochastic Processes and Orthogonal Polynomials. xiv, 163 pp., 2000.

147: Jürgen Franke, Wolfgang Härdle, and Gerhard Stahl, Measuring Risk in Complex Stochastic Systems. xvi, 272 pp., 2000.

148: S.E. Ahmed and Nancy Reid, Empirical Bayes and Likelihood Inference. x, 200 pp., 2000.

149: D. Bosq, Linear Processes in Function Spaces: Theory and Applications. xv, 296 pp., 2000.

150: Tadeusz Caliński and Sanpei Kageyama, Block Designs: A Randomization Approach, Volume I: Analysis. ix, 313 pp., 2000.

151: Håkan Andersson and Tom Britton, Stochastic Epidemic Models and Their Statistical Analysis. ix, 152 pp., 2000.

152: David Ríos Insua and Fabrizio Ruggeri, Robust Bayesian Analysis. xiii, 435 pp., 2000.

153: Parimal Mukhopadhyay, Topics in Survey Sampling. x, 303 pp., 2000.

154: Regina Kaiser and Agustin Maravall, Measuring Business Cycles in Economic Time Series. vi, 190 pp., 2000.

155: Leon Willenborg and Ton de Waal, Elements of Statistical Disclosure Control. xvii, 289 pp., 2000.

156: Gordon Willmot and X. Sheldon Lin, Lundberg Approximations for Compound Distributions with Insurance Applications. xi, 272 pp., 2000.

157: Anne Boomsma, Marijtje A. J. van Duijn, and Tom A.B. Snijders (Editors), Essays on Item Response Theory. xv, 448 pp., 2000.

158: Dominique Ladiray and Benoît Quenneville, Seasonal Adjustment with the X-11 Method. xxii, 220 pp., 2001.

159: Marc Moore (Editor), Spatial Statistics: Methodological Aspects and Some Applications. xvi, 282 pp., 2001.

160: Tomasz Rychlik, Projecting Statistical Functionals. viii, 184 pp., 2001.

161: Maarten Jansen, Noise Reduction by Wavelet Thresholding, xxii, 224 pp., 2001.

162: Constantine Gatsonis, Bradley Carlin, Alicia Carriquiry, Andrew Gelman, Robert E. Kass, Isabella Verdinelli, and Mike West (Editors), Case Studies in Bayesian Statistics, Volume V. xiv, 448 pp., 2001.

163: Erkki P. Liski, Nripes K. Mandal, Kirti R. Shah, and Bikas K. Sinha, Topics in Optimal Design. xii, 164 pp., 2002.

164: Peter Goos, The Optimal Design of Blocked and Split-Plot Experiments. xiv, 244 pp., 2002.

165: Karl Mosler, Multivariate Dispersion, Central Regions and Depth: The Lift Zonoid Approach. xii, 280 pp., 2002.

166: Hira L. Koul, Weighted Empirical Processes in Dynamic Nonlinear Models, Second Edition. xiii, 425 pp., 2002.

167: Constantine Gatsonis, Alicia Carriquiry, Andrew Gelman, David Higdon, Robert E. Kass, Donna Pauler, and Isabella Verdinelli (Editors), Case Studies in Bayesian Statistics, Volume VI. xiv, 376 pp., 2002.

168: Susanne Rässler, Statistical Matching: A Frequentist Theory, Practical Applications and Alternative Bayesian Approaches. xviii, 238 pp., 2002.

169: Yu. I. Ingster and Irina A. Suslina, Nonparametric Goodness-of-Fit Testing Under Gaussian Models. xiv, 453 pp., 2003.

170: Tadeusz Caliński and Sanpei Kageyama, Block Designs: A Randomization Approach, Volume II: Design. xii, 351 pp., 2003.

171: D.D. Denison, M.H. Hansen, C.C. Holmes, B. Mallick, B. Yu (Editors), Nonlinear Estimation and Classification. x, 474 pp., 2002.

172: Sneh Gulati, William J. Padgett, Parametric and Nonparametric Inference from Record-Breaking Data. ix, 112 pp., 2002.

173: Jesper Møller (Editor), Spatial Statistics and Computational Methods. xi, 214 pp., 2002.

174: Yasuko Chikuse, Statistics on Special Manifolds. xi, 418 pp., 2002.

175: Jürgen Gross, Linear Regression. xiv, 394 pp., 2003.

176: Zehua Chen, Zhidong Bai, Bimal K. Sinha, Ranked Set Sampling: Theory and Applications. xii, 224 pp., 2003.

177: Caitlin Buck and Andrew Millard (Editors), Tools for Constructing Chronologies: Crossing Disciplinary Boundaries, xvi, 263 pp., 2004.

178: Gauri Sankar Datta and Rahul Mukerjee, Probability Matching Priors: Higher Order Asymptotics, x, 144 pp., 2004.

179: D.Y. Lin and P.J. Heagerty (Editors), Proceedings of the Second Seattle Symposium in Biostatistics: Analysis of Correlated Data, vii, 336 pp., 2004.

180: Yanhong Wu, Inference for Change-Point and Post-Change Means After a CUSUM Test, xiv, 176 pp., 2004.

181: Daniel Straumann, Estimation in Conditionally Heteroscedastic Time Series Models, x, 250 pp., 2004.

182: Lixing Zhu, Nonparametric Monte Carlo Tests and Their Applications, xi, 192 pp., 2005.

183: Michel Bilodeau, Fernand Meyer, and Michel Schmitt (Editors), Space, Structure and Randomness, xiv, 416 pp., 2005.

184: Viatcheslav B. Melas, Functional Approach to Optimal Experimental Design, vii., 352 pp., 2005.

185: Adrian Baddeley, Pablo Gregori, Jorge Mateu, Radu Stoica, and Dietrich Stoyan, (Editors), Case Studies in Spatial Point Process Modeling, xiii., 324 pp., 2005.

186: Estela Bee Dagum and Pierre A. Cholette, Benchmarking, Temporal Distribution, and Reconciliation Methods for Time Series, xiv., 410 pp., 2006.

187: Patrice Bertail, Paul Doukhan and Philippe Soulier, Dependence in Probability and Statistics, viii., 504 pp., 2006.

Benchmarking, Temporal Distribution, and Reconciliation Methods for Time Series

Estela Bee Dagum and Pierre A. Cholette

This book discusses the statistical methods most often applied for adjustments, ranging from ad hoc procedures to regression-based models. The latter are emphasized, because of their clarity, ease of application, and superior results. Each topic is illustrated with many real case examples. In order to facilitate understanding of their properties and limitations of the methods discussed, a real data example, the Canada Total Retail Trade Series, is followed throughout the book.

2006. 410 p. (Lecture Notes in Statistics, Vol. 186) Softcover
ISBN 0-387-31102-5

Case Studies in Spatial Point Process Modeling

A. Baddeley, P. Gregori, J. Mateu, R. Stoica and D. Stoyan (Editors)

Point process statistics is successfully used in fields such as material science, human epidemiology, social sciences, animal epidemiology, biology, and seismology. Its further application depends greatly on good software and instructive case studies that show the way to successful work. This book satisfies this need by a presentation of the spatstat package and many statistical examples.

2005. 312 p. (Lecture Notes in Statistics, Vol. 185) Softcover
ISBN 0-387-28311-0

Functional Approach to Optimal Experimental Design

V.B. Melas

The book presents a novel approach for studying optimal experimental designs. The functional approach consists of representing support points of the designs by Taylor series. It is thoroughly explained for many linear and nonlinear regression models popular in practice including polynomial, trigonometrical, rational, and exponential models. Using the tables of coefficients of these series included in the book, a reader can construct optimal designs for specific models by hand.

2005. 336 p. (Lecture Notes in Statistics, Vol. 184) Softcover
ISBN 0-387-98741-X